T0188814

Communications in Computer and Information Science 855

More information about this series at http://www.springer.com/series/7899

Jesús Medina · Manuel Ojeda-Aciego
José Luis Verdegay · Irina Perfilieva
Bernadette Bouchon-Meunier · Ronald R. Yager (Eds.)

Information Processing and Management of Uncertainty in Knowledge-Based Systems

Applications

17th International Conference, IPMU 2018
Cádiz, Spain, June 11–15, 2018
Proceedings, Part III

Editors
Jesús Medina
Universidad de Cádiz
Cádiz, Cadiz
Spain

Manuel Ojeda-Aciego
Universidad de Málaga
Málaga, Málaga
Spain

José Luis Verdegay
Universidad de Granada
Granada
Spain

Irina Perfilieva
Institute for Research and Applications
University of Ostrava
Ostrava
Czech Republic

Bernadette Bouchon-Meunier
LIP6
Université Pierre et Marie Curie, CNRS
Paris
France

Ronald R. Yager
Iona College
New Rochelle, NY
USA

ISSN 1865-0929 ISSN 1865-0937 (electronic)
Communications in Computer and Information Science
ISBN 978-3-319-91478-7 ISBN 978-3-319-91479-4 (eBook)
https://doi.org/10.1007/978-3-319-91479-4

Library of Congress Control Number: 2018944294

Printed on acid-free paper

This Springer imprint is published by the registered company Springer International Publishing AG part of Springer Nature
The registered company address is: Gewerbestrasse 11, 6330 Cham, Switzerland

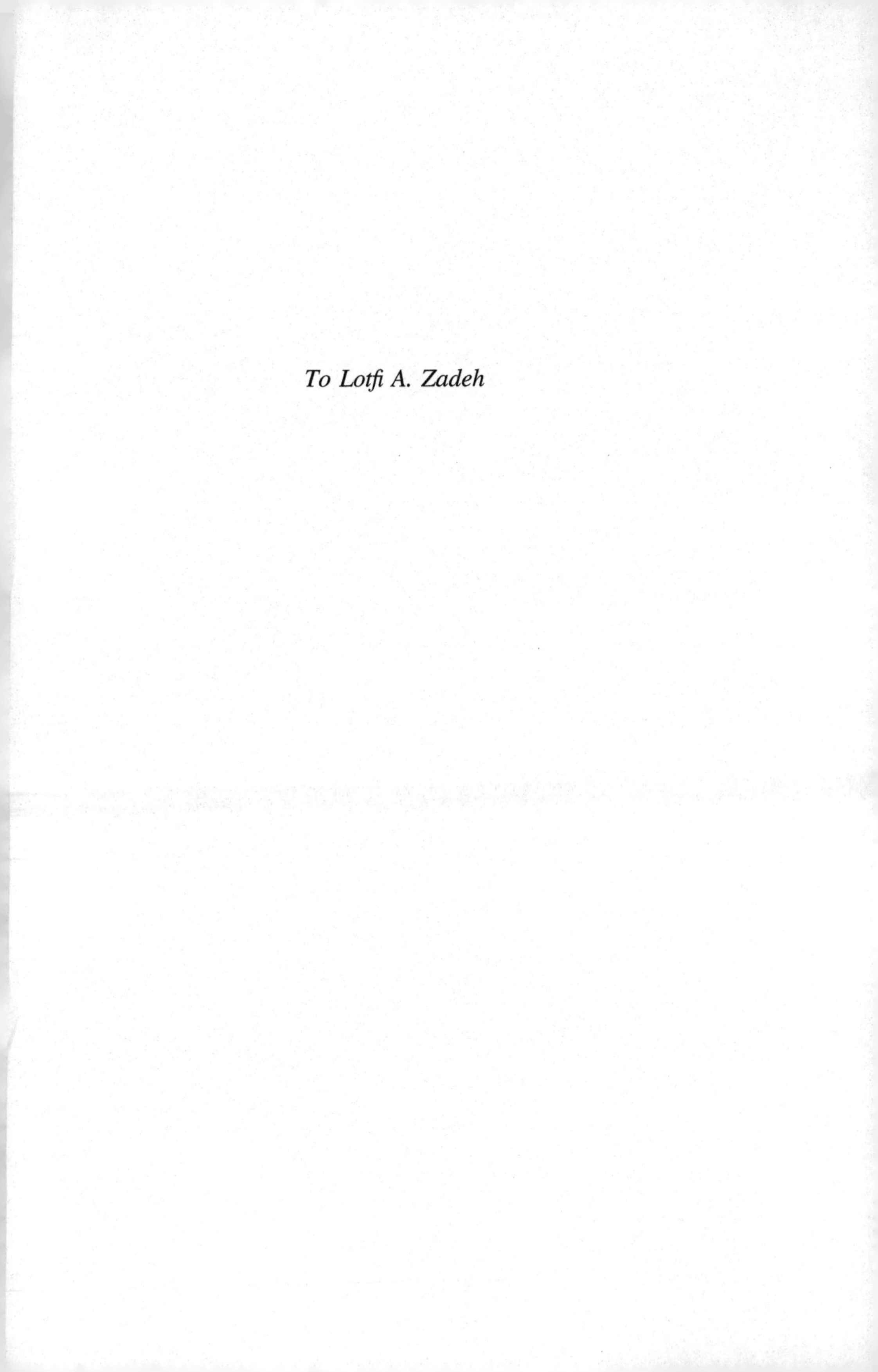

To Lotfi A. Zadeh

Preface

These are the proceedings of the 17th International Conference on Information Processing and Management of Uncertainty in Knowledge-Based Systems, IPMU 2018. The conference was held during June 11–15, in Cádiz, Spain.

The IPMU conference is organized every two years with the aim of bringing together scientists working on methods for the management of uncertainty and aggregation of information in intelligent systems. Since 1986, the IPMU conference has been providing a forum for the exchange of ideas between theoreticians and practitioners working in these areas and related fields.

This IPMU edition held special meaning since one of its co-founders, Lotfi A. Zadeh, passed away on September 6, 2017. To pay him a well-deserved tribute, and in memory of his long relationship with IPMU participants, a special plenary panel was organized to discuss the scientific legacy of his ideas. Renowned researchers and Lotfi's good friends made up the panel: it was chaired by Ronald Yager, while Bernadette Bouchon-Meunier, Didier Dubois, Janusz Kacprzyk, Rudolf Kruse, Rudolf Seising, and Enric Trillas acted as panelists. Besides this, a booklet of pictures with Lotfi Zadeh and friends was compiled and distributed at the conference.

Following the IPMU tradition, the Kampé de Fériet Award for outstanding contributions to the field of uncertainty and management of uncertainty was presented. Past winners of this prestigious award were Lotfi A. Zadeh (1992), Ilya Prigogine (1994), Toshiro Terano (1996), Kenneth Arrow (1998), Richard Jeffrey (2000), Arthur Dempster (2002), Janos Aczel (2004), Daniel Kahneman (2006), Enric Trillas (2008), James Bezdek (2010), Michio Sugeno (2012), Vladimir N. Vapnik (2014), and Joseph Y. Halpern (2016). In this 2018 edition, the award was given to Glenn Shafer (Rutgers University, Newark, USA) for his seminal contributions to the mathematical theory of evidence and belief functions as well as to the field of reasoning under uncertainty. The so-called Dempster–Shafer theory, an alternative to the theory of probability, has been widely applied in engineering and artificial intelligence.

The program consisted of the keynote talk of Glenn Shafer, as recipient of the Kampé de Feriet Award, five invited plenary talks, two round tables, and 30 special sessions plus a general track for the presentation of the 190 contributed papers that were authored by researchers from more than 40 different countries. The plenary presentations were given by the following distinguished researchers: Gloria Bordogna (IREA CNR – Institute for the Electromagnetic Sensing of the Environment of the Italian National Research Council), Lluis Godo (Artificial Intelligence Research Institute of the Spanish National Research Council, Barcelona, Spain), Enrique Herrera-Viedma (Department of Computer Science and Artificial Intelligence, University of Granada, Spain), Natalio Krasnogor (School of Computing Science at Newcastle University, UK), and Yiyu Yao (Department of Computer Science, University of Regina, Canada).

The conference followed a single-blind review process, respecting the usual conflict-of-interest standards. The contributions were reviewed by at least three reviewers. Moreover, the conference chairs further checked the contributions in those cases were conflicting reviews were obtained. Finally, the accepted papers are published in three volumes: Volumes I and II focus on "Theory and Foundations," while Volume III is devoted to "Applications."

The organization of the IPMU 2018 conference was possible thanks to the assistance, dedication, and support of many people and institutions. In particular, this renowned international conference owes its recognition to the great quality of the contributions. Thank you very much to all the participants for their contributions to the conference and all the authors for the high quality of their submitted papers. We are also indebted to our colleagues, members of the Program Committee, and the organizers of special sessions on hot topics, since the successful organization of this international conference would not have been possible without their work. They and the additional reviewers were fundamental in maintaining the excellent scientific quality of the conference. We gratefully acknowledge the local organization for the efforts in the successful development of the multiple tasks that a great event like IPMU involves.

We also acknowledge the support received from different areas of the University of Cádiz, including the Department of Mathematics, the PhD Program in Mathematics, the Vice-Rectorate of Infrastructures and Patrimony, and the Vice-Rectorate for Research; the International Global Campus of Excellence of the Sea (CEI·Mar) led by the University of Cádiz and composed of institutions of three different countries; the European Society for Fuzzy Logic and Technology (EUSFLAT); and the Springer team who managed the publication of these proceedings. Finally, J. Medina, M. Ojeda-Aciego, J. L. Verdegay, I. Cabrera, and D. Pelta acknowledge the support of the following research projects: TIN2016-76653-P, TIN2015-70266-C2-P-1, TIN2014-55024-P, TIN2017-86647-P, and TIN2017-89023-P (Spanish Ministery of Economy and Competitiveness, including FEDER funds).

June 2018

Jesús Medina
Manuel Ojeda-Aciego
Irina Perfilieva
José Luis Verdegay
Bernadette Bouchon-Meunier
Ronald R. Yager
Inma P. Cabrera
David A. Pelta

Organization

General Chair

Jesús Medina	Universidad de Cádiz, Spain

Program Chairs

Manuel Ojeda-Aciego	Universidad de Málaga, Spain
Irina Perfilieva	University of Ostrava, Czech Republic
José Luis Verdegay	Universidad de Granada, Spain

Executive Directors

Bernadette Bouchon-Meunier	LIP6 - Université Pierre et Marie Curie, CNRS, Paris, France
Ronald R. Yager	Iona College, USA

Sponsors and Publicity Chair

Martin Stepnicka	University of Ostrava, Czech Republic

Special Sessions Chair

Inma P. Cabrera	Universidad de Málaga, Spain

Publication Chair

David A. Pelta	Universidad de Granada, Spain

Local Organizing Committee

María José Benítez-Caballero	Universidad de Cádiz, Spain
María Eugenia Cornejo	Universidad de Cádiz, Spain
Juan Carlos Díaz-Moreno	Universidad de Cádiz, Spain
David Lobo	Universidad de Cádiz, Spain
Óscar Martín-Rodríguez	Universidad de Granada, Spain
Eloísa Ramírez-Poussa	Universidad de Cádiz, Spain

International Advisory Board

Anne Laurent
Benedetto Matarazzo
Christophe Marsala
Enric Trillas
Eyke Hüllermeier
Giulianella Coletti
João Paulo Carvalho
José Luis Verdegay
Julio Gutierrez-Rios
Laurent Foulloy
Llorenç Valverde
Lorenza Saitta
Luis Magdalena
Manuel Ojeda-Aciego
Maria Amparo Vila
Maria Rifqi
Marie-Jeanne Lesot
Mario Fedrizzi
Miguel Delgado
Olivier Strauss
Salvatore Greco
Uzay Kaymak

Program Committee

Michal Baczynski, Poland
Rafael Bello-Pérez, Cuba
Jim Bezdek, USA
Isabelle Bloch, France
Ulrich Bodenhofer, Austria
Bernadette Bouchon-Meunier, France
Humberto Bustince, Spain
Inma P. Cabrera, Spain
Joao Carvalho, Portugal
Giulianella Coletti, Italy
Oscar Cordon, Spain
Inés Couso, Spain
Keeley Crockett, UK
Fabio Cuzzolin, UK
Bernard De Baets, Belgium
Guy De Tré, Belgium
Sébastien Destercke, France
Antonio Di Nola, Italy
Didier Dubois, France
Fabrizio Durante, Italy
Francesc Esteva, Spain
Juan C. Figueroa-García, Colombia
Sylvie Galichet, France
Lluis Godo, Spain
Fernando Gomide, Brazil
Gil González-Rodríguez, Spain
Michel Grabisch, France
Przemysław Grzegorzewski, Poland
Lawrence Hall, USA
Francisco Herrera, Spain
Enrique Herrera-Viedma, Spain
Ludmilla Himmelspach, Germany
Janusz Kacprzyk, Poland
Uzay Kaymak, The Netherlands
James Keller, USA
Laszlo Koczy, Hungary
Vladik Kreinovich, USA
Tomas Kroupa, Italy
Rudolf Kruse, Germany
Christophe Labreuche, France
Weldon A. Lodwick, USA
Jean-Luc Marichal, Luxembourg
Trevor Martin, UK
Sebastian Massanet, Spain
Gaspar Mayor, Spain
Jesús Medina, Spain
Radko Mesiar, Slovakia
Ralf Mikut, Germany
Enrique Miranda, Spain
Javier Montero, Spain
Jacky Montmain, France
Serafín Moral, Spain
Zbigniew Nahorski, Poland
Pavel Novoa, Ecuador
Vilem Novak, Czech Republic
Hannu Nurmi, Finland
Manuel Ojeda-Aciego, Spain
Gabriella Pasi, Italy

Witold Pedrycz, Canada
David A. Pelta, Spain
Irina Perfilieva, Czech Republic
Fred Petry, USA
Vincenzo Piuri, Italy
Olivier Pivert, France
Henri Prade, France
Marek Reformat, Canada
Daniel Sánchez, Spain
Mika Sato-Ilic, Japan
Ricardo C. Silva, Brazil
Martin Stepnicka, Czech Republic
Umberto Straccia, Italy
Eulalia Szmidt, Poland
Settimo Termini, Italy
Vicenc Torra, Sweden/Spain
Linda van der Gaag, The Netherlands
Barbara Vantaggi, Italy
José L. Verdegay, Spain
Thomas Vetterlein, Austria
Susana Vieira, Portugal
Slawomir Zadrozny, Poland

Additional Reviewers

Jesús Alcalá-Fernández
José Carlos R. Alcantud
Svetlana Asmuss
Laszlo Aszalos
Mohammad Azad
Cristobal Barba Gonzalez
Gleb Beliakov
María J. Benítez-Caballero
Pedro Bibiloni
Alexander Bozhenyuk
Michal Burda
Ana Burusco
Camilo Alejandro Bustos Téllez
Francisco Javier Cabrerizo
Yuri Cano
Andrea Capotorti
J. Manuel Cascón
Dagoberto Castellanos
Francisco Chicano
María Eugenia Cornejo
Susana Cubillo
Martina Dankova
Luis M. de Campos
Robin De Mol
Yashar Deldjoo
Pedro Delgado-Pérez
Juan Carlos Díaz
Susana Díaz
Marta Disegna
Alexander Dockhorn
Paweł Drygaś
Talbi El Ghazali
Javier Fernández
Joao Gama
José Gámez
M. D. García Sanz
José Luis García-Lapresta
José García Rodríguez
Irina Georgescu
Manuel Gómez-Olmedo
Adrián González
Antonio González
Manuel González-Hidalgo
Jerzy Grzymala Busse
Piotr Helbin
Daryl Hepting
Michal Holcapek
Olgierd Hryniewicz
Petr Hurtik
Atamanyuk Igor
Esteban Induráin
Vladimir Janis
Andrzej Janusz
Sándor Jenei
Pascual Julian-Iranzo
Aránzazu Jurío
Katarzyna Kaczmarek
Martin Kalina
Gholamreza Khademi
Margarita Knyazeva
Martins Kokainis
Galyna Kondratenko

Oleksiy Korobko
Piotr Kowalski
Oleksiy Kozlov
Anna Król
Sankar Kumar Roy
Angelica Leite
Tianrui Li
Ferenc Lilik
Nguyen Linh
Hua Wen Liu
Bonifacio Llamazares
David Lobo
Marcelo Loor
Ezequiel López-Rubio
Gabriel Luque
Rafael M. Luque-Baena
M. Aurora Manrique
Nicolás Marín
Ricardo A. Marques-Pereira
Stefania Marrara
Davide Martinetti
Víctor Martínez
Raquel Martínez España
Miguel Martínez-Panero
Tamás Mihálydeák
Arnau Mir
Katarzyna Miś
Miguel A. Molina-Cabello
Michinori Nakata
Bac Nguyen
Joachim Nielandt
Wanda Niemyska
Juan Miguel Ortiz De Lazcano Lob
Sergio Orts Escolano
Iván Palomares
Esteban José Palomo
Manuel Pegalajar-Cuéllar
Barbara Pękala
Renato Pelessoni
Tomasz Penza
Davide Petturiti
José Ramón Portillo
Cristina Puente
Eloísa Ramírez-Poussa
Ana Belén Ramos Guajardo
Jordi Recasens
Juan Vicente-Riera
Rosa M. Rodríguez
Luis Rodríguez-Benítez
Estrella Rodríguez-Lorenzo
Maciej Romaniuk
Jesús Rosado
Clemente Rubio-Manzano
Pavel Rusnok
Hiroshi Sakai
Sancho Salcedo-Sanz
José María Serrano
Ievgen Sidenko
Gerardo Simari
Julian Skirzynski
Andrzej Skowron
Grégory Smits
Marina Solesvik
Anna Stachowiak
Sebastian Stawicki
Michiel Stock
Andrei Tchernykh
Jhoan S. Tenjo García
Luis Terán
Karl Thurnhofer Hemsi
S. P. Tiwari
Joan Torrens
Gracian Trivino
Matthias Troffaes
Shusaku Tsumoto
Diego Valota
Matthijs van Leeuwen
Sebastien Varrette
Marco Viviani
Pavel Vlašánek
Yuriy Volosyuk
Gang Wang
Anna Wilbik
Andrzej Wójtowicz

Special Session Organizers

Stefano Aguzzoli	University of Milan, Italy
José M. Alonso	University of Santiago de Compostela, Spain
Michal Baczynski	University of Silesia, Poland
Isabelle Bloch	University Paris-Saclay, France
Reda Boukezzoula	LISTIC- Université de Savoie Mont-Blanc, France
Humberto Bustince	Universidad Pública de Navarra, Spain
Inma P. Cabrera	Universidad de Málaga, Spain
Tomasa Calvo	Universidad de Alcalá, Spain
Ciro Castiello	University of Bari Aldo Moro, Italy
Juan Luis Castro	University of Granada, Spain
Yurilev Chalco-Cano	Universidad de Tarapacá, Chile
Davide Ciucci	University of Milano-Bicocca, Italy
Didier Coquin	LISTIC- Université de Savoie Mont-Blanc, France
Pablo Cordero	University of Málaga, Spain
Rocío de Andrés Calle	University of Salamanca, Spain
Bernard De Baets	Ghent University, Belgium
Juan Carlos de la Torre	University of Cádiz, Spain
Graçaliz Dimuro	Institute of Smart Cities and Universidade Federal do Rio Grande, Brazil
Enrique Domínguez	University of Málaga, Spain
Bernabe Dorronsoro	University of Cádiz, Spain
Krzysztof Dyczkowski	Adam Mickiewicz University in Poznań, Poland
Ali Ebrahimnejad	Islamic Azad University, Iran
Tommaso Flaminio	Artificial Intelligence Research Institute (CSIC), Barcelona, Spain
Pilar Fuster-Parra	Universitat de les Illes Balears, Spain
M. Socorro Garcia Cascales	Polytechnic University of Cartagena, Spain
Brunella Gerla	University of Insubria, Italy
Juan Gómez Romero	University of Granada, Spain
Teresa González-Arteaga	University of Valladolid, Spain
Balasubramaniam Jayaram	Indian Institute of Technology Hyderabad, India
László Kóczy	Budapest University of Technology and Economics, Hungary
Yuriy Kondratenko	Petro Mohyla Black Sea National University, Ukraine
Weldon Lodwick	University of Colorado, USA
Vincenzo Loia	University of Salerno, Italy
Nicolás Madrid	University of Málaga, Spain
Luis Magdalena	Universidad Politécnica de Madrid, Spain
María J. Martín-Bautista	University of Granada, Spain
Sebastián Massanet	University of the Balearic Islands, Spain
Corrado Mencar	University of Bari Aldo Moro, Italy
Radko Mesiar	University of Technology, Slovakia
Enrique Miranda	University of Oviedo, Spain
Ignacio Montes	University of Oviedo, Spain

Contents – Part III

Logical Methods in Mining Knowledge from Big Data

Metaheuristics and Machine Learning

Uncertainty in Medicine

Uncertainty in Video/Image Processing (UVIP)

General Track

Decision Making Modeling and Applications

On the Structure of Acyclic Binary Relations

José Carlos R. Alcantud[1], María J. Campión[2], Juan C. Candeal[3], Raquel G. Catalán[4], and Esteban Induráin[4](✉)

[1] Departamento de Economía e Historia Económica, Universidad de Salamanca, Campus Miguel de Unamuno, Salamanca, Spain
jcr@usal.es

[2] Inarbe (Institute for Advanced Research in Business and Economics) and Departamento de Matemáticas, Universidad Pública de Navarra, Pamplona, Spain
mjesus.campion@unavarra.es

[3] Departamento de Análisis Económico, Universidad de Zaragoza, Zaragoza, Spain
candeal@unizar.es

[4] InaMat (Institute for Advanced Materials) and Departamento de Matemáticas, Universidad Pública de Navarra, Campus Arrosadía, Pamplona, Spain
{raquel.garcia,steiner}@unavarra.es

Abstract. We investigate the structure of acyclic binary relations from different points of view. On the one hand, given a nonempty set we study real-valued bivariate maps that satisfy suitable functional equations, in a way that their associated binary relation is acyclic. On the other hand, we consider acyclic directed graphs as well as their representation by means of incidence matrices. Acyclic binary relations can be extended to the asymmetric part of a linear order, so that, in particular, any directed acyclic graph has a topological sorting.

Keywords: Acyclic binary relations · Functional equations
Acyclic directed graphs · Arborescences · Incidence matrices
Total preorders · Numerical representability
Topological sorting algorithms

AMS Subject Class. (2010): 06A06 · Secondary: 54F05 · 39B52
39B22 · 05C20 · 05C38 · 05C62 · 65F30 · 91B16

1 Introduction

Acyclic binary relations are crucial in the mathematical analysis of Decision Making and Social Choice, as well as in Theoretical Computer Science. To put

This work has been partially supported by the research projects MTM2012-37894-C02-02, TIN2013-47605-P, ECO2015-65031-R, MTM2015-63608-P (MINECO/FEDER), TIN2016-77356-P and the Research Services of the Public University of Navarre (Spain).

J. Medina et al. (Eds.): IPMU 2018, CCIS 855, pp. 3–15, 2018.
https://doi.org/10.1007/978-3-319-91479-4_1

only an example, binary relations that model preferences of agents are often asked to be compulsorily acyclic, in order to avoid incoherences. By this reason, theoretical studies on the structure, main properties and scope in possible applications of acyclic binary relations should be welcome as the grounds that support many aspects of Decision Making.

The origin of the study addressed in the present paper comes from an analysis of those binary relations $\mathcal{R}$ on a nonempty set X that appear through a bivariate real-valued function $F : X \times X \to \mathbb{R}$ such that $x\mathcal{R}y \Leftrightarrow F(x,y) > 0$. In some appealing particular cases, the special kind of binary relation considered is characterized by the fact of the function F being the solution of some functional equation (e.g. the Sincov's one $F(x,y) + F(y,z) + F(z,x) = 0 \quad (x,y,z \in X)$, see [10], closely related to representable total preorders). Surprisingly as it may appear at first glance, the types of binary relations that have already been characterized this way correspond either to very simple situations (namely, reflexivity, irreflexivity and asymmetry) or to sophisticated ones as representable total preorders, interval orders and semiorders. Intermediate situations as transitivity or acyclicity among others remain as open problems. At that stage, we did not have at hand yet any characterization of acyclicity by means of suitable functional equations. Nor we had characterized binary relations that give rise to an acyclic graph, or to a tree –that is also a directed graph– or to a finite union of trees among others. Nevertheless, in some particular situations (e.g., on countable sets) a few characterizations of acyclicity can actually be encountered in the literature (see [3,9]). Also, there are techniques that detect if a binary relation on a finite set is actually an arborescence, as the well-known Kruskal's algorithm (see [8]). However, they have not been built in terms of functional equations but using other techniques (see e.g. [1,3]).
The structure of the manuscript goes as follows: We analyze the relationship between functional equations and acyclicity in Sect. 3. Next we study particular situations where the set on which the binary relations are defined is finite. In that case, alternative mathematical tools to deal with binary relations are graph theory and incidence matrices (see Sect. 4).

2 Preliminaries

Definition 1. A binary relation $\mathcal{R}$ on a nonempty set X is a subset of the Cartesian product $X^2 = X \times X$. Given two elements $x, y \in X$, we will use the standard notation $x\mathcal{R}y$ to express that the pair (x,y) belongs to $\mathcal{R}$.

Naturally associated to a binary relation $\mathcal{R}$ on a set X, we will also deal with the binary relations $\mathcal{R}^c$ and $\mathcal{R}^{-1}$ on X, respectively given by $\mathcal{R}^c = X^2 \setminus \mathcal{R}$, and by $x\mathcal{R}^{-1}y \iff y\mathcal{R}x, \quad (x,y \in X)$.

A binary relation $\mathcal{R}$ defined on a set X is said to be

(i) *reflexive* if $\Delta \subseteq \mathcal{R}$, with $\Delta = \{(x,x) : x \in X\}$ (here Δ stands for the *diagonal* of X^2),
(ii) *irreflexive* if $\mathcal{R} \cap \Delta = \emptyset$,
(iii) *symmetric* if $\mathcal{R}$ and $\mathcal{R}^{-1}$ coincide,

(iv) *antisymmetric* if $\mathcal{R} \cap \mathcal{R}^{-1} \subseteq \Delta$,
(v) *asymmetric* if $\mathcal{R} \cap \mathcal{R}^{-1} = \emptyset$,
(vi) *total* (or *complete*) if $\mathcal{R} \cup \mathcal{R}^{-1} = X^2$,
(vii) *transitive* if $x\mathcal{R}y \wedge y\mathcal{R}z \Rightarrow x\mathcal{R}z$ for every $x, y, z \in X$,
(viii) *negatively transitive* if $\mathcal{R}^c$ is transitive.

Given two binary relations $\mathcal{R}$, $\mathcal{S}$ on X, its *composition* $\mathcal{R} \circ \mathcal{S}$ is a new binary relation on X, defined as follows: For any pair $(x, y) \in X^2$, we declare that $x\ (\mathcal{R} \circ \mathcal{S})\ y$ holds true -equivalently, we say that the pair (x, y) belongs to $\mathcal{R} \circ \mathcal{S} \subseteq X \times X$- whenever there exists $z \in X$ such that (x, z) belongs to $\mathcal{R} \subseteq X \times X$, whereas (z, y) belongs to $\mathcal{S} \subseteq X \times X$. The composition of binary relations is associative. Given a natural number n, we will use the standard notation $\mathcal{R}^n$ to denote the composition $\mathcal{R} \circ \ldots$ (n-times) $\ldots \circ \mathcal{R}$.

The binary relation $\mathcal{R}$ is said to be *acyclic* if $\mathcal{R}^n \cap \Delta = \emptyset$ holds true for every natural number n. The *transitive closure* $\bar{\mathcal{R}}$ of a binary relation $\mathcal{R}$ is defined as $\bar{\mathcal{R}} = \bigcup_{n=1}^{\infty} \mathcal{R}^n$. It is plain that $\bar{\mathcal{R}}$ is transitive.

In the particular case of dealing with *orderings* on X, the standard notation is different. We include it here for sake of completeness.

Definition 2. A *preorder* $\precsim$ on a nonempty set X is a binary relation on X which is reflexive and transitive. An antisymmetric preorder is said to be a *partial order.* A *total preorder* $\precsim$ on a set X is a preorder such that if $x, y \in X$ then $x \precsim y$ or $y \precsim x$ holds. An antisymmetric total preorder is said to be a *total order.* A total order is also called a *linear order.*

If $\precsim$ is a preorder on X, then as usual we denote the associated *asymmetric* relation by $\prec$ and the associated *equivalence* relation by $\sim$ and these are defined, respectively, by $x \prec y \iff (x \precsim y) \wedge \neg (y \precsim x)$ and by $x \sim y \iff (x \precsim y) \wedge (y \precsim x)$. The asymmetric part of a linear order (respectively, of a partial order, of a total preorder) is said to be a *strict linear order* (respectively, a *strict partial order*, a *strict total preorder*).

A total preorder $\precsim$ on a set X is said to be *representable* if there exists a real-valued map $u : X \to \mathbb{R}$ such that, for any $x, y \in X$, we have $x \precsim y \Leftrightarrow u(x) \leq u(y)$. The map u is said to be a *utility function* or an *order-isomorphism.*

Definition 3. Let X be a nonempty set. Let $F : X \times X \to \mathbb{R}$ be a real-valued bivariate function defined on X. The function F satisfies the *Sincov functional equation* if $F(x, y) + F(y, z) = F(x, z)$ holds for every $x, y, z \in X$ (see [4,10]).

The following easy result arises (see e.g. [10]).

Proposition 1. *A bivariate function* $F : X \times X \to \mathbb{R}$ *satisfies the Sincov functional equation if and only if there exists a real-valued function* $G : X \to \mathbb{R}$ *such that* $F(x, y) = G(y) - G(x)$ *holds for all* $x, y \in X$.

Given a binary relation $\mathcal{R}$ on a nonempty set X, we may immediately interpret $\mathcal{R}$ through a bivariate real-valued function $F : X \times X \to \mathbb{R}$. To do so, it is enough to consider the characteristic function of the binary relation $\mathcal{R} \subseteq X \times X$, namely $F(x,y) = 1 \Leftrightarrow (x,y) \in \mathcal{R}$ and $F(x,y) = 0$ otherwise. However, this F may fail to satisfy suitable additional properties, as, for instance, to be the solution of some classical functional equation. Paying attention to the converse situation, we begin with a bivariate map $F : X \times X \to \mathbb{R}$, and we define its associated binary relation $\mathcal{R}_F$ by declaring that $(x,y) \in \mathcal{R}_F$ holds true if and only if $F(x,y) > 0$. It is clear that if F satisfies certain additional properties, its associated binary relation $\mathcal{R}_F$ will a fortiori feature some related special characteristics. To put an obvious example, we may notice that if F vanishes on the diagonal Δ, then $\mathcal{R}_F$ is irreflexive. In this direction, the following result arises. Its proof is straightforward and follows from the corresponding definitions.

Proposition 2. *Let X denote a nonempty set and $F : X \times X \to \mathbb{R}$ a bivariate map. Let $\mathcal{R}_F$ the binary relation defined on X by means of F, as follows: $x\mathcal{R}_F y \Leftrightarrow F(x,y) > 0 \quad (x,y \in X)$. The following statements hold true:*

(i) If $F(x,x) > 0$ holds for every $x \in X$ then $\mathcal{R}_F$ is reflexive.
(ii) If $F(x,x) \leq 0$ holds for every $x \in X$ then $\mathcal{R}_F$ is irreflexive.
(iii) If $F(x,y) + F(y,x) = 0$ holds for every $x,y \in X$ then $\mathcal{R}_F$ is asymmetric.
(iv) If F satisfies the Sincov functional equation, then $\mathcal{R}_F$ is asymmetric and negatively transitive. It is actually a strict total preorder.

For the particular case of representable total preorders, the following well-know result stated in Proposition 3 above plays a crucial role (see e.g. [4]).

Proposition 3. *Let X be a nonempty set. Let $\precsim$ be a total preorder on X. Then the following statements are equivalent:*

(i) The total preorder $\precsim$ is representable by means of a utility function $u : X \to \mathbb{R}$ such that $x \precsim y \Leftrightarrow u(x) \leq u(y) \quad (x,y \in X)$.
(ii) There exists a real-valued bivariate map $F : X \times X \to \mathbb{R}$ that satisfies the Sincov functional equation and, in addition, $x \prec y \Leftrightarrow F(x,y) > 0$ holds true for every $x,y \in X$.

3 Acyclic Binary Relations vs. Functional Equations

Definition 4. Given a nonempty set X endowed with a binary relation $\mathcal{R}$, we say that another binary relation $\mathcal{Q}$ is an *extension* of $\mathcal{R}$ if $x\mathcal{R}y \Rightarrow x\mathcal{Q}y$ holds true for every $x,y \in X$. In other words, as subsets of the Cartesian product $X \times X$, this means that $\mathcal{R} \subseteq \mathcal{Q} \subseteq X \times X$.

In this direction, a classical extension theorem was obtained by E. Szpilrajn in 1930. That theorem will be an important key in this Sect. 3.

Lemma 1 *(Szpilrajn extension theorem, 1930). Let X be a nonempty set. Let $\prec$ stand for an irreflexive and transitive binary relation defined on X. Then $\prec$ can be extended to a strict linear order.*

Proof. See [12]. For some related results, see also [11]. □

Using Szpilrajn extension theorem as a tool, we may prove now, as a direct consequence of it, the following result on extension of acyclic binary relations.

Theorem 1. *Let X be a nonempty set. Let $\mathcal{R}$ be an acyclic binary relation defined on X. Then $\mathcal{R}$ can be extended to a strict linear order.*

Proof. Let $\bar{\mathcal{R}}$ be the transitive closure of the given relation $\mathcal{R}$. It is plain that $\bar{\mathcal{R}}$ is transitive, by its own definition, and it is also irreflexive because $\mathcal{R}$ is acyclic. Moreover, $\bar{\mathcal{R}}$ is an extension of $\mathcal{R}$. Since $\bar{\mathcal{R}}$ is irreflexive and transitive, by Lemma 1 (Szpilrajn extension theorem), it can actually be extended to a linear order defined on X. Obviously, such linear order is also an extension of the former acyclic binary relation $\mathcal{R}$. □

Parallel to Szpilrajn extension theorem, the following result is also classical.

Theorem 2 *(Hansson extension theorem, 1968). Let X be a nonempty set. Let $\precsim$ be a preorder defined on X. Then $\precsim$ can be extended to a total preorder defined on X, so that the asymmetric part of that total preorder is also an extension of $\prec$, the asymmetric part of $\precsim$.*

Proof. See [5]. For generalizations, see [11]. □

Remark 1. Matching Hansson extension theorem and Lemma 1 (Szpilrajn extension theorem) we can prove again Theorem 1. To do so, we may observe that given an acyclic binary relation $\mathcal{R}$, and $\bar{\mathcal{R}}$ its transitive closure, the binary relation $\mathcal{Q} = \Delta \cup \bar{\mathcal{R}}$ is a preorder whose asymmetric part is $\bar{\mathcal{R}}$. By Theorem 2, $\mathcal{Q}$ can be extended to a total preorder $\precsim$ whose asymmetric part $\prec$ extends $\bar{\mathcal{R}}$ and consequently $\mathcal{R}$. Finally, by Lemma 1, $\prec$ can be extended to a linear order.

Definition 5. Let X be a nonempty set. Let $\mathcal{S}$ be a binary relation defined on X. Associated to $\mathcal{S}$, let $\mathcal{T}$ be the binary relation defined as $x\mathcal{T}y \Leftrightarrow x\mathcal{S}y \wedge y\mathcal{S}^c x \quad (x, y \in X)$. Given a natural number $n \geq 2$, a n-tuple $(x_1, x_2, \dots, x_n) \in X^n$ is called a *$\mathcal{TS}$-cycle of order n* if we have $x_1\mathcal{T}x_2\mathcal{S}\dots\mathcal{S}x_n\mathcal{S}x_1$. Then we say that $\mathcal{S}$ is *consistent* if no $\mathcal{TS}$-cycle of order n appears, for any natural number $n \geq 2$.

Theorem 3 *(Suzumura extension theorem, 1976). Let X be a nonempty set. Let $\mathcal{S}$ be a binary relation defined on X. Associated to $\mathcal{S}$, let $\mathcal{T}$ be the binary relation defined as $x\mathcal{T}y \Leftrightarrow x\mathcal{S}y \wedge y\mathcal{S}^c x \quad (x, y \in X)$. Then, there exists a total preorder $\precsim$ on X that extends $\mathcal{S}$, and with its asymmetric part $\prec$ extending $\mathcal{T}$ too, if and only if the binary relation $\mathcal{S}$ is consistent.*

Proof. See Theorem 3 in [11]. □

Remark 2. A weaker version of Theorem 1 appears now as a corollary of Suzumura extension theorem. As a matter of fact, if $\mathcal{P}$ is an acyclic binary relation on X, the associated binary relation $\mathcal{T}$ defined as $x\mathcal{T}y \Leftrightarrow x\mathcal{P}y \wedge y\mathcal{P}^c x \quad (x, y \in X)$ coincides with $\mathcal{P}$ since $\mathcal{P}$ is acyclic, hence asymmetric. Therefore, for the relation $\mathcal{P}$, the condition of being consistent directly follows from acyclicity. So $\mathcal{P}$ can be extended to the asymmetric part of a total preorder.

Moreover, if we use now Szpilrajn extension theorem (Lemma 1) again, it follows that from this weaker version of Theorem 1 we may retrieve the whole version, since every asymmetric part of a total preorder is indeed irreflexive and transitive.

Definition 6. An acyclic binary relation $\mathcal{R}$ defined on a nonempty set X is said to be *representable* if it is extendable to the asymmetric part of a representable total preorder.

Definition 7. Given a binary relation $\mathcal{R}$ defined on a nonempty set X, a real-valued function $u : X \to \mathbb{R}$ is said to be a *pseudoutility* for $\mathcal{R}$ if $x\mathcal{R}y \Rightarrow u(x) < u(y)$ holds true for every $x, y \in X$.

Representable acyclic binary relations can be characterized in terms of a suitable modification of Sincov functional equation, as follows.

Theorem 4. *Let X be a nonempty set. Let $\mathcal{R}$ be an acyclic binary relation defined on X. The following statements are equivalent:*

(i) $\mathcal{R}$ is representable,
(ii) there exist bivariate functions $F : X \times X \to \mathbb{R}$ and $G : X \times X \to \{0, 1\}$ such that F satisfies the Sincov functional equation and $x\mathcal{R}y \Leftrightarrow F(x, y) \cdot G(x, y) > 0$ holds true for every $x, y \in X$,
(iii) there exists a pseudoutility function u for the given binary relation $\mathcal{R}$.

Proof. To prove that (i) $\Rightarrow$ *(ii)* we take a representable total preorder $\precsim$ on X, whose asymmetric part $\prec$ extends $\mathcal{R}$. By Proposition 3, there is a function $F : X \times X \to \mathbb{R}$ that satisfies the Sincov functional equation and $x \prec y \Leftrightarrow F(x, y) > 0 \quad (x, y \in X)$. Define now $G : X \times X \to \{0, 1\}$ as $G(x, y) = 1 \Leftrightarrow x\mathcal{R}y$ and $G(x, y) = 0$ otherwise $\quad (x, y \in X)$. We have that $x\mathcal{R}y \implies x \prec y \implies F(x, y) > 0$. Also $x\mathcal{R}y \implies G(x, y) = 1$. Therefore $x\mathcal{R}y \Rightarrow F(x, y) \cdot G(x, y) > 0$ holds true for every $x, y \in X$. Conversely, given $x, y \in X$, if $F(x, y) \cdot G(x, y) > 0$ it follows that $G(x, y) = 1$ by definition of G, so that $x\mathcal{R}y$ holds true. Hence $x\mathcal{R}y \Leftrightarrow F(x, y) \cdot G(x, y) > 0 \quad (x, y \in X)$.

To prove that (ii) $\Rightarrow$ *(iii)*, let $F : X \times X \to \mathbb{R}$ and $G : X \times X \to \{0, 1\}$ be such that F satisfies the Sincov functional equation and $x\mathcal{R}y \Leftrightarrow F(x, y) \cdot G(x, y) > 0 \quad (x, y \in X)$. Consider the binary relation $\precsim$ defined on X by declaring that $x \precsim y \Leftrightarrow F(x, y) \geq 0$. Since F satisfies the Sincov functional equation, we have that $F(x, x) = 0 = F(x, y) + F(y, x)$ holds true for every $x, y \in X$. Thus $\precsim$ is reflexive and total. Moreover, the fact $F(x, y) + F(y, z) = F(x, z) \quad (x, y, z \in X)$ immediately implies that $\precsim$ is transitive, hence it is indeed a total preorder. By definition, its asymmetric part $\prec$ satisfies that for any $x, y \in X$, $x \prec y$ holds if and only if $y \precsim x$ does not hold. Equivalently, $x \prec y$ if and only if $F(y, x) < 0$. This last fact, jointly with $F(x, y) + F(y, x) = 0$, is equivalent to say that $F(x, y) > 0$. By Proposition 3, the total preorder $\precsim$ is representable by a utility function $u : X \to \mathbb{R}$. Therefore, for any $x, y \in X$ we have that $x\mathcal{R}y \implies x \prec y \implies u(x) < u(y)$.

Finally, to prove that (iii) $\Rightarrow$ (i), we consider a pseudoutility u for $\mathcal{R}$. Observe now that the binary relation $\precsim$ on X given by $x \precsim y \Leftrightarrow u(x) \leq u(y)$ $(x, y \in X)$ is a total preorder, whose asymmetric part $\prec$ satisfies that $x \prec y \Leftrightarrow u(x) < u(y)$ $(x, y \in X)$. Hence $\prec$ is actually an extension of the given binary relation $\mathcal{R}$. Thus $\mathcal{R}$ is representable. □

Remark 3. An acyclic binary relation $\mathcal{R}$ defined on a nonempty set X may fail to be representable. A clear example is the asymmetric part of a non-representable linear order. By the way, the structure of non-representable linear orders has been analyzed in depth in [2]. Whenever X is finite or countable, any acyclic binary relation defined on X is representable because any total preorder on a finite or countable set is actually representable (see e.g. Theorem 1.4.8 in [3], or else [2] for further details).

Consider now a nonempty *finite* set X.

Definition 8. Let $\mathcal{R}$ be a binary relation on X. We say that $\mathcal{R}$ is an *arborescence* if the following conditions hold:

(i) $\mathcal{R}$ is irreflexive,
(ii) there exists a unique element $x_0 \in X$, called *root*, such that $x\mathcal{R}x_0$ does not hold for any $x \in X$,
(iii) for any $x \in X$ with $x \neq x_0$, there exists a *unique* $(k+1)$-tuple $(x_0, x_1, \ldots, x_k = x) \in X^{k+1}$, for some suitable $k \in \mathbb{N}$, such that $x_0\mathcal{R}x_1\mathcal{R}\ldots\mathcal{R}x_k$ holds true.

Remark 4. Notice that the uniqueness restriction arising in condition (iii), with respect to the $(k+1)$-tuple $(x_0, x_1, \ldots, x_k = x) \in X^{k+1}$, avoids that a given point x could be reached from x_0 by two different "sequences of branches".

Proposition 4. *Any arborescence is acyclic.*

Proof. Let $\mathcal{R}$ be an arborescence on X. Suppose that there is a n-cycle $y_1\mathcal{R}y_2\mathcal{R}\ldots\ \mathcal{R}y_n\mathcal{R}y_1$ in X as regards $\mathcal{R}$. Then, the condition iii) for x_0 and $x = y_1$ is no longer true, since for any $(k+1)$-tuple $(x_0, x_1, \ldots, x_k = y_1)$ with $x_0\mathcal{R}x_1\mathcal{R}\ldots\mathcal{R}x_k = y_1$ we have, repeating now the cycle, that the $(k+n+1)$-tuple $(x_0, x_1, \ldots, x_k = y_1, y_2, \ldots, y_n, y_1)$ also satisfies $x_0\mathcal{R}x_1\mathcal{R}\ldots\mathcal{R}x_k = y_1\mathcal{R}y_2\mathcal{R}\ldots\mathcal{R}y_n\mathcal{R}y_1$, in contradiction with the hypothesis of uniqueness. □

We introduce another equivalent way to define the notion of arborescence.

Proposition 5. *Let $\mathcal{R}$ be a binary relation on a nonempty finite set X with at least two elements. Then $\mathcal{R}$ is an arborescence if and only if the following conditions hold:*

(i) $\mathcal{R}$ is irreflexive,
(ii) there exists a unique element $x_0 \in X$ such that $x\mathcal{R}x_0$ does not hold for any $x \in X$ (in particular, the relation $\mathcal{R}$ is nonvoid),
(iii) for any $x, y, z \in X$, it holds true that $(y\mathcal{R}x \wedge z\mathcal{R}x) \Rightarrow y = z$.

Proof. Assume that $\mathcal{R}$ is an arborescence. If there exist $x, y, z \in X$ such that $y\mathcal{R}x \wedge z\mathcal{R}x$ holds true with $y \neq z$, then taking a $(k+1)$-tuple $(x_0, x_1, \ldots, x_k = y)$ with $x_0\mathcal{R}x_1\mathcal{R}\ldots\mathcal{R}x_k = y$ and another $l+1$-tuple $(x_0, y_1, \ldots, y_l = z)$ with $x_0\mathcal{R}y_1\mathcal{R}\ldots\mathcal{R}y_l = z$, we may construct two *different* tuples from x_0 to x, namely the $(k+2)$-tuple $(x_0, x_1, \ldots, x_k = y, x)$ for which we have $x_0\mathcal{R}x_1\mathcal{R}\ldots\mathcal{R}x_k = y\mathcal{R}x$ and the $(l+2)$-tuple $(x_0, y_1, \ldots, y_l = z, x)$ satisfying that $x_0\mathcal{R}y_1\mathcal{R}\ldots\mathcal{R}y_l = z\mathcal{R}x$. But this contradicts condition (iii) in Definition 8.

Conversely, suppose now that $\mathcal{R}$ satisfies the conditions in the statement of Proposition 5. Given $x \in X$ with $x \neq x_0$, by conditions (ii) and (iii) there exists a unique element $y \in X$ such that $y\mathcal{R}x$ holds true. If $y = x_0$ we are done. And if $y \neq x_0$, then with the same argument, there exists a unique element $z \in X$ for which $z\mathcal{R}y$ holds true. Again if $z = x_0$ we are done. Also, if $z \neq x_0$, there exists a unique element $t \in X$ for which $t\mathcal{R}z$ holds true. This process goes on until we arrive at x_0. This must compulsorily happen by condition (ii) and the fact of X being finite. So it is clear that condition (iii) in Definition 8 must hold true, too. This concludes the proof. □

Definition 9. Let X be a finite set and $\mathcal{R}$ a binary relation on X. Then $\mathcal{R}$ is said to be a *forest* if X can be split as a union of pairwise disjoint subsets, say $X = \bigcup_{i=1}^{n} X_n$, accomplishing the following conditions:

(i) The restriction of $\mathcal{R}$ to X_i is an arborescence, por any $i \in \{1, \ldots, n\}$,
(ii) If $i \neq j$ then $x_i\mathcal{R}x_j$ does not hold, for any $x_i \in X_i$, $x_j \in X_j$.

Proposition 6. *Any forest –and, in particular, any arborescence– is a representable acyclic binary relation.*

Proof. The fact of being acyclic is a direct consequence of Proposition 4 and Definition 9 (of the concept of a forest). Since the support set X is finite, the results follows now from Remark 3. In addition, a different alternative argument to prove the representability follows from Theorem 1, since $\mathcal{R}$ can be extended to the asymmetric part of a linear order on X. That linear order is a fortiori representable because X is finite (see e.g. Theorem 1.2.1 in [3]). Therefore $\mathcal{R}$ is also representable, by Definition 6. □

Now we analyze which conditions should be added to those in the statement of Theorem 4 in order to characterize arborescences and forests among acyclic binary relations, by using some functional equation.

Theorem 5. *Let X be a nonempty finite set. Let $\mathcal{R}$ be an acyclic binary relation defined on X. The following statements are equivalent:*

(i) $\mathcal{R}$ is an arborescence,
(ii) there exist bivariate functions $F : X \times X \to \mathbb{R}$ and $G : X \times X \to \{0, 1\}$ such that the following conditions are met:
- *(a) F satisfies the Sincov functional equation and $x\mathcal{R}y \Leftrightarrow F(x, y) \cdot G(x, y) > 0$ holds true for every $x, y \in X$,*
- *(b) there exists a unique $x_0 \in X$ such that $F(x, x_0) \cdot G(x, x_0) \leq 0$ for every $x \in X$,*

– *(c) for every $x, y, z \in X$ we have that $F(y,x) \cdot F(z,x) \cdot G(y,x) \cdot G(z,x) = [F(y,x) \cdot G(y,x)]^2 \cdot \delta(y,z)$, where δ stands here for the Kronecker delta function, that is, given $(a,b) \in X^2$, we have that $\delta(a,b) = 1$ if $a = b$, whereas $\delta(a,b) = 0$ otherwise.*

Proof. To prove that (i) $\implies$ (ii) we argue as in Theorem 4, so that again we consider a representable total preorder $\precsim$ on X, whose asymmetric part $\prec$ extends $\mathcal{R}$. Once more, by Proposition 3, there is a function $F : X \times X \to \mathbb{R}$ that satisfies the Sincov functional equation and $x \prec y \Leftrightarrow F(x,y) > 0 \quad (x, y \in X)$. Let now $G : X \times X \to \{0,1\}$ be given as $G(x,y) = 1 \Leftrightarrow x\mathcal{R}y$ and $G(x,y) = 0$ otherwise $(x, y \in X)$. Since $\mathcal{R}$ is acyclic, condition (ii)-(a) directly follows from the proof of Theorem 4. In addition, since $\mathcal{R}$ is an arborescence, there exists $x_0 \in X$ such that $x\mathcal{R}x_0$ never holds, for any $x \in X$. In other words, by definition of G, we have $G(x, x_0) = 0$ for every $x \in X$, so that condition (ii)-(b) is also satisfied. Finally, given $x, y, z \in X$, if $y = z$ the condition (ii)-(c) trivially follows. If $y \neq z$, we have that $\delta(y,z) = 0$. By condition (iii) in the statement of Proposition 5 we have that $y\mathcal{R}x$ or $z\mathcal{R}x$ fails to be true, so that $G(y,x).G(z,x) = 0$. Therefore the condition (ii)-(c) is always met.

Let us prove now that (ii) $\implies$ (i): The binary relation $\mathcal{R}$ is obviously irreflexive, since it is acyclic. By condition (ii)-(b) we have that there exists a unique x_0 such that $F(x,x_0) \cdot G(x,x_0) \leq 0$ holds for every $x \in X$. Equivalently, there exists a unique element $x_0 \in X$, for which $x\mathcal{R}x_0$ does not hold for any $x \in X$.

Finally, given any $x, y, z \in X$ such that both $y\mathcal{R}x$ and $z\mathcal{R}x$ hold true, we have that $F(y,x) \cdot G(y,x) > 0$ and also $F(z,x) \cdot G(z,x) > 0$. Hence, by condition (ii)-(c) it follows that $F(y,x) \cdot F(z,x) \cdot G(y,x) \cdot G(z,x) = [F(y,x) \cdot G(y,x)]^2 \cdot \delta(y,z)$, so that by simplifying we arrive at $F(z,x) \cdot G(z,x) = F(y,x) \cdot G(y,x)\delta(y,z)$. Thus $\delta(y,z) = 1$ a fortiori, since $F(z,x) \cdot G(z,x) > 0$. So we conclude that $y = z$. Therefore $\mathcal{R}$ is an arborescence by Proposition 5. □

4 Directed Acyclic Graphs and Incidence Matrices

Each result on binary relations of a finite set can immediately be interpreted in terms of Graph Theory, a branch of Discrete Mathematics. Basically, a *graph* consists of a finite set of vertices or points –also known as *nodes*– that are connected by arcs or lines –also known as *edges*–. In fact, some of the nodes can be pairwise related (or not), and we say that each pair of related nodes constitutes an edge of the graph. In addition, the edges may be directed or undirected, giving rise to the so-called *directed graphs*, where the edges have an orientation and are also said to be *directed edges* or *arrows*, as well as to *undirected graphs*, in which edges have no orientation at all.

Now we may observe that if X is a nonempty finite set and $\mathcal{R}$ is a binary relation on X, we can schematically represent $\mathcal{R}$ as a graph in which each node corresponds to each element in X, and an arrow is drawn from the node that represents the element $x \in X$ to the node that corresponds to $y \in X$ if and only if $x\mathcal{R}y$ holds true. Conversely, if we are given a directed graph, we immediately can interpret it as a binary relation on a finite set.

Definition 10. A *cycle* in a directed graph is an ordered tuple of nodes $(x_1, \ldots, x_k)$ such that there is an arrow from x_i to x_{i+1} for every $i < k$ and there is also one arrow from x_k to x_1. In the particular case in which $k = 1$ a 1-cycle is said to be a *loop*. A *directed acyclic graph* is a directed graph with no cycles.

(Notice that every directed acyclic graph can be interpreted as an acyclic binary relation on a nonempty finite set, and viceversa).

Definition 11. We say that a directed graph *admits a topological ordering* (also known as a *topological sorting* in this literature) if there exists a suitable linear order $\prec$ on the nodes of the graph such that it preserves the existing arrows. That is, if there is in the graph an arrow from the node x_i to the node x_j, then $x_i \prec x_j$ must hold true.

The following classical theorem is just a rephrasal of Theorem 1. It is a classical in Graph Theory, where several *sorting algorithms* have been introduced to get a topological sorting on a directed acyclic graph (see e.g. [6,7]). We should notice that the topological sorting on a directed acyclic graph *is not unique*, in general.

Theorem 6. *Any directed acyclic graph admits a topological sorting.*

It is a classical in Graph Theory, where *sorting algorithms* have been introduced to get a topological sorting on a directed acyclic graph (see e.g. [6,7]).

Another alternative way to deal with binary relations on nonempty finite sets comes from Matrix Theory. Thus, given a binary relation $\mathcal{R}$ on a set $X = \{x_1, \ldots, x_n\}$, we can visualize $\mathcal{R}$ by means of a suitable square matrix $n \times n$, called its incidence matrix. Needless to say, from such a matrix we can retrieve the binary relation $\mathcal{R}$ as well as its corresponding directed graph, already considered above. Conversely, from the graph we can easily get the corresponding matrix.

Definition 12. A $n \times n$ square matrix each of whose entries is either 0 or 1 is said to be an *incidence matrix*. Given a nonempty finite set X and a binary relation $\mathcal{R}$ on X, the *incidence matrix relative to the binary relation* $\mathcal{R}$ is the $n \times n$ matrix $M_{\mathcal{R}} = (m_{ij})$ with $m_{ij} = \chi_{\mathcal{R}}(x_i, x_j) \quad (i, j \in \{1, \ldots, n\})$. (Here $\chi_{\mathcal{R}}(x_i, x_j) = 1 \Leftrightarrow x_i \mathcal{R} x_j$. Otherwise $\chi_{\mathcal{R}}(x_i, x_j) = 0$.)

Let us analyze now how some properties of a binary relation $\mathcal{R}$ defined on a nonempty set can directly be observed by looking at its corresponding incidence matrix $M_{\mathcal{R}}$.

Proposition 7. *Let $\mathcal{R}$ be a binary relation defined on a nonempty finite set X. Let $M_{\mathcal{R}}$ be incidence matrix relative to $\mathcal{R}$. The following properties hold true.*

(i) $\mathcal{R}$ *is reflexive if and only if* $m_{ii} = 1$ *for every* $1 \leq i \leq n$.
(ii) $\mathcal{R}$ *is irreflexive if and only if* $m_{ii} = 0$ *for every* $1 \leq i \leq n$.
(iii) $\mathcal{R}$ *is asymmetric if and only if all the entries in the main diagonal of* $M_{\mathcal{R}}^2$ *are zeroes.*
(iv) If the cardinality of X *(henceforward denoted* $\#X$*) is* n, $\mathcal{R}$ *is acyclic if and only if for any natural number* k *with* $1 \leq k \leq n$, *all the entries in the main diagonal of* $M_{\mathcal{R}}^k$ *are zeroes.*
(v) If $\#X = n$, *and* $\mathcal{R}$ *is acyclic, then* $M_{\mathcal{R}}^n$ *is the null matrix.*
(vi) If $\#X = n$ *and* $\mathcal{R}$ *is acyclic, then* $I - M_{\mathcal{R}}$ *is a regular matrix such that* $(I - M_{\mathcal{R}})^{-1} = I + M_{\mathcal{R}} + \ldots + M_{\mathcal{R}}^{n-1}$.

Proof. Parts (i) and (ii) directly follow from the corresponding definitions.

Let us prove part (iii). Assume first that $\mathcal{R}$ is asymmetric. Then for every $i, j \in \{1, \ldots, n\}$ we have that $\chi_{\mathcal{R}}(x_i, x_j) \cdot \chi_{\mathcal{R}}(x_j, x_i) = m_{ij}m_{ji} = 0$. Hence the sum $\Sigma_{j=1}^n m_{ij}m_{ji} = 0$. But this sum is the i-th element in the main diagonal of $M_{\mathcal{R}}^2$. Conversely, if $\Sigma_{j=1}^n m_{ij}m_{ji} = 0$ then it is plain that $\chi_{\mathcal{R}}(x_i, x_j) \cdot \chi_{\mathcal{R}}(x_j, x_i) = 0$ for every $i, j \in \{1, \ldots, n\}$ so that $x_i \mathcal{R} x_j$ forces the negation of $x_j \mathcal{R} x_i$. Hence $\mathcal{R}$ is asymmetric.

To prove part (iv), first we assume that $\mathcal{R}$ is acyclic. Let $k \in$ be such that $1 \leq k \leq n$. Observe now that the $i-th$ term in the main diagonal of $M_{\mathcal{R}}^k$ consists of sums of products of the kind $m_{i_1 i_2} \cdot m_{i_2 i_3} \cdot \ldots \cdot m_{i_k i_{k+1}}$ with $i = i_1$ and also $i_{k+1} = i$. But, being $\mathcal{R}$ is acyclic, it is clear that all these products are null. Conversely, we may notice that the existence of a cycle on k elements, where $1 \leq k \leq n$, $\{x_{i1}, \ldots, x_{ik}\} \subseteq X^k$ such that $x_{i1} \mathcal{R} x_{i2} \mathcal{R} \ldots \mathcal{R} x_{ik} \mathcal{R} x_{i1}$ forces the i_1-th element in the main diagonal of $M_{\mathcal{R}}^k$ to be different from zero, in contradiction with the hypothesis of the statement.

To prove (v), notice that any entry in $M_{\mathcal{R}}^n$ consists of sums of products of the type $m_{i_1 i_2} \cdot m_{i_2 i_3} \cdot \ldots \cdot m_{i_n i_{n+1}}$. Since $\#X = n$ in the tuple $(i_1, i_2, \ldots, i_{n+1})$ a repetition occurs, so giving rise to a part of that tuple of the kind $(j_1, j_2, \ldots, j_k)$ with $k \leq n$ and $j_1 = j_k$. Therefore the product $m_{j_1 j_2} \cdot m_{j_2 j_3} \cdot \ldots \cdot m_{j_{k-1} j_k}$ is zero, and so is $m_{i_1 i_2} \cdot m_{i_2 i_3} \cdot \ldots \cdot m_{i_n i_{n+1}}$.

Let us conclude by proving part (vi). Since $M_{\mathcal{R}}^n$ is the null matrix by part (v), it follows that $(I - M_{\mathcal{R}}) \cdot (I + M_{\mathcal{R}} + \ldots + M_{\mathcal{R}}^{n-1}) = (I + M_{\mathcal{R}} + \ldots + M_{\mathcal{R}}^{n-1}) - (M_{\mathcal{R}} + \ldots + M_{\mathcal{R}}^{n-1} + M_{\mathcal{R}}^n) = I - M_{\mathcal{R}}^n = I$. So $I - M_{\mathcal{R}}$ is a regular matrix whose inverse equals $I + M_{\mathcal{R}} + \ldots + M_{\mathcal{R}}^{n-1}$. □

Theorem 7. *Let* $\mathcal{R}$ *be an acyclic binary relation defined on a nonempty finite set* X. *Then* $\mathcal{R}$ *is an arborescence if an only if there is a unique* $i \in \{1, \ldots, n\}$ *such that all the entries in the* i*-th column of* $M_{\mathcal{R}}$ *are zeroes, while all the entries in the* i*-th row of* $(I - M_{\mathcal{R}})^{-1}$ *equal* 1.

Proof. Assume first that $\mathcal{R}$ is an arborescence. Let $X = \{x_1, \ldots, x_n\}$. By Definition 8, there exists an element $x_i \in X$ such that $x_j \mathcal{R} x_i$ does not hold for any $x_j \in X$. Therefore $m_{ji} = 0$ for every $1 \leq j \leq n$, so that the i-th column of $M_{\mathcal{R}}$ is null. Moreover, given $j \neq i$, there is a unique $k + 1$-tuple $(x_i = x_{i0}, x_{i1}, \ldots, x_{ik} = x_j) \in X^{k+1}$, for some suitable $k \in \mathbb{N}$, such that $x_i = x_{i0} \mathcal{R} x_{i1} \mathcal{R} \ldots \mathcal{R} x_{ik} = x_j$ holds true. Therefore the entry in the row i and column j of $M_{\mathcal{R}}^k$ must be 1 by the uniqueness hypothesis. Since that k is also unique, we have that all the entries in the i-th row of $(I + M_{\mathcal{R}} + \ldots + M_{\mathcal{R}}^{n-1})$ are 1, so that by parts (v) and (vi) of Proposition 7 we conclude that the all the terms in the i-th row of $(I - M_{\mathcal{R}})^{-1}$ equal 1.

To prove the converse, first we notice that condition (i) in Definition 8 is trivially met because $\mathcal{R}$ is acyclic and, in particular, irreflexive. In addition, since all the entries in the i-th column of $M_{\mathcal{R}}$ are zeroes for a unique $i \in \{1, \ldots, n\}$, the condition (ii) in Definition 8 is accomplished by taking $x_0 = x_i$. Moreover, because all the entries in the in the i-th row of $(I - M_{\mathcal{R}})^{-1}$ equal 1, and taking into account that, by part vi) in Proposition 7, the equality $(I - M_{\mathcal{R}})^{-1} = I + M_{\mathcal{R}} + \ldots + M_{\mathcal{R}}^{n-1}$ holds true, we observe that being $x_0 = x_i$ and $x_j = x$, there exists a unique $1 \leq k \leq n$ such that the entry in the i-th row and j-th column of $M_{\mathcal{R}}^k$ equals 1, whereas for any other $l \neq k$ the entry in the i-th row and j-th column of $M_{\mathcal{R}}^k$ equals 0. Hence, there exists a unique $(k+1)$-tuple $(x_i = x_{i0}, x_{i1}, \ldots, x_{ik} = x_j) \in X^{k+1}$, with $x_{i0} \mathcal{R} x_{i1} \mathcal{R} \ldots \mathcal{R} x_{ik}$ holding true. So the condition iii) in Definition 8 is also accomplished. □

5 Concluding Remarks

Acyclic binary relations have been considered under different points of view, paying an special attention to the use of some suitable functional equation. When the set on the relations are considered is finite, the parallelism between binary relations, graphs and incidence matrix has been shown. Here, any result arising in one of those approaches –namely: abstract binary relations, directed acyclic graphs, and incidence matrices– immediately has a translation into any other one of those settings.

Among open problems within this theory, we point out that, as far as we know, the question of characterizing all those acyclic binary relations on a set that fail to admit a pseudoutility representation has not been solved yet.

References

1. Alcantud, J.C.R.: Weak utilities from acyclicity. Theor. Decis. **47**(2), 185–196 (1999)
2. Beardon, A.F., Candeal, J.C., Herden, G., Induráin, E., Mehta, G.B.: The non-existence of a utility function and the structure of non-representable preference relations. J. Math. Econom. **37**, 17–38 (2002)
3. Bridges, D.S., Mehta, G.B.: Representations of Preference Orderings. Springer, Berlin (1995). https://doi.org/10.1007/978-3-642-51495-1
4. Campión, M.J., Catalán, R.G., Induráin, E., Ochoa, G.: Reinterpreting a fuzzy subset by means of a Sincov's functional equation. J. Intell. Fuzzy Syst. **27**, 367–375 (2014)
5. Hansson, B.: Choice structures and preference relations. Synthese **18**, 443–458 (1968)
6. Kahn, A.B.: Topological sorting of large networks. Commun. ACM **5**(11), 558–562 (1962)

7. Knuth, D.E., Szwarcfiter, J.L.: A structured program to generate all topological sorting arrangements. Inform. Process. Lett. **2**(6), 153–157 (1974)
8. Kruskal, J.B.: On the shortest spanning subtree of a graph and the traveling salesman problem. Proc. Amer. Math. Soc. **7**, 48–50 (1956)
9. Rodríguez-Palmero, C.: A representation of acyclic preferences. Econom. Lett. **54**, 143–146 (1997)
10. Sincov, D.M.: Über eine Funktionalgleichung. Arch. Math. Phys. **6**(3), 216–227 (1903)
11. Suzumura, K.: Remarks on the theory of collective choice. Economica **43**(172), 381–390 (1976)
12. Szpilrajn, E.: Sur l' extension de l' ordre partiel. Fund. Math. **16**, 386–389 (1930)

A Comparison Between NARX Neural Networks and Symbolic Regression: An Application for Energy Consumption Forecasting

Ramón Rueda Delgado(✉), Luis G. Baca Ruíz, Manuel Pegalajar Cuéllar(✉), Miguel Delgado Calvo-Flores, and María del Carmen Pegalajar Jiménez

Department of Computer Science and Artificial Intelligence, University of Granada, C/. Pdta. Daniel Saucedo Aranda s.n., Granada, Spain
{ramonrd,bacaruiz}@ugr.es, {manupc,mdelgado,mcarmen}@decsai.ugr.es

Abstract. Energy efficiency in public buildings has become a major research field, due to the impacts of the energy consumption in terms of pollution and economic aspects. For this reason, governments know that it is necessary to adopt measures in order to minimize the environmental impact and saving energy. Technology advances of the last few years allow us to monitor and control the energy consumption in buildings, and become of great importance to extract hidden knowledge from raw data and give support to the experts in decision-making processes to achieve real energy saving or pollution reduction among others. Prediction techniques are classical tools in machine learning, used in the energy efficiency paradigm to reduce and optimize the energy using. In this work we have used two prediction techniques, symbolic regression and neural networks, with the aim of predict the energy consumption in public buildings at the University of Granada. This paper concludes that symbolic regression is a promising and more interpretable results, whereas neural networks lack of interpretability take more computational time to be trained. In our results, we conclude that there are no significant differences in accuracy considering both techniques in the problems addressed.

Keywords: Energy efficiency · Symbolic regression
Neural networks · Genetic programming

1 Introduction

Growth in population and the necessity of saving the energy consumption in the building sector have become a major concern in different governments. The increase of energy consumption is related not only with the economy of each country, but also with the environmental impact. Thus, the literature gathers plenty of works that discuss about the high number of studies that show the

J. Medina et al. (Eds.): IPMU 2018, CCIS 855, pp. 16–27, 2018.
https://doi.org/10.1007/978-3-319-91479-4_2

large energy consumption registered in public buildings. For example, in USA [24] the HVAC (heating, ventilation, and air conditioning) is the 20% of the total consumption. Besides, other countries such as China [9] or Iran [30] show that their energy consumption has been increased in more than 10% from China and in 1.54 times from Iran in the past 20 years. On the other hand, Diakiki et al. [12] argue that the building sector in the European Union (EU) has increased it consumption until a 56%. As a consequence, European energy policy focuses its interests in the preservation and efficient usage of energy in public buildings, to optimize energy consumption or minimize the environmental impact.

In this way, there are different works that relate the energy consumption with global warming. Jenkins et al. [18] use a model in order to investigate how the climate change will affect the energy demand for heating and cooling in 2030 in different locations of UK, concluding that the heating consumption is dominant regarding energy demand. Besides, Braun et al. [8] summarize that global warming will be directly related with the consumption in offices, buildings and home. More specifically, in future years global warming will imply a reduction in the energy used for colder climates and an electricity consumption increase in warmer climates. Researchers from different countries have been aware that having energy consumption data and external information could be used in order to minimize environmental impacts or saving energy.

Nowadays, the use of different technologies to monitor and control the energy consumption in buildings is widely spread. Despite of the existence of several types of sensors, the selection of the most appropriate has become in a hard task [12]. The objective of the study should motivate the technology selection considering environmental, energy, financial or social factors. For this reason, the installation of suitable technologies in buildings to monitor the energy consumption has provided useful and descriptive information [1]. This data can furnish knowledge about: consumed energy, temperature, humidity, wind speed, etc. Nevertheless, the diversity of these data implies a previous application of machine learning techniques in order to preprocess and extract hidden knowledge of the data. Once the data are stored and available to be used, a big amount of applications could be applied, such as: creation of consumption profiles to estimate the hourly and seasonal energy needed in the building sector [17], methods for identifying anomalies in energy consumption [15] or making recommendation systems to save household electricity [2].

The computational advance together with the development of diverse algorithms in the machine learning paradigm provide us an optimal exploitation of the energy data, in terms of knowledge extraction and data behaviour understanding. In the literature there are several techniques that make to achieve this purpose easily. Ruiz et al. [29] use neural networks and evolutionary algorithms to predict the energy consumption in public buildings, Zhao et al. [34] show statistical and artificial intelligence methods such as regression analysis to predict energy consumption. Ekici and Aksoy [13] describe the use of artificial neural network to predict the heating energy consumption in different buildings using different data sources such as orientation, building transparency ratio and insulation thickness. Besides, Kumar et al. [20] apply three kind of time series models to forecast the energy consumption in India, and many other works [4,10,19,33]. In this work, we focus on the use of prediction techniques such as neural networks and symbolic regression to infer, model and predict the energy consumption in public buildings of the University of Granada, combining information from energy consumption together with exogenous variables such as ambient temperature. We are interested in the estimation of the energy consumption from the weekdays using the own information of the consumption registered in a specific building and also considering the temperature, with the goal of verify whether the energy consumption registered in a specific building is related with external variables. On the other hand, a comparison between both neural networks and symbolic regression techniques will be carried out.

This paper is organized as follows: Sect. 2 describes the methodologies used in this work: symbolic regression and neural networks. Section 3 introduces the experiments carried out in order validate each technique and discusses the advantages of each one. Finally, Sect. 4 concludes with a discussion and future works.

2 Prediction Paradigm

Prediction techniques are classical tools used in data science since several years ago in different areas, such as: *medicine* –using hemodynamic prediction try to predict and reduce abdominal aortic aneurysm diseases [23] or to predict drug responses in cancer research using regressions or support vector machine techniques [6]–, *economy* – whereas Lam et al. [21] use neural networks techniques in order to predict the financial performance of different companies, White [32] uses neural network techniques and learning techniques so as to search non-linear regularities in asset price movements– or *environmental sciences* –Holger and Graeme [22] use neural networks models to predict water resources variables –.

As it has been mentioned before, we can verify that in the literature exists several applications of prediction tools and there are two basic pillars in order to solve prediction problems: neural networks techniques and regression.

2.1 NARX and NARX-GA Neural Networks

Artificial neural networks (ANN) are a kind of methodology inspired by the biological neural networks that simulate animal brains. An ANN is based on a set of connected artificial neurons where each connection sends signals to other neuron. The connection developed between all of these neurons take different weights during the learning process and will determine the quality of the solution found. Besides, there are different types of neural networks techniques that include hidden layers or back-propagation.

In the real world, there are several kind of problems, which can involve different static or dynamic parameters that model a certain pattern of behaviour; so the selection of the more affordable method will be fundamental in the search solution. Particularly as the problem addressed in this work make use of historical data of energy consumption and external data, such as temperature, to predict energy consumption in public buildings. Motivated by this fact, a NARX neural network is used in this work due to its nonlinear model is able to work with dynamic inputs represented by time series, using the p past values of the time series and another external series, that can be single or multidimensional. The whole structure of NARX neural network and the interactions between the neurons in the hidden layer is explained in detail in [29]. As first approximation we have used the *Levenberg-Marquardt backpropagation* (LMBP) method in order to estimate the weights of the models. Despite being a faster approximation, the results achieved are not promising. For this reason, a genetic algorithm is used such as alternative in order to optimize the weights of the models, instead of LMBP, which is explained with detail in [28].

The evolutionary process used in the NARX model using a genetic algorithm, is based on the CHC algorithm [14] and consists of a hybrid algorithm, using a genetic algorithm and local search in order to create a kind of evolutionary algorithm commonly known as memetic algorithm. As it is known, in a genetic algorithm it is necessary to find out a representation of the individual; in our case, an individual represents the structure of a neural network, encoded by a vector which stores the inputs, outputs and the hidden layer neuron weights. Once the individual representation is created, the algorithm begins creating a random set of individuals. After that, a local search take place in order to improve the individual characteristics based on LM method. After that, each individual in the population are evaluated by a MSE measure –see formula 2– as fitness function. Once the value of the solution found is calculated, a quality criterium is carried out to determine if it is necessary to apply an incest prevention –like is used in the CHC algorithm (see [14] for more detail)– in order to delay convergence. Finally, the memetic algorithm will return the best neural network configuration ready to predict the energy consumption.

2.2 Symbolic Regression

There are different methods in the literature that solve modelling [11] or prediction [25,31] problems, but the most used is known as regression analysis. Regression analysis is a mathematical methodology used by researchers in order to fit a functional model between the dependent and independent variables. An example of a regression problem can be seen in formula 1, where we can identify the following elements: a set of input data $\bar{x} = (x_1, x_2, ..., x_n)$, a set of output data $\bar{y} = (y_1, y_2, ..., y_m)$ and a set of parameters that depend of the model hypothesis f $\bar{w} = (w_1, w_2, ..., w_n)$. In order to find out the best combination between the dependent and independent variables of the problem, regression analysis uses an error that try to minimize; usually this error is the Mean Square Error (MSE). Once regression coefficients are obtained, a prediction equation can be used in order to predict the value of a continuous target as a function.

$$\bar{y} = f(\bar{x}, \bar{w}) \tag{1}$$

Furthermore, the literature shows different kind of regression problems: Whereas in some issues all the parameters are known in advance and the problem consists of combining the set of variables until find out the optimal solution [5,8], other works [3,16] argue that singularities could appear in the problem definition, such as the unawareness of some variables of the problem.

On the other hand, Symbolic Regression is a kind of regression that generalize the process of regression analysis, assuming that the parameters $\bar{w}$ and f are unknown in advance, and try to find these parameters using a set of atomic operators as for example $(+, -, *, /, cos, sin, tan)$, and the goal consist of combining the input data $\bar{x}$ with the mathematical operators in order to approximate a function $\tilde{f}$ that minimize an error measure such as $||\bar{y} - \tilde{f}(\bar{x}, \bar{w})||$. Despite being a suitable method to find the algebraic expression that better models the output data $\bar{y}$ from input data $\bar{x}$, it is a NP-hard problem that needs the use of meta-heuristics to be solved. The classic algorithm used in the literature is genetic programming [3]. Genetic programming is a supervised learning method based on biological evolution that try to imitate the behaviour of the natural selection, using a population of individuals –traditionally encoded by tree structures– that should interact between them in order to combine the best genes of each candidate and obtain an individual that represent the best characteristics of all of them. In our case, each individual represents an algebraic expression encoded by a Straight Line Program and the goal is to combine the population using a set of operators.

Straight Line Programs (SLP) are a kind of grammar based on Straight Line Grammars [7] –which are a type of non-recursive grammar that allow to generate a unique expression– that are able to encode algebraic expressions in a linear representation. This linear representation can be seen as a table where each row of the table is a production rule where the antecedent is a non-terminal symbol and the consequent is a concatenation of mathematical operators, terminal symbols (such as constants or variables) or non-terminal symbols (references to other row of the table). In this way, a SLP structure is able to encode a set of grammar rules that can generate an algebraic expression and will be the representation used to encode algebraic expression in the genetic procedure.

Thereupon, the main components of a genetic algorithm are presented below (the paper [26] explains in detail each operator, for each tree and SLP representation scheme):

- The process start out selecting a set of individuals, that will behave as parents. A **tournament selection** is used as operator that select the best individuals of the population.
- Once the best individuals have been selected, a **crossover operator** is applied over each pair of individuals. This operator tries to combine the genetic material of the two parents in order to obtain a new individual that encode the best characteristics of each parents.
- After that, in order to imitate the behaviour of the nature and establish a balance between the convergence and the exploration of our algorithm, a **mutation operator** is carried out with a low probability.

- Finally, the new population replaces the previous individuals. Whereas in a stationary scheme, the individual with better fitness value is included in the previous population, in a generational scheme all the new population will replace the previous population. We emphasize that the fitness value in the genetic procedure is the Mean Square Error (see formula 2).

$$MSE = \frac{1}{m}\sum_{i=1}^{m}(\hat{y}_i - y_i)^2 \tag{2}$$

In addition, it is necessary to notice that usually the algebraic expressions are encoded by trees in genetic algorithms, but in [26, 27] we show that there are other promising alternatives to encode algebraic expression like Straight Line Programs. The main potential of SLP is its linear structure, that allows our genetic algorithm to avoid a premature convergence and explore efficiently the search space.

3 Experiments

In this section we describe the experiments carried out in order to compare the results obtained with Symbolic Regression and a NARX Neural Network with or without use a genetic algorithm as training method. The main goal of this experimentation consists on predicting the energy consumption of a specific day of the week using the energy used in the past 4 days and the temperature registered for the last day. Combining all of this information we are interested on estimating the energy consumption from the following days and, on the other hand, comparing and establishing differences considering accuracy, interpretability in the solutions found and training computational time.

In order to compare these algorithms, we have used a set of real consumption datasets of different buildings of the University of Granada. These data come from an automated system installed in the University of Granada that registers the energy consumption from different buildings measured hourly. In this work we have used the data of 8 different buildings. For confidentiality reasons we will name these buildings as b_1 to b_8. Despite having consumption data of these buildings, we cannot apply our algorithm over these datasets because the database saves raw data which could contains noise or incomplete data. For this reason, as a previous step of the experimentation, we need to preprocess the data as follows:

- On one hand, we need to remove breaks or irregular patterns in the data, using a linear interpolation.
- After that, we are interested in the aggregation of the energy consumption daily, in order to apply the methods explained in this work.
- Finally, we have normalized the data between $[0, 1]$ –using the formula 3– because we want to guarantee that there are no variables which may shrink the search.

$$y_{normalized} = \frac{y - y_{min}}{y_{max} - y_{min}} \tag{3}$$

After the previous preprocessing, we have available around 1000 instances from each building. After that, we use a 70% of the data as train data and the remaining 30% is used as test data. In this way, our model is trained with 700 instances and is able to predict the energy consumption from 300 days.

An example of the data available to use in this experimentation for each building, once the data has been preprocessed, is shown in Fig. 1. Thereupon, the application of both Symbolic Regression and Neural Networks algorithms are carried out with the experimental configuration shown in Tables 1 and 2:

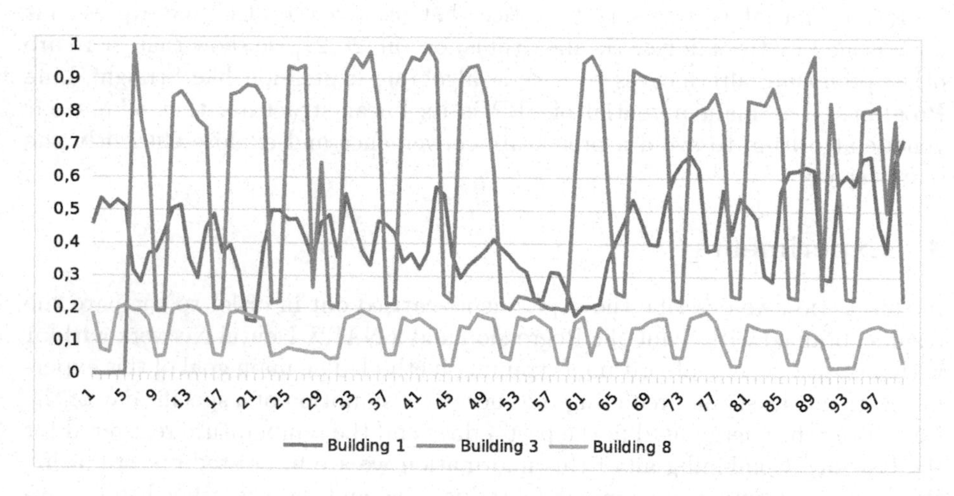

Fig. 1. Example of the energy consumption registered from 100 days from 3 buildings.

In order to verify the models robustness, we have divided each dataset into two subsets: training (70%) in order to build the model and test (30%) for validation, running each experimentation 5 times. Table 3 gathers the results achieved by NARX Neural Network with LMBP and using the genetic algorithm, and Symbolic Regression using the generational genetic algorithm. The first column of the table shows the *building id* and the following columns show the average of mean square error for each algorithm –represented as *AF (Average Fitness)*– the computational time measured in seconds –shown as *AT (Average Time)*– and the number of neurons and delay used in both neural networks: NARX with LMBP and NARX using GA. Finally, the expression size –contains the number of mathematical operators used in the expression– found by the symbolic regression algorithm is shown in the last column in order to present the interpretability of the model.

Considering the results shown in Table 3 we can verify that, although NARX neural networks need a small amount of time to find a solution, it shows a worse performance than Symbolic Regression using SLP in all cases. On the other hand, attending to NARX using genetic algorithms, these results are better in terms of precision. Nevertheless, the MSE found by symbolic regression is near

Table 1. Experimental configuration for NARX and NARX-GA.

	NARX	NARX-GA
Hidden neurons	[2, 20]	[2, 20]
Memory/Delay	[1,21]	5
Function	LM backpropagation	LM backpropagation
Minimum gradient	10^{-7}	10^{-7}
Learning rate	$[10^{-3}, 10^{10}]$	$[10^{-3}, 10^{10}]$
ν decrease	0.1	0.1
ν increase	10	10
Crossover rate	–	90%
Mutation rate	–	10%
N. evaluations	–	100
Population size	–	25

Table 2. Experimental configuration for genetic algorithm using SLP - Symbolic Regression

Symbolic regression - GGASLP	
Population size	180
SLP size	15
Crossover rate	90%
Mutation rate	10%
Function set F	$\{+, -, *, /, sqrt, pow, \exp, \sin, \cos, \tan, \log, \min, \max\}$
Constants	$\{k_1, k_2, k_3, k_4, k_5\}$
Variables	$\{x_1, x_2, x_3, x_4, t_4\}$
Number of evaluation	40000

Table 3. Results of 8 buildings using NARX, NARX-GA and SLP.

	NARX		NARX-GA		Delay	Neurons	SLP		
	AF	AT	AF	AT			AF	AT	Size
1	1.33	88.29	0.011	1206.54	7	4	0.028	554.01	14
2	1.31	91.26	0.008	1582.4	7	2	0.037	491.44	14
3	1.24	84.68	0.01	1095.07	7	2	0.025	477.24	15
4	1.4	85.77	0.01	1523.57	5	14	0.026	511.56	13
5	0.71	87.83	0.005	1040.45	5	2	0.008	489.66	11
6	0.79	88.28	0.003	1556.36	5	2	0.023	389.03	10
7	1.17	91.45	0.007	1582.33	7	2	0.024	467.68	12
8	0.88	82.69	0.007	1128.63	8	2	0.024	549.43	12

to 0 and the execution time in all cases is 2 times less in symbolic regression than with neural networks. Indeed, neural networks have a better performance than regression techniques in prediction problems, but is necessary to emphasize that in some cases the precision goes to the background, overall when the aim of the solution found should have the purpose of helping to the expert in the decision-making paradigm. In this way, the black-box model of neural networks, implies a low interpretability of the results. On the other hand, symbolic regression is able to find good approaches in a considerably less time than using neural networks, and also the algebraic expressions found, in contrast with the black-box model of the neural networks, are more interpretable; as is shown in formulas 4 to 6. We emphasize that the maximum size of an algebraic expression equals 15 operators, so that these SLP solutions involve a high interpretability in the decision-making problem to assess the expert.

$$\begin{aligned} &y = max(x_4, max(x_4, ({}^{x_1}/_{1.35})))^{a} \\ &where\, a = (x_3 + 0.63) * exp(x_3) * ({}^{x_1}/_{0.91})^{(max(x_4,({}^{x_1}/_{1.35}))^{max(x_4,1.67)})} \end{aligned} \tag{4}$$

$$y = \frac{((max(x_5, x_4)^{((((x_4^{0.21})^{(((x_4^{0.21})/0.18)^{x_2})})/0.18)^{x_2})})^{(x_3^{x_2})})^{((x_3^{x_2})^{x_2})}}{0.95} \tag{5}$$

$$y = \frac{tan(max(sqrt((x_4 * 0.38)), sin(x_4)))/min((x_3 * (1.35/x_2)), 1.01)}{(1.06 + x_1)^{x_2}} \tag{6}$$

4 Conclusions

In this paper we have studied the use of different alternatives to predict the energy consumption of the working days in different public buildings of the University of Granada: symbolic regression and NARX neural networks trained with LMBP and with genetic algorithms. In order to achieve this prediction we have been interested in the use of both internal (energy consumption) and exogenous (temperature) information.

Our experimentation shows that the use of the NARX model with LMBP is not as promising as the other alternatives. On the other hand, NARX-GA and SLP methods seem to be able to reach good performance. Indeed, NARX neural networks, using the genetic algorithm, work better than SLP in terms of precision, but SLP has achieved similar solutions in 2 times less than neural networks. Nevertheless, the main advantage of symbolic regression over NARX Neural Networks is the interpretability of the solutions found. In this way, whereas it is difficult to understand the solution of a neural network –due to its black-box model performance–, symbolic regression brings us the solution in terms of an algebraic expression, which will be of assistance to the expert in the decision-making problem.

Acknowledgements. This work has been supported by the project TIN201564776-C3-1-R. We thank reviewers for their constructive comments and useful ideas for future works.

References

1. Ahmad, M.W., Mourshed, M., Mundow, D., Sisinni, M., Rezgui, Y.: Building energy metering and environmental monitoring a state-of-the-art review and directions for future research. Energy Build. **120**, 85–102 (2016)
2. de Almeida, A., Fonseca, P., Schlomann, B., Feilberg, N.: Characterization of the household electricity consumption in the eu, potential energy savings and specific policy recommendations. Energy Build. **43**, 1884–1894 (2011). http://www.sciencedirect.com/science/article/pii/S0378778811001058
3. Alonso, C.L., Montaa, J.L., Puente, J., Borges, C.E.: A new linear genetic programming approach based on straight line programs: some theoretical and experimental aspects. Int. J. Artif. Intell. Tools **18**, 757–781 (2009). http://www.worldscientific.com/doi/abs/10.1142/S0218213009000391
4. Arghira, N., Ploix, S., Făgărăşan, I., Iliescu, S.S.: Forecasting energy consumption in dwellings. In: Dumitrache, L. (ed.) Advances in Intelligent Control Systems and Computer Science. AISC, vol. 187, pp. 251–264. Springer, Heidelberg (2013). https://doi.org/10.1007/978-3-642-32548-9_18
5. Arregi, B., Garay, R.: Regression analysis of the energy consumption of tertiary buildings. Energy Procedia **122**, 9–14 (2017). http://www.sciencedirect.com/science/article/pii/S1876610217328886
6. Azuaje, F.: Computational models for predicting drug responses in cancer research. Brief. Bioinform. **18**, 820–829 (2017). http://dx.doi.org/10.1093/bib/bbw065
7. Benz, F., Kötzing, T.: An effective heuristic for the smallest grammar problem. In: Genetic and Evolutionary Computation Conference, GECCO 2013, Amsterdam, The Netherlands, 6–10 July 2013, pp. 487–494 (2013). http://doi.acm.org/10.1145/2463372.2463441
8. Braun, M., Altan, H., Beck, S.: Using regression analysis to predict the future energy consumption of a supermarket in the UK. Appl. Energy **130**, 305–313 (2014). http://www.sciencedirect.com/science/article/pii/S0306261914005674
9. Cai, W., Wu, Y., Zhong, Y., Ren, H.: China building energy consumption: situation, challenges and corresponding measures. Energy Policy **37**, 2054–2059 (2009). http://www.sciencedirect.com/science/article/pii/S0301421508007398
10. Daut, M.A.M., Hassan, M.Y., Abdullah, H., Rahman, H.A., Abdullah, M.P., Hussin, F.: Building electrical energy consumption forecasting analysis using conventional and artificial intelligence methods: a review. Renew. Sustain. Energy Rev. **70**, 1108–1118 (2017). http://www.sciencedirect.com/science/article/pii/S1364032116310619
11. Cochrane, D., Orcutt, G.H.: Application of least squares regression to relationships containing auto-correlated error terms. J. Am. Stat. Assoc. **44**, 32–61 (1949)
12. Diakaki, C., Grigoroudis, E., Kolokotsa, D.: Towards a multi-objective optimization approach for improving energy efficiency in buildings. Energy Build. **40**, 1747–1754 (2008). http://www.sciencedirect.com/science/article/pii/S0378778808000649
13. Ekici, B.B., Aksoy, U.T.: Prediction of building energy consumption by using artificial neural networks. Adv. Eng. Softw. **40**, 356–362 (2009). http://www.sciencedirect.com/science/article/pii/S0965997808001105
14. Eshelman, L.J.: The CHC adaptive search algorithm: how to have safe search when engaging in nontraditional genetic recombination. Found. Genet. Algorithms **1**, 265–283 (1991). http://www.sciencedirect.com/science/article/pii/B9780080506845500203

15. Fontugne, R., Ortiz, J., Tremblay, N., Borgnat, P., Flandrin, P., Fukuda, K., Culler, D., Esaki, H.: Strip, bind, and search: a method for identifying abnormal energy consumption in buildings. In: Proceedings of the 12th International Conference on Information Processing in Sensor Networks, New York, NY, USA, pp. 129–140 (2013)
16. Haeri, M.A., Ebadzadeh, M.M., Folino, G.: Statistical genetic programming for symbolic regression. Appl. Soft Comput. **60**, 447–469 (2017). http://www.sciencedirect.com/science/article/pii/S1568494617303939
17. Heiple, S., Sailor, D.J.: Using building energy simulation and geospatial modeling techniques to determine high resolution building sector energy consumption profiles. Energy Build. **40**, 1426–1436 (2008). http://www.sciencedirect.com/science/article/pii/S0378778808000200
18. Jenkins, D., Liu, Y., Peacock, A.: Climatic and internal factors affecting future UK office heating and cooling energy consumptions. Energy Build. **40**, 874–881 (2008). http://www.sciencedirect.com/science/article/pii/S0378778807001880
19. Kalogirou, S.A., Bojic, M.: Artificial neural networks for the prediction of the energy consumption of a passive solar building. Energy **25**, 479–491 (2000)
20. Kumar, U., Jain, V.: Time series models (grey-markov, grey model with rolling mechanism and singular spectrum analysis) to forecast energy consumption in India. Energy **35**, 1709–1716 (2010). http://www.sciencedirect.com/science/article/pii/S0360544209005416
21. Lam, M.: Neural network techniques for financial performance prediction: integrating fundamental and technical analysis. Decis. Support Syst. **37**, 567–581 (2004). http://www.sciencedirect.com/science/article/pii/S0167923603000885
22. Maier, H.R., Dandy, G.C.: Neural networks for the prediction and forecasting of water resources variables: a review of modelling issues and applications. Environ. Model. Softw. **15**, 101–124 (2000). http://www.sciencedirect.com/science/article/pii/S1364815299000079
23. Paramasivam, V., Yee, T.S., Dhillon, S.K., Sidhu, A.S.: A methodological review of data mining techniques in predictive medicine: an application in hemodynamic prediction for abdominal aortic aneurysm disease. Biocyber. Biomed. Eng. **34**, 139–145 (2014). http://www.sciencedirect.com/science/article/pii/S0208521614000266
24. Prez-Lombard, L., Ortiz, J., Pout, C.: A review on buildings energy consumption information. Energy Build. **40**, 394–398 (2008). http://www.sciencedirect.com/science/article/pii/S0378778807001016
25. Ranjan, M., Jain, V.: Modelling of electrical energy consumption in Delhi. Energy **24**, 351–361 (1999)
26. Rueda, R., Cuéllar, M.P., Delgado, M., Pegalajar, M.: Preliminary evaluation of symbolic regression methods for energy consumption modelling. In: Proceedings of the 6th International Conference on Pattern Recognition Applications and Methods, ICPRAM 2017, Porto, Portugal, 24–26 February, pp. 39–49 (2017). https://doi.org/10.5220/0006108100390049
27. Rueda Delgado, R., Ruiz, L.G.B., Jimeno-Sáez, P., Cuellar, M.P., Pulido-Velazquez, D., Del Carmen Pegalajar, M.: Experimental evaluation of straight line programs for hydrological modelling with exogenous variables. In: Martínez de Pisón, F.J., Urraca, R., Quintián, H., Corchado, E. (eds.) HAIS 2017. LNCS (LNAI), vol. 10334, pp. 447–458. Springer, Cham (2017). https://doi.org/10.1007/978-3-319-59650-1_38

28. Ruiz, L., Rueda, R., Cullar, M., Pegalajar, M.: Energy consumption forecasting based on Elman neural networks with evolutive optimization. Expert Syst. Appl. **92**, 380–389 (2018). http://www.sciencedirect.com/science/article/pii/S0957417417306565
29. Ruiz, L.G.B., Cuéllar, M.P., Calvo-Flores, M.D., Jiménez, M.D.C.P.: An application of non-linear autoregressive neural networks to predict energy consumption in public buildings. Energies **9**, 684 (2016)
30. Sadeghi, H., Zolfaghari, M., Heydarizade, M.: Estimation of electricity demand in residential sector using genetic algorithm approach. Int. J. Ind. Eng. Prod. Res. **22**, 43–50 (2011)
31. Tso, G.K., Yau, K.K.: Predicting electricity energy consumption: a comparison of regression analysis, decision tree and neural networks. Energy **32**, 1761–1768 (2007). http://www.sciencedirect.com/science/article/pii/S0360544206003288
32. White, H.: Economic prediction using neural networks: the case of IBM daily stock returns. In: IEEE 1988 International Conference on Neural Networks, pp. 451–458, July 1988
33. Zeng, Y.R., Zeng, Y., Choi, B., Wang, L.: Multifactor-influenced energy consumption forecasting using enhanced back-propagation neural network. Energy **127**(Suppl. C), 381–396 (2017). http://www.sciencedirect.com/science/article/pii/S0360544217304759
34. Zhao, H., Magouls, F.: A review on the prediction of building energy consumption. Renew. Sustain. Energy Rev. **16**, 3586–3592 (2012). http://www.sciencedirect.com/science/article/pii/S1364032112001438

Fuzzy Relational Compositions Can Be Useful for Customers Credit Scoring in Financial Industry

Soheyla Mirshahi(✉) and Nhung Cao(✉)

Centre of Excellence IT4Innovations,
Institute for Research and Applications of Fuzzy Modeling,
University of Ostrava, 30. dubna 22, 701 03 Ostrava 1, Czech Republic
{soheyla.mirshahi,nhung.cao}@osu.cz
http://irafm.osu.cz/

Abstract. Fuzzy relational compositions is an important topic in fuzzy mathematics and many researchers have applied that in various fields which the classification problem was more and less accounted for the significant part. Related to this problem, in this paper, we will show that fuzzy relational compositions assist in evaluating customers creditability (credit scoring) which is one of the most important problems in the financial industry. The purpose is to classify a given customer into two classes of accepted or rejected and to help loan officers to make a better decision. We will illustrate an experimental example with initial values provided by an bank expert and use LFL R-package as the practical tool to calculate the compositions for our application. The concept of so-called generalized quantifiers and excluding features incorporating in the compositions will be employed as well.

Keywords: Fuzzy relational compositions
Bandler-Kohout products · Excluding features
Generalized quantifiers · Credit assessment · Credit scoring
Credit evaluation

1 Introduction

The credit scoring is a risk evaluation task considered as a critical decision for financial institutions in order to avoid wrong decision that may result in huge amount of losses [1]. This process includes collecting, analysing and classifying different credit elements and variables to assess the credit decisions. The quality of bank loans is the key determinant of competition, survival and profitability [2].

Evaluation of banking system customers risk in Iranian banks relies on experts judgement and fingertip rule. This type of evaluation resulted in high rate of postponed claims, therefore, designing new intelligent model for credit risk evaluation can be helpful [3]. Hence, in this paper, we consider credit scoring process based on one of the private banks in Iran which we had access to its data. This bank has a lot of branches around the country quite far from each

J. Medina et al. (Eds.): IPMU 2018, CCIS 855, pp. 28–39, 2018.
https://doi.org/10.1007/978-3-319-91479-4_3

other, therefore, it is important for the bank to assist young loan officers who are far from experts with their decision making process. For this purpose, first, we need to collect the knowledge of experienced loan experts along with analyzed information of previous customers and second, to apply fuzzy relational compositions based on LFL R-package [4] to support young loan officers making the decision.

Fuzzy relational compositions have been firstly introduced by Bandler and Kohout in the late 70s and the early 80s [5]. Since then, they have become an important topic in fuzzy mathematics. This can be substantiated by numerous studies deeply elaborating the topic on various aspects, see e.g. [6–8]. Its application has been presented in many areas, including the formal constructions of fuzzy inference systems [9,10], related systems of fuzzy relational equations, see e.g. [11–13] and recently [14], medical diagnosis [5], architectures of information processing [15] or in flexible queries to relational databases [16].

Apart from the numerous recent works such as flexible query answering systems [17], inference systems [18] or modeling monotone fuzzy rule bases [19], there are two other interesting directions extending the topic: the incorporation of excluding features in the compositions [20] and the compositions based on generalized quantifiers [21,22]. It is worth mentioning that both approaches may be combined together in order to obtain a more flexible and effective tool, see [23]. The real potential of the approaches was demonstrated on the classification problem of Odonata (dragonflies) [20,24].

In this paper, we will apply the approach of the compositions based on generalized quantifiers combining the concept of excluding features to the customers credit scoring as this application is close to the classification one. As we show, the final result is positive and illustrate that this method would help to classify a customer to the suitable category.

2 Background of the Approach

This section recalls some basic facts about fuzzy relational compositions combining together with two their recently extensions on the concept of excluding features [20] and the employment of intermediate generalized quantifiers [21]. From now, let us fix a residuated lattice $\mathcal{L} = \langle [0,1], \wedge, \vee, \otimes, \rightarrow, 0, 1\rangle$ as the underlying algebraic structure and define two more additional operations: the biresiduation (bi-implication) $\leftrightarrow$ given by $a \leftrightarrow b = (a \rightarrow b) \wedge (b \rightarrow a)$ and the negation $\neg$ defined by $\neg a = a \rightarrow 0$, for all $a; b \in L$. Furthermore, by $\mathcal{F}(U)$ we denote the set of all fuzzy sets on a given universe U.

2.1 Fuzzy Relational Compositions

Definition 1 [21]. *Let X, Y, Z be non-empty finite universes, and let $R \in \mathcal{F}(X \times Y)$, $S \in \mathcal{F}(Y \times Z)$ then the basic composition $\circ$, BK (Bandler-Kohout)-subproduct $\lhd$, BK-superproduct $\rhd$ and BK-square product $\square$ of R and S are fuzzy relations on $X \times Z$ defined as follows:*

$$(R \circ S)(x,z) = \bigvee_{y \in Y} (R(x,y) \otimes S(y,z)),$$

$$(R \lhd S)(x,z) = \bigwedge_{y \in Y} (R(x,y) \to S(y,z)),$$

$$(R \rhd S)(x,z) = \bigwedge_{y \in Y} (S(y,z) \to R(x,y)),$$

$$(R \square S)(x,z) = \bigwedge_{y \in Y} (R(x,y) \leftrightarrow S(y,z)),$$

for all $x \in X$ and $z \in Z$.

Let us illustrate the meaning of these compositions on the following example in the financial problem. Assume that X is a finite set of customers, Y is a finite set of attributes, and Z is a finite set of classes of credit decisions. Then $(R \circ S)(x,z)$ means that the customer x has at least one attribute belonging to the class z. The two other BK products and the square product provide a sort of strengthening of the initial suspicion. The subproduct $(R \lhd S)(x,z)$ means that all the attributes of the customer x belong to the class z, and $(R \rhd S)(x,z)$ expresses that the customer x has all attributes of the class z. The meaning of $(R \square S)(x,z)$ is that the customer x has all attributes of the class z and all attributes of the customer belong to this class.

2.2 Excluding Features in Fuzzy Relational Compositions

Now we follow [20] and recall the concept of *excluding features* incorporated in the fuzzy relational compositions for our problem which is motivated by the existence of excluding symptoms for some particular diseases in the medical diagnosis problem.

Definition 2 [20]. *Let X, Y, Z be non-empty finite universes, and let $R \in \mathcal{F}(X \times Y)$, $S, E \in \mathcal{F}(Y \times Z)$. Then the composition of R and S incorporating E is a fuzzy relation $R \circ S^{\backprime} E \in \mathcal{F}(X \times Z)$ defined by:*

$$(R \circ S^{\backprime} E)(x,z) = \bigvee_{y \in Y} (R(x,y) \otimes S(y,z)) \otimes \neg \bigvee_{y \in Y} (R(x,y) \otimes E(y,z)). \quad (1)$$

The semantics of $E(y,z)$ is that y is an excluding feature for the class z, and the semantics of the composition $(R \circ S^{\backprime} E)(x,z)$ is that customer x has at least one feature belonging to the class z and at the same time there is no excluding feature related to the class and carried by the customer.

Formula (1) can be rewritten into a more comprehensible form as the product of the two basic compositions:

$$(R \circ S^{\backprime} E)(x,z) = (R \circ S)(x,z) \otimes \neg (R \circ E)(x,z).$$

2.3 Compositions Based on Generalized Quantifiers

Another extension of the fuzzy relational compositions are *the compositions based on generalized quantifiers* [21]. The usual quantifiers are e.g. "Many", "A few", or "Majority". Before giving the main definition of the extension, let us recall the concept of fuzzy measures [25].

Definition 3 [21]. *Let $U = \{u_1, \ldots, u_n\}$ be a finite universe, let $\mathcal{P}(U)$ be the power set of U. A mapping $\mu : \mathcal{P}(U) \to [0,1]$ is called a fuzzy measure on U if $\mu(\emptyset) = 0$ and $\mu(U) = 1$ and, if for all $C, D \in \mathcal{P}(U), C \subseteq D$ then $\mu(C) \leq \mu(D)$.*

Fuzzy measure μ is called symmetric if for all $C, D \in \mathcal{P}(U) : |C| = |D| \Rightarrow \mu(C) = \mu(D)$ where $|\cdot|$ denotes the cardinality of a set.

Example 1. The fuzzy measure μ^f given by $\mu^f(C) = f(\frac{|C|}{|U|})$ is symmetric, where $f : [0,1] \to [0,1]$ is a non-decreasing mapping with $f(0) = 0$ and $f(1) = 1$. Note that fuzzy sets modeling the evaluative linguistic expressions of the type `Big` (and modified by any linguistic hedges, e.g. `Roughly`, `Very`, `Significantly` etc.) [26,27] are satisfy the conditions of the function f.

Definition 4 [21,22]. *A mapping $Q : \mathcal{F}(U) \to [0,1]$ defined by*

$$Q(C) = \bigvee_{D \in \mathcal{P}(U) \setminus \{\emptyset\}} \left(\left(\bigwedge_{u \in D} C(u) \right) \otimes \mu(D) \right), \qquad C \in \mathcal{F}(U)$$

is called generalized (fuzzy) quantifier determined by a fuzzy measure μ on U.

If μ is a symmetric fuzzy measure then the quantifier can be reduced to a computational form:

$$Q(C) = \bigvee_{i=1}^{n} \left(C(u_{\pi(i)}) \otimes \mu(\{u_1, \ldots, u_i\}) \right), \qquad C \in \mathcal{F}(U) \tag{2}$$

where π is a permutation on $\{1, \ldots, n\}$ such that $C(u_{\pi(1)}) \geq C(u_{\pi(2)}) \geq \cdots \geq C(u_{\pi(n)})$.

Definition 5 [21]. *Let X, Y, Z be non-empty finite universes, let $R \in \mathcal{F}(X \times Y)$, $S \in \mathcal{F}(Y \times Z)$. Let Q be a quantifier on Y determined by a fuzzy measure μ. Then, the compositions $R@^Q S$ where $@ \in \{\circ, \lhd, \rhd, \square\}$ are defined as follows:*

$$(R@^Q S)(x, z) = \bigvee_{D \in \mathcal{P}(Y) \setminus \{\emptyset\}} \left(\left(\bigwedge_{y \in D} R(x, y) \circledast S(y, z) \right) \otimes \mu(D) \right) \tag{3}$$

for all $x \in X$, $z \in Z$ and for $\circledast \in \{\otimes, \to, \leftarrow, \leftrightarrow\}$ respectively where $R(x, y) \leftarrow S(y, z)$ is the same as $S(y, z) \to R(x, y)$.

Note that if the used fuzzy measure is symmetric then formula (3) can be rewritten into a form similar to formula (2).

It is worth mentioning that incorporation of excluding features in fuzzy relational compositions and the compositions based on generalized quantifiers may be combined together, see [23].

3 Customer Credit Scoring Problem

3.1 Overview of the Problem

In the financial industry, the consolidation of using classification models occurred in the 90s. Changes in the world scene, such as deregulation of interest rates and exchange rates, leads to an increase in bank competition and made financial institutions more and more worried about credit risk, i.e., the risk they were running when accepting someone as their customer. The granting of credit started to be more important in the profitability of companies in the financial sector, becoming one of the main sources of revenue for banks and financial institutions in general. Due to this fact, the sector of the economy realized that it was highly recommended to increase the number of allocated resources without losing the agility and quality [28]. The overall idea of credit scoring is to compare the features or characters of a customer with other earlier period customers, whose loans they have already paid back. If a customer's characters are adequately similar to those, who have been granted credit, and have consequently defaulted, the application will normally be rejected. If the customer's features are satisfactorily like those, who have not defaulted, the application will normally be granted [29]. In many researches the goal of credit scoring models is defined as to classify loan customers to either good credit or bad credit [30,31].

Many methods have been used to solve this problem such as discriminant analysis, regression analysis, probity analysis, neural networks, logistic regression and case-based reasoning [2,32]. Despite different methods we mentioned before, it seems that fuzzy logic modelling methods have been neglected in this area, although various researchers tried to bring fuzzy techniques into the credit scoring problem [27,33–36]. Moreover, one should keep in mind that the credit scoring models are not standardized, they differ from one market to another [2]. They can differ on the basis of the country situation, policies of the banks, culture and, of course, availability of data.

In this research we use the concept called Five C's of credit (Character, Capacity, Capital, Conditions, and Collateral). "the five C's of credit" is a pedagogical concept used in lending training. Loan officers use the five C's of credit to classify loan information and consider relationships among "categories of information" or one can say knowledge structure [37]. Decision making for credit scoring based on this knowledge structure enable loan officers to recall more judgemental and consistent information because loan officers frequently perform follow-up work on loans they originated, and they must recall loan information during conversations with borrowers and loan committee meetings.

The following section describes the five C's of credit as being the foundation of loan officers decision making.

3.2 Loan Officers' Decision Making Process

As mentioned before, loan officers are generally taught to seek and organize information using a framework called the five C's of credit. This provides a common structure to their judgements of loan applications and help them to make the final decision to accept or reject the borrowers request [37]. In Iran, this model is very common and experts are using it very frequently.

The following brief description of the five C's is based on [32,37].

- **Character:** The credit history of the borrower. It is about the borrower's reputation for repaying debts and appears on the credit reports which contain detailed information about how much the customer has borrowed from banking systems in the past and whether he or she has repaid loans on time or had some problems (delay, default). They also contain information of bankruptcies, etc. Also variables like length of living in the current address, occupation, length of current employment, payment status of previous credit, and further running credits, can show integrity, stability, and honesty of the customer.
- **Capacity:** Ability to operate a business or having a job capable of repaying debt. Loan officer compares income(s) against recurring debts of the customer and assess his or her debt-to-income (DTI) ratio.
- **Capital:** The funds available to operate a business or task. Loan officer considers any capital the customer puts toward a potential investment and it can be found in financial statements. Usually, large contribution by the borrower decreases the chance of default.
- **Condition:** The economic conditions (e.g., recession, growth) and also the condition of the loan, such as its interest rate and amount of principal, affect the loan officer's desire to finance the customer. It refers to how a borrower is going to pay back the loan and how he or she intends to use the money. For example, if a customer applies for a car loan or a home improvement loan, the lender may be more likely to approve those loans because of their specific purpose, rather than a signature loan that could be used for anything.
- **Collateral:** An alternative source of repayment of the loan. Collateral can help loan officer to secure the loan. Usually, an explicit pledge is required when there are some weaknesses in other C's although, collateral alone should not be used to make the decision. In other words, it assures the loan officer if the borrower defaults on the loan, the bank can repossess the collateral. For example, car loans are secured by cars and mortgages are secured by homes.

Instruction of the five C's serves four functions. First, it ensures that loan officers acquire data in categories that are critical to the success or failure of loans. Second, it helps them develop internal standards or reference points for customers information in those categories. Third, it forces loan officers to consider relationships among these categories. However, this information is influenced by borrowers character; borrowers of poor character may provide non-credible accounting information, such as unrealistically high valuations of their property. Since borrowers character influences their presentation of financial information, their responses to economic conditions, and even the security of their pledges of collateral, loan officers combine social judgement of character with judgments regarding the other C's of credit to obtain overall judgments of loan applications [37]. And the last one is that this model enable them to develop shared language with upper managers, different loan committee and even with the customer [32].

4 Experiment on the Credit Scoring and Discussion

In the previous section, we explained how the decision to accept or reject the loan is made by loan officers. In this section, we will show how the approach of fuzzy relational compositions based generalized quantifiers together combining with the excluding features can be useful for the problem of credit scoring in banks which are using the 5C's model. For the influence of the approach, an experimental example will be demonstrated while the membership degrees in fuzzy relations are determined by an bank expert. For the sake of simplicity, let us change the name of these 5C's to make them more clear and understandable. We need to determine the relationship between customers and features and similarly, features and outputs therefore, We should divide some of the features into more specific ones. Following, we specify how to extract the features:

- First C is Character and we will call it positive reputation.
- Second C is Capacity, we use the same name but we divide it into three categories, low capacity, medium capacity and high capacity.
- Third C is Capital and we will keep the name.
- Fourth C is Condition and again we divide this feature into three categories, easy condition, challenging condition and difficult condition.
- Fifth C is Collateral and we will rename it as security of the loan and divide it into three categories, low secured, medium secured and high secured.

Therefore totally out of these 5C's we extract 11 features which we will use in the next step. Let $Y = \{y_1, y_2, \ldots, y_{11}\}$ be a set of features, $Z = \{z_1, z_2\}$ be a set of classes (categories), and let $X = \{x_1, x_2, \ldots, x_{12}\}$ be a set of particular customers chosen randomly by a bank expert, where z_1 – accepted, z_2 – rejected; y_1 – positive reputation, y_2 – low capacity, y_3 – medium capacity, y_4 – high capacity, y_5 – sufficient capital, y_6 – low secured loan, y_7 – medium secured loan, y_8 – high secured loan, y_9 – easy condition, y_{10} – challenging condition, y_{11} – difficult condition.

The task is to classify a particular customer into a suitable category of the set Z to see that up to a certain degree, a given customer will be eligible to receive a loan. Although some of features might have low-degree connection with a given class, they should not be excluded from the possibility of having the class. The reason is that there might be particular customers who have this kind of connection and in this situation, the excluding features can be applied as well. For example, the "sufficient capital" feature has low degree 0.5 for accepted class, but it should not be excluded from this class since there might be a customer who has that feature in medium (or high) degree and he could be considered as accepted. On the other hand, there exist features directly excluded from having class no matter how many other features of that class linking to given customer. For example, if a customer has a high degree on low secured loan feature, since this feature is excluded from the accepted class therefore, the customer does not have to be in the consideration of accepted any more, no matter how many other features carried by him linking to the accepted class. Thus, the fuzzy relations $S, E \in \mathcal{F}(Y \times Z)$ and $R \in \mathcal{F}(X \times Y)$ can be given as follows:

R	y_1	y_2	y_3	y_4	y_5	y_6	y_7	y_8	y_9	y_{10}	y_{11}
x_1	0.9	0.2	0.8	0.0	0.7	0.0	0.2	0.8	0.4	0.6	0.0
x_2	0.3	0.0	0.6	0.4	0.8	0.0	0.7	0.3	0.0	0.5	0.5
x_3	0.8	0.8	0.2	0.0	0.5	0.0	0.4	0.6	0.0	0.3	0.7
x_4	0.7	0.1	0.9	0.0	0.9	0.0	0.8	0.2	0.6	0.4	0.0
x_5	0.8	0.8	0.2	0.0	0.7	0.6	0.4	0.0	0.0	0.6	0.4
x_6	0.9	0.2	0.8	0.0	0.9	0.0	0.1	0.9	0.4	0.6	0.0
x_7	0.6	0.6	0.4	0.0	0.8	0.3	0.7	0.0	0.0	0.6	0.4
x_8	0.3	0.6	0.4	0.0	0.9	0.2	0.8	0.0	0.0	0.5	0.5
x_9	0.9	0.0	0.3	0.7	0.8	0.3	0.7	0.0	0.3	0.7	0.0
x_{10}	0.9	0.0	0.3	0.7	0.9	0.0	0.2	0.8	0.1	0.9	0.0
x_{11}	0.8	0.3	0.7	0.0	0.5	0.0	0.1	0.9	0.6	0.4	0.0
x_{12}	0.9	0.2	0.8	0.0	0.9	0.0	0.2	0.8	0.8	0.2	0.0

S	z_1	z_2
y_1	0.7	0.3
y_2	0.2	0.8
y_3	0.5	0.5
y_4	0.7	0.3
y_5	0.5	0.5
y_6	0.2	0.8
y_7	0.5	0.5
y_8	0.7	0.3
y_9	0.7	0.3
y_{10}	0.5	0.5
y_{11}	0.3	0.7

E	z_1	z_2
y_1	0	0.5
y_2	0.8	0
y_3	0.2	0.2
y_4	0	0.6
y_5	0.1	0.3
y_6	0.8	0
y_7	0.2	0.2
y_8	0.1	0.2
y_9	0	0.6
y_{10}	0.5	0.4
y_{11}	0.6	0

The semantics of the membership degrees in the fuzzy relations R, S, E can be expressed as follows: $R(x, y)$ means that how much it is true, that the customer x has the feature y; $S(y, z)$ stands for the truth degree of the prediction, that y is a feature of the class z; and $E(y, z)$ means how much the truth degree should be, that y is an excluding feature of the class z. For example, $R(x_1, y_1) = 0.9$ means that the customer x_1 is evaluated to have positive reputation at 90%. This value is obtained on the basis of the fact that the customer x_1 delayed repaying her/his borrowed loans one or two times. As we have mentioned in the preliminaries section, that the role of the basic composition $R \circ S$ is to give the initial guess of the relationship between a given customer and a given class, and it is strengthened by the other products i.e., $R \lhd S$, $R \rhd S$ and $R \square S$, we make a computation on all these compositions to compare the results. If we fix the standard Łukasiweicz MV-algebra as the underlying algebraic structure then we obtain:

$R \circ S$	z_1	z_2
x_1	0.6	0.3
x_2	0.3	0.3
x_3	0.5	0.6
x_4	0.4	0.4
x_5	0.5	0.6
x_6	0.6	0.4
x_7	0.3	0.4
x_8	0.4	0.4
x_9	0.6	0.3
x_{10}	0.6	0.4
x_{11}	0.6	0.2
x_{12}	0.6	0.4

$R \lhd S$	z_1	z_2
x_1	0.7	0.4
x_2	0.7	0.7
x_3	0.4	0.5
x_4	0.6	0.6
x_5	0.4	0.5
x_6	0.6	0.4
x_7	0.6	0.7
x_8	0.6	0.6
x_9	0.7	0.4
x_{10}	0.6	0.4
x_{11}	0.8	0.4
x_{12}	0.6	0.4

$R \rhd S$	z_1	z_2
x_1	0.3	0.2
x_2	0.3	0.2
x_3	0.3	0.2
x_4	0.3	0.2
x_5	0.3	0.7
x_6	0.3	0.2
x_7	0.3	0.5
x_8	0.3	0.4
x_9	0.3	0.2
x_{10}	0.4	0.2
x_{11}	0.3	0.2
x_{12}	0.3	0.2

$R \square S$	z_1	z_2
x_1	0.3	0.2
x_2	0.3	0.2
x_3	0.3	0.2
x_4	0.3	0.2
x_5	0.3	0.5
x_6	0.3	0.2
x_7	0.3	0.5
x_8	0.3	0.4
x_9	0.3	0.2
x_{10}	0.4	0.2
x_{11}	0.3	0.2
x_{12}	0.3	0.2

As we can see, the basic composition gives us the initial suspicion which is a nice information where the difference between the acceptance and rejection membership degree is adequate to distinguish the class of customers. However, it has some weaknesses itself because it may have high membership degrees for both classes. On the other hand, the membership degree in two classes may be equal. In this situation, the second obstacle occurs as one may see for customer 2, customer 4 and customer 8. This does not provide enough information to solve the classification problem. In this case, one might suggest that the strengthening products may help however, because of the use of the universal quantifier, they often drop the membership degrees too much, for example, customer $1, 2, 3, 4$. Consequently, the square product should give a similar problem with too much dropping the membership degrees. Nevertheless, based on our experience in most cases, the use of the square composition based on generalized quantifier can help. In this context, let us consider fuzzy set modelling the meaning of the linguistic expression `Roughly Big` (abbr. `RoBi`) (cf. [27]) which enables us to construct a generalized quantifier $Q =$ "Majority". In a standard context (cf. Chap. 5 in [27]), this fuzzy set takes values $\texttt{RoBi}(1/11) = \texttt{RoBi}(2/11) = \cdots = \texttt{RoBi}(7/11) = 0$, $\texttt{RoBi}(8/11) = 0.113$, and $\texttt{RoBi}(9/11) = \texttt{RoBi}(10/11) = \texttt{RoBi}(1) = 1$. Thus, based on the computational form of the compositions based on generalized quantifiers, one may easily compute the composition $R \,\square^Q\, S$. The result, as we may see in the following table, eliminates the equality of the membership degrees related the acceptance and rejection classes. In details, the chance of "rejection" for the customer x_2 is more likely than getting "acceptance", the customer x_4 is vice-versa as the percentage of having the acceptance is greater than the other one, and finally, the customer x_8 has more risk of rejection. Furthermore, the differentiation of the acceptance degree and the rejection degree is more distinguished for the rest of the customers which undoubtedly proceeds the classification problem easier. As we have raised reason for combining the approach of excluding features in the compositions, we have applied the product $R \,\square^Q\, S \backslash E$ which serves as a better tool. Indeed, it eliminates the false assigned membership degrees without losing the possibility of having the correct ones. We describe the results in the following tables:

$R \,\square^Q\, S$	z_1	z_2
x_1	0.7	0.4
x_2	0.6	0.7
x_3	0.4	0.7
x_4	0.6	0.3
x_5	0.3	0.7
x_6	0.6	0.4
x_7	0.3	0.7
x_8	0.3	0.7
x_9	0.7	0.4
x_{10}	0.6	0.3
x_{11}	0.7	0.4
x_{12}	0.7	0.4

,

$R \,\square^Q\, S \backslash E$	z_1	z_2
x_1	0.6	0.0
x_2	0.5	0.6
x_3	0.0	0.4
x_4	0.5	0.1
x_5	0.0	0.4
x_6	0.5	0.0
x_7	0.0	0.6
x_8	0.0	0.5
x_9	0.5	0.0
x_{10}	0.2	0.0
x_{11}	0.6	0.1
x_{12}	0.7	0.0

.

Let us mention two points regarding our final results:

- First, we should point out in our research, we had access to small but real data sample from one of the private Iranian banks, and the outcome matched with the expected outputs.
- Furthermore, we found an interesting point during our research which we will explain by an example. As we mentioned before, z_1 shows the membership degree of the accepted class. In the table above, it is clear that even among the customers who already categorized as accepted, some of them have higher accepted membership degree compare to other ones. For instance, customer x_{12}'s membership degree is 0.7 and it is greater than membership degree of customer x_{10}. This is exactly based on our expert's expectation. In another word, even though two customers might be accepted, the creditability of one might be different than the other. It inspires us to consider later research and instead of measuring the acceptance or rejection of customers, we will measure the risk of giving a loan to each customer. For example, the output can be categorized into low-risk, medium-risk, and high-risk classes.

One of the common problems regarding working in the area of credit scoring is the sample data. Usually, researchers in this field are using the German credit assessment dataset (donated by Professor Dr. Hans Hofmann) but we did not have access to experts to evaluate the five C's based on this dataset therefore, we have performed our experiment on a small amount of data based on one of the private Iranian banks. We are aware that though it gives us an idealistic result for the classification, it might not seem enough convincing for readers regarding its real potential. However, we noticed that the approach behaves similarly and keep the performance while we increased the number of samples.

Another issue should be kept in mind is the dependency among features which may affect the credit scoring.

5 Conclusion and Further Work

We have proposed that fuzzy relational compositions can be used as a useful tool to apply for customers creditability assessment. Its real potential was demonstrated on a simple yet real example determining from the expert knowledge. The application has contributed for the extension of the applied files of fuzzy relational compositions, especially the compositions combining together with recent its extensions on the excluding features and the employment of intermediate generalized quantifiers such as "Most", "Many" or "Majority" which somehow provide us with a wider choice of models that may better fit for particular problems consistent with human reasoning using natural language. For the further work, a system of fuzzy rule base using linguistic expressions seems to be another appropriate direction toward solving the assessment of customer creditability as well.

References

1. Khashei, M., Rezvan, M.T., Hamadani, A.Z., Bijari, M.: A bi-level neural-based fuzzy classification approach for credit scoring problems. Complexity **18**, 46–57 (2013)
2. Abdou, H.A., Pointon, J.: Credit scoring, statistical techniques and evaluation criteria: a review of the literature. Intell. Syst. Account. Finance Manag. **18**(2–3), 59–88 (2011)
3. Salehi, M., Mansoury, A.: An evaluation of iranian banking system credit risk: Neural network and logistic regression approach. Int. J. Phys. Sci. **6**(25), 6082–6090 (2011)
4. Burda, M.: Linguistic fuzzy logic in R. In: Proceedings of IEEE International Conference on Fuzzy Systems, Istanbul, Turkey (2015)
5. Bandler, W., Kohout, L.J.: Semantics of implication operators and fuzzy relational products. Int. J. Man Mach. Stud. **12**(1), 89–116 (1980)
6. De Baets, B., Kerre, E.: Fuzzy relational compositions. Fuzzy Sets Syst. **60**, 109–120 (1993)
7. Belohlavek, R.: Sup-t-norm and inf-residuum are one type of relational product: unifying framework and consequences. Fuzzy Sets Syst. **197**, 45–58 (2012)
8. Běhounek, L., Daňková, M.: Relational compositions in fuzzy class theory. Fuzzy Sets Syst. **160**(8), 1005–1036 (2009)
9. Pedrycz, W.: Applications of fuzzy relational equations for methods of reasoning in presence of fuzzy data. Fuzzy Sets Syst. **16**, 163–175 (1985)
10. Štěpnička, M., Jayaram, B.: On the suitability of the Bandler-Kohout subproduct as an inference mechanism. IEEE Trans. Fuzzy Syst. **18**(2), 285–298 (2010)
11. Di Nola, A., Sessa, S., Pedrycz, W., Sanchez, E.: Fuzzy Relation Equations and Their Applications to Knowledge Engineering. Kluwer, Boston (1989)
12. De Baets, B.: Analytical solution methods for fuzzy relational equations. In: Dubois, D., Prade, H. (eds.) The Handbook of Fuzzy Set Series vol. 1, pp. 291–340. Academic Kluwer Publishers, Boston (2000)
13. Pedrycz, W.: Fuzzy relational equations with generalized connectives and their applications. Fuzzy Sets Syst. **10**, 185–201 (1983)
14. Cao, N., Štěpnička, M.: Fuzzy relation equations with fuzzy quantifiers. In: Kacprzyk, J., Szmidt, E., Zadrożny, S., Atanassov, K.T., Krawczak, M. (eds.) IWIFSGN/EUSFLAT -2017. AISC, vol. 641, pp. 354–367. Springer, Cham (2018). https://doi.org/10.1007/978-3-319-66830-7_32
15. Bandler, W., Kohout, L.J.: Relational-product architectures for information processing. Inf. Sci. **37**, 25–37 (1985)
16. Dubois, D., Prade, H.: Semantics of quotient operators in fuzzy relational databases. Fuzzy Sets Syst. **78**, 89–93 (1996)
17. Pivert, O., Bosc, P.: Fuzzy preference queries to relational databases. Imperial College Press, London (2012)
18. Mandal, S., Jayaram, B.: SISO fuzzy relational inference systems based on fuzzy implications are universal approximators. Fuzzy Sets Syst. **277**, 1–21 (2015)
19. Štěpnička, M., Jayaram, B.: Interpolativity of at-least and at-most models of monotone fuzzy rule bases with multiple antecedent variables. Fuzzy Sets Syst. **297**, 26–45 (2016)
20. Cao, N., Štěpnička, M., Burda, M., Dolný, A.: Excluding features in fuzzy relational compositions. Expert Syst. Appl. **81**, 1–11 (2017)

21. Cao, N., Štěpnička, M., Holčapek, M.: Extensions of fuzzy relational compositions based on generalized quantifer. Fuzzy Sets Syst. (in Press)
22. Štěpnička, M., Holčapek, M.: Fuzzy relational compositions based on generalized quantifiers. In: Laurent, A., Strauss, O., Bouchon-Meunier, B., Yager, R.R. (eds.) IPMU 2014. CCIS, vol. 443, pp. 224–233. Springer, Cham (2014). https://doi.org/10.1007/978-3-319-08855-6_23
23. Cao, N., Štěpnička, M.: Incorporation of excluding features in fuzzy relational compositions based on generalized quantifiers. In: Kacprzyk, J., Szmidt, E., Zadrożny, S., Atanassov, K.T., Krawczak, M. (eds.) IWIFSGN/EUSFLAT-2017. AISC, vol. 641, pp. 368–379. Springer, Cham (2018). https://doi.org/10.1007/978-3-319-66830-7_33
24. Cao, N., Štěpnička, M.: Fuzzy relational compositions based on grouping features. In: The 9th International Conference on Knowledge and Systems Engineering (KSE 2017), Hue, Vietnam, pp. 94–99. IEEE (2017)
25. Dvořák, A., Holčapek, M.: L-fuzzy quantifiers of type ⟨1⟩ determined by fuzzy measures. Fuzzy Sets Syst. **160**(23), 3425–3452 (2009)
26. Novák, V.: A comprehensive theory of trichotomous evaluative linguistic expressions. Fuzzy Sets Syst. **159**(22), 2939–2969 (2008)
27. Novák, V., Perfilieva, I., Dvorák, A., et al.: Insight Into Fuzzy Modeling. Wiley (2016)
28. Ferreira, P.H., Louzada, F., Diniz, C.: Credit scoring modeling with state-dependent sample selection: a comparison study with the usual logistic modeling. Pesquisa Operacional **35**(1), 39–56 (2015)
29. Crook, J.: Credit Scoring: An Overview. British Association, Festival of Science, University of Birmingham and the University of Edinburgh (1996)
30. Lee, T.S., Chiu, C.C., Lu, C.J., Chen, I.F.: Credit scoring using the hybrid neural discriminant technique. Expert Syst. Appl. **23**(3), 245–254 (2002)
31. Lim, M.K., Sohn, S.Y.: Cluster-based dynamic scoring model. Expert Syst. Appl. **32**(2), 427–431 (2007)
32. Khadivar, A., Mirshahi, S., Aghababaei, S.: Modeling and knowledge acquisition processes using case-based inference. Iran. J. Inf. Process. Manag. **32**(2), 467–490 (2017)
33. Darwish, N.R., Abdelghany, A.S.: A fuzzy logic model for credit risk rating of egyptian commercial banks. Int. J. Comput. Sci. Inf. Secur. **14**(2), 11–18 (2016)
34. Nosratabadi, H.E., Nadali, A., Pourdarab, S.: Credit assessment of bank customers by a fuzzy expert system based on rules extracted from association rules. Int. J. Mach. Learn. Comput. **2**(5), 662–666 (2012)
35. Bazmara, A., Donighi, S.S.: Bank customer credit scoring by using fuzzy expert system. Int. J. Intell. Syst. Appl. **6**(11), 29–35 (2014)
36. Abdulrahman, U.F.I., Panford, J.K., Hayfron-acquah, J.B.: Fuzzy logic approach to credit scoring for micro finance in ghana: a case study of KWIQPLUS money lending. Int. J. Comput. Appl. **94**(8), 11–18 (2014)
37. Beaulieu, P.R.: Commercial lenders' use of accounting information in interaction with source credibility. Contemp. Account. Res. **10**(2), 557–585 (1994)

Two-Sample Dispersion Tests for Interval-Valued Data

Przemysław Grzegorzewski[1,2](✉)

[1] Systems Research Institute, Polish Academy of Sciences, Newelska 6, 01-447 Warsaw, Poland
pgrzeg@ibspan.waw.pl
[2] Faculty of Mathematics and Information Science, Warsaw University of Technology, Koszykowa 75, 00-662 Warsaw, Poland

Abstract. The two-sample dispersion testing problem is considered. Two generalizations of the Sukhatme test for interval-valued data are proposed. These two versions correspond to different possible views on the interval outcomes of the experiment: the epistemic or the ontic one. Each view yields its own approach to data analysis which results in a different test construction and the way of carrying on the statistical inference.

Keywords: Interval-valued data · Nonparametric test · p-value Random interval · Tests for dispersion · The Sukhatme test

1 Introduction

Two-sample tests for dispersion belong to a basic toolbox of statistical procedures. Having two random samples one may be interested in detecting a difference in variability of distributions corresponding to these samples. In this paper we consider the two-sample dispersion testing problem for interval-valued data.

Analysts sometimes do not realize that interval-valued data which appear in applications may deliver two different types of information: the imprecise description of a point-valued quantity or the precise description of a set-valued entity (see [4]). Firstly, quite often the results of an experiment cannot be observed precisely or are so uncertain that they are recorded just as intervals containing the precise outcomes. Sometimes even having precise data the exact value of some variables are hidden deliberately for confidentiality reasons. These are examples of the *epistemic view* on intervals, where an *epistemic interval* A contains an ill-known actual value of a point-valued quantity x, so we can write $x \in A$. However, since it represents only the epistemic state of an agent, it does not exist per se. Secondly, there are situations when the outcomes of an experiment appear as essentially interval-valued data describing a precise information. Typical example are ranges of fluctuations of some physical measurements or time intervals spanned by some activity. Here we face the *ontic view* on intervals. Thus an *ontic interval* is the precise representation of an objective entity, i.e. A is an actual value of a set-valued variable X, so we can write $X = A$.

J. Medina et al. (Eds.): IPMU 2018, CCIS 855, pp. 40–51, 2018.
https://doi.org/10.1007/978-3-319-91479-4_4

Our goal is to generalize a suitable test for interval-valued data perceived both from the epistemic and ontic perspective. Usually to verify whether the dispersion of two populations differ one compares their sample variances. Unfortunately, a generalization of such statistical procedures into the interval-valued framework may cause considerable computational problems, especially if a sample is large enough. Indeed, a sample variance computation for the epistemic intervals is NP-hard task (see [7]). Moreover, a classical F-test for comparing variances assumes that populations are normally distributed. Therefore, to avoid problems in verifying assumptions on the underlying distributions we consider nonparametric tests which are not based on variances (see [5]). Thus the first advantage of our tests is their distribution-free nature. The second advantage of the proposed procedures is their low computational costs.

The paper is organized as follows. In Sect. 2 we discuss the two-sample dispersion problem and recall some facts on the Sukhatme test. In Sect. 3 we introduce basic notation and concepts related to the interval-valued data. Next, we propose two generalizations of the Sukhatme test adequate to each type of data: for the epistemic sets in Sect. 4 and for the ontic sets (random intervals) in Sect. 5.

2 Two-Sample Dispersion Problem

2.1 Detecting a Difference in Variability

Suppose $\mathcal{X}_1, \ldots, \mathcal{X}_n$ and $\mathcal{Y}_1, \ldots, \mathcal{Y}_m$ denote two independent samples of independent and identically distributed real random variables. We are interested in detecting whether there is a difference in variability (dispersion) between these two underlying distributions. The classical test applied in this context is the test for equality of variances, $H_0 : \sigma_x^2 = \sigma_y^2$, against the one- or two-sided alternative. Assuming that both populations are normally distributed we can use the so called F-test with a test statistic

$$T_F = \frac{\frac{1}{n-1}\sum_{i=1}^{n}(\mathcal{X}_i - \overline{\mathcal{X}})^2}{\frac{1}{m-1}\sum_{i=1}^{m}(\mathcal{Y}_i - \overline{\mathcal{Y}})^2},$$

which has the F-Snedecor distribution with $n-1$ and $m-1$ degrees of freedom. The aforementioned test is not particularly robust with respect to the normality assumption. Therefore, we are interested in nonparametric tests that could be used if the population distributions are not normal or unknown.

Thus, let us assume that $\mathcal{X}_1, \ldots, \mathcal{X}_n$ are i.i.d. random variables with c.d.f. $F(x) = H\left(\frac{x-\xi_x}{\eta_x}\right)$ and $\mathcal{Y}_1, \ldots, \mathcal{Y}_m$ are i.i.d. random variables with c.d.f. $G(x) = H\left(\frac{x-\xi_y}{\eta_y}\right)$, where H is a continuous c.d.f., ξ_x and ξ_y are population medians, while η_x and η_y denote scale parameters of the $\mathcal{X}$'s and $\mathcal{Y}$'s, distributions, respectively. Moreover, we assume that $\mathcal{X}$'s and $\mathcal{Y}$'s are mutually independent.

Please, note that the F-test does not require any assumption regarding the locations of the distributions. Indeed, variances are directly comparable, because they are each computed as measures of deviations around the respective sample means. In general, before comparing variability one should ascertain that both distributions under study do not differ in location since possible location differences may mask dispersion differences. Otherwise, if $\xi_x \neq \xi_y$, the sample observations should be adjusted by $\mathcal{X}_i' := \mathcal{X}_i - \xi_x$ and $\mathcal{Y}_j' := \mathcal{Y}_j - \xi_y$, for $i = 1, \ldots, n$ and $j = 1, \ldots, m$. Then the $\mathcal{X}'$ and $\mathcal{Y}'$ populations both have zero medians and the arrangement of $\mathcal{X}'$ and $\mathcal{Y}'$ random variables in the combined sample indicates dispersion differences as unaffected by location differences. Hence, further on, we assume either that the medians of the two populations are equal or that the sample observations can be adjusted to have equal locations, by subtracting the respective location parameters.

If we denote the ratio of the scale parameters by $\beta := \frac{\eta_x}{\eta_y}$ then the null hypothesis corresponding to the assertion of no difference in dispersion is given by

$$H_0 : \beta = 1, \tag{1}$$

and might be considered against one- or two-sided alternatives.

Several nonparametric tests for the two-sample dispersion problem have been proposed in the literature. The Siegel-Tukey test is the most frequently used procedure because it has the same null hypothesis distribution as the Wilcoxon rank-sum test and therefore it does not require a new set of critical values. Other popular tests are the Freund test, the Ansari-Bradley test and the David-Barton test, which are statistically equivalent although described by different formulas. One can also apply here the Mood test, the Klotz normal-scores test or the Sukhatme test. For more details on the nonparametric test for scale we refer the reader to [5]. Further on we focus our attention on the Sukhatme test because it seems to be the most suitable one for adopting to interval-valued data. Let us briefly recall how it works.

2.2 The Sukhatme Test

As it was stated above, we assume that both $\mathcal{X}$ and $\mathcal{Y}$ observations have or can be adjusted to have equal medians. Moreover, without loss of generality we can assume that this common median is zero, i.e. $\xi_x = \xi_y = 0$. Then, obviously, if the $\mathcal{X}$'s have a larger spread than $\mathcal{Y}$'s those $\mathcal{X}$ observations which are positive should be larger than most of the positive $\mathcal{Y}$ observations, while the negative $\mathcal{X}$ observations should be arranged so that most of the $\mathcal{Y}$ observations are larger than the $\mathcal{X}$'s. In other words, most of the negative $\mathcal{X}$'s should precede negative $\mathcal{Y}$'s, and most of the positive $\mathcal{X}$'s should follow positive $\mathcal{Y}$'s. Hence, we may define the Sukhatme test statistic [12] as follows

$$T = \sum_{i=1}^{n} \sum_{j=1}^{m} D_{ij}, \tag{2}$$

where

$$D_{ij} = \begin{cases} 1 & \text{if } \mathcal{X}_i < \mathcal{Y}_j < 0 \text{ or } 0 < \mathcal{Y}_j < \mathcal{X}_i, \\ 0 & \text{otherwise.} \end{cases} \tag{3}$$

We reject the null hypothesis (1) and conclude that the distribution of $\mathcal{X}$ population is more dispersed than $\mathcal{Y}$'s if the p-value $p = \mathbb{P}_{H_0}(T \geqslant t)$ is small enough (say $p < \alpha$, where α is the given significance level) and t stands for the actual value of the test statistic (2).

One may notice that the indicator variable (3) resembles that which is applied for the Mann-Whitney test statistic. The exact null distribution of (2) can be also found by a method similar to that for the Mann-Whitney test. If the null hypothesis holds than the mean and the variance of the test statistic are $\mathbb{E}_{H_0}(T) = \frac{1}{4}mn$ and $\mathrm{Var}_{H_0}(T) = \frac{1}{48}(mn(m+n+7))$, respectively. Thus, for sample sizes large enough, the distribution of the statistic

$$T^* = \frac{T - \mathbb{E}_{H_0}(T)}{\sqrt{\mathrm{Var}_{H_0}(T)}} = \frac{4\sqrt{3}(T - \frac{1}{4}mn)}{\sqrt{mn(m+n+7)}} \tag{4}$$

is asymptotically normal under H_0, i.e. $T^* \overset{H_0}{\sim} N(0,1)$ as $n \longrightarrow \infty$.

If the medians of the $\mathcal{X}$ and $\mathcal{Y}$ populations are unknown, we can firstly estimate the sample medians $\widehat{\xi}_x = \mathrm{Me}(\mathcal{X}_1, \ldots, \mathcal{X}_n)$ and $\widehat{\xi}_y = \mathrm{Me}(\mathcal{Y}_1, \ldots, \mathcal{Y}_m)$, and then define the adjusted observations $\mathcal{X}_i' := \mathcal{X}_i - \widehat{\xi}_x$ for $i = 1, \ldots, n$ and $\mathcal{Y}_j' := \mathcal{Y}_j - \widehat{\xi}_y$ for $j = 1, \ldots, m$. Next, we calculate a value of the test statistic T' for these adjusted $\mathcal{X}'$ and $\mathcal{Y}'$ observations and apply the large-sample approximation to the modified statistic T'.

3 Interval-Valued Data

Let $\mathcal{K}_c(\mathbb{R})$ denote a family of all non-empty closed and bounded intervals in the real line $\mathbb{R}$. Each interval $A \in \mathcal{K}_c(\mathbb{R})$ can be expressed by means of a two-dimensional value, defined in terms of its endpoints, $(\inf A, \sup A) \in \mathbb{R}^2$ with $\inf A \leq \sup A$. Other way for describing interval, which is in some situations more operative, is based on the point $(\mathrm{mid}\, A, \mathrm{spr}\, A) \in \mathbb{R} \times \mathbb{R}^+$, where $\mathrm{mid}\, A = (\sup A + \inf A)/2$ is the mid-point (center) of the interval A, and $\mathrm{spr}\, A = (\sup A - \inf A)/2$ denotes the spread (radius). Thus, A can be represented as $A = [\inf A, \sup A] = [\mathrm{mid}\, A \pm \mathrm{spr}\, A]$.

When dealing with intervals, a natural arithmetic is defined on $\mathcal{K}_c(\mathbb{R})$ based on the Minkowski addition and the product by scalars. These operations are settled as $A + B = \{a + b : a \in A, b \in B\}$ and $\lambda A = \{\lambda a : a \in A\}$, for all $A, B \in \mathcal{K}_c(\mathbb{R})$ and $\lambda \in \mathbb{R}$, respectively. Using the mid/spr notation the above operations can be jointly expressed as follows $A + \lambda B = [(\mathrm{mid}\, A + \lambda \mathrm{mid}\, B) \pm (\mathrm{spr}\, A + |\lambda| \mathrm{spr}\, B)]$.

It should be noted that the space $(\mathcal{K}_c(\mathbb{R}), +, \cdot)$ is not linear but semilinear, due to the lack of the inverse element with respect to the Minkowski addition: in general, $A + (-1)A \neq \{0\}$, unless $A = \{a\}$ is a singleton. To overcome this problem, sometimes it is possible to consider the so-called Hukuhara difference $A -_H B$ between the intervals A and B, defined by such interval $C \in \mathcal{K}_c(\mathbb{R})$ that $B + C = A$. Unfortunately, the Hukuhara difference does not exist for any two intervals $A, B \in \mathcal{K}_c(\mathbb{R})$ but only for such $A, B \in \mathcal{K}_c(\mathbb{R})$ that $\mathrm{spr}\, A \geqslant \mathrm{spr}\, B$.

Although we use the same notation and basic operations on intervals both for the epistemic and ontic approach, there are significant differences in statistics of interval-valued data perceived from those two perspectives.

4 Two-Sample Dispersion Tests for Epistemic Data

4.1 General Remarks on Statistical Reasoning with Epistemic Data

Let us consider a sequence of intervals $X_1, \ldots, X_n$, where $X_i = [\underline{x}_i, \overline{x}_i]$, which are interval-valued perceptions of the unknown true outcomes of the experiment $x_1, \ldots, x_n$, where $x_i \in X_i$, for $i = 1, \ldots, n$. These true, but not observed outcomes x_i are realizations of some real-valued random variables $\mathcal{X}_i : \Omega \longrightarrow \mathbb{R}$ defined on a probability space $(\Omega, \mathcal{A}, P)$, i.e. $\mathcal{X}_i(\omega) = x_i$.

Usually to make any statistical inference we have to determine a value of some statistic $T = T(\mathcal{X}_1, \ldots, \mathcal{X}_n)$, e.g. an estimator or a test statistic. However, having interval data only we may consider different possible values of that statistic, i.e. we obtain

$$T_I = \{T(x_1, \ldots, x_n) : x_1 \in X_1, \ldots, x_n \in X_n\}.$$

It is worth noting that in general it is not always possible to find the actual range of T_I. Thus we try to compute its *enclosure*, i.e. an interval $\widetilde{T}_I$ such that $\widetilde{T}_I \supseteq T_I$. If $\widetilde{T}_I = T_I$ we say that the enclosure is *exact*. When a statistic under study is in some sense regular (e.g., continuous or monotonic), it is usually enough to identify the smallest and largest value of T denoted by $T_{\min}$ and $T_{\max}$, respectively, to determine $\widetilde{T}_I$. However, finding the exact (or even satisfactory) enclosure is not easy in general. Moreover, in some cases, it is even impossible in a reasonable time. For instance, computing the largest value of the sample variance $\overline{S^2}$ for arbitrary interval-valued data perceived from the epistemic perspective is the NP-hard problem (see [7]). Keeping this in mind it is clear that a generalization of the F-test for arbitrary epistemic intervals may have no practical sense.

4.2 The Sukhatme Test for Epistemic Interval-Valued Data

Suppose we observe two sequences of intervals $X_1, \ldots, X_n$ and $Y_1, \ldots, Y_m$, where $X_i = [\underline{x}_i, \overline{x}_i]$ and $Y_j = [\underline{y}_j, \overline{y}_j]$, are perceptions of the unknown true outcomes $x_1, \ldots, x_n$ and $y_1, \ldots, y_m$ of the experiment, respectively, where $x_i \in X_i$ and $y_j \in Y_j$ for $i = 1, \ldots, n$ and $j = 1, \ldots, m$. We assume that the samples and all observations are independent. Actually, we deal with the interval perceptions of the real-valued random samples $\mathcal{X}_1, \ldots, \mathcal{X}_n$ and $\mathcal{Y}_1, \ldots, \mathcal{Y}_m$. We assume that these samples satisfy all the assumptions discussed in Sect. 2.1. In particular, we assume that the medians of both distributions are equal, i.e. $\xi_x = \xi_y$. Without loss of generality we can assume that $\xi_x = \xi_y = 0$. Moreover, let us assume that $0 \notin X_i$ for $i = 1, \ldots, n$ and $0 \notin Y_j$ for $j = 1, \ldots, m$ (otherwise, interval observations containing zero are thrown out).

We verify $H_0 : \beta = 1$ vs. $H_1 : \beta > 1$. Keeping in mind formulae (2)–(3) and the considerations given in Sect. 4.1, it can be shown that the upper and the lower bound of T_I are given as follows

$$T_{\max} = \max\left\{T(x_1,\ldots,x_n,y_1,\ldots,y_m):\ x_i \in X_i, y_j \in Y_j\right\} = \sum_{i=1}^{n}\sum_{j=1}^{m} D^*_{ij},$$

where

$$D^*_{ij} = \begin{cases} 1 & \text{if } \underline{x}_i < \overline{y}_j < 0 \text{ or } 0 < \underline{y}_j < \overline{x}_i, \\ 0 & \text{otherwise;} \end{cases}$$

$$T_{\min} = \min\left\{T(x_1,\ldots,x_n,y_1,\ldots,y_m):\ x_i \in X_i, y_j \in Y_j\right\} = \sum_{i=1}^{n}\sum_{j=1}^{m} D^{**}_{ij},$$

where

$$D^{**}_{ij} = \begin{cases} 1 & \text{if } \overline{x}_i < \underline{y}_j < 0 \text{ or } 0 < \overline{y}_j < \underline{x}_i, \\ 0 & \text{otherwise.} \end{cases}$$

Thus we conclude immediately that $T_I \subseteq \left\{T_{\min}, T_{\min}+1,\ldots,T_{\max}\right\}$. The aforementioned set implies that instead of a real-valued p-value, typical for the classical test, we obtain a set of possible p-values corresponding to our interval-valued testing problem, i.e. $p_I = \{\mathbb{P}_{H_0}(T \geqslant t) : t \in T_I\}$, which is a subset of the interval $[\underline{p}, \overline{p}]$, where

$$\underline{p} = \min\{\mathbb{P}_{H_0}(T \geqslant t) : t \in T_I\} = \mathbb{P}_{H_0}(T \geqslant T_{\max}),$$
$$\overline{p} = \max\{\mathbb{P}_{H_0}(T \geqslant t) : t \in T_I\} = \mathbb{P}_{H_0}(T \geqslant T_{\min}).$$

If sample sizes are large enough then substituting $T_{\min}$ and $T_{\max}$ into (4) we obtain $T^*_{\min}$ and $T^*_{\max}$, respectively and we can utilize the normal approximation to determine p-values.

Now, having a set of possible p-values we need a new decision making criteria. In particular, we may behave as follows:

- if $\overline{p} < \alpha$ then reject H_0,
- if $\alpha < \underline{p}$ then do not reject (accept) H_0,
- otherwise (i.e. if $\underline{p} \leqslant \alpha \leqslant \overline{p}$) we abstain.

The last option might be interpreted as a demand for more observations to make a well-based decision or the abstention because of too imprecise measurements that do not allow to make a final binary decision. However, if one requires just a binary decisions – either reject or accept H_0 – we may apply an appropriate randomization (see [6]).

5 Two-Sample Dispersion Tests for Ontic Data

5.1 Random Interval

In the ontic approach, contrary to the classical statistical analysis, we deal no longer with usual real-valued random variables but with random intervals defined as follows.

Definition 5.1. *Given a probability space $(\Omega, \mathcal{A}, P)$, a mapping $X : \Omega \longrightarrow \mathcal{K}_c(\mathbb{R})$ is said to be a random interval (interval-valued random set) if it is Borel-measurable with the Borel σ-field generated by the topology associated with by the Hausdorff metric on $\mathcal{K}_c(\mathbb{R})$.*

Equivalently, a mapping $X : \Omega \longrightarrow \mathcal{K}_c(\mathbb{R})$ is a random interval if $\operatorname{mid} X : \Omega \to \mathbb{R}$ and $\operatorname{spr} X : \Omega \to \mathbb{R}^+$ are (real-valued) random variables defined as the mid-point and the spread of the interval $X(\omega)$, respectively, for each $\omega \in \Omega$. The mid/spr characterization of random intervals has appeared very valuable for different statistical purposes (see, e.g., [2,11]).

If X is a random interval and $\operatorname{mid} X, \operatorname{spr} X \in L^1(\Omega, \mathcal{A}, P)$, then the *Aumann mean of* X [1] is given by the set $\mathbb{E}[X] = \{\int_\Omega f dP : f \in X \ a.s.[P]\}$, which leads to the following interval $\mathbb{E}[X] = [\mathbb{E}(\operatorname{mid} X) \pm \mathbb{E}(\operatorname{spr} X)]$. It has been shown that the expected value of a random interval is linear and it is coherent with the arithmetic considered for finite populations in the sense of the Strong Law of Large Numbers.

The dispersion of a random interval can be measured by means of a distance between X and $\mathbb{E}[X]$. Consider a generalized family of L_2-type metric between intervals, introduced by Trutschnig et al. [13] for any $A, B \in \mathcal{K}_c(\mathbb{R})$ as

$$d_\theta = \sqrt{(\operatorname{mid} A - \operatorname{mid} B)^2 + \theta(\operatorname{spr} A - \operatorname{spr} B)^2}, \tag{5}$$

were $\theta > 0$ determines the relative weight of the distance between the spreads against the distance between the mids. It should be noted that a value of θ closer to 0 gives more importance to the midpoint, while a high value of θ gives more importance to the spread of the interval.

Such metric appears useful for defining the Fréchet-variance of a random interval X as $\sigma_X^2 = \mathbb{E}\left(d_\theta^2(X, \mathbb{E}[X])\right)$. As it is detailed in [3] this variance can be also expressed in terms of the classical variances of the mid and spr variables, namely, $\sigma_X^2 = \sigma^2_{\operatorname{mid} X} + \theta\sigma^2_{\operatorname{spr} X}$.

Procedures to test hypotheses for equality of variances of random intervals using the distance-based methodology can be deduced from tests for equality of variances of random fuzzy sets proposed in [8,10]. The crucial difficulty here is to find the distribution of the test statistic, so the advised way-out is to use a bootstrap or an asymptotic approach, provided a sample is large enough.

It seems that a reasonable remedy for problems with sampling distributions is to develop nonparametric techniques. Below we suggest a simple generalization of the Sukhatme test that avoids the above mentioned problem. We make use of the idea utilized in metrics (5) and connected with the aforementioned prior settlement on the relative importance of the difference in location and imprecision. We'll achieve it by a suitable projection, so the proposed approach may be called the *projection-based methodology.*

5.2 The Sukhatme Test for Random Intervals

Let us now consider the null hypothesis H_0 that both samples of random intervals $X_1, \ldots, X_n$ and $Y_1, \ldots, Y_m$, where $X_i = [\text{mid}\, X_i \pm \text{spr}\, X_i]$ and $Y_i = [\text{mid}\, Y_i \pm \text{spr}\, Y_i]$, have the same distributions against the alternative hypothesis that the population of X's is more dispersed than the population of Y's.

Each interval-valued observation $X_i = [\text{mid}\, X_i \pm \text{spr}\, X_i]$ can be perceived as a point $(\text{mid}\, X_i, \text{spr}\, X_i) \in \mathbb{R} \times \mathbb{R}^+$. Let us consider the following line:

$$l_x : \gamma(\text{spr}\, X - \text{spr}\, M_X) = (1 - \gamma)(\text{mid}\, X - \text{mid}\, M_X),$$

where $\gamma \in [0, 1]$ and $M_X = [\text{Me}(\text{mid}\, X) \pm \text{Me}(\text{spr}\, X)]$ is a median of X. Now, by representing a mid-point and spread of each interval-valued observation as a point of the two-dimensional Euclidean half-space we may find a projection of this point onto the line l_x as follows

1. if $\gamma = 0$, then
$$l_x : \text{mid}\, X = \text{mid}\, M_X,$$
and the projection of $(\text{mid}\, X_i, \text{spr}\, X_i)$ onto the line l_x is given by
$$X_i^{\mathbf{P}} = \mathbf{P}_{l_x}(\text{mid}\, X_i, \text{spr}\, X_i) = (\text{mid}\, M_X, \text{spr}\, X_i).$$

2. if $\gamma \neq 0$, then
$$l_x : \text{spr}\, X = \theta(\text{mid}\, X - \text{mid}\, M_X) + \text{spr}\, M_X,$$
where $\theta = \frac{1-\gamma}{\gamma} \in [0, \infty)$, and the projection of $(\text{mid}\, X_i, \text{spr}\, X_i)$ onto the line l_x can be written as
$$X_i^{\mathbf{P}} = \mathbf{P}_{l_x}(\text{mid}\, X_i, \text{spr}\, X_i) = (U_i^X, V_i^X),$$
where
$$U_i^X = \frac{\theta^2 \text{mid}\, M_X + \text{mid}\, X_i + \theta(\text{spr}\, X_i - \text{spr}\, M_X)}{\theta^2 + 1}, \tag{6}$$
$$V_i^X = \frac{\theta(\text{mid}\, X_i - \text{mid}\, M_X) + \theta^2 \text{spr}\, X_i + \text{spr}\, M_X}{\theta^2 + 1}. \tag{7}$$

Obviously, the projection of $(\text{mid}\, M_X, \text{spr}\, M_X)$ is $\mathbf{P}_l(\text{mid}\, M_X, \text{spr}\, M_X) = (\text{mid}\, M_X, \text{spr}\, M_X)$. Further on, to simplify notation, the situation $\gamma = 0$ will be denoted by $\theta = \infty$. As we see soon such convention would have a natural interpretation which also corresponds nicely with that applied in the distance-based methodology.

Some examples of this projection for different θ are given in Fig. 1. By choosing θ we specify the relative importance of the difference in location and imprecision of the interval data. In particular, by taking $\theta = 0$ we restrict our attention to the location of intervals. For any $\theta > 0$ we take into account both the difference in location and in imprecision of our interval-valued observations. By increasing

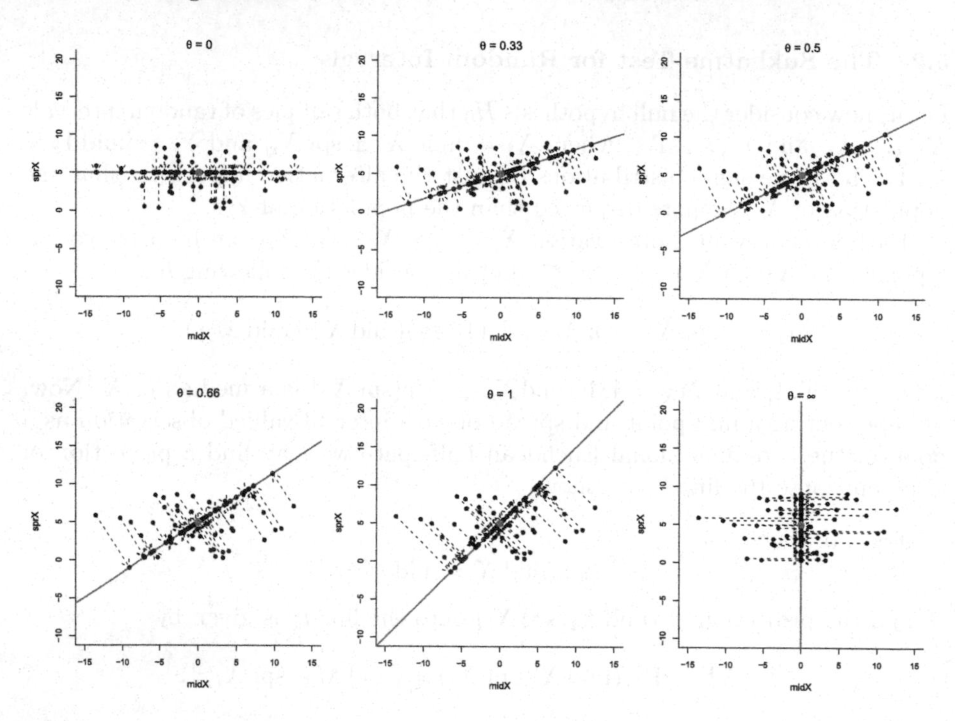

Fig. 1. The scatter plot of the points $(\operatorname{mid} X_i, \operatorname{spr} X_i)$ and their projection on the line l_x.

θ we also increase the relative importance of the interval imprecision. Finally, for $\theta = \infty$ our whole attention is restricted to the difference between imprecision of the considered intervals. Note, that this interpretation agrees with that related to the Trutschnig distance (5).

Besides projecting observations, we have to perform two more transformations so that the testing problem originally expressed for the two-dimensional objects (described by the mid-points and spreads) could be treated as a one-dimensional case. Firstly, we shift coordinates by a translation vector $\boldsymbol{w}_x$, where

$$\boldsymbol{w}_x = \begin{cases} [0, -\operatorname{spr} M_X] & \text{if} \quad \theta \in [0, \infty), \\ [-\operatorname{mid} M_X, 0] & \text{if} \quad \theta = \infty. \end{cases}$$

We will denote this transformation by $\mathbf{T}_{\boldsymbol{w}_x}$. Secondly, we rotate coordinates so that the abscissae coincides with the line l_x. Here we use the rotation matrix R_ϕ such that

$$R_\phi = \begin{bmatrix} \cos\phi & \sin\phi \\ -\sin\phi & \cos\phi \end{bmatrix},$$

where ϕ stands for the desired angle. In our case $\phi = \arctan\theta$.

As a result we obtain a new real-valued sample $\Psi_1, \ldots, \Psi_n$, where

$$\Psi_i = \Psi(\operatorname{mid} X_i, \operatorname{spr} X_i) = R_\phi \circ \mathbf{T}_{\boldsymbol{w}_x} \circ \mathbf{P}_{l_x}(\operatorname{mid} X_i, \operatorname{spr} X_i),$$

for $i = 1, \dots, n$. More precisely, $\Psi : \mathbb{R} \times \mathbb{R}^+ \to \mathbb{R}$, i.e. the composition of the projection on the line l_x, translation by the vector $\boldsymbol{w}_x$ and rotation through an angle ϕ, results in

$$\Psi_i = \Psi(\operatorname{mid} X_i, \operatorname{spr} X_i) = \begin{cases} \cos\phi \cdot U_i^X + \sin\phi \cdot (V_i^X - \operatorname{spr} M_X) & \text{if } \theta \in [0, \infty), \\ \operatorname{spr} X_i & \text{if } \theta = \infty, \end{cases}$$

where U_i^X and V_i^X are defined by (6) and (7), respectively.

Applying the same reasoning to the second sample of interval-valued observations, i.e. $Y_1, \dots, Y_m$, where $Y_j = [\operatorname{mid} Y_j \pm \operatorname{spr} Y_j]$, and performing similar transformations (i.e. a composition of the projection onto the line l_y, translation $\mathbf{T}_{\boldsymbol{w}_y}$ and rotation R_ϕ), we finally obtain a second real-valued sample $\Upsilon_1, \dots, \Upsilon_m$, where

$$\Upsilon_j = \Upsilon(\operatorname{mid} Y_j, \operatorname{spr} Y_j) = R_\phi \circ \mathbf{T}_{\boldsymbol{w}_y} \circ \mathbf{P}_{l_y}(\operatorname{mid} Y_j, \operatorname{spr} Y_j),$$

for $j = 1, \dots, m$. More precisely,

$$\Upsilon_j = \Upsilon(\operatorname{mid} Y_j, \operatorname{spr} Y_j) = \begin{cases} \cos\phi \cdot U_j^Y + \sin\phi \cdot (V_j^Y - \operatorname{spr} M_Y) & \text{if } \theta \in [0, \infty), \\ \operatorname{spr} Y_j & \text{if } \theta = \infty, \end{cases}$$

where U_j^Y and V_j^Y are defined by

$$U_j^Y = \frac{\theta^2 \operatorname{mid} M_Y + \operatorname{mid} Y_j + \theta(\operatorname{spr} Y_j - \operatorname{spr} M_Y)}{\theta^2 + 1},$$

$$V_j^Y = \frac{\theta(\operatorname{mid} Y_j - \operatorname{mid} M_Y) + \theta^2 \operatorname{spr} Y_j + \operatorname{spr} M_Y}{\theta^2 + 1}.$$

The last thing to do before testing our hypothesis is to adjust samples $\Psi_1, \dots \Psi_n$ and $\Upsilon_1, \dots, \Upsilon_m$ so they both have medians equal zero. Therefore, we define the adjusted observations $\widetilde{\Psi}_i := \Psi_i - \operatorname{Me}(\Psi_1, \dots \Psi_n)$ for $i = 1, \dots, n$ and $\widetilde{\Upsilon}_j := \Upsilon_j - \operatorname{Me}(\Upsilon_1, \dots, \Upsilon_m)$ for $j = 1, \dots, m$.

Now we are able to define the generalized Sukhatme test statistic for interval-valued observations

$$\widetilde{T} = \sum_{i=1}^{n} \sum_{j=1}^{m} \widetilde{D}_{ij}, \tag{8}$$

where

$$\widetilde{D}_{ij} = \begin{cases} 1 & \text{if } \widetilde{\Psi}_i < \widetilde{\Upsilon}_j < 0 \text{ or } 0 < \widetilde{\Upsilon}_j < \widetilde{\Psi}_i, \\ 0 & \text{otherwise.} \end{cases}$$

Thus, by the appropriate transformation of the original interval-valued observations into the one-dimensional problem, the rejection criteria remain as it is described in Sect. 2.2: we state that the distribution of X population is more dispersed than the distribution of Y population if $p = \mathbb{P}_{H_0}(\widetilde{T} \geqslant \widetilde{t})$ is small enough, where t stands for the actual value of the test statistic (8). If sample sizes are large enough we may substitute T in (4) by $\widetilde{T}$ and compute the p-value applying the normal approximation.

Remark 5.1. If is worth noting that we may project both samples onto the same straight line l instead of using l_x and l_y. Indeed, both l_x and l_y have the same slope determined by θ, while their intercepts are negligible. However, using l_x and l_y for $X_1, \ldots, X_n$ and $Y_1, \ldots, Y_m$, respectively, might be convenient for a graphical illustration or an interpretation.

Remark 5.2. Transformations Ψ and Υ are the compositions of the projection onto the straight line, translation and rotation. Although rotation depends on θ, the only operation influencing the final result is the projection. Actually, neither translation nor rotation change the relative position of the projected data.

5.3 Illustrative Example

Data in Table 1 correspond to the observed ranges over a day of the systolic and diastolic blood pressures of two independent samples of hospitalized patients, supplied by the Nephrology Unit of the Hospital Valle del Naln in Asturias, Spain (see [9,11]). We verify the null hypothesis that the dispersions of the ranges of the systolic and diastolic blood pressures do not differ significantly against the alternative hypothesis that the dispersion of the ranges of the systolic blood pressure exceeds the dispersion of the and diastolic blood pressure.

Table 1. Data on the ranges of systolic (X) and diastolic (Y) blood pressure.

X				Y			
118–173	111–192	109–174	101–194	56–121	47–93	55–97	64–121
104–161	116–201	128–210	112–162	50–94	73–105	59–101	53–109
131–186	102–167	94–145	116–168	52–95	74–125	59–101	37–94
105–157	104–161	136–201	148–201	63–118	52–112	60–98	55–85
120–179	106–167	90–177		57–113	69–133	55–121	

Let us assume the equal importance of the location and the interval spreads, i.e. $\gamma = \frac{1}{2}$, which leads to $\theta = 1$. As a result of all calculations we obtain $\widetilde{T} = 100$. In our case $n = m = 19$, so $\mathbb{E}_{H_0}(T) = 95$, $\text{Var}_{H_0}(T) = 16245$. Since both sample sizes are large enough to use the normal approximation, then by (4) we obtain $\widetilde{T}^* = 0.5399$. Finally, the desired p-value equals $p = \mathbb{P}_{H_0}(\widetilde{T}^* \geqslant 0.5399) = 0.2981$. Its value is large enough to state that there is no reason to reject the null hypothesis. In other words, we may conclude that there is no significant difference between the dispersions of the ranges of the systolic and diastolic blood pressures.

6 Conclusions

We have proposed two generalizations of the Sukhatme nonparametric two-sample test for dispersion designed for two different views on interval-valued data. Both ontic and epistemic view yield different approach to data analysis and statistical inference and generates its own problems. In particular, epistemic data may lead to situations when no definite decision can be made, especially if the intervals are too broad. This problem does not concern the ontic data modeled by random intervals. Their analysis yields a desired binary decision with respect to the null hypothesis. However, it requires a coefficient of the relative importance of the difference in location and imprecision in the considered data set, which leaves some place for subjectivity.

The common advantage of both generalizations is their computational simplicity. Moreover, general ideas presented in this contribution could be also applied for adopting other nonparametric two-sample dispersion tests for interval-valued data.

References

1. Aumann, R.J.: Integrals of set-valued functions. J. Math. Anal. Appl. **12**, 1–12 (1965)
2. Blanco-Fernández, A., Casals, M.R., Colubi, A., Corral, N., García-Bárzana, M., Gil, M.A., González-Rodríguez, G., López, M.T., Lubiano, M.A., Montenegro, M., Ramos-Guajardo, A.B., de la Rosa de Sáa, S., Sinova S.: A distance-based statistic analysis of fuzzy number-valued data. Int. J. Approx. Reason. **55**, 1487–1501, 1601–1605 (2014)
3. Blanco-Fernández, A., Corral, N., González-Rodríguez, G.: Estimation of a flexible simple linear model for interval data based on set arithmetic. Comput. Stat. Data Anal. **55**, 2568–2578 (2011)
4. Couso, I., Dubois, D.: Statistical reasoning with set-valued information: ontic vs. epistemic views. Int. J. Approx. Reason. **55**, 1502–1518 (2014)
5. Gibbons, J.D., Chakraborti, S.: Nonparametric Statistical Inference. Marcel Dekker, New York (2003)
6. Grzegorzewski, P.: Fuzzy tests – defuzzification and randomization. Fuzzy Sets Syst. **118**, 437–446 (2001)
7. Nguyen, H.T., Kreinovich, V., Wu, B., Xiang, G.: Computing Statistics under Interval and Fuzzy Uncertainty. Springer, Heidelberg (2012). https://doi.org/10.1007/978-3-642-24905-1
8. Ramos-Guajardo, A.B., Colubi, A., González-Rodríguez, G., Gil, M.A.: One-sample tests for a generalized Fréchet variance of a fuzzy random variable. Metrika **71**, 185–202 (2010)
9. Ramos-Guajardo, A.B., Colubi, A., González-Rodríguez, G.: Inclusion degree tests for the Aumann expectation of a random interval. Inf. Sci. **288**, 412–422 (2014)
10. Ramos-Guajardo, A.B., Lubiano, M.A.: K-sample tests for equality of variances of random fuzzy sets. Comput. Stat. Data Anal. **56**, 956–966 (2012)
11. Sinova, B., Colubi, A., Gil, M.A., González-Rodríguez, G.: Interval arithmetic-based linear regression between interval data: discussion and sensitivity analysis on the choice of the metric. Inf. Sci. **199**, 109–124 (2012)
12. Sukhatme, B.V.: On certain two sample nonparametric tests for variances. Ann. Math. Stat. **28**, 188–194 (1957)
13. Trutschnig, W., González-Rodríguez, G., Colubi, A., Gil, M.A.: A new family of metrics for compact, convex (fuzzy) sets based on a generalized concept of mid and spread. Inf. Sci. **179**, 3964–3972 (2009)

Ideal and Real Party Positions in the 2015–2016 Spanish General Elections

M. D. García-Sanz[1(✉)], I. Llamazares[2], and M. A. Manrique[3]

[1] BORDA Research Unit and Department of Economics and Economic History, University of Salamanca, 37007 Salamanca, Spain
dgarcia@usal.es

[2] Department of General Public Law, University of Salamanca, 37007 Salamanca, Spain
illamaz@usal.es

[3] Department of Economics and Economic History, University of Salamanca, 37007 Salamanca, Spain
amg@usal.es

Abstract. In this paper, using data from the pre-electoral and post-electoral Spanish surveys conducted by the Centro de Investigaciones Sociológicas (CIS) in the 2015 and 2016 general elections and assuming that parties maximize votes, we use an iterative algorithm to derive the optimal party positions (as predicted by spatial competition models based on proximity and directional models of voting). These optimal policy positions constitute a Nash equilibria, in which no party can increase its vote share by changing unilaterally its policy position. Then we compare the actual ideological positions of Spanish parties (as perceived by all voters) to their ideological party positions. Our aims are to examine the predictive power of proximity and directional models in the two Spanish electoral processes, to explore the degree to which parties deviate from their ideal positions and to examine the evolution of party positions from December 2015 to June 2016.

Keywords: Spatial models of voting · Directional, proximity models
Party competition · Nash equilibrium

1 Introduction

This work deals with an application of decision making modeling in Political Science. We compare the actual ideological positions of Spanish parties (as perceived by all voters) to their ideal ideological party positions (as predicted by spatial competition models based on proximity and directional models of voting). We do that by analyzing the pre-electoral and post-electoral surveys conducted by the Centro de Investigaciones Sociológicas (CIS) in the 2015 and 2016 general elections (Survey numbers 3117, 3126, 3141 and 3145). Our analysis restricts the

J. Medina et al. (Eds.): IPMU 2018, CCIS 855, pp. 52–62, 2018.
https://doi.org/10.1007/978-3-319-91479-4_5

sample to the respondents who voted for the main Spanish-wide parties: Partido Popular (PP), Partido Socialista Obrero Español (PSOE), Podemos (Ps), Ciudadanos (Cs) and Izquierda Unida (IU) in 2015, and PP, PSOE, Unidos Podemos (coalition of Podemos and Izquierda Unida, UPs) and Ciudadanos in 2016. Our work has several goals. In the first place, we examine the predictive power of proximity and directional models in these two Spanish electoral processes. In the second place, we intend to examine the degree to which parties deviate from their ideal positions. This analysis can pave the way for future work on the ideological, organizational and strategic factors conditioning theoretically non-vote maximizing positions. Finally, our work allows us to examine the evolution of party positions from December 2015 to June 2016.

For this analysis we follow the unified model of party competition in [1] to predict ideal party positions in the sense of Nash equilibrium. We use their unified model of voting, but we consider both spatial proximity and directional voting [11]. In Nash equilibrium, parties adopt ideological positions from which none of the vote-maximizing parties has incentives to deviate [1] if the others remain at their positions. That is, in such situations, no party would improve its electoral share by unilaterally modifying its position. We consider in our analysis both ideology and non-policy characteristics and attitudes (including here party identification). We first estimate conditional logit models for each survey using proximity and directional models, and then, following [1], we calculate the Nash equilibrium for party positions using the estimated parameters. We derive Nash equilibrium by using the iterative algorithm developed by [8], as implemented in the nopp R package (Nash Optimal Party Positions) that has been developed by [6]. The original package only deals with the proximity model, therefore we have developed a new R-project to implement the directional model.

The results of our analysis can be compared to those obtained on other cases by [1,5,9]. [10] has contributed a spatial analysis of voting in Spain that includes both proximity and directional models. However, whereas [10] focused on the receptivity of Spanish voters to positions that are distant from the status quo, a goal for which both proximity and directional models are used and tested in his work, our paper uses the unified proximity model of party competition in order to predict optimal policy positions. For that reason, although our interests and findings partially overlap with those of [10], our analytical approach and focus diverge from the ones he developed.

Our findings confirm the importance of ideological voting and the stability of party ideological positions. They also reveal the well-established centripetal bias of spatial models. This bias is particularly intense in the cases of Podemos, Ciudadanos, and, to a much larger extent, the PP, a fact that is entirely consistent with the findings in [10] about the importance of directional considerations among the PP voters. This result suggests the need to refine our analytical tools, test new models of party competition, and deepen our understanding of the organizational, strategic, and ideological factors conditioning the ideological positions of political parties.

The paper is organised as follows. In Sect. 2 we analyse the real positions of the Spanish political parties (as perceived by all the voters) and predict their optimal ideological positions in the sense of Nash equilibrium. We finish with some concluding remarks in Sect. 3.

2 Real and Optimal Ideological Positions

The first step in our analysis consists in the prediction of vote-choices through a conditional logit model. We use the Survey numbers 3117, 3126, 3141 and 3145[1] conducted by Centro de Investigaciones Sociológicas (CIS) in the 2015 and 2016 general elections. We have restricted our analysis to Spanish-wide parties, and have left out of the analysis voters for subnational parties, on the grounds that just voters from specific territories had the possibility to vote for them[2].

The conditional logit model of vote choice assumes that voter is probability of voting for party j is given by

$$P_i(j) = \frac{\exp(U_i(j))}{\sum_{l=1}^{n} \exp(U_i(l))} \tag{1}$$

where n stands for the number of political parties.

The conditional logit model can not determine absolute utility. The utility for an individual must be specified with respect to a base value. We have chosen PP, the government party, as the reference value.

The normalized utility is given by $U_i(j) - U_i(1)$, where 1 stands for the reference level, and $j = 2, \ldots, n$.

Our conditional logit model is the unified model of voting of party competition (see [1]).

The utility of voter i for voting party j, $U_i(j)$, is given by

$$U_i(j) = \alpha V_{ij} + \beta p_{ij} + \sum_{k=1}^{4} \gamma_{jk} Z_{ik} + \epsilon_{ij} \tag{2}$$

where ϵ_{ij} have standard Type 1 extreme value distributions.

We conduct models using both ideological proximity (quadratic proximity utility given by the negative of the squared distance between the voters and the partys location in the left-right dimension, scale 1 to 10) and directional (product of the difference between the respondent's position and the status quo by the difference between the mean party position and the status quo).

The variables in model (2) can be grouped into two types:

- Alternative specific variables, which vary with alternative, V_{ij} and p_{ij}, where p_{ij} equals to 1 if i identifies with party j and 0 elsewhere and V_{ij} has different expressions depending on the model we are working with.

[1] Survey numbers 3117 and 3141 are included in a two stage panel data study (panel 7715).

[2] We have considered as voters of Podemos respondents who voted for the alliances in which Podemos participated in Cataluña, Galicia and Valencia.

Proximity model:

$$V_{ij} = -(x_i - s_j)^2 \tag{3}$$

Directional model:

$$V_{ij} = (x_i - sq)(s_j - sq) \tag{4}$$

(x_i stands for i's location, s_j for j's location (mean party positions) and sq for the neutral point –status quo–.

- Individual specific variables which do not vary with alternative, Z_{ik}. Table 1 reports these variables in our analysis (k from 1 to 4).

Table 1. Individual specific variables in the model.

Variable	Description
Z_{i1}	Sex of voter i
Z_{i2}	Age of voter i
Z_{i3}	Education of voter i
Z_{i4}	Evaluation of government performance of voter i

We follow the existing literature and use individually perceived party positions in both models. As status quo, we take the center of the policy space. Both directional and proximity models include non-ideological variables as predictors of vote choices. Our dependent variable is vote intention for the pre-electoral surveys and vote choices as reported by the respondents for the post-electoral surveys. As [1] have shown, parties have incentives to present policies distant from the center in the direction of voters leaning towards them for non-policy reasons. Party identification is a critical variable in this respect. In addition to party identification we use cultural and territorial identifications as measured by the Linz-Moreno question[3], evaluations of the current economic situation, and controls for education, gender, and age. Full statistical results for these models are available on demand. As expected, ideological variables (based on proximity or direction) and partisanship carry the bulk of the models explanatory power. We report in Table 2 the number of voters selected and in Table 3 ideological and party identification impact coefficients in these models. The vote shares for the surveys are given in Table 4.

Our analyses show that the coefficients for ideology are larger in the post-electoral than in pre-electoral surveys (very considerably so in the case of directional models). They also show, interestingly, that directional and proximity

[3] Based on the Linz-Moreno question, the standard CIS question on the balance of Spanish and regional identities, which asks people if they feel only Spanish; more Spanish than from their autonomous community; both Spanish and from their autonomous community; more from their autonomous community than Spanish; only from their autonomous community.

Table 2. Number of voters selected.

	Pre-electoral 2015	Post-electoral 2015	Pre-electoral 2016	Post-electoral 2016
Survey number	3117	3126	3141	3145
Number of selected voters	8479	2241	8562	1870

Table 3. Ideological and party identification coefficients in conditional logistic models (all coefficients are significant at the 0.001 level).

	Pre-electoral 2015	Post-electoral 2015	Pre-electoral 2016	Post-electoral 2016
Survey number	3117	3126	3141	3145
Number of selected voters	8479	2241	8562	1870
Ideology				
Proximity model	0.08	0.10	0.08	0.12
Directional model	0.10	0.20	0.15	0.23
Party Id				
Proximity model	3.19	4.47	3.49	4.44
Directional model	3.36	4.5	3.46	4.46

coefficients have become larger in 2016 (if we compare pairwise pre-electoral and post-electoral surveys of 2015 and 2016 elections). The only coefficient that remains identical is the proximity indicator in the 2015 and 2016 pre-electoral models. Party id coefficients are also larger in post-electoral surveys. They have become larger from 2015 to 2016 in the proximity models, but they have remained almost identical in directional models. In general, our data reveal that from 2015 to 2016 a trend towards the intensification of the effects of ideological orientations took place, and that party identifications either increased their effects (in proximity models) or remained stable (in directional models).

Table 4. Vote shares.

	Pre-electoral 2015	Post-electoral 2015	Pre-electoral 2016	Post-electoral 2016
PP	31.2%	31.1%	32.6%	30%
Cs	20.1%	10.3%	14.6%	7.6%
PSOE	27.3%	29.5%	26.5%	31.8%
Ps	16.5%	22.9%	-	-
IU	4.9%	6.2%	-	-
UPs	-	-	26.3%	30.6%

Based on the coefficients estimated by our conditional logit model we have inferred the Nash equilibria of party ideological positions. This equilibrium leads a system involving different participants to a stable state, in which none of them can gain by a unilateral change of strategy (position), if the strategies of the others do not change. We refer to the positions given by the Nash equilibrium (NE) as ideal –or optimal– positions. To compute it, we implement the iterative algorithm developed by [8]. Assuming that parties maximize vote-shares, in each step of the algorithm each partys position is shifted to its vote-maximizing position holding the other parties positions constant. This leads to a new vector of party positions and eventually converges to a unique NE.

We compare then those ideal positions to the actual positions of political parties as perceived by all voters in the sample. But before showing the results of this analysis it must be taken into consideration that the perceptions of voters on party positions have remained extremely stable. As Table 5 shows, the Pearson correlation coefficients between the average perceptions of party positions are never lower than +0.99.

Table 5. Ideological and party id coefficients in conditional logistic models (all coefficients are significant at the 0.001 level).

	Pre-electoral 2015	Post-electoral 2015	Pre-electoral 2016	Post-electoral 2016
Pre-electoral 2015	1			
Post-electoral 2015	0.999	1		
Pre-electoral 2016	0.999	0.998	1	
Post-electoral 2016	0.998	0.992	0.999	1

Figures 1, 2, 3 and 4 display the real and ideal positions (as estimated by both proximity and directional models) for each of these surveys. As it is the case in similar analyses [5,9], ideal positions have a strong centripetal bias. The magnitude of this centripetal bias can be better grasped by examining the spread between extreme parties in actual and ideal positions. Whereas actual distances are never lower than 6 points, distances between ideal positions are always lower than 2 points. Directional models perform better in this respect in the 2015 surveys. They are also always better at predicting the positions of the PP. However, in terms of general spread, in 2016 there is almost no difference between the predictions derived from directional and proximity models. Also interestingly, although the number of players moved from 5 in 2015 to 4 in 2016, the spread of actual ideological did not diminish but in fact slightly increased. Also in this case, parties (at least in the perception of citizens) chose to intensify their ideological messages in the face of second general electoral contest in a short time span.

Table 6 below reports the ideological spread given by the respondents and estimated by both models.

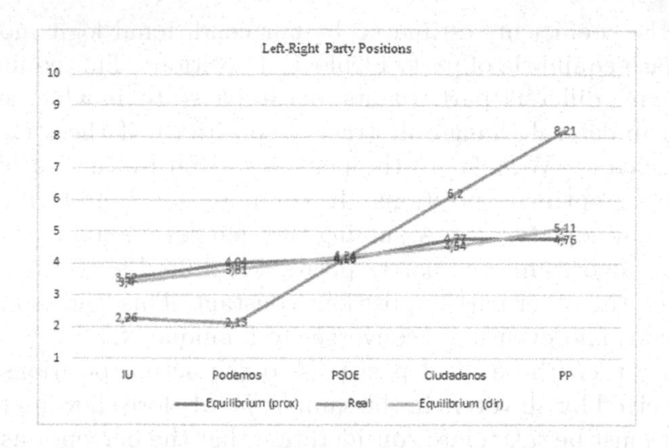

Fig. 1. Ideal and actual party positions (2015 pre-electoral survey).

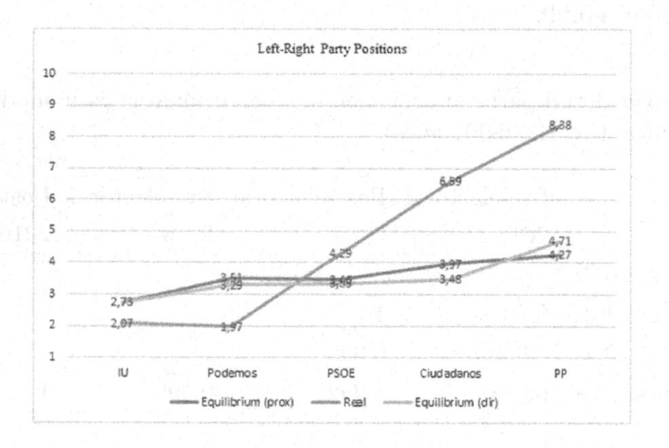

Fig. 2. Ideal and actual party positions (2015 post-electoral survey).

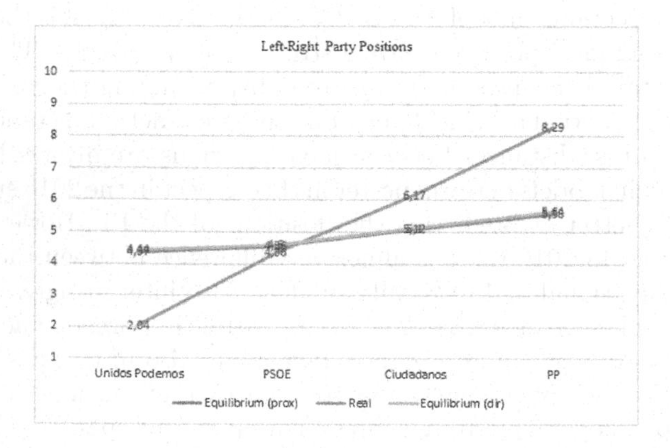

Fig. 3. Ideal and actual party positions (2016 pre-electoral survey).

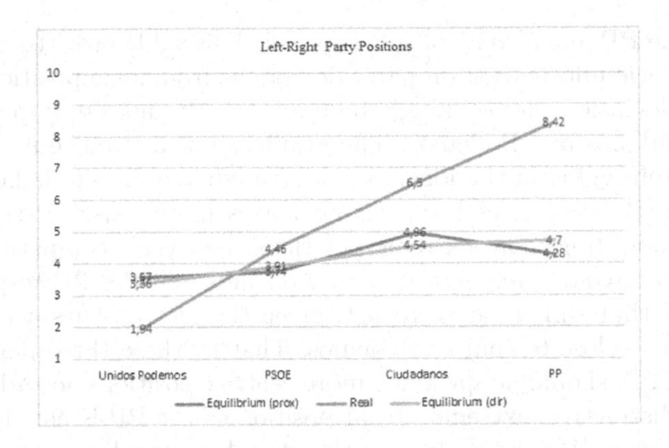

Fig. 4. Ideal and actual party positions (2016 post-electoral survey).

Table 6. Actual and predicted ideological spread.

	Pre-electoral 2015	Post-electoral 2015	Pre-electoral 2016	Post-electoral 2016
Ideological spread				
Actual	6.08	6.41	6.25	6.48
Proximity model	1.25	1.54	1.21	1.39
Directional model	1.71	1.98	1.2	1.34

Figures 1, 2, 3 and 4 show that the highest deviations from ideal positions are to be found, quite consistently, in the PP. But they tend to be high also in the case of Ciudadanos, and they are particularly high in the case of Unidos Podemos in 2016. Table 7 reports the average values for actual and ideal positions in the directional model[4], which is the one that tends to do better at predicting the

Table 7. Average actual (as perceived by all voters) and ideal party positions (directional model).

	Actual	Ideal	Average distance
PP	8.33	5.04	3.29
Cs	6.37	4.42	1.95
PSOE	4.32	4.02	0.30
Ps	2.02	3.73	1.71
IU	2.17	3.07	0.9

[4] IU averages are based on just the 2015 values. Podemos values are based on its values in 2015 and the values of Unidos Podemos in 2016. This decision was based on the relative electoral size of Podemos and IU in the 2016 Unidos Podemos coalition.

positions of the PP and also ideological spread. This table provides us thus with a more systematic information on party deviations from ideal positions. It shows that average deviations are strikingly high for the PP and also important in the cases of Ciudadanos and Podemos. The graphs show as well some instances of ideological leapfrogging in the ideal positions predicted for Ciudadanos and IU. In general, as [1], ideological leapfrogging makes more sense for small parties. As shown by adopting more extreme positions than their competitors, IU and Ciudadanos can avoid being squeezed by Podemos and the PP respectively. In the case of IU that shift is more realist, given the fact that its actual position was already very close to that of Podemos. That is, these three parties, and in particular the PP, should adopt much more centrist positions according to these models. The distinctive, extreme actual position of the PP is clearly consistent with the findings in [10] about the prevalence of directional over proximity components in the voting for the PP[5].

3 Concluding Remarks

We can draw several main inferences from our analysis. The first one refers to the powerful explanatory role of ideological voting in the Spanish party system. Ideological coefficients (in proximity or directional models) carry the bulk of explanatory power in vote-choice models in Spain. The second one concerns the stability of actual ideological positions in the 2015–2016 period in Spain, despite the complexity of the institutional scenario from 2015 to 2016 and in spite of the presence of significant differences in the party offer in this period (from 5 to 4 national parties due to the electoral coalition between IU and Podemos). Our contrast of ideal and actual ideological positions reveals also the strong centripetal bias of the ideal positions estimated through both proximity and directional models. This bias is strong in the cases of Ciudadanos, Unidos Podemos and, in particular, the PP. The extreme position of the PP is consistent with the strength of directional components for PP voters in the analysis in [10]. At least part of this bias could be accounted for by the role played by discount factors in party system competition and voting decisions. Now, this explanation leaves us with new unanswered questions. In the first place, since the PP government enjoyed an absolute majority in the Spanish parliament from 2011 to 2015, we can only assume that the status quo will be much closer to the preferences of this party than to those of left and center-left parties. And in the second place, even if discount factors are generally strong and PP voters are strongly directional, the question remains as to why the PP does not shift to more centrist positions in order to improve its electoral results.

Different tentative answers can be advanced as to the reasons of the PP positions. Following [9] we could speculate on the interdependence of valence considerations and ideological positions: by moving to the center the PP party

[5] The prevalence of directional voting among rightist voters has been also identified by [3] in several Latin American party systems.

could lose general credibility among voters (and not only among its conservative sympathizers). Still, the association of strongly conservative positions and credibility already depends on a continuous reassertion of very rightist positions. In contrast to policy switches, slow-paced and incremental ideological changes are not incompatible with maintaining general credibility, even if such moves displease very extreme or very ideologically committed voters.

Perhaps more importantly, the adoption of extreme positions may result from the preferences of intensive policy-demanders [4] among both core party constituents and social groups and civil actors endorsing the party. Internally, the fact that the PP is a strongly hierarchical and centralized organization should favor strategic maneuverability. However, this does not preclude the possibility that ideological shifts are penalized by core constituents and social actors with strong conservative leanings. Furthermore, to the extent that party leaders fear new party entries in the right side of the political spectrum (a realistic development that has already taken place in many other European countries), they may feel strongly compelled to adopt staunch conservative platforms, even if they risk leading the party, at least in the short term, to suboptimal electoral results.

Future developments of this work will demand testing the unified discount model of party competition. As it has been shown by previous analyses [1,9], this model can render more realistic predictions than proximity and directional models. Given data availability, our analyses will also have to assess the influence of new dimensions of political competition not mapped by the CIS survey we used for this work. This is particularly the case of populist attitudes, whose influence on voting decisions in Spain has already been established, from different perspectives, by [2] and by [7].

Acknowledgment. The authors acknowledge financial support by the Spanish Ministerio de Economía e Innovación Project CSO2013-47667-P.

References

1. Adams, J.F., Merrill, S., Grofman, B.: A Unified Theory of Party Competition. Cambridge University Press, Cambridge (2005)
2. Andreadis, I., Hawkins, K.A., Llamazares, I., Singer, M.: The conditional effects of populist attitudes on voter choices in four democracies. In: Littway, L., Hawkins, K.A., Carlin, R., Rovira Kaltwasser, C. (eds.) The Ideational Approach to Populism: Concept, Theory and Method. Routledge, Abingdon (2018, forthcoming)
3. Boscán, G.: Voto y competición electoral en América Latina. Ph.D. thesis, Universidad de Salamanca (2016)
4. Cohen, M., Karol, D., Noel, H.: The Party Decides Presidential Nominations Before and After Reform. The University of Chicago Press, Chicago (2008)
5. Curini, L.: Explaining party ideological stances. Public Choice **162**(1), 79–96 (2015).https://doi.org/10.1007/s11127-014-0199-6
6. Curini, L., Iacus, S.: Nash optimal party positions: the nopp R package. J. Stat. Softw. **81**(11), 311–320 (2017). https://doi.org/10.18637/jss.v081.i11
7. Lavezzolo, S., Ramiro, L.: Stealth democracy and the support for new and challenger parties. Eur. Polit. Sci. **1**, 1–23 (2017). https://doi.org/10.1109/JPROC.2010.2070470

8. Merill, S., Adams, J.F.: Computing nash equilibria in probabilistic, multiparty spatial models with nonpolicy components. Polit. Anal. **9**, 347–361 (2001)
9. Meyer, T.M., Müller, W.C.: Testing theories of spatial competition. the Austrian case. Party Polit. **20**(5), 802–813 (2014)
10. Queralt, D.: Spatial voting in spain. S. Eur. Soc. Polit. **17**, 356–392 (2012).https://doi.org/10.1080/13608746.2012.701890
11. Rabinowitz, G., Macdonald, S.E.: A directional theory of issue voting. Am. Polit. Sci. Rev. **83**, 93–121 (1989)

DNBMA: A Double Normalization-Based Multi-Aggregation Method

Huchang Liao[1,2], Xingli Wu[1(✉)], and Francisco Herrera[2,3]

[1] Business School, Sichuan University, Chengdu 610064, China
liaohuchang@163.com, xingliwusly@foxmail.com
[2] Department of Computer Science and Artificial Intelligence,
University of Granada, 18071 Granada, Spain
herrera@decsai.ugr.es
[3] Faculty of Computing and Information Technology,
King Abdulaziz University, Jeddah, Saudi Arabia

Abstract. We propose a double normalization-based multi-aggregation method to deal with the multi-criteria decision making problems considering the benefit, cost and target criterion values. To do so, we introduce an enhanced target-based linear normalization formula and a target-based vector normalization formula. Given that different normalization techniques maintain special advantages and disadvantages, we combine them with three aggregation models to describe the alternatives' performance from different aspects. Then, a new integration approach is developed to integrate three types of subordinate utility values and ranks to derive the final ranking. The selected alternative not only has a comprehensive performance but does not perform badly under each criterion. Finally, the proposed method is highlighted by a case study of selecting an optimal innovation enterprise.

Keywords: Multi-criteria decision making
Double Normalization-Based Multi-Aggregation Method
Target-based linear normalization · Target-based vector normalization

1 Introduction

Multi-Criteria Decision Making (MCDM) is a process of ranking a finite set of alternatives based on multiple criteria. There are basically two types of techniques to handle the MCDM problems: the outranking methods and the multi-criterion value methods [1]. The former is based on pairwise comparisons of alternatives under each criterion, which is limited in dealing with massive alternatives due to the complicated calculation. The latter composes a simple process of aggregating the criterion values to rank alternatives, which includes the normalization process and the aggregation process. There are mainly three kinds of normalization techniques: the linear normalization model, the vector normalization model and the non-normalization model. Jahan and Edwards [2] illustrated that different results can be derived by different normalization models. The multi-criterion value methods are various from different normalization and aggregation tools. Simple Multi-Attribute Rating Technique (SMART) is a rather

J. Medina et al. (Eds.): IPMU 2018, CCIS 855, pp. 63–73, 2018.
https://doi.org/10.1007/978-3-319-91479-4_6

simple MCDM method which uses the linear normalization model to eliminate the different dimensions among criteria and employs the weighted arithmetic aggregation operator to integrate the normalized criterion values [3]. Considering that the weighted arithmetic aggregation operator, the geometric weighted aggregation operator and the weighted maximum operator have different effects in representing the performance of objects, the MULTIplicative Multi-Objective Optimization by Ratio Analysis (MULTIMOORA) method [4], based on the vector normalization, applies these operators respectively to derive three kinds of utility values and then to yield the subordinate rankings of alternatives. Based on the vector normalization and the weighted arithmetic aggregation, the Technique for Order Preference by Similarity to Ideal Solution (TOPSIS) method [5] determines the compromise solution which is nearest to the positive ideal solution by calculating the distance of each alternative from the reference point. Opricovic and Tzeng [6] claimed that the solution selected by TOPSIS may not be closest to the ideal one since it ignores the relative importance between the distance from the ideal point and that from the negative-ideal point, and then proposed the VlseKriterijumska Optimizacija I Kompromisno Resenje (VIKOR) method [7] to further compute the "individual regret value" by the weighted maximum formula after deriving the "group utility" based on linear normalization and weighted arithmetic aggregation. However, the subordinate ranks are not taken into consideration in the VIKOR method when integrating two types of utility values, which makes the result with low robustness. Given that the criteria include benefit, cost, and target values in practice, Jahan et al. [8, 9] extended the linear normalization to the target-based linear normalization. On this basis, the target-based TOPSIS method [8], the target-based VIKOR method [9] and the target-based MULTIMOORA method [10] were proposed.

In conclusion, the common defect of the existing methods is that they eliminate the criterion dimensions only based on one normalization method, which may bias the results since all the normalization methods loss the original information more or less from different aspects. Furthermore, calculating the utility values by different aggregation operators is useful, but there still is a challenge to integrate the subordinate utility values and the ranks of alternatives at the same time to derive the final ranking.

This paper aims to propose a new MCDM method, named Double Normalization-Based Multi-Aggregation (DNBMA), to solve these problems. The paper is highlighted by the following innovative work:

(1) We introduce an improved target-based linear normalization formula and a target-based vector normalization formula.
(2) After analyzing the advantages and disadvantages of the target-based linear and vector normalization techniques, respectively, we make a suitable combination on two kinds of normalized values and three types of aggregation models to derive the subordinate utility values and ranks. It can reduce the information loss caused by one normalization technique.
(3) We propose a new aggregation formula to derive the final ranking of alternatives. It considers the subordinate utility values and the ranks of alternative simultaneously, and their relative importance is also taken into consideration. In this way, the result is more robustness than the ranking which is only integrated by subordinate utility values or subordinate ranks.

The paper is organized as follows: Sect. 2 presents the target-based linear and vector normalization formulas. The DNBMA method is proposed in Sect. 3. Section 4 illustrates the method by an example. Final concluding remarks are pointed out in Sect. 5.

2 The Target-Based Normalization Techniques

A general MCDM problem contains a set of alternatives $A = \{a_1, a_2, \ldots, a_m\}$ and a set of criteria $C = \{c_1, c_2, \ldots, c_n\}$ with the weight vector $W = (\omega_1, \omega_2, \ldots, \omega_n)^T$. The decision matrix is composed as $X = (x_{ij})_{m \times n}$ where x_{ij} is the value of alternative a_i with respect to criterion c_j. Given that the non-normalization tool can only be applied to special problems, in the following, we start our investigation by introducing two commonly used linear and vector normalization formulas. Afterwards, we illustrate their advantages and disadvantages based on some examples.

2.1 The Target-Based Linear Normalization

Considering cost, benefit, and target-based criteria at the same time for a MCDM problem, based on the distance of each judgment to target value, Jahan et al. [9] proposed a linear normalization formula as:

$$y_{ij}'^{1} = 1 - \frac{|x_{ij} - r_j|}{\max\left\{\max\limits_i x_{ij}, r_j\right\} - \min\left\{\min\limits_i x_{ij}, r_j\right\}} \tag{1}$$

where r_j is the target value on criterion c_j, especially, $r_j = \max\limits_i x_{ij}$ if c_j is in benefit type, $r_j = \min\limits_i x_{ij}$ if c_j is in cost type.

Motivated by the simplified form of linear normalization [10], we improve the target-based linear normalization formula as:

$$y_{ij}^{1} = 1 - \frac{|x_{ij} - r_j|}{\max\limits_i |x_{ij} - r_j|} \tag{2}$$

The linear normalization given as Eq. (2) can reflect the closeness between each alternative and the target solution under each criterion. The normalization values are the same for different convertible units ($\phi_{ij} = \alpha x_{ij} + \beta$, $\alpha > 0$) with the same criterion function given that

$$\phi_{ij}^{1} = 1 - \frac{|\phi_{ij} - \phi_j|}{\max\limits_i |\phi_{ij} - \phi_j|} = \frac{\left|(\alpha x_{ij} + \beta) - (\alpha r_j + \beta)\right|}{\max\limits_i \left|(\alpha x_{ij} + \beta) - (\alpha r_j + \beta)\right|} = y_{ij}^{1}$$

where ϕ_j is the target value on criterion c_j and ϕ_{ij}^{1} is the linear normalized value of ϕ_{ij}.

Thus, it is reasonable to aggregate the values of all criteria of an alternative directly because the normalized values only represent the normalized distances between the judgments of alternatives and the ideal solution. However, it loses the size of value itself, which would bias the result. This defect can be illustrated by Example 1.

Example 1. Suppose that there are three projects a_1, a_2 and a_3 against the internal rate of return c_1 (in %) and the payback period c_2 (in years), and the decision matrix is given as:

$$D_1 = \begin{bmatrix} 1 & 5 \\ 6 & 5.5 \\ 11 & 6 \end{bmatrix}$$

By Eq. (2), we get $y_{11}^1 = 0$, $y_{21}^1 = 0.5$, $y_{31}^1 = 1$, $y_{12}^1 = 1$, $y_{22}^1 = 0.5$ and $y_{32}^1 = 0$. If the weight vector of criteria is $w = (0.5, 0.5)^T$, based on the weighted arithmetic aggregation operator, we obtain $y_1 = 0.5$, $y_2 = 0.5$ and $y_3 = 0.5$, then $a_1 \sim a_2 \sim a_3$. However, we could not accept this result. There are great differences on the values of c_1 and x_{11} is so inferior that we cannot select a_1, while there are small differences on the values of c_2 and x_{32} is not so bad. Thus, the linear normalization is unable to describe the real differences between different data.

2.2 The Target-Based Vector Normalization

The vector normalization, employed in the MULTIMOORA method [4] and the TOPSIS method [5], is shown as

$$y_{ij}^{\prime 2} = \begin{cases} x_{ij} \Big/ \sqrt{\sum_{i=1}^{m} (x_{ij})^2} & \text{if } c_j \text{ is a benefit criterion} \\ 1 - x_{ij} \Big/ \sqrt{\sum_{i=1}^{m} (x_{ij})^2} & \text{if } c_j \text{ is a cost criterion} \end{cases} \tag{3}$$

The vector normalization aims to normalize the values of all alternatives with respect to criteria to the interval $[0, 1]$. The dimensionless number $y_{ij}^{\prime 2}$ can maintain the size of the original value x_{ij} compared with the linear normalization formulas. Brauers et al. [11] proved that the vector normalization formula is a robust option. But it fails to eliminate the evaluation units of criteria essentially in two aspects: (1) On the one hand, it cannot eliminate the influences of different convertible units of the same criterion function on the result of a MCDM method, such as the length x_{ij} $[m]$ or ϕ_{ij} $[km]$, and the temperature x_{ij} $[C^0]$ or ϕ_{ij} $[F^0]$. These "convertible" units are related as $\phi_{ij} = \alpha x_{ij} + \beta$, $\alpha > 0$. The normalized value $y_{ij}^{\prime 2N}$ is different with respect to different evaluation units of a criterion function [6]. That is to say, $y_{ij}^{\prime 2} = x_{ij} \Big/ \sqrt{\sum_{i=1}^{m} (x_{ij})^2}$ but $\phi_{ij}^2 = \phi_{ij} \Big/ \sqrt{\sum_{i=1}^{m} (\phi_{ij})^2} = (\alpha x_{ij} + \beta) \Big/ \sqrt{\sum_{i=1}^{m} (\alpha x_{ij} + \beta)^2}$, and if $\beta \neq 0$, then $y_{ij}^{\prime 2} \neq \phi_{ij}^2$.

(2) On the other hand, it is unable to eliminate the influence of different units of different criteria on the result of a MCDM method which integrates the information based on the fully compensated aggregation operator. This defect can be verified by Example 2.

Example 2. Suppose that there are three production lines a_1, a_2 and a_3 against the cost c_1 (million) and the production c_2 (number of packages), and the decision matrix is given as:

$$D_2 = \begin{bmatrix} 43 & 1100 \\ 42 & 1050 \\ 41 & 900 \end{bmatrix}$$

By Eq. (3), we get $y_{11}^{\prime 2} = 0.41$, $y_{21}^{\prime 2} = 0.42$, $y_{31}^{\prime 2} = 0.44$, $y_{12}^{\prime 2} = 0.62$, $y_{22}^{\prime 2} = 0.59$ and $y_{32}^{\prime 2} = 0.51$. If the weight vector of criteria is $w = (0.5, 0.5)^T$, based on the weighted arithmetic aggregation function, we obtain $y_1 = 0.515$, $y_2 = 0.505$ and $y_3 = 0.475$. Then $a_1 \succ a_2 \succ a_3$. However, the fact is that the performance of a_2 is the best and a_1 is as bad as a_3. The result is misleading since the differences of alternatives on cost are decreased by the vector normalization which is only able to measure the differences between numbers but ignores the unit differences. In fact, there is a big separation between 42 million and 41 million of cost but a small division between 1100 and 1050 of production packages. Thus, a_2 is superior to a_1. In conclusion, the vector normalization is not suitable to aggregate the values of an alternative under all criteria based on the completely compensated arithmetic aggregation operator.

To fill the gap of normalizing all the benefit, cost and target-based criterion values by the vector normalization, we introduce a target-based vector normalization formula based on the distance of each judgment to the target value, shown as:

$$y_{ij}^{2} = 1 - \frac{|x_{ij} - r_j|}{\sqrt{\sum_{i=1}^{m} (x_{ij})^2 + (r_j)^2}} \tag{4}$$

3 The Double Normalization-Based Multi-aggregation Method

The MULTIMOORA method employs three different aggregation methods to calculate the utility values respectively based on the vector normalization. However, it does not take into account the matching of normalization and aggregation methods. Consequently, the selected alternative is not always nearest to the ideal one. Given that both the target-based linear and vector normalization methods have their advantages and disadvantages, we combine them with different aggregation operators to obtain the different utility values of alternatives. This section aims to propose a DNBMA method to deal with the defects of the existing multi-criterion value methods and obtain a decision result with high reliability and robustness.

3.1 The Subordinate Aggregation Models

In the following, we develop three kinds of aggregation operators based on the two target-based normalization techniques.

3.1.1 The Complete Compensatory Model (CCM)

To measure the closeness to the ideal solution, Zeleny [12] proposed a measurement $r(x;\ p)$, used as an aggregation function, to measure the regret from alternative a_i to the ideal solution a^*.

$$R_{i,p} = \left\{ \sum_{j=1}^{n} \left(\omega_j \left| x_{ij} - r_j \right| \right)^p \right\}^{1/p}, 1 \le p \le \infty \tag{5}$$

where ω_j is the weight of c_j.

As we know, with the increase of p, the weight of the larger value $\omega_j \left| x_{ij} - r_j \right|$ becomes greater and greater. The measurements $r(x;\ p)$ of $p = 1$ and $p = \infty$ are used in the VIKOR method, and the measurement $r(x;\ p)$ of $p = 2$ is used in the classical TOPSIS method as well. Since each criterion has a weight, there is no reason to add a weight to a bigger one. Thus, we employ the measurement $r(x;\ p)$ of $p = 1$ as the first aggregation function of the proposed method. From Sect. 2, we can find that the target-based linear normalization is superior to the target-based vector normalization to combine with the linear aggregation operator to aggregate the values of an alternative under all criteria. Thus, we define the CCM based on the arithmetic weighted aggregation operator as:

$$u_1(a_i) = \sum_{j=1}^{n} \omega_j y_{ij}^1 \tag{6}$$

The alternatives are ranked by $u_1(a_i)$ $(i = 1, 2, \ldots, m)$ in descending order and we get the first type of ranks $r_1(a_i)$ $(i = 1, 2, \ldots, m)$.

Here we let the ranks obtained in this paper be the Besson's mean ranks [13]: If an object a_i ranks the u th position, then $r(a_i) = u$; if both a_i and a_t rank the u th position, then $r(a_i) = r(a_t) = (u + u + 1)/2 = u + 0.5$. For example, if a_1 prefers to a_2, and a_2 is indifferent to a_3, then $r(a_1) = 1$ and $r(a_2) = r(a_3) = 2.5$.

3.1.2 The Un-Compensatory Model (UCM)

To avoid the selected solution having an extremely poor performance under a criterion, we employ the measurement $r(x;\ p)$ with $p = \infty$ and the linear normalized values to compose the second aggregation function, shown as Eq. (7).

$$u_2(a_i) = \max_j \omega_j(1 - y^1_{ij}) \tag{7}$$

The alternatives are ranked by $u_2(a_i)$ $(i = 1, 2, \ldots, m)$ in ascending order and we get the second type of ranks $r_2(a_i)$ $(i = 1, 2, \ldots, m)$.

3.1.3 The Incomplete Compensatory Model (ICM)

Since the linear normalization is unable to reflect the quality of original values, the results would be misleading by the above two aggregation functions in some cases as illustrated by Example 3.

Example 3. Suppose that there are two decision matrices of two MCDM problems with three types of products a_1, a_2 and a_3 against the reliability c_1 and the price c_2 given as:

$$D_3 = \begin{bmatrix} 90\% & 100 \\ 94\% & 105 \\ 98\% & 110 \end{bmatrix}, \; D_4 = \begin{bmatrix} 75\% & 100 \\ 85\% & 105 \\ 95\% & 110 \end{bmatrix}$$

By Eqs. (6) and (7), we obtain the same results regarding to D_3 and D_4 that a_2 is the optimal solution and $a_1 \sim a_3$. It is easy to accept the result of D_3 but hard to accept the result of D_4 since the alternatives associated to D_4 have a big gap in quality, but a small gap in price. There is a fact that the alternative a_1 is of extremely poor quality but not good price, the alternative a_2 is of medium quality and slightly bad price, and the alternative a_3 is of extremely good quality and not so bad price. According to our intuition, we may select a_3 and deem a_1 as the worst.

The above two aggregation functions fail to consider the size of the value itself. To solve this defect and make the result more reliable, we employ the vector normalized values combined with the multiplicative form to propose the third aggregation function as Eq. (8).

$$u_3(a_i) = \prod_j (y^2_{ij})^{\omega_j} \tag{8}$$

The multiplicative formula can reflect people's preferences that the former case is superior to the latter case. That is, the good performance of an alternative cannot fully compensate for poor performance. The alternatives are ranked by $u_2(a_i)$ $(i = 1, 2, \ldots, m)$ in ascending order and we obtain the third type of ranks $r_3(a_i)$ $(i = 1, 2, \ldots, m)$.

3.2 The Subordinate Utilities and Ranks Integration

In the final phase, we need to obtain a comprehensive ranking of the alternatives by integrating the results of the above three models. The three models can be deemed as three criteria: CCM (denoted by C_1), UCM (denoted by C_2) and ICM (denoted by C_3). Each alternative a_i have two kinds of evaluation values: the utility value $u_y(a_i)$ and the rank $r_y(a_i)$ with respect to each criterion $C_y (y = 1, 2, 3)$. Obviously, this is a MCDM problem composed by two decision matrixes: the utility value decision matrix $D(u) = \left[u_y(a_i)\right]_{m\times3}$

and the ranking decision matrix $D(r) = \left[r_y(a_i)\right]_{m \times 3}$. $u_y(a_i)$ can be normalized by the vector normalization formula:

$$u_y^N(a_i) = \frac{u_y(a_i)}{\sqrt{\sum_{i=1}^{m} \left(u_y(a_i)\right)^2}}, y = 1, 2, 3 \tag{9}$$

We then define the integrated score by a weighted Euclidean distance formula as Eq. (10).

$$S_i = \sqrt{\varphi\left(u_1^N(a_i)\right)^2 + (1-\varphi)\left(\frac{m - r_1(a_i) + 1}{m(m+1)/2}\right)^2} - \sqrt{\varphi\left(u_2^N(a_i)\right)^2 + (1-\varphi)\left(\frac{r_2(a_i)}{m(m+1)/2}\right)^2} + \sqrt{\varphi\left(u_3^N(a_i)\right)^2 + (1-\varphi)\left(\frac{m - r_3(a_i) + 1}{m(m+1)/2}\right)^2} \tag{10}$$

where φ is the coefficient to highlight the importance between the subordinate utility value and the subordinate rank. The final rank set $R = \{r(a_1), r(a_2), \ldots, r(a_m)\}$ is determined in descending order of S_i $(i = 1, 2, \ldots, m)$.

The DNBMA method is summarized as follows:

Step 1. Calculate the target-based linear normalization values by Eq. (2) and the target-based vector normalization values by Eq. (4). Go to next step.
Step 2. Compute the utility values $u_1(a_i)$, $u_2(a_i)$ and $u_3(a_i)$ $(i = 1, 2, \ldots, m)$ based on the CCM (as Eq. (6)), the UCM (as Eq. (7)) and the ICM (as Eq. (8)), respectively, and then determine the three types of subordinate ranks $r_y(a_i)$ $(y = 1, 2, 3;\ i = 1, 2, \ldots, m)$. Go to next step.
Step 3. Normalize the utility values $u_y(a_i)$ $(y = 1, 2, 3;\ i = 1, 2, \ldots, m)$ by Eq. (9). Go to next step.
Step 4. Integrate the subordinate normalized utility values and the subordinate ranks by Eq. (10). Determine the final ranking and ends the algorithm.

4 A Case Study and Some Comparative Analyses

To promote the innovation of small and medium iron and steel enterprises, one city decides to choose an optimal green enterprise to reward. There are four candidates a_1, a_2, a_3, a_4. The evaluation criteria are R & D investment accounting for the proportion of total investment (c_1, target criterion in %), the number of developers (c_2, benefit criterion in number), sales revenue of new products (c_3, benefit criterion in 10^6 million), and comprehensive energy consumption (c_4, cost criterion in number). Suppose that the target value of criterion c_1 is 6%. The decision matrix D is determined as:

$$D = \begin{bmatrix} 15.4 & 10 & 0.8 & 1 \\ 10.9 & 25 & 16.5 & 1.1 \\ 1.8 & 14 & 5.7 & 0.9 \\ 4.5 & 8 & 10.3 & 0.95 \end{bmatrix}$$

We solve the case by the DNBMA method. The target-based linear normalized values are computed by Eq. (2), shown in Table 1 and the target-based vector normalized values are calculated by Eq. (4), shown in Table 2.

Table 1. The target-based linear normalized values

	c_1	c_2	c_3	c_4
a_1	0	0.12	0	0.5
a_2	0.48	1	1	0
a_3	0.55	0.35	0.31	1
a_4	0.84	0	0.61	0.75

Table 2. The target-based vector normalized values

	c_1	c_2	c_3	c_4
a_1	0.54	0.63	0.4	0.95
a_2	0.76	1	1	0.91
a_3	0.79	0.73	0.59	1
a_4	0.93	0.58	0.76	0.98

Suppose that the criterion weights are the same as $\omega_j = 0.25$, $j = 1, 2, 3, 4$. By Eq. (6), we obtain $u_1(a_1) = 0.16$, $u_1(a_2) = 0.62$, $u_1(a_3) = 0.55$ and $u_1(a_4) = 0.55$. Then $R_1 = \{4, 1, 2.5, 2.5\}$. By Eq. (7), we obtain $u_2(a_1) = 0.25$, $u_2(a_2) = 0.25$, $u_2(a_3) = 0.17$ and $u_2(a_4) = 0.25$. Then $R_2 = \{3, 3, 1, 3\}$. By Eq. (8), we obtain $u_3(a_1) = 0.6$, $u_3(a_2) = 0.92$, $u_3(a_3) = 0.76$ and $u_3(a_4) = 0.8$. Then $R_3 = \{4, 1, 3, 2\}$. Let $\varphi = 0.5$. According to Eqs. (9) and (10), we obtain the values $S_1 = -0.02$, $S_2 = 0.59$, $S_3 = 0.53$ and $S_4 = 0.41$. Thus we get the ranking relation $a_2 \succ a_3 \succ a_4 \succ a_1$, which shows that a_2 is an optimal innovation enterprise. a_2 have good performances on all criteria except criterion c_4. The government can encourage a_2 to reduce the comprehensive energy consumption. The enterprise a_1 invests much money for innovation, but the sales revenue of new products is low. It can introduce more developers to obtain the advanced technology.

To make comparison, we also solve the case by other MCDM methods. The results are shown in Table 3.

Comparative Analysis: We obtain different results derived by the target-based VIKOR method. From calculation process, we find that the "individual regret value" has a huge impact on the utility values. When using the linear normalization formula, the "individual regret value" of a_3 is significantly superior to a_1, which leads to the

Table 3. The results derived by different Target-based MCDM methods

Methods	Utility values				Rankings
	c_1	c_2	c_3	c_4	
Target-based TOPSIS [8]	0.25	0.58	0.55	0.51	$a_2 \succ a_3 \succ a_4 \succ a_1$
Target-based VIKOR [9]	0	0.5	0.87	0.36	$a_3 \succ a_2 \succ a_4 \succ a_1$
Target-based MULTIMOORA [10]	0.47,0.16,0.46	0.77,0.16,0.71	0.69,0.13,0.66	0.68,0.16,0.65	$a_2 \succ a_3 \succ a_4 \succ a_1$
The proposed DNBMA method	–0.02	0.59	0.53	0.41	$a_2 \succ a_3 \succ a_4 \succ a_1$

final result $a_3 \succ a_1$. Besides, the result of the target-based VIKOR is sensitive to the threshold depicting the relative importance between the "group utility" and "individual regret value", and it is hard for decision maker to select a suitable threshold. Despite that the same ranks are derived by the target-based TOPSIS method, the target-based MULTIMOORA method and the proposed DNBMA method, but there are different utility values since different normalization and aggregation techniques are employed. The target-based TOPSIS method only calculates the "group utility" by the arithmetic weighted aggregation operator based on the linear normalization but ignores the "individual regret value". There is a fact that the same results are obtained by the ratio system model and the full multiple form model of the target-based MULTIMOORA method due to the same normalization values are utilized. In DNBMA method, we obtain different normalized valued from the linear and the vector normalization formula. Besides, the different subordinate utility values are calculated by three aggregation models. After integrating these utility values and subordinate ranks, we obtain a robust ranking result.

5 Conclusions

We proposed a new MCDM method named DNBMA, which can handle the benefit, cost and target-based criteria at the same time. The proposed method is based on two normalization tools: the target-based linear and the target-based vector normalization formulas, and consists three aggregation models: the CCM with arithmetic weighted aggregation, the UCM with weighted maximization formula and ICM with the geometric weighted aggregation. These subordinate methods depict the performance of alternatives from different aspects, which make the DNBMA method robust. After making comparative analysis with other MCDM methods based on a case study, the advantages of the proposed method were highlighted. As future studies, the DNBMA method can be combined with different fuzzy information, such as the hesitant fuzzy linguistic term set to deal with the subjective MCDM problems.

Acknowledgements. The work was supported by the National Natural Science Foundation of China (71501135, 71771156), the Fundamental Research Funds for the central Universities (No. YJ201535), and the Scientific Research Foundation for Excellent Young Scholars at Sichuan University (No. 2016SCU04A23).

References

1. Segura, M., Maroto, C.: A multiple criteria supplier segmentation using outranking and value function methods. Expert Syst. Appl. **69**, 87–100 (2017). https://doi.org/10.1016/j.eswa.2016.10.031
2. Jahan, A., Edwards, K.L.: A state-of-the-art survey on the influence of normalization techniques in ranking: Improving the materials selection process in engineering design. Mater. Des. **65**, 335–342 (2015). https://doi.org/10.1016/j.matdes.2014.09.022
3. Risawandi, R.R.: Study of the simple multi-attribute rating technique for decision support. Int. J. Sci. Res. Sci. Technol. **2**(6), 491–494 (2016)
4. Brauers, W.K.M., Zavadskas, E.K.: Project management by MULTIMOORA as an instrument for transition economies. Ukio Technologinis Ir Ekonominis Vystymas. **16**(1), 5–24 (2010). https://doi.org/10.3846/tede.2010.01
5. Chen, S.J., Hwang, C.L.: Fuzzy multiple attribute decision making methods. In: Fuzzy Multiple Attribute Decision Making. Lecture Notes in Economics and Mathematical Systems, vol. 375, pp. 289–486. Springer, Heidelberg (1992). https://doi.org/10.1007/978-3-642-46768-4_5
6. Opricovic, S., Tzeng, G.H.: Compromise solution by MCDM methods: a comparative analysis of VIKOR and TOPSIS. Eur. J. Oper. Res. **156**(2), 445–455 (2004). https://doi.org/10.1016/S0377-2217(03)00020-1
7. Liao, H.C., Xu, Z.S., Zeng, X.J.: Hesitant fuzzy linguistic VIKOR method and its application in qualitative multiple criteria decision making. IEEE Trans. Fuzzy Syst. **23**(5), 343–1355 (2015). https://doi.org/10.1109/TFUZZ.2014.2360556
8. Jahan, A., Bahraminasab, M., Edwards, K.L.: A target-based normalization technique for materials selection. Mater. Des. **35**, 647–654 (2012). https://doi.org/10.1016/j.matdes.2011.09.005
9. Jahan, A., Mustapha, F., Ismail, M.Y., Sapuan, S.M., Bahraminasab, M.: A comprehensive VIKOR method for material selection. Mater. Des. **32**(3), 1215–1221 (2011). https://doi.org/10.1016/j.matdes.2010.10.015
10. Hafezalkotob, A., Hafezalkotob, A.: Comprehensive MULTIMOORA method with target-based attributes and integrated significant coefficients for materials selection in biomedical applications. Mater. Des. **87**, 949–959 (2015). https://doi.org/10.1016/j.matdes.2015.08.087
11. Brauers, W.K.M., Zavadskas, E.K.: The MOORA method and its application to privatization in a transition economy. Control Cybern. **35**(2), 445–469 (2006)
12. Zeleny, M.: Multiple Criteria Decision Making. McGraw-Hill, New York (1982). https://doi.org/10.1007/978-3-642-45486-8
13. Wu, X.L., Liao, H.C.: An approach to quality function deployment based on probabilistic linguistic term sets and ORESTE method for multi-expert multi-criteria decision making. Inf. Fusion **43**, 13–26 (2018). https://doi.org/10.1016/j.inffus.2017.11.008

Design of a Decision Support System for Buried Pipeline Corrosion Assessment

Laurence Boudet[1(✉)], Jean-Philippe Poli[1], Alicia Bel[2], François Castillon[3], Frédéric Gaigne[3], and Olivier Casula[2]

[1] CEA, LIST, Data Analysis and System Intelligence Laboratory, 91191 Gif-sur-Yvette cedex, France
{laurence.boudet,jean-philippe.poli}@cea.fr
[2] CEA Tech Aquitaine, 33600 Pessac, France
{alicia.bel,olivier.casula}@cea.fr
[3] TIGF, 64000 Pau, France
{francois.castillon,frederic.gaigne}@tigf.fr

Abstract. Maintaining the level of integrity of pipeline networks to guarantee at least a reliable and safe service is a challenge operators of such networks are facing everyday. TIGF is one of the French operator which manages 5000 km of pipelines in the south-west quarter of France. This paper presents a decision-making tool which automatically ranks the pipeline sections regarding the risk of deterioration (damages and corrosion) and the gravity of the consequences, indicating which pipeline sections should be excavated. The tool relies on a fuzzy expert system which gathers 26 input variables, processes more than 300 rules, classifies the risk of deterioration into 7 classes and estimates the gravity. The rules are a formalization of human expertise: the fuzzy logic helps to tackle the vagueness of their knowledge and the measurement inaccuracy of some of the 26 input variables. The method has been tested on past excavations to assess its performances.

Keywords: Corrosion · Pipeline networks · Risk assessment
Decision-making · Fuzzy expert system

1 Introduction

Everywhere in the world, high-pressure pipelines are used to transport gaz from the production and storage sites to the customers. The major challenge with such networks of pipelines is to maintain their level of integrity to guarantee at least a reliable and safe service. Regarding the conditions of the pipelines (ground, buried, subsea, etc.), their environment (soil, etc.) and their features (coating, etc.), different kinds of corrosion may damage them [6,8,14]. The phenomenon is too complex and too little understood to be modeled analytically, or only in specific context (coating material, soil type and features, pipe age, ...).

However, different measures can help to identify which pipeline sections are affected by a certain type of corrosion. We can distinguish several approaches

J. Medina et al. (Eds.): IPMU 2018, CCIS 855, pp. 74–85, 2018.
https://doi.org/10.1007/978-3-319-91479-4_7

to exploit these measures. Bayesian techniques have been used to incorporate uncertainty and measurement errors [4,5,14] and to either compute the probabilities of corrosion occurrence and consequences or to assess the size of the corrosion. Other papers introduce a probabilization of physical models like [9] which predicts the corrosion remaining life. All those methods are also related to Monte-Carlo simulation to overcome the lack of data and suffer from the difficulty to choose the most proper distributions. Another way of taking the uncertainty and inaccuracy into account is performed by fuzzy logic [10,11,15]: various qualitative and quantitative factors are considered in assessing the security of pipeline network. The authors fuzzify classical models used in risk assessment. We can also cite [1] in which the authors introduce an expert system to suggest the adequate coating regarding several quantitative and qualitative parameters. It ranks all the available materials and the most suitable one is chosen. In a more anecdotal way, other papers apply Multi-Attribute Utility Theory (MAUT) [3] and machine learning [7] to risk assessment.

In this article, we present an application of fuzzy expert system to tackle both the recognition of the type of deterioration and the risk assessment. Fuzzy expert systems can handle a cold start, i.e. the lack of data at the start of the project, knowledge vagueness and measurement uncertainty. Moreover, as corrosion depends on environmental conditions, the decision making tool can be easily adapted from a region to another.

We focus on the network managed by our partner, TIGF, responsible of the pipeline network in the south-west quarter of France. The whole network is buried and TIGF is facing corrosion of different natures. Ground inspection must help them to ensure a section of the network is affected by corrosion. The difficulty resides in the fact that several criteria have to be merged to make the decision. Our approach is based on knowledge modeling and testing on data.

The paper is structured as follows. The next section introduces the context of corrosion of non-piggable pipelines. Then, Sect. 3 motivates the choices made for the decision support system (DSS) which has been built for TIGF in order to assess risk and corrosion types. Section 4 describes the knowledge modeling of human experts with fuzzy logic and introduces the user interface of the system. The results of the application of the tool are presented in Sect. 5. Finally, we draw a conclusion and some perspectives to this work.

2 The Case-Study

Pipeline pigging is an effective way to accurately locate steel defects and metal loss. Unfortunately, this method is inappropriate for some pipe configurations, such as small diameter or multi-diameter pipes. Therefore, above ground inspection surveys are conducted to gather information about the whole network condition. TIGF mainly uses the Direct Current Voltage Gradient (DCVG) technique.

We explain in this paragraph the principle of the DCVG technique. If the pipe is exposed at holidays in its protective coating, the current impressed by the cathodic protection system will flow from the soil into the bare steel. It results in voltage gradients in the soil surrounding the defect. The DCVG method consists in pulsing the input current signal and detecting associated voltage gradients in the soil above the pipeline, that betray the presence of a soil-metal interface. To this end, an operator performs regular measures with a milli-voltmeter of the voltage drop between two electrodes placed on the soil surface at a distance that remains constant (about 1.5 m). As the operator approaches a coating defect, he observes an increasing pulsing signal. This signal finally stabilizes then decreases as the defect is passed.

Each defect severity is characterized by its value of %IR, which is computed from DCVG measures. Then, thanks to calibrated references, the size of the steel surface exposed can be estimated.

DCVG surveys return alerts on the pipeline protective coating. However, it does not inform about the cathodic protection state, nor does it imply a real metal deterioration. To better assess the risk of a pipeline defect, TIGF gathers additional data:

- the pipeline specific features (age, type of coating, ...);
- the pipeline environment (presence of stray currents, soil resistivity, soil bacteria, ...);
- the history of the pipeline cathodic protection.

All of this information is carefully examined with multiple risks in mind. In addition to mechanical attacks and high-voltage damages, a typology of 5 different types of corrosion is considered (see Fig. 1): stray current corrosion, corrosion caused by alternating current (AC corrosion), corrosion under a disbonded shielding coating, bacterial corrosion and insufficient cathodic protection. Finally, the gravity of the consequences that a severe metal defect would have is a crucial parameter that is also carefully taken into consideration. A pivotal factor is the pipeline proximity from any public location or infrastructure.

Conducting a systematic analysis of the thousands of coating defects that are detected each year by DCVG surveys is a real challenge. Indeed, since the excavation of a pipeline section is very expensive, only few anticipated defects can be checked. Thus, it requires to apply a wide expertise in a consistent fashion to a large variety of configurations. In this context, a decision support tool offers clear benefits.

3 Decision Support System Design

In this section, we first motivate the choices made for the DSS according to the context and the constraints. Then, we explain the different phases of its life cycle.

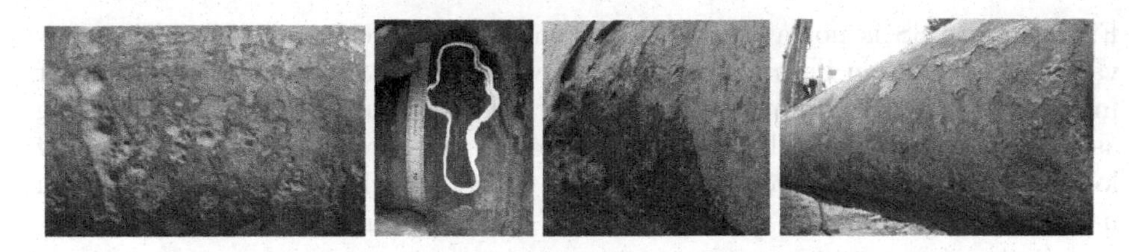

Fig. 1. Example of different corrosion shapes and surfaces.

3.1 Technology Choice

The proposed DSS assists human experts in identifying the most critical coating defects for excavation. There are quite a lot of available measurements or information (26 variables per coating defect): some of them are qualitative, others are quantitative. There are a lot of individuals (several thousands of defects) in the first data set but very few of them were labeled at the beginning of the study (around 50 for one DCVG survey). There are 7 risks of metal deterioration. This data set is unbalanced since one deterioration type represents two thirds of actual defects and some are missing. Moreover, each defect should not be considered independent from the others because a set of defects (even a lot of them sometimes) can be detected at different locations of the same pipeline. Their features are then correlated. Thus, it was not possible to learn the features of the defects because of the small number of labeled individuals available at the beginning of the study nor to apply clustering techniques because of the difficulty to find a distance dealing with both quantitative and qualitative features and separating pipeline sections with or without deterioration.

Despite these observations, the DSS should be usable immediately because it is not conceivable to wait a series of yearly acquisition campaigns. Thus, it has to exploit the expert background and experience at TIGF. It will allow to formalize and to structure this knowledge on the one hand, and to treat all the coating defects in an homogeneous and systematic way on the other hand. Unfortunately, underground corrosion phenomena comprehension is much more a matter of vague and uncertain knowledge and understanding than a precise and definitive knowledge. Indeed, there are multiple factors implied in corrosion formation and it is a local phenomenon because it depends on soil features. Moreover, coating used for pipes as well as the type of protections changed over the last 70 years because of acquired experience on coating ageing and corrosion and improvements of materials.

Obviously, the DSS should improve the effectiveness of realized excavations compared to the ones decided without it. Moreover, a higher success rate will allow TIGF to reduce the number of excavations following DCVG surveys.

Taking into account these different points, we proposed to use a fuzzy expert system to model expert knowledge. It may be designed from expert interviews before the availability of a data set. Fuzziness is helpful to deal with vagueness and uncertainty. Two other features pleaded for a fuzzy expert system.

Firstly, the DSS is not a black box: its suggestions are justified thank to activated rules and can be understood by the user. Secondly, if we found out that modeled knowledge is imprecise or incorrect, it can be updated by only changing associated fuzzy rules. These ones are saved in external files and automatically loaded at the next start of the DSS. Thus, expert knowledge can be refined with new experiences and the virtuous circle of knowledge improvement is possible.

3.2 DSS Life Cycle

There are five steps in the design and exploitation of the DSS:

1. **System design**: the DSS is designed with human experts at TIGF. Knowledge is gathered and modeled with fuzzy rules. A dedicated graphic user interface (GUI) is specified and implemented. During this phase, we decided to use some features that were available in the databases but not extracted for this purpose yet. Then, the first annual DCVG survey with actual values was formatted more or less automatically to be used by the DSS.
2. **Manual corrections**: The DSS results are evaluated and analyzed with the first data set. Some rule improvements are identified by comparing the actual values of the defects to what was modeled by knowledge elicitation. Then, more data are collected and formatted during this step. Several annual DCVG surveys are automatically extracted. There are 330 labeled metal defects in this second data set (including the first one). Knowledge update is natural when using fuzzy rules, and is very easy by using the GUI [12] of our fuzzy expert system. We followed two ways to discover corrections to apply. Firstly, parameters used for defining fuzzy sets of some variables are refined by a statistical analysis when sufficient data are available. Secondly, this data set enables to question about the influence of some parameters on some risks.
3. **First exploitation**: The DSS is used to evaluate the excavations to select for the next yearly DCVG survey. The software is used to navigate between coating defects, visualize, analyze and understand possible risks. Moreover, reports are automatically generated on most possible risks.
4. **Automatic improvements**: Data registered in case of swabbing of pipelines at accessible location (in particular, at manholes) are automatically extracted with their actual values. The latter are much more numerous than the labeled ones obtained by DCVG survey (several thousands). Optimization of fuzzy rules is then possible with the second and the third data sets for the most frequent observed risks. After validation by human experts at TIGF, correction of fuzzy sets are made with the dedicated GUI [12]. This step is not introduced in this article.
5. **Full exploitation**: After fine tuning of fuzzy rules by optimization, the DSS is used to choose the defects to excavate for the next year. Efficiency and usefulness of the DSS can be assessed by different considerations: success rate of realized excavations, spent time to select defects to be excavated, justification of decisions and evolution of repair costs. The latter depend on the number of defects selected for excavation which will decrease in several years if the success rate raises.

4 Knowledge Modeling and Visual Tool

This section introduces the approach used to gather and model human expertise in this domain and shows the user interface created for this application.

4.1 Human Expert Knowledge Modeling

The DSS evaluates the criticity of each coating defect in a large variety of configurations in a structured and systematic way. The work conducted to make the knowledge of human experts fully explicit. First of all, we identified nine indicators including seven risks of deterioration which must be distinguished. Some works close to the pipe may simply damage it if there is no mechanical protection. Another pipe damage may arise because of a close high-voltage line. Moreover, five corrosion types are considered: stray current corrosion, corrosion caused by alternating current, corrosion under a disbonded shielding coating, bacterial corrosion and insufficient cathodic protection. The development of each corrosion type is multi-factorial: it depends on intrinsic factors of the pipe (age, type of coating, ...) and extrinsic factors given by environmental features (presence of stray currents, soil resistivity, soil bacteria, ...). The %IR based on DCVG measures gives an information about the estimated surface of the coating defect of the pipe. The estimated risks are considered with a factor that indicates a higher risk of deterioration based on background (history of the pipeline) and a factor that assesses the severity of a potential metal defect (its proximity from any public location or infrastructure). All this information is available in knowledge management systems at TIGF in different forms: either quantitative, categorical or boolean. Only one part of this information was already used to assess the risk of each coating defect. In some cases, only the simplest raw information was used such as the presence of bacteria in the soil while a concentration was known.

After having identified the different inputs and outputs of the DSS, we have to understand the influence of each input on the outputs. During the interviews, a lot of expressed knowledge was in the form "the higher the value of X, the higher the risk of A" like in the rule "the higher the age of the pipe, the higher the risk of the corrosion of type A", or at the contrary the form "the lesser the resistivity, the higher the risk of the corrosion of type A". We could have used powerful and synthetic fuzzy rules like fuzzy gradual rules or those based on "all the more" clauses [2] for instance. However, there were generally several factors that influence each risk and we did not know which relation between the inputs and the outputs should be chosen. Thus, we decided to adopt a grid structure for modeling the input - output relations for each module of fuzzy rules. Indeed, it is very easy to understand for the corrosion domain experts who were not used to fuzzy logic. Moreover, it is simple to update a conclusion according to one (or several) influencing parameter(s). However, we did not follow a strict flat grid structure approach. When possible, intermediate variables were introduced in order to synthesize pieces of information at intermediate level. The latter ones are used into other modules of fuzzy rules thanks to chaining. This provides a hierarchical structure of the fuzzy system [13] that makes it more concise and understandable. It is used for instance for establishing a bacterial risk due to the soil characteristics. This risk depends on 4 input variables of pipe environment. Then, the bacterial risk is considered with the coating material to determine the bacterial corrosion risk of the pipe.

Thus, we modeled each variable (either input or output) by a linguistic variable and built fuzzy rules to make the expert knowledge more explicit. We chose to model each risk of pipe deterioration with the same set of fuzzy sets (see Fig. 2): the risks come from null to very high with 4 intermediate levels. The risk suggested by the pipeline history is rated with the same levels. With the Mamdani inference system, the numerical defuzzified value of each risk belongs to [0, 100]. A boolean alert about the history is given by coupling the potential corrosion risk with the one implied in the history of the pipe close to the coating defect when known. Indeed, an old mechanical attack does not tell anything on a possible corrosion due a bacteria for instance. Finally, the severity of the consequences is evaluated with four fuzzy terms from low to critical.

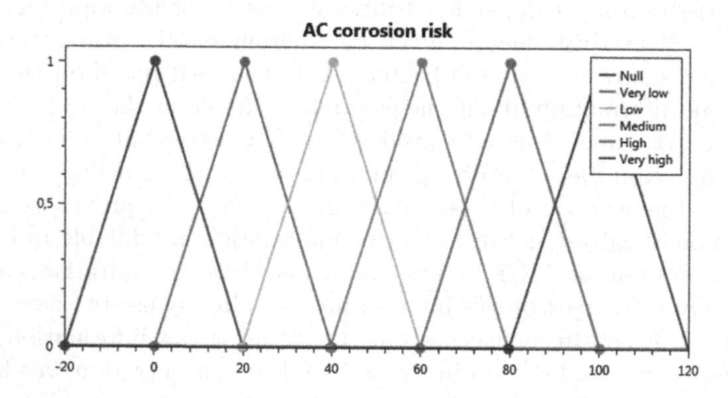

Fig. 2. Fuzzy sets for risk level, in this case for alternating current corrosion

The Table 1 shows the risk level of one type of corrosion according to four input variables for one specific coating. There are 24 rules for this example. It represents a gradual influence of each parameter on the risk level. The worst case occurs when the diameter of the pipe is large, the pipe is old (more than 50 years old), the resistivity is low and the estimated defect surface is large.

On the one hand, the number of possible items of qualitative variables is given by the specifications of the database system. They are often described by two or four terms, but sometimes by many more like in the case of the coating type. We often had to use the same granularity to build the fuzzy rules. Sometimes, we were able to group some items to build fuzzy sets, like the one shown in Fig. 3 telling which coating type is sensitive to stray currents.

On the other hand, quantitative variables description is more flexible because of their continuous domain. We often restrict their description to two or three fuzzy sets due to the number of combinations to consider, like in the case of the

Table 1. Example of risk level determination of corrosion under a disbanded shielding coating for pipeline with bituminous coating according to four input variables. For Diameter: S = small, L = large; for Resistivity: L = low, H = high; for surface: S = small, M = medium, L = large

Level of risk	Diameter is S & age is low	Diameter is S & age is high	Diameter is L & age is low	Diameter is L & age is high
Resistivity is H & surface is S	Null	Null	Very low	Low
Resistivity is L & surface is S	Null	Very low	Low	Medium
Resistivity is H & surface is M	Very low	Low	Low	Medium
Resistivity is L & surface is M	Very low	Low	Medium	High
Resistivity is H & surface is L	Low	Medium	Medium	High
Resistivity is L & surface is L	Low	Medium	High	Very high

estimated defect surface (see Fig. 4). Indeed, the number of fuzzy rules grows exponentially according to the number of inputs, and it becomes difficult to tell if a parameter is more important than another for each corrosion risk. We ended the modeling with a little more than 300 rules for assessing 9 indicators.

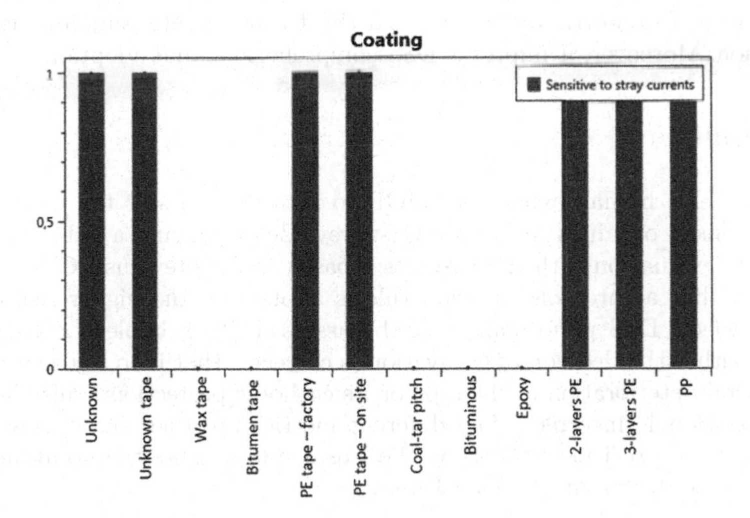

Fig. 3. Coating types that are sensitive to stray currents

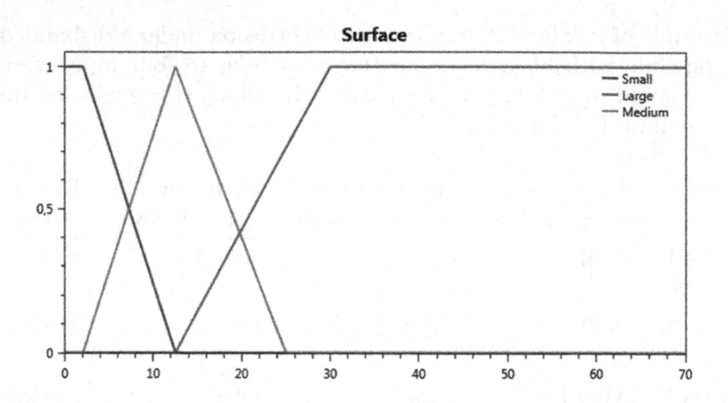

Fig. 4. Fuzzy sets defined for the estimated defect surface

4.2 Visual Tool for Case Assessment

A GUI (Fig. 5) has been designed for the end-users according to their specifications. It is dedicated to the analysis and prioritization of DCVG defects. It embeds the inference engine that applies fuzzy rules onto all DCVG defects loaded for analysis. It offers a synthetic view of them with 9 visual indicators. Seven of them stand for an assessed deterioration risk. The two last ones consider an alert about the history of the pipe and its surroundings and the severity level of the consequences of a potential pipeline failure. The list of DCVG defects can be sorted by these scores so that the most critical cases stand out in a straightforward manner. When a particular defect is selected, its parameter values, the defuzzified values of each indicator and associated activated rules are displayed in specific areas. It is worth mentioning that this feature greatly simplifies feedback integration. Moreover, it improves DSS comprehension and adoption.

5 Results

The DSS helps human experts at TIGF to identify and sort the most critical coating defects obtained by a DCVG survey. Before having a field experience return, the evaluation of the DSS results is based on the previous DCVG surveys. Assuming that a threshold decision rule is applied on the higher risk of each coating defect, DSS performance can be assessed like a boolean classification method: either the decision of excavation is correct – that is to say that there is a real metal deterioration of the pipe or its cathodic protection is insufficient –, or the decision is incorrect. The different metrics are then the sensibility (or recall) and the precision defined by the Eqs. 1 and 2 using the notation of the well known confusion matrix (see Table 2):

$$\mathbf{Sensibility} = TP/(TP + FP) \tag{1}$$

$$\mathbf{Precision} = TP/(TP + FN). \tag{2}$$

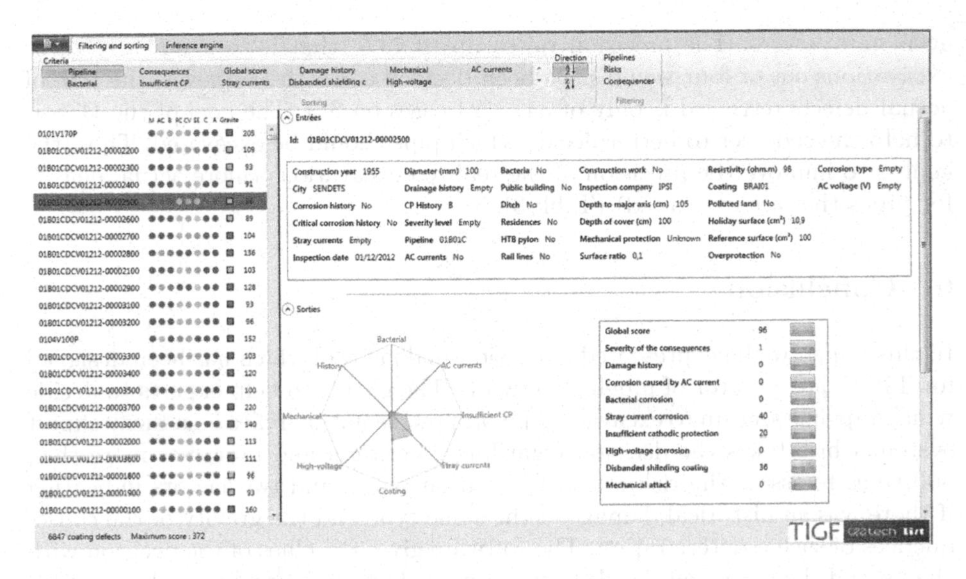

Fig. 5. Dedicated user interface with a synthetic view of all the DCVG defects (left part), and a detailed view of a particular defect with its parameter values and the inferred corrosion risks (right part).

Table 2. Confusion matrix

		System decision	
		Positive	Negative
Actual value	Real defect	True Positive (TP)	False Negative (FN)
	No defect	False Positive (FP)	True Negative (TN)

Table 3. Results according to a threshold T

Metrics	Threshold		
	T = 0	T = 70	T = 89
Precision	20%	35%	73%
Sensibility	100%	55%	34%

Applying a null threshold on the highest risk of each defect is the situation of reference, before having the DSS. In that case, only 20% of excavations are correctly realized (65 cases on 330 excavations). When we use a high threshold ($T = 70$) on the risks estimated by the DSS, 35% of the excavated pipes would have really been damaged. A little bit more than half of actual defects are retrieved (36 on 65). With a higher threshold ($T = 89\%$), the selection is even more severe. The precision raises up to 73%, that is to say nearly three excavations out of four would have been decided correctly. However, the rate of actual defects retrieved is only of 34% (24 cases on 65). The goal of the DSS is to help the end-user to better decide which pipe should be excavated. Thus, the goal is to improve the precision of the realized excavations (more actual defects for pipes that are excavated) (Table 3).

6 Conclusion

In this paper, we have presented a decision-making tool which has been designed for TIGF, an operator of pipeline network. The goal is to help experts to decide which pipeline section to excavate. The proposed solution relies on a fuzzy expert system which takes the decision regarding 26 criteria and processes more than 300 rules to assess the risk of 7 deterioration types, and two factors indicating if there was an historical damage of the same type and the gravity of the consequences of such a deterioration. The software gives the clues of the decision with charts and the activated rule list, which allow human experts to make the final decision from this explanation of the ranking. Moreover, the software is able to automatically generate reports.

Fuzzy logic is used here to take into account the vagueness of the knowledge of human experts, gathered by iterative interviews: the rules are close to natural language thanks to the linguistic variables and the fuzzy sets avoid crisp thresholds which have low sense for an experience-based knowledge. Moreover, the ground measures are performed by different workers with different tools and thus come with measurement uncertainty which are easily handled with fuzzy logic.

The results show an improvement of the decision thanks to the tool while it is only a formalization of the knowledge of experts at TIGF: up to 73% of the decisions are accurate and reveal a real defect.

Contrary to other methods, this approach is adaptable to other pipeline network operators: since it relies on a fuzzy expert system, the rule base can be easily adapted to another region of the world if human experts are available to share their knowledge.

The perspectives of this work consist in improving the tool to make it capable of taking advantage of past results to parameterize the rule base and learn new rules.

References

1. Athanasopoulos, G., Riba, C.R., Athanasopoulou, C.: A decision support system for coating selection based on fuzzy logic and multi-criteria decision making. Expert Syst. Appl. **36**(8), 10848–10853 (2009)
2. Bouchon-Meunier, B., Laurent, A., Lesot, M.J., Rifqi, M.: Strengthening fuzzy gradual rules through "all the more" clauses. In: IEEE International Conference on Fuzzy Systems (FUZZ-IEEE), Barcelona, Spain, pp. 2940–2946 (2010)

3. Brito, A., de Almeida, A.: Multi-attribute risk assessment for risk ranking of natural gas pipelines. Reliab. Eng. Syst. Saf. **94**(2), 187–198 (2009)
4. Cagno, E., Caron, F., Mancini, M., Ruggeri, F.: Using AHP in determining the prior distributions on gas pipeline failures in a robust bayesian approach. Reliab. Eng. Syst. Saf. **67**, 275–284 (2000)
5. Caleyo, F., Valor, A., Alfonso, L., Vidal, J., Perez-Baruch, E., Hallen, J.: Bayesian analysis of external corrosion data of non-piggable underground pipelines. Corros. Sci. **90**(Suppl. C), 33–45 (2015)
6. Cole, I., Marney, D.: The science of pipe corrosion: a review of the literature on the corrosion of ferrous metals in soils. Corros. Sci. **56**(Suppl. C), 5–16 (2012)
7. El-Abbasy, M., Senouci, A., Zayed, T., Parvizsedghy, L., Mirahadi, F.: Unpiggable oil and gas pipeline condition forecasting models. J. Perform. Constr. Facil. **30**(1) (2014). https://doi.org/10.1061/(ASCE)CF.1943-5509.0000716
8. Kishawy, H.A., Gabbar, H.A.: Review of pipeline integrity management practices. Int. J. Press. Vessels Pip. **87**(7), 373–380 (2010)
9. Li, S.X., Yu, S.R., Zeng, H.L., Li, J.H., Liang, R.: Predicting corrosion remaining life of underground pipelines with a mechanically-based probabilistic model. J. Petrol. Sci. Eng. **65**(3), 162–166 (2009)
10. Markowski, A.S., Mannan, M.S.: Fuzzy logic for piping risk assessment (pfLOPA). J. Loss Prev. Process Ind. **22**(6), 921–927 (2009)
11. Martins, A.G., Nicholson, E.: A fuzzy logic model designed for quantitative risk analysis based on ECDA data. In: CORROSION, Dallas (2015)
12. Poli, J.P., Laurent, J.P.: Touch interface for guided authoring of expert systems rules. In: IEEE International Conference on Fuzzy Systems (FUZZ-IEEE), pp. 1781–1788 (2016)
13. Torra, V.: A review of the construction of hierarchical fuzzy systems. Int. J. Intell. Syst. **17**(5), 531–543 (2002)
14. Yang, Y., Khan, F., Thodi, P., Abbassi, R.: Corrosion induced failure analysis of subsea pipelines. Reliab. Eng. Syst. Saf. **159**(Suppl. C), 214–222 (2017)
15. You, Q., Fan, J., Zhu, W., Bai, Y.: Risk assessment of urban gas pipeline network based on fuzzy multi-attribute method. J. Appl. Sci. **14**, 955–959 (2014)

How Potential BLFs Can Help to Decide Under Incomplete Knowledge

Florence Dupin de Saint-Cyr and Romain Guillaume(✉)

IRIT, Université de Toulouse, Toulouse, France
{florence.bannay,romain.guillaum}@irit.fr

Abstract. In a Bipolar Leveled Framework (BLF) [7], the comparison of two candidates is done on the basis of the decision principles and inhibitions which are validated given the available knowledge-bases associated with each candidate. This article defines a refinement of the rules for comparing candidates by using the potential-BLFs which can be built according to what could additionally be learned about the candidates. We also propose a strategy for selecting the knowledge to acquire in order to better discriminate between candidates.

Keywords: Qualitative decision making · Bipolarity · Arguments
Incomplete knowledge

1 Introduction

Making decision is both one of the most current task and one of the most difficult problem that human beings should face. Hence designing an intelligent system able to help people to make decisions is a very important challenge. Tchangani et al. [13] recall that decision analysis is a process requiring first to formulate the decision goals, then to identify the attributes that characterize the potential alternatives and then decide. In classical approaches about decision making (see e.g. the introductory book of [9]), the standard way is to use a utility function that evaluates the quality of each decision hence that helps to select the one that has the best utility. This utility function should be designed in order to take into account the uncertainty and the multi-criteria aspects of the problem.

Studies in Psychology (see e.g. [12]) have shown that decision making is often guided by affect. Even more, Slovic et al. [11] argue that "affect is essential to rational action" where affect is defined as "the specific quality of "goodness" and "badness", as felt consciously or not by the decision maker, and demarcating a positive or negative quality of a stimulus". Then it is natural to use a scale going from negative (bad) to positive (good) values, including a central neutral value, to encode the bipolarity of the affect. And even use a bipolar scale, indeed, it is often the case that human people evaluate the possible alternatives considering positive and negative aspects separately [5].

J. Medina et al. (Eds.): IPMU 2018, CCIS 855, pp. 86–98, 2018.
https://doi.org/10.1007/978-3-319-91479-4_8

In Artificial Intelligence literature some models have already been proposed based on a bipolar view of alternatives (see [8] for an aggregation function approach and [3] for a pairwise comparison approach). In this paper we further explore the Bipolar Leveled Framework (BLF), which is a new representation framework for decision making, first introduced in [2] and extended in [6,7]. The BLF is a bipolar structure that enables the human decision maker to *visualize the attributes and goals that are involved in the decision problem, together with their links and their importance levels.* The structure is bipolar in the sense that the goals are either positive (i.e. wished to be achieved, a decision that achieves that goal is good) or negative (i.e. dreaded to be achieved, a decision that achieves that goal, is bad). Information in a BLF is encoded under the form of "decision principles" (DP). A DP is a kind of argument linking a description of a factual situation (here the situation is a candidate, described by some attributes) to the achievement of a goal. Informally a BLF may be viewed as a kind of qualitative utility function with some extra features: (1) the links between attributes and goals are made explicit into the decision principles, (2) the fact that a decision principle can be inhibited in the presence of some attribute is represented by an arrow from the attribute to the DP, (3) the importance levels of decision principles are represented by the height of their position in the structure. For more details on the link between BLF and qualitative decision Theory see [6]. When an alternative is known, the attributes of this alternative define what is called a "Valid BLF" which is an instance of a generic BLF.

The problem addressed is the pairwise comparison of alternatives given a bipolar utility representation. More precisely, the aim of this paper is to study how the BLF can take into account the awareness of the user about the completeness of her knowledge. We propose a measure called "sensibility" that evaluates this awareness in terms of what is known about the alternative versus what could be known (given the generic BLF). Then we propose two different ways to deal with this *sensibility*. The first one aims at refining the comparison that could be done with the valid BLFs associated to the alternatives, by taking into account the potential BLFs that could be obtained if we had more available information on each alternative. The second one aims at helping the decision maker to choose which information is relevant in order to make the more reliable choice between two decisions, hence what information should be obtained before deciding.

In the next section we recall the BLF definitions introduced in [6,7]. The third section describes how to take into account the actual knowledge with regard to information that could be learned given the generic BLF, leading to define a potential BLF. The last section proposes two ways to take into account the potential BLF either for refining the ordering of candidates or to select which information has to be acquired in order to be more accurate in the comparison.

2 BLF: A Structure Encoding Decision Criteria

We consider a set $\mathscr{C}$ of candidates[1] about which some information is available. We propose two distinct languages in order to clearly differentiate beliefs (coming from observations) from desires (goals to be achieved when selecting a candidate): $\mathscr{L}_F$ (a propositional language based on a vocabulary $\mathcal{V}_F$) represents information about some features that are believed to hold for a candidate and $\mathscr{L}_G$ (another propositional language based on a distinct vocabulary $\mathcal{V}_G$) represents information about the achievement of some goals when a candidate is selected. In the propositional languages used here, the logical connectors "or", "and", "not" are denoted respectively by $\vee$, $\wedge$, and $\neg$. A *literal* is a propositional symbol x or its negation $\neg x$, the set of literals of $\mathscr{L}_G$ are denoted by LIT_G. Classical inference, logical equivalence and contradiction are denoted respectively by $\models$, $\equiv$, $\perp$.

In the following we denote by K a set of formulas representing the beliefs of an agent about the features that hold: hence $K \subseteq \mathscr{L}_F$ is the available information. Using the inference operator $\models$, the fact that a formula $\varphi \in \mathscr{L}_F$ holds[2] in K is written $K \models \varphi$.

2.1 BLF: Definitions [7]

The BLF is a structure that contains two kinds of information: decision principles and inhibitors. A decision principle can be viewed as a defeasible reason enabling to reach a conclusion about the achievement of a goal. More precisely, a decision principle is a pair (φ, g), it represents the default rule meaning that "if the formula φ is believed to hold for a candidate then the goal g is a priori believed to be achieved by selecting this candidate":

Definition 1 (decision principle (DP)). *A decision principle p is a pair $(\varphi, g) \in \mathscr{L}_F \times LIT_G$, where φ is the reason of p, denoted $reas(p)$ and g the conclusion of p, denoted $concl(p)$. $\mathcal{P}$ denotes the set of decision principles.*

Depending on whether the achievement of its goal is wished or dreaded, a decision principle may have either a positive or a negative polarity. Moreover some decision principles are more important than others because their goal is more important.

Definition 2 (polarity and importance). *A function $pol : \mathcal{V}_G \to \{\oplus, \ominus\}$ gives the polarity of a goal $g \in \mathcal{V}_G$, this function is extended to goal literals by $pol(\neg g) = -pol(g)$ with $-\oplus = \ominus$ and $-\ominus = \oplus$. A decision principle p is polarized accordingly to its goal: $pol(p) = pol(concl(p))$. The set of positive and negative goals are abbreviated $\overline{\oplus}$ and $\overline{\ominus}$ respectively: $\overline{\oplus} = \{g \in LIT_G : pol(g) = \oplus\}$ and $\overline{\ominus} = \{g \in LIT_G : pol(g) = \ominus\}$.*

[1] Candidates are also called alternatives in the literature.

[2] The agent's knowledge K being considered to be certain, we write "φ holds" instead of "φ is believed to hold".

LIT_G is totally ordered by the relation $\preceq$ ("less or equally important than"). Decision principles are ordered accordingly: $(\varphi, g) \preceq (\psi, g')$ iff $g \preceq g'$.

The polarities and the relative importances of the goals in $\mathcal{V}_G$ are supposed to be given by the decision maker, *e.g.*, he may want to avoid to select an expensive hotel (hence "expensive hotel" can be a negative goal), while selecting a hotel where it is possible to swim can be a positive goal, moreover he may give more importance to swim than to pay less.

A decision principle (φ, g) is a defeasible piece of information because sometimes there may exist some reason φ' to believe that it does not apply in the situation, this reason is called an *inhibitor*.

The fact that ψ inhibits a decision principle (φ, g) is interpreted as follows: "when the decision maker only knows $\varphi \wedge \psi$ then he is no longer certain that g is achieved". In that case, the inhibition is represented with an arc towards the decision principle. The decision principles and their inhibitors are supposed to be given by the decision maker. An interpretation of decision principles in terms of possibility theory is described in [6].

We are now in position to define the BLF structure.

Definition 3 (BLF). *Given a set of goals $\mathcal{V}_G$, a BLF is a triplet $(\mathcal{P}, \mathcal{R}, pol, \preceq)$ where $\mathcal{P}$ is a set of decision principles ordered[3] accordingly to their goals by $\preceq$ and with a polarity built on pol as defined in Definition 2, $\mathcal{R} \subseteq (\mathscr{L}_F \times \mathcal{P})$ is an inhibition relation.*

The four elements of the BLF are supposed to be available prior to the decision and to be settled for future decisions as if it was a kind of utility function. A graphical representation of a BLF is given below, it is a tripartite graph represented in three columns, the DPs with a positive level are situated on the left column, the inhibitors are in the middle, and the DPs with a negative polarity are situated on the right. The more important (positive and negative) DPs are in the higher part of the graph, equally important DPs are drawn at the same horizontal level. By convention the highest positive level is at the top left of the figure and the lowest negative level is at the bottom right. The height of the inhibitors is not significant only their existence is used.

Example 1. *Let us imagine an agent who wants to find an inexpensive hotel in which he can swim. This agent would also be happy to have free drinks but it is less important for him. $\mathcal{V}_G = \{swim, free\_drinks, expensive, crowded\}$, with $pol(swim) = pol(free\_drinks) = \oplus$ and $pol(expensive) = pol(crowded) = \ominus$ and $swim \simeq expensive \succ free\_drinks \succ crowded$. The possible pieces of information concern the following attributes: $\mathcal{V}_F = \{pool, open\_bar, four\_stars, fine\_weather, special\_offer\}$. The agent considers the following principles: $\mathcal{P} = \{p_1 = (pool, swim), p_2 = (open\_bar, free\_drinks), p_3 = (four\_stars, expensive), p_4 = (fine\_weather, crowded)\}$. When the weather is not fine then*

[3] The equivalence relation associated to $\preceq$ is denoted $\simeq$ ($x \simeq y \Leftrightarrow x \preceq y$ and $y \preceq x$) and the strict order is denoted $\prec$ ($x \prec y \Leftrightarrow x \preceq y$ and not $y \preceq x$).

the fact that there is a pool is not sufficient to ensure that the agent can swim, it means that there is an inhibition on p_1 *by* $\neg fine\_weather$*, and the DP* p_4 *that expresses that "if the weather is fine the hotel will be crowded" is inhibited when its a four stars hotel, and the DP* p_3 *is inhibited when the agent have a special offer, i.e.* $\mathcal{R} = \{(\neg fine\_weather,\, p_1),\, (four\_stars,\, p_4),\, (special\_offer,\, p_3)\}$.

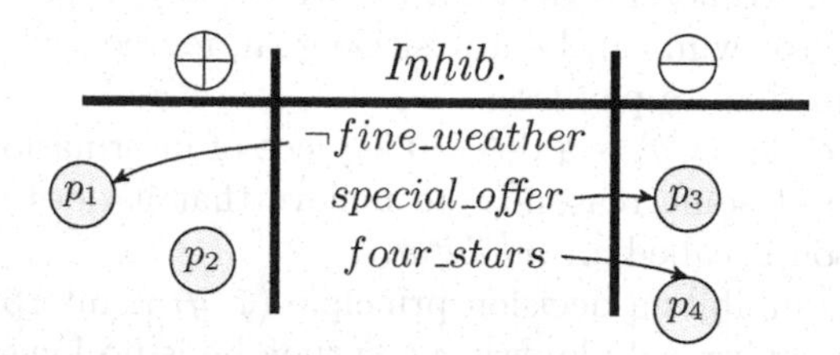

In the following, the BLF $(\mathcal{P}, \mathcal{R}, pol, \preceq)$ is set and we show how it can be used for comparing candidates. First, we present the available information and the notion of instantiated BLF, called valid-BLF.

Given a candidate $c \in \mathscr{C}$, we consider that the knowledge of the decision maker about c has been gathered in a knowledge base K_c with $K_c \subseteq \mathscr{L}_F$. K_c is **supposed to be consistent**. Given a formula φ describing a configuration of features ($\varphi \in \mathscr{L}_F$), the decision maker can have three kinds of knowledge about c: φ holds for candidate c (i.e., $K_c \models \varphi$), or φ does not hold ($K_c \models \neg\varphi$) or the feature φ is unknown for c ($K_c \not\models \varphi$ and $K_c \not\models \neg\varphi$). When there is no ambiguity about the candidate c, K_c is denoted K.

Definition 4 (K-Valid-BLF). *Given a base* K*, a* K*-Valid-BLF is a quadruplet* $(\mathcal{P}_K, \mathcal{R}_K, pol, \preceq)$ *where*

- $\mathcal{P}_K = \{(\varphi, g) \in \mathcal{P},\ s.t.\ K \models \varphi\}$ *is the set of DPs in* $\mathcal{P}$ *whose reason* φ *holds in* K*, called* valid-DP*s.*
- $\mathcal{R}_K = \{(\psi, p) \in \mathcal{R},\ s.t.\ K \models \psi\}$ *is the set of* valid inhibitions *wrt to* K*.*

When there is no ambiguity, we simply use "valid-BLF" instead of "K-Valid-BLF". The validity of a DP only depends on whether the features that constitute its reason φ hold or not, it does not depend on its goal g since the link between the reasons and the goal is given in the BLF (hence it is not questionable).

Example 2. *The agent has information about a hotel situated in a place where the weather will not be fine and that has a pool* ($reas(p_1)$) *and an open bar* ($reas(p_2)$)*:* $K_1 = \{\neg fine\_weather, pool, open\_bar\}$*. The* K_1*-Valid-BLF corresponding to what is known about this hotel is on the left. Now, we can consider another knowledge base* $K_2 = \{fine\_weather,\ four\_stars,\ open\_bar\}$ *describing a hotel that has an open-bar and that is located somewhere where the weather is nice but with no information about the existence of a pool, its associated* K_2*-Valid BLF is on the right.*

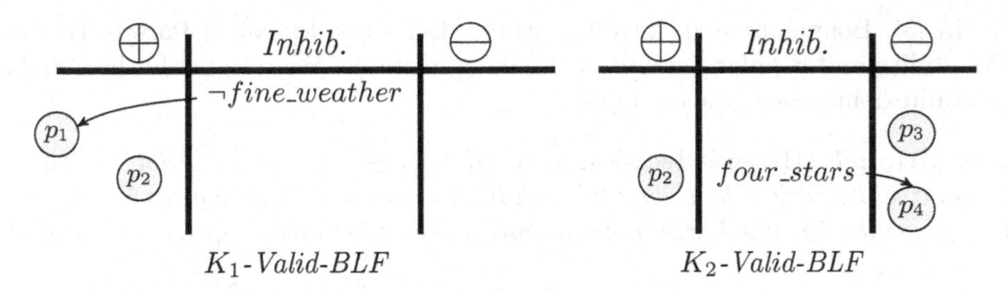

Now in the valid-BLF the principles that are not inhibited are the ones that are going to be trusted. A goal in $\mathcal{V}_G$ is said to be "realized" if there is a valid-DP that is not inhibited by any valid-inhibitor.

Definition 5 (realized goal). *Let g be a goal in LIT_G, g is* realized *w.r.t. a K-Valid-BLF $(\mathcal{P}_K, \mathcal{R}_K, pol, \preceq)$ iff $\exists(\varphi, g) \in \mathcal{P}_K$ and (φ, g) not inhibited in $\mathcal{R}_K$. The set of realized goals is denoted $\mathtt{R}_K$ (and simply $\mathtt{R}$ when there is no ambiguity about K) the positive and negative realized goals are denoted by $\mathtt{R}^{\oplus} = \mathtt{R} \cap \overline{\oplus}$ and $\mathtt{R}^{\ominus} = \mathtt{R} \cap \overline{\ominus}$ respectively.*

Example 3. *In the BLF with the knowledge K_1, $concl(p_2)$ is the only realized goals. K_2 with the same initial BLF allows us to conclude that both $concl(p_2)$ and $concl(p_3)$ are realized. To summarize, the first valid-BLF has one positive realized goal: $\mathtt{R}_{K_1} = \{free\_drinks\}$, while the second valid-BLF has a positive and a negative realized goal, $\mathtt{R}_{K_2} = \{free\_drinks, expensive\}$. But the negative goal that is realized has greater importance for the agent than the positive one, hence he should prefer the first hotel.*

In the next section we show how to use a BLF in order to compare several candidates based on the goals that are realized in their corresponding valid-BLF.

2.2 Decision Rules for Comparing Candidates

In order to compare candidates we have to compare the levels of DPs that are valid, hence we are going to define an absolute scale of the levels of the goals in the BLF (this definition is straightforward from the BLF). We start by attributing levels to the goals starting from the least important ones that are assigned a level 1 and stepping by one each time the importance grows.

Definition 6 (levels of goals wrt a BLF). *Given a BLF $B = (\mathcal{P}, \mathcal{R}, pol, \prec)$ the levels of the goals of the BLF are defined by induction:*

- $L(B)_1 = \{g \in Goals(B) : \nexists g' \in Goals(B) \text{ s.t. } g' \prec g\}$
- $L(B)_{i+1} = \{g \in Goals(B) : \nexists g' \in Goals(B) \setminus (\bigcup_{k=1}^{i} L(B)_k) \text{ s.t. } g' \prec g\}$

where $Goals(B) = \bigcup_{p \in \mathcal{P}} concl(p)$

Given a set of goals $G \in Goals(B)$, we write $G_k = G \cap L(B)_k$, and the level of a goal $g \in Goals(B)$ is defined by $level(g) = k$ iff $g \in G_k$.

In [3], Bonnefon et al. introduce three decision rules called Pareto, Bipolar Possibility and Bipolar Leximin dominance relations. We recall only the Bipolar Leximin dominance relation below:

Definition 7 (BiLexi decision rule of [3]). *Given two candidates c and c' respectively described by K and K' with their associated realized goals $\mathtt{R} = \mathtt{R}_K$ and $\mathtt{R}' = \mathtt{R}_{K'}$, the Bipolar Leximin dominance relation denoted $\succeq_{BiLexi}$ is defined by:*

$$c \succeq_{BiLexi} c' \quad \mathit{iff}\ |\mathtt{R}_\delta^{\oplus}| \geq |\mathtt{R}_\delta'^{\oplus}| \ \mathit{and}\ |\mathtt{R}_\delta^{\ominus}| \leq |\mathtt{R}_\delta'^{\ominus}|$$

where $\delta = argmax_\lambda(\{|\mathtt{R}_\lambda^{\oplus}| \neq |\mathtt{R}_\lambda'^{\oplus}| \ or\ |\mathtt{R}_\lambda^{\ominus}| \neq |\mathtt{R}_\lambda'^{\ominus}|\})$.

Example 4. *As we expected, the hotel described by K_1 is preferred to the one described by K_2, wrt $\succeq_{Bilexi}$, since*
$\mathtt{R}_{K_1}^{\oplus} = \{free\_drinks\}\, \mathtt{R}_{K_1}^{\ominus} = \varnothing$
$\mathtt{R}_{K_2}^{\oplus} = \{free\_drinks\}\, \mathtt{R}_{K_2}^{\ominus} = \{expensive\}$
and $free\_drinks \in L(B)_2$, $expensive \in L(B)_3$. Hence, we have the same realized goals at level 1 and 2, hence $\delta = 3$.

3 Awareness and K-Potential-BLF

In [6], only the features that the agent knows are used to compare candidates, i.e., the decision is based on the K-Valid-BLF. It means that the knowledge about the potential existence of a DP or of an inhibition is not taken into account. Hence, the quality of the agent knowledge is not taken into account (see example below).

Example 5. *In Example 1 with $K_1 = \{\neg fine\_weather, pool, open\_bar\}$, the agent believes that the weather is not fine and that the hotel has a pool and has an open bar. If we compare this state of belief with the belief $K_3 = \{open\_bar, fine\_weather, special\_offer\}$, then the realized goal is the same ($concl(p2)$). However, knowing that there could be no possibility to swim in the first hotel while there could be a pool in the second one may incline the agent to prefer the second hotel.*

To refine the comparison of candidates, we propose to improve the evaluation of a candidate by evaluating the goals that *could be* realized under the actual knowledge K.

3.1 K-Potential-BLF

Potential DPs and potential inhibition relations are the ones that could belong to the valid BLF if we had more information. Hence they are the DPs and inhibition relations that can be consistently assumed to be valid. In other words, they are not proven to be not valid wrt to the agent knowledge about the candidate. A DP is proven not valid when the agent knows that its reason does not hold ($K \models \neg reas(p)$). An inhibition on a DP cannot be valid if the agent knows that the inhibitor does not hold. Hence a potential inhibition is an inhibition that is not proven impossible when the reason of the DP holds.

A K-Potential-BLF is made of potential DPs and potential inhibitions with respect to a knowledge base K.

Definition 8 (K-Potential-BLF). *Given a base K and a BLF $(\mathcal{P}, \mathcal{R}, pol, \preceq)$, the K-Potential-BLF associated to K is the quadruplet $(\widehat{\mathcal{P}}_K, \widehat{\mathcal{R}}_K, pol, \preceq)$ where*

- *$\widehat{\mathcal{P}}_K$ is the set of potential DPs defined by:*

$$\widehat{\mathcal{P}}_K = \{p \in \mathcal{P} \quad | \quad K \cup \{reas(p)\} \text{ is consistent}\}$$

- *$\widehat{\mathcal{R}}_K$ is the set of potential inhibition relations defined by:*

$$\widehat{\mathcal{R}}_K = \{(\psi, p) \in \mathcal{R} \mid K \cup \{reas(p) \wedge \psi\} \text{ is consistent}\}$$

Example 6. *The K_1-Potential-BLF associated to K_1 contains $\widehat{\mathcal{P}}_{K_1} = \{p_1, p_2, p_3\}$ and $\widehat{\mathcal{R}}_{K_1} = \{(\neg fine\_weather, p_1), (special\_offer, p_3)\}$.*

Now according to whether the BLF considered is the K-Valid-BLF or the K-Potential-BLF, some goals can be simply realized (we recall Definition 5) or necessarily/possibly/potentially realized.

Definition 9 (Potential realization). *A goal g in LIT_G can have eight statuses w.r.t. a knowledge base K and a BLF $(\mathcal{P}, \mathcal{R}, pol, \preceq)$:*

status	***Realized***	***Not realized***
notation	$g \in \mathrm{R}_K$	$g \in \overline{\mathrm{R}}_K$
definition	$\exists p \in \mathcal{P}_K$ $concl(p) = g$ and $\nexists(\psi, p) \in \mathcal{R}_K$	$\forall p \in \mathcal{P}_K$ either $concl(p) \neq g$ or $\exists(\psi, p) \in \mathcal{R}_K$
status	***Necessarily realized***	***Possibly not realized***
notation	$g \in N\mathrm{R}_K$	$g \in \Pi\overline{\mathrm{R}}_K$
definition	$\exists p \in \mathcal{P}_K$ $concl(p) = g$ and $\nexists(\psi, p) \in \widehat{\mathcal{R}}_K$	$\forall p \in \mathcal{P}_K$ either $concl(p) \neq g$ or $\exists(\psi, p) \in \widehat{\mathcal{R}}_K$
status	***Potentially realized***	***Potentially not realized***
notation	$g \in P\mathrm{R}_K$	$g \in P\overline{\mathrm{R}}_K$
definition	$\exists p \in \widehat{\mathcal{P}}_K$ $concl(p) = g$ and $\nexists(\psi, p) \in \widehat{\mathcal{R}}_K$	$\forall p \in \widehat{\mathcal{P}}_K$ either $concl(p) \neq g$ or $\exists(\psi, p) \in \widehat{\mathcal{R}}_K$
status	***Possibly realized***	***Necessarily not realized***
notation	$g \in \Pi\mathrm{R}_K$	$g \in N\overline{\mathrm{R}}_K$
definition	$\exists p \in \widehat{\mathcal{P}}_K$ $concl(p) = g$ and $\nexists(\psi, p) \in \mathcal{R}_K$	$\forall p \in \widehat{\mathcal{P}}_K$ either $concl(p) \neq g$ or $\exists(\psi, p) \in \mathcal{R}_K$

In other words, a *necessarily realized* goal is realized in the K-Valid BLF and has no potential inhibitor (i.e. no valid inhibitor and more information cannot bring anymore inhibitor). A *necessarily not realized* goal is either not achieved by any potential DP or it has a K-Valid inhibitor. A *possibly realized* goal is the conclusion of a DP whose reason could hold and for which no inhibition is known to hold.

Example 7. *We have already seen that* $\mathtt{R}_{K_1} = \{free\_drinks\}$. *We have also:* $N\mathtt{R}_{K_1} = \{free\_drinks\}$, $P\mathtt{R}_{K_1} = \{free\_drinks\}$, $\Pi\mathtt{R}_{K_1} = \{free\_drinks, expensive\}$.

3.2 Link Between K-Potential-BLF and K-Valid-BLF

In the following proposition we show that we have upper and lower bounds of the set of realized goals according to the potential knowledge.

Proposition 1. *For any BLF* $(\mathcal{P}, \mathcal{R}, pol, \preceq)$ *and any knowledge base* K

$$N\mathtt{R}_K \subseteq \mathtt{R}_K \subseteq \Pi\mathtt{R}_K \qquad \Pi\overline{\mathtt{R}}_K \subseteq \overline{\mathtt{R}}_K \subseteq N\overline{\mathtt{R}}_K$$
$$N\mathtt{R}_K \subseteq P\mathtt{R}_K \subseteq \Pi\mathtt{R}_K \qquad \Pi\overline{\mathtt{R}}_K \subseteq P\overline{\mathtt{R}}_K \subseteq N\overline{\mathtt{R}}_K$$

Since we are able to give an interval containing the realized goals, the confidence in the decision can be defined with respect to the size of this interval: the smallest the interval the surest the evaluation of the candidate (since learning the values of unknown features cannot change this evaluation). Hence we are going to define a measure that evaluates the size of this interval which is called sensibility of the BLF wrt knowledge.

Definition 10 (Sensitivity). *The sensitivity of a BLF* $B = (\mathcal{P}, \mathcal{R}, pol, \preceq)$ *wrt a knowledge base* K *is*

$$s(B, K) = |\Pi\mathtt{R}_K \setminus N\mathtt{R}_K|$$

The sensitivity is the number of goals that are possibly realized but not necessarily realized. The aim is to take this sensitivity into account while making a decision. In our example, $s(B, K_1) = 1$.

Definition 11. K *is a perfect knowledge wrt a BLF* $(\mathcal{P}, \mathcal{R}, pol, \preceq)$ *iff* $\forall\varphi \in \bigcup_{p\in P}\{reas(p)\} \cup \bigcup_{(\psi,p)\in\mathcal{R}}\{\psi\}$, *either* $K \models \varphi$ *or* $K \models \neg\varphi$.

Note that it is not necessary to have perfect knowledge in order to have perfect information about the set of realized goals. In case of perfect knowledge the Valid BLF and the Potential BLF are equal, then all the eight statuses reduced to two, each goal is either necessarily realized or necessarily not realized.

Proposition 2. *For all BLF* $B = (\mathcal{P}, \mathcal{R}, pol, \preceq)$,
(K *is a perfect knowledge wrt* B) $\Rightarrow$ ($\widehat{\mathcal{P}}_K = \mathcal{P}_K$ *and* $\widehat{\mathcal{R}}_K = \mathcal{R}_K$) $\Rightarrow s(B, K) = 0$.
But the converse does not necessarily hold.

4 *K*-Potential-BLF and Decision Making

In the BLF framework, a goal $g \in \mathtt{R}_K$ induces that "g is achieved" is the nominal conclusion. In other words, it is the conclusion drawn under the available knowledge K. Nevertheless, the ranking on goals obtained with $OM(\mathtt{R}_K)$ or $|\mathtt{R}_K|$ could be challenged when the quality of the knowledge is not the same for the candidates that we want to compare. In this section, we explore two different ways to exploit the K-Potential-BLF. The first way is to use the K-Potential-BLFs in case of equality or incomparability of two candidates wrt to their K-Valid-BLFs. Indeed, it allows us to use the three sets: the set of Necessarily (resp. Possibly and Potentially) realized goals $N\mathtt{R}_K$ (resp. $\Pi\mathtt{R}_K$ and $P\mathtt{R}_K$) additionally to the set of realized goals $\mathtt{R}_K$.

The second way aims at helping the decision maker to choose which information is relevant in order to make the more reliable choice between two candidates. When it is possible to acquire more information, it is fairer to obtain nearly the same level of sensitivity in the knowledge bases of the candidates to be compared.

4.1 Refining the Ordering of Candidates

In order to compare two candidates we should use one of the comparison operator recalled in Definition 7 on the Valid-BLFs of the candidates (hence on their respective realized goals). In case of equality or incomparability between two candidates the decision maker can use its awareness of the possible DPs given in the generic BLF. More precisely according to the decision maker's profile he may choose to use either $N\mathtt{R}_K$ (if "skeptic") or $\Pi\mathtt{R}_K$ (if "believer")[4]. The decision maker is called *skeptic* when he considers that a DP is valid and not inhibited only if this DPs remains valid and not inhibited whatever the missing information is, in accordance to the definition of Necessary realized goals (Definition 9). Similarly a *believer* considers that a DP is valid and not inhibited if there is a way to complete the missing information in order to make it possible.

4.2 Acquisition of Knowledge in Order to Discriminate Candidates

In this section, we consider that the decision maker is able to increase her knowledge K when she considers that this knowledge is not pertinent enough. After this acquisition K is increased into $K \cup \varphi$, the DM can compare the candidates by using the rules applied to the set $\mathtt{R}_{K\cup\varphi}$. In order to evaluate the quality of the knowledge available for each candidate, we can compare the sensitivity associated to their different knowledge bases. We may compare the candidates only if the knowledge about them has approximately the same sensitivity:

Definition 12. *Given a BLF $B = (\mathcal{P}, \mathcal{R}, pol, \preceq)$, a given constant ε and two knowledge bases K and K' describing two candidates c and c',*

- *c is* ε-sensitivity-comparable *to c' iff* $|s(B,K) - s(B,K')| \leq \varepsilon$.
- *c is $BiLexi$*-preferred to c' with ε-sensitivity awareness *iff they are ε-sensitivity-comparable and $c \succeq_{BiLexi} c'$*

If the candidates are ε-sensitivity comparable but are equal wrt to BiLexi preference, or if we want to decrease the sensitivity then we have to choose the subject on which we have to increase our knowledge. The question to answer is

[4] Note that the set $P\mathtt{R}$ is not meaningful in this context.

"What is the most important goal that could be necessarily realized by adding only one formula φ to K which would not be possibly realized by adding $\neg\varphi$? and what is the simplest formula φ that could do that?"

Definition 13. *Given a BLF* $B = (\mathcal{P}, \mathcal{R}, pol, \preceq)$ *and a knowledge base* K*, a best-discriminating formula* (φ^*) *is a formula* $\varphi \in \mathcal{L}_F$ *such that:*

$$\varphi^* = \arg\max_{\varphi \in \mathcal{L}_F}\{k : level(g) = k \text{ and } g \in N\mathrm{R}_{K\cup\{\varphi\}} \text{ and } g \notin \Pi\mathrm{R}_{K\cup\{\neg\varphi\}}\}$$

The simplest-best-discriminating formula is a DNF[5] *best-discriminating formula that is not subsumed by any other DNF best-discriminating formula.*

We illustrate the two ways to use the K-Potential-BLF in the next section.

4.3 Example

We would like to compare 4 hotels: the three hotels described by K_1, K_2 and K_3, and a new one described by $K_4 = \{open\_bar; fine\_weather\}$:

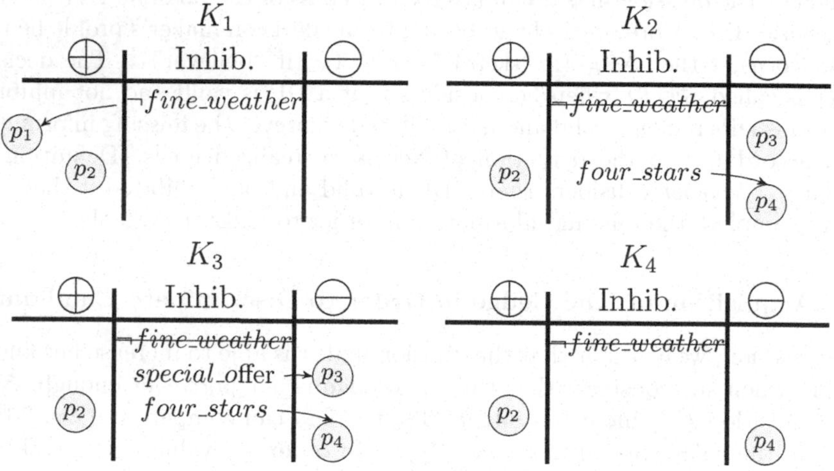

If we apply the *BiLexi* rule, we obtain $1 \sim_{BiLexi} 3 \succ_{BiLexi} 2 \succ_{BiLexi} 4$. The order relation between K_1 and K_2 can be refined by using the *BiLexi* rule either on the set $N\mathrm{R}_K$ or on $\Pi\mathrm{R}_K$. The choice depends on the DM's profile: $N\mathrm{R}_K$ is taken if she is a skeptic and $\Pi\mathrm{R}_K$ if she is a believer. Using $N\mathrm{R}_K$, the ordering between candidate 1 and 3 remains the same but with $\Pi\mathrm{R}_K$, we get $1 \succ_{BiLexi} 3$.

Now, in case we can increase our knowledge. We can take into account the sensibility of the BLF associated to each candidate, which are $s(B, K_1) = 1$, $s(B, K_2) = 2$, $s(B, K_3) = 1$ and $s(B, K_4) = 3$. Note that Candidate 4 is very sensitive since its sensitivity is close to the maximal possible value of sensitivity (the number of goals). Hence, before concluding on the ordering the DM should increase her knowledge about candidate 4. She can investigate the feature *pool*: if the answer is Yes, then $swim \in N\mathrm{R}_{K_4\cup\{pool\}}$ else $swim \notin \Pi\mathrm{R}_{K_4\cup\{\neg pool\}}$. In the first case, 4 becomes the most preferred hotel otherwise it is the worst hotel.

[5] DNF: Disjunctive Normal Form.

5 Conclusion

The BLF is a visual tool made to help human decision makers in their tasks. Note that once the BLF is defined the decision is automatically computed, hence BLF can be used by artificial or human agents. In [6] we have already studied the comparison of two candidates a and b on the basis of the K_a-Valid-BLF and the K_b-Valid-BLF, that gathers the decision principles and inhibitions which are validated given the available knowledge-bases K_a and K_b associated with each candidate. In this paper, we have proposed a refinement of the rules for comparing candidates by using the *potential*-BLFs which can be built according to *what could additionally be learned* about the candidates.

We can consider that our approach is of the kind "compare then aggregate" in the sense that when we want to select one candidate among a set of candidates, we can do a pairwise comparison and then decide which candidate to elect. This last step is a kind of aggregation. Bonnefon et al. approach [3], and classical decision making methods like Electre [10], Promethee [4] and Condorcet [1] could be assigned to the same category of approaches where only [3] also uses bipolarity. Another approach of decision making is "aggregate then compare", it means that first candidates are given an absolute value and then the best one is selected. In this family of approaches we can find the weighted average method, Choquet integral-based methods (like [8]), the uninorm aggregation operators [14]. An interesting direction could be to override the pairwise comparisons done with the valid BLFs towards defining an absolute scale for ranking the candidates based on an aggregation function defined on BLFs.

References

1. Austen-Smith, D., Banks, J.: Information aggregation, rationality, and the condorcet jury theorem. Am. Polit. Sci. Rev. **90**(1), 34–45 (1996)
2. Bannay, F., Guillaume, R.: Towards a transparent deliberation protocol inspired from supply chain collaborative planning. In: Laurent, A., Strauss, O., Bouchon-Meunier, B., Yager, R.R. (eds.) IPMU 2014. CCIS, vol. 443, pp. 335–344. Springer, Cham (2014). https://doi.org/10.1007/978-3-319-08855-6_34
3. Bonnefon, J.F., Dubois, D., Fargier, H.: An overview of bipolar qualitative decision rules. In: Della Riccia, G., Dubois, D., Kruse, R., Lenz, H.J. (eds.) Preferences and Similarities. CISM International Centre for Mechanical Sciences, vol. 504, pp. 47–73. Springer, Vienna (2008). https://doi.org/10.1007/978-3-211-85432-7_3
4. Brans, J.-P., Vincke, Ph.: Note-a preference ranking organisation method: (the promethee method for multiple criteria decision-making). Manage. Sci. **31**(6), 647–656 (1985)
5. Cacioppo, J., Berntson, G.: Relationship between attitudes and evaluative space: a critical review, with emphasis on the separability of positive and negative substrates. Psychol. Bull. **115**(3), 401 (1994)
6. de Saint-Cyr, F.D., Guillaume, R.: Analyzing a bipolar decision structure through qualitative decision theory. KI - Künstliche Intelligenz **31**(1), 53–62 (2017)
7. de Saint-Cyr, F.D., Guillaume, R.: Group decision making in a bipolar leveled framework. In: An, B., Bazzan, A., Leite, J., Villata, S., van der Torre, L. (eds.)

PRIMA 2017. LNCS (LNAI), vol. 10621, pp. 34–52. Springer, Cham (2017). https://doi.org/10.1007/978-3-319-69131-2_3
8. Labreuche, C., Grabisch, M.: Generalized choquet-like aggregation functions for handling bipolar scales. Eur. J. Oper. Res. **172**(3), 931–955 (2006)
9. Raiffa, H.: Decision Analysis: Introductory Lectures on Choices Under Uncertainty. Addison-Wesley, Reading (1970)
10. Roy, B.: The outranking approach and the foundations of electre methods. In: Bana e Costa, C.A. (eds.) Readings in Multiple Criteria Decision Aid, pp. 155–183. Springer, Heidelberg (1990). https://doi.org/10.1007/978-3-642-75935-2_8
11. Slovic, P., Finucane, M., Peters, E., MacGregor, D.G.: Rational actors or rational fools: implications of the affect heuristic for behavioral economics. J. Socio-Econ. **31**(4), 329–342 (2002)
12. Suci, G.J., Tannenbaum, P.H.: The Measurement of Meaning, vol. 2. University of Illinois Press, Urbana (1957)
13. Tchangani, A., Bouzarour-Amokrane, Y., Pérès, F.: Evaluation model in decision analysis: bipolar approach. Informatica **23**(3), 461–485 (2012)
14. Yager, R.R., Rybalov, A.: Uninorm Aggregation Operators. Fuzzy Sets Syst. **80**(1), 111–120 (1996)

A Proposal to Measure Human Group Behaviour Stability

Teresa González-Artega[1], José Manuel Cascón[2], and Rocio de Andrés Calle[3](✉)

[1] BORDA and PRESAD Research Groups and Multidisciplinary Institute of Enterprise (IME), University of Valladolid, 47011 Valladolid, Spain
teresag@eio.uva.es

[2] Department of Economics and Economic History, Institute on Fundamental Physics and Mathematics, University of Salamanca, 37007 Salamanca, Spain
casbar@usal.es

[3] BORDA Research Unit, PRESAD Research Group and Multidisciplinary Institute of Enterprise (IME), University of Salamanca, 37007 Salamanca, Spain
rocioac@usal.es

Abstract. A non-traditional approach on the measurement of agents behaviour is presented. This contribution focus on measuring stability of agents' preferences on an intertemporal context under the assumption of considering uncertainty opinions. To this aim, the concept of behaviour stability measure is defined as well as a particular one, the sequential behaviour stability measure. Finally and in order to highlight the good behaviour of novel measure, some properties are also provided.

Keywords: Human behaviour stability · Uncertainty opinions
Behaviour stability measure

1 Introduction

Human behaviour involves intertemporal decisions. In these choices, person must evaluate costs and benefits of doing something at different points of time. Every day, humans makes intertemporal decisions - when they select between eating snacks now or complete meal later or between going on holidays or increasing their pension fund contribution and so on. Therefore, intertemporal choice has been obtaining attention from several research fields such as Economics, Psychology, Decision Analysis and Neuroscience.

Generally speaking, two different approaches exist in the literature regarding intertemporal decision behaviour. The first one is focused on exploring behaviour causes by means of general optimization criterion (see [2, 7, 8] amog other). In this line it is possible to frame the contemporary economics models where humans takes intertemporal decision maximizing an exponentially-discount utility function to make temporal consistency choices. The second approach, meanly provided by psychologists, is the empirical one. The human behaviour in intertemporal choices is studied by means of empirical data collected from laboratories (see [1, 4–6]).

J. Medina et al. (Eds.): IPMU 2018, CCIS 855, pp. 99–110, 2018.
https://doi.org/10.1007/978-3-319-91479-4_9

In order to extend from another different perspective the intertemporal decision topic, the aim of this contribution is to develop a new tool capable of analysing intertemporal decision-making human behaviour and measuring the stability of their decisions. This work is inspired in the methodology proposed by González-Arteaga, de Andrés Calle and Peral [3] where the notion of intertemporal decision stability is considered in the same vein that the notion of cohesiveness. In [3] agents must choose between to approve or disapprove an alternative at diverse points of time (opinion takes value 1 if agent approves or 0 if agent disapproves it).

The assumption of dichotomous opinions in this particular context could limit and disturb the results of the behaviour analysis due to evidence suggests that humans may experience difficulties in expressing uncertain knowledge in a dichotomous way [9]. Consequently, this research is focused on an intertemporal decision-making problem under a general framework, i.e., agents can express their opinions on an alternative in the unit interval along different moments of time and then overcomes the aforementioned approach. Thus, the paper objective is to determine how much stability agents' opinions conveys to the group over time. In order to analyse intertemporal human behaviour by means of measuring such stability, a new general methodology is defined, the *behaviour stability measure*. Moreover, an specific formulation of the behaviour stability measure is introduced, the *sequential behaviour stability measure* as well as a study of its analytic properties. Under this approach, the stability of human behaviour is understood like the probability that for a randomly chosen moment of time, two randomly chosen agents have the same opinion at such a time and its consecutive.

The overall structure of the study takes the form of three sections. Section 2 introduces some notation as well as our proposal to measure preference stability: the behaviour stability measure. Moreover, an specific type of this measure, the sequential behaviour stability measure, is presented. In Sect. 3, the main properties of the sequential behaviour stability measure are defined and explained by an illustrative example. Finally, some concluding remarks and further researches are provided.

2 Behaviour Stability Measure: Notation and Definitions

Let $\mathbf{N} = \{1, 2, ..., N\}$ a set of agents or experts that expresses their opinions on an alternative, x, at different time moments $\mathbf{T} = \{t_1, \ldots, t_T\}$. A *behaviour profile* of a set of agents $\mathbf{N}$ on an alternative x at T different time moments is an $N \times T$ matrix

$$\mathbf{P} = \begin{pmatrix} P_{1t_1} & \ldots & P_{1t_T} \\ \vdots & \ddots & \vdots \\ P_{Nt_1} & \ldots & P_{Nt_T} \end{pmatrix}_{N \times T}$$

where $P_{it_j} \in [0, 1]$ is the opinion of the agent i over alternative x at t_j moment, in the sense:

- If agent i disapproves x at the t_j time, then $P_{it_j} = 0$ (in occasion for simplicity notation, this option is denoted by d),
- If agent i mostly disapproves x at the t_j time, then $0 < P_{it_j} < 0.5$ (in the same vein, this option is denoted by md),
- If agent i is undecided on x at the t_j time, then $P_{it_j} = 0.5$ (similarly, this option is denoted by u),
- If agent i mostly approves x at the t_j time, then $0.5 < P_{it_j} < 1$ (likewise, this option is denoted by ma),
- If agent i approves x at the t_j time, then $P_{it_j} = 1$ (analogously, this option is denoted by a).

For abbreviation, let $\mathbf{O} = \{d, md, u, ma, a\}$ stand for the label set associated to all possible opinions. Moreover, the behaviour profile arising from replacing the values in $[0, 1]$ by the labels in $\mathbf{O}$, is also called in the same way.

Let $\mathbb{P}_{N \times T}$ denote the set of all such $N \times T$ matrices. For simplicity of notation, $(1)_{N\times T}$, $(0.5)_{N\times T}$ and $(0)_{N\times T}$ are the $N \times T$ matrices whose cells are universally equal to 1, 0.5 and 0, respectively.

A behaviour profile $\mathbf{P}$ is *unanimous* if all agents have the same opinion on alternative x over $\mathbf{T}$. In matrix terms, if behaviour profile $\mathbf{P} \in \mathbb{P}_{N\times T}$ is constant. Any permutation σ of the agents $\{1, 2, ..., N\}$ determines a behaviour profile $\mathbf{P}^\sigma$ by permutation of the rows of $\mathbf{P}$, that is, row i of the profile $\mathbf{P}^\sigma$ is row $\sigma(i)$ of the profile $\mathbf{P}$.

For each behaviour profile $\mathbf{P}$, $\mathbf{P}_S$ is the restriction to a subset of agents, an *agent-subprofile* on the agents in $S \subseteq \mathbf{N}$, and it emerges from selecting the rows of $\mathbf{P}$ that are associated with the respective agents in S. For each behaviour profile $\mathbf{P}$, $\mathbf{P}^I$ is the restriction to a subset of moments of time, a *time-behaviour subprofile* on the moments of time in $I \subseteq \mathbf{T}$, and it emerges from selecting consecutive columns of $\mathbf{P}$ that are associated with the respective moments of time in I. Any partition $\{I_1, \dots, I_p\}$ of $\mathbf{T}$ generates a decomposition of $\mathbf{P}$ into behaviour subprofiles $\mathbf{P}^{I_1}, \dots, \mathbf{P}^{I_p}$ where $\mathbf{P}^{I_1} \cup \dots \cup \mathbf{P}^{I_p} = \mathbf{P}$.

An *extension* of a behaviour profile $\mathbf{P}$ of a group of agents $\mathbf{N}$ at $\mathbf{T} = \{t_1, \dots, t_T\}$ is a behaviour profile $\overline{\mathbf{P}}$ at $\overline{\mathbf{T}} = \{t_1, \dots, t_T, t_{T+1}, \dots, t_{T+q}\}$ such that the restriction of $\overline{\mathbf{P}}$ to the first T moments of time of $\overline{\mathbf{T}}$ coincides with $\mathbf{P}$. A *replication* of a behaviour profile $\mathbf{P}$ of a group of agents $\mathbf{N}$ on alternative x is the behaviour profile $\mathbf{P} \uplus \mathbf{P} \in \mathbb{P}_{2N\times T}$ obtained by duplicating each row of $\mathbf{P}$, in the sense that rows r and $N + r$ of $\mathbf{P} \uplus \mathbf{P}$ are row r of $\mathbf{P}$, for each $r = 1, ..., N$.

For each behaviour profile $\mathbf{P}$ on alternative x $n_a^{t_j}$ is the number of agents that approve x at t_j, $n_{ma}^{t_j}$ is the number of agents that mostly approve x at t_j, $n_u^{t_j}$ is the number of agents that are undecided on x at t_j, $n_{md}^{t_j}$ is the number of agents that mostly disapprove x at t_j and $n_d^{t_j}$ is the number of agents that disapprove x at t_j. Therefore, $N = n_a^{t_j} + n_{ma}^{t_j} + n_u^{t_j} + n_{md}^{t_j} + n_d^{t_j}$ for each $t_j \in \mathbf{T}$.

In addition, $n_{d,d}^{t_j,t_{j+1}}$ denotes the number of agents that disapprove alternative x at t_j and keep their opinion at the following point of time t_{j+1}. Similarly, $n_{a,a}^{t_j,t_{j+1}}$ denotes the number of agents that approve alternative x at t_j and keep their opinion at the following point of time t_{j+1} and so on. See Table 1 for improving understanding.

Table 1. Notation summary table.

$t_j \backslash t_{j+1}$	d	md	u	ma	a	
d	$n_{d,d}^{t_j,t_{j+1}}$	$n_{d,md}^{t_j,t_{j+1}}$	$n_{d,u}^{t_j,t_{j+1}}$	$n_{d,ma}^{t_j,t_{j+1}}$	$n_{d,a}^{t_j,t_{j+1}}$	$n_d^{t_j}$
md	$n_{md,d}^{t_j,t_{j+1}}$	$n_{md,md}^{t_j,t_{j+1}}$	$n_{md,u}^{t_j,t_{j+1}}$	$n_{md,ma}^{t_j,t_{j+1}}$	$n_{md,a}^{t_j,t_{j+1}}$	$n_{md}^{t_j}$
u	$n_{u,d}^{t_j,t_{j+1}}$	$n_{u,md}^{t_j,t_{j+1}}$	$n_{u,u}^{t_j,t_{j+1}}$	$n_{u,ma}^{t_j,t_{j+1}}$	$n_{u,a}^{t_j,t_{j+1}}$	$n_u^{t_j}$
ma	$n_{ma,d}^{t_j,t_{j+1}}$	$n_{ma,md}^{t_j,t_{j+1}}$	$n_{ma,u}^{t_j,t_{j+1}}$	$n_{ma,ma}^{t_j,t_{j+1}}$	$n_{ma,a}^{t_j,t_{j+1}}$	$n_{ma}^{t_j}$
a	$n_{a,d}^{t_j,t_{j+1}}$	$n_{a,md}^{t_j,t_{j+1}}$	$n_{a,u}^{t_j,t_{j+1}}$	$n_{a,ma}^{t_j,t_{j+1}}$	$n_{a,a}^{t_j,t_{j+1}}$	$n_a^{t_j}$
	$n_d^{t_{j+1}}$	$n_{md}^{t_{j+1}}$	$n_u^{t_{j+1}}$	$n_{ma}^{t_{j+1}}$	$n_a^{t_{j+1}}$	N

Definition 1. An *behaviour stability measure* for a group of agents $\mathbf{N} = \{1, ..., N\}$ on an alternative x is a mapping

$$\varphi : \mathbb{P}_{N\times T} \to [0, 1]$$

that assigns a number $\varphi(\mathbf{P}) \in [0, 1]$ to each behaviour profile $\mathbf{P}$, with the following properties:

(i) $\varphi(\mathbf{P}) = 1$ if and only if $\mathbf{P}$ is unanimous (full stable behaviour).
(ii) $\varphi(\mathbf{P}^\sigma) = \varphi(\mathbf{P})$ for each permutation σ of the agents and $\mathbf{P} \in \mathbb{P}_{N\times T}$ (anonymous behaviour).

A behaviour stability measure is a collection of behaviour stability measures for each group of agents $\mathbf{N}$.

Now a particular behaviour stability measure is introduced. Formally:

Definition 2. Let $\mathbf{N} = \{1, ..., N\}$ be group of agents that give their opinion on an alternative x by $\mathbf{O} = \{d, md, u, ma, a\}$ at different moments of time $\mathbf{T}$. The *sequential behaviour stability measure* is the mapping $\varphi_S : \mathbb{P}_{N\times T} \to [0, 1]$ given by

$$\varphi_S(\mathbf{P}) = \frac{1}{T-1} \cdot \frac{\sum_{o\in\mathbf{O}} \sum_{j=1}^{j=T-1} n_{o,o}^{t_j,t_{j+1}} \cdot (n_{o,o}^{t_j,t_{j+1}} - 1)}{N(N-1)}.$$

Intuitively, it measures the probability that for a randomly chosen moment of time, two randomly chosen agents of a group have the same opinion upon an alternative at the moment of time selected and its consecutive. It is easy to check that Definition 2 provides a behaviour stability measure.

3 Sequential Behaviour Stability Measure: Properties

In this section some desirable properties of the sequential behaviour stability measure are defined but firstly, an illustrative example is included in order to improve the understanding of the notation and the properties.

Example 1. Suppose a set of twelve experts $\mathbf{N} = \{1, 2, \ldots, 12\}$ that express their opinions on alternative x over four consecutive moments of time $\mathbf{T} = \{t_1, t_2, t_3, t_4\}$. Their opinions are collected in the following behaviour profile:

$$\mathbf{P} = \begin{pmatrix} P_{1t_1} & \ldots & P_{1t_4} \\ \vdots & \ddots & \vdots \\ P_{12t_1} & \ldots & P_{12t_4} \end{pmatrix}_{12\times 4} = \begin{pmatrix} a & a & a & d \\ d & u & md & d \\ md & u & md & d \\ d & a & d & a \\ ma & u & u & a \\ a & a & a & a \\ d & d & d & d \\ md & md & md & d \\ d & md & ma & a \\ d & u & u & a \\ d & d & md & md \\ a & ma & md & d \end{pmatrix}$$

This behaviour profile is summarized in Table 2. Using Definition 2, the value of the sequential behaviour stability measure is $\varphi_S(\mathbf{P}) = 0.02$.

Henceforth, the properties of the sequential behaviour stability measure are defined and explained by means of the example.

Reversal Invariance. This property shows that the main aspect of the sequential behaviour stability measure is the stability of agents' opinions more than an specific value. If agents' opinions totally change in opposite course, then the sequential behaviour stability measure reminds equal. Formally:

Let $\mathbf{P}^c$ be the complementary behaviour profile of $\mathbf{P}$ defined by $\mathbf{P}^c = (1)_{N\times T} - \mathbf{P}$. If φ_S verifies reversal invariance then $\varphi_S(\mathbf{P}^c) = \varphi_S(\mathbf{P})$.

Focusing on the following matrices (from Example 1), this property means that those agents whose opinions coincide at t_j and t_{j+1} in $\mathbf{P}$ have also coincident opinions at t_j and t_{j+1} in $\mathbf{P}^c$ although those opinions are totally different than in $\mathbf{P}$, then the sequential behaviour stability measure is the same in both cases, $\varphi_S(\mathbf{P}) = \varphi_S(\mathbf{P}^c) = 0.02$.

Table 2. Notation summary table for Example 1

$t_1 \backslash t_2$	d	md	u	ma	a	
d	$n_{d,d}^{t_1,t_2}=2$	$n_{d,md}^{t_1,t_2}=1$	$n_{d,u}^{t_1,t_2}=2$	$n_{d,ma}^{t_1,t_2}=0$	$n_{d,a}^{t_1,t_2}=1$	$n_d^{t_1}=6$
md	$n_{md,d}^{t_1,t_2}=0$	$n_{md,md}^{t_1,t_2}=1$	$n_{md,u}^{t_1,t_2}=1$	$n_{md,ma}^{t_1,t_2}=0$	$n_{md,a}^{t_1,t_2}=0$	$n_{md}^{t_1}=2$
u	$n_{u,d}^{t_1,t_2}=0$	$n_{u,md}^{t_1,t_2}=0$	$n_{u,u}^{t_1,t_2}=0$	$n_{u,ma}^{t_1,t_2}=0$	$n_{u,a}^{t_1,t_2}=0$	$n_u^{t_1}=0$
ma	$n_{ma,d}^{t_1,t_2}=0$	$n_{ma,md}^{t_1,t_2}=0$	$n_{ma,u}^{t_1,t_2}=1$	$n_{ma,ma}^{t_1,t_2}=0$	$n_{ma,a}^{t_1,t_2}=0$	$n_{ma}^{t_1}=1$
a	$n_{a,d}^{t_1,t_2}=0$	$n_{a,md}^{t_1,t_2}=0$	$n_{a,u}^{t_1,t_2}=0$	$n_{a,ma}^{t_1,t_2}=1$	$n_{a,a}^{t_1,t_2}=2$	$n_a^{t_1}=3$
	$n_d^{t_2}=2$	$n_{md}^{t_2}=2$	$n_u^{t_2}=4$	$n_{ma}^{t_2}=1$	$n_a^{t_2}=3$	$N=12$

$t_2 \backslash t_3$	d	md	u	ma	a	
d	$n_{d,d}^{t_2,t_3}=1$	$n_{d,md}^{t_2,t_3}=1$	$n_{d,u}^{t_2,t_3}=0$	$n_{d,ma}^{t_2,t_3}=0$	$n_{d,a}^{t_2,t_3}=0$	$n_d^{t_2}=2$
md	$n_{md,d}^{t_2,t_3}=0$	$n_{md,md}^{t_2,t_3}=1$	$n_{md,u}^{t_2,t_3}=0$	$n_{md,ma}^{t_2,t_3}=1$	$n_{md,a}^{t_2,t_3}=0$	$n_{md}^{t_2}=2$
u	$n_{u,d}^{t_2,t_3}=0$	$n_{u,md}^{t_2,t_3}=2$	$n_{u,u}^{t_2,t_3}=2$	$n_{u,ma}^{t_2,t_3}=0$	$n_{u,a}^{t_2,t_3}=0$	$n_u^{t_2}=4$
ma	$n_{ma,d}^{t_2,t_3}=0$	$n_{ma,md}^{t_2,t_3}=1$	$n_{ma,u}^{t_2,t_3}=0$	$n_{ma,ma}^{t_2,t_3}=0$	$n_{ma,a}^{t_2,t_3}=0$	$n_{ma}^{t_2}=1$
a	$n_{a,d}^{t_2,t_3}=1$	$n_{a,md}^{t_2,t_3}=0$	$n_{a,u}^{t_2,t_3}=0$	$n_{a,ma}^{t_2,t_3}=0$	$n_{a,a}^{t_2,t_3}=2$	$n_a^{t_2}=3$
	$n_d^{t_3}=2$	$n_{md}^{t_3}=5$	$n_u^{t_3}=2$	$n_{ma}^{t_3}=1$	$n_a^{t_3}=2$	$N=12$

$t_3 \backslash t_4$	d	md	u	ma	a	
d	$n_{d,d}^{t_3,t_4}=1$	$n_{d,md}^{t_3,t_4}=0$	$n_{d,u}^{t_3,t_4}=0$	$n_{d,ma}^{t_3,t_4}=0$	$n_{d,a}^{t_3,t_4}=1$	$n_d^{t_3}=2$
md	$n_{md,d}^{t_3,t_4}=4$	$n_{md,md}^{t_3,t_4}=1$	$n_{md,u}^{t_3,t_4}=0$	$n_{md,ma}^{t_3,t_4}=0$	$n_{md,a}^{t_3,t_4}=0$	$n_{md}^{t_3}=5$
u	$n_{u,d}^{t_3,t_4}=0$	$n_{u,md}^{t_3,t_4}=0$	$n_{u,u}^{t_3,t_4}=0$	$n_{u,ma}^{t_3,t_4}=0$	$n_{u,a}^{t_3,t_4}=2$	$n_u^{t_3}=2$
ma	$n_{ma,d}^{t_3,t_4}=0$	$n_{ma,md}^{t_3,t_4}=0$	$n_{ma,u}^{t_3,t_4}=0$	$n_{ma,ma}^{t_3,t_4}=0$	$n_{ma,a}^{t_3,t_4}=1$	$n_{ma}^{t_3}=1$
a	$n_{a,d}^{t_3,t_4}=1$	$n_{a,md}^{t_3,t_4}=0$	$n_{a,u}^{t_3,t_4}=0$	$n_{a,ma}^{t_3,t_4}=0$	$n_{a,a}^{t_3,t_4}=1$	$n_a^{t_3}=2$
	$n_d^{t_4}=6$	$n_{md}^{t_4}=1$	$n_u^{t_4}=0$	$n_{ma}^{t_4}=0$	$n_a^{t_4}=5$	$N=12$

$$\mathbf{P} = \begin{pmatrix} a & a & a & d \\ d & u & md & d \\ md & u & md & d \\ d & a & d & a \\ ma & u & u & a \\ a & a & a & a \\ d & d & d & d \\ md & md & md & d \\ d & md & ma & a \\ d & u & u & a \\ d & d & md & md \\ a & ma & md & d \end{pmatrix} \quad \mathbf{P}^c = \begin{pmatrix} d & d & d & a \\ a & u & ma & a \\ ma & u & ma & a \\ a & d & a & d \\ md & u & u & d \\ d & d & d & d \\ a & a & a & a \\ ma & ma & ma & a \\ a & ma & md & d \\ a & u & u & d \\ a & a & ma & ma \\ d & md & ma & a \end{pmatrix}$$

Time-Reducibility. It means that the stability of a behaviour profile is the average of the sequential behaviour stability measures of all its consecutive behaviour-subprofiles of two consecutive moments of time. This says that we can first compute the sequential behaviour in the opinions of each time and the sequential one by the agents (the proportion of pairs of agents whose opinions keep the same for both times), and then aggregate these values by taking their average. Formally:

Let $\mathbf{P} \in \mathbb{P}_{N\times T}$ be a behaviour profile. We say that φ_S verifies time-reducibility if

$$\varphi_S(\mathbf{P}) = \frac{1}{T-1}\sum_{j=1}^{T-1} \varphi_S(\mathbf{P}^{I_{j,j+1}})$$

where $\mathbf{P}^{I_{j,j+1}} \in \mathbb{P}_{N\times 2}$ is the behaviour-subprofile of $\mathbf{P}$ containing the columns corresponding to times t_j and t_{j+1}.

Staying on Example 1 it means that:

$$\varphi_S(\mathbf{P}) = \frac{1}{3}\left(\varphi_S(\mathbf{P}^{I_{1,2}}) + \varphi_S(\mathbf{P}^{I_{2,3}}) + \varphi_S(\mathbf{P}^{I_{3,4}})\right) = \frac{1}{3}(0.03 + 0.03 + 0) = 0.02$$

where

$$\mathbf{P}^{I_{1,2}} = \begin{pmatrix} a & a \\ d & u \\ md & u \\ d & a \\ \vdots & \vdots \\ d & d \\ a & ma \end{pmatrix}, \; \mathbf{P}^{I_{2,3}} = \begin{pmatrix} a & a \\ u & md \\ u & md \\ a & d \\ \vdots & \vdots \\ d & md \\ ma & md \end{pmatrix}, \; \mathbf{P}^{I_{3,4}} = \begin{pmatrix} a & d \\ md & d \\ md & d \\ d & a \\ \vdots & \vdots \\ md & md \\ md & d \end{pmatrix}.$$

Replication Monotonicity. When a non-unanimous behaviour profile is replicated, its sequential behaviour stability measure increases. Formally:

Let $\mathbf{P} \in \mathbb{P}_{N \times T}$ be a non unanimous behaviour profile then

$$\varphi_S(\mathbf{P} \uplus \mathbf{P}) > \varphi_S(\mathbf{P}).$$

In addition, for an unanimous behaviour profile $\mathbf{P} \in \mathbb{P}_{N \times T}$, by Definition 2, φ_S verifies

$$\varphi_S(\mathbf{P} \uplus \mathbf{P}) = \varphi_S(\mathbf{P}) = 1.$$

Keeping up with Example 1:

$$\varphi_S(\mathbf{P} \uplus \mathbf{P}) = 0.035 > \varphi_S(\mathbf{P}) = 0.02$$

being

$$\mathbf{P} = \begin{pmatrix} a & a & a & d \\ d & u & md & d \\ md & u & md & d \\ d & a & d & a \\ ma & u & u & a \\ a & a & a & a \\ d & d & d & d \\ md & md & md & d \\ d & md & ma & a \\ d & u & u & a \\ d & d & md & md \\ a & ma & md & d \end{pmatrix}, \quad \mathbf{P} \uplus \mathbf{P} = \begin{pmatrix} a & a & a & d \\ d & u & md & d \\ md & u & md & d \\ d & a & d & a \\ ma & u & u & a \\ a & a & a & a \\ d & d & d & d \\ md & md & md & d \\ d & md & ma & a \\ d & u & u & a \\ d & d & md & md \\ a & ma & md & d \\ a & a & a & d \\ d & u & md & d \\ md & u & md & d \\ d & a & d & a \\ ma & u & u & a \\ a & a & a & a \\ d & d & d & d \\ md & md & md & d \\ d & md & ma & a \\ d & u & u & a \\ d & d & md & md \\ a & ma & md & d \end{pmatrix}.$$

Minimum Time Stability. If all agents express their opinions at a moment of time and change their opinions at the next moment of time, that is, all agents change their opinions over two successive moments of time, then the sequential behaviour stability measure takes a zero value. It also happens when there are at most two agents that keep their opinion at two consecutive moments of time but their opinions do not coincide each other. Formally:

Let $\mathbf{P} \in \mathbb{P}_{N \times T}$ be a behaviour profile such that there is at most one agent who has the same opinion at t_j and t_{j+1} for $j \in \{1, \ldots, T-1\}$, that is, $n_{a,a}^{t_j,t_{j+1}} \leq 1$, $n_{ma,ma}^{t_j,t_{j+1}} \leq 1$, $n_{u,u}^{t_j,t_{j+1}} \leq 1$, $n_{d,d}^{t_j,t_{j+1}} \leq 1$ and $n_{md,md}^{t_j,t_{j+1}} \leq 1$ Then, $\varphi_S(\mathbf{P}) = 0$.

Following an illustrative example of a profile showing minimum time stability is given.

$$\mathbf{P_0} = \begin{pmatrix} d & u & md & d \\ a & a & a & d \\ md & u & ma & d \\ d & a & d & a \\ ma & u & ma & a \\ a & ma & a & ma \\ d & d & d & d \\ md & md & md & d \\ d & md & ma & a \\ d & u & u & a \\ u & d & md & d \\ a & ma & md & d \end{pmatrix} \rightarrow \mathbf{P_0} = \begin{pmatrix} d & u & md & d \\ a & a & a & d \\ md & u & ma & d \\ d & a & d & a \\ ma & u & ma & a \\ a & ma & a & ma \\ d & d & d & d \\ md & md & u & d \\ d & md & ma & a \\ d & md & u & a \\ u & d & md & d \\ a & ma & md & d \end{pmatrix} \rightarrow \mathbf{P_0} = \begin{pmatrix} d & u & md & d \\ a & a & a & d \\ md & u & ma & d \\ d & a & d & a \\ ma & u & ma & a \\ a & ma & a & ma \\ d & d & d & d \\ md & md & md & d \\ d & md & ma & a \\ d & u & u & a \\ u & d & md & d \\ a & ma & md & d \end{pmatrix}$$

Leaving Minimum Time Stability. In order to leave the minimum time stability it is needed that at least the opinions of two agents coincide at the same moment of time and the next one. Formally:

Let $\mathbf{P} \in \mathbb{P}_{N \times T}$ be a behaviour profile such that there exists at least a j, $j \in \mathbf{T}$, such that $n_{a,a}^{t_j,t_{j+1}} > 1$ or $n_{ma,ma}^{t_j,t_{j+1}} > 1$ or $n_{u,u}^{t_j,t_{j+1}} > 1$ or $n_{d,d}^{t_j,t_{j+1}} > 1$ or $n_{md,md}^{t_j,t_{j+1}} > 1$, then $\varphi_S(\mathbf{P}) > 0$.

Taking the matrix exposes in the previous property $\mathbf{P_0}$, if the expert in line 4 of the profile gives "approve", instead of "disapprove", the sequential behaviour stability measure is greater than zero, $\varphi_S(\mathbf{P}) = 0.005 > 0$.

$$\mathbf{P_1} = \begin{pmatrix} d & u & md & d \\ a & a & a & d \\ md & u & ma & d \\ d & a & a & a \\ ma & u & ma & a \\ a & ma & a & ma \\ d & d & d & d \\ md & md & md & d \\ d & md & ma & a \\ d & u & u & a \\ u & d & md & d \\ a & ma & md & d \end{pmatrix} \rightarrow \mathbf{P_1} = \begin{pmatrix} d & u & md & d \\ a & a & a & d \\ md & u & ma & d \\ d & a & a & a \\ ma & u & ma & a \\ a & ma & a & ma \\ d & d & d & d \\ md & md & md & d \\ d & md & ma & a \\ d & md & u & a \\ u & d & md & d \\ a & ma & md & d \end{pmatrix} \rightarrow \mathbf{P_1} = \begin{pmatrix} d & u & md & d \\ a & a & a & d \\ md & u & ma & d \\ d & a & a & a \\ ma & u & ma & a \\ a & ma & a & ma \\ d & d & d & d \\ md & md & md & d \\ d & md & ma & a \\ d & u & u & a \\ u & d & md & d \\ a & ma & md & d \end{pmatrix}$$

Time Monotonicity. Consider two behaviour profiles, $\mathbf{P}$ and $\mathbf{P}'$, that coincide in all their elements excepting the opinion of an agent $m \in \mathbf{N}$, at t_k and t_{k+1}. Concretely, this agent has different opinion at t_k and t_{k+1} in $\mathbf{P}$: $P_{mt_k} \neq P_{mt_{k+1}}$, and the agent's opinion is the same at t_k and t_{k+1} in $\mathbf{P}'$: $P'_{mt_k} = P'_{mt_{k+1}}$. In this case, the sequential behaviour stability measure verifies $\varphi_S(\mathbf{P}') \geq \varphi_S(\mathbf{P})$. Formally:

Let $\mathbf{P}, \mathbf{P}' \in \mathbb{P}_{N \times T}$ be behaviour profiles such that:

(a) $P_{it_j} = P'_{it_j}$, $i \in \{\mathbf{N} \setminus \{m\}\}$,

(b) $P_{mt_k} \neq P_{mt_{k+1}}$, $m \in \mathbf{N}$, $t_k, t_{k+1} \in \mathbf{T}$,
(c) $P'_{mt_k} = P'_{mt_{k+1}}$, $m \in \mathbf{N}$, $t_k, t_{k+1} \in \mathbf{T}$.

Then, $\varphi_S(\mathbf{P}') \geq \varphi_S(\mathbf{P})$.
Focusing on Example 1, the matrices $\mathbf{P}$ and $\mathbf{P}'$ are:

$$\mathbf{P} = \begin{pmatrix} a & a & a & d \\ d & u & md & d \\ md & u & md & d \\ d & a & d & a \\ ma & u & u & a \\ a & a & a & a \\ d & d & d & d \\ md & md & md & d \\ d & md & ma & a \\ d & u & u & a \\ d & d & md & md \\ a & ma & md & d \end{pmatrix}, \mathbf{P}' = \begin{pmatrix} a & a & a & d \\ d & u & md & d \\ md & md & md & d \\ d & a & d & a \\ ma & u & u & a \\ a & a & a & a \\ d & d & d & d \\ md & md & md & d \\ d & md & ma & a \\ d & u & u & a \\ d & d & md & md \\ a & ma & md & d \end{pmatrix}$$

and then $\psi_S(\mathbf{P}) = 0.02 \leq \psi_S(\mathbf{P}') = 0.03$.

Convergence to Full Stability. If new moments of times are repeatedly introduced into the problem and all agents have the same opinion at them, then the sequential behaviour stability measure approaches 1. Formally:

Suppose that q moments of time $t_{T+1}, \ldots t_{T+q}$ are added to $\mathbf{T}$, and at these new moments of time the alternative x is unanimous. If the introduction of new moments of time does not affect agents' opinions in past times, then the sequential time cohesiveness measure of the extended behaviour profile $\overline{\mathbf{P}}^{(q)} \in \mathbb{P}_{N\times(T+q)}$ approaches 1 when q tends to infinity.

$$\lim_{q\to\infty} \varphi_S(\overline{\mathbf{P}}^{(\mathbf{q})}) = 1$$

Following Example 1 the consecutive matrices are:

$$\overline{\mathbf{P}}^{(1)} = \begin{pmatrix} a & a & a & ma \\ d & u & md & ma \\ md & u & md & ma \\ d & a & d & ma \\ \vdots & \vdots & \vdots & \vdots \\ d & d & md & ma \\ a & ma & md & ma \end{pmatrix}, \quad \overline{\mathbf{P}}^{(2)} = \begin{pmatrix} a & a & a & ma & ma \\ d & u & md & ma & ma \\ md & u & md & ma & ma \\ d & a & d & ma & ma \\ \vdots & \vdots & \vdots & \vdots & \vdots \\ d & d & md & ma & ma \\ a & ma & md & ma & ma \end{pmatrix},$$

$$\overline{\mathbf{P}}^{(3)} = \begin{pmatrix} a & a & a & ma & ma & ma \\ d & u & md & ma & ma & ma \\ md & u & md & ma & ma & ma \\ d & a & d & ma & ma & ma \\ \vdots & \vdots & \vdots & \vdots & \vdots & \vdots \\ d & d & md & ma & ma & ma \\ a & ma & md & ma & ma & ma \end{pmatrix}, \quad \ldots \quad, \overline{\mathbf{P}}^{(q)} = \begin{pmatrix} a & a & a & ma & ma \ldots & ma \\ d & u & md & ma & ma \ldots & ma \\ md & u & md & ma & ma \ldots & ma \\ d & a & d & ma & ma \ldots & ma \\ \vdots & \vdots & \vdots & \vdots & \vdots & \vdots \\ d & d & md & ma & ma \ldots & ma \\ a & ma & md & ma & ma \ldots & ma \end{pmatrix}.$$

Convexity. It means the sequential behaviour stability measure of a behaviour profile is a weighted average of the measures of any decomposition of **P** into consecutive time-subprofiles. Formally:

For each behaviour profile $\mathbf{P} \in \mathbb{P}_{N\times T}$, and each decomposition of **P** into two consecutive time-subprofiles, $\mathbf{P}^{I_1} \in \mathbb{P}_{N\times(k_1+1)}$ and $\mathbf{P}^{I_2} \in \mathbb{P}_{N\times(T-k_1)}$ with $I_1 = \{t_1, \ldots, t_{k_1+1}\}$ and $I_2 = \{t_{k_1+1}, \ldots, t_T\}$, and $(\mid I_1 \mid -1)+(\mid I_2 \mid -1) = T-1$

$$\varphi_S(\mathbf{P}) = \frac{(\mid I_1 \mid -1) \cdot \varphi_S(\mathbf{P}^{I_1}) + (\mid I_2 \mid -1) \cdot \varphi_S(\mathbf{P}^{I_2})}{T-1}.$$

In our example:

$$\mathbf{P} = \begin{pmatrix} a & a & a & d \\ d & u & md & d \\ md & u & md & d \\ d & a & d & a \\ \vdots & \vdots & \vdots & \vdots \\ d & d & md & md \\ a & ma & md & d \end{pmatrix}, \quad \mathbf{P}^{I_1} = \begin{pmatrix} a & a \\ d & u \\ md & u \\ d & a \\ \vdots & \vdots \\ d & d \\ a & ma \end{pmatrix}, \quad \mathbf{P}^{I_2} = \begin{pmatrix} a & a & ma \\ u & md & ma \\ u & md & ma \\ a & d & ma \\ \vdots & \vdots & \vdots \\ d & md & ma \\ ma & md & ma \end{pmatrix}$$

$$\varphi_S(\mathbf{P}) = \frac{(2-1) \cdot \varphi_S(\mathbf{P}^{I_1}) + (3-1) \cdot \varphi_S(\mathbf{P}^{I_2})}{3} = \frac{(2-1) \cdot 0.03 + (3-1) \cdot 0.015}{3}$$

4 Conclusion and Futher Research

Research in the subject of preference stability has made progress mostly in Economics. The aim of this paper is to manage the problem of measuring human behaviour stability from a non-traditional perspective. In order to set forth the context of our research a framework is established where agents express their opinions on an alternative at different moments by uncertainty information. The general notion of behaviour stability measure is introduced. Then, a specific formulation is developed with particular regard to any two successive time moments. In this way, the sequential behaviour stability measure is proposed. Moreover, some meaningful properties which make our proposal compelling are also provided as well as an illustrative example. Overall, the proposals of this contribution have a range of implications for future research. Many problems on human behaviour stability from a diversity of fields can be faced by our approach such as the consumers preferences, risk preference, medical preferences, and so on.

Acknowledgment. The authors acknowledge financial support by the Spanish Ministerio de Economía y Competitividad under Project ECO2016-77900-P (T. González-Arteaga and R. de Andrés Calle) and by the Conserjería de Educación of the Junta de Castilla y León under Project SA020U16 (J.M. Cascón).

References

1. Alcock, J., Sherman, P.: The utility of the proximate-ultimate dichotomy in ethology. Ethology **96**(1), 58–62 (1994)
2. Bateson, M., Kacelnik, A.: Rate currencies and the foraging starling: the fallacy of the averages revisited. Behav. Ecol. **7**(3), 341–352 (1996)
3. González-Artega, T., de Andrés Calle, R., Peral, M.: Preference stability along time: the time cohesiveness measure. Progress. Artif. Intell. **6**(3), 235–244 (2017)
4. Kim, S., Hwang, J., Lee, D.: Prefrontal coding of temporally discounted values during intertemporal choice. Neuron **59**(1), 161–172 (2008)
5. Kim, S., Lee, D.: Prefrontal cortex and impulsive decision making. Biological Psychiatry **69**(12), 1140–1146 (2011). Prefrontal Cortical Circuits Regulating Attention, Behavior and Emotion
6. Peters, J., Büchel, C.: Overlapping and distinct neural systems code for subjective value during intertemporal and risky decision making. J. Neurosci. **29**(50), 15727–15734 (2009)
7. Samuelson, P.A.: A note on measurement of utility. Rev. Econ. Stud. **4**(2), 155–161 (1937)
8. Stephens, D.W., Anderson, D.: The adaptive value of preference for immediacy: when shortsighted rules have farsighted consequences. Behav. Ecol. **12**(3), 330–339 (2001)
9. Zadeh, L.: Fuzzy sets. Inf. Control **8**, 338–375 (1965)

Identifying Criteria Most Influencing Strategy Performance: Application to Humanitarian Logistical Strategy Planning

Cécile L'Héritier[1(✉)], Abdelhak Imoussaten[1], Sébastien Harispe[1], Gilles Dusserre[1], and Benoît Roig[2]

[1] LGI2P, IMT Mines Ales, Univ Montpellier, Ales, France
cecile.lheritier@mines-ales.fr

[2] EA7352 CHROME, Université de Nîmes, Nîmes, France

Abstract. A growing interest is expressed by organizations for the development of approaches enabling to take advantage of past experiences to improve their decision processes; they may be referred to as Lessons Learned (LL) processes. Within the LL processes implementation framework, the development of semi-automatic approaches able to distinguish criteria having major influence on the evaluation of experiences is crucial for identifying relevant recommendations and performing efficient prescriptive analysis. In this paper, we propose to contribute to LL study by focusing on the definition of an approach enabling, in a specific setting, to identify the criteria most influencing the decision process regarding the overall performance evaluation of a reduced set of experiences. The proposed approach is framed on Multi-Criteria Decision Analysis, and specifically is based on the Electre tri method. In this paper, an illustration of the proposed approach is provided studying the evaluation of logistical response strategies in humanitarian emergency situations.

Keywords: Multi-Criteria Decision Analysis (MCDA)
Lessons Learned · Electre tri · Influencing criteria identification
Humanitarian domain

1 Introduction

Lessons Learned process (LL)[1] refers to a general Knowledge Management approach aiming at ensuring and improving proper functioning of organizations by collecting, analysing, disseminating and reusing tacit experiential knowledge. One of the main focuses of LL is therefore to study how to improve techniques enabling to discover and further take advantage of expert knowledge that is not

[1] Sometimes referred to as Experience feedback.

J. Medina et al. (Eds.): IPMU 2018, CCIS 855, pp. 111–123, 2018.
https://doi.org/10.1007/978-3-319-91479-4_10

explicitly formalized into organizations' Information Systems. In this context, LL is particularly interested in taking advantage of knowledge that is most often difficult if not impossible to formalize nowadays - in particular because of the technical difficulties to systematically and efficiently formalize such knowledge considering every-day practices used in organizations, e.g. domain experts often take decisions based on complex bodies of professional histories. Importance of LL is thus critical to fight against sensitive losses of expertise in organizations, and to achieve well-supported and understood decision-making processes. LL can indeed help improving both the performance and the quality of organizations by analysing past experiences and by using problem-solving methods – narrow links with well-established related domains such as Case Based Reasoning, Multiple-Criteria Decision Analysis and Knowledge Management/Representation exist; some of them will further be discussed.

In this paper, we contribute to LL study by focusing on the definition of an approach enabling, in a specific setting, to identify the criteria most influencing the decision process regarding the overall performance evaluation of a reduced set of experiences. Identifying the criteria most impacting performance is indeed of major importance for numerous applications related to LL; it can also be central for the global implementation of the LL process, e.g. it indeed helps a lot to identify relevant recommendations and to perform efficient prescriptive analyses. Identifying such criteria in real-world problems can however be challenging. We indeed here consider - as it is most often the case in numerous complex decision processes - that no explicit and formal definition of performance based on criteria analysis can be provided by domain experts. Experts can indeed evaluate performance but are not able to explicitly define the way it is evaluated, i.e. to define the model on which the evaluation is based, if any. We also consider the following common constraints: only a reduced set of observations is available[2] and only limited interactions with domain experts can be expected. Additionally, we consider that it is most often unthinkable to perform automatic deep analyses of all pieces of information that could be related to each criterion to further distinguish relevant teachings. Therefore, based on the analysis of a reduced set of evaluated past experiences characterized by their performance on criteria, we propose a general approach framed on Multiple-Criteria Decision Analysis (MCDA) in order to distinguish the minimal set of criteria most impacting performance. The proposed approach is built upon Electre tri, a MCDA method enabling classifying a set of alternatives into predefined categories.

An application of the proposed approach to the humanitarian field is presented; it is used for analysing logistical response strategies in emergency situations. Our case study rests upon a corpus of past missions carried out by a Non-Governmental Organization (NGO); the core mission of the NGO is to deliver emergency aid to people affected by disasters (earthquake, tsunami, conflict, famine, epidemic...). Our general goal is therefore to exploit knowledge that can be extracted from prior missions in order to formulate relevant recommendations that could be used for better defining future logistical response

[2] This *de facto* prevents the use of traditional Machine Learning approaches.

strategies. To this aim, this paper focuses on identifying the set of criteria most impacting performance evaluation of logistical response strategies. The paper is structured as follows: Sect. 2 briefly introduces state of the art related to LL and motivates the use of MCDA for extracting knowledge from past experiences. Section 3 presents our contribution for identifying the criteria most impacting performance evaluation; an introduction to Electre tri, the MCDA method used in our work, is also given in that section. Section 4 illustrates the use of the proposed approach for studying the logistical response in humanitarian emergency situations. Section 5 concludes this paper and highlights future work.

2 Related Works and Positioning

Whatever the field of application, learning from experience is a growing concern for numerous organizations. In this context, several approaches have emerged to fulfill the different LL needs expressed by organizations; they differ both in (i) the nature and the heterogeneity of information to collect, as well as in (ii) the expected purpose: exploitation, reuse, capitalization, and dissemination of knowledge. Various types of LL approaches can therefore be identified in the literature. Some of them rely on databases composed of past experiences and take advantage of domain-specific querying strategies, e.g. REX method [1]. Other approaches take advantage of knowledge models - using techniques related to Knowledge Management and Knowledge Representation (KM/KR) such as CommonKads [2], or MKSM (Methodology for Knowledge System Management) [3]. Another category of LL approaches also considers the use of problem-solving methods. These methods, in particular Case-Based Reasoning (CBR) [4], rely on the use of knowledge extracted from previously experienced cases to solve similar new cases. A significant number of cases is however generally required to distinguish analogies using these methods.

In our study, we are considering LL towards an application to the logistical emergency response carried out by humanitarian organizations. In this domain, data collection is even more difficult, e.g. oral testimony often prevails, analyses most often rest upon reduced corpora of past missions. With this in mind, another challenge arises: to exploit and reason on a small number of complex past experiences. These experiences are composed of successions of decisions being influenced by different factors such as: the intuition, the context-dependent nature of the missions - instable environment, unpredictable events affecting the system - the multiple, and sometimes conflicting objectives. Nevertheless, despite decision-making complexity, according to the collaborating NGO, the logistical strategy of several missions could be, to a certain extent, generalized and reproduced. In that context, the NGO has established a procedure to learn lessons from past experiences, particularly from choices made in missions. It aims at providing an *a posteriori* evaluation of singular cases entailing a noteworthy success or a significant failure. To do so, a board of multidisciplinary experts conducts an analysis of the case based on related data and complementary interviews; then a conclusive report is provided. However, according to the NGO stakeholders, provided outcomes often face major weaknesses - e.g. hardly exploitable, lack of sharpness - the time-consuming nature of the process is also pointed out.

An open challenge is thus offered to LL automation for knowledge extraction from past experiences in NGOs. Recall however that in the humanitarian domain, relevant and accurate data is difficult to obtain, and cannot easily be automatically analysed - data access and analysis often require costly staff and experts involvement. A further constraint is therefore, while maintaining interactions with domain experts and decision-makers (DMs), to minimize information requests in order to lessen their involvement (both in terms of time and cognitive effort). All these constraints underline the need for a prior identification of relevant criteria in order to reduce the scope of analysis for applying deeper data analyses. In practice, DMs always - at least implicitly - consider a set of objectives that the mission has to reach in order to meet success. In addition, a set of relevant criteria can be considered to measure, for a specific mission, the degree of success of each objective, *i.e.*, performance. Such a performance evaluation can however be a complex process. Indeed, to be successful, for DMs (i) a mission does not necessarily require maximal performance with respect to each criterion, and (ii) the criteria may not have the same importance. We therefore propose to take advantage of MCDA techniques to identify the subset of criteria most impacting the overall performance assessment of a specific mission.

The best known MCDA approaches are those derived from Multi-Attribute Value/Utility Theory (MAVT/MAUT) [5] and outranking methods [6]. Since every MCDA method differently combines performance and criteria importance to define the overall performance of a mission, the identification of the subset of criteria of interest will necessarily depend on the selected MCDA method. In this work, we have chosen to use Electre tri, a method from the well-known ELECTRE family methods [7]. It has been selected according to three main criteria related to NGOs context: it is simple to reuse; non-compensatory, and techniques exist to reduce the amount of information asked to DMs while determining model parameters - agreement among DMs is considered here. The non-compensatory condition means that an excellent performance on specific criteria (e.g. cost) cannot compensate for a bad performance on other prevailing criteria (e.g. degree of achievement of the medical objectives). In order to reduce the amount of information that has to be provided, Electre tri relies on indirect identification of the parameters. In addition, Electre tri is an outranking method: instead of scoring alternatives, outranking methods are based on pair comparisons. Indeed, it is more natural for DMs to compare the strength and the weakness of two alternatives rather than assigning an arbitrary value to each alternative, and comparing them subsequently. Interestingly, outranking methods are based on relatively weak mathematical assumptions [8]. Nevertheless, obtaining the model parameters is a very difficult issue because of the non-linearity due to the thresholds induced in the model. However some recent works have proposed simplified assumptions and a procedure to determine the parameters of the Electre tri method has been proposed [9].

3 Determining Criteria Contribution Within Electre Tri

The following section first introduces the Electre tri method and is then dedicated to the presentation of our approach: the identification of the subset criteria having the major contribution on the overall performance of a mission.

3.1 General Approach

The approach proposed in this paper aims to determine the criteria on which LL process should focus when a mission is analysed. To facilitate DMs' involvement, the information requirement is minimized and the cognitive effort reduced as much as possible. For instance, the parameters of the MCDA model used for the analysis will be transparent for DMs. They will only be asked to assign examples of fictive missions to the predefined categories. These examples of missions are defined by their performance on each evaluation criterion, such that DMs are able to easily classify them. The general approach can be decomposed as follows:

1. Determine the set of criteria on which missions will be evaluated;
2. Automatic identification of Electre tri parameters using a training set of fictive examples of missions;
3. Classify/Assign the mission to one of the predefined categories using this Electre tri model;
4. Determine the criteria strongly contributing to that classification.

3.2 Classifying Alternatives Using Electre Tri

We introduce the technical details related to Electre Tri, the specific MCDA method used in our work.

Let $A = \{a_1, \ldots, a_m\}$ a set of alternatives, and $\mathcal{F}$ a set of criteria satisfying consistency conditions, i.e. completeness (all relevant criteria are considered), monotonicity (the better the evaluation of an alternative on considered criteria, the more it is preferable to another), and non–redundancy (no superfluous criteria are considered) [6]. According to traditional MCDA outranking methods notations, we denote $g_j(a)$ the score of alternative $a \in A$ on the criterion $j \in \mathcal{F}$. For some pairs of alternatives $a, b \in A$, Electre tri builds an outranking relation aSb meaning that 'a is at least as good as b'. To this aim, the method uses concordance, comprehensive concordance, discordance and credibility indices [7]. The concordance index is defined using the indifference $q_j(g_j(a))$ and preference $p_j(g_j(a))$ thresholds; these thresholds allow to take the imprecision of the evaluations $g_j(a)$ into account.[3]

Let's define three subsets $\mathcal{F}_1^c(a,b)$, $\mathcal{F}_2^c(a,b)$ and $\mathcal{F}_3^c(a,b)$ that divide criteria of $\mathcal{F}$ with regard to the comparison of a and b performances:

[3] Note that Electre tri non-compensatory behaviour is defined by the fact that whenever $g_j(a) - g_j(b)$ is greater than $p_j(g_j(a))$ no distinction is made computing the concordance index; a big difference thus cannot compensate any negative difference on another criterion j with $j \neq i$.

- $\mathcal{F}_1^c(a,b) = \{j \in \mathcal{F} : g_j(a) \leq g_j(b) - p_j(g_j(b))\}$,
- $\mathcal{F}_2^c(a,b) = \{j \in \mathcal{F} : g_j(b) - p_j(g_j(b)) < g_j(a) < g_j(b) - q_j(g_j(b))\}$,
- $\mathcal{F}_3^c(a,b) = \{j \in \mathcal{F} : g_j(b) - q_j(g_j(b)) \leq g_j(a)\}$.

The partial concordance index on criterion j is defined using the previous subsets as follows:

$$\forall a,b \in A, \quad c_j(a,b) = \begin{cases} 0 & \text{if } j \in \mathcal{F}_1^c(a,b), \\ \alpha_j(a,b) & \text{if } j \in \mathcal{F}_2^c(a,b), \\ 1 & \text{if } j \in \mathcal{F}_3^c(a,b). \end{cases} \tag{1}$$

where $\alpha_j(a,b) = \frac{g_j(a)-g_j(b)+p_j(g_j(b))}{p_j(g_j(b))-q_j(g_j(b))}$. Criteria in $\mathcal{F}_3^c(a,b)$ vote without reservation for alternative a; in $\mathcal{F}_2^c(a,b)$ only the proportion $w_j \cdot c_j$ is considered voting for alternative a.

The comprehensive concordance index is then defined using the relative importance w_j of each criterion $j \in \mathcal{F}$:

$$\forall a,b \in A, \quad c(a,b) = \sum_{j\in\mathcal{F}} w_j \cdot c_j(a,b) = \sum_{j\in\mathcal{F}_3^c(a,b)} w_j + \sum_{j\in\mathcal{F}_2^c(a,b)} w_j \cdot \alpha_j(a,b) \tag{2}$$

The discordance index is defined by introducing a veto threshold $v_j(g_j(b))$:

$$\forall a,b \in A, \quad d_j(a,b) = \begin{cases} 1 & \text{if } j \in \mathcal{F}_1^d(a,b), \\ \frac{g_j(a)-g_j(b)+p_j(g_j(b))}{p_j(g_j(b))-v_j(g_j(b))} & \text{if } j \in \mathcal{F}_2^d(a,b), \\ 0 & \text{if } j \in \mathcal{F}_3^d(a,b). \end{cases} \tag{3}$$

where:

- $\mathcal{F}_1^d(a,b) = \{j \in \mathcal{F} : g_j(a) \leq g_j(b) - v_j(g_j(b))\}$,
- $\mathcal{F}_2^d(a,b) = \{j \in \mathcal{F} : g_j(b) - v_j(g_j(b)) < g_j(a) < g_j(b) - p_j(g_j(b))\}$,
- $\mathcal{F}_3^d(a,b) = \{j \in \mathcal{F} : g_j(b) - p_j(g_j(b)) \leq g_j(a)\}$.

The credibility index is further defined as follows:

$$\rho_S(a,b) = c(a,b) \cdot \prod_{j\in\mathcal{F}^d(a,b)} \frac{1-d_j(a,b)}{1-c(a,b)} \tag{4}$$

where $\mathcal{F}^d(a,b) = \{j \in \mathcal{F} : d_j(a,b) > c(a,b)\}$. Finally, the outranking relation aSb is considered when $\rho_S(a,b) \geq \lambda$, $\lambda \in [0.5, 1]$ [9].

In Electre tri, the credibility index is used to assign alternatives to predefined categories. Suppose that alternatives should be assigned to p categories $\{C_1, ...C_p\}$ defined by $p-1$ profiles $\{b_1, ...b_{p-1}\}$. Each profile b_h, $h \in \{1, \ldots, p-1\}$, is the upper limit of C_h and the lower limit of C_{h+1}. Let a be an alternative to assign to a category C_h, $h \in \{1, \ldots, p\}$. Two assignment procedures are possible:

- Pessimistic procedure: a is compared first to profiles defining best categories until meeting the first b_h such that aSb_h, then a is assigned to C_{h+1}.

- Optimistic procedure: a is compared first to profiles defining worst categories until meeting the first b_h such that $b_h S a$, then a is assigned to C_h.

Note that Electre tri requires determining the following parameters:

- the profiles of the categories, defined by their evaluations $g_j(b_h)$, $\forall j \in \mathcal{F}$, $\forall h \in \{1, \ldots, p\}$,
- the weight coefficients, $w_j, \forall j \in \mathcal{F}$,
- indifference, preference and veto thresholds, $q_j(g_j(b_h))$,$p_j(g_j(b_h))$, $v_j(g_j(b_h))$, $\forall j \in \mathcal{F}$, $\forall h \in \{1, \ldots, p\}$,
- the cutting level λ.

These parameters define the preference model of the DMs. The direct elicitation of these parameters values, i.e. asking directly the DMs to provide them, remains difficult since they do not correspond to the usual language or indicators that DMs use to express their opinion and expertise. To overcome this issue, procedures have been proposed to infer the parameters of Electre tri model. Indeed, in [9] this inference is based on the assignment of a set of examples of alternatives A^* for which DMs have a clear preference. This is done by resolving a non-linear programming problem, given in Eq. (5), where Electre tri parameters are the variables. The slack variables x_k and y_k are introduced to transform the inequality constraints $\rho(a_k, b_{h_k-1}) \geq \lambda$ and $\rho(a_k, b_{h_k}) \leq \lambda$ into equality constraints. Further details about variables and constants involved in Eq. (5) are given in [9].

$$\left(\alpha + \epsilon \sum_{a_k \in A^*} (x_k + y_k)\right) \rightarrow max, \quad \text{such that:} \tag{5}$$

$$\begin{aligned}
&\alpha \leq x_k \text{ and } \alpha \leq y_k, \quad \forall a_k \in A^* \\
&\frac{\sum_{j=1}^{m} w_j c_j(a_k, b_{h_k-1})}{\sum_{j=1}^{m} w_j} - x_k = \lambda, \quad \forall a_k \in A^* \\
&\frac{\sum_{j=1}^{m} w_j c_j(a_k, b_{h_k})}{\sum_{j=1}^{m} w_j} + y_k = \lambda, \quad \forall a_k \in A^* \\
&\lambda \in [0.5, 1] \\
&g_j(b_{h+1}) \geq g_j(b_h) + p_j(b_h) + p_j(b_{h+1}), \quad \forall j \in \mathcal{F}, \quad \forall h \in B \\
&p_j(b_j) \geq q_j(b_h), \quad \forall j \in \mathcal{F}, \quad \forall h \in B \\
&w_j \geq 0, \quad w_j \leq \frac{1}{2} \sum_{j=1}^{m} w_j, \quad q_j(b_h) \geq 0, \quad \forall j \in \mathcal{F}, \quad \forall h \in B
\end{aligned}$$

3.3 Determining Criteria of Interest

Let consider a the alternative defined by the performance of the currently evaluated mission on all evaluation criteria. The alternative a is classified as "good", "medium" or "bad" by DMs. The aim of this section is to propose a method to determine which subset of criteria strongly contributes to the overall result of a. In our setting, this alternative is assigned to one of the categories using Electre tri; this problem is therefore equivalent to identifying the subset of criteria strongly contributing to the credibility index of a when it is compared to the profiles determining the category finally assigned to a.

As explained above, when an outranking method is used, the contribution of criteria to the overall score is the portion of their weights considered in the comprehensive concordance index. This contribution depends on the belonging of criteria to one of the three subsets $\mathcal{F}_1^c$, $\mathcal{F}_2^c$ and $\mathcal{F}_3^c$. Naturally, when comparing a to a profile b, we define the contribution of a subset of criteria $I \subseteq \mathcal{F}$ to the comprehensive concordance index, as the quantity $c_I(a,b)$:

$$c_I(a,b) = \sum_{j \in I} w_j \cdot c_j(a,b) = \sum_{j \in \mathcal{F}_3^c(a,b) \cap I} w_j + \sum_{j \in \mathcal{F}_2^c(a,b) \cap I} w_j \cdot \alpha_j(a,b) \quad (6)$$

By convention $c_{\mathcal{F}}(a,b)$ is $c(a,b)$.

Let define two indices $r, s \in \{1, ..., p\}$ obtained by the two procedures of Electre tri classification:

1. Pessimistic procedure: $\forall h \in \{r, ..., p\}$, $not(aSb_h)$ and aSb_r: a is classified in category C_r;
2. Optimistic procedure: $\forall h \in \{1, ..., s-1\}$, aSb_h and b_sSa and $not(aSb_s)$: a is classified in category C_s.

Two cases have to be distinguished according to the quality of C_s:

1. The category C_s is considered as a "good" category. Then we have to determine criteria strongly contributing to the strength of a;
2. The category C_s is considered as a "bad" category. Then we have to determine criteria strongly contributing to the weakness of a.

Knowing that the criteria contributing strongly to the weakness of a are those strongly contributing to the strength of b_s, we only present the first case.

When a is assigned to a "good" category, we have aSb_r. Then for the cutting level $\lambda \in [0.5, 1]$ defined by the DMs, $\rho_S(a, b_r) > \lambda$ means:

$$c(a,b_r) \cdot \prod_{j \in \mathcal{F}^d(a,b_r)} \frac{1 - d_j(a,b_r)}{1 - c(a,b_r)} > \lambda \quad (7)$$

Let denote $\lambda(a,b) = \frac{\lambda}{\prod_{j \in \mathcal{F}^d(a,b)} \frac{1-d_j(a,b)}{1-c(a,b)}}$. The most contributing subset of criteria $I \subseteq \mathcal{F}$ is the one guaranteeing $c_I(a,b_r) > \beta\lambda(a,b_r)$, $\beta \in]0,1]$. The criteria of I are those having a large voting weight, i.e. $c_{\{j\}}(a,b_r)$, and the larger β is, the stronger the contribution of I is. Moreover the individual contribution of each criterion of I should exceed a minimal value $\gamma \in]0,1]$. This threshold prevents the selection of criteria that could contribute to state $c_I(a,b_r) > \beta\lambda(a,b_r)$ but in an insignificant proportion. Since the aim is to minimize the subset of criteria to analyse, we have to focus on the smallest subsets $I \subseteq \mathcal{F}$ such that the previous conditions are verified. Finally, we can set:

$$I^* = arg \min_{I \subseteq \mathcal{F}} \{|I| : c_I(a, b_r) > \beta^* \lambda(a, b_r) \;\; and \;\; \forall j \in I, c_{\{j\}}(a, b_r) > \gamma\} \quad (8)$$

where β^* is the biggest $\beta \in]0, 1]$ such that $\{I \subseteq \mathcal{F} : c_I(a, b_r) > \beta^* \lambda(a, b_r)\} \neq \emptyset$ and γ is chosen by the DMs. In case of existence of several subsets I^*, the choice of I^* is made considering performance of a on the criteria of I^*.

4 Case Study

This section illustrates the use of the proposed approach in a practical case aiming at studying the logistical response in humanitarian emergency situations.

We consider the context of LL in a humanitarian organisation case (see Sect. 2), it aims at determining, *a posteriori*, if choices were good or bad w.r.t objectives of an analysed mission, and to explain why. We deal with this problem by (i) analysing a real mission using Electre tri method which assigns the mission to a predefined category characterizing its overall success, (ii) identifying the criteria contributing to the mission success and those most impacting its performance assessment. In this case study, among the diversity of missions carried out by the collaborating NGO, we are focusing on distribution missions - distribution of medicines, food and shelters - which clearly entail similar logistics. For this illustrative example, we consider a mission a, corresponding to a mission largely inspired – but slightly simplified for the example – from a real food distribution mission carried out by the NGO. We define three categories:

- C_1 includes failed missions for which objectives have not been reached,
- C_2 covers missions with met objectives but moderate success,
- C_3 covers successful missions: those exceeding the objectives.

The performance of a mission is evaluated according to a set of eight criteria, *i.e.* $\mathcal{F} = \{1, 2, ..., 8\}$, listed in Table 1. Criterion g_1 assesses deadline compliance, criterion g_3 assesses the portion of the logistical costs w.r.t the total costs and criterion g_4 the number of enlisted human resources. The preference decreases on these criteria. Criterion g_2 assesses the percentage of achievement of the targeted population: preference increase on it. Criteria g_5, g_6, g_7 and g_8 are assessed in a qualitative way using an ordinal scale with four levels: the scale is $\{Small, Medium, High, Very\ High\}$ for g_5 and g_6, and $\{Bad, Moderate, Good, Very\ Good\}$ for g_7 and g_8. We use the scale $\{1, 2, 3, 4\}$ to encode these two scales. Preferences on the criteria will further be characterized by the definition of pseudo-criteria using thresholds q_j, p_j and v_j.

4.1 Determination of Parameters

As a starting point, several parameters have to be fixed to determine the Electre tri model that best fits DMs' preferences. As mentioned in Sect. 3.2, an approach

Table 1. Evaluation criteria.

	Criterion	Preference
g_1	Deadline	↘
g_2	Achievement %	↗
g_3	Portion of the logistics costs	↘
g_4	Human resources	↘
g_5	Added value for the organization	↗
g_6	Added value for the beneficiary country	↗
g_7	Environmental impact	↗
g_8	Security level	↗

can be used to determine those parameters by solving a non-linear optimisation problem [9]. Similarly to the work proposed in that paper [9], to facilitate the optimisation problem setting and resolution, veto thresholds are not inferred, we assume that they are directly given by DMs (see Table 3). Therefore, the variables are: $g_j(b_h)$, w_j, $q_j(g_j(b_h))$, $p_j(g_j(b_h))$, $\forall j \in \{1, \ldots, 8\}$, $\forall h \in \{1, 2\}$, λ. The problem constraints are defined from parameters definition and the assignment of a training set:

- $p_j(g_j(b_h)) \geq q_j(g_j(b_h)) \geq 0$, $\lambda \in [0.5, 1]$,
- a_k assigned to C_h means that $a_k S b_{h-1}$ and not$(a_k S b_h)$.

Regarding the profiles of the categories, a training set of ten alternatives has been used (see Table 2); each alternative a_k is assigned to one of the three categories by the DMs. These fictive missions a_k, used as a training set to infer model parameters are distinct from the alternative a, a real mission that we aim at analysing here. Then in this context, we have to consider 58 variables and 90 constraints. Table 3 gives the values of parameters obtained from the resolution of the non-linear optimisation problem resulting from the training set analysis.

4.2 Criteria Contribution

Let denote the analysed mission by a. DMs assess the performance of a on the criteria of $\mathcal{F}$ as shown in Table 4. We apply the pessimistic procedure of Electre tri using the identified model parameters (Table 3), and we consider the credibility indices $\rho(a, b_1) = 0.995$ and $\rho(a, b_2) = 0.2$. Then the mission a is assigned to category C_2, since not(aSb_2), and aSb_1 with b_1 the lower profile of C_2. Knowing that aSb_1, we are looking for the subset of criteria $I \subseteq \mathcal{F}$ that most contributes to the establishment of this outranking relation, both because of the importance of the criteria, and the performance of a on these criteria. Table 4 shows the individual contribution $c_{\{j\}}(a, b_1)$ of each criterion g_j to the global performance of a when it is compared to b_1.

Then in our Electre tri model, the cutting level is $\lambda = 0.93$ and the threshold $\gamma = \frac{c(a,b_1)}{2^n} = 0.004$ is considered, $n = |\mathcal{F}|$. Thus, solving the equation

Table 2. Training set.

	g_1	g_2	g_3	g_4	g_5	g_6	g_7	g_8	Category
a_1	5	90	38	30	3	4	3	3	C_3
a_2	10	97	30	37	3	3	3	4	C_3
a_3	12	90	40	50	4	2	4	4	C_3
a_4	23	85	40	60	2	2	3	2	C_2
a_5	10	74	48	55	3	3	2	3	C_2
a_6	20	60	50	65	3	2	3	2	C_2
a_7	16	80	55	80	2	2	2	3	C_2
a_8	23	55	70	120	1	2	2	1	C_1
a_9	60	55	60	125	2	1	1	2	C_1
a_{10}	27	40	50	100	2	3	2	1	C_1

Table 3. Parameters values obtained from the optimisation problem resolution and given veto thresholds.

	g_1	g_2	g_3	g_4	g_5	g_6	g_7	g_8
$g_j(b_1)$	26	62	53	80	2	1	2	2
$g_j(b_2)$	6.5	97	40	48	4	3	3	4
$q_j(g_j(b_1))$	3	3.5	2	7	0	0	0	0
$q_j(g_j(b_2))$	4.6	8	4	3	0	0	0	0
$p_j(g_j(b_1))$	10	5	6	14	1	0	1	1
$p_j(g_j(b_2))$	9.5	9	7	5	1	0	1	1
w_j	2.4	37.5	14	18.3	3.3	0.5	3.7	2.3
v_j	80	40	70	200	3	3	2	3
λ	0.93							

Table 4. Individual criteria contribution.

	g_1	g_2	g_3	g_4	g_5	g_6	g_7	g_8
$g_j(a)$	22	88	52	30	2	4	2	3
$c_{\{j\}}(a, b_1)$	0.03	0.46	0.17	0.22	0.038	0.006	0.04	0.03

proposed in (8), the subset of criteria $I^* = \{2, 3, 4\}$ is identified, with a maximal $\beta^* = 0.914$. This identified subset has the contribution $c_I(a, b_1) = 0.85$, stating that I^* contributes to 91.4% of the global performance of a when it is compared to the reference profile b_1. In the example, the subset of identified criteria I^* corresponds to the subset of all criteria having a large weight ($w_2 = 37.5, w_3 = 14, w4 = 18.3$) w.r.t the others criteria having weights lower than 4. Thus, a outranks b_1 on all of the criteria of I^*. Conversely, the performance of a compared to b_2 on the important criteria gives $I^* \subset \mathcal{F}_2^c(a, b_2)$ with $\alpha_2(a, b_2) = 0.06$, $\alpha_3(a, b_2) = 6.10^{-6}$ and $\alpha_4(a, b_2) = 0.99$: the bad performance of a compared to b_2 is then explained.

In the presented solution, the selection of criteria in I^* is not affected by the defined threshold γ. Indeed, the coalition of criteria $\{2, 3, 4\}$ having significant contributions is enough to exceed λ. In order to highlight the role of γ, lets consider $\beta = 0.92$, such that $c_I(a, b_1)$ has to exceed 0.86. It means that at least one criterion with a weak contribution ($\geq$0.01) has to be selected in I^*, e.g. $c_{\{1,2,3,4\}}(a, b_1) = c_{\{2,3,4,8\}}(a, b_1) = 0.88$ or $c_{\{2,3,4,5\}}(a, b_1) = c_{\{2,3,4,7\}}(a, b_1) = 0.89$. If a small threshold $\gamma_{small} = \frac{c(a,b_1)}{2^n} = 0.004$ is considered, it does not prevent the selection of any criterion with an insignificant contribution. Conversely, a large $\gamma_{large} = \frac{c(a,b_1)}{n} = 0.124$, refuses the selection of criteria $\{1, 5, 6, 7, 8\}$, then the subset I^* cannot be built. The higher the value of γ is, the more restrictive it is on the final identified subset of criteria.

It is important to mention here that we have applied the approach according to one selected combination of values for the parameters – local solution. Indeed, the programming problem solved to determine the parameters of the model has infinity of solutions, and thus the determined subset I^* could change. This solution also depends on the initial values provided to the program. In addition, for illustration purpose a small training set has been built. Consequently, a throughout robustness analysis should be carried out in our future works.

This application aimed at analysing, *a posteriori*, the performance of a food distribution mission. The proposed approach allowed to identify three criteria: "Percentage of population reached w.r.t the targeted one", "Number of human resources enlisted", "Invested logistical costs". It means that, in similar contexts, a good performance on these three criteria could guaranty a successful mission. It is then relevant to focus on these elements for learning lessons. Furthermore, identifying this subset of criteria is of major interest to restrain the search space to perform deeper analyses. Indeed, in a further step our work aims at identifying, between similar distribution missions, the shared features explaining the success or failure of the mission to then infer general lessons. Thus, knowing the criteria mainly responsible for the achievement of a *good* or *bad* overall performance, allows to focus searching on properties influencing these outlined criteria. In more comprehensive approaches that we are targeting, dealing with up to twenty criteria, the use of such a procedure will be required.

5 Conclusion

Organizations today express a growing interest for semi-automatic approaches enabling to take advantage of past experiences to improve their decision processes. In this paper, based on Electre tri - a well-known and established MCDA method -, we have presented an approach for identifying the criteria most influencing the decision process regarding the overall performance evaluation of a reduce set of experiences. A case study applying the approach to the humanitarian sector has also been proposed. Such an approach will further be used to enable searching for recommendations in large and highly dimensional search spaces. Future works are envisaged. First, as mentioned above a robustness analysis should be made. Furthermore, we plan to present an extended application, and to sophisticate the proposed model on several aspects: (i) to include veto thresholds into the procedure used to infer model parameters from assignment examples; (ii) to take into consideration the possible interactions/dependences between criteria; (iii) to consider potential disagreements of DMs. Finally, we are also interested by improving the procedure for minimizing the DM's cognitive load (i.e. information requirement), and to both manage and reduce experts' subjectivity.

References

1. Malvache, P., Prieur, P.: Mastering corporate experience with the REX method. In: Proceedings of ISMICK 1993, pp. 33–41 (1993)
2. De How, R., Benus, B., Vogler, M., Metselaar, C.: The commonKADS organization model: content, usage and computer support. Expert Syst. Appl. **11**(1), 29–40 (1996)
3. Ermine, J.L., Chaillot, M., Bigeon, P., Charreton, B., Malavieille, D.: Méthode pour la gestion des connaissances. Ingénierie des systèmes d?information, AFCET-Hermès **4**(4), 541–575 (1996)
4. Aamodt, A., Plaza, E.: Case-based reasoning: foundational issues, methodological variations, and system approaches. AI Commun. **7**(1), 39–59 (1994)
5. Dyer, J.S.: Multiattribute utility theory (MAUT). In: Greco, S., Ehrgott, M., Figueira, J. (eds.) Multiple Criteria Decision Analysis, vol. 233, pp. 285–314. Springer, New York (2016). https://doi.org/10.1007/978-1-4939-3094-4_8
6. Figueira, J.R., Mousseau, V., Roy, B.: ELECTRE methods. In: Greco, S., Ehrgott, M., Figueira, J.R. (eds.) Multiple Criteria Decision Analysis, vol. 233, pp. 155–185. Springer, New York (2016). https://doi.org/10.1007/978-1-4939-3094-4_5
7. Figueira, J.R., Greco, S., Roy, B., Słowiński, R.: An overview of ELECTRE methods and their recent extensions. J. Multi-Criteria Decis. Anal. **20**(1–2), 61–85 (2013)
8. Vincke, P.: Multicriteria Decision-Aid. Wiley, Chichester (1992)
9. Mousseau, V., Slowinski, R.: Inferring an ELECTRE TRI model from assignment examples. J. Glob. Optim. **12**(2), 157–174 (1998)

Combining Weighted Description Logic with Fuzzy Logic for Decision Making

Nadine Mueller(✉), Klemens Schnattinger, and Heike Walterscheid

Baden-Wuerttemberg Cooperative State University, Hangstr. 46-50, 79539 Lörrach, Germany
{muellnad, schnattinger, walterscheid}@dhbw-loerrach.de

Abstract. In this paper we present a consensus-theoretic framework based on weighted description logic and on a consensus modelling approach, which is used to retrieve a consistent decision among experts along multi-attributes. We will show that the integration of these two approaches is best suited for consensus building between (human) experts, especially when their preferences are not easily found or disturbed by coincidental influences. As an application of our methodology, we interviewed experts (in our case students) on the choice of means of transport. One time we asked them directly about their preferences and another time we asked them about their attitudes towards ecology, economy, and others. We will show how these two approaches of gathering data lead to different constructed hypothetical consensus and how the additional use of weighted description logic reveals other diverse insights. Our consensus-theoretical methodology begins with the modelling of basic attribute characteristics, mapping them into fuzzy preference relations and thus supports the decision-making process with respect to consensus.

Keywords: Weighted description logic · Group decision making
Fuzzy preference relations · Consensus

1 Introduction

When groups of experts need to find consensus upon different choices, there are multiple possibilities that model and support this process. One which is only able to model preferences and rankings using weighted description logic [1] is given by [2]. Another one which contains a consistent as well as consensus focused approach is described by [3] and uses simple fuzzy preferences relations. Papers around basic fuzzy decision making start with a known [4] or a known but incomplete [5] preference matrix, no matter whether it is about 1-type or 2-type fuzzy. Thus, all of them start with more or less well-articulated preferences (see also [6–10]), which in real world problems are often impossible to obtain from experts, receiving benefits from leaving preferences unrevealed [11]. Preferences as well can be superimposed in the context of decision making if people decide subjectively, but not objectively rational. So-called subjective rationality describes a biased individual perception of the subject matter of a decision, which only becomes apparent in critical exchange with neutral third parties [12].

J. Medina et al. (Eds.): IPMU 2018, CCIS 855, pp. 124–136, 2018.
https://doi.org/10.1007/978-3-319-91479-4_11

Reasons for a bias may be a lack of knowledge of benefits [13] or psychological manipulation, e.g. nudging [14, 15]. To neutralize or even eliminate the distorting influences, we will use weighted description logic on easy to vote attributes to retrieve objective or rather the true non-biased preferences, instead of prompting those directly from the experts. In this way we combine two existing approaches for group decision making and consensus retrieving and build up a framework which constructs a hypothetical end-to-end process, when true preferences are inferred or not detectable.

2 Preliminaries

2.1 Opinion and Consensus Mining Architecture OMA

The original architecture OMA (Opinion Mining Architecture) is part of a project of the same name. OMA was first used for sentiment analysis from tweets for the financial sector [2]. To achieve an automated calculation of sentiment scores from texts, traditional approaches of natural language processing and machine learning from texts were used for the preprocessing of the texts [16]. In addition an extension of description logic [17], so-called weighted description logic according to [1], is used to calculate the sentiment scores in an automated way (for more details, see Sect. 2.2). The idea of separating the "text processing" task (pre-processing, filtering out relevant phrases) and the decision support task (evaluating extracted phrases) derives from the text understanding system SYNDIKATE [18] and its qualitative calculus [19, 20]. In order to introduce the extension of OMA to include consensus mining, we must first clarify the essential components of OMA: the *TBox*, which accommodates models via compliances, judgements, etc., the *ABox*, which contains unweighted statements on the model of the *TBox* and U_i*Box*es, that contain weighted attribute models of experts.

From a technical point of view, the models of the *TBox* are entirely expressed in description logic by means of terminological concepts, roles and is-a-relations. The elements of the *ABox* are terminological assertions that enter into an instance-of-relationship with concepts of the *TBox*. The weighted individual attributes per expert e_i are shown in the U_i*Box*. These models consist of a subjective-rationality based a priori rating over attributes of concepts and represent the individual utility function of an expert e_i. With these a first a posteriori preference order for each expert's choice can be derived. Next, the preference relations of each expert are used to build consensus. This is done by means of incomplete fuzzy preference relations for group decision making [5], which repeatedly adapts the preference relations of all experts until a satisfying consistent consensus is achieved. The theoretical basis of this approach comes from [21] and its IOWA operator. For more details see Sect. 2.3. Finally, the consistent consensus is represented by a preference order.

2.2 Weighted Description Logic

The signature of description logic [17] is usually specified as a triple (N_C, N_R, N_I), where N_C represents the set of atomic concepts, N_R the set of role names and N_I the set of atomic instances. We denote concepts or classes C and D, roles R and S and instances a and b. Concept descriptions such as $\neg C$, $C \sqcap D$, and $C \sqcup D$ can be derived from N_C if C and D are concept descriptions. Further, $\exists r.C$ and $\forall r.C$ exist if $r \in N_R$ and C is a concept description. The top concept $\top$ is an abbreviation for $C \sqcup \neg C$ and $\bot$ for $\neg \top$. To specify a semantics, an interpretation for the given syntax is introduced. An interpretation is a pair $\mathcal{J} := (\Delta^{\mathcal{J}}, \cdot^{\mathcal{J}})$, where the domain $\Delta^{\mathcal{J}}$ is a non-empty set and $\cdot^{\mathcal{J}}$ is a so-called interpretation function. Further details can be found in [17].

The description logic distinguishes between terminological knowledge (the so-called *TBox*) and assertional knowledge (the so-called *ABox*). A *TBox* contains concept inclusions of type $C \sqsubseteq D$ which have the semantics $C^{\mathcal{J}} \subseteq D^{\mathcal{J}}$ and concept definitions of type $C \equiv D$, where $C \sqsubseteq D$ and $D \sqsubseteq C$. An *ABox* is a set of assertions about concepts $C(a)$, where $a \in N_I$ and $C(a)^{\mathcal{J}} := a^{\mathcal{J}} \in C^{\mathcal{J}}$, as well as role assertions $R(a, b)$, where $(a, b) \in N_I \times N_I$ and $R(a, b)^{\mathcal{J}} := (a^{\mathcal{J}}, b^{\mathcal{J}}) \in R^{\mathcal{J}}$. In the following, only a coherent *TBox* $\mathcal{T}$ and a consistent *ABox* $\mathcal{A}$ is considered (see [17] for details). The pair $\mathcal{K} := \langle \mathcal{T}, \mathcal{A} \rangle$ is then called a knowledge base.

To automate decision making processes, it is necessary to rank the available choices with respect to a set of weighted attributes that have been specified by a user. To take advantage of the reasoning capabilities of description logic the user's preferences over attributes have to be modelled into the artefacts of the description logic. Weighted description logic is an ontological approach to decision making and can be considered as a generic framework, so-called DL decision base [1]. For this purpose, an a priori preference relation via attributes (so-called ontological classes correspond concepts) is used. From this relation, an a posteriori preference relation can be derived (so-called ontological individuals correspond instances). This relationship is then used to make a choice for decision making. In formal terms, an a priori utility function U is defined by the attribute set $\mathcal{X}$ ($U : \mathcal{X} \rightarrow \mathbb{R}$). In addition, a utility function u, which is defined by choices that use logical entailment, extends the utility function U to the subset of attributes. The utility function u is used because it allows a choice to be defined as an instance and its outcome as a set of concepts. Another reason is that U can take various forms, e.g., *max*, *mean*. Modelling attributes has two steps:

1. Each attribute is modelled by a concept.
2. For every value of an attribute a new (sub)concept is introduced.

For instance, if *attitude* is an attribute of an expert to be modelled, it is simply represented by the concept *Attitude* (i.e., *Attitude* $\in \mathcal{X}$). Then, an attitude can be regarded as a value, as if it were a concept of its own. If ecological minded is a value of the attribute *attitude*, the attribute set $\mathcal{X}$ is simply extended by adding the concept *EcologicalM* as a sub-concept of *Attitude*. It should be noted, that axioms have to be introduced to guarantee the disjointedness from all other attitudes.

Assuming a total preference relation (i.e., $\succcurlyeq_{\mathcal{X}}$) over a set of attributes $\mathcal{X}$, and a function $U : \mathcal{X} \rightarrow \mathbb{R}$ that represents $\succcurlyeq$ (i.e., $U(X_1) \geq U(X_2)$ **iff** $X_1 \succcurlyeq_{\mathcal{X}} X_2$ for X_1, $X_2 \in \mathcal{X}$), the function U assigns an *a priori* weight to each concept $X \in \mathcal{X}$. The utility of a concept $X \in \mathcal{X}$ is denoted by $U(X)$. In addition, the greater the utility of an attribute, the more preferable the attribute is. As mentioned above, a *choice* is an instance $c \in N_I \cdot \mathcal{C}$ denotes the finite set of choices. To determine a preference relation (*a posteriori*) over $\mathcal{C}$ (i.e., $\succcurlyeq_{\mathcal{C}}$), which respects $\succcurlyeq_{\mathcal{X}}$, a utility function $u : \mathcal{C} \rightarrow \mathbb{R}$ is introduced. $u(c)$ indicates *the utility of a choice c* relative to the attribute set $\mathcal{X}$. Also, a utility function U over attributes is introduced. For simplicity, the symbol $\succcurlyeq$ is used for both choices and attributes whenever it is evident from the context. The σ-utility is a particular u and is defined as $u_\sigma(c) := \sum\{U(X)|X \in \mathcal{X} \text{ and } \mathcal{K} \models X(c)\}$ and is called the *sigma utility of a choice* $c \in \mathcal{C} \cdot u_\sigma$ triggers a preference relation over $\mathcal{C}$ i.e., $u_\sigma(c_1) \geq u_\sigma(c_2)$ **iff** $c_1 \succcurlyeq_{\mathcal{C}} c_2$. Each choice corresponds to a set of attributes, which is logically *entailed* e.g., $\mathcal{K} \models X(c)$. Due to the criterion additivity, each selection c corresponds to a result. Putting things (DL, U and u) together, a generic *UBox* (so-called *Utility Box*) is defined as a pair $\mathcal{U} := (u_\sigma, U)$, where U is a utility function over $\mathcal{X}$ and u is the utility function over $\mathcal{C}$. Also, a *decision base* can be defined as a triple $D = (\mathcal{K}, \mathcal{C}, \mathcal{U})$ where $\mathcal{K} := \langle \mathcal{T}, \mathcal{A} \rangle$ is a consistent knowledge base, $\mathcal{T}$ is a *TBox* and $\mathcal{A}$ is an *ABox*, $\mathcal{C} \subseteq N_I$ is the set of choices, and $\mathcal{U} = (u, U)$ is an *UBox* (all definitions are due to [1]).

Example

A student would like to decide which transportation to use to get to university. Five alternatives are considered, which fit the original purpose. The student's decision base $D = (\mathcal{K}, \mathcal{C}, \mathcal{U})$ is given as follows:

$$\begin{aligned}\mathcal{T} = \{ & \mathit{EcologicalM} \sqsubseteq \mathit{Attitude}, \mathit{EconomicalM} \sqsubseteq \mathit{Attitude},\\ & \mathit{MobilityM} \sqsubseteq \mathit{Attitude}, \mathit{SpontaneousM} \sqsubseteq \mathit{Attitude}, \mathit{VelocityM} \sqsubseteq \mathit{Attitude},\\ & \mathit{EcologicalM} \sqcap \mathit{EconomicalM} \sqsubseteq \bot, \mathit{EconomicalM} \sqcap \mathit{MobilityM} \sqsubseteq \bot,\\ & \mathit{MobilityM} \sqcap \mathit{SpontaneousM} \sqsubseteq \bot, \mathit{SpontaneousM} \sqcap \mathit{VelocityM} \sqsubseteq \bot\}\end{aligned}$$

$$\begin{aligned}\mathcal{A} = \{ & \mathit{UniTransfer}(car), \mathit{UniTransfer}(motorcycle), \mathit{UniTransfer}(ride),\\ & \mathit{UniTransfer}(publicTrans), \mathit{UniTransfer}(byFoot), \mathit{MobilityM}(car),\\ & \mathit{VelocityM}(car), \mathit{SpontaneousM}(car), \mathit{EconomicalM}(motorcycle),\\ & \mathit{VelocityM}(motorcycle), \mathit{SpontaneousM}(motorcycle), \mathit{EconomicalM}(ride),\\ & \mathit{EcologicalM}(ride), \mathit{EcologicalM}(publicTrans), \mathit{EconomicalM}(publicTrans),\\ & \mathit{VelocityM}(publicTrans), \mathit{EcologicalM}(byFoot),\\ & \mathit{EconomicalM}(byFoot), \mathit{SpontaneousM}(byFoot)\}\end{aligned}$$

$$\begin{aligned}\mathcal{C} &= \{car, motorcycle, publicTrans, ride, byFoot\}\\ \mathcal{U} &= \{(\mathit{EcologicalM}, 60), (\mathit{EconomicalM}, 30), (\mathit{MobilityM}, 20), (\mathit{VelocityM}, 10),\\ &\quad (\mathit{SpontaneousM}, 20)\}\end{aligned}$$

Considering $\mathcal{U}$, the attitude of the student is more ecologically minded than all other attitudes, more economically minded than e.g. velocity minded, etc. The utility scores can be calculated by

$$\begin{array}{ll} u_\sigma(car) = 20 + 20 + 10 = 50 & u_\sigma(motorcycle) = 60 + 30 + 20 = 110\\ u_\sigma(ride) = 60 + 30 = 90 & u_\sigma(publicTrans) = 60 + 30 = 90\\ u_\sigma(byFoot) = 60 + 20 = 80 & \end{array}$$

2.3 Fuzzy Group Decision Making

To obtain a consensus and select the most common preference relation a modelling technique is needed, that respects consistency and at the same time finds consensus. In our work the consensus modelling along [5] is used, as it fulfills all the underlying criteria. In the following, the basic ideas of the procedure are given.

Initial position is a set of alternatives $X = \{x_1, x_2, \ldots, x_n\}$ and a group of experts $E = \{e^1, e^2, \ldots, e^m\}$ who have preferences between some of those alternatives. These preferences are synthesized in matrices (one for every expert). They reflect the underlying preference relation for each combination of compared alternatives. The representation of the preference relation is a numerical fuzzy value, based on its membership function $\mu_P : X \times X \rightarrow [0, 1]$ with $X \times X$ being the Cartesian product of the alternatives defined above. This means that in the cell ik of matrix l the preference degree between the two alternatives x_i and $x_k (p^l_{ik} = \mu^l_P(x_i, x_k))$ of expert e^l is denoted. The higher the value, the stronger expert e^l prefers alternative x_i to x_k. If the preference degree is unknown or undefined an 'x' will be set. In order to carry out the consensus modelling process along [5], various measures and values are required, which are introduced below.

Consistency Measures

A consistent preference relation matrix fulfills the transitivity property. Therefore, this property is used as base to create conditions each matrix should satisfy. Transitivity means that if two alternatives are directly compared with each other then this value should be at least as great as all other preference values using an indirect path. To articulate this in formulas, the additive transitivity is given by [9]:

$$(p_{ij} - 0.5) + (p_{jk} - 0.5) = (p_{ik} + 0.5) \quad \forall i, j, k \in \{1, \ldots, n\}$$

where $p_{i/j/k}$ indicate the preference values of an arbitrary expert e^l.

Or rewritten and reordered to obtain a condition for p_{ik} :

$$p_{ik} = p_{ij} + p_{jk} - 0.5 \quad \forall i, j, k \in \{1, \ldots, n\} \tag{1}$$

The preference relations are considered to be additive consistent if they satisfy for every possible triple $x_i, x_j, x_k \in X$ the additive transitivity according to (1). This leads to the following three equations which need to be fulfilled:

$$\begin{aligned}
p_{ik} = p_{ij} + p_{jk} - 0.5 \quad &\Rightarrow \quad (cp_{ik})^{j1} \overset{\text{def}}{=} p_{ij} + p_{jk} - 0.5 \quad \forall i, j, k \in \{1, \ldots, n\} \\
p_{jk} = p_{ji} + p_{ik} - 0.5 \quad &\Rightarrow \quad (cp_{ik})^{j2} \overset{\text{def}}{=} p_{jk} - p_{ji} + 0.5 \quad \forall i, j, k \in \{1, \ldots, n\} \\
p_{ij} = p_{ik} + p_{kj} - 0.5 \quad &\Rightarrow \quad (cp_{ik})^{j3} \overset{\text{def}}{=} p_{ij} - p_{kj} + 0.5 \quad \forall i, j, k \in \{1, \ldots, n\}
\end{aligned}$$

A preference matrix is called fully additive consistent if the arithmetic mean is the value itself across all possible constellations:

$$cp_{ik} \stackrel{\text{def}}{=} \frac{\sum_{j=1;i \neq k \neq j}^{n} (cp_{ik})^{j1} + (cp_{ik})^{j2} + (cp_{ik})^{j3}}{3(n-2)} \tag{2}$$

where $(cp_{ik})^{j1}$ is an additive transitivity estimation for p_{ik} according to the three equations above (note $cp_{ik} \in [-0.5, 1.5]$). Therefore, the normalized difference is:

$$\varepsilon p_{ik} \stackrel{\text{def}}{=} \frac{2}{3} \cdot |cp_{ik} - p_{ik}|$$

The consistency level is defined as

$$cl_{ik} \stackrel{\text{def}}{=} 1 - \varepsilon p_{ik} \tag{3}$$

which indicates a high consistency if the value is close to 1.

To obtain an overall consistency measure the consistency level for each alternative $x_i \in X$ is first calculated by building the arithmetic mean over all consistency measures:

$$cl_i \stackrel{\text{def}}{=} \frac{\sum_{k=1;i \neq k}^{n} (cl_{ik} + cl_{ki})}{2(n-1)}$$

The overall consistency measure is then accordingly determined:

$$cl = \frac{\sum_{i=1}^{n} cl_i}{n} \stackrel{\text{def}}{=} CL \tag{4}$$

Again, the closer this value is to 1, the more consistent the preference relation matrix.

Not going further into details, the above Eq. (2) is used to compute missing preference values, in which all related existing values are incorporated. A more detailed explanation can be found in [5]. To be able to distinguish the initial preference matrix with values p_{jk} given explicitly by an expert and the estimated ones obtained along the concept described above, the latter ones are formally noted as $\bar{p}_{jk}$.

The Consensus Measures

In order to assess the degree of consensus among experts, two different kinds of degree are calculated: the consensus degree and the proximity measure. The consensus degree indicates how close the preferences of the experts are, while the proximity measure shows how close the different experts are to a common consensus.

When calculating the consensus degree, the first step is to consider the distance between each pair of experts. In order to not double count pairs of experts, they are ordered sequentially $(h < l)$ and the similarity value is then:

$$sm_{ik}^{hl} = 1 - \left|\bar{p}_{ik}^{h} - \bar{p}_{ik}^{l}\right| \tag{5}$$

The collective similarity matrix is obtained by building the arithmetic mean as aggregation function ϕ over all calculated similarity matrices according to (5):

$$sm_{ik} = \phi\left(sm_{ik}^{hl}\right) \stackrel{\text{def}}{=} cop_{ik}$$

In this way cop_{ik} is a measure across all experts on how close they are on the specific preference relation between alternative x_i and x_k.

Consensus degree on an alternative x_i and analogously the overall consensus degree are determined through building the arithmetic mean over all related cop_{ik}/ca_i values:

$$ca_i \stackrel{\text{def}}{=} \frac{\sum_{k=1;i\neq k}^{n}(cop_{ik} + cop_{ki})}{2(n-1)}$$

$$CR \stackrel{\text{def}}{=} \frac{\sum_{i=1}^{n} ca_i}{n} \tag{6}$$

To calculate the proximity measure, first a collective fuzzy preference matrix is needed. Therefore, an IOWA operator is used, which is defined according to [22]:

$$\Phi_W(\langle u_1, p_1\rangle, \ldots, \langle u_n, p_n\rangle) = \sum_{i=1}^{n} w_i \cdot p_{\sigma(i)} \stackrel{\text{def}}{=} p_{ik}^{c} \tag{7}$$

where w_i is a weighting vector with $\sum_{i=1}^{n} w_i = 1$ and $p_{\sigma(i)}$ a permutation, so that the following condition for the inducing variables u_i is fulfilled: $u_{\sigma(i)} > u_{\sigma(i+1)}$. In this work the weighting vector is calculated with the help of the linguistic quantifier '*most of*' (function Q) and Yager's idea on a quantifier guided aggregation [23]:

$$w_h = Q\left(\frac{\sum_{j=1}^{h} u_{\sigma(j)}}{T}\right) - Q\left(\frac{\sum_{j=1}^{h-1} u_{\sigma(j)}}{T}\right)$$

being $T = \sum_{j=1}^{n} u_j$. Applied to several alternatives and experts and taking into account a balance between consistency and consensus, the values for the different u's are given by the following equation:

$$u_{ik}^{h} = (1-\delta) \cdot cl_{ik}^{h} + \delta \cdot co_{ik}^{h}$$

The parameter δ controls the influence of consistency vs. consensus, cl_{ik}^{h} is the consistency level per expert according to (3) and co_{ik}^{h} is the expert's e^h degree of proximity to all other experts:

$$co_{ik}^{h} = \frac{\sum_{l=h+1}^{n} sm_{ik}^{hl} + \sum_{l=1}^{h-1} sm_{ik}^{lh}}{n-1}$$

using the similarity values defined in (5). In this way a consistent consensus is specified. The proximity measure is now defined first on the level of pairs of alternatives:

$$pp_{ik}^{h} \stackrel{\text{def}}{=} 1 - \left|\bar{p}_{ik}^{h} - p_{ik}^{c}\right|$$

The proximity measure of an expert on an alternative x_i and accordingly his overall proximity measure is then defined by using the arithmetic mean:

$$pa_{i}^{h} \stackrel{\text{def}}{=} \frac{\sum_{k=1;i\neq k}^{n} \left(pp_{ik}^{h} + pp_{ki}^{h}\right)}{2(n-1)} \qquad pr^{h} \stackrel{\text{def}}{=} \frac{\sum_{i=1}^{n} pa_{i}^{h}}{n}$$

This results in the following formula for a global consistency and consensus measure:

$$CCL \stackrel{\text{def}}{=} (1-\delta) \cdot CL + \delta \cdot CR \tag{8}$$

At this point all necessary values and measures are defined to perform the consensus model process based on consistency and consensus criteria, which runs in four steps:

1. **Computing Missing Information**
2. **Computing Consistency and Consensus Measures**
   All relevant measures according to Eqs. (4), (6), (8) are calculated
3. **Consistency/Consensus Control**
   The CCL (8) is compared against a defined threshold. If it is reached, the process stops, otherwise step 4 is executed
4. **Feedback Process**
   Some values of some experts need to be changed to reach a better level of consistency and consensus (for more details see [5]). After these preference relations are adjusted, the consensus modelling routine returns to step 2 and is executed again.

The consensus model process in this work uses the following parameters: 0.75 as the parameter controlling weight against consensus, 0.3 and 0.8 for the quantifier used in the IOWA operator and 0.85 as the threshold to reach for global consistency and consensus.

2.4 Inquiry Setup and Elicitation of Attributes and Attitudes

As an example for the application of the framework, a survey among five students (experts) will be presented which covers aspects of transportation. The three aspects addressed in the survey are:

- How do the students actually come to university?
- How would they ideally like to come to the university?
- Ratings of some personal attitude values that are strongly related to transportation.

The goal of this inquiry is to reveal a constructed hypothetical consensus on the means of transportation for students. As directly involved persons these students can be considered as experts of this question.

The first two questions had to be filled out in percentages using a table as form

Car	Motorcycle	Ride	Public transport	By foot
p_1	p_2	...		

To compress the representation of the results of the survey only summarized tables are shown in the following:

Actual	Car	Motorcycle	Ride	Public transport	By foot
$\sum$ *Experts*	310	0	80	110	0
Ideal	Car	Motorcycle	Ride	Public transport	By foot
$\sum$ *Experts*	350	0	100	0	50

For the third aspect the personal attitudes on ecology, economics, velocity, mobility, and spontaneity were rated on the basis of the values important, rather important, rather unimportant, unimportant, not applicable. The following table shows the results:

Attributes	Important	Rather important	Rather unimportant	Unimportant	Not applicable
Ecology		3	1	1	
Economics	2	2	1		
Velocity	5				
Mobility	4	1			
Spontaneity	5				

2.5 Consensus Mining on Actual and Ideal Situation

For the first two questions of the inquiry setup, where the preferences of the experts were gathered directly, the consensus modelling process runs without incorporated weighted description logic.

After the first three steps the relevant measures for the actual situation are: CL = 0.96, CR = 0.72 and CCL = 0.78. This means that the necessary threshold could not be reached. Therefore, a feedback loop is needed. In this scenario, expert 3 and 5 need to change some of their preferences because their individual measures were not satisfying. Assuming that these two experts would change their preferences according to the recommendations, after a first feedback round the CCL = 0.88 and the process ends.

The feedback loop only affects some minor preferences and the overall ranking of means of transport stays the same: 1. Car; 2. Ride; 3. Public transport; 4. Motorcycle; 5. By foot. This ranking is quite interesting, as looking into the preference matrices public transport has a sum of 110% and Ride only 80%. But taking the consensus and the consistency into account, also the distances among the different preferences of the experts are respected.

Looking now to the ideally preferred means of transport, the calculation returns the measures: CL = 0.99, CR = 0.77 and CCL = 0.83. This Global CCL is already significantly higher than the one of the actual world, which is due to the fact that the ideal world is most often less complex and therefore a higher degree of consensus can be expected.

In this scenario only expert 1 does not reach the necessary threshold and needs to adjust his individual preferences.

After this is done the CCL = 0.9 and the ranking is: 1. Car; 2. By foot; 3. Ride; 4. Public transport & Motorcycle.

2.6 Combination of Weighted Description Logic with Fuzzy Logic

For the third aspect weighted description logic will be incorporated to handle and process the attitudes of each expert. The underlying knowledge base is the same as introduced in Sect. 2.3, with additional U_i*Boxes* per expert (depending on his ratings).

To create the relevant *UBox*-es the four linguistic fuzzy labels (important, rather important, …) need to be transferred into discrete fuzzy numbers. As there is no special logic behind those labels a normalization up to 100 is used and divided into 3 regular ranges to obtain weights. Summary of weights:

	Ecology	Economics	Velocity	Mobility	Spontaneity
$\sum$ *Experts*	433.33	400	400	366.67	466.67

For each expert one separate *UBox* is with the respective weights is created:

$$\mathcal{U}^{Expert_1} = \{(EcologicalM, 66.67), (EconomicalM, 66.67), \ldots\}$$
$$\mathcal{U}^{Expert_2} = \{(EcologicalM, 0), (EconomicalM, 100), \ldots\}\ldots$$

Then every choice gets a utility value by expert:

$$u_{\sigma}^{Expert_1}(car) = 100 + 100 + 100$$
$$u_{\sigma}^{Expert_1}(motorcycle) = 66.67 + 66.67 + 100 + 100 + 100 = 433.33$$
$$\ldots$$
$$u_{\sigma}^{Experts_5}(ride) = 66.67 + 100 = 166.67$$
$$u_{\sigma}^{Experts_5}(byFoot) = 66.67 + 100 + 100 = 266.67$$

With this we are now able to determine the fuzzy preference relation matrix of each expert. E.g.:

$$A_{Expert_1} = \begin{pmatrix} - & 0,53 & 0,69 & 0,56 & 0,56 \\ 0,47 & - & 0,67 & 0,53 & 0,53 \\ 0,31 & 0,33 & - & 0,36 & 0,36 \\ 0,44 & 0,47 & 0,64 & - & 0,5 \\ 0,44 & 0,47 & 0,64 & 0,5 & - \end{pmatrix}$$

Based upon those preference relation matrices the consensus modelling process as introduced in Sect. 2.3 is run.

After one loop the measures are CL = 0.99, CR = 0.97 and CCL = 0.97. As anticipated the relevant measures are very high, because there were no controversial opinions among the experts regarding their attitudes. A feedback loop is therefore not necessary.

The overall ranking of means of transport is: 1. Car; 2. Motorcycle; 3. Public transport & By foot and 4. Ride.

The combination of weighted description logic with consensus modelling returns a different ranking. The first choice is still car, but the second choice is already different. One reason for this could be, that some of the experts are not aware, that motorcycle is also a feasible option for them use to commute to university. In the context of decision making and consensus mining this is an interesting outcome as the combination of the two approaches reveals a different result.

3 Results and Conclusion

As shown above, there are several ways to explore a constructed hypothetical common sense and therefore a hypothetical decision within a group. The direct way of asking experts about their preferences could reveal a rather superficial opinion and thus lead to a consensus that is accepted "only" as common sense. Asking about an ideal world does change some results compared with the actual situation. But asking for attitudes reveals a different picture again. This proofs that even abstracted preferences in an ideal world, do not necessarily conform to results obtained from given attitudes. Changing the way of how preferences are captured has a high impact on common sense. This leads to the conclusion, that a priori the way of how opinions/preferences are revealed should be carefully selected. The combination of weighted description logic and consensus modeling is a powerful construct for common sense mining, mainly in complex real-world situations.

From a decision theoretical point of view, it can be shown to what extent subjective rationality distorts the decision results. At the same time, however, it can also be shown how, theoretically, the objective rationality could be stretched to create a consensus in a group. Thereby, we assume an individual willingness to adapt all preferences (loss of identity) in order to achieve consensus.

4 Outlook

Our next step will be to combine consensus modelling with weighted description logic for different application scenarios with the preprocessing NLP component in OMA to build an enriched opinion & consensus mining system for different applications. In order to adapt the model to different realistic decision-making scenarios we will check for example, how the results will turn out if some of the decision-makers are not willing to give up their identity as an expression of absolutely stable preferences which could be e.g. ethic values. This can be modeled with weightings not supposed to change, which can be the case when it comes to social preferences [24] or when decision makers are under duress, maybe because any compromises would massively harm e.g. their individual legal or economic status. Another option to model parts of a decision-making process is to start the automated feedback process not before a primal adapting feedback loop within the group of deciders took place. This would describe dynamic adaptive decision-making, which becomes relevant within e.g. political voting behavior tweeted in advance of the official announcement of the election results.

References

1. Acar, E., Fink, M., Meilicke, C., Thome, C., Stuckenschmidt, H.: Multi-attribute decision making with weighted description logics. IFCoLog J. Log. Its Appl. **4**, 1973–1995 (2017)
2. Schnattinger, K., Walterscheid, H.: Opinion mining meets decision making: towards opinion engineering. In: Fred, A., Filipe, J. (eds.) IC3K17 – Proceedings of the 9th International Joint Conference on Knowledge Discovery, Knowledge Engineering and Knowledge Management, vol. 1, pp. 334–341 (2017)
3. Alonso, S., Cabrerizo, F., Chiclana, F., Herrera, F.: Group decision making with incomplete fuzzy linguistic preference relations. Int. J. Intell. Syst. **24**, 201–222 (2009). https://doi.org/10.1002/int.20332
4. Xu, K., Liao, S.S., Li, J., Song, Y.: Mining comparative opinions from customer reviews for competitive intelligence. Decis. Support Syst. **50**, 743–754 (2011)
5. Herrera-Viedma, E., Alonso, S., Chiclana, F., Herrera, F.: A consensus model for group decision making with incomplete fuzzy preference relations. IEEE Trans. Fuzzy Syst. **15**(5), 863–877 (2007)
6. Hsu, H.-M., Chen, C.-T.: Aggregation of fuzzy opinions under group decision making. Fuzzy Sets Syst. **79**, 279–285 (1996)
7. Kacprzyk, J., Fedrizzi, M., Nurmi, H.: Group decision making and consensus under fuzzy preferences and fuzzy majority. Fuzzy Sets Syst. **49**, 21–31 (1992). https://doi.org/10.1016/0165-0114(92)90107-f
8. Nurmi, H.: Approaches to collective decision making with fuzzy preference relations. Fuzzy Sets Syst. **6**, 249–259 (1981)
9. Tanino, T.: Fuzzy preference orderings in group decision making. Fuzzy Sets Syst. **12**, 117–131 (1984). https://doi.org/10.1016/0165-0114(84)90032-0
10. Xu, Z.: A method based on linguistic aggregation operators for group decision making with linguistic preference relations. Inf. Sci. **166**, 19–30 (2004)
11. Hardin, G.: The tragedy of the commons. Science **162**(13), 1243–1248 (1968)
12. Walterscheid, H.: Who owns Digital Data? Working and Discussion Paper DHBW Loerrach (4) (2017)

13. Stiglitz, J.: Economics of the Public Sector, 3rd edn. WW Norton & Co., New York (2000)
14. Thaler, R., Sunstein, C.: Nudge: Improving Decisions About Health, Wealth, and Happiness. Penguin (2009). https://doi.org/10.1007/s10602-008-9056-2
15. Aprem, A., Krishnamurthy, V.: Online social media: a revealed preference framework. IEEE Trans. Sig. Process. **65**(7), 1869–1880 (2017)
16. Sun, S., Luo, C., Chen, J.: A review of natural language processing techniques for opinion mining systems. Inf. Fusion **36**, 10–25 (2017)
17. Baader, F., McGuinness, D., Narci, D., Patel-Schneider, P.: The Description Logic Handbook: Theory, Implementation, and Applications. Cambridge University Press, New York (2003). https://doi.org/10.1017/CBO9780511711787
18. Hahn, U., Schnattinger, K.: Towards text knowledge engineering. In: AAAI 1998 – Proceedings of the 15th National Conference on Artificial Intelligence, pp. 524–531 (1998)
19. Schnattinger, K., Hahn, U.: A sketch of a qualification calculus. In: FLAIRS – Proceedings of the 9th Florida Artificial Intelligence Research Symposium, pp. 198–203 (1996)
20. Schnattinger, K., Hahn, U.: Quality-based learning. In: ECAI 1998 – Proceedings of the 13th Biennial European Conference on Artificial Intelligence, pp. 160–164 (1998)
21. Yager, R.: Quantifier guided aggregation using OWA operators. Int. J. Intell. Syst. **11**(11), 49–73 (1996)
22. Yager, R.R., Filev, D.P.: Operations for granular computing: mixing words and numbers. In: IEEE International Conference on Fuzzy Systems, vol. 2, no. 1, pp. 123–128 (1998). https://doi.org/10.1109/fuzzy.1998.687470
23. Zadeh, L.A.: A computational approach to fuzzy quantifiers in natural languages. Comput. Math. Appl. **9**(1), 149–184 (1983)
24. Chuan, Y., Schechter, L.: Stability of experimental and survey measures of risk, time, and social preferences: a review and some new results. J. Dev. Econ. **117**, 151–170 (2015)

The Use of Fuzzy Linguistic Information and Fuzzy Delphi Method to Validate by Consensus a Questionnaire in a Blended-Learning Environment

Jeovani Morales[1], Rosana Montes[1,2](✉), Noe Zermeño[1], Jeronimo Duran[1], and Francisco Herrera[1,3]

[1] Andalusian Research Institute Data Science and Computational Intelligence, DaSCI, University of Granada, Granada, Spain
{jeovani,nzermeno,jeronimoduran}@correo.ugr.es,
rosana@ugr.es, herrera@decsai.ugr.es
[2] Software Engineering Department, School of Informatics and Telecommunications Engineering, University of Granada, Granada, Spain
[3] Computer Science and Artificial Intelligence Department, School of Informatics and Telecommunications Engineering, University of Granada, Granada, Spain

Abstract. The virtual learning landscapes have created complex environments when evaluating an educational experience. The Fuzzy Delphi method, which incorporates the theory of fuzzy sets, takes the opinions issued by judges, from a linguistic perspective, to validate a questionnaire that will measure the degree of success of an educational experience. The judges have to reach a consensus on the validity and applicability of the instrument. This work contributes to the validation of questionnaires by enabling linguistic assessments and not only binary answers and with a calculus of consistency and consensus degrees for each item, which contributes to consensus reaching. It has been used as a practical experience to define, with the consensus of nine experts, a questionnaire that measures the virtual communication and the satisfaction with in a Blended-Learning pilot experience in the subject of Software Fundamentals, 1st semester of the Degree in Computer Engineering of the University of Granada.

Keywords: Linguistic decision making · Fuzzy Delphi method B-Learning · Instrument validation

1 Introduction

Newly emerging educational methodologies tend to encourage the creation of virtual learning environments. In higher education they are promoted with different technological tools for self-regulation of learning, as well as collaborative

J. Medina et al. (Eds.): IPMU 2018, CCIS 855, pp. 137–149, 2018.
https://doi.org/10.1007/978-3-319-91479-4_12

and cooperative learning [19]. Examples of this are the Flipped Classroom (FC) [18] and Mobile-Learning (ML) [11] methodologies. In addition, in the context of e-Learning, interactions between participants have been defined through the concept of Community of Inquiry (CoI) [8]. On the other hand, the application of in-person interactions combined with e-learning support is also an educational methodology that integrates the advantages offered by each of the above, and is known as Blended-Learning (B-Learning).

The application of those environments altogether become complex to evaluate as there are many constructs and different ways of structuring a measuring instrument such as the questionnaire. Moreover, specialized literature has developed each area separately, that is questionnaires for FC or questionnaires for ML or questionnaires for CoI. So there is little basis for taking a validated questionnaire to be applied in combination of the above methodologies.

Based on a proposal to evaluate a teaching experience that combines ML and FC [7] in a B-Learning environment, our aim is to know the degree of applicability of that questionnaire in a pilot educational experience, by checking the robustness of the instrument through the evaluation of judges.

In this paper, content validity of a questionnaire has been checked by the Fuzzy Delphi (FD) method, which is based on obtaining the opinion of judges in an iterative process for assessing consistency and consensus among the items of the instrument. Given that experts usually evaluate on a binary linguistic or numeric scale, the aim of this work is to use an enriched linguistic term set. In this way, we take advantage of the expert's knowledge in the assessments issued. To this end, we had the support of 9 expert judges in the area of Education Sciences and Information and Communication Technologies (ICT). At the end of the application of this methodology, a consensual questionnaire is obtained.

In the following section, a descriptive overview of the preliminaries relating to an educational experience and the validation of a questionnaire is provided. In Sect. 3, the FD method is applied to the research context. Finally, Sect. 4 presents the conclusions.

2 Preliminaries

This section describes the educational experience to be evaluated in addition to the criteria and steps required to validate data collection instrument together with the FD method.

2.1 A Blended-Learning Experience in Higher Education

B-learning is a flexible approach to course design supported by the combination of different learning moments (face-to-face blended with online activities). Thanks to technological advances that promote interaction between students, traditional focus of education shifts from individual to collaborative approaches. Collaboration and virtual communication are fundamental aspects of e-Learning because of the effect they have on learning and satisfaction [12], so it has long been sought to analyze the characteristics necessary to improve learning outcomes in higher education environments. A theoretical and analytical model is the CoI [8] model, which is based on a collaborative-constructivist perspective of education that conceptualizes the learning and virtual communication. Thus representing the process of creating a deep and meaningful learning experience that develops through three interdependent core elements:

- *Cognitive Presence*: Through a series of phases, it allows the student to construct new educational experiences.
- *Social Presence:* Develops interpersonal relationships through the media available in the learning environment.
- *Teaching Presence:* Integrates the above elements through design, direct teaching and resource facilitation.

A relatively new and popular pedagogical methodology for B-Learning is known as *Flipped Classroom* (FC) [1]. It is based on flipping moments of learning, conceptual acquisition and application of knowledge allowing students to learn theory outside the classroom, through resources provided by the teacher, mainly videos. And also through the application of knowledge within the classroom in a collaborative and meaningful way with the support of the teacher and/or peers, promoting more active and responsible learning by students [13].

In the same sense, the use of mobile devices such as laptops or smartphones, being highly individualized and collaborative tools, has allowed the incorporation of Mobile Learning (ML) which is a methodology that intersects mobile computing with e-Learning, offering benefits for the learning environment, such as flexibility (to develop anywhere, anytime). The combined use of ML and FC methodologies enable teachers to easily provide B-Learning environments [4].

In the subject of Software Fundamentals of the 1st semester of the Degree in Computer Engineering at the University of Granada, 9 Telegram groups have been used to work with 70 students. This communication tool has made it possible to carry out synchronous meetings and asynchronous teamwork, arising from the viewing of videos and the proposal of group activities. FC has therefore been combined with ML and it is desired to evaluate the underlying CoI model in the virtual community. In order to accomplish our aim, we validate the questionnaire [7] that contains the necessary characteristics to evaluate this specificity in blended learning situations. Table 1 shows the distribution of 45

Table 1. Blocks, dimensions and items corresponding to the questionnaire to evaluate virtual communication and students' satisfaction in FC and ML methodologies.

Assessment instrument in combined environments							
Blocks	Virtual communication			Students' satisfaction			
Dimensions	Cognitive presence	Social presence	Teaching presence	Cognitive presence	Social presence	Teaching presence	General satisfaction
Items	1–8	9–14	15–21	22–28	29–35	36–41	42–45

items in 7 dimensions covering the two blocks that we want to evaluate: Virtual Communication and Students' Satisfaction.

2.2 Instrument Validation: Questionnaire

There are several methodologies for data collection, among them are the use of surveys and questionnaires. These instruments are cost-effective and time-efficient, allowing for the initiation of more developed research [14].

In order for an instrument to be valid, it must meet three requirements:

1. *Reliability*: Consisting of consistency and stability.
2. *Validity*: It is the capacity of an instrument to measure the variable for which it was designed, it contains three dimensions: *criteria*, *construct* and *content*.
3. *Objectivity*: It is the degree to which this is or is not permeable to the influence of the biases and tendencies of the researcher or researchers who administer, qualify and interpret it.

Consensus is the agreement produced by consent between all members of a group or between several groups. Therefore, judgmental review process [2] is a method which reports agreement among judges regarding the evaluation of a questionnaire. According to Lynn [16] at least three judges are required for the validation of an instrument, although this is not a specific figure, it depends on the complexity of the work. In addition to, a moderator figure collects the judges' suggestions and redefine the proposal for the next iteration until a consensus is reached. Then, the instrument can be applied. Consensus methods for questionnaire validation include the Delphi method.

2.3 Fuzzy Delphi Method

The Delphi method is an iterative process [5], where participants express their opinion as many times as necessary until consensus is reached; it has the characteristic of being anonymous, thus avoiding that they are influenced by the group. The sequential process that defines the Delphi method includes three phases: (1) identify the problem and its characteristics, (2) create a coordinating group that elaborates a pilot instrument and, (3) choose the group of judges that values the instrument during iterations. Once these have been carried out, the method must go through a series of stages:

1. Disseminate the instrument to judges.
2. Sort, assess and compare the responses obtained in the first iteration.
3. Modify the instrument items according to the judges' suggestions.
4. Feedback to the judges in each iteration, at least three are recommended, or until they have a positive consistency.

At the end a report is issued describing each of the elements and stages that made up the study, the development of each iteration and the degree of consensus reached.

Ishikawa et al. [10], who introduced the FD method, argued that the classic Delphi Method requires time and high costs to achieve an efficient consensus of judges' judgments as it requires several iterations in the instrument's responses. In addition, according to Gupta [9], in expert judgments, there is ambiguity about the different meaning or interpretation each one has of what it evaluates, so that neither real situations nor personal interpretations are usually adequately reflected by quantitative values.

Therefore the use of the FD method, which is a combination of the Delphi method and the theory of fuzzy sets proposed by Zadeh [20], solves some of the drawbacks of the classical method. It avoids confusion of common understanding between expert opinions [17] or interpretation of the responses by involving diffuse numbers and taking these opinions from a linguistic perspective, providing more reasonable results.

3 Application of Fuzzy Delphi Method with Linguistic Assessments to Get the Validation of a Questionnaire

A measurement instrument for B-Learning is used in Sect. 3.1. As exemplification for our proposal. This consist in the application of two iterations of the FD method with fuzzy linguistic information provided by our experts, as it is described in Sect. 3.2.

3.1 Proposal Design

There are few occasions when linguistic decision-making (LDM) has been associated with the validation of a questionnaire, although it has been used to normalize the results of several questionnaires to the one given as a reference [3].

A questionnaire is defined as a set of r items $Q = \{Q_1, Q_2, \ldots, Q_r\}$ which are evaluated over q criteria $C = \{C_1, C_2, \ldots, C_q\}$ of equal weights by p judges $J = \{J_1, J_2, \ldots, J_p\}$. Judges decide if each item is valid to represent the construct for which it is designed, or should be discarded for not doing so (binary answer). To validate a questionnaire, the judges face r different decision-making problems. The assessment matrix for each item $Q_l (l = 1, \ldots, r)$, is represented by a $p \times q$ matrix. Elements are the valuation of the item over criterion C_j by the expert J_k. The full problem of questionnaire validation stores a $p \times q \times r$ matrix.

The judges answered the questionnaire using a scale of 7 linguistic terms, $S = \{s_0 = Lousy, s_1 = Very\ Wrong, s_2 = Wrong, s_3 = Moderate, s_4 = Correct, s_5 = Very\ Correct, s_6 = Excellent\}$ to express their opinion. So a single assessment over item Q_l is $s_{i_{jk}} \in S (i = 0, \ldots, 6)$. This setting completely differs from the usual binary answer (accept or discard) in the assessment matrix.

Example 1 (An Instrument to apply for B-Learning methodology). Our aim is to use a tool that assesses the quality of virtual communication and satisfaction in higher education in combined methodologies. Thus, our problem of LDM proposes a variant of the questionnaire [7] with $r = 45$ items that are evaluated according to $q = 4$ criteria (clarity, writing, belonging and scale). For the purposes of this research, $p = 9$ judges are selected considering various aspects such as: teaching experience in blended/mobile/flipped methodologies, seniority and academic degree. The semantic for the linguistic labels is $S = \{s_0 = 0, s_1 = 0.10, s_2 = 0.25, s_3 = 0.50, s_4 = 0.75, s_5 = 0.90, s_6 = 1\}$.

The instrument design is part of the FD steps as shown in Fig. 1. The following sections detail the iterative processes.

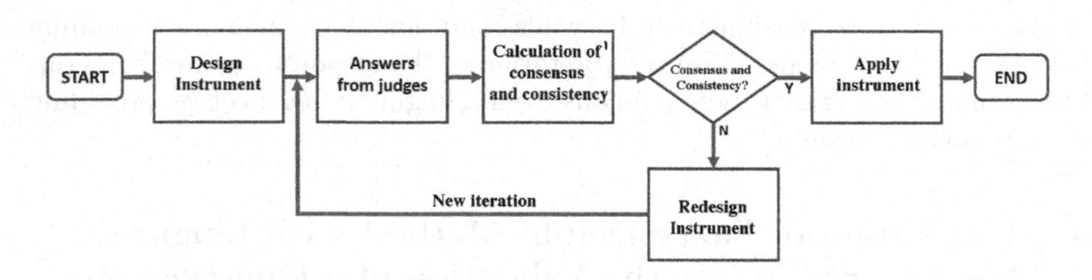

Fig. 1. Flowchart of Fuzzy Delphi methodology.

3.2 The Iterative Process of the Fuzzy Delphi Method

This section describes the process of each iteration within the FD method used in this document. The first iteration collects the judges' opinion, calculates the consistency index and evaluates the level of consensus reached. In the second iteration, modifications to the questionnaire are made according to the judges' suggestions and disseminated together with the average response of each item so that each judge can reassess their opinion. Finally, discussion of results is made in a comparison between iteration 1 and 2.

First Iteration of Fuzzy Delphi Method. Once the questionnaire has been defined, it is sent to the judges for their opinion. Opinions are then represented by a family of parametric functions. Each linguistic valuation $s_{i_{jk}}$ in S is processed using a triangular function by defining a triangular fuzzy number (TFN) $s_{i_{jk}} = (a_{i_{jk}}, b_{i_{jk}}, c_{i_{jk}})$.

We consider consensus as the agreement between several members of a group. Let us note it by a boolean value CS that takes the value of T if there is consensus or F in other case. The Consistency Index $CI \in [0, 1]$ measures the degree of consensus that judges have. The closer it is to 1, the more consistent the judges' opinions are.

In our model, consistency is a boolean value noted as CC that allows us to tag the agreement as above a minimum accepted value, set as $s_4 = Correct$ within our scale. So CC take its value T when $CI \geq \varepsilon$, where $\varepsilon = 0.75$ and $\varepsilon \in [0, 1]$, since this value numerically corresponds to the s_4 label.

Table 2. Judges' evaluation matrix for dimension 7 of the questionnaire (general satisfaction).

	Criteria															
	Clarity				Writing				Belonging				Scale			
	Q_{42}	Q_{43}	Q_{44}	Q_{45}	Q_{42}	Q_{43}	Q_{44}	Q_{45}	Q_{42}	Q_{43}	Q_{44}	Q_{45}	Q_{42}	Q_{43}	Q_{44}	Q_{45}
J_1	s_4	s_6	s_6	s_6	s_4	s_6	s_6	s_6	s_6	s_6	s_6	s_4	s_6	s_6	s_6	s_6
J_2	s_3	s_5	s_4	s_4	s_3	s_5	s_4	s_3	s_3	s_4	s_5	s_4	s_3	s_5	s_3	s_6
J_3	s_5	s_4	s_4	s_5	s_5	s_4	s_4	s_5	s_4	s_4	s_4	s_4	s_4	s_4	s_4	s_5
J_4	s_6	s_6	s_6	s_6	s_6	s_6	s_6	s_6	s_6	s_6	s_6	s_4	s_6	s_6	s_6	s_6
J_5	s_2	s_4	s_3	s_2	s_2	s_4	s_3	s_3	s_3	s_4	s_3	s_2	s_2	s_2	s_2	s_2
J_6	s_3	s_4	s_4	s_4	s_3	s_4	s_4	s_4	s_3	s_4	s_4	s_4	s_4	s_4	s_4	s_6
J_7	s_3	s_4	s_4	s_4	s_3	s_4	s_4	s_4	s_3	s_4	s_4	s_4	s_4	s_4	s_4	s_5
J_8	s_4	s_6	s_4	s_5	s_4	s_6	s_4	s_5	s_4	s_4	s_4	s_4	s_4	s_4	s_4	s_5
J_9	s_6	s_6	s_6	s_6	s_6	s_6	s_6	s_6	s_6	s_6	s_6	s_5	s_6	s_6	s_6	s_6

Example 2 (Valuations). Following our example, we represent data gather for Dimension 7 that ranges from items Q_{42} to Q_{45}. Assessments are shown in Table 2.

The triangular function is applied to each valuation. Table 3 shows the valuations of item Q_{45} and its corresponding TFNs.

Table 3. Triangular Fuzzy Numbers matrix represented as $s_{i_{jk}} = (a_{i_{jk}}, b_{i_{jk}}, c_{i_{jk}})$.

	Item Q_{45}: "I have a positive impression of the course"							
	Clarity		Writing		Belonging		Scale	
	Label	TFN	Label	TFN	Label	TFN	Label	TFN
J_1	s_6	(0.9, 1.0, 1.0)	s_6	(0.9, 1.0, 1.0)	s_4	(0.5, 0.75, 0.9)	s_6	(0.9, 1.0, 1.0)
J_2	s_4	(0.5, 0.75, 0.9)	s_3	(0.25, 0.5, 0.75)	s_4	(0.5, 0.75, 0.9)	s_3	(0.25, 0.5, 0.75)
J_3	s_5	(0.75, 0.9, 1.0)	s_5	(0.75, 0.9, 1.0)	s_4	(0.5, 0.75, 0.9)	s_4	(0.5, 0.75, 0.9)
J_4	s_6	(0.9, 1.0, 1.0)	s_6	(0.9, 1.0, 1.0)	s_4	(0.5, 0.75, 0.9)	s_6	(0.9, 1.0, 1.0)
J_5	s_2	(0.1, 0.25, 0.5)	s_3	(0.25, 0.5, 0.75)	s_2	(0.1, 0.25, 0.5)	s_2	(0.1, 0.25, 0.5)
J_6	s_4	(0.5, 0.75, 0.9)	s_4	(0.5, 0.75, 0.9)	s_4	(0.5, 0.75, 0.9)	s_4	(0.5, 0.75, 0.9)
J_7	s_4	(0.5, 0.75, 0.9)	s_4	(0.5, 0.75, 0.9)	s_4	(0.5, 0.75, 0.9)	s_4	(0.5, 0.75, 0.9)
J_8	s_5	(0.75, 0.9, 1.0)	s_5	(0.75, 0.9, 1.0)	s_4	(0.5, 0.75, 0.9)	s_4	(0.5, 0.75, 0.9)
J_9	s_6	(0.9, 1.0, 1.0)	s_6	(0.9, 1.0, 1.0)	s_5	(0.75, 0.9, 1.0)	s_6	(0.9, 1.0, 1.0)

In order to find the values of consistence and consensus, we establish a conservative valuation with the triangular number $t = (l, m, u)$, set by Eq. (1). For each criterion, and with respect to the lower of the TFN experts opinions, l is the minimum value, m is the geometric mean and u the maximum value.

$$l = \min\{a_{i_{jk}}, ..., a_{p_{jk}}\}, \quad m = (\prod_{k=1}^{p} a_{i_{jk}})^{(1/p)}, \quad u = \max\{a_{i_{jk}}, ..., a_{p_{jk}}\} \tag{1}$$

Subsequently, the optimistic valuation $T = (L, M, U)$ is calculated using the expression given in Eq. (2) where L is the minimum value, M is the geometric mean and U the maximum value, with respect to the TFN upper values considered for each criterion.

$$L = \min\{c_{i_{jk}}, ..., c_{p_{jk}}\}, \quad M = (\prod_{k=1}^{p} c_{i_{jk}})^{(1/p)}, \quad U = \max\{c_{i_{jk}}, ..., c_{p_{jk}}\} \tag{2}$$

The consistency index CI is then calculated for each criterion of each item using Eq. (3). For this purpose, certain elements and requirements must be met.

In Fig. 2, the elements related to the calculus of CI are appreciated and explained below.

In this work we use a combination of the methods described by Dong et al. [6] and Lin [15] with certain modifications, based on the following cases:

(a) If $L \geq u$, the item has an excellent consensus according to the scale used, where $CS = T$, and the value of CI is:

$$CI = \frac{M + m}{2} \tag{3}$$

(b) If $L \leq u$, there is a grey interval, defined as $GI = (L, u)$, one of the following 2 cases may occur:
   (i) If this interval lies between the range of mean values of optimistic and conservative valuation (HI), defined as $HI = (m, M)$, consensus exists and $CS = T$, as shown in Fig. 2a. In this case, CI is determined by:

$$CI = \frac{(M \times u) - (L \times m)}{(u - m) + (M - L)} \tag{4}$$

   (ii) If the GI interval is not within the HI range, there is no consistency and no consensus between the judges' valuations, as shown in Fig. 2b. Hence $CS = F$, and it is necessary to perform another iteration until all the items are consistent.

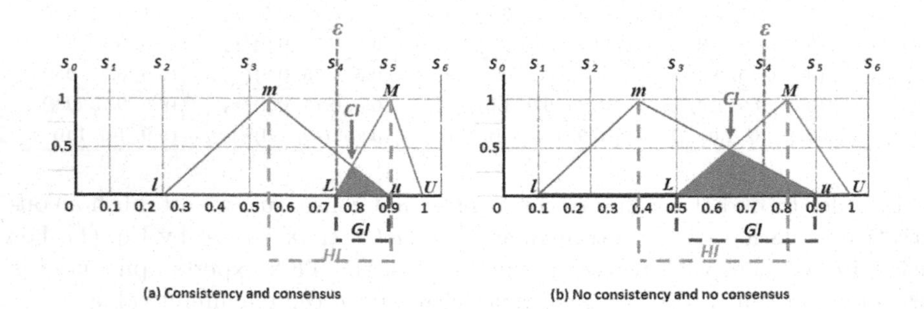

Fig. 2. Requirements for consensus and consistency. (a) Representation of consensus and consistency. (b) Representation of lack of consensus and consistency.

Table 4. Content validation of Q_{45} in the first round.

Item Q_{45}: I have a positive impression of the course											
Criteria	l	m	u	L	M	U	GI	HI	CI	CC	CS
Clarity	0.10	0.56	0.90	0.50	0.89	1.00	(0.50, 0.90)	(0.56, 0.89)	0.76	T	*F*
Writing	0.25	0.57	0.90	0.75	0.92	1.00	(0.75, 0.90)	(0.57, 0.92)	0.80	T	T
Belonging	0.10	0.44	0.75	0.50	0.85	1.00	(0.50, 0.75)	(0.44, 0.85)	0.63	*F*	T
Scale	0.10	0.47	0.90	0.50	0.86	1.00	(0.50, 0.90)	(0.47, 0.86)	0.68	*F*	*F*

Example 3 (Consistency and Consensus). Continuing with the example of item 45, Table 4 shows the results of the application of Eq. (1) and Eq. (2).

The results for the criteria in item Q_{45} are as follows:

- *Clarity:* meets consistency but not consensus guidelines as the GI interval is not within the HI range, see Fig. 3a.
- *Writing*: is the only one that achieves consensus with average value of $s_4 = Correct$ and an adequate consistency, having a $CI \geq \varepsilon$, see Fig. 3.b.
- *Belonging:* achieves a consensus $s_3 = Moderate$ but a consistency below the accepted value with $CI = 0.65$, see Fig. 3c.
- *Scale:* where there is no consistency or consensus, being $CI \leq \varepsilon$ and the interval GI is out of range HI, see Fig. 3d.

Concluding that the item does not have validity of content because not all criteria are satisfactorily validated. As shown in Fig. 3, the consensus is between s_3 and s_4, so we proceed to make appropriate modifications based on the judges' suggestions.

The moderator takes the suggestions made by the judges. In the case of item Q_{45} some of them are literally as follows: (1) *"After using the phrase* 'I am satisfied...' *in the previous questions, it change to* 'I have a positive impression...' *it breaks the dynamics of the questions of the dimension"*, (2) *"The scale I am satisfied does not agree with having a positive impression or not, it is better to have a binary scale"*. Based on these suggestions, the moderator modifies item Q_{45} *"I am satisfied with the course development"* for the second iteration.

Second Iteration of Fuzzy Delphi Method. Once the first iteration is completed and modifications to the first instrument have been made, the judges are provided with the average of each item obtained from the consensus reached during the first iteration. The GI and HI indices are recalculated to obtain the degree of consensus from the judges. Continuing with the example of item Q_{45}, we can see in the Table 5 the results.

The new calculations obtained from the second iteration, considerably improve what reveals that item Q_{45} has satisfactory content validity because it achieves consistency and consensus in each of its four criteria, providing an overall average for the CI of 0.91, indicating that there is a final consistency for

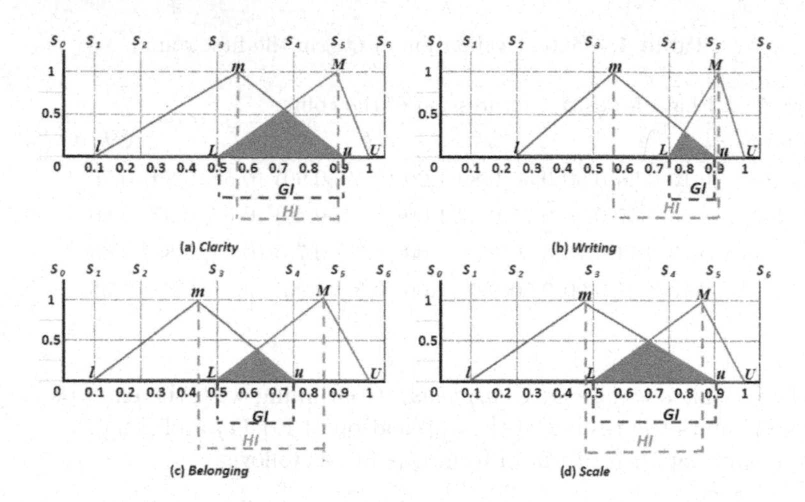

Fig. 3. The graphs represent the values for CC and CS of item Q_{45} for each criterion: (a) Clarity, (b) Writing, (c) Belonging and (d) Scale.

Table 5. Validation of the content of item Q_{45} in the second round.

Item Q_{45}: I am satisfied with the course development											
Criteria	l	m	u	L	M	U	GI	HI	CI	CC	CS
Clarity	0.90	0.90	0.90	1.00	1.00	1.00	(1.00, 0.90)	(0.90, 1.00)	0.95	T	T
Writing	0.75	0.88	0.9	1.00	1.00	1.00	(1.00, 0.90)	(0.88, 1.00)	0.94	T	T
Belonging	0.75	0.88	0.9	1.00	1.00	1.00	(1.00, 0.90)	(0.88, 1.00)	0.94	T	T
Scale	0.25	0.64	0.9	0.75	0.93	1.00	(0.75, 0.90)	(0.64, 0.93)	0.78	T	T

the item of 91% agreement between the judges, thus recognizing that the FD process has been successful in this item.

3.3 Discussion of Results

As can be seen in the FD process, the possibility of obtaining a consensus to validate the content of an instrument in a single iteration is complicated, whether numerically or linguistically evaluated, since each judge has their own perception of the clarity, writing, membership and scale of each item. Using linguistic terms gives the judge the ability to choose the rating that really suits his or her expertise.

The second iteration provides very significant information for judges and educational promoters of the pilot experience. It does so by highlighting the modifications, average assessments and suggestions of the judges themselves. Thus, in spite of subjectivity, it is achieved that opinions are directed towards a point in common for all, the final consensus.

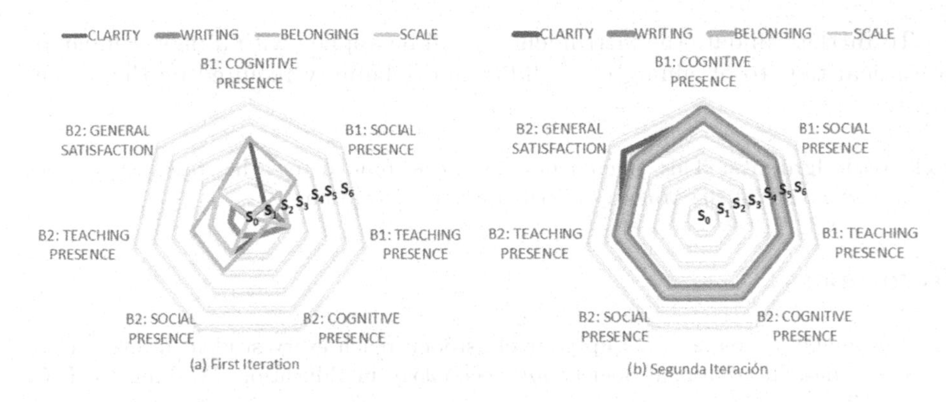

Fig. 4. Final assessment for all criteria and for all dimensions: (a) First iteration. (b) Second iteration.

Example 4 (Comparison of first and second iteration of FD method). We analyze, using a radial chart, the agreement over all the dimension of the questionnaire by the application of the FD method. Results are linguistic values in S. Corresponding to the first iteration (see Fig. 4a there is no consensus neither consistency. Figure 4b illustrates the second iteration where both criteria are satisfactorily met by having values of at least s_4.

4 Conclusions

Higher Education makes use of ICT, so it is very important to evaluate the quality of every b-Learning experience. All questionnaires should be validated prior its use. Currently, if we apply in combination the methodologies of FC and ML is hard to find in the literature a questionnaire already validated.

A questionnaire can be validated by judgmental review such as the Delphi method. This is an iterative process that tries to find consensus in the judgment opinions. When binary scales are used, much of the expert information is lost. Our proposal is to use fuzzy linguistic information to account for situations with imprecision and subjectivity. Then the use of the FD method is proposed to validate by consensus a questionnaire centered in the B-Learning environments. Our proposal incorporates the computation of the consensus status CS and consistency CC, which are fundamental to consensus reaching.

We put in practice a Fuzzy Delphi method with linguistic assessments to get the validation of a questionnaire of 45 items. The acceptance or rejection of a given item is defined as LDM problem that uses 9 judges, to assess the item considering 4 criteria. The method also uses a moderator who collects the suggestions and makes the pertinent changes for the next iteration. In our practical case, consensus was obtained with minimal loss of information due to the application of the fuzzy linguistic model.

To further validate the instrument, it will be applied with a pilot sample for statistical test, to obtaining the validity and reliability required by the instrument.

Acknowledgments. This document has been funded by the research project TIN2017-89517-P of the Ministry of Science and Innovation.

References

1. Bergmann, J., Sams, A.: Flip your classroom: reach every student in every class every day. International Society for Technology in Education, Washington, D.C. (2012)
2. Berk, R.: Importance of expert judgment in content-related validity evidence. West. J. Nurs. Res. **12**(5), 659–671 (1990). https://doi.org/10.1177/019394599001200507
3. Carrasco, R.A., et al.: A linguistic multi-criteria decision making model applied to the integration of education questionnaires. Int. J. Comput. Intell. Syst. **4**(5), 946–959 (2011). https://doi.org/10.1080/18756891.2011.9727844
4. Cornelius, S., Gordon, C.: Providing a flexible, learner-centred programme: challenges for educators. Internet High. Educ. **11**(1), 33–41 (2008). https://doi.org/10.1016/j.iheduc.2007.11.003
5. Dalkey, N.: An experimental study of group opinion: the delphi method. Futures **1**(5), 408–426 (1969). https://doi.org/10.1016/S0016-3287(69)80025-X
6. Dong, J., Huo, H.: Identification of financing barriers to energy efficiency in small and medium-sized enterprises by integrating the fuzzy delphi and fuzzy dematel approaches. Energies **10**(8), 1172 (2017). https://doi.org/10.3390/en10081172
7. García-Lira, K., Gutiérrez-Santiuste, E., Montes-Soldado, R.: Cuestionarios para la evaluacion de la comunicacion y la satisfaccion al aplicar metodologias flipped classroom combinadas con m-learning en educacion superior. In: III Congreso Internacional de Educación Mediática y Competencia Digital, pp. 1145–1163 (2017)
8. Garrison, D., Akyol, Z.: Toward the development of a metacognition construct for communities of inquiry. Internet High. Educ. **17**(Suppl. C), 84–89 (2013). https://doi.org/10.1016/j.iheduc.2014.10.001
9. Gupta, R., et al.: Selection of 3PL service provider using integrated fuzzy delphi and fuzzy topsis. In: Proceedings of the World Congress on Engineering and Computer Science, vol. 2, pp. 20–22 (2010). https://doi.org/10.1504/IJISE.2012.048862
10. Ishikawa, A., et al.: The max-min delphi method and fuzzy delphi method via fuzzy integration. Fuzzy Sets Syst. **55**(3), 241–253 (1993). https://doi.org/10.1016/0165-0114(93)90251-C
11. Jaldemark, J., et al.: Editorial introduction: collaborative learning enhanced by mobile technologies. Br. J. Educ. Technol. **49**, 201–206 (2017). https://doi.org/10.1111/bjet.12596
12. Kim, J.: Developing an instrument to measure social presence in distance higher education. Br. J. Educ. Technol. **42**(5), 763–777 (2011). https://doi.org/10.1111/j.1467-8535.2010.01107.x
13. Lage, M.J., et al.: Inverting the classroom: a gateway to creating an inclusive learning environment. J. Econ. Educ. **31**(1), 30–43 (2000). https://doi.org/10.1080/00220480009596759

14. Lensing, S.Y., et al.: Encouraging physicians to respond to surveys through the use of fax technology. Eval. Health Prof. **23**(3), 348–359 (2000). https://doi.org/10.1177/01632780022034642
15. Lin, C.: Application of fuzzy delphi method (FDM) and fuzzy analytic hierarchy process (FAHP) to criteria weights for fashion design scheme evaluation. Int. J. Clothing Sci. Technol. **25**(3), 171–183 (2013). https://doi.org/10.1108/09556221311300192
16. Lynn, M.R.: Determination and quantification of content validity. Nurs. Res. **35**(6), 382–386 (1986)
17. Noorderhaven, N.G.: Strategic Decision Making. Addison-Wesley, Wokingham (1995)
18. Tucker, B.: The flipped classroom. Educ. Next **12**(1) (2012)
19. Valtonen, T., et al.: Perspectives on personal learning environments held by vocational students. Comput. Educ. **58**(2), 732–739 (2012). https://doi.org/10.1016/j.compedu.2011.09.025
20. Zadeh, L.: Fuzzy sets. Inf. Control **8**(3), 338–353 (1965). https://doi.org/10.1016/S0019-9958(65)90241-X

On the Interaction Between Feature Selection and Parameter Determination in Fuzzy Modelling

Peipei Chen[1,2](✉), Caro Fuchs[2], Anna Wilbik[2], Tak-Ming Chan[3], Saskia van Loon[4], Arjen-Kars Boer[4], Xudong Lu[1,2], Volkher Scharnhorst[4,5], and Uzay Kaymak[1,2]

[1] College of Biomedical Engineering and Instrument Science, Zhejiang University, Hangzhou, Zhejiang, China
[2] Information Systems, School of Industrial Engineering, Eindhoven University of Technology, Eindhoven, The Netherlands
{p.chen,c.e.m.fuchs}@tue.nl
[3] Health Systems, Philips Research China, Shanghai, China
[4] Clinical Chemistry, Catharina Hospital, Michelangelolaan 2, 5623 EJ Eindhoven, The Netherlands
[5] Faculty of Biomedical Engineering, Eindhoven University of Technology, Eindhoven, The Netherlands

Abstract. Nowadays the amount of data that is collected in various settings is growing rapidly. These elaborate data records enable the training of machine learning models that can be used to extract insights and for making better informed decisions. When doing the data mining task, on one hand, feature selection is often used to reduce the dimensionality of the data. On the other hand, we need to decide the structure (parameters) of the model when building the model. However, feature selection and the parameters of the model may interact and affect the performance of the model. Therefore, it is difficult to decide the optimal parameter and the optimal feature subset without an exhaustive search of all the combination of the parameters and the feature subsets which is time-consuming. In this paper, we study how the interaction between feature selection and the parameters of a model affect the performance of the model through experiments on four data sets.

Keywords: Feature selection · Model parameters
Number of clusters · Fuzzy models

1 Introduction

Nowadays the amount of data that is collected in various settings is growing rapidly [6]. These elaborate data records enable the training of machine learning models that can be used to extract insights and for making better informed decisions [3]. One type of these machine learning models are fuzzy inference systems (FIS).

J. Medina et al. (Eds.): IPMU 2018, CCIS 855, pp. 150–161, 2018.
https://doi.org/10.1007/978-3-319-91479-4_13

Fuzzy inference systems are an interpretable and transparent type of machine learning models, whose parameters can partly be estimated from data. However, some of its parameters, the so-called hyperparameters, should be provided.

Machine learning models are trained to discover true underlying relations in data and so they should have good generalization capabilities. Since fuzzy inference systems are interpretable and transparent, they function as white box or grey box models (in opposition to for example (deep) artificial neural networks or random forests), which makes it possible to study their underlying reasoning mechanism. The simpler the model, the better the comprehension of these white box models.

One way of making models simpler and therefore easier to understand is by employing feature selection. Feature selection methods aim to select a subset of the original variables that can efficiently describe the input data while reducing effects from noise or irrelevant variables [8]. Feature selection reduces the computational load when training a model and lowers the chances of overfitting the model to the data. Therefore feature selection enhances the model's generalizability [18]. However, when features are correlated, feature selection is not an easy task. Results of feature selection might also depend on the structure of the model class that has been chosen. Machine learning therefore has a dual goal: identifying both the optimal model structure and the optimal model parameters. Since these two are dependent on each other, the subset of informative features might change for different model parameters. Therefore, it is difficult to decide the optimal parameter and the optimal feature subset without an exhaustive search of all the combination of the parameters and the feature subsets, which is time-consuming. In this paper, we attempt to figure out the relationship between feature selection and model parameters with which we may be able to simplify the process of feature selection and parameter searching.

In fuzzy systems model structure is highly related to its hyperparameters such as the number of rules in the system. Therefore, the selected features for a FIS might depend on the number of rules in the system and vice versa. In this paper we investigate this relationship between feature selection and number of rules in a fuzzy system. We will follow a clustering-based approach for modeling, which means that the number of rules of the system is determined by the number of clusters in the feature space.

The remainder of the paper is structured as follows. Section 2 defines the problem. In Sect. 3, we introduce the background information on feature selection, the Takagi-Sugeno model and hyperparameter Setting. Next, the methodology is described in Sect. 4 and the results are shown in Sect. 5. The discussions of our work are presented in Sect. 6. Finally, Sect. 7 concludes this work.

2 Problem Definition

Currently, feature selection is often seen as a pre-processing step which is executed before model building. However, the most relevant feature set is not only inherent to the problem context. Also the structure of the model is of importance.

For example, consider a process in which data are generated according to

$$x_1 = r * cos(t),$$
$$x_2 = r * sin(t).$$

where t is uniformly distributed in $[0, 2\pi]$ and r is uniformly distributed in $[0.99, 1.01]$. The data belong to class 1 if $r > 1$ and to class 0 otherwise. The variables r and t can not be observed directly. Consider now that a data set is available consisting of variables x_1, x_2, x_1^2 and x_2^2.

As can be seen in Fig. 1a, the instances in this data set show a circular pattern in the space $x_1 \times x_2$, where both output categories could be separated by a circle. In this original feature space, a linear classifier would not be able to reach a satisfactory performance. However, both categories can be separated linearly when using features x_1^2 and x_2^2 as can be seen in Fig. 1b. Hence, when using a linear model, x_1^2 and x_2^2 would be the optimal feature subset while other features might be relevant for nonlinear models.

This example shows the dependency between the type of model and the selected feature set, but the problem is even more complex. Not only the type of model, but also the structure of the chosen model affects which feature subset is optimal. For example, the behaviour of a (first-order) Takagi-Sugeno (TS) model is heavily dependent on its hyperparameters. A TS model with only one rule behaves like a linear classifier, while adding rules makes them universal approximators that can implement a non-linear mapping between inputs and output. Also other hyperparameters (such as the type of membership functions (MFs) and aggregation function) change the structure of the model and are therefore likely to affect which features are relevant. Choosing the wrong hyperparameter setting and choosing the optimal feature set sequently (or the other way around) might therefore lead to sub-optimal model performance. This suggests that feature selection has to be an integral part of model selection, instead of being a pre-processing step.

3 Constructing Takagi-Sugeno Fuzzy Models

3.1 Feature Selection

Not all features present in a data set are necessarily informative. Including these irrelevant or redundant features in the model causes a high computational load and may lower the model performance due to overfitting [18]. To avoid this, feature selection is a crucial step in the modelling process [8,15]. Feature selection methods aim to select a subset of the original variables that can efficiently describe the input data while reducing effects from noise or irrelevant variables [8].

Feature selection methods can be divided into three categories. Filter methods use a fast computable proxy measure (like Fisher's score [18] or mutual information [20]) to score a feature set. This measure should reflect the usefulness of a given feature subset for modeling purposes. When using wrapper

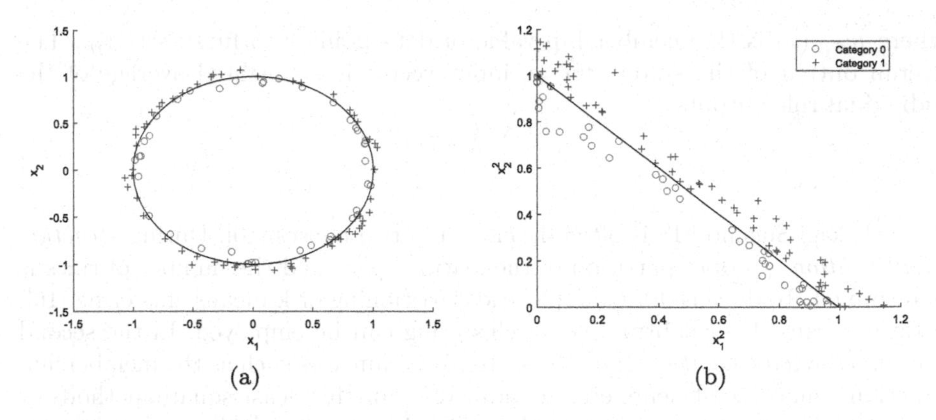

Fig. 1. Artificial data with (a) x_1 and x_2 features and (b) x_1^2 and x_2^2 features.

methods, for every feature subset a prediction model is trained and tested using a hold-out set. The subset resulting in the model with the highest performance is then selected. For embedded methods (such as SVM-RFE [9] and FS-P [17]) feature selection is performed integrated in the training process while developing the final model.

Fuzzy modeling techniques can not deal with large number of features due to 'the curse of dimensionality' [12]. Therefore, feature selection is an indispensable step. Filter methods are often used because of their simplicity and generality [2]. These methods are less computationally intensive than wrapper methods, because wrapper methods have to train and test the classifier for each feature subset candidate. However, wrapper methods usually provide the best performing features subset [13].

3.2 The Takagi-Sugeno Model

A Takagi-Sugeno (TS) fuzzy inference system [19] consists of fuzzy rules where each rule describes a local input-output relation. For a first order system, these rules are of the type:

$$R_j: \ \textbf{If} \ x_1 \text{ is } A_{j1} \text{ and } \dots \text{ and } x_N \text{ is } A_{jN} \ \textbf{then} \ y_j = \mathbf{a}_j^T \mathbf{x} + b_j \tag{1}$$

where, $j = 1, \dots J$ corresponds to the rule number, $\mathbf{x} = (x_1, \dots x_N)$ is the input vector, N is the total number of input variables, A_{jn} is the fuzzy set for rule Rj and n^{th} feature, and y_j is the consequent function for rule R_j. The degree of fulfillment of rule j is given by:

$$\beta_j = \prod_{n=1}^{N} \mu_{A_{jn}}(\mathbf{x}), \tag{2}$$

where $\mu_{A_{jn}}(x)$ is the membership value of data point x to fuzzy set A_{jn}. The overall output of the system for an input vector is a weighted average of the individual rule outputs:

$$y^* = \frac{\sum_{j=1}^{J} \beta_j y_j}{\sum_{j=1}^{J} \beta_j}. \tag{3}$$

A Takagi-Sugeno FIS is often developed in two phases [10]. During *structure identification* a proper partition of the feature space and the number of rules is determined. To do this, for example grid partitioning or k-means clustering [16], fuzzy c-means [1] or subtractive [5] clustering can be employed. In the second phase, *parameter identification*, the system's parameters such as the membership functions, linear coefficients etc. are adjusted. For this, least-square methods or derivative-based optimization techniques can be used (see [10]).

3.3 Hyperparameter Setting

A model hyperparameter is a tuning parameter of a machine learning algorithm that is external to the model and whose value cannot be estimated from data. Therefore, hyperparameters must be specified by the data scientist. This can be done by trial and error, or using (meta) heuristics, for example evolutionary algorithms (e.g. [7,21]).

When using a clustering-based modelling approach, the most important hyperparameter of the FIS is the number of clusters, since it influences the partitioning of the data in the feature space and determines the number of rules in the FIS. Therefore, we will study the interaction of this hyperparameter with feature selection in this paper. Other hyperparameters, such as the t-(co)norms and shape of the membership functions also influence the performance of the FIS, but more subtly. When these hyperparameters have been provided by the data scientist, other model parameters, such as the membership functions and linear coefficients can be estimated from data.

4 Methodology

The purpose of this study is to explore the interaction between feature selection and the structure of the model. There are various feature selection methods and also multiple parameters related to the structure of the models. In this study, to make a start, we study the interaction between the simple filter feature selection and one parameter of the model. Specifically, we design experiments to check the interaction between the results of filter feature selection based on Fisher's score and the number of clusters of TS model based on fuzzy c-means.

4.1 The Data

One artificial data set and three real world data sets are used to study the interaction between selected features and hyperparameter settings:

The artificial data set is generated according to the process described in Sect. 2. There are 100 records in total and each record contains 4 features: x_1, x_2, x_1^2 and x_2^2. The output has two categories: 0 ($r \leq 1$) and 1 ($r > 1$).

The first real world data set contains data from 569 breast cancer patients. Based on 30 features, a prediction model is trained to classify whether these patients suffer from a malignant (37.3% of the cases) or benign (62.7%) tumor. This data is from the UCI Repository [14].

The second real world data set is also from the UCI Repository [14]. Toms Hardware is a worldwide forum network focusing on new technology with more conservative dynamics but distinctive features [11]. The data set contains 7905 records and 96 features which provide time-windows showing an upward trend. The objective of the prediction model is to determine whether or not these time-windows are followed by buzz events. There are 61% records followed by 'buzz events' and 39% records followed by 'non buzz events'.

The last real world data set has data from 187 patients who underwent cardiac resynchronization therapy (CRT) between January 1, 2008 and July 1, 2015. This therapy involves implanting a biventricular pacemaker that detects rate irregularities and emits pulses of electricity to correct them. However, CRT only improves the condition of a fraction of the patients. This data set is used to build a predictor to identify patients that will be responsive to this therapy. The set contains 137 features: gender, age, surgery type, 11 lab variables and 123 ultra sound variables. There are two output classes, which are 'responsive' (22%) and 'non-responsive' (78%).

4.2 The Experiment

For each data sets described above, we do the following experiments shown in Fig. 2. In this study, we rank the features through Fisher's score and build Takagi-Sugeno models using fuzzy c-means. Without prior knowledge about the data, the number of clusters needs to be decided. Therefore, for each number of clusters from 1 to 10 (when number of cluster equals 1, it is linear model), we build TS fuzzy models with from 1 to 30 features. We build models with 1 to 4 features for the artificial data set because it only has four features. For each data set, the experiment uses ten-fold cross validation and is repeated 5 times. The average AUC of the five runs is used to assess the model performance.

For the artificial data set, we also build TS fuzzy models with the 15 combinations of the features for each number of clusters from 1 to 10. The experiment still uses ten-fold cross validation and is repeated 5 times. The average AUC of the five runs is used as the final performance indicator. The 15 combinations of the features is shown in Table 2.

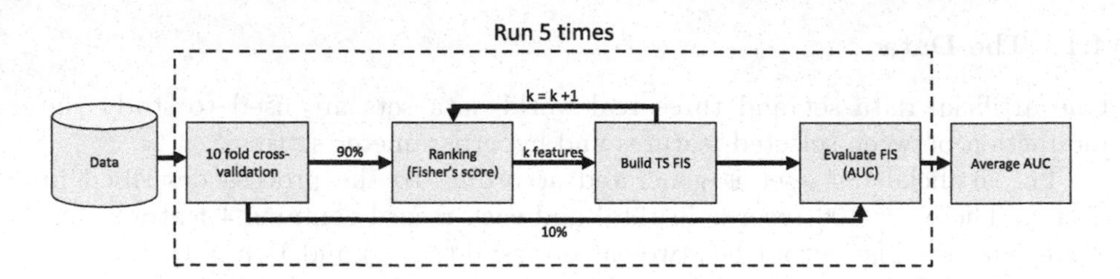

Fig. 2. Experimental framework.

Table 1. The average rank of each number of clusters.

Data set	No. of clusters	1	2	3	4	5	6	7	8	9	10
Artificial data	Average rank	3.25	**1.50**	2.75	3.75	4.50	6.00	7.75	7.50	9.50	8.50
CRT	Average rank	**1.40**	1.73	3.27	5.33	5.67	6.20	7.40	7.77	8.00	8.23
Breast	Average rank	**1.63**	2.70	2.80	5.10	4.70	5.13	6.50	7.83	8.80	9.80
Tomshardware	Average rank	9.63	8.83	6.37	3.97	**2.57**	3.07	4.33	5.07	5.73	5.43

Table 2. The 15 feature combinations of the artificial data set.

No. of features	Feature subsets
1	$\{x_1\}$, $\{x_2\}$, $\{x_1^2\}$, $\{x_2^2\}$
2	$\{x_1, x_2\}$, $\{x_1, x_1^2\}$, $\{x_1, x_2^2\}$, $\{x_2, x_1^2\}$, $\{x_2, x_2^2\}$, $\{x_1^2, x_2^2\}$
3	$\{x_1, x_2, x_1^2\}$, $\{x_1, x_2, x_2^2\}$, $\{x_1, x_1^2, x_2^2\}$, $\{x_2, x_1^2, x_2^2\}$
4	$\{x_1, x_2, x_1^2, x_2^2\}$

5 Results

Figure 3 shows the average AUC of the models with 15 feature combinations for each number of clusters from 1 to 10 on the artificial data. It shows that when building fuzzy models with different feature subsets, the optimal numbers of clusters are different and vice versa. For example, fuzzy models with 10 clusters performs the best when the feature subset is $\{x_1, x_2\}$, while the models with 1 cluster (linear models) have the best performance when the feature subset is $\{x_1^2, x_2^2\}$ as expected. Hence, the parameters of the model and the feature subsets affect each other.

Figures 4, 5, 6 and 7 show the average AUC of the models for each number of clusters from 1 to 10 on the artificial data, CRT data, Breast data and Tomshardware data respectively. To compare the results more explicitly, we also rank the models with different number of clusters for each number of features and calculate the average rank for each number of clusters, which is shown in Table 1.

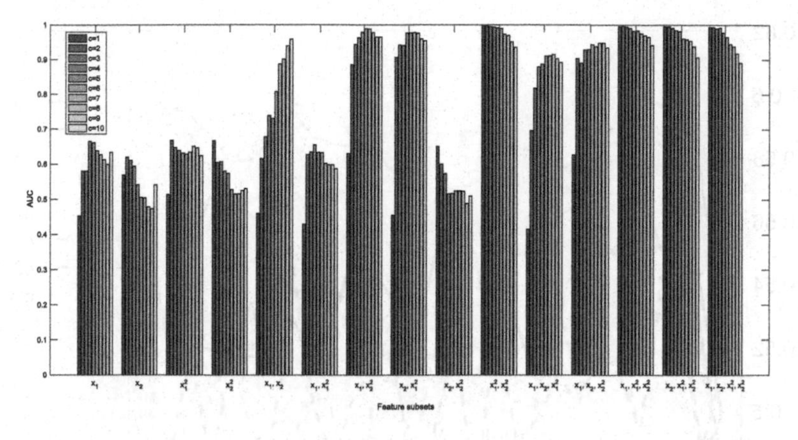

Fig. 3. AUC of fuzzy models with all 15 feature combinations with 1 to 10 clusters on the artificial data.

For the artificial data, model with two rules outperforms other models, except for when building models with one feature. In general, the models with 2 clusters perform better with the average rank 1.50. From Fig. 5, we can see that for most of the number of features, the linear models perform better and have the best average rank 1.4 on CRT data. Figure 6 shows that the average AUC of models with 1, 2, 3 and 5 clusters are similar for number of features from 1 to 15. However, the linear models perform the best when there are more than 15 features. In general, the linear models performs the best on the Breast data with the average rank 1.63. In Fig. 7, there are slightly difference among the models

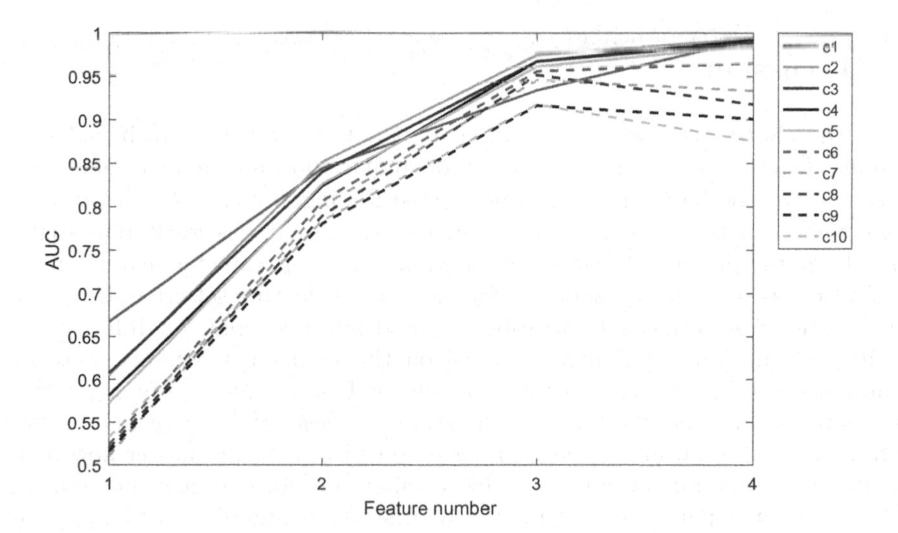

Fig. 4. Average AUC vs. number of features using fuzzy models with 1 to 10 clusters on the artificial data.

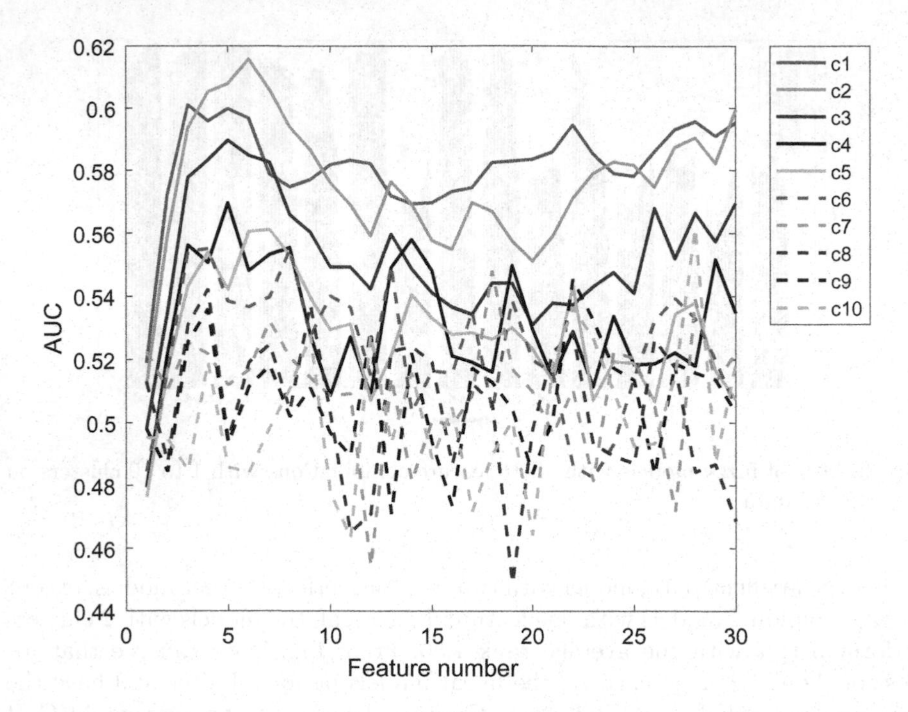

Fig. 5. Average AUC vs. number of features using fuzzy models with 1 to 10 clusters on CRT data.

with all the number of clusters, but models with 5 clusters have the best average rank 2.57.

6 Discussion

In all the above cases we have seen that models built with certain values of number of clusters are performing better or similarly for most number of features we select. On one hand this is an unexpected finding, on the other hand if this observation can be verified in more general settings, it has great potential of simplifying the process of feature selection and hyper-parameter searching.

This can be especially important for the wrapper feature selection. Currently, ideally feature selection with wrapper method is carried out for different possibilities of the hyper-parameter. Based on the model performance, the best combination of features and hyper-parameter is found. This is very costly from the computational point of view. Our study suggests that there may exist a certain value of a parameter performing better in general no matter how many features we select for filter method in certain conditions. It may be sufficient to test the hyper-parameter values for the first several number of features, and based of that, optimal hyper-parameter values can be determined. It will be significant if this can be generalized to wrapper method. However, we can not make

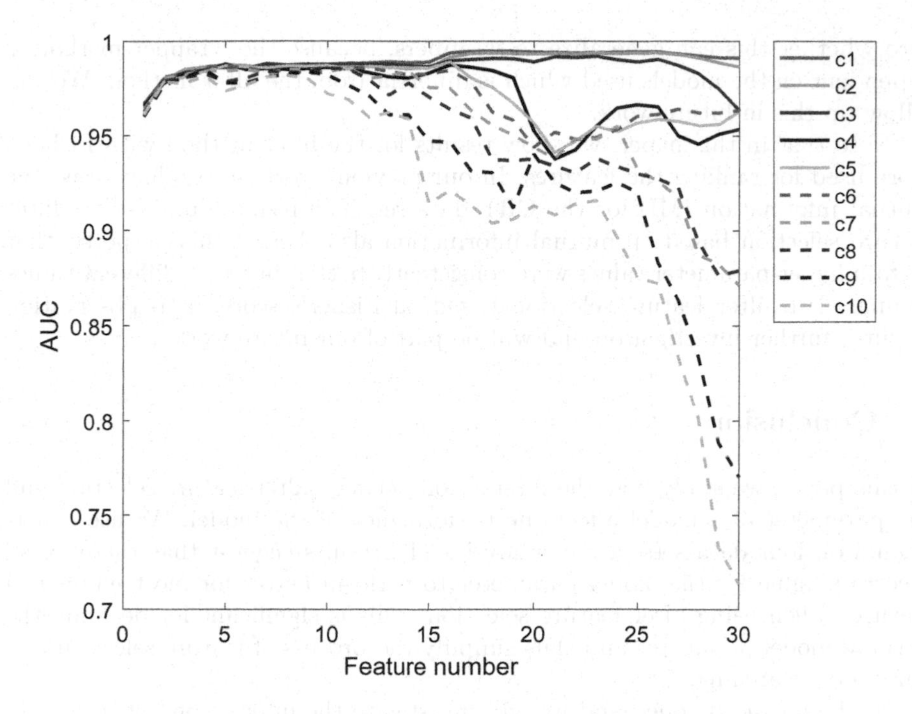

Fig. 6. Average AUC vs. number of features using fuzzy models with 1 to 10 clusters on Breast data.

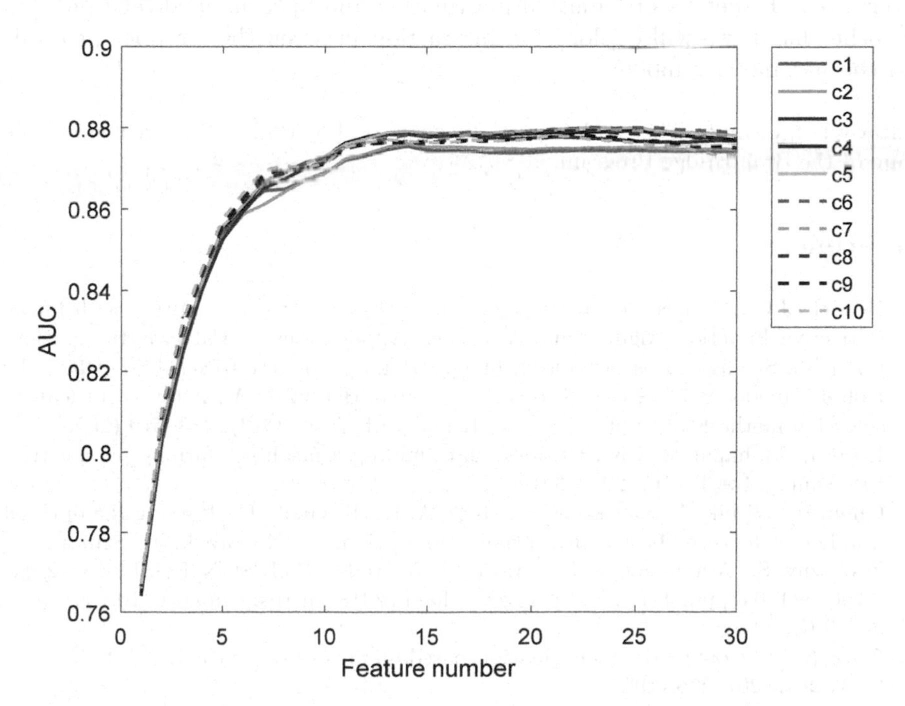

Fig. 7. Average AUC vs. number of features using fuzzy models with 1 to 10 clusters on Tomshardware data.

sure whether this can generalize to wrappers, because the wrapper method is dependent on the models used which is different from the filter method. We will follow up this in future work.

Note that in this paper, we show results for the filter method with Fisher's score used for ranking the features. In our previous work [4], we have used the mutual information (MI) for the CRT data set. The results obtained for filter feature selection based on mutual information also showed this property, that certain hyper-parameter values were consistently better, but with different values compared to filter feature selection based on Fisher's score. This observation requires further investigation and will be part of our future work.

7 Conclusion

In this paper, we study how the interaction between filter feature selection and the parameter of a model affect the performance of the model. We use experiments on four data sets in our analysis. The results suggest that there exists a certain value for the model parameter to perform better for most number of features when using filter feature selection. This is significant for deciding the optimal model parameter and thus simplify the process of feature selection and parameter searching.

In the future, on one hand we will investigate the interaction between filter method and the parameter of the model when using different feature ranking criteria (e.g. Fisher's score, mutual information and Spearman correlation). On the other hand, we will explore the interaction between the wrapper methods and the parameter a model.

Acknowledgement. This work is partially supported by Philips Research within the scope of the BrainBridge Program.

References

1. Bezdek, J.C.: Models for pattern recognition. In: Pattern Recognition with Fuzzy Objective Function Algorithms. Advanced Applications in Pattern Recognition, pp. 1–13. Springer, Boston (1981). https://doi.org/10.1007/978-1-4757-0450-1_1
2. Bolón-Canedo, V., Sánchez-Maroño, N., Alonso-Betanzos, A.: A review of feature selection methods on synthetic data. Knowl. Inf. Syst. **34**(3), 483–519 (2013)
3. Bose, I., Mahapatra, R.K.: Business data mining a machine learning perspective. Inf. Manag. **39**(3), 211–225 (2001)
4. Chen, P., Wilbik, A., van Loon, S., Boer, A.-K., Kaymak, U.: Finding the optimal number of features based on mutual information. In: Kacprzyk, J., Szmidt, E., Zadrożny, S., Atanassov, K.T., Krawczak, M. (eds.) IWIFSGN/EUSFLAT -2017. AISC, vol. 641, pp. 477–486. Springer, Cham (2018). https://doi.org/10.1007/978-3-319-66830-7_43
5. Chiu, S.L.: Fuzzy model identification based on cluster estimation. J. Intell. Fuzzy Syst. **2**(3), 267–278 (1994)

6. Cisco Visual Networking Index: The Zettabyte Era-trends and Analysis (2013). https://www.cisco.com/c/en/us/solutions/collateral/service-provider/visual-networking-index-vni/vni-hyperconnectivity-wp.pdf
7. Claesen, M., Simm, J., Popovic, D., Moreau, Y., De Moor, B.: Easy hyperparameter search using optunity. arXiv preprint arXiv:1412.1114 (2014)
8. Guyon, I., Elisseeff, A.: An introduction to variable and feature selection. J. Mach. Learn. Res. **3**(Mar), 1157–1182 (2003)
9. Guyon, I., Weston, J., Barnhill, S., Vapnik, V.: Gene selection for cancer classification using support vector machines. Mach. Learn. **46**(1), 389–422 (2002)
10. Jang, J.S.R., Sun, C.T., Mizutani, E.: Neuro-fuzzy and Soft Computing, a Computational Approach to Learning and Machine Intelligence. Prentice-Hall Inc., Upper Saddle River (1997)
11. Kawala, F., Douzal-Chouakria, A., Gaussier, E., Dimert, E.: Prédictions d'activité dans les réseaux sociaux en ligne. In: 4ième Conférence sur les Modèles et l'Analyse des Réseaux: Approches Mathématiques et Informatiques, p. 16 (2013)
12. Keogh, E., Mueen, A.: Curse of dimensionality. In: Liu L., Özsu, M.T. (eds.) Encyclopedia of Machine Learning, pp. 257–258. Springer, Boston (2011). https://doi.org/10.1007/978-0-387-39940-9
13. Kohavi, R., John, G.H.: Wrappers for feature subset selection. Artif. Intell. **97**(1–2), 273–324 (1997)
14. Lichman, M.: UCI Machine Learning Repository (2013). http://archive.ics.uci.edu/ml
15. Liu, H., Motoda, H.: Feature Extraction, Construction and Selection: A Data Mining Perspective, vol. 453. Springer, New York (1998). https://doi.org/10.1007/978-1-4615-5725-8
16. Lloyd, S.: Least squares quantization in PCM. IEEE Trans. Inf. Theor. **28**(2), 129–137 (1982)
17. Mejía-Lavalle, M., Sucar, E., Arroyo, G.: Feature selection with a perceptron neural net. In: Proceedings of the International Workshop on Feature Selection for Data Mining, pp. 131–135 (2006)
18. Pehro, D., Stork, D.: Pattern Classification. Wiley, New York (2001)
19. Takagi, T., Sugeno, M.: Fuzzy identification of systems and its applications to modeling and control. IEEE Trans. Syst. Man Cybern. **1**, 116–132 (1985)
20. Vergara, J.R., Estévez, P.A.: A review of feature selection methods based on mutual information. Neural Comput. Appl. **24**(1), 175–186 (2014)
21. Young, S.R., Rose, D.C., Karnowski, T.P., Lim, S.H., Patton, R.M.: Optimizing deep learning hyper-parameters through an evolutionary algorithm. In: Proceedings of the Workshop on Machine Learning in High-Performance Computing Environments, p. 4. ACM (2015)

Logical Methods in Mining Knowledge from Big Data

Fuzzy Association Rules on Data with Undefined Values

Petra Murinová(✉), Viktor Pavliska, and Michal Burda

Institute for Research and Applications of Fuzzy Modeling,
Centre of Excellence IT4Innovations, Division University of Ostrava,
30. dubna 22, 701 03 Ostrava, Czech Republic
{petra.murinova,viktor.pavliska,michal.burda}@osu.cz

Abstract. Handling of missing values is a very common in data processing. However, data values may be missing not only because of lack of information, but also because of undefinedness (such as asking for the age of non-married person's spouse). The aim of this paper is to propose an extension of fuzzy association rules framework for data with undefined values.

Keywords: Association rules · Undefined values · Fuzzy sets
Support · Confidence

1 Introduction

Searching for association rules is a tool for explanatory analysis of large data sets. Association rule is a formula of the form $A \rightharpoonup C$, where A is called an antecedent and C is a consequent, and which denotes some interesting relationship between A and C. There exist many different types of association rules. In this paper, we focus on implicative rules.

Association rules were firstly introduced by Hájek et al. in the late 1960s [1] by formulating a GUHA (General Unary Hypotheses Automaton) method [2]. Independently on them, a similar framework was developed by Agrawal [3] in 1993. Many different authors extended the association rules framework for fuzzy data, see [4] for a recent survey. A framework for a construction of linguistic summaries is also very closely related to fuzzy association rules. It was proposed by Yager in [5] and later further developed by Kacprzyk [6]. Another approach [7] introduces intermediate quantifiers to interpret association rules in natural language.

In real-world applications, data being analyzed are sometimes missing. Non-availability comes very often from the fact that some values are unknown or concealed. Handling of missing values is very common in data processing. There were developed many techniques for missing values imputation, and many existing methods were extended to directly work with unknown values. Hájek et al. proposed within their GUHA method [2] an extension capable of searching for

J. Medina et al. (Eds.): IPMU 2018, CCIS 855, pp. 165–174, 2018.
https://doi.org/10.1007/978-3-319-91479-4_14

association rules on data with missing values. However, their approach is applicable on *binary* (or categorical) data only. Moreover, our approach focuses on handling undefined values, whereas GUHA deals with unknown values.

Data values may be missing not only because of lack of information, but also because of undefinedness. For instance, asking for a spouse's age of a single person or for an error-rate of an air conditioning unit in cars with no such unit results in missing values that cannot be treated as unknown. It makes no sense to impute any averages at their places nor use any existing techniques suitable for handling unknown values. In this paper, we focus on searching for fuzzy association rules on data with undefined values.

The objective of handling undefined values is not new. The fundamental grounds in mathematical logic were established by Kleene, Bochvar, Sobociński and others, who studied the properties of three-valued logics $0/1/*$, which was also studied by Łukasiewicz in 1920 in [8]. These authors showed that the third value $*$ may represent an unknown, undefined or indeterminate truth value. An overview of main contributions can be found e.g. in [9].

Their work is being generalized to fuzzy propositional partial logic and later extended to predicate partial fuzzy logic by Novák, Běhounek and Daňková in [10,11]. They propose a fuzzy logic (based on expansions of a well-known fuzzy logic MTL_{Δ} of left-continuous t-norms (see [12]) that handles a special truth value $*$ with several types of fuzzy logical connectives that each treat the $*$ value in a different way. In [13], the author proposed a study of fuzzy type theory (FTT) with partial functions which are used for a characterization of the undefined values. In [14], several truth values representing different kinds of unavailability together with a single type of fuzzy logical connectives were introduced.

The rest of the paper is organized as follows. Section 2 recalls basic definitions from fuzzy set theory and Sect. 3 provides a mathematical background for computing with non-existent truth value. Section 4 presents a framework generalizing association rules for data with undefined values. In Sect. 5, we discuss some other ideas of how to handle missing values, and Sect. 6 concludes the paper by drawing some directions for future work.

2 Mathematical Background

The main goal of this section is to introduce a mathematical background which will be used for fuzzy association rules.

2.1 Basic Logical Operations

Convention

Please recall that Zadeh [15] defined a fuzzy set as a mapping from universe of discourse U to a real interval $[0,1]$, i.e. $F : U \to [0,1]$. Unlike crisp sets, where an object fully belongs or does not belong to a set, fuzzy sets enable an object $u \in U$ to belong partially to a set F in a degree $F(u)$. We will denote it by

$F \underset{\sim}{\subseteq} U$. We work with a finite universe U in this paper, $|U| = n$. A fuzzy set X is a subset of a fuzzy set Y, $X \subseteq Y$, if $X(u) \leq Y(u)$, for all $u \in U$. A size of a fuzzy set X is $|X| = \sum_{u \in U} X(u)$.

In [11], the authors proposed an algebraic structure for an interpretation of "undefined", "meaningless" or "non-applicable" values as an extension of MTL_{Δ}-algebra of left-continuous t-norms.

We do not limit this approach by assuming of a concrete algebra. The proposal concept will be associated with t-norms, which represent the general class of multiplications. These are binary operations $\otimes : [0,1]^2 \longrightarrow [0,1]$ which have been mainly studied by Klement et al. in [16] and later elaborated by many others. A concept associated with t-norm is the triangular conorm (t-conorm) $\oplus : [0,1]^2 \longrightarrow [0,1]$. The other two operations which correspond to these operations are the residuation operation and the negation operation.

Definition 1. *A t-norm is a binary operation $\otimes : [0,1]^2 \longrightarrow [0,1]$ such that the following axioms are satisfied for all $a, b, c \in [0,1]$:*

(a) commutativity: $a \otimes b = b \otimes a$,
(b) associativity: $a \otimes (b \otimes c) = (a \otimes b) \otimes c$,
(c) monotonicity: $a \leq b$ implies $a \otimes c \leq b \otimes c$,
(d) boundary condition: $1 \otimes a = a$.

Example 1. Typical examples of t-norms are *minimum* "$\wedge$", '*Łukasiewicz conjunction* "$\otimes_L$", *drastic product* and *nilpotent minimum.*

Definition 2. *A t-conorm is a binary operation $\oplus : [0,1]^2 \longrightarrow [0,1]$ which fulfils the axioms (a)–(c) from Definition 1 and for all $a \in [0,1]$ it fulfils the following boundary condition:*

(e) boundary condition: $0 \oplus a = a$.

Every t-conorm is dual to the given t-norm $\otimes$ if

$$a \oplus b = 1 - ((1 - a) \otimes (1 - b)) \tag{1}$$

Example 2. The most important t-conorms dual to the t-norms from Example 1 are the following. Dual to minimum is *maximum* "$\vee$", dual to product is *probabilistic sum.* Dual to Łukasiewicz conjunction is *Łukasiewicz disjunction.* Dual to drastic product is *drastic sum* and finally, dual to nilpotent minimum is *nilpotent maximum.*

Definition 3. *A* generalized implication *is a binary operation $\rightsquigarrow: [0,1]^2 \longrightarrow [0,1]$ that is monotone decreasing in the first and monotone increasing in the second argument and that satisfies the boundary conditions as follows:*

(a) $a \rightsquigarrow 1 = 1$,
(b) $0 \rightsquigarrow b = 1$,
(c) $1 \rightsquigarrow b = b$.

The typical example of that kind is Łukasiewicz implication. Other generalized implications were discussed in [17].

Definition 4. *The* negation *is a non-increasing operation* $\neg : [0,1] \longrightarrow [0,1]$ *such that* $\neg(0) = 1$ *and* $\neg(1) = 0$. *The negation is* involutive *if* $\neg(\neg(a)) = a$ *holds for every* $a \in [0,1]$.

For simplicity, we will work with an algebra $\mathcal{A} = \langle [0,1], \otimes, \oplus, \rightsquigarrow, \neg \rangle$ where all the axioms from Definitions 1, 2, 3 and 4 are fulfilled. Please recall that we will work with the involutive negation, which is defined as $\neg a = 1 - a$.

3 Representation of "non-Sense"

Now let us extend the support of $[0,1]$ by adding an additional truth value $*$ that represents a non-sense (i.e. non-existence of a truth value).

Definition 5. *Let* $\mathcal{A} = \langle [0,1], \otimes, \oplus, \rightsquigarrow, \neg \rangle$ *be an algebra. An* extended algebra $\mathcal{A}^* = \langle [0,1]^*, \otimes^*, \oplus^*, \rightsquigarrow^*, \neg^* \rangle$ *is defined as follows:*

$$[0,1]^* = [0,1] \cup \{*\} \tag{2}$$

where for all $a, b \in [0,1]$, $a \otimes^* b = a \otimes b$, $a \oplus^* b = a \oplus b$, $a \rightsquigarrow^* b = a \rightsquigarrow b$, *and* $\neg^*(a) = \neg(a)$, *with the following extension for* $*$ *(where* $*$ *acts as an* annihilator*):*

(a) $a \bigcirc * = * \bigcirc a = * \bigcirc * = *$, *for* $\bigcirc \in \{\otimes^*, \oplus^*, \rightsquigarrow^*\}$ *and* $a \in [0,1]$,
(b) $\neg^*(*) = *$.

Furthermore, the following unary connectives can be defined. All these unary connectives were introduced in [11]. We start with the definitions of unary operators $\uparrow$ and $\downarrow$ which reinterpret $*$ to 1 respectively to 0.

Definition 6. *Let* $[0,1]^* = [0,1] \cup \{*\}$ *be a support. Then* $\uparrow, \downarrow : [0,1]^* \longrightarrow [0,1]$ *such that the following is true:*

(a) $\uparrow a = \downarrow a = a$, *for all* $a \in [0,1]$;
(b) $\uparrow * = 1$, $\downarrow * = 0$.

Moreover, we can define the unary connective ! for the crisp modality "is defined" and similarly the unary connective ? for the crisp modality "is undefined". For the detail see ([10]).

Definition 7. *Let* $[0,1]^* = [0,1] \cup \{*\}$ *be a support. Then* $?, ! : [0,1]^* \longrightarrow \{0,1\}$ *such that the following is true:*

(a) $!a = 1$ *and* $?a = 0$, *for any* $a \in [0,1]$;
(b) $!* = 0$ *and* $?* = 1$.

4 Association Rules

4.1 Background

Let $\mathcal{O} = \{o_1, o_2, \ldots, o_N\}$, $N > 0$, be a finite set of abstract elements called objects and $\mathcal{A} = \{a_1, a_2, \ldots, a_M\}$, $M > 0$, be a finite set of attributes. Within the association rules framework, a dataset D is a mapping that assigns to each object $o \in \mathcal{O}$ and attribute $a \in \mathcal{A}$ a truth degree $D(a, o) \in [0, 1]$, which represents the intensity of assignment of attribute a to object o.

For fixed D, we can treat the attribute a as a predicate, which assigns a truth value $a(o) \in [0, 1]$ to each object $o \in \mathcal{O}$. Similarly, for each subset $X \subseteq \mathcal{A}$ of attributes, we define a predicate $X(o)$ for a selected t-norm $\otimes$ as follows:

$$X(o) = \bigotimes_{a \in X} a(o). \tag{3}$$

Association rule is a formula $A \rightharpoonup C$, where $A \subseteq \mathcal{A}$ is the *antecedent* and $C \subseteq \mathcal{A}$ is the *consequent*. It is natural to assume $A \cap C = \emptyset$ and also $|C| = 1$.

As each combination of predicates in antecedent and consequent form a well-formed association rules, an important problem is to identify such rules that are relevant to the given dataset D. So far, there exist a large number of measures of such relevance. An overview can be found in [18].

Perhaps the most commonly known indicators of a rule quality are the *support* and *confidence*. Dubois et al. [19] define them on the basis of a partition of $\mathcal{O}$: they argue that a rule $A \rightharpoonup C$ is a three valued entity, which partitions the objects from $\mathcal{O}$ into three (fuzzy) subsets, namely, into a set of *positive examples* S_+ that verify the rule, *negative examples* S_- that falsify the rule, and *irrelevant examples* $S_\pm$ that do not contribute in either direction. For any $o \in \mathcal{O}$, Dubois et al. [19] provide the following formal definitions for a fixed t-norm $\otimes$ and a generalized implication $\rightsquigarrow$:

$$\begin{aligned} S_+(o) &= A(o) \otimes C(o); \\ S_-(o) &= \neg\big(A(o) \rightsquigarrow C(o)\big); \\ S_\pm(o) &= \neg A(o). \end{aligned} \tag{4}$$

They argue that for $\langle S_+, S_-, S_\pm \rangle$ to be a proper fuzzy partition, all o should satisfy Ruspini condition:

$$S_+(o) + S_-(o) + S_\pm(o) = 1. \tag{5}$$

As noted in [19], Eqs. (4) and (5) lead to the admissible operator problem. [19] selects three pairs of $\otimes$ and $\rightsquigarrow$, which together satisfy both (4) and (5), see Table 1. Dubois et al. [19] assumes that $\neg(a) = 1 - a$, which together with conditions (4) and (5) results in

$$a \rightsquigarrow c = \neg a \oplus (a \otimes b). \tag{6}$$

Based on (4), the *support* and *confidence* of a fuzzy association rule $A \rightharpoonup C$ may be defined as follows [19].

Table 1. Admissible operators induced by (4) and (5) accordingly to [19]

$\otimes$	$\rightsquigarrow$
$\min\{a, b\}$	$\min\{1, 1-a+b\}$
$a \cdot b$	$1 - a(1-b)$
$\max\{a+b-1, 0\}$	$\max\{1-a, b\}$

Definition 8. *Let $R = A \rightharpoonup C$ be a rule and $S = \langle S_+, S_-, S_\pm \rangle$ be a partition of $\mathcal{O}$ with respect to R. Then*

$$\mathrm{supp}_S(A \rightharpoonup C) = |S_+|, \tag{7}$$

$$\mathrm{conf}_S(A \rightharpoonup C) = \frac{|S_+|}{|S_+| + |S_-|}. \tag{8}$$

4.2 Extension for Undefined Values

In order to extend the association rules framework for data containing *undefined* membership degrees, one has to switch the range of membership degrees from $[0, 1]$ to $[0, 1]^*$. In other words, dataset D becomes a mapping such that $D(a, o) \in [0, 1]^*$ for each $a \in \mathcal{A}$ and each $o \in \mathcal{O}$. Similarly to (3), an attribute a may be treated as a predicate with truth value $a(o) \in [0, 1]^*$ and each subset $X \subseteq \mathcal{A}$ may be used to define a predicate $X(o) \in [0, 1]^*$ by applying an extended t-norm $\otimes^*$ in (3):

$$X(o) = \bigotimes_{a \in X}^{*} a(o). \tag{9}$$

In order to extend Dubois [19] definitions of S_+, S_-, $S_\pm$ for undefined values $(*)$, let us consider an association rule $A \rightharpoonup C$ and examine all variants of evaluations of A and C where $*$ may occur.

If both $A(o)$ and $C(o)$ are defined ($A(o) \neq *$, $C(o) \neq *$) for some $o \in \mathcal{O}$, the definition (4) is directly applicable.

For such $o \in \mathcal{O}$ that $A(o)$ is undefined, then, regardless of $C(o)$, o should be considered as irrelevant to $A \rightharpoonup C$ since this object o neither supports nor falsifies the rule (o does not have the property A at all).

Finally, if $A(o) \neq *$ and $C(o) = *$, there are two alternatives possible. We can treat $A \rightharpoonup C$ either as a rule about defined values only, which means that any undefined values have to be ignored. Therefore, o with $C(o) = *$ can be considered as an example *irrelevant* to $A \rightharpoonup C$. The corresponding partition of $\mathcal{O}$ will be called the *I-partition* and denoted by $\langle S_+^I, S_-^I, S_\pm^I \rangle$.

On the other hand, o with C being undefined ($C(o) = *$) may be treated as the example that o does not have the attribute C at all and therefore, o is a *negative example* for $A \rightharpoonup C$. The corresponding partition of $\mathcal{O}$ will be called the *N-partition* and denoted by $\langle S_+^N, S_-^N, S_\pm^N \rangle$.

Definition 9 (I-partition). *Let* $\mathcal{A} = \langle [0,1], \otimes, \oplus, \rightsquigarrow, \neg \rangle$ *be an algebra with* $\otimes$ *and* $\rightsquigarrow$ *being operators satisfying the admissible operator problem [19] (see Table 1) and let* $\mathcal{A}^* = \langle [0,1]^*, \otimes^*, \oplus^*, \rightsquigarrow^*, \neg^* \rangle$ *be its extension (as in Definition 5). The* I-partition *of* $\mathcal{O}$ *with respect to association rule* $A \rightharpoonup C$ *can be defined as follows:*

$$\begin{aligned} S^I_+(o) &= \downarrow\big(A(o) \otimes^* C(o)\big); \\ S^I_-(o) &= \downarrow\neg^*\big(A(o) \rightsquigarrow^* C(o)\big); \\ S^I_\pm(o) &= \uparrow\neg^* A(o) \oplus^* !C(o). \end{aligned} \tag{10}$$

Let us check that the definition of I-partition satisfies Ruspini condition.

Proposition 1. $S^I_+(o) + S^I_-(o) + S^I_\pm(o) = 1$ *for any* $o \in \mathcal{O}$.

Proof. For $A(o) \neq *$ and $C(o) \neq *$, the arrow operators $(\uparrow, \downarrow)$ can be ignored and hence $S^I_+(o) = S_+(o)$, $S^I_-(o) = S_-(o)$. Also $!(C(o)) = 0$ and therefore $S^I_\pm(o) = \neg A(o) = S_\pm(o)$. Therefore, by (5) we see that the proposition holds.

If $A(o) = *$ or $C(o) = *$ then $S^I_+(o) = S^I_-(o) = 0$. Evidently also either $\uparrow\neg^* A(o) = 1$ or $!C(o) = 1$ so that $S^I_\pm = 1$.

Definition 10 (N-partition). *Let* $\mathcal{A} = \langle [0,1], \otimes, \oplus, \rightsquigarrow, \neg \rangle$ *be an algebra with* $\otimes$ *and* $\rightsquigarrow$ *being operators satisfying the admissible operator problem [19] (see Table 1),* $\neg(a) = 1 - a$*, and let* $\mathcal{A}^* = \langle [0,1]^*, \otimes^*, \oplus^*, \rightsquigarrow^*, \neg^* \rangle$ *be its extension (as in Definition 5). The* N-partition *of* $\mathcal{O}$ *with respect to association rule* $A \rightharpoonup C$ *can be defined as follows:*

$$\begin{aligned} S^N_+(o) &= \downarrow\big(A(o) \otimes^* C(o)\big); \\ S^N_-(o) &= \downarrow\neg^*\big(A(o) \rightsquigarrow^* \downarrow C(o)\big); \\ S^N_\pm(o) &= \uparrow\neg^* A(o). \end{aligned} \tag{11}$$

Proposition 2. $S^N_+(o) + S^N_-(o) + S^N_\pm(o) = 1$ *for any* $o \in \mathcal{O}$.

Proof. For $A(o) \neq *$ and $C(o) \neq *$, the proof is similar to the proof of Proposition 1. If $A(o) = *$ then evidently $S^N_+(o) = S^N_-(o) = 0$ and $S^N_\pm(o) = 1$.

If $A(o) \neq *$ and $C(o) = *$ then $S^N_+(o) = 0$, $S^N_-(o) = \neg(A(o) \rightsquigarrow 0)$, which equals (by using (6)) to $\neg\neg A(o) = A(o)$ and $S^N_\pm(o) = \neg A(o)$. Therefore, $S^N_+(o) + S^N_-(o) + S^N_\pm(o) = 0 + A(o) + (1 - A(o)) = 1$.

Note that the support and confidence slightly differs based on the selected partitioning method.

Proposition 3. *Let* $S^I = \langle S^I_+, S^I_-, S^I_\pm \rangle$ *and* $S^N = \langle S^N_+, S^N_-, S^N_\pm \rangle$*. Then:*

(a) $\mathrm{supp}_{S^N}(A \rightharpoonup C) = \mathrm{supp}_{S^I}(A \rightharpoonup C)$;
(b) $\mathrm{conf}_{S^N}(A \rightharpoonup C) \leq \mathrm{conf}_{S^I}(A \rightharpoonup C)$.

Proof. (a) is evidently fulfilled. (b) $S^I_-(o)$ differs from $S^N_-(o)$ only if $A(o) \neq *$ and $C(o) = *$, then $0 = S^I_-(o) \leq S^N_-(o)$ and therefore evidently (b) holds too.

Table 2. Sobociński $\otimes_S$, Bochvar $\otimes_B$ and Kleene $\otimes_K$ variants for handling of $*$

$\otimes_S$	0	b	1	$*$
0	0	0	0	0
a	0	$a \otimes b$	a	a
1	0	b	1	1
$*$	0	b	1	$*$

$\otimes_B$	0	b	1	$*$
0	0	0	0	$*$
a	0	$a \otimes b$	a	$*$
1	0	b	1	$*$
$*$	$*$	$*$	$*$	$*$

$\otimes_K$	0	b	1	$*$
0	0	0	0	0
a	0	$a \otimes b$	a	$*$
1	0	b	1	$*$
$*$	0	$*$	$*$	$*$

5 Discussion

In Definition 5, $*$ representing non-existence of a truth-value (i.e. non-sense, undefinedness) is defined as an *annihilator*. That is, the whole antecedent gets undefined if any of its predicates obtains $*$. Such behaviour is sometimes called Bochvar's accordingly to [9,11]. However, there exist other variants for handling of $*$. See Table 2 for some examples. It is therefore natural to ask for their benefits or drawbacks with respect to association analysis.

Kleene's $\otimes_K$ preserves 0 as an annihilator, while other values get annihilated by $*$. Such behaviour supports interpretation of $*$ in the sense of "unknown" truth value: if $a = 0$ and b is unknown then surely $a \otimes_K b = 0$. For any other value of a, $a \otimes_K b$ remains unknown.

Unfortunately, defining $\otimes^*$ like $\otimes_K$ without any changes in our proposal is not recommended, since all objects $o \in \mathcal{O}$ resulting in "unknown" values for $A(o)$ or $C(o)$ would be treated as irrelevant (resp. negative) examples of the association rule $A \rightharpoonup C$, which does not correspond to reality. In our opinion, a better approach for handling of "unknown" truth values would be to use e.g. *interval-valued fuzzy logic* [20,21] and compute intervals of possible supports and confidences. We would like to address that topic in more detail in our future research.

Sobociński's $\otimes_S$ treats $*$ as something that has to be ignored. That is, for $a = *$, $a \otimes_S b = b$ for any b. However, ignoring predicates with $*$ value may be sometimes favorable, but sometimes disadvantageous. For example, let us consider the following fuzzy attributes:

- X: children's age is high
- Y: family educational expenses are high
- Z: parents have a lot of free time

In Sobociński style, $*$ is in a conjunction treated as 1. That is, for o being a family without any children ($X(o) = *$) is treated as a family with grown children ($X(o) = 1$). Such interpretation positively affects association rule $X \rightharpoonup Z$, which fortunately corresponds with reality (indeed, parents with grown children as well as non-children families have both more free time). On the other hand, association rule $X \rightharpoonup Y$, i.e. families with grown children ($X(o) = 1$) have high educational expenses ($Y(o) = 1$), is negatively affected by families with no children ($X(o) = *$), which have low or nil educational expenses ($Y(o) = 0$) in reality.

Based on the above discussed example, Sobociński style seems not to be generally useful in association analysis as ignorance of $*$ sometimes favorably corresponds with positive cases and sometimes not.

6 Conclusion and Future Work

In this paper, an association analysis framework was developed that allows to process data with undefined values. A mathematical background from [10] was selected, which extends a set of truth values with $*$, which represents undefined truth values (non-sense). Based on that, an extended definition of association rule's support and confidence was proposed by generalizing Dubois et al. approach [19].

A future work will address the processing of missing ("unknown") values, which may be represented e.g. with fuzzy interval-valued logic. Also a combination of missing and non-existent truth values may be interesting: for instance, an attribute "spouse's age is high" may have known truth value, non-existing (in case the respondent is single), unknown (concealed age), or unknown whether non-sense (knowing neither whether the spouse exists nor what age he/she is).

Acknowledgements. Authors acknowledge support by project "LQ1602 IT4Innovations excellence in science" and by GAČR 16-19170S.

References

1. Hájek, P.: The question of a general concept of the GUHA method. Kybernetika **4**, 505–515 (1968)
2. Hájek, P., Havránek, T.: Mechanizing Hypothesis Formation (Mathematical Foundations for a General Theory). Springer-Verlag, Heidelberg (1978). https://doi.org/10.1007/978-3-642-66943-9
3. Agrawal, R., Srikant, R.: Fast algorithms for mining association rules. In: Proceedings of 20th International Conference on Very Large Databases, Chile, pp. 487–499. AAAI Press (1994)
4. Ralbovský, M.: Fuzzy *GUHA*. PhD thesis, University of Economics, Prague (2009)
5. Yager, R.R.: A new approach to the summarization of data. Inf. Sci. **28**(1), 69–86 (1982)
6. Kacprzyk, J., Yager, R.R., Zadrożny, S.: A fuzzy logic based approach to linguistic summaries of databases. Int. J. Appl. Math. Comput. Sci. **10**(4), 813–834 (2000)
7. Murinová, P., Burda, M., Pavliska, V.: An Algorithm for Intermediate Quantifiers and the Graded Square of Opposition Towards Linguistic Description of Data. In: Kacprzyk, J., Szmidt, E., Zadrożny, S., Atanassov, K.T., Krawczak, M. (eds.) IWIFSGN/EUSFLAT -2017. AISC, vol. 642, pp. 592–603. Springer, Cham (2018). https://doi.org/10.1007/978-3-319-66824-6_52
8. Lukasiewicz, J.: O logice trojwartosciowej. Ruch filozoficzny **5**, 170–171 (1920)
9. Malinowski, G.: The Many Valued and Nonmonotonic Turn in Logic. North-Holland, Amsterdam (2007)
10. Běhounek, L., Novák, V.: Towards fuzzy partial logic. In: Proceedings of the IEEE 45th International Symposium on Multiple-Valued Logics (ISMVL 2015), pp. 139–144 (2015)

11. Běhounek, L., Daňková, M.: Towards fuzzy partial set theory. In: Carvalho, J.P., Lesot, M.-J., Kaymak, U., Vieira, S., Bouchon-Meunier, B., Yager, R.R. (eds.) IPMU 2016, Part II. CCIS, vol. 611, pp. 482–494. Springer, Cham (2016). https://doi.org/10.1007/978-3-319-40581-0_39
12. Hájek, P.: Metamathematics of Fuzzy Logic. Kluwer, Dordrecht (1998). https://doi.org/10.1007/978-94-011-5300-3
13. Novák, V.: Towards Fuzzy Type Theory with Partial Functions. In: Kacprzyk, J., Szmidt, E., Zadrożny, S., Atanassov, K.T., Krawczak, M. (eds.) IWIFSGN/EUSFLAT -2017. AISC, vol. 643, pp. 25–37. Springer, Cham (2018). https://doi.org/10.1007/978-3-319-66827-7_3
14. Murinová, P., Burda, M., Pavliska, V.: Undefined Values in Fuzzy Logic. In: Kacprzyk, J., Szmidt, E., Zadrożny, S., Atanassov, K.T., Krawczak, M. (eds.) IWIFSGN/EUSFLAT -2017. AISC, vol. 642, pp. 604–610. Springer, Cham (2018). https://doi.org/10.1007/978-3-319-66824-6_53
15. Zadeh, L.A.: Fuzzy sets. Inf. Control **8**, 338–353 (1965)
16. Klement, E.P., Mesiar, R., Pap, E.: Triangular Norms. Kluwer, Dordrecht (2000). https://doi.org/10.1007/978-94-015-9540-7
17. Baczynski, M., Jayaram, B.: Fuzzy Implications. Studies in Fuzziness and Soft Computing, vol. 231. Springer, Heidelberg (2008). https://doi.org/10.1007/978-3-540-69082-5
18. Geng, L., Hamilton, H.J.: Interestingness measures for data mining: a survey. ACM Comput. Surv. (CSUR) **38**(3), 9 (2006)
19. Dubois, D., Hüllermeier, E., Prade, H.: A systematic approach to the assessment of fuzzy association rules. Data Min. Knowl. Disc. **13**(2), 167–192 (2006)
20. Deschrijver, G., Cornelis, C., Kerre, E.E.: Advances and challenges in intervalvalued fuzzy logic. Fuzzy Sets Syst. **157**, 622–627 (2005)
21. Van Gasse, B., et al.: A fuzzy formal logic for interval-valued residuated lattices. In: Proceedings of the 5th EUSFLAT Conference, Ostrava, Czech Republic, 11–14 September 2007, vol. 2 (2007)

On the Use of Subproduct in Fuzzy Relational Compositions Based on Grouping Features

Nhung Cao[1(✉)], Martin Štěpnička[1(✉)], Michal Burda[1], and Aleš Dolný[2]

[1] Institute for Research and Applications of Fuzzy Modeling, Centre of Excellence IT4Innovations, University of Ostrava, 30. dubna 22, 701 03 Ostrava 1, Czech Republic
{nhung.cao,martin.stepnicka,michal.burda}@osu.cz

[2] Department of Biology and Ecology, Faculty of Sciences, University of Ostrava, Chittussiho 10, 710 00 Slezská Ostrava, Ostrava, Czech Republic
ales.dolny@osu.cz

http://irafm.osu.cz/, http://prf.osu.cz/kbe/

Abstract. Fuzzy relational compositions have been extended and studied from distinct perspectives, and their use on the classification problem has been already demonstrated too. One of the recent approaches foreshadowed the positive influence of the so-called grouping features. When this improvement is being applied, the universe of features is partitioned into a number of groups of features and then the relevant composition is applied. The use of the concept was demonstrated on the real classification of Odonata (dragonflies). This paper shows that the Bandler-Kohout subproduct may appropriately serve as the chosen compositions in order to obtain an effective tool. The concepts of excluding features and generalized quantifiers will be employed in the constructed method as well. Some interesting properties will be introduced and a real example of the influence of the new concept will be provided.

Keywords: Fuzzy relational compositions
Fuzzy relational products · Bandler-Kohout products
Grouping features · Excluding features · Generalized quantifiers
Classification

1 Introduction

Fuzzy relational compositions became a crucial topic in fuzzy relational calculus [1] and in fuzzy mathematics in general. The biggest development in the late 70's has been provided due to Bandler and Kohout and later on, the compositions (or also fuzzy relational products) have been applied in a large number of applications, for instance: medical diagnosis [2], formal constructions of fuzzy inference

This research was partially supported by the NPU II project LQ1602 "IT4Innovations excellence in science" provided by the MŠMT.

J. Medina et al. (Eds.): IPMU 2018, CCIS 855, pp. 175–186, 2018.
https://doi.org/10.1007/978-3-319-91479-4_15

systems [3,4] and related systems of fuzzy relational equations [5–8] and recently [9], modeling monotone fuzzy rule bases [10,11], fuzzy concept analysis [1], data mining [12] or in flexible query answering systems [13–15]. The development and improvement on this topic have been updated during the time by numerous authors, let us refer to some of them only [16–18].

There are several recent directions significantly contributing to the extension of the topic, for example, the compositions of partial fuzzy relations [19]. We will recall two other extensions, namely the concept of excluding features incorporated in the compositions [20,21] and the compositions based on generalized quantifiers [22–24]. It is worth mentioning that the both approaches may be combined together in order to obtain a more flexible and effective tool [25].

As we have shown in [26], for some types of features, the introduced approaches do not provide sufficiently appropriate results. This may be caused by a specific yet not unnatural way of the construction of the original features. So, we propose a preprocessing grouping features first and then applied the chosen composition, particularly Bandler-Kohout (BK) superproduct in [26], incorporating both excluding features and fuzzy quantifiers.

The application potential of the grouping features combined with the both above-mentioned extensions of fuzzy relational compositions was demonstrated on a real example of classification of dragonflies. Due to the effect of the grouping of features into appropriate groups before the application of the composition, we could classify a given sample (observed dragonfly) with the BK superproduct with a different semantics. In particular, instead of classifying a sample into a class of dragonflies in case of "all the features" belonging to the given class are observed, it was sufficient to observe "at least one feature for any group of features" related to the given class to be carried by a given sample. This approach serves as motivation to show that there is a possibility for applying the BK subproduct jointly with grouping of features as well.

2 Preliminaries

We recall some basic definitions of fuzzy relation compositions and their extensions, namely of the incorporation of excluding features in fuzzy relational composition [20,21] and the compositions based on generalized quantifiers [22–24], and their combination [25]. For the whole paper, all the used operations will be from a residuated lattice $\mathcal{L} = \langle [0,1], \wedge, \vee, \otimes, \rightarrow 0, 1\rangle$ with the negation $\neg$ and biresiduation (bi-implication) $\leftrightarrow$ defined in standard ways for the residuated lattices, i.e., $\neg a = a \rightarrow 0$ and $a \leftrightarrow b = (a \rightarrow b) \wedge (b \rightarrow a)$ for any $a, b \in [0,1]$. Furthermore, we will denote the set of all fuzzy sets on a given universe U by $\mathcal{F}(U)$. Finally, sets X, Y and Z will be non-empty finite universes of objects (samples), features and classes, respectively.

2.1 Fuzzy Relational Compositions

We shortly recall four fundamental fuzzy relational compositions.

Definition 1. *Consider* $R \in \mathcal{F}(X \times Y)$ *and* $S \in \mathcal{F}(Y \times Z)$*. Then the* basic composition $\circ$, BK-subproduct $\lhd$, BK-superproduct $\rhd$ *and* BK-square product $\Box$ *of* R *and* S *are fuzzy relations on* $X \times Z$ *defined as follows:*

$$(R \circ S)(x,z) = \bigvee_{y \in Y} (R(x,y) \otimes S(y,z)),$$

$$(R \lhd S)(x,z) = \bigwedge_{y \in Y} (R(x,y) \to S(y,z)),$$

$$(R \rhd S)(x,z) = \bigwedge_{y \in Y} (R(x,y) \leftarrow S(y,z)),$$

$$(R \Box S)(x,z) = \bigwedge_{y \in Y} (R(x,y) \leftrightarrow S(y,z)),$$

respectively.

In order to explain the semantics of the fuzzy relations as well as their compositions, let us consider, e.g., the context of the medical diagnosis. $R(x,y)$ stands for the truth degree of the predicate *"patient x carries symptom y"*, the meaning of $S(y,z)$ similarly expresses the relationship between symptom y and disease z. The value $(R \circ S)(x,z)$ expresses the truth-degree of the predicate: *"patient x has at least one symptom belonging to disease z"* and thus, it expresses an initial suspicion. The two other BK products and the square product provide a sort of a strengthening of the initial suspicion. In particular, the subproduct $(R \lhd S)(x,z)$ expresses the truth-degree of the predicate: *"all symptoms of the patient x belong to disease z"*, and $(R \rhd S)(x,z)$ expresses the truth-degree of the predicate: *"patient x has all symptoms belonging to the diseases z"*. The meaning of $(R \Box S)(x,z)$ is clear as it is a conjunction of both previously described triangle BK products, i.e., *"patient x has all symptoms of disease z and at the same time all symptoms of the patient x belong to the disease z"*.

2.2 Excluding Features in Fuzzy Relational Compositions

This approach was motivated by the existence of excluding symptoms for some particular diseases in the medical diagnosis problem [20] and it has been successfully applied to the classification of Odonata (dragonflies) in biology [21].

Definition 2 [21]**.** *Let* X, Y, Z *be non-empty finite universes, let* $R \in \mathcal{F}(X \times Y)$*,* $S, E \in \mathcal{F}(Y \times Z)$*. Then the composition* $R \circ S^{\backprime} E \in \mathcal{F}(X \times Z)$ *is defined:*

$$(R \circ S^{\backprime} E)(x,z) = \bigvee_{y \in Y} (R(x,y) \otimes S(y,z)) \otimes \neg \bigvee_{y \in Y} (R(x,y) \otimes E(y,z)).$$

The definition provided above may be rewritten into the following comprehensible form:

$$(R \circ S^{\backprime} E)(x,z) = (R \circ S)(x,z) \otimes \neg (R \circ E)(x,z).$$

We recall that the membership degree of the pair (x, y) to the fuzzy relation E expresses how much it is true that y is an excluding feature for class (animal species) z. In other words, fuzzy relation $(R \circ S^{\backslash} E)(x, z)$ expresses the truth-degree of the predicate: *"patient x has at least one symptom of disease z and at the same time does not have any symptom that would be excluding the disease z"*.

2.3 Compositions Based on Generalized Quantifiers

This approach has been proposed with the motivation to fill in the big gap between basic composition $\circ$, based on the existential quantifier, and the BK products based on the universal quantifier. This gap is naturally caused by the use of existential quantifier in the construction of $\circ$ and by the use of the universal quantifier in the construction of the BK products. Indeed, the use of the existential quantifiers will cause that too many classes will be assigned to any sample and vice-versa, due to some potential mistakes in the determination of the features, the use of the universal quantifier may cause that for many samples, no class will be assigned a high degree.

The use of generalized quantifiers such as *Most*, *Many* or *A Few* in the construction of the fuzzy relational compositions proved its potential [24]. We recall the concept of fuzzy measures [27] first and then revisit the concept of fuzzy relational compositions based on generalized (fuzzy) quantifiers [24].

Definition 3 [24]. Let $U = \{u_1, \ldots, u_n\}$ be a finite universe, let $\mathcal{P}(U)$ denotes the power set of U. A mapping $\mu : \mathcal{P}(U) \to [0, 1]$ is called a fuzzy measure on U if $\mu(\emptyset) = 0$ and $\mu(U) = 1$ and, if $\forall C, D \in \mathcal{P}(U), C \subseteq D$ then $\mu(C) \leq \mu(D)$.

Fuzzy measure μ is called *symmetric* if $\forall C, D \in \mathcal{P}(U) : |C| = |D| \Rightarrow \mu(C) = \mu(D)$ where $|\cdot|$ denotes the cardinality of a set.

Example 1. The measure

$$\mu(A) = f(|A|/|U|) ,$$

where $f : [0, 1] \to [0, 1]$ is a non-decreasing mapping with $f(0) = 0$ and $f(1) = 1$ is a symmetric fuzzy measure.

Definition 4 [24]. Let μ be a fuzzy measure on a finite universe U. A mapping $Q : \mathcal{F}(U) \to [0, 1]$ defined by

$$Q(C) = \bigvee_{D \in \mathcal{P}(U) \setminus \{\emptyset\}} \left(\left(\bigwedge_{u \in D} C(u) \right) \otimes \mu(D) \right), \; C \in \mathcal{F}(U)$$

is called *generalized quantifier determined by μ*.

If μ is a symmetric fuzzy measure, the quantifier can be calculated with help of the following computationally unexpensive form that is actually nothing but a Sugeno integral [28]:

$$Q(C) = \bigvee_{i=1}^{n} C(u_{\pi(i)}) \otimes \mu(\{u_1, \ldots, u_i\}), \; C \in \mathcal{F}(U)$$

where π is a permutation on $\{1, \ldots, n\}$ such that $C(u_{\pi(1)}) \geq C(u_{\pi(2)}) \geq \cdots \geq C(u_{\pi(n)})$.

Now, we recall how the fuzzy quantifiers may be directly used in the definition of the fuzzy relational compositions such that the four classical compositions recalled above are only special cases of the below provided general frame.

Definition 5 [24]. Let Q be a quantifier determined by a fuzzy measure μ on Y. Then, the compositions $R@^Q S$ where $@ \in \{\circ, \lhd, \rhd, \square\}$ are defined as follows:

$$(R@^Q S)(x,z) = \bigvee_{D \in \mathcal{P}(Y) \setminus \{\emptyset\}} \left(\left(\bigwedge_{y \in D} R(x,y) \circledast S(y,z) \right) \otimes \mu(D) \right)$$

for all $x \in X$, $z \in Z$ and for $\circledast \in \{\otimes, \rightarrow, \leftarrow, \leftrightarrow\}$ corresponding to the composition @.

It is worth mentioning that both approaches, namely the incorporation of excluding features and the use of the generalized quantifiers may be combined together, see [25].

2.4 Compositions Based on Grouping Features

Let us briefly recall some basic facts on the way of getting the composition based on grouping features, see [26].

Let X, Y, Z be non-empty finite universes of cardinalities I, J and K, respectively. Let $R \in \mathcal{F}(X \times Y)$ and $S, E \in \mathcal{F}(Y \times Z)$. Assume that Y can be partitioned into M disjoint sets

$$Y = G_1 \cup \cdots \cup G_M$$

such that each set contains all features of the "same type"[1].

We define M fuzzy relations $S_m \in \mathcal{F}(Y \times Z)$ by:

$$S_m(y,z) = \begin{cases} S(y,z), & y \in G_m \\ 0, & \text{otherwise} \end{cases} \tag{1}$$

and define a new universe of features of some types:

$$Y_Z = \{y_1^1, y_1^2, \ldots, y_1^K, y_2^1, \ldots, y_2^K, \ldots, y_M^1, \ldots, y_M^K\},$$

in which the subscript of each element stands for the group number and the superscript of each element stands for the class number. Each component y_m^k expresses how much it is true, that any of the features in the m-th group corresponds to the k-th class.

[1] By the type we mean, e.g., a color or a morphological type and the features are particular colors (green, black, blue etc.) or particular morphological types (Anisoptera, Zygoptera), see [21].

Based on the new universe Y_Z, we define the fuzzy relation $R' \in \mathcal{F}(X \times Y_Z)$ as follows:

$$R'(x_i, y_m^k) = (R \circ S^{\backslash}_m E)(x_i, z_k)$$

and the fuzzy relation $S' \in \mathcal{F}(Y_Z \times Z)$ as follows:

$$S'(y_m^{k_1}, z_{k_2}) = \begin{cases} 1, & k_1 = k_2, \\ 0, & \text{otherwise.} \end{cases}$$

Then, using the BK superproduct, the composition $R' \rhd S' \in \mathcal{F}(X \times Z)$ is correctly defined and it has the following natural semantics. The value $(R' \rhd S')(x, z)$ expresses the degree of truth of the predicate:

"If class z is related to some features from a certain group of features then at least one of the features from the group is carried by object x and x does not carry any feature that would be excluding for the classification to z.

The above-provided semantics may be easily comprehended on the real example of the classification of dragonflies. Let the group of features is "a color". Then if there are some colors pre-determining that a given sample could be of a given family of dragonflies z then $R' \rhd S'(x, z)$ expresses how much it is true that the sample x carries at least one of these colors. Observing the semantics or its first part, it is clear, that it could be comprehended also in the following way:

"Object x carries **at least a single** *feature that relates to class z from* **all groups** *of features and at the same time, it carries no features that would be excluding for the classification to z"*

which will be recalled later on.

The proposed BK superproduct composition $R' \rhd S' \in \mathcal{F}(X \times Z)$ of fuzzy relations R' and S' may be expressed as an intersection of m circlet compositions incorporating excluding features.

Proposition 1 [26]. *Let $\cap$ denotes the min-intersection. Then*

$$R' \rhd S' = \bigcap_{m=1}^{M} (R \circ S^{\backslash}_m E) \,. \tag{2}$$

3 BK Subproduct and Grouping Features

3.1 Motivation

We have studied the new method of grouping features and applied the BK superproduct to the grouped features. Naturally, one may ask whether a similar approach for the BK subproduct makes sense too and what would be the resulting composition appropriate for. And in the case of positive answer, what is the semantics of such a composition.

3.2 Formal Part

Let us use the same notation, i.e., we consider finite universes X, Y, Z with $|X| = I$, $|Y| = J$ and $|Z| = K$ as in the Sect. 2.4, fuzzy relations $R \in \mathcal{F}(X \times Y)$, $S, E \in \mathcal{F}(Y \times Z)$. The universe Y is again partitioned into M disjoint sets $Y = G_1 \cup \cdots \cup G_M$, and the definition of S_m is given by (1).

Furthermore, similarly to the case of Y_Z, we again define a new universe of features Y_X, however, unlike in the case of Y_Z this universe will not partition features from the perspective of classes but from the perspective of samples:

$$Y_X = \{y_1^1, y_1^2, \ldots, y_1^I, y_2^1, \ldots, y_2^I, \ldots, y_M^1, \ldots, y_M^I\}\ .$$

The subscript $m \in \{1, \ldots, M\}$ of each element y_m^i stands for the group number and the superscript $i \in \{1, \ldots, I\}$ stands for the sample number.

By this setting up, the universe Y_X has $M \cdot I$ elements. Each component y_m^i expresses how much it is true, that any of the features in the m-th group relates to the i-th sample.

Furthermore, let us define the fuzzy relation $R^* \in \mathcal{F}(X \times Y_X)$ as follows:

$$R^*(x_{i_1}, y_m^{i_2}) = \begin{cases} 1, & i_1 = i_2, \\ 0, & \text{otherwise.} \end{cases}$$

and the fuzzy relation $S^* \in \mathcal{F}(Y_X \times Z)$ as follows:

$$S^*(y_m^i, z_k) = (R \circ S^{\backslash}_m E)(x_i, z_k).$$

So, in the matrix form, the fuzzy relation S^* is a matrix of type $M \cdot I \times K$ that looks as follows:

$$S^* = \begin{pmatrix} [R \circ S^{\backslash}_1 E] \\ \cdots \\ [R \circ S^{\backslash}_M E] \end{pmatrix}$$

and the fuzzy relation R^* may be represented in a matrix form as a matrix of the type $I \times M \cdot I$ that is constituted by M diagonal identity matrices of the type $I \times I$ ordered horizontally next to each other:

$$R^* = \left([\mathrm{Id}^1_{I \times I}]\ \ldots\ [\mathrm{Id}^M_{I \times I}]\right)\ .$$

Then we may introduce the composition $R^* \lhd S^* \in \mathcal{F}(X \times Z)$ with the following semantics. $(R^* \lhd S^*)(x, z)$ expresses the truth-degree of the following predicate:

> *"If object x carries some feature(s) from a certain group of features then at least one of the carried features from the group relates to class z and x does not carry any of the features that would be excluding for the classification to z.*

Let us again demonstrate the semantics on the real example of the classification of dragonflies. Let the group of features is "a color". Then if there are some colors carried by a given sample x then $(R^* \lhd S^*)(x, z)$ expresses how much it is true that the sample x carries at least one of the colors related to the class of dragonflies z.

The following interesting proposition again shows that the approach of grouping features based on BK subproduct can be presented as an intersection of basic compositions (with excluding features) of fuzzy relation R and fuzzy relations S_m.

Proposition 2

$$R^* \lhd S^* = \bigcap_{m=1}^{M} (R \circ S_m^{\backslash} E) \tag{3}$$

Proof. Using the definitions of R^* and S^*, we obtain

$$\begin{aligned}
&(R^* \lhd S^*)(x_i, z_k)\\
&= \bigwedge_{h=1}^{I} (R^*(x_i, y_1^h) \to S_E^*(y_1^h, z_k) \wedge \cdots \wedge \bigwedge_{h=1}^{I} (R^*(x_i, y_M^h) \to S_E^*(y_M^h, z_k)\\
&=(1 \to S_E^*(y_1^i, z_k)) \wedge (1 \to S_E^*(y_2^i, z_k)) \wedge \cdots \wedge (1 \to S_E^*(y_M^i, z_k))\\
&=(R \circ S_1^{\backslash} E)(x_i, z_k) \wedge (R \circ S_2^{\backslash} E)(x_i, z_k) \wedge \cdots \wedge (R \circ S_M^{\backslash} E)(x_i, z_k)\\
&= \left(\bigcap_{m=1}^{M} (R \circ S_m^{\backslash} E) \right) (x_i, z_k).
\end{aligned}$$

Observing the result provided by Proposition 2, one may immediately notice that the right hand side of (3) is equivalent to the right hand side of (2) provided in Proposition 1 and so, the left hand sides are consequently equivalent too. This means, that both compositions are actually identical.

Indeed, standard BK triangle products differ in the direction of the used implication. BK subproduct assumes that "*all feature*" of the a given object x belong to a class z and BK superproduct assumes that "*all feature*" of the a given class z are carried by a given object x. These are natural meanings, but in many situations too much lowering the degree as the universal quantifiers operates in both cases over all features which are in many situations not only hard to expect (human mistake in the determination of some characteristic feature) but even impossible (features build in a specific way – e.g. altitudes in which a given dragonfly appears in nature) and all features related to a given class cannot be observed at the same time (a single sample is caught only in a single altitude, never in all the possible ones related to a given class).

As soon as we downgrade from "all features" to "all types of features" represented by "at least one" representative of the group/type of features, we approach a solution to this, say, lowering problem, but we also change the character of the compositions. The universal quantifiers is not operating on the features, only on the groups, and within the groups of features, we deal with the existential quantifier. Therefore, we naturally come to the conclusion that the newly introduced products may be constructed as intersections of the circlet products. These circlet products are using a commutative t-norm and thus, do not consider the order of the composed elements. Thus, both approaches necessarily lead to the same fuzzy relation.

Indeed, observing the above introduced semantics of the BK subproduct or its first part, it is clear, that it could be comprehended also in the following way:

> *"Object x carries* **at least a single** *feature that relates to class z from* **all groups** *of features and at the same time, it carries no features that would be excluding for the classification to z"*

which is equivalent to the semantics of the BK superproduct.

Consequently, if defined a sort of BK square product, it would be an intersection of both triangle products and as both are equivalent, we would come again to the same fuzzy relation. Thus, the concept of grouping features helps to deal with the required tolerance to some missing features and ignores the direction of the composition. So, it serves universally for both types of discussed problems. As both BK triangle products (and thus the square product too) may be equipped with a generalized quantifier ($R' \rhd^Q S'$ or $R^* \lhd^Q S^*$, respectively), the quantifier is the only preserved choice of freedom we have above the originally given matrices and the choice of the particular underlying algebraic structure.

Corollary 1. *Let Q be a generalized quantifier determined by a fuzzy measure μ on the set $\{1, \ldots, M\}$. Then*

$$R' \rhd^Q S' = R^* \lhd^Q S^* \, .$$

3.3 Experiment and Conclusions

Most of the recent as well as historical direct applications of fuzzy relational compositions were more or less of the classification type, no matter we talk about very original medical diagnosis problems studied by Bandler and Kohout, recent dragonfly classification problem [21] or a sort of pattern recognition [29].

On the other hand, the area of fuzzy relational databases and flexible query answering systems[2] recognizes a task that is sort of inverse to the classification. That task is a database querying. Unlike in the classification problems where for a given sample we search for a single class, the querying a database does not expect to get back a single object. Indeed, for a given patient we seek for a single disease in the classification problem[3]. And although (multi-label) classification methods often provide us with sets of more diseases with some assigned numbers expressing how much it can be true or probable, that each of the assigned diseases is the ground truth one carried by a given patient. But in the querying problem, we formulate an opposite question to the database seeking for "all patients" having a particular disease, so the uniqueness makes no sense at all.

[2] These systems are formally based mainly on compositions and fuzzy quantifiers.

[3] Here, we abstract from the more complicated cases with more illnesses at the same time as usually there is always "the one" we seek for.

In general, both tasks have their unquestionable importance but each of them is based on a different approach and requires a different accuracy evaluation. But due to the equivalence of $(R' \rhd^Q S')(x, z)$ with $(R^* \lhd^Q S^*)(x, z)$ for any pair (x, z) provided by Corollary 1, both tasks may be obtained by the same calculation and simply by substituting the known instance to the particular variable x or z.

For the experimental purposes, let us adopt the dragonfly data used already for the evaluation of the classification task in [21]. This experimental dataset consists of $I = 105943$ presence records containing a description of an observed dragonfly (morphological category, date, altitude, and up to 6 colors). Each record has assigned also the correct dragonfly species and gender. The total number of possible dragonfly species is 70, thus each record falls into one of 140 categories.

The objective of the experiment is to find all records corresponding to a selected species/gender category based on the observed description of dragonflies. For that purpose, odonatology experts defined the following features: 6 colors encoded using three modalities (0/0.5/1 = cannot/may/must have); 14 intervals of altitudes (encoded using 6 intensities); 36 decades in the year (encoded using 6 intensities); and 4 morphological categories. These features formed feature-category matrices S and E (60 rows and 140 columns).

Within the experiment, several types of compositions were performed and evaluated by means of ratios of correct selections (C_k), false positives (FP_k) and false negatives (FN_k), over all categories ($k = 1, \ldots, K = 140$), where

$$C_k = \frac{\sum_{i=1}^{I} r_{ik} g_{ik}}{\sum_{i=1}^{I} r_{ik}}, \quad FP_k = \frac{\sum_{i=1}^{I} (1 - r_{ik}) g_{ik}}{\sum_{i=1}^{I} (1 - r_{ik})}, \quad FN_k = \frac{\sum_{i=1}^{I} r_{ik} (1 - g_{ik})}{\sum_{i=1}^{I} r_{ik}},$$

and where r_{ik} encodes the correctness of selection (0 or 1), i.e., $r_{ik} = 1$ if x_i is a sample of z_k and $r_{ik} = 0$ in the opposite cases; and g_{ik} is the value of the resulting composition evaluated for the pair (x_i, z_k). The results are then averaged over all K categories and presented in Table 1.

Table 1. The results of the experiment (C = ratio of correct selections, FP = ratio of false positives, FN = ratio of false negatives)

Method	C	FP	FN
$R \circ S$	1.0000	0.9755	0.0000
$R \lhd S$	0.7408	0.0740	0.2592
$R \rhd S$	0.0000	0.0000	1.0000
$R \square S$	0.0000	0.0000	1.0000
$R^* \lhd S^*$	0.7949	0.0790	0.2051
$R^* \lhd^{75\%} S^*$	0.9476	0.1106	0.0524
$R^* \lhd^{50\%} S^*$	0.9672	0.1209	0.0328

As we may see, the basic composition $\circ$ provides us with a maximal score $C = 1$ which expresses, that each sample of a given particular category was really assigned to this category in the query task. However, this high score is painfully redeemed by the too high FP rate equal to 0.9755, which expresses the average membership degree of a sample to a given particular category, to which it does not belong to. In other words, nearly all samples belong to all categories. The BK subproducts nicely reduces the FP error however, it lowers the C score (and thus increases the FN score) as well. This is the effect of the usage of the universal quantifier over all features. The BK superproduct naturally leads to a zero success, as no sample may carry all features created in such a way (e.g. all decades). Grouping the features already itself brings some improvement compared to the BK subproduct and in the combination with the generalized quantifiers *at least 3 out of 4* ($>= 75\%$) and *at least 2 out of 4* ($>= 50\%$) operating on the groups of the features, we are obtaining very promising results. The choice of the quantifier is the parameter of freedom that is highly context and particular application dependent.

References

1. Bělohlávek, R.: Fuzzy Relational Systems: Foundations and Principles. Kluwer Academic, Plenum Press, Dordrecht, New York (2002)
2. Bandler, W., Kohout, L.J.: Semantics of implication operators and fuzzy relational products. Int. J. Man Mach. Stud. **12**(1), 89–116 (1980)
3. Pedrycz, W.: Applications of fuzzy relational equations for methods of reasoning in presence of fuzzy data. Fuzzy Sets Syst. **16**, 163–175 (1985)
4. Štěpnička, M., De Baets, B., Nosková, L.: Arithmetic fuzzy models. IEEE Trans. Fuzzy Syst. **18**, 1058–1069 (2010)
5. Sanchez, E.: Resolution of composite fuzzy relation equations. Inf. Control **30**, 38–48 (1976)
6. Di Nola, A., Sessa, S., Pedrycz, W., Sanchez, E.: Fuzzy Relation Equations and Their Applications to Knowledge Engineering. Kluwer, Boston (1989)
7. De Baets, B.: Analytical solution methods for fuzzy relational equations. In: Dubois, D., Prade, H. (eds.) The Handbook of Fuzzy Set Series, vol. 1, pp. 291–340. Academic Kluwer Publ, Boston (2000)
8. Pedrycz, W.: Fuzzy relational equations with generalized connectives and their applications. Fuzzy Sets Syst. **10**, 185–201 (1983)
9. Cao, N., Štěpnička, M.: Fuzzy relation equations with fuzzy quantifiers. In: Kacprzyk, J., Szmidt, E., Zadrożny, S., Atanassov, K.T., Krawczak, M. (eds.) IWIFSGN/EUSFLAT -2017. AISC, vol. 641, pp. 354–367. Springer, Cham (2018). https://doi.org/10.1007/978-3-319-66830-7_32
10. Štěpnička, M., De Baets, B.: Interpolativity of at-least and at-most models of monotone single-input single-output fuzzy rule bases. Inf. Sci. **234**, 16–28 (2013)
11. Štěpnička, M., Jayaram, B.: Interpolativity of at-least and at-most models of monotone fuzzy rule bases with multiple antecedent variables. Fuzzy Sets Syst. **297**, 26–45 (2016)
12. Kohout, L., Kim, E.: The role of BK-products of relations in soft computing. Soft Comput. **6**, 92–115 (2002)

13. Dubois, D., Prade, H.: Semantics of quotient operators in fuzzy relational databases. Fuzzy Sets Syst. **78**, 89–93 (1996)
14. Pivert, O., Bosc, P.: Fuzzy Preference Queries to Relational Databases. Imperial College Press, London (2012)
15. Delgado, M., Sánchez, D., Vila, M.A.: Fuzzy cardinality based evaluation of quantified sentences. Int. J. Approx. Reason. **23**, 23–66 (2000)
16. Belohlavek, R.: Sup-t-norm and inf-residuum are one type of relational product: unifying framework and consequences. Fuzzy Sets Syst. **197**, 45–58 (2012)
17. Běhounek, L., Daňková, M.: Relational compositions in fuzzy class theory. Fuzzy Sets Syst. **160**(8), 1005–1036 (2009)
18. De Baets, B., Kerre, E.: Fuzzy relational compositions. Fuzzy Sets Syst. **60**, 109–120 (1993)
19. Daňková, M.: Fuzzy relations and fuzzy functions in partial fuzzy set theory. In: Kacprzyk, J., Szmidt, E., Zadrożny, S., Atanassov, K.T., Krawczak, M. (eds.) IWIFSGN/EUSFLAT -2017. AISC, vol. 641, pp. 563–573. Springer, Cham (2018). https://doi.org/10.1007/978-3-319-66830-7_50
20. Cao, N., Štěpnička, M.: How to incorporate excluding features in fuzzy relational compositions and what for. In: Carvalho, J.P., Lesot, M.-J., Kaymak, U., Vieira, S., Bouchon-Meunier, B., Yager, R.R. (eds.) IPMU 2016. CCIS, vol. 611, pp. 470–481. Springer, Cham (2016). https://doi.org/10.1007/978-3-319-40581-0_38
21. Cao, N., Štěpnička, M., Burda, M., Dolný, A.: Excluding features in fuzzy relational compositions. Expert Syst. Appl. **81**, 1–11 (2017)
22. Štěpnička, M., Holčapek, M.: Fuzzy relational compositions based on generalized quantifiers. In: Laurent, A., Strauss, O., Bouchon-Meunier, B., Yager, R.R. (eds.) IPMU 2014. CCIS, vol. 443, pp. 224–233. Springer, Cham (2014). https://doi.org/10.1007/978-3-319-08855-6_23
23. Cao, N., Štěpnička, M., Holčapek, M.: An extension of fuzzy relational compositions using generalized quantifiers. In: Proceedings of 16th World Congress of the International Fuzzy Systems Association (IFSA) and 9th Conference of the European Society for Fuzzy-Logic and Technology (EUSFLAT). Advances in Intelligent Systems Research, vol. 89, pp. 49–58. Atlantis press, Gijón (2015)
24. Cao, N., Štěpnička, M., Holčapek, M.: Extensions of fuzzy relational compositions based on generalized quantifer. Fuzzy Sets Syst. **339**, 73–98 (2018)
25. Cao, N., Štěpnička, M.: Incorporation of excluding features in fuzzy relational compositions based on generalized quantifiers. In: Kacprzyk, J., Szmidt, E., Zadrożny, S., Atanassov, K.T., Krawczak, M. (eds.) IWIFSGN/EUSFLAT -2017. AISC, vol. 641, pp. 368–379. Springer, Cham (2018). https://doi.org/10.1007/978-3-319-66830-7_33
26. Cao, N., Štěpnička, M.: Fuzzy relational compositions based on grouping features. In: The 9th International Conference on Knowledge and Systems Engineering (KSE 2017), Hue, Vietnam, pp. 94–99. IEEE (2017)
27. Dvořák, A., Holčapek, M.: L-fuzzy quantifiers of type $\langle 1 \rangle$ determined by fuzzy measures. Fuzzy Sets Syst. **160**(23), 3425–3452 (2009)
28. Klement, E.P., Mesiar, R., Pap, E.: A universal integral as common frame for Choquet and Sugeno integral. IEEE Trans. Fuzzy Syst. **18**, 178–187 (2010)
29. Peeva, K., Kyosev, Y.: Fuzzy Relational Calculus: Theory, Applications and Software. World Scientific, Singapore (2005)

Compositions of Partial Fuzzy Relations

Nhung Cao(✉) and Martin Štěpnička(✉)

Institute for Research and Applications of Fuzzy Modeling,
Centre of Excellence IT4Innovations, University of Ostrava,
30. Dubna 22, 701 03 Ostrava 1, Czech Republic
{nhung.cao,martin.stepnicka}@osu.cz
http://irafm.osu.cz/

Abstract. The aim of this contribution is to study compositions of partial fuzzy relational compositions, i.e., of fuzzy relations with membership degrees not defined on the whole universe. This is motivated by the possibility of existence of the relationships which are "undefined", "unknown", "meaningless", "non-applicable", "irrelevant", etc. We introduce definitions for the new concept based on suitable operations used in the framework of the partial fuzzy set theory. The preservations of well-known interesting properties of compositions are studied for the compositions of partial fuzzy relations as well. An illustrative example is provided.

Keywords: Partial fuzzy relations · Fuzzy relational compositions
Excluding features · Classification

1 Introduction

Fuzzy relational compositions became a crucial topic in fuzzy mathematics. This concept was firstly introduced by Bandler and Kohout in late 70's [1–3]. Afterward, the topic has been continuously developed on various aspects and led to an extensive amount of results. For more details on its theory, we refer readers to some sources [4–6], and for many fields of applications including formal constructions of fuzzy inference systems, related systems of fuzzy relational equations or modeling monotone fuzzy rule bases, we refer to [7–13]. Apart from others, two directions extending the topic have been recently proposed, namely the incorporation of excluding features in the fuzzy relational compositions [14] and the compositions based on generalized quantifiers [15]. We assume that readers are familiar with the topic and let we deal with another extension of the topic, namely, with partial fuzzy relations and their compositions.

Partial fuzzy set theory has been recently studied in [16–18] elaborating a new theory of fuzzy sets with partially undefined membership degrees. These undefined degrees appear mainly values in the cases of "unknown (missing)",

This research was partially supported by the NPU II project LQ1602 "IT4Innovations excellence in science" provided by the MŠMT.

J. Medina et al. (Eds.): IPMU 2018, CCIS 855, pp. 187–198, 2018.
https://doi.org/10.1007/978-3-319-91479-4_16

"meaningless", "non-applicable" or "irrelevant" values. We may accept this motivation as a natural one for the fuzzy relational compositions because having no information or having missing part of data is a natural situation as well and it should be treated theoretically correctly. Compared to [18], the domain of partial fuzzy relations will be not explicitly treated in the notation and we will keep the same fixed universes. A natural question is, what are suitable operations that should be used for the meaningful definitions. The answer will be partly provided in this work. Similar to the standard fuzzy relational compositions, where the concept of excluding feature is incorporated to get an effective tool [14,19], the new partial compositions are allowed to involve this concept as well. We will provide a list of properties of the new concept of the compositions of partial fuzzy relations and demonstrate the concept on an illustrative example.

2 Partial Fuzzy Relations

We fix a residuated lattice $\mathcal{L} = \langle L, \wedge, \vee, \otimes, \rightarrow 0, 1\rangle$ where the support $L = [0, 1]$ as the underlying algebraic structure and let sets X, Y and Z will be non-empty finite universes of objects (samples), features and classes, respectively, and let us denote $\mathcal{F}(U)$ the set of all fuzzy sets on the universe U.

The appearance of undefined (e.g. missing) values for some relationships of some pairs of elements will be represented by a dummy element $\bullet$, the support L will be extended to $L^\bullet = L \cup \{\bullet\}$. Following the previous works on partial fuzzy set theory [17,18,20], we assume that $L^\bullet$ is a poset consisting of a chain $[0, 1]$ and a single element $\bullet$ that is not comparable to any element from the chain $[0, 1]$. So, the ordering relation $\leq$ on $L^\bullet$ is then defined standardly for the values from $[0, 1]$ and $\bullet \leq \bullet$ but $\bullet \not\leq a$ and $a \not\leq \bullet$ for any $a \in [0, 1]$. The equality $a = b$ of two elements $a, b \in L^\bullet$ is then defined standardly as the conjunction: $a \leq b$ and $b \leq a$.

Definition 1. *Partial fuzzy relation* R is a fuzzy set $R : X \times Y \rightarrow L^\bullet$.

The set of all partial fuzzy sets on U will be denoted as $\mathcal{F}^\bullet(U)$ and thus, we can freely write $R \in \mathcal{F}^\bullet(X \times Y)$ and $S, E \in \mathcal{F}^\bullet(Y \times Z)$. For the interpretation of the meaning of partial fuzzy relations, for example in the classification problem of animals, a defined value $R(x, y)$ express that the sample x has the feature y up to some degrees and $S(y, z)$ means that how much it is true y is a feature of the class z. In case of having no information of the relationship of the sample x and the feature y or of the feature y and the class z, we replace the relationships by the dummy value $\bullet$.

3 Connective Operations for Undefined Degrees

This section recalls suitable connective operations for dealing with the undefined values. There are several useful families of connectives extended from three-valued logic [16,21,22] such as Bochvar-style connectives, Sobociński-style connectives, Kleene-style connectives, or McCarthy-style sequential binary connectives. In this paper, we pick the first two well-known ones, i.e., Bochvar-style and Sobociński-style connectives.

Definition 2 [16,18]. The Bochvar operation $c_B \in \{\wedge_B, \vee_B, \otimes_B, \rightarrow_B\}$, $c_B : L^\bullet \times L^\bullet \rightarrow L^\bullet$ is represented by the following truth table:

c_B	β	$\bullet$
α	$\alpha\, c\, \beta$	$\bullet$
$\bullet$	$\bullet$	$\bullet$

Definition 3 [16,18]. The Sobociński operation $c_S \in \{\wedge_S, \vee_S, \otimes_S\}$, $c_S : L^\bullet \times L^\bullet \rightarrow L^\bullet$ and the Sobociński residuum $\rightarrow_S$ residuated with $\otimes_S$ are represented by the following truth tables:

c_S	β	$\bullet$
α	$\alpha\, c\, \beta$	α
$\bullet$	β	$\bullet$

$\rightarrow_S$	β	$\bullet$
α	$\alpha \rightarrow \beta$	$\neg\alpha$
$\bullet$	β	$\bullet$

As one can see, the Bochvar operations treat the value $\bullet$ as the annihilator and the Sobociński operations treat it as the neutral element. The two respective equivalence connectives (bi-implications) of the Bochvar and Sobociński can be defined on $L^\bullet$ as follows [16]:

$$\alpha \leftrightarrow_B \beta = (\alpha \rightarrow_B \beta) \wedge_B (\beta \rightarrow_B \alpha);$$
$$\alpha \leftrightarrow_S \beta = (\alpha \rightarrow_S \beta) \wedge_S (\beta \rightarrow_S \alpha).$$

Let us recall in the following the infimum and supremum of a set of elements from $L^\bullet$ according to the Bochvar-style and Sobociński-style.

Definition 4 [18]. Let $\alpha_i \in L^\bullet$. Then, for arbitrary index set I we define:

(i) The Bochvar infimum: $\bigwedge^B_{i\in I} \alpha_i = \begin{cases} \inf\limits_{i\in I} \alpha_i & \text{if } \alpha_i \neq \bullet \text{ for each } i \in I, \\ \bullet & \text{otherwise.} \end{cases}$

(ii) The Bochvar supremum: $\bigvee^B_{i\in I} \alpha_i = \begin{cases} \sup\limits_{i\in I} \alpha_i & \text{if } \alpha_i \neq \bullet \text{ for each } i \in I, \\ \bullet & \text{otherwise.} \end{cases}$

(iii) The Sobociński infimum: $\bigwedge^S_{i\in I} \alpha_i = \begin{cases} \inf\limits_{\substack{i\in I\\ \alpha_i\neq\bullet}} \alpha_i & \text{if } \alpha_i \neq \bullet \text{ for some } i \in I, \\ \bullet & \text{otherwise.} \end{cases}$

(iv) The Sobociński supremum: $\bigvee^S_{i\in I} \alpha_i = \begin{cases} \sup\limits_{\substack{i\in I\\ \alpha_i\neq\bullet}} \alpha_i & \text{if } \alpha_i \neq \bullet \text{ for some } i \in I, \\ \bullet & \text{otherwise.} \end{cases}$

The four operations are self-explainable and thus, we avoid commenting on them.

Given two partial fuzzy relations on the same domain, their Bochvar and Sobociński intersection and union can be defined as follows.

Definition 5. Let R_1, R_2 be partial fuzzy relations on $X \times Y$. Then we define

(i) Bochvar union: $(R_1 \cup_B R_2)(x,y) = R_1(x,y) \vee_B R_2(x,y)$;
(ii) Bochvar intersection: $(R_1 \cap_B R_2)(x,y) = R_1(x,y) \wedge_B R_2(x,y)$;
(iii) Strong Bochvar intersection: $(R_1 \Cap_B R_2)(x,y) = R_1(x,y) \otimes_B R_2(x,y)$;
(iv) Sobociński union: $(R_1 \cup_S R_2)(x,y) = R_1(x,y) \vee_S R_2(x,y)$;
(v) Sobociński intersection: $(R_1 \cap_S R_2)(x,y) = R_1(x,y) \wedge_S R_2(x,y)$;
(vi) Strong Sobociński intersection: $(R_1 \Cap_S R_2)(x,y) = R_1(x,y) \otimes_S R_2(x,y)$.

The equivalence and the inclusions of partial fuzzy relations are defined in a standard way no matter we deal on $L^\bullet$.

Definition 6. For any $R_1, R_2 \in \mathcal{F}^\bullet(X \times Y)$ we say that $R_1 \subseteq R_2$ if $R_1(x,y) \leq R_2(x,y)$ for $(x,y) \in X \times Y$ and, $R_1 = R_2$ if $R_1(x,y) = R_2(x,y)$ for $(x,y) \in X \times Y$.

4 Compositions of Partial Fuzzy Relations and Excluding Features

4.1 Compositions of Partial Fuzzy Relations

We have recalled partial fuzzy relations and suitable operations for dealing with their membership degrees. This section investigates possibilities of forming their compositions. We propose four kinds of compositions following the idea of the standard ones. Mainly, the basic composition with the inner operation representing the multiplication and the outer operation stands for the supremum will be considered as an initial guess of the relationship between elements from X and Z. The three other ones will be proposed as a strengthening of the first guess and they are set by the infimum operation standing outside, and implication-based operations for the inner ones. A natural question is which style of connectives should be used for formalizing the compositions. We will address this question below and demonstrate the proposed definitions on a demonstrative example.

Definition 7. Let $R \in \mathcal{F}^\bullet(X \times Y), S \in \mathcal{F}^\bullet(Y \times Z)$ then we define compositions $R \circ_{SB} S$, $R \lhd_{BS} S$, $R \rhd_{BS} S$ and $R \square_{BS} S$ as follows

$$(R \circ_{SB} S)(x,z) = \bigvee_{y \in Y}^{S} (R(x,y) \otimes_B S(y,z)) \; ; \tag{1}$$

$$(R \lhd_{BS} S)(x,z) = \bigwedge_{y \in Y}^{B} (R(x,y) \rightarrow_S S(y,z)) \; ; \tag{2}$$

$$(R \rhd_{BS} S)(x,z) = \bigwedge_{y \in Y}^{B} (R(x,y) \leftarrow_S S(y,z)) \; ; \tag{3}$$

$$(R \square_{BS} S)(x,z) = \bigwedge_{y \in Y}^{B} (R(x,y) \leftrightarrow_S S(y,z)) \; . \tag{4}$$

The semantics of the compositions can be described in the same way to the standard ones. For instance, considering the classification of animals, $(R \circ_{SB} S)(x, z)$ expresses the truth degrees of the predicate: "sample x has at least one feature belonging to class z". The strengthening product $(R \lhd_{BS} S)(x, z)$ expresses the truth degrees of the predicate: "all the features of sample x belong to class z" and, similarly, $(R \rhd_{BS} S)(x, z)$ expresses the truth degrees of the predicate: "sample x has all features of class z". The last square product $(R \,\square_{BS}\, S)(x, z)$ represents the conjunction of the both triangle products: "sample x carries all features of class z and all features of the sample are related to the class z". The difference, compared to the compositions of (non-partial) fuzzy relations is that the above-defined compositions are correctly defined even for the case of missing values.

An important question is why the Sobociński and Bochvar connectives are used for the outer and inner operation in the composition $R \circ_{SB} S$, respectively, while the three other ones apply Bochvar-style to the outer operations and Sobociński for the inner ones? Let us first focus on the circlet composition. Consider the hypothetic use of the Sobociński inner operator (conjunction) for the animal classification example. An animal without any feature that would connect it with a certain class (family of animals) would be potentially classified to such a class. Therefore, the use of the Bochvar conjunction seems reasonable. Vice-versa, using the Bochvar outer operator could lead to a loss of potential initial suspicion even in the case of more connecting features only due to a single dummy value, which is undoubtedly not desirable.

However, the case of the triangle and the square BK products is different. Consider, e.g., the BK subproduct. It should express that for all considered features it holds that if a feature is possessed by the sample then the feature should be related to a given class. So, the Sobociński-style outer operation (infimum) could be too tolerant to the frequent appearance of the dummy values still allowing to evaluate the membership degree of a certain pair (x, z) to the composed fuzzy relation equal to 1. Therefore, we opt for the Bochvar infimum. On the other hand, if the Bochvar-style of the operation would be chosen also for the inner operation, this would significantly increase the occurrence of the dummy values in the resulting composition. Indeed, actually any pair (y, z) for which $S(y, z)$ is equal to $\bullet$ would cause that $(R \lhd_{BB} S)(x, z) = \bullet$ for any x. And similarly, any pair (x, y) for which $R(x, y) = \bullet$ would cause that $(R \lhd_{BB} S)(x, z) = \bullet$ for any z and thus, actually, such samples x's could be easily deleted from the sample file as unclassifiable ones, and this is surely not the goal of the partial fuzzy relational compositions. Therefore, we opt for $R @_{BS} S$ for any $@ \in \{\lhd, \rhd, \square\}$. On the other hand, any other combination of connectives is also formally correct and a study of BK products with the opposite connectives, i.e. SB instead of BS, is provided in [20].

4.2 The Compositions Incorporating Excluding Features

Similarly to the standard basic fuzzy relational composition which may be extended with the concept of excluding features [14], we allow to employ this concept to the circlet composition of partial fuzzy relations as well. We introduce the partial fuzzy relation $E \in \mathcal{F}^{\bullet}(Y \times Z)$ where $E(y,z) \in L^{\bullet}$ stands for the truth-degree of the predicate: y is an excluding feature of z and, $E(y,z) = \bullet$ means that we cannot decide (or do not know) whether y is an excluding feature of z or not.

Definition 8. Let $R \in \mathcal{F}^{\bullet}(X \times Y)$ and let $S, E \in \mathcal{F}^{\bullet}(Y \times Z)$ then the composition $R \circ_{SB} S' E$ is defined as follows

$$(R \circ_{SB} S' E)(x,z) = \bigvee_{y \in Y}^{S} (R(x,y) \otimes_B S(y,z)) \otimes_B \neg_S \bigvee_{y \in Y}^{S} (R(x,y) \otimes_B E(y,z))$$

where $\neg_S a = a \rightarrow_S 0$.

The semantic of the composition is the same to the standard one. The membership degree of (x,z) to the composed fuzzy relation $R \circ_{SB} S' E$ expresses how much it is true, that the sample x has at least one feature belonging to the class z and at the same time, there is no excluding feature of the family z carried by the sample x.

The above-proposed composition can be written in a comprehensible form which given as follows

$$(R \circ_{SB} S' E)(x,z) = (R \circ_{SB} S)(x,z) \otimes_B \neg_S (R \circ_{SB} E)(x,z)$$

We propose to use the Bochvar connective $\otimes_B$ between the two basic compositions as we find unintuitive to assign non-dummy value to a pair (x,z) in the case of not a single known connection of the sample x and the class z only due to the fact that x has no excluding features related to the class z, which could happen in the case of the use of the Sobociński conjunction. On the other hand, we also agree that making the overall result unknown (dummy value) in the case of a significant suspicion provided by $(R \circ_{SB} S)(x,z)$ and an unknown result of the other part $(R \circ_{SB} E)(x,z)$. This can serve as a a motivation for the use of non-commutative conjunctors which we leave for the future consideration.

5 Properties

Let us provide a list of useful properties for the proposed concept of compositions of partial fuzzy relations. Let $R, R_1, R_2 \in \mathcal{F}^{\bullet}(X \times Y)$, let $S, S_1, S_2 \in \mathcal{F}^{\bullet}(Y \times Z)$, and let $T \in \mathcal{F}^{\bullet}(Z \times U)$.

5.1 Basic Partial Composition

Proposition 1. (Associativity property)

$$(R \circ_{SB} S) \circ_{SB} T = R \circ_{SB} (S \circ_{SB} T) \quad (5)$$

Sketch of the proof: The proof of (5) is based on the fact that $(a \vee_S b) \otimes_B c = (a \otimes_B c) \vee_S (b \otimes_B c)$ and that $(a \otimes_B b) \otimes_B c = a \otimes_B (b \otimes_B c)$ for $a, b, c \in L^\bullet$. □

Proposition 2. (Monotonicity)

$$R_1 \subseteq R_2 \Rightarrow R_1 \circ_{SB} S \subseteq R_2 \circ_{SB} S \tag{6}$$

$$S_1 \subseteq S_2 \Rightarrow R \circ_{SB} S_1 \subseteq R \circ_{SB} S_2 \tag{7}$$

Sketch of the proof: Property (6) and (7) are proved using the fact that $a \leq b \Rightarrow a \otimes_B c \leq b \otimes_B c$ for $a, b, c \in L^\bullet$. □

Proposition 3. (Sobociński union)

$$(R_1 \cup_S R_2) \circ_{SB} S = (R_1 \circ_{SB} S) \cup_S (R_2 \circ_{SB} S) \tag{8}$$

$$R \circ_{SB} (S_1 \cup_S S_2) = (R \circ_{SB} S_1) \cup_S (R \circ_{SB} S_2) \tag{9}$$

Sketch of the proof: Let us sketch only the proof of the equality (8) and the other one is analogous. Fix some $x \in X$ and some $z \in Z$ and let us partition the set Y into five subsets $Y_1, Y_2, \ldots, Y_5$ as follows:

$$\begin{aligned}
Y_1 &= \{y \in Y \mid S(y,z) = \bullet\}, \\
Y_2 &= \{y \in Y \mid R_1(x,y) = \bullet, R_2(x,y) = \bullet, S(y,z) \neq \bullet\}, \\
Y_3 &= \{y \in Y \mid R_1(x,y) = \bullet, R_2(x,y) \neq \bullet, S(y,z) \neq \bullet\}, \\
Y_4 &= \{y \in Y \mid R_1(x,y) \neq \bullet, R_2(x,y) = \bullet, S(y,z) \neq \bullet\}, \\
Y_5 &= \{y \in Y \mid R_1(x,y) \neq \bullet, R_2(x,y) \neq \bullet, S(y,z) \neq \bullet\}.
\end{aligned}$$

Then, for each $Y_i, i = 1, \ldots, 5$, we easily check that the left hand side and the right hand side of (8) are equal.

□

Proposition 4. (Convertibility)

$$(R \circ_{SB} S)^T = S^T \circ_{SB} R^T \tag{10}$$

where R^T denotes the transposition of R, i.e., $R^T(x,y) = R(y,x)$.

Sketch of the proof: The proof is analogous to the proof for standard fuzzy relational compositions. □

Proposition 5. (Associativity – the case of excluding features)

$$(R \circ_{SB} S) \circ_{SB} T^{\backprime} E = R \circ_{SB} (S \circ_{SB} T)^{\backprime}(S \circ_{SB} E). \tag{11}$$

Sketch of the proof: The proof is based on Proposition 1. □

Proposition 6. (Monotonicity – the case of excluding features)

$$S_1 \subseteq S_2 \Rightarrow R \circ_{SB} S_1^{\backprime} E \subseteq R \circ_{SB} S_2^{\backprime} E \tag{12}$$

$$E_1 \subseteq E_2 \Rightarrow R \circ_{SB} S^{\backprime} E_1 \supseteq R \circ_{SB} S^{\backprime} E_2 \tag{13}$$

Sketch of the proof: Property (12) can be proved similarly to the proof of (7) from Proposition 2 and property (13) is derived from the fact that $a \leq b \Rightarrow \neg_S a \geq \neg_S b$ for $a, b \in L^\bullet$. □

Proposition 7. (Sobociński union – the case of excluding features)

$$R \circ_{SB} (S_1 \cup_S S_2)^{\backprime} E = (R \circ_{SB} S_1^{\backprime} E) \cup_S (R \circ_{SB} S_2^{\backprime} E) \quad (14)$$

Sketch of the proof: The proof uses the equality: $(a \vee_S b) \otimes_B c = (a \otimes_B c) \vee_S (b \otimes_B c)$. □

5.2 The Bandler-Kohout Products

Let us provide some useful properties for the Bandler-Kohout products of partial fuzzy relations which strengthen the basic partial compositions.

Proposition 8. (Interdefinability)

$$R \square_{BS} S = (R \lhd_{BS} S) \cap_S (R \rhd_{BS} S) \quad (15)$$

Sketch of the proof: The proof is based on the definition of $\leftrightarrow_S$ and the fact that $\bigwedge^B_{i \in I}(a_i \wedge_S b_i) = \bigwedge^B_{i \in I} a_i \wedge_S \bigwedge^B_{i \in I} b_i$ for $a_i, b_i \in L^\bullet$. □

Proposition 9. (Monotonicity)

$$R_1 \subseteq R_2 \Rightarrow R_1 \lhd_{BS} S \supseteq R_2 \lhd_{BS} S, \quad (16)$$
$$S_1 \subseteq S_2 \Rightarrow R \lhd_{BS} S_1 \subseteq R \lhd_{BS} S_2, \quad (17)$$
$$R_1 \subseteq R_2 \Rightarrow R_1 \rhd_{BS} S \subseteq R_2 \rhd_{BS} S, \quad (18)$$
$$S_1 \subseteq S_2 \Rightarrow R \rhd_{BS} S_1 \supseteq R \rhd_{BS} S_2. \quad (19)$$

Sketch of the proof: The proof of (16)–(19) uses the antitonicity and the isotonicity of $\rightarrow_S$ that are $a \leq b \Rightarrow a \rightarrow_S c \geq b \rightarrow_S c$ and $c \rightarrow_S a \leq c \rightarrow_S b$ for $a, b, c \in L^\bullet$. □

Proposition 10. (Bochvar union and the BK subproduct)

$$(R_1 \cup_B R_2) \lhd_{BS} S \subseteq (R_1 \lhd_{BS} S) \cap_B (R_2 \lhd_{BS} S). \quad (20)$$

Sketch of the proof: The proof of the proposition uses the same technique the proof of Proposition 3 where we consider all possible cases. □

Proposition 11. (Bochvar intersection and the BK subproduct)

$$R \lhd_{BS} (S_1 \cap_B S_2) \subseteq (R \lhd_{BS} S_1) \cap_B (R \lhd_{BS} S_2). \quad (21)$$

Sketch of the proof: The proof of the proposition uses the same technique the proof of Proposition 3 where we consider all possible cases. □

Proposition 12. (Bochvar union and intersection and the BK superproduct)

$$(R_1 \cap_B R_2) \rhd_{BS} S \subseteq (R_1 \rhd_{BS} S) \cap_B (R_2 \rhd_{BS} S). \tag{22}$$

$$R \rhd_{BS} (S_1 \cup_B S_2) \subseteq (R \rhd_{BS} S_1) \cap_B (R \rhd_{BS} S_2). \tag{23}$$

Sketch of the proof: The proof of the proposition uses the same technique the proof of Proposition 3 where we consider all possible cases. □

Proposition 13. (Convertibility of the BK products)

$$(R \lhd_{BS} S)^T = S^T \rhd_{BS} R^T\,, \tag{24}$$

$$(R \rhd_{BS} S)^T = S^T \lhd_{BS} R^T\,, \tag{25}$$

$$(R \square_{BS} S)^T = S^T \square_{BS} R^T. \tag{26}$$

where R^T denotes the transposition of R, i.e., $R^T(x,y) = R(y,x)$.

Sketch of the proof: The proof is analogous to the proof for standard fuzzy relational compositions. □

Proposition 14. (Exchange property)

$$R \lhd_{BS} (S \rhd_{BS} T) = (R \lhd_{BS} S) \rhd_{BS} T. \tag{27}$$

Sketch of the proof: Using the property $a \to_S (b \to_S)c = b \to_S (a \to_S c)$ and the property $a \to_S (b \wedge_B c) = (a \to_S b) \wedge_B (a \to_S c)$ we obtain the proof of (27). □

6 Illustrative Example, Conclusions and Future Work

We provide readers with an illustrative example that demonstrates the influence of the use of partial fuzzy relational compositions. For this purpose, let us consider a simple example of classification of animals in biology that is similar yet slightly modified example from [14]. Let $Z = \{z_1, z_2, \ldots, z_7\}$ be a set of families of animals, $Y = \{y_1, y_2, \ldots, y_{10}\}$ be a set of animal features and let $X = \{x_1, x_2, \ldots, x_6\}$ be a set of particular animals, where z_1 – Bird, z_2 – Fish, z_3 – Dog, z_4 – Equidae, z_5 – Mosquito, z_6 – Monotreme, and z_7 – Reptile; y_1 – animal flies, y_2 – animal has feathers, y_3 – animal has fins, y_4 – animal has claws, y_5 – animal has hair, y_6 – animal has teeth, y_7 – animal has a beak, y_8 – animal has scales, y_9 – animal swims, y_{10} – animal is warm blooded; x_1 – Platypus, x_2 – Emu, x_3 – Hairless dog, x_4 – Aligator, x_5 – Parrotfish, x_6 – Puffin.

The task is to classify the animals from X to their classes. It may happen, that for some of given animals, we do not know the membership degrees assigning to some features. For instance, a decision whether a dog swims or not may be missing in the fuzzy relation R as some dog breeds do not swim (e.g. Pugs or Bulldogs) or swim very badly (e.g. Basset hounds or Dachshunds) and thus, the data provider does not fill in this information. Similarly, for some providers is might not be easy to determine whether an animal is warm-blooded or cold-blooded. Thus, at some cases, the truth degrees of the relationships will be assigned to the dummy element •. Also, in case of a given feature cannot be determined to be excluded from possibility of belonging the classes, its respective truth degree should turn to the unknown value • as well. Thus, fuzzy relations $S, E \in \mathcal{F}(Y \times Z)$ and $R \in \mathcal{F}(X \times Y)$ can be given as follows

R	y_1	y_2	y_3	y_4	y_5	y_6	y_7	y_8	y_9	y_{10}
Platypus	0	0	0	1	1	0	1	0	0.9	•
Emu	0	1	0	1	0	0	1	0.5	•	1
Hairless dog	0	0	0	1	0.2	1	0	0	•	1
Aligator	0	0	0	1	0	1	0	1	0.9	0
Parrotfish	0	0	1	0	0	0.9	0.8	1	1	0
Puffin	1	1	0	1	0	0	1	0.4	0.9	1

,

S	z_1	z_2	z_3	z_4	z_5	z_6	z_7
y_1	0.8	0	0	0	1	0	0
y_2	1	0	0	0	0	0	0
y_3	0	1	0	0	0	0.5	0
y_4	0.9	0	1	0	0	0.8	0.3
y_5	0	0	0.8	1	0	0.9	0
y_6	0	0.6	1	1	0	0	0.7
y_7	1	0.1	0	0	0	0.5	0
y_8	0.7	0.9	0	0	0	0	1
y_9	0.5	1	0.8	0.6	0.1	0.7	0.8
y_{10}	1	0.1	1	1	0	0.7	0

,

E	z_1	z_2	z_3	z_4	z_5	z_6	z_7
y_1	0	1	1	1	0	1	1
y_2	0	1	1	1	1	1	1
y_3	1	0	1	1	1	0	1
y_4	0	1	0	1	1	0	0
y_5	0.8	1	0	0	1	0	1
y_6	1	0	0	0	1	1	0
y_7	0	•	1	1	1	0	1
y_8	0	0	1	0	1	1	0
y_9	0	0	0	0	0.8	0	0
y_{10}	0	0.8	0	0	1	•	1

.

If we use the Łukasiewicz algebra as the underlying algebraic structure then we get the following compositions.

$R \circ_{SB} S$	z_1	z_2	z_3	z_4	z_5	z_6	z_7
Platypus	1	0.9	1	1	0	0.9	0.7
Emu	1	0.4	1	1	0	0.8	0.5
Hairless dog	1	0.6	1	1	0	0.8	0.7
Aligator	0.9	0.9	1	1	0	0.8	1
Parrotfish	0.8	1	0.9	0.9	0.1	0.7	1
Puffin	1	0.9	1	0.5	1	0.8	0.7

,

$R \lhd_{BS} S$	z_1	z_2	z_3	z_4	z_5	z_6	z_7
x_1	0	0.1	0	0	0	0.5	0
x_2	0.5	0	0	0	0	0	0
x_3	0	0	0.8	0	0	0	0
x_4	0	0	0	0	0	0	0.3
x_5	0	0.3	0	0	0	0	0
x_6	0.6	0	0	0	0	0	0

,

$R \rhd_{BS} S$	z_1	z_2	z_3	z_4	z_5	z_6	z_7
x_1	0	0.1	0	0	0	0.3	0
x_2	0.2	0	0	0	0	0.1	0.2
x_3	0	0	0.2	0.2	0	0.3	0
x_4	0	0	0	0	0	0.1	1
x_5	0	0.9	0	0	0	0.1	0.7
x_6	0.9	0	0	0	1	0.1	0.3

.

As we may see that the initial partial relation provided by $R \circ_{SB} S$ gives too much pairs of elements in a high degree which does not help for the classification. The strengthening partial subproduct and superproduct improve the results as well as making the results to be more precise however, they are somehow strengthening too much the assigned values for some pairs of relationships. The proposed method of partial composition incorporating excluding features would help as, in this situation, it can eliminate the false suspicions from the initial ones. This is given by the following table.

$R \circ_{SB} S^{\backprime} E$	z_1	z_2	z_3	z_4	z_5	z_6	z_7
Platypus	0	0	0	0	0	0.9	0
Emu	1	0	0	0	0	0	0
Hairless dog	0	0	1	0	0	0	0
Aligator	0	0	0	0	0	0	1
Parrotfish	0	1	0	0	0	0	0
Puffin	1	0	0	0	0	0	0

All the results demonstrated that the proposed compositions of partial fuzzy relations may successfully deal with unknown (missing) values. In the case of an appropriate choice of the operations, the compositions may even eliminate the occurrence of the unknown values. The last result shows that a positive result could be obtained if the number of known features is large enough for the classification. This example, though it is simple and only illustrative, it provides with a demonstration of the performance of compositions of partial fuzzy relations. A real example on a real data is left for future work.

Furthermore, the results motivate us to think of studying the concept of partial fuzzy relational compositions based on generalized quantifiers in future work.

References

1. Bandler, W., Kohout, L.J.: Fuzzy relational products and fuzzy implication operators. In: Proceedings of International Workshop on Fuzzy Reasoning Theory and Applications, London, Queen Mary College (1978)
2. Bandler, W., Kohout, L.J.: Fuzzy power sets and fuzzy implication operators. Fuzzy Sets Syst. **4**, 183–190 (1980)
3. Bandler, W., Kohout, L.J.: Semantics of implication operators and fuzzy relational products. Int. J. Man-Mach. Stud. **12**(1), 89–116 (1980)
4. Belohlavek, R.: Sup-t-norm and inf-residuum are one type of relational product: unifying framework and consequences. Fuzzy Sets Syst. **197**, 45–58 (2012)
5. Běhounek, L., Daňková, M.: Relational compositions in fuzzy class theory. Fuzzy Sets Syst. **160**(8), 1005–1036 (2009)
6. De Baets, B., Kerre, E.: Fuzzy relational compositions. Fuzzy Sets Syst. **60**, 109–120 (1993)
7. Štěpnička, M., De Baets, B., Nosková, L.: Arithmetic fuzzy models. IEEE Trans. Fuzzy Syst. **18**, 1058–1069 (2010)
8. Mandal, S., Jayaram, B.: SISO fuzzy relational inference systems based on fuzzy implications are universal approximators. Fuzzy Sets Syst. **277**, 1–21 (2015)
9. Sanchez, E.: Resolution of composite fuzzy relation equations. Inf. Control **30**, 38–48 (1976)

10. Di Nola, A., Sessa, S., Pedrycz, W., Sanchez, E.: Fuzzy Relation Equations and Their Applications to Knowledge Engineering. Kluwer, Boston (1989)
11. Cao, N., Štěpnička, M.: Fuzzy relation equations with fuzzy quantifiers. In: Kacprzyk, J., Szmidt, E., Zadrożny, S., Atanassov, K.T., Krawczak, M. (eds.) IWIFSGN/EUSFLAT -2017. AISC, vol. 641, pp. 354–367. Springer, Cham (2018). https://doi.org/10.1007/978-3-319-66830-7_32
12. Štěpnička, M., Jayaram, B.: Interpolativity of at-least and at-most models of monotone fuzzy rule bases with multiple antecedent variables. Fuzzy Sets Syst. **297**, 26–45 (2016)
13. Dubois, D., Prade, H.: Semantics of quotient operators in fuzzy relational databases. Fuzzy Sets Syst. **78**, 89–93 (1996)
14. Cao, N., Štěpnička, M., Burda, M., Dolný, A.: Excluding features in fuzzy relational compositions. Expert Syst. Appl. **81**, 1–11 (2017)
15. Cao, N., Štěpnička, M., Holčapek, M.: Extensions of fuzzy relational compositions based on generalized quantifer. Fuzzy Sets Syst. **339**, 73–98 (2018)
16. Běhounek, L., Novák, V.: Towards fuzzy partial logic. In: 2015 IEEE International Symposium on Multiple-Valued Logic (ISMVL), pp. 139–144. IEEE (2015)
17. Běhounek, L., Daňková, M.: Towards fuzzy partial set theory. In: Carvalho, J.P., Lesot, M.-J., Kaymak, U., Vieira, S., Bouchon-Meunier, B., Yager, R.R. (eds.) IPMU 2016. CCIS, vol. 611, pp. 482–494. Springer, Cham (2016). https://doi.org/10.1007/978-3-319-40581-0_39
18. Daňková, M.: Fuzzy relations and fuzzy functions in partial fuzzy set theory. In: Kacprzyk, J., Szmidt, E., Zadrożny, S., Atanassov, K.T., Krawczak, M. (eds.) IWIFSGN/EUSFLAT -2017. AISC, vol. 641, pp. 563–573. Springer, Cham (2018). https://doi.org/10.1007/978-3-319-66830-7_50
19. Cao, N., Štěpnička, M.: How to incorporate excluding features in fuzzy relational compositions and what for. In: Carvalho, J.P., Lesot, M.-J., Kaymak, U., Vieira, S., Bouchon-Meunier, B., Yager, R.R. (eds.) IPMU 2016. CCIS, vol. 611, pp. 470–481. Springer, Cham (2016). https://doi.org/10.1007/978-3-319-40581-0_38
20. Běhounek, L., Daňková, M.: Variable-domain fuzzy set theory–part II: apparatus. Fuzzy Sets and Systems (submitted)
21. Sobociński, B.: Axiomatization of a partial system of three-value calculus of propositions. J. Comput. Syst. **1**, 23–55 (1952)
22. Ciucci, D., Dubois, D.: A map of dependencies among three-valued logics. Inf. Sci. **250**, 162–177 (2013)

Quantification over Undefined Truth Values

Martina Daňková(✉)

Institute for Research and Applications of Fuzzy Modeling, University of Ostrava, NSC IT4Innovations, 30. dubna 22, 701 03 Ostrava, Czech Republic
martina.dankova@osu.cz

Abstract. We will recall three prominent families of quantifiers in first-order partial fuzzy logic and study their properties. The background fuzzy logic is the monoidal t-norm based logic MTL. First we will overview the semantics of partial fuzzy propositional logic, recall the basic notions, comment on axiomatization and present its first-order variant. Then we will present results on the properties of quantifiers from different families.

Keywords: Partial fuzzy logic · Undefinedness · Generalized quantifier

1 Introduction

Undefined values that are present in a common practice have usually different characters given by their sources, e.g., an argument that is out of a domain of definition such as square root or division by zero (coded as NaN in programing languages); a machine halt state of a recursive function which loops forever [8]; ill-posed questions in questionnaires; false presumptions [10] etc. Generally, they are a source of bugs and thus needed to be represented in order to be correctly handled [11].

It has been shown that partial fuzzy logic introduced in [4] can capture undefinedness of specific types, i.e., those that behave functionally. The important point to note here is that unknown or uncertain data cannot be handled functionally and thus, partial fuzzy logic is unable to cover undefinedness in a sense of unknown or uncertain. More on distinctions between those phenomena can be found in [2, page 17].

The advantage of using this logic lies in the fact that it combines graduality and undefinedness into the common framework. A simple system of partial fuzzy propositional logic is based on (any implicative expansion of) the well-known fuzzy logic $\mathrm{MTL}_\triangle$ of left-continuous t-norms [7]. The first-order variant of partial fuzzy logic has been proposed in [1]. There were introduced three basic families of quantifiers further used to develop partial fuzzy set theory [1,6], fuzzy type theory with partial functions [12] and to identify non-denoting terms [3].

Recently, it is of the main interest to know how and which properties of fuzzy logic (relativized) quantifiers can be transmitted to partial fuzzy logic

J. Medina et al. (Eds.): IPMU 2018, CCIS 855, pp. 199–208, 2018.
https://doi.org/10.1007/978-3-319-91479-4_17

(relativized) quantifiers from various families. The main aim of this paper is to show some basic properties of the prominent families of quantifiers in partial fuzzy logic.

2 Partial Fuzzy Propositional Logic

Let the logic L be any implicative [13] expansion of $\mathrm{MTL}_\triangle$, i.e., an expansion of $\mathrm{MTL}_\triangle$ such that $\varphi \leftrightarrow \psi \models_{\mathrm{L}} c(\chi_1, \ldots, \varphi, \ldots, \chi_n) \leftrightarrow c(\chi_1, \ldots, \psi, \ldots, \chi_n)$ each connective c. Since L is assumed to expand $\mathrm{MTL}_\triangle$, the language $\mathscr{S}$ of L contains at least the connectives $\wedge$, $\vee$, $\&$, $\rightarrow$, $\leftrightarrow$, 0, 1, and $\triangle$.

The partial fuzzy propositional logic L^* based on L is defined as follows [4]:

- The *language* (or signature) $\mathscr{S}^*$ of L^* extends the language $\mathscr{S}$ of L by the truth constant $*$ (representing the undefined truth degree of propositions), the unary connective ! (for the crisp modality "is defined"), and the binary connective $\wedge_{\mathrm{K}}$ (for Kleene-style min-conjunction).
- The intended L^*-*algebras* are defined by expanding the L-algebras by a dummy element $*$ (interpreting the truth constant $*$). In the intended L^*-algebra $\boldsymbol{L}_* = \boldsymbol{L} \cup \{*\}$ (where $\boldsymbol{L}$ is an L-algebra), the connectives of L^* are interpreted as described by the following truth tables, for all unary connectives $u \in \mathscr{S}$, binary connectives $c \in \mathscr{S}$ (and similarly for higher arities), $\alpha, \beta \in \boldsymbol{L}$ and $\gamma, \delta \in \boldsymbol{L} \setminus \{0\}$:

	!
α	1
$*$	0

	u_{B}
α	$u\alpha$
$*$	$*$

c_{B}	β	$*$
α	$\alpha\, c\, \beta$	$*$
$*$	$*$	$*$

$\wedge_{\mathrm{K}}$	0	δ	$*$
0	0	0	0
γ	0	$\gamma \wedge \delta$	$*$
$*$	0	$*$	$*$

(1)

- *Tautologies* of L^* are defined as formulae that are evaluated to 1 under all evaluations in all intended L^*-algebras (notation: $\models_{\mathrm{L}^*} \varphi$).
- An *axiomatic system* for L^* extends the (suitably modified) axioms and rules of L by 4 additional derivation rules and 10 additional axiom schemata. The general, linear, and (if enjoyed by L) standard *completeness theorems* (respectively w.r.t. L^*-algebras $\boldsymbol{L}_*$ over all, linear, or standard L-algebras $\boldsymbol{L}$) can be proved for this axiomatic system. However, since in this paper we only deal with the semantics of partial fuzzy logic, we leave the axiomatic system for L^* aside.

The connectives of $\mathscr{S}^*$ make a broad class of derived connectives available in L^*. This includes several useful families of connectives, analogous to those well-known from three-valued logic (see, e.g., [5]):

- The *Bochvar-style* connectives treat $*$ as the absorbing element. Recall that in L^*, the connectives of the original language $\mathscr{S}$ of the underlying fuzzy logic L are actually interpreted Bochvar-style: see the truth tables (1) above.

– The *Sobociński-style* connectives treat $*$ as the neutral element; and the Sobociński-style implication is $\to_S$ residuated with $\&_S$; for $c \in \{\wedge, \vee, \&\}$ we have

c_S	β	$*$
α	$\alpha\, c\, \beta$	α
$*$	β	$*$

$\to_S$	β	$*$
α	$\alpha \to \beta$	$\neg\alpha$
$*$	β	$*$

(2)

– The *Kleene-style* connectives keep the absorbing elements of the corresponding connectives of L and are evaluated Bochvar-style otherwise, i.e., for $c \in \{\wedge, \&\}$:

c_K	0	β	$*$
0	0	0	0
α	0	$\alpha\, c\, \beta$	$*$
$*$	0	$*$	$*$

$\vee_K$	δ	1	$*$
γ	$\gamma \vee \delta$	1	$*$
1	1	1	1
$*$	$*$	1	$*$

$\to_K$	δ	1	$*$
0	1	1	1
α	$\alpha \to \delta$	1	$*$
$*$	$*$	1	$*$

(3)

Moreover, the following useful unary connectives are L*-definable:

x	$?x$	$\downarrow x$	$\uparrow x$	$\circledast x$
γ	0	γ	γ	0
1	0	1	1	$*$
$*$	1	0	1	0

(4)

The connective ? expresses the modality of being defined; $\downarrow$ and $\uparrow$ shift $*$ to 0 and 1, respectively; and $\circledast$ outputs $*$ provided that the input is evaluated to 1 and it is intended for designing definable connectives (see (7) below).

The following binary connectives are L*-definable:

$\to_*$	β	$*$
α	$\alpha \to \beta$	0
$*$	1	1

$\to^*$	β	$*$
α	$\alpha \to \beta$	1
$*$	0	1

(5)

$\sim$	β	$*$
α	$\alpha \leftrightarrow \beta$	0
$*$	0	1

$\precsim$	β	$*$
α	$\alpha \leftrightarrow \beta$	0
$*$	1	1

(6)

for $\alpha, \beta \neq *$ and $\gamma \notin \{1, *\}$.

Moreover, we define the following crisp ($\{0,1\}$-valued) binary connectives:

$$x \leq_* y =_{\mathrm{df}} \triangle(x \to_* y) \qquad x \equiv y =_{\mathrm{df}} \triangle(x \sim y)$$
$$x \leq^* y =_{\mathrm{df}} \triangle(x \to^* y) \qquad x \sqsubseteq y =_{\mathrm{df}} \triangle(x \precsim y)$$

The connective $\equiv$ expresses the identity of truth values; $\sqsubseteq$ the 'information' order of truth values; $\leq_*$ the matrix order w.r.t. which L* is implicative; and $\leq^*$ the dual matrix order; and connectives defined by tables (5) and (6) are their respective graded variants.

Remark 1. Observe that $\sim$ and $\underset{\sim}{\sqsubset}$ are definable from $\rightarrow_*$ and $\rightarrow^*$ as follows:

$$x \sim y =_{\mathrm{df}} (x \rightarrow_* y) \wedge (y \rightarrow^* x)$$
$$x \underset{\sim}{\sqsubset} y =_{\mathrm{df}} (x \rightarrow_* y) \wedge (y \rightarrow_* x)$$

And analogously, $\equiv$ and $\sqsubseteq$ are definable from $\leq_*$ and $\leq^*$.

Remark 2. It follows from properties of $\triangle$ that for an arbitrary formulas φ, ψ:

$$\models_{\mathrm{L}^*} \varphi \rightarrow_* \psi \quad \text{if and only if} \quad \models_{\mathrm{L}^*} \varphi \leq_* \psi$$
$$\models_{\mathrm{L}^*} \varphi \rightarrow^* \psi \quad \text{if and only if} \quad \models_{\mathrm{L}^*} \varphi \leq^* \psi$$
$$\models_{\mathrm{L}^*} \varphi \sim \psi \quad \text{if and only if} \quad \models_{\mathrm{L}^*} \varphi \equiv \psi$$
$$\models_{\mathrm{L}^*} \varphi \underset{\sim}{\sqsubset} \psi \quad \text{if and only if} \quad \models_{\mathrm{L}^*} \varphi \sqsubseteq \psi$$

Due to this fact, it is a matter of taste which connective is used in formulas of the above listed forms. We choose $\{\leq_*, \leq^*, \equiv, \sqsubseteq\}$ because of their close connection to style of proving tautologies in models of L* where we check validity of sequence of inequalities or identities.

By means of unary connectives from table (4), we can define e.g. Kleene-style connectives $c \in \{\wedge, \&\}$:

$$x \, c_{\mathrm{K}} \, y =_{\mathrm{df}} \downarrow(x \, c_{\mathrm{B}} \, y) \vee_{\mathrm{B}} \circledast((?x \wedge (y \not\equiv 0)) \vee (?y \wedge (x \not\equiv 0))) \tag{7}$$

where $x \not\equiv y =_{\mathrm{df}} \neg(x \equiv y)$.

Let us give a few examples of logical laws valid in L*. Obviously,

$$\models_{\mathrm{L}^*} (\varphi \, c_{\mathrm{B}} \, \psi) \sqsubseteq (\varphi \, c_{\mathrm{K}} \, \psi)$$
$$\models_{\mathrm{L}^*} (\varphi \, c_{\mathrm{K}} \, \psi) \sqsubseteq (\varphi \, c_{\mathrm{S}} \, \psi)$$

for arbitrary $c \in \{\&, \wedge, \vee, \rightarrow\}$. And the distributive laws in L* work as follows: if the connectives (c, c') are distributive in the logic L, then $(c_{\mathrm{B}}, c'_{\mathrm{B}})$ and $(c_{\mathrm{B}}, c'_{\mathrm{S}})$ are distributive[1] in L*; however, this is not so for $(c_{\mathrm{B}}, c'_{\mathrm{K}})$. Indeed, non of $\{(c_{\mathrm{B}}, c'_{\mathrm{K}})\}$ is distributive, e.g., $\not\models_{\mathrm{L}^*} \varphi \,\&_{\mathrm{B}} (\psi \vee_{\mathrm{K}} \chi) \equiv (\varphi \,\&_{\mathrm{B}} \psi) \vee_{\mathrm{K}} (\varphi \,\&_{\mathrm{B}} \chi)$, but it can be proved that

$$?\varphi \vee [((\psi \not\equiv 1) \vee !\chi) \wedge ((\chi \not\equiv 1) \vee !\psi)] \models_{\mathrm{L}^*} [\varphi \,\&_{\mathrm{B}} (\psi \vee_{\mathrm{K}} \chi)] \equiv [(\varphi \,\&_{\mathrm{B}} \psi) \vee_{\mathrm{K}} (\varphi \,\&_{\mathrm{B}} \chi)]$$

3 Quantifiers in Partial Fuzzy First-Order Logic

As a natural next step, a semantical approach to the first-order variant L∀* of partial fuzzy propositional logic L* has been given in [1], where the Bochvar,

[1] A complete characterization of distributive pairs of Sobociński and Bochvar connectives has been provided in [4, Theorem 3.5].

Sobociński and Kleene quantifiers were defined. In this paper, we will recall these $\mathrm{L}\forall^*$-extensions of the quantifiers of $\mathrm{L}\forall$ and list some of their properties.

The models for $\mathrm{L}\forall^*$ are the same as those for $\mathrm{L}\forall$, only using L^*-algebras $\boldsymbol{L}_*$ instead of L-algebras $\boldsymbol{L}$. The primitive quantifiers $\forall_{\mathrm{B}}$ and $\exists_{\mathrm{B}}$ of $\mathrm{L}\forall^*$ are interpreted Bochvar-style, yielding $*$ whenever there is an undefined instance of the quantified formula:

$$\|(\forall_{\mathrm{B}} x)\varphi\|^{\mathbf{M}}_{e} =_{\mathrm{df}} \begin{cases} * & \text{if } \|\varphi\|^{\mathbf{M}}_{e[x\mapsto a]} = * \text{ for some } a \in D_{\mathbf{M}} \\ \inf_{a\in D_{\mathbf{M}}} \|\varphi\|^{\mathbf{M}}_{e[x\mapsto a]} & \text{otherwise} \end{cases}$$

$$\|(\exists_{\mathrm{B}} x)\varphi\|^{\mathbf{M}}_{e} =_{\mathrm{df}} \begin{cases} * & \text{if } \|\varphi\|^{\mathbf{M}}_{e[x\mapsto a]} = * \text{ for some } a \in D_{\mathbf{M}} \\ \sup_{a\in D_{\mathbf{M}}} \|\varphi\|^{\mathbf{M}}_{e[x\mapsto a]} & \text{otherwise} \end{cases}$$

Similarly, the Sobociński quantifiers $\forall_{\mathrm{S}}$ and $\exists_{\mathrm{S}}$ (which, like $\wedge_{\mathrm{S}}$ and $\vee_{\mathrm{S}}$, ignore the undefined instances) are introduced by the following Tarski conditions:

$$\|(\forall_{\mathrm{S}} x)\varphi\|^{\mathbf{M}}_{e} =_{\mathrm{df}} \begin{cases} * & \text{if } \|\varphi\|^{\mathbf{M}}_{e[x\mapsto a]} = * \text{ for all } a \in D_{\mathbf{M}} \\ \inf_{a\in D_{\mathbf{M}}} \|{\uparrow}\varphi\|^{\mathbf{M}}_{e[x\mapsto a]} & \text{otherwise.} \end{cases}$$

$$\|(\exists_{\mathrm{S}} x)\varphi\|^{\mathbf{M}}_{e} =_{\mathrm{df}} \begin{cases} * & \text{if } \|\varphi\|^{\mathbf{M}}_{e[x\mapsto a]} = * \text{ for all } a \in D_{\mathbf{M}} \\ \sup_{a\in D_{\mathbf{M}}} \|{\downarrow}\varphi\|^{\mathbf{M}}_{e[x\mapsto a]} & \text{otherwise.} \end{cases}$$

Sobociński- and Kleene-style quantifiers can be defined by means of $\forall_{\mathrm{B}}, \exists_{\mathrm{B}}$ and the connectives of L^* as follows:

$$(\forall_{\mathrm{S}} x)\varphi \equiv_{\mathrm{df}} (\forall_{\mathrm{B}} x){\uparrow}\varphi \vee_{\mathrm{B}} \circledast(\forall_{\mathrm{B}} x)?\varphi \qquad (\forall_{\mathrm{K}} x)\varphi \equiv_{\mathrm{df}} (\forall_{\mathrm{B}} x)\varphi \wedge_{\mathrm{K}} (\forall_{\mathrm{S}} x)\varphi$$
$$(\exists_{\mathrm{S}} x)\varphi \equiv_{\mathrm{df}} (\exists_{\mathrm{B}} x){\downarrow}\varphi \vee_{\mathrm{B}} \circledast(\forall_{\mathrm{B}} x)?\varphi \qquad (\exists_{\mathrm{K}} x)\varphi \equiv_{\mathrm{df}} (\exists_{\mathrm{B}} x)\varphi \vee_{\mathrm{K}} (\exists_{\mathrm{S}} x)\varphi\,.$$

Convention 1. *For simplicity of notation, we write* $\varphi_1 \leq_* \varphi_2 \leq_* \varphi_3 \leq_* \dots \leq_* \varphi_{n-1} \leq_* \varphi_n$ *instead of the sequence of tautologies of* $\mathrm{L}\forall^*$ $\models_{\mathrm{L}\forall^*} \varphi_1 \leq_* \varphi_2$, $\models_{\mathrm{L}\forall^*} \varphi_2 \leq_* \varphi_3, \dots, \models_{\mathrm{L}\forall^*} \varphi_{n-1} \leq_* \varphi_n$.

Analogously, we shorten sequences with $\equiv, \leq^*$, *or combinations all three relations.*

Let us give some examples of tautologies of $\mathrm{L}\forall^*$. Assume that χ does not contain x freely. Then:

$$(\forall_{\mathrm{B}} x)\varphi \mathbin{\&_{\mathrm{B}}} (\forall_{\mathrm{B}} x)\psi \leq_* (\forall_{\mathrm{B}} x)(\varphi \mathbin{\&_{\mathrm{B}}} \psi) \leq_* (\forall_{\mathrm{B}} x)(\varphi \mathbin{\&_{\mathrm{K}}} \psi) \leq_* (\forall_{\mathrm{B}} x)(\varphi \mathbin{\&_{\mathrm{S}}} \psi)\,,$$
$$(\forall_{\mathrm{S}} x)\varphi \mathbin{\&_{\mathrm{B}}} (\forall_{\mathrm{S}} x)\psi \leq_* (\forall_{\mathrm{S}} x)(\varphi \mathbin{\&_{\mathrm{B}}} \psi) \leq_* (\forall_{\mathrm{S}} x)(\varphi \mathbin{\&_{\mathrm{K}}} \psi) \leq_* (\forall_{\mathrm{S}} x)(\varphi \mathbin{\&_{\mathrm{S}}} \psi)\,,$$

$(\forall_{\mathrm{S}} x)\varphi \mathbin{\&_{\mathrm{K}}} (\forall_{\mathrm{S}} x)\psi \leq_* (\forall_{\mathrm{S}} x)(\varphi \mathbin{\&_{\mathrm{K}}} \psi)$ and $(\forall_{\mathrm{S}} x)\varphi \mathbin{\&_{\mathrm{S}}} (\forall_{\mathrm{S}} x)\psi \leq_* (\forall_{\mathrm{S}} x)(\varphi \mathbin{\&_{\mathrm{S}}} \psi)$ are tautologies of $\mathrm{L}\forall^*$.

Since the focus of this paper is on the semantics, we leave the axiomatization of $\mathrm{L}\forall^*$ aside. Let us just hint that $\mathrm{L}\forall^*$ turns out to be implicative (in the sense of Rasiowa) w.r.t. the connective $\leq_*$ of (5) above. Therefore it can be axiomatized straightforwardly by adding Rasiowa's axioms for quantifiers [13] to the axiomatic system for L^* (and, optionally, Hájek's axiom (∀3) of [9] to ensure completeness w.r.t. safe models over *linear* intended L^*-algebras). This axiomatizes the quantifiers $\forall_B$ and $\exists_S$, which correspond, respectively, to inf and sup w.r.t. the order generated by $\leq_*$ (the primitive Bochvar-style quantifier $\exists_B$ is definable from $\forall_B, \exists_S$ and the connectives of L^*) and dually, $\forall_S, \exists_B$ are inf, sup, respectively, w.r.t. $\leq^*$.

Due to the above arguments, we obtain that particular tautologies of $L\forall$ are tautologies of $L\forall^*$ in the following manner: the universal quantifier is taken as Bochvar-style; the existential quantifier as Sobociński-style; the principal implication as $\leq_*$; and the principal equivalence $\leftrightarrow$ as $\equiv$.

Proposition 1. *Assume that χ does not contain x freely, and x' is a variable not occurring in φ. Then:*

$$\models_{L\forall^*} \chi \equiv (\forall_B x)\chi \tag{8}$$
$$\models_{L\forall^*} (\exists_S x)\chi \equiv \chi \tag{9}$$
$$\models_{L\forall^*} (\exists_S x)\varphi(x) \equiv (\exists_S x')\varphi(x') \tag{10}$$
$$\models_{L\forall^*} (\forall_B x)(\forall_B y)\varphi \equiv (\forall_B y)(\forall_B x)\varphi \tag{11}$$
$$\models_{L\forall^*} (\exists_S x)(\exists_S y)\varphi \equiv (\exists_S y)(\exists_S x) \tag{12}$$
$$\models_{L\forall^*} (\exists_S x)(\forall_B y)\varphi \leq_* (\forall_B y)(\exists_S x)\varphi\,. \tag{13}$$

and dually

$$\models_{L\forall^*} \chi \equiv (\forall_S x)\chi \tag{14}$$
$$\models_{L\forall^*} (\exists_B x)\chi \equiv \chi \tag{15}$$
$$\models_{L\forall^*} (\exists_B x)\varphi(x) \equiv (\exists_B x')\varphi(x') \tag{16}$$
$$\models_{L\forall^*} (\forall_S x)(\forall_S y)\varphi \equiv (\forall_S y)(\forall_S x)\varphi \tag{17}$$
$$\models_{L\forall^*} (\exists_B x)(\exists_B y)\varphi \equiv (\exists_B y)(\exists_B x)\varphi \tag{18}$$
$$\models_{L\forall^*} (\exists_B x)(\forall_S y)\varphi \leq^* (\forall_S y)(\exists_B x)\varphi\,. \tag{19}$$

Formulas (8)–(12) work for Kleene-style universal and existential quantifiers as well.

Taking a closer look to the specification axiom and its dual of $L\forall$, we obtain the following results in $L\forall^*$:

$$\begin{array}{rlrl} & \models_{L\forall^*} (\forall_B x)\varphi \leq_* \varphi(t) & & \models_{L\forall^*} f(t) \leq^* (\exists_B x)\varphi \\ & \models_{L\forall^*} (\forall_S x)\varphi \leq^* \varphi(t) & & \models_{L\forall^*} f(t) \leq_* (\exists_S x)\varphi \\ !(\forall_K x)\varphi & \models_{L\forall^*} (\forall_K x)\varphi \leq^* \varphi(t) & !\varphi(t) & \models_{L\forall^*} \varphi(t) \leq_* (\exists_K x)\varphi \\ !(\exists_K x)\varphi & \models_{L\forall^*} (\forall_K x)\varphi \leq_* \varphi(t) & !\varphi(t) & \models_{L\forall^*} \varphi(t) \leq^* (\exists_K x)\varphi. \end{array}$$

If we would like to have formulas with an implication of L^* instead of the above inequalities, e.g. $(\forall_K x)\varphi \rightarrow_K \varphi(t)$, then such formula is not a tautology of $L\forall^*$. But we can fix this problem by excluding cases in which the considered formula does not evaluate to 1:

$$\{!(\forall_K x)\varphi \vee (\varphi(t) \equiv 1), ((\forall_K x)\varphi \equiv 0) \vee !\varphi(t)\} \models_{L\forall^*} (\forall_K x)\varphi \rightarrow_K \varphi(t).$$

It means that $(\forall_{\mathrm{K}} x)\varphi \to_{\mathrm{K}} \varphi(t)$ is true provided that ($\|(\forall_{\mathrm{K}} x)\varphi\|$ is defined or $\|\varphi(t)\| = 1$) and ($\|(\forall_{\mathrm{K}} x)\varphi\| = 0$ or $\|\varphi(t)\|$ is defined).

Convention 2. *If* $\models_{\mathrm{L}\forall^*} \varphi \leq^* \psi$ *and* $\models_{\mathrm{L}\forall^*} \varphi \leq_* \psi$ *then we write* $\models_{\mathrm{L}\forall^*} \varphi \leq_*^* \psi$.

Let us finish with some examples of provable formulas with various quantifiers and Bochvar-style implication. The subsequent semantical proofs provide a general manual for proving first order formulas in $\mathrm{L}\forall^*$.

Proposition 2

$$\models_{\mathrm{L}\forall^*} (\forall_{\mathrm{B}} x)(\varphi \to_{\mathrm{B}} \psi) \leq_*^* (\forall_{\mathrm{B}} x)\varphi \to_{\mathrm{B}} (\forall_{\mathrm{B}} x)\psi \tag{20}$$

$$!\varphi \models_{\mathrm{L}\forall^*} (\forall_{\mathrm{S}} x)(\varphi \to_{\mathrm{B}} \psi) \leq_*^* (\forall_{\mathrm{S}} x)\varphi \to_{\mathrm{B}} (\forall_{\mathrm{S}} x)\psi \tag{21}$$

$$\psi \not\equiv 0 \models_{\mathrm{L}\forall^*} (\forall_{\mathrm{K}} x)(\varphi \to_{\mathrm{B}} \psi) \leq_* (\forall_{\mathrm{K}} x)\varphi \to_{\mathrm{B}} (\forall_{\mathrm{K}} x)\psi \tag{22}$$

Proof (20). In this case we have to check a validity w.r.t. the both orderings $\leq_*$ and $\leq^*$. If φ and ψ are defined everywhere then both inequalities are trivially valid. Let us consider that $\|\varphi\|^{\mathbf{M}}_{e[x\mapsto a]} = *$ for some $a \in D_{\mathbf{M}}$ then $\|(\forall_{\mathrm{B}} x)\varphi\|^{\mathbf{M}} = *$ and also $\|\varphi \to_{\mathrm{B}} \psi\|^{\mathbf{M}}_{e[x\mapsto a]} = *$ which implies $\|(\forall_{\mathrm{B}} x)(\varphi \to_{\mathrm{B}} \psi)\|^{\mathbf{M}} = *$. Hence, both sides of inequalities are $*$ at the same time and the validity of the inequalities follows.

(21) Let us denote $\mathrm{Dom}_x(\psi) = \{a \in D_{\mathbf{M}} \mid \|\psi\|^{\mathbf{M}}_{e[x\mapsto a]} \neq *\}$. The conditional requirement $\|!\varphi\|^{\mathbf{M}} = 1$ implies $\mathrm{Dom}_x(\psi) \subseteq \mathrm{Dom}_x(\varphi) = D_{\mathbf{M}}$ and $\mathrm{Dom}_x(\psi) \cap \mathrm{Dom}_x(\varphi) = \mathrm{Dom}_x(\psi)$. Let $\mathrm{Dom}_x(\psi) \neq \emptyset$ (ψ is not undefined everywhere) then we have

$$\begin{aligned}
\|(\forall_{\mathrm{S}} x)(\varphi \to_{\mathrm{B}} \psi)\|^{\mathbf{M}} &= \inf_{a\in D_{\mathbf{M}}} \|{\uparrow}(\varphi \to_{\mathrm{B}} \psi)\|^{\mathbf{M}} = \inf_{a\in \mathrm{Dom}_x(\psi)} \|\varphi \to_{\mathrm{B}} \psi\|^{\mathbf{M}} \\
&\leq \inf_{a\in \mathrm{Dom}_x(\psi)} \|\varphi\|^{\mathbf{M}} \to \inf_{a\in \mathrm{Dom}_x(\psi)} \|\psi\|^{\mathbf{M}} \leq \inf_{a\in D_{\mathbf{M}}} \|\varphi\|^{\mathbf{M}} \to \inf_{a\in \mathrm{Dom}_x(\psi)} \|\psi\|^{\mathbf{M}} \\
&= \inf_{a\in D_{\mathbf{M}}} \|\varphi\|^{\mathbf{M}} \to \inf_{a\in D_{\mathbf{M}}} \|{\uparrow}\psi\|^{\mathbf{M}} = \|(\forall_{\mathrm{S}} x)\varphi\|^{\mathbf{M}} \to \|(\forall_{\mathrm{S}} x)\psi\|^{\mathbf{M}}
\end{aligned}$$

Otherwise, $\|(\forall_{\mathrm{S}} x)(\varphi \to_{\mathrm{B}} \psi)\|^{\mathbf{M}} = *$ and

$$\|(\forall_{\mathrm{S}} x)\varphi\|^{\mathbf{M}} \to_{\mathrm{B}} \|(\forall_{\mathrm{S}} x)\psi\|^{\mathbf{M}} = \|(\forall_{\mathrm{S}} x)\varphi\|^{\mathbf{M}} \to_{\mathrm{B}} * = *$$

(22) $\psi \not\equiv 0$ implies $(\forall_{\mathrm{K}} x)\psi \equiv (\forall_{\mathrm{B}} x)\psi \not\equiv 0$ and also $(\forall_{\mathrm{K}} x)(\varphi \to_{\mathrm{B}} \psi) \equiv (\forall_{\mathrm{B}} x)(\varphi \to_{\mathrm{B}} \psi) \not\equiv 0$. Due to (20), we have that

$$\begin{aligned}
\underbrace{(\forall_{\mathrm{K}} x)(\varphi \to_{\mathrm{B}} \psi)}_{L} &\leq_*^* (\forall_{\mathrm{B}} x)\varphi \to_{\mathrm{B}} (\forall_{\mathrm{B}} x)\psi \equiv (\forall_{\mathrm{B}} x)\varphi \to_{\mathrm{B}} (\forall_{\mathrm{K}} x)\psi \\
&\leq_* ((\forall_{\mathrm{B}} x)\varphi \wedge_{\mathrm{K}} (\forall_{\mathrm{S}} x)\varphi) \to_{\mathrm{B}} (\forall_{\mathrm{K}} x)\psi \equiv \underbrace{(\forall_{\mathrm{K}} x)\varphi \to_{\mathrm{B}} (\forall_{\mathrm{K}} x)\psi}_{R}.
\end{aligned}$$

We have to check two cases to verify the above inequality:

1. If $\|(\forall_{\mathrm{K}} x)\varphi\|^{\mathbf{M}} \neq 0$ then Kleene-style quantifiers in the formula reduce to Bochvar-style one and the inequality follows from (20).
2. If $\|(\forall_{\mathrm{K}} x)\varphi\|^{\mathbf{M}} = 0$ then either $R \equiv *$ or $R \equiv 1$. In the first case, it follows that there exists some $a \in D_{\mathbf{M}}$ such that $\|(\forall_{\mathrm{K}} x)\psi\|^{\mathbf{M}} = *$ and consequently, $L \equiv *$. In the second case, it follows that ψ is defined everywhere this is not so for φ. Hence, it may happen that $L \equiv *$ and therefore only $\leq_*$ is applicable in this case.

Proposition 3

$$\models_{\mathrm{L}\forall^*} (\forall_{\mathrm{B}} x)(\varphi \to_{\mathrm{S}} \psi) \leq^* (\forall_{\mathrm{B}} x)\varphi \to_{\mathrm{S}} (\forall_{\mathrm{B}} x)\psi \tag{23}$$

$$\models_{\mathrm{L}\forall^*} (\forall_{\mathrm{S}} x)(\varphi \to_{\mathrm{S}} \psi) \leq^*_* (\forall_{\mathrm{S}} x)\varphi \to_{\mathrm{S}} (\forall_{\mathrm{S}} x)\psi \tag{24}$$

$$(\exists_{\mathrm{S}} x)\neg\varphi \models_{\mathrm{L}\forall^*} (\forall_{\mathrm{K}} x)(\varphi \to_{\mathrm{S}} \psi) \leq_* (\forall_{\mathrm{K}} x)\varphi \to_{\mathrm{S}} (\forall_{\mathrm{K}} x)\psi \tag{25}$$

Proof. Let us denote $\mathrm{Dom}_x(\varphi) = \{a \in D_{\mathbf{M}} \mid \|\varphi\|^{\mathbf{M}}_{e[x\mapsto a]} \neq *\}$ and analogously for $\mathrm{Dom}_x(\psi)$. Moreover, let $U = \mathrm{Dom}_x(\varphi) \cup \mathrm{Dom}_x(\psi)$.

(23) If $\|(\forall_{\mathrm{B}} x)(\varphi \to_{\mathrm{S}} \psi)\|^{\mathbf{M}}$ is undefined then both $(\forall_{\mathrm{B}} x)\varphi, (\forall_{\mathrm{B}} x)\psi$ evaluate to $*$ and the inequality is fulfilled. Consider $\|(\forall_{\mathrm{B}} x)(\varphi \to_{\mathrm{S}} \psi)\|^{\mathbf{M}} \neq *$ then

$$\begin{aligned}
\|(\forall_{\mathrm{B}} x)(\varphi \to_{\mathrm{S}} \psi)\|^{\mathbf{M}} &= \inf_{a\in D_{\mathbf{M}}} \|{\uparrow}\varphi \to_{\mathrm{S}} {\downarrow}\psi\|^{\mathbf{M}} \\
&\leq \inf_{a\in D_{\mathbf{M}}} \|{\uparrow}\varphi\| \to_{\mathrm{S}} \inf_{a\in D_{\mathbf{M}}} \|{\downarrow}\psi\|^{\mathbf{M}} \leq^* \|(\forall_{\mathrm{B}} x)\varphi\| \to_{\mathrm{S}} \|(\forall_{\mathrm{B}} x)\psi\|^{\mathbf{M}}
\end{aligned}$$

In case $\|(\forall_{\mathrm{B}} x)\varphi\| = *$ and defined value of $\|(\forall_{\mathrm{B}} x)\psi\|$, we have that

$$\begin{aligned}
\|(\forall_{\mathrm{B}} x)\varphi\| \to_{\mathrm{S}} \|(\forall_{\mathrm{B}} x)\psi\|^{\mathbf{M}} &= * \to_{\mathrm{S}} \|(\forall_{\mathrm{B}} x)\psi\|^{\mathbf{M}} \\
&= 1 \to_{\mathrm{S}} \|(\forall_{\mathrm{B}} x)\psi\|^{\mathbf{M}} = \inf_{a\in D_{\mathbf{M}}} \|{\uparrow}\varphi\| \to_{\mathrm{S}} \inf_{a\in D_{\mathbf{M}}} \|{\downarrow}\psi\|^{\mathbf{M}}
\end{aligned}$$

Analogously, we proceed for $\|(\forall_{\mathrm{B}} x)\psi\| = *$ and defined value of $\|(\forall_{\mathrm{B}} x)\varphi\|$. The case when both $(\forall_{\mathrm{B}} x)\varphi, (\forall_{\mathrm{B}} x)\psi$ evaluate to $*$ is trivial.

(24) As in the above case, if $\|(\forall_{\mathrm{S}} x)(\varphi \to_{\mathrm{S}} \psi)\|^{\mathbf{M}}$ is undefined then both $(\forall_{\mathrm{S}} x)\varphi, (\forall_{\mathrm{S}} x)\psi$ evaluate to $*$ because $\mathrm{Dom}_x(\varphi) = \mathrm{Dom}_x(\psi) = \emptyset$ and both inequalities ar fulfilled. Consider $\|(\forall_{\mathrm{S}} x)(\varphi \to_{\mathrm{S}} \psi)\|^{\mathbf{M}} \neq *$ and $\mathrm{Dom}_x(\varphi), \mathrm{Dom}_x(\psi)$ are non-empty then

$$\begin{aligned}
\|(\forall_{\mathrm{S}} x)(\varphi \to_{\mathrm{S}} \psi)\|^{\mathbf{M}} &= \inf_{a\in U} \|{\uparrow}\varphi \to_{\mathrm{S}} {\downarrow}\psi\|^{\mathbf{M}} \leq \inf_{a\in U} \|{\uparrow}\varphi\| \to_{\mathrm{S}} \inf_{a\in U} \|{\downarrow}\psi\|^{\mathbf{M}} \\
&= \inf_{a\in \mathrm{Dom}_x(\varphi)} \|\varphi\| \to_{\mathrm{S}} \inf_{a\in U} \|{\downarrow}\psi\|^{\mathbf{M}} \leq \inf_{a\in \mathrm{Dom}_x(\varphi)} \|\varphi\| \to_{\mathrm{S}} \inf_{a\in \mathrm{Dom}_x(\psi)} \|\psi\|^{\mathbf{M}} \\
&= \|(\forall_{\mathrm{S}} x)\varphi\| \to_{\mathrm{S}} \|(\forall_{\mathrm{S}} x)\psi\|^{\mathbf{M}}
\end{aligned}$$

In the case $\|(\forall_{\mathrm{S}} x)\varphi\| = *$ and defined value of $\|(\forall_{\mathrm{S}} x)\psi\|$, we have that

$$\begin{aligned}
\|(\forall_{\mathrm{S}} x)(\varphi \to_{\mathrm{S}} \psi)\|^{\mathbf{M}} &= \inf_{a\in \mathrm{Dom}_x(\psi)} \|{\uparrow}\varphi \to_{\mathrm{S}} {\downarrow}\psi\|^{\mathbf{M}} \\
&\leq \inf_{a\in \mathrm{Dom}_x(\psi)} \|{\uparrow}\varphi\| \to_{\mathrm{S}} \inf_{a\in \mathrm{Dom}_x(\psi)} \|{\downarrow}\psi\|^{\mathbf{M}} = 1 \to_{\mathrm{S}} \inf_{a\in \mathrm{Dom}_x(\psi)} \|{\downarrow}\psi\|^{\mathbf{M}} \\
&= \|(\forall_{\mathrm{S}} x)\varphi\| \to_{\mathrm{S}} \|(\forall_{\mathrm{S}} x)\psi\|^{\mathbf{M}}
\end{aligned}$$

And analogously, we proceed for $\|(\forall_S x)\psi\| = *$ and defined value of $\|(\forall_S x)\varphi\|$.

(25) The conditional $\|(\exists_S x)\neg\varphi\| = 1$ implies

$$\mathrm{Dom}_x(\varphi) \neq \emptyset \qquad \text{and} \qquad \inf_{x \in \mathrm{Dom}_x(\varphi)} \|\varphi\| = 0.$$

Hence, $\|(\forall_K x)\varphi \rightarrow_S (\forall_K x)\psi\| = 1$ and $v \leq_* 1$ for an arbitrary $v \in \boldsymbol{L}_*$.

4 Conclusions

In this paper, some selected properties of fuzzy logic (relativized) quantifiers has been studied in partial fuzzy logic for the three prominent families of quantifiers. On this selection we explain a way of designing provable formulas so that their sets of requirements are as small as possible. This is achieved by choosing $\{0, 1\}$-valued implications $\leq_*$ and $\leq^*$ (represent orderings on a L^*-algebra) as principal implications. It was shown that formulas in a form of implication from any family of extended connectives do not lead directly to tautologies of L^* and there are tautological consequences of quite complex sets of formulas which indeed ensure only that the respective formula does not evaluate to $*$.

A detailed investigation is left for a future work and it will serve as a prerequisite for further research in fields mentioned in the introduction.

Acknowledgments. The work was supported by grant No. 16–19170S 'Fuzzy partial logic' of the Czech Science Foundation and the programme NPU II, project LQ1602 'IT4I XS' of the MŠMT ČR.

References

1. Běhounek, L., Daňková, M.: Towards fuzzy partial set theory. In: Carvalho, J.P., Lesot, M.-J., Kaymak, U., Vieira, S., Bouchon-Meunier, B., Yager, R.R. (eds.) IPMU 2016. CCIS, vol. 611, pp. 482–494. Springer, Cham (2016). https://doi.org/10.1007/978-3-319-40581-0_39
2. Běhounek, L., Daňková, M.: Variable-domain fuzzy sets - part 1. Fuzzy Sets Syst. (Submitted 2017)
3. Běhounek, L., Dvořák, A.: Non-denoting terms in fuzzy logic: an initial exploration. In: Kacprzyk, J., Szmidt, E., Zadrożny, S., Atanassov, K.T., Krawczak, M. (eds.) IWIFSGN/EUSFLAT -2017. AISC, vol. 641, pp. 148–158. Springer, Cham (2018). https://doi.org/10.1007/978-3-319-66830-7_14
4. Běhounek, L., Novák, V.: Towards fuzzy partial logic. In: Proceedings of the IEEE 45th International Symposium on Multiple-Valued Logics (ISMVL 2015), pp. 139–144 (2015)
5. Ciucci, D., Dubois, D.: A map of dependencies among three-valued logics. Inf. Sci. **250**, 162–177 (2013)
6. Daňková, M.: Fuzzy relations and fuzzy functions in partial fuzzy set theory. In: Kacprzyk, J., Szmidt, E., Zadrożny, S., Atanassov, K.T., Krawczak, M. (eds.) IWIFSGN/EUSFLAT -2017. AISC, vol. 641, pp. 563–573. Springer, Cham (2018). https://doi.org/10.1007/978-3-319-66830-7_50

7. Esteva, F., Godo, L.: Monoidal t-norm based logic: towards a logic for left-continuous t-norms. Fuzzy Sets Syst. **124**(3), 271–288 (2001)
8. Falai, C., Dongming, W.: Geometric Computation. Lecture Notes Series On Computing, World Scientific Publishing Company (2004). https://books.google.cz/books?id=SgfJCgAAQBAJ
9. Hájek, P.: Metamathematics of Fuzzy Logic. Kluwer, Dordrecht (1998)
10. Little, W.: Applied Logic. University college. University of Florida, Houghton Mifflin (1955). https://books.google.cz/books?id=9ThSaurNmsEC
11. Marlet, R.: Program Specialization. ISTE, Wiley (2013). https://books.google.cz/books?id=XTlio_HSJ0cC
12. Novák, V.: Towards fuzzy type theory with partial functions. In: Kacprzyk, J., Szmidt, E., Zadrożny, S., Atanassov, K.T., Krawczak, M. (eds.) IWIFSGN/EUSFLAT -2017. AISC, vol. 643, pp. 25–37. Springer, Cham (2018). https://doi.org/10.1007/978-3-319-66827-7_3
13. Rasiowa, H.: An Algebraic Approach to Non-Classical Logics. North-Holland, Amsterdam (1974)

Metaheuristics and Machine Learning

FS4RV$_{DD}$: A Feature Selection Algorithm for Random Variables with Discrete Distribution

Fiorella Cravero[1], Santiago Schustik[1], María Jimena Martínez[2], Mónica Fátima Díaz[1,3], and Ignacio Ponzoni[2,4(✉)]

[1] Planta Piloto de Ingeniería Química – PLAPIQUI (Universidad Nacional del Sur-CONICET), Bahía Blanca, Argentina

[2] Instituto de Ciencias e Ingeniería de LA Computación (UNS-CONICET), Bahía Blanca, Argentina
ip@cs.uns.edu.ar

[3] Departamento de Ingeniería Química, Universidad Nacional del Sur (UNS), Bahía Blanca, Argentina

[4] Departamento de Ciencias e Ingeniería de la Computación, Universidad Nacional del Sur, Bahía Blanca, Argentina

Abstract. Feature Selection is a crucial step for inferring regression and classification models in QSPR (Quantitative Structure–Property Relationship) applied to Cheminformatics. A particularly complex case of QSPR modelling occurs in Polymer Informatics because the features under analysis require the management of uncertainty. In this paper, a novel feature selection method for addressing this special QSPR scenario is presented. The proposed methodology assumes that each feature is characterized by a probabilistic distribution of values associated with the polydispersity of the polymers included in the training dataset. This new algorithm has two sequential steps: ranking of the features, generated by correlation analysis, and iterative subset reduction, obtained by feature redundancy analysis. A prototype of the algorithm has been implemented in order to conduct a proof of concept. The method performance has been evaluated by using synthetic datasets of different sizes and varying the cardinality of the feature selected sub-sets. These preliminary results allow concluding that the chosen mathematical representation and the proposed method is suitable for managing the uncertainty inherent to the polymerization. Nevertheless, this research constitutes a piece of work in progress and additional experiments should be conducted in the future in order to assess the actual benefits and limitations of this methodology.

Keywords: Feature selection · QSPR · Polymer informatics

1 Introduction

In Computer Science and Statistics, Feature Selection (FS), also known as variable selection, is the process of selecting a subset of relevant features (or variables) for using in the design of a computational model [1]. FS techniques can be applied in predictive

J. Medina et al. (Eds.): IPMU 2018, CCIS 855, pp. 211–222, 2018.
https://doi.org/10.1007/978-3-319-91479-4_18

modeling for several reasons as: to avoid the curse of dimensionality, to reduce the computational efforts of the model training step, to reduce overfitting and to improve the interpretability of the models. The main hypothesis behind the application of a FS technique is that the data usually contains many features that are either redundant or irrelevant, which can be removed without suffering a significant loss of information. Irrelevant or redundant variables are key concepts in FS, because it could be the case where a relevant feature is redundant in the presence of another relevant feature with which it has a strong correlation.

In Cheminformatics, FS methods are commonly used as preprocessing step in QSAR/QSPR (Quantitative Structure–Activity/Property Relationship) modelling [2]. QSAR/QSPR models are regression or classification models used for predicting a target property from molecular descriptors (MDs), where each MD characterizes a piece of information encoded in the structure of the chemical compounds. Therefore, these models can be used for estimating relevant biological and physicochemical properties as virtual screening methods for drug design [3, 4]. In this context, the selection of the MDs subset more correlated with the target property constitutes an instance of the FS problem.

A particularly challenging case of QSPR modelling occurs in Polymer Informatics [5, 6]. Polymer Informatics is an interdisciplinary field that requires knowledge and tools from polymer chemistry, computer science and information science. The key goal behind Polymer Informatics is to progress on the design and understanding of polymer systems. An expert in this field develops in silico approaches for polymer research by means of systematic computational studies based on knowledge acquisition methodologies and machine learning algorithms [7]. Polymer Informatics requires a judicious management of macromolecules, which are chain-like molecules consisting of one or more structural repeat units (SRUs). Therefore, Polymer Informatics, as also occurs with Cheminformatics, is a mostly design-oriented discipline but the chemical structures studies for polymer informaticians are more complex and computationally demanding than the compounds modelled in drug design [8].

A polymeric material is made of several polymer chains with different lengths and molecular weights. Therefore, in contrast with a typical drug molecule, a polymeric material is characterized by a distribution of molecular weights instead of a single molecular weight value. This distinctive aspect of polymeric materials is known as polydispersion and, as a consequence of this issue, each molecular descriptor of a polymeric material has also associated a discrete distribution of values that it is obtained by calculating the molecular descriptor for the different polymeric chains and its frequencies.

FS algorithms had been extensively proposed in the field of combinatorial optimization for different application areas [1, 9, 10], including the selection of molecular descriptors [11–14]. Nevertheless, none method was designed for dealing with the uncertainty introduced by the polydispersion of materials described before. For this reason, a novel feature selection algorithm is proposed in this paper as a strategy for addressing the selection of molecular descriptors in QSPR modelling applied to material sciences.

2 Proposed Computational Approach

2.1 Polydispersity and Molecular Descriptors with Uncertainty

Polymers consist of repeat units (monomers) chemically bonded into long chains [15]. Understanding the physical properties of a polymer (such as mechanical strength, solubility and brittleness) requires knowledge of the length of the polymer chains. When the polymer chain length is defined, Molar mass of all monomers included in the molecule is considered. However, all synthetic polymers are polydisperse because they contain polymer chains of unequal length [16]. For this reason, the molecular weight of a synthetic polymeric material is not a single value, and must be represented as a distribution of chain lengths or molecular weights (see Fig. 1).

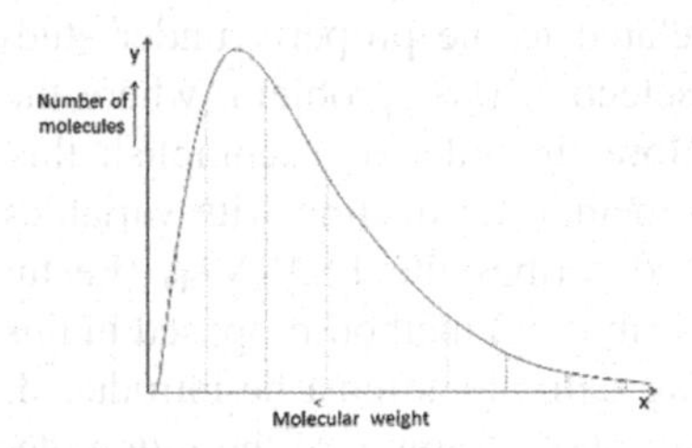

Fig. 1. The typical molecular weight distribution of a polymeric material. The x-axis represents the different weights of molecular chains polymerized in the material and the y-axis represents the numbers of molecular chains of each weight present in the material.

As polymer properties are dependent on the molecular-weight distribution of the material, the molecular descriptors used for inferring a QSPR model for these properties should be characterized taking into account the polydispersity of the materials. In order to address this problem, a novel approach is proposed in this paper. The key idea consists of obtaining a discrete distribution of values for each molecular descriptor. This distribution can be obtained by computing the molecular descriptor values for a subset of molecular chains with different weights present in the material. Figure 2 schematizes this procedure for any molecular descriptor.

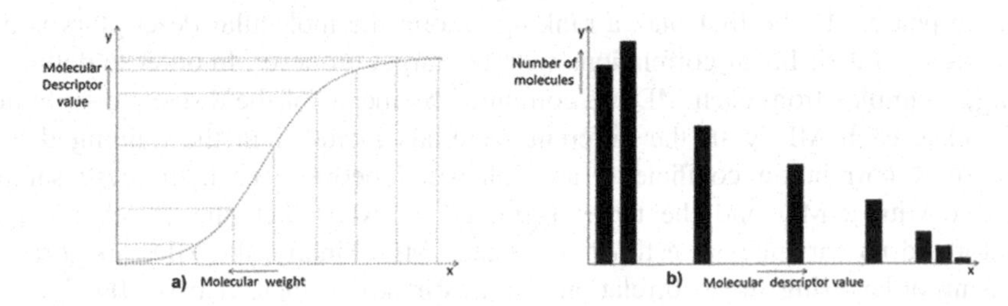

Fig. 2. **(a)** For each sample, the molecular descriptor value that corresponds to this polymer chain is computed. **(b)** Finally, the discrete distribution of the molecular descriptor can be obtained by matching the descriptor value of each sampled molecular chain with its corresponding number of molecules present in the material.

At this point, the study of polymer properties can be mathematically defined as a QSPR modeling problem under uncertainty. In order to address this problem some notations must be introduced.

Let *Mat* be a matrix with m rows (number of materials) and n columns (number of molecular descriptors), where each entry Mat_{ij} contains the discrete distribution DD_{ij} associated with *i–th* material and *j–th* molecular descriptor, and let Y be target property vector of length m, such that y_i is the experimental value of the target property for the *i–th* material. Therefore, a QSPR model can be defined as a regression function f, such that $f(Mat) \cong Y$.

2.2 Feature Selection for Variables with Uncertainty

The first step for inferring a QSPR model involves choosing the subset of molecular descriptors (MDs) more related to the property under study. This task constitutes a particular case of feature selection (FS) problem, where the variables under analysis present uncertainty. Therefore, in order to accomplish this step, it is mandatory to propose a feature selection method for dealing with variables characterized by discrete distributions. Algorithm 1 describes the $FS4RV_{DD}$ (Feature Selection for Random Variables with Discrete Distribution) method proposed in this work. In order to explain this algorithm, the following definitions must be introduced.

Sampled molecular descriptor: a sample of the *j–th* molecular descriptor, $s_{MD}j$, is a m-dimensioned vector such as $s_{MD_j}[i]$ is a random value sampled from DD_{ij}. In other words, s_{MD_j} vector represents a sample of the *j–th* molecular descriptor obtained from the discrete distributions that characterize this descriptor for the different materials available in the training dataset.

Similarity Ratio between molecular descriptors: let MD_A and MD_B two molecular descriptors that belong to *Mat* training database. These molecular descriptors are similar for the *i–th* material of database if only if the Bhattacharyya Distance (BD) [17] between $Mat[i, MD_A]$ and $Mat[i, MD_B]$ is lower than a predefined threshold, θ_{BD}. Taking into account that BD measures the similarity of two discrete probability distributions, therefore, the similarity ratio between MD_A and MD_B can be defined as the proportion between the number of materials where these MDs have a BD below θ_{BD} and the number of materials, m, available in *Mat*.

Using these definitions, the $FS4RV_{DD}$ algorithm can be explained. The method has two main phases. In the first one, a ranking among the molecular descriptors is generated based on their linear correlations with the target property. In order to define this ranking, k samples from each MD are computed by means of the discrete distributions which relate each MD with the different materials included in the training dataset. Therefore, k correlation coefficients are calculated between each of these samples associated with a MD and the target property, Y. After that, the correlation point estimator and its variance are estimated for each MD. Finally, the MDs are sorted by decreasing order using their correlation point estimator values. If two MDs have the same value, the second criteria for sorting is based on lower variance; and if two MDs have the same point estimator and variance values, they are sorted in alphabetic order.

In the second phase of the algorithm, all the MDs with a correlation point estimator value below a predefined threshold is removed from the ranking. Therefore, the remaining MDs in the ranking are contrasted in a pair-wise fashion in order to detect MDs with similar discrete distributions. For this task, the Bhattacharyya Distance is used for comparing the distributions associated with each pair of MDs for each material in the training dataset. If two MD have a similarity ratio higher than some predefined threshold, the MD with a lower position in the ranking is removed. The goal is to achieve a subset of MDs with a low degree of redundancy. As the last step, the final subset of selected features is cutoff to the maximum cardinality, defined by the user as a constraint, if it is necessary.

Algorithm 1: FS4VDD Method

```
Input:  Mat: training dataset, Y: target property values,
        max.card: maximum cardinality,
        min.corr: minimum correlation threshold,
        sampling.size: number of values sampled from the dis-
crete distributions,
        θBD: Bhattacharyya Distance threshold,
        θSR: Similarity ratio threshold.

Output: SSF: subset of selected features.
```

Phase 1: Correlation Ranking

```
point.estimators <- array [1..n] of null;
point.estimator.variances <- array [1..n] of null;
For each j varying from 1 to n do:
  correlations <- array [1..k] of null;
  For each k varying from 1 to sampling.size do:
    SMDj <- array [1.. sampling.size] of null;
    For each i varying from 1 to m do:
      SMDj[i] <- random.sampling (Matij)
    end-for
    correlations[k] <- correlation(SMDj, Y)   # Kendall's tau
  end-for
  point.estimators[j] <- 1/sampling.size Σ_{k=1}^{sampling.size} correlati[k]
  point.estimator.variances[j] <- 1/sampling.size*(sampling.size−1)
                       Σ_{k=1}^{sampling.size}(correlati[k] − point.estimator [j])²
end-for
```

```
ranking.MD <- sort MDs by decreasing value of their point
estimators for correlations, if two MD have the same
point estimator value, put first in the ranking the MD
with lower point estimator variance.
```

Phase 2: Iterative Subset Reduction

```
Delete from ranking.MD all MDs with correlation point es-
timator below the min.corr threshold;
stop <- false;
size.subset <- length(ranking.MD)
For each r1 varying from 1 to (size.subset-1) do:
  While (r2 <= size.subset) do:
    MDA <- ranking.MD[r1];  # the molecular descriptor in the r1
                            # position of the ranking is assigned
    MDB <- ranking.MD[r2];  # the molecular descriptor in the r2
                            #position of the ranking is assigned
    Similar.q <- 0;
      For each i varying from 1 to m do:
        DDAi <- [i,MDA];     # the discrete distribution associated to
                             # MDA in the material i-th is assigned
        DDBi <- [i,MDB];     # the discrete distribution associated to
                             # MDB in the material i-th is assigned
        BDi <- BD(DDAi, DDBi)  # the Bhattacharyya distance between
                               # DDAi and DDBi is assigned
        If (BDi < θBD) then similar.q <- similar.q+1;
      end-for
      similarity.ratio <- similar.q/m;
      if (similarity.ratio > θSR) then remove MDB from
ranking.MD;
  end- while
  r2 <- r2+1;
  size.subset <- length(ranking.MD);
end- for
if (size.subset > max.card) then SSF <- ranking.MD[1..max.card]
else SSF <- ranking.MD;
end-algorithm
```

2.3 Inferring QSPR Models from Variables with Uncertainty

In order to evaluate the performance of the $FS4RV_{DD}$ algorithm, it is necessary to define a computational approach for inferring QSPR models from molecular descriptors represented as discrete distributions. Let Mat_{SSF} a submatrix of *Mat*, which only includes the columns corresponding to the molecular descriptors selected by $FS4RV_{DD}$. A real-valued matrix $sMat_{SSF}$ can be obtained by random sampling the discrete distributions included as entries of Mat_{SSF}. In this way, the sampled $sMat_{SSF}$ matrix can be considered a particular instance of the information represented by Mat_{SSF}. Thus, from $sMat_{SSF}$ and Y, a QSPR model can be inferred for this instance using a traditional regression method. Therefore, if several instances of Mat_{SSF} are obtained by random sampling, a consensus QSPR model can be defined from the distribution of outcomes, DD_Y, generated by the QSPR models inferred by the different instances. Finally, the $\hat{Y}$ value estimated by consensus will be the mean of DD_Y. Figure 3 outlines this approach.

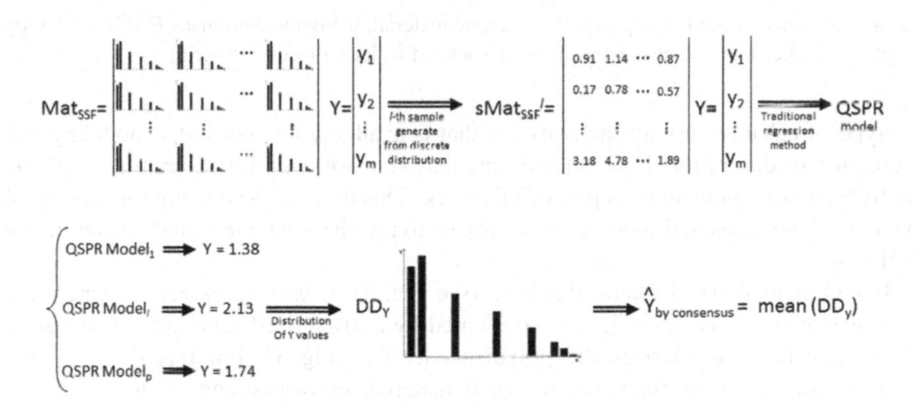

Fig. 3. Inference of a QSPR model from a matrix obtained by sampling p-times the discrete distributions associated to the selected subset of molecular descriptors.

Applying a QSPR Model Based on Variables with Uncertainty to a New Material
As a final methodological point, it is necessary to define how the consensus QSPR model is applied to a new material, mat_{new}, during the validation step. This new material, not included in the training dataset (*Mat*), can be represented as a vector. Each entry of mat_{new} is a discrete distribution associated with one of the molecular descriptors included in *SSF*. Using these distributions, it is possible to define several instances of this vector by random sampling. These instances of mat_{new} can be used as inputs for each $QSPR\,Model_l$ and, consequently, a discrete distribution, DD_{Y_l}, of Y values can be generated from the outcomes of each model. As before, the mean of DD_{Y_l} can be reported of the Y value estimated for $QSPR\,Model_l$. Finally, the Y consensus value can be estimated as the mean of DD_Y generated from the mat_{new} estimations. Figure 4 schematizes this procedure.

3 Results

3.1 Synthetic Data

Generate a database with real values of molecular descriptors for high molecular weight polymers has a high computational cost. This is because the molecules must exceed the average molecular weight that characterizes them. Today, it is not possible to perform

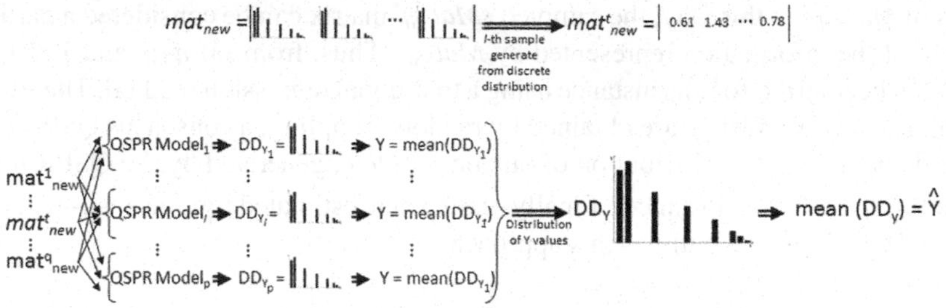

Fig. 4. Inference of the target property for a new material, using the consensus QSPR model, by sampling *q*-times the discrete distributions associated to its molecular descriptors.

this type of calculation with the software that is available for molecular modeling and calculation of descriptors [18, 19]. For this reason, a software for generating synthetic databases was implemented as part of this work. This tool allows developing a proof of concept of the proposed methodology, for studying the soundness and scalability of $FS4RV_{DD}$.

For obtaining the synthetic database (see Fig. 5), it was necessary to generate a table of materials (see 1 in Fig. 5) represented by a discretized lognormal distribution which typically characterizes the polydispersity (see Fig. 1). For this purpose, two random numbers were generated for each material, corresponding to the *mean* and *standard deviation* of the distribution of molecular weights. Thus, a vector of coordinates is saved in the table of materials for each material, with a length equal to *sampling.size*, where the first value represents the different sampled molecular weights (x-axis) and the second one the frequencies (y-axis).

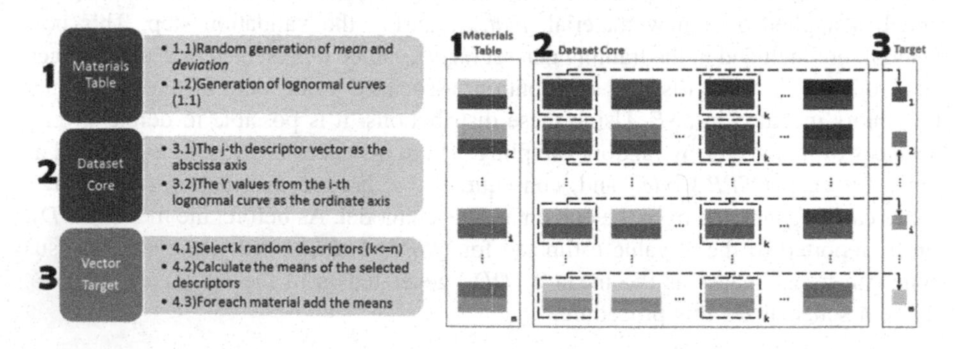

Fig. 5. Methodology for synthetic dataset generation.

In a second step, the increasing values that the descriptors take for the different instances of each material are simulated (2 in Fig. 5). These vectors correspond to the y-axis in Fig. 2.**a**. In order to construct each one of these vectors, two random numbers were generated, *start* and *step*. In this way, the lowest value associated to the descriptor will coincide with *start*, while the subsequent values will be defined sequentially adding *step* until all the observations are completed. Then, the dataset core generation starts (2 in Fig. 5). Each cell (i, j) of the core corresponds to the distribution of values for the *j*-th descriptor for the *i*-th material, like the plot represented in Fig. 2.**b**. These cells are represented as coordinates vectors in a manner that the y-axis of the cell (i, j) coincides with the y-axis of the *i*-th material, while the x-axis is built from the values randomly generated for each cell, *start* and *step*. Hence, the lowest value associated with the descriptor will coincide with *start*, while the subsequent values will be defined by adding *step* until all the observations are completed.

The final step is the generation of the target (see 3 in Fig. 5), for which it is necessary to randomly select as many descriptors as one desires to correlate with the output variable. Finally, the target vector is defined by fitting a linear correlation to the means of the distributions associated to the chosen descriptors.

3.2 Experiments

Four dataset cores with a different number of materials (200, 400, 800 and 1600) had been generated using the procedure described in Sect. 3.1. In all cases, the number of molecular descriptors per table was fixed in 100. For each dataset core, four target vectors were computed by linear correlation assuming different cardinalities for the selected subset of molecular descriptors (5, 10, 15 and 20). Therefore, a total amount of sixteen synthetic datasets were created for the experiments in order to evaluate the performance scalability of the $FS4RV_{DD}$ algorithm.

The performance metrics calculated in these experiments have been choosing in order to assess the algorithm skills for recovering the correct subset of features. In other words, as an exercise of reverse engineering, the goal is to determine if $FS4RV_{DD}$ can trace back the subset of features artificially correlated with the target during the creation of the synthetic datasets. The computed metrics are four: Accuracy, which represents the percentage of features correctly classified as *selected* or *not selected*; Mean Absolute Percentage Error (MAPE), which is a measure of prediction accuracy of a statistical forecasting method (0% and 100% represent the best and worst performance respectively); Sensitivity, which measures the proportion of positives cases that are correctly identified as such; and Specificity, which measures the proportion of negatives cases that are correctly identified as such. In the context of this work, the positive cases correspond to features that belong to the selected subset, whereas the remaining features constitute negative cases.

Table 1 and Fig. 6 show the metrics values obtained by the experiments. The accuracy does not present significant variations when the size of the data set is increased. Nonetheless, the values decline with increment in the cardinality of the selected subsets. In general terms, the accuracy values are moderately high although this observation is tricky because they are a consequence of the strong unbalance between the number of positive and negative cases. This can be concluded from the inspection of the sensitivity and specificity values.

On the other hand, MAPE values trend to improve when the dataset size and the selected subset cardinality are incremented. This is an expected result considering that when bigger the number of materials and the selected features are, its simpler to detect the linear correlations with the target.

Table 1. Metrics values obtained for each data set (200, 400, 800 and 1600 materials) with the different cardinalities for the selected subset of molecular descriptors (5, 10, 15 and 20).

# Materials	#MD Selected	Accuracy	MAPE	Sensitivity	Specificity
200	5	96.00%	13.30%	60.00%	97.89%
	10	86.00%	12.28%	30.00%	92.22%
	15	82.00%	11.97%	40.00%	89.41%
	20	78.00%	11.23%	45.00%	86.25%
400	5	96.00%	13.68%	60.00%	97.89%
	10	90.00%	13.79%	50.00%	94.44%
	15	80.00%	9.45%	33.33%	88.24%
	20	78.00%	9.93%	45.00%	86.25%
800	5	96.00%	13.26%	60.00%	97.89%
	10	90.00%	12.73%	50.00%	94.44%
	15	88.00%	11.36%	60.00%	92.94%
	20	82.00%	9.36%	55.00%	88.75%
1600	5	100.00%	13.23%	100.00%	100.00%
	10	98.00%	10.50%	90.00%	98.89%
	15	90.00%	10.08%	66.67%	94.12%
	20	80.00%	8.95%	50.00%	87.50%

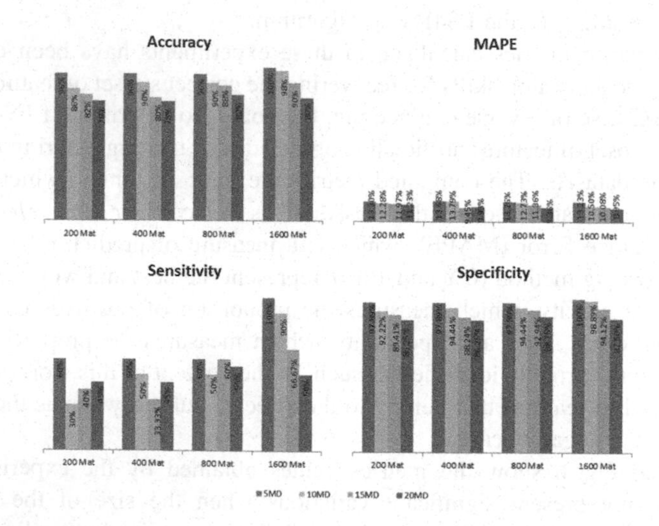

Fig. 6. Graphical representation of the metrics computed for each experiment.

4 Conclusions

Feature selection (FS) is a well-known combinatorial optimization problem extensively studied for different real-world environments, which usually request the design of novel approaches for dealing with new emerging fields such as Big Data or Data Streaming applications [10]. In particular, a challenging feature selection problem in the area of

Quantitative Structure–Property Relationship (QSPR) for Polymer Informatics has been addressed in this paper. The main characteristic of this scenario is the uncertainty introduced for the polymerization phenomenon intrinsic to these industrial materials, which makes necessary to represent the molecular descriptor values (feature values) as probabilistic distributions.

Following this probabilistic representation a new feature selection method, called with the acronym FS4RV$_{DD}$, has been proposed in this work. This new FS approach has two main steps: feature ranking definition by correlation analysis and iterative subset reduction by redundancy analysis based on the Bhattacharyya Distance as a similarity measure between molecular descriptors. The performance of the method has been assessed by using synthetic datasets of different sizes and varying the number of features to be selected in the experiments.

From these preliminary results, it is possible to conclude that the mathematical representation and FS method proposed in this paper are suitable for handling the uncertainty inherent to this particular QSPR problem. Nevertheless, more trials are indispensable for testing this approach under several experimental conditions, by incorporating different levels of noise and more type of correlations among features and target variables in the synthetic data generation procedure.

In summary, as far as we know, the FS4RV$_{DD}$ algorithm constitutes the first feature selection method proposed for selecting subsets of molecular descriptors in the context of polymeric material design. Therefore, the main contribution of this paper is to present a proof of concept of this general idea for discussion in our scientific community, as a previous step for more exhaustive validations and potential improvements of the proposed approach.

Acknowledgments. This work is kindly supported by CONICET, grant PIP 112-2012-0100471 and UNS, grants PGI 24/N042 and PGI 24/ZM17.

References

1. Li, Y., Li, T., Liu, H.: Recent advances in feature selection and its applications. Knowl. Inf. Syst. **53**, 551–577 (2017)
2. Eklund, M., Norinder, U., Boyer, S., Carlsson, L.: Choosing feature selection and learning algorithms in QSAR. J. Chem. Inf. Model. **54**, 837–843 (2014)
3. Li, J., Fong, S., Siu, S., Mohammed, S., Fiaidhi, J., Wong, K.K.L.: WITHDRAWN: improving classification of protein binders for virtual drug screening by novel swarm-based feature selection techniques. Comput. Med. Imaging Graph. (2016, in press)
4. Ponzoni, I., Sebastián-Pérez, V., Requena-Triguero, C., Roca, C., Martínez, M.J., Cravero, F., Díaz, M.F., Páez, J.A., Gómez Arrayás, R., Adrio, J., Campillo, N.E.: Hybridizing feature selection and feature learning approaches in QSAR modeling for drug discovery. Sci. Rep. **7**, Article number 2403 (2017)
5. Adams, N.: Polymer informatics. In: Meier, M., Webster, D. (eds.) Polymer Libraries. Advances in Polymer Science, vol. 225, pp. 107–149 (2010)
6. Audus, D.J., De Pablo, J.J.: Polymer informatics: opportunities and challenges. ACS Macro Lett. **6**, 1078–1082 (2017)

7. Liu, Y., Zhao, T., Ju, W., Shi, S.: Materials discovery and design using machine learning. J. Materiomics **3**, 159–177 (2017)
8. Huan, T.D., Mannodi-Kanakkithodi, A., Kim, C., Sharma, V., Pilania, G., Ramprasad, R.: A polymer dataset for accelerated property prediction and design. Sci. Data **3**, Article number 160012 (2016)
9. Singh, R.K., Sivabalakrishnan, M.: Feature selection of gene expression data for cancer classification: a review. Procedia Comput. Sci. **50**, 52–57 (2015)
10. Tommasel, A., Godoy, D.: A Social-aware online short-text feature selection technique for social media. Inf. Fusion **40**, 1–17 (2018)
11. Soto, A.J., Cecchini, R.L., Vazquez, G.E., Ponzoni, I.: A wrapper-based feature selection method for ADMET prediction using evolutionary computing. In: Marchiori, E., Moore, J.H. (eds.) EvoBIO 2008. LNCS, vol. 4973, pp. 188–199. Springer, Heidelberg (2008). https://doi.org/10.1007/978-3-540-78757-0_17
12. Soto, A.J., Cecchini, R.L., Vazquez, G.E., Ponzoni, I.: Multi-objective feature selection in QSAR using a machine learning approach. Mol. Inf. **28**, 1509–1523 (2009)
13. Martínez, M.J., Ponzoni, I., Díaz, M.F., Vazquez, G.E., Soto, A.J.: Visual analytics in cheminformatics: user-supervised descriptor selection for QSAR methods. J. Cheminform. **7**, 39 (2015)
14. Cravero, F., Martínez, M.J., Vazquez, G.E., Díaz, M.F., Ponzoni, I.: Feature learning applied to the estimation of tensile strength at break in polymeric material design. J. Integr. Bioinf. **13**, 286 (2016)
15. McCrum, N.G., Buckley, C.P., Bucknall, C.B.: Principles of Polymer Engineering. Oxford University Press, Oxford; New York (1997)
16. Sheu, W.-S.: Molecular weight averages and polydispersity of polymers. J. Chem. Educ. **78**, 554–555 (2001)
17. Bhattacharyya, A.: On a measure of divergence between two statistical populations defined by probability distributions. Bull. Calcutta Math. Soc. **35**, 99–109 (1943)
18. Cravero, F., Schustik, S., Martínez, M.J., Ponzoni, I., Díaz, M.F.: Macro approach to molecular modelling of linear polymers applied to estimation of tensile modulus for new materials development. In: VIII International Symposium on Materials (Materias2017), Aveiro, Portugal (2017)
19. Cravero, F., Martínez, M.J., Vazquez, G.E., Ponzoni, I., Díaz, M.F.: Representación de la Estructura Molecular de Polímeros Sintéticos de Alto Peso. In: XXXI Congreso Argentino de Química, Buenos Aires, Argentina (2016)

A Comparative Analysis of Accurate and Robust Bi-objective Scheduling Heuristics for Datacenters

Sergio Nesmachnow[1] and Bernabé Dorronsoro[2(✉)]

[1] Universidad de la República, Montevideo, Uruguay
sergion@fing.edu.uy
[2] Universidad de Cádiz, Cádiz, Spain
bernabe.dorronsoro@uca.es

Abstract. This article presents and evaluates twenty-four novel bi-objective efficient heuristics for the simultaneous optimization of makespan and robustness in the context of the static robust tasks mapping problem for datacenters. The experimental analysis compares the proposed methods over realistic problem scenarios. We study their accuracy, as well as the regions of the search space they explore, by comparing versus state-of-the-art Pareto fronts, obtained with four different specialized versions of well-known multi-objective evolutionary algorithms.

Keywords: Independent tasks scheduling · Robustness optimization

1 Introduction

The scheduling problem proposes assigning resources to a set of processes (usually modeled by a set of *tasks*), such that the resources are effectively shared or a given quality of service metric is optimized. There are many different variants of the problem, depending on how resources and processes are defined [3,16].

All effective scheduling algorithms consider that the time to perform a task is known beforehand. This assumption is realistic in some versions of the problem (e.g., manufacturing processes [10]), but not in the case of scheduling computing tasks in datacenters, considered in this article. The reason is that accurately predicting the time a process takes in a given processor is an open problem yet. Therefore, estimations are used to approximate the execution time [4]. Because estimations are often not accurate, they lead to considerable performance loss of the system. Looking for solutions that are robust against such inaccuracies may help alleviating, or even neglecting, the performance decrease they produce.

Scheduling independent tasks is common in datacenters [4,20,22], where independent users submit their jobs to distributed computing infrastructures. The performance of a schedule is measured in terms of the makespan or total execution time. It is an NP-complete problem [11].

J. Medina et al. (Eds.): IPMU 2018, CCIS 855, pp. 223–235, 2018.
https://doi.org/10.1007/978-3-319-91479-4_19

This article addresses a multi-objective version of the problem, considering the robustness of the schedule against inaccuracies in the execution time estimation of tasks, together with the makespan. The objective of this problem is to find schedules that not only execute all tasks as quickly as possible, but are resilient to inaccuracies in the execution time estimations of the tasks. These two objectives are in conflict. This article proposes designing a number of fast heuristics, suitable for real environments, and analyze their results.

The main contributions of this article are: (i) the formulation and implementation of twenty-four bi-objective efficient greedy scheduling heuristics for the simultaneous optimization of makespan and robustness ratio in datacenters, (ii) the experimental analysis comparing the proposed methods, performed over realistic problem scenarios, and (iii) the study of the accuracy of the methods and the regions of the search space they explore by comparing versus state-of-the-art Pareto fronts, obtained with four different specialized versions of well-known multi-objective evolutionary algorithms (MOEAs).

The article is organized as follows. Next section reviews the main related works in literature about the robust scheduling problem and similar variants. Section 3 presents the problem formulation. The greedy scheduling algorithms proposed in this work are introduced in Sect. 4. The experimental analysis of the proposed methods is reported in Sect. 5. Finally, Sect. 6 summarizes the main conclusions and formulates the main lines for future work.

2 Related Work

Only a few works proposing mathematical metrics to quantify the robustness of schedules exist, and they cannot guarantee the correct behavior of the schedule against uncertainties. Leon et al. [15] studied different robustness measures and analyzed the impact on improving the average performance of the schedules after several breakdowns. Jensen [13] defined the robustness of a schedule with task dependencies in terms of the variation in the makespan when rescheduling after a breakdown. Carretero et al. [5] compared robustness metrics for scheduling parallel applications, defined in terms of the makespan variation (standard deviation) of different realizations of the same schedules. They proposed several algorithms to optimize makespan and robustness, and they claim that solutions offer a *smaller* variation in makespan, compared to regular schedulers. Other works focused on optimizing the worst case, thus obtaining highly robust solutions [14], but with low makespan.

Ali et al. [1] proposed a robustness metric for allocating independent tasks in heterogeneous computing systems guaranteeing that, if the differences between actual and estimated task execution times is within a certain range, then a given makespan requirement is met. This metric was used in our previous work [8] using different state-of-the-art MOEAs. Iturriaga et al. [12] studied the energy-aware scheduling problem in heterogeneous computing systems considering uncertainty. Versions of well-known scheduling heuristics were proposed for applying a realistic power consumption model [21]. A model for uncertainty on power consumption was determined through empirical evaluations using three CPU-bound benchmarks. Results confirmed that uncertainly has a significant impact in the accuracy of the scheduling algorithms.

In [9,23], authors focus on finding schedules that are somehow *flexible* to uncertainties. A *flexible* schedule is expected to be less affected by those uncertainties (e.g., by some delay) than a regular schedule. De Falco et al. [7] applied a true multi-objective approach in which some kind of robustness is considered by optimizing, together with the resource utilization, the resources reliability by assigning some static reliability values to every resource in the problem definition. However, none of these techniques provide any exact metric to measure the flexibility (how they defined the robustness) of a solution, and therefore no guarantee can be assured on the robustness level of the solution.

Several works deal with robustness in other scheduling domains, as the robust software planning and scheduling [2,17] or robust flight schedules to minimize the impact of unplanned events as delays or disruptions [6].

3 The Robust Static Mapping Problem

This section presents the Robust Static Mapping Problem (RSMP) for robust assignment of independent tasks on heterogeneous computing systems.

Consider the following elements:

- A set of n independent tasks $T = \{t_1, t_2, \ldots, t_n\}$ to be scheduled. Each task t_i has a *workload* w_i (in millions of instructions).
- A pool of k heterogeneous machines $M = \{m_1, m_2, \ldots, m_k\}$ available for executing tasks. Each machine has a *computing capacity* (in millions of instructions per second).
- The ready time ($ready_j$) indicates when machine j will finish the previously assigned tasks. Without loss of generality, the problem formulation assumes $ready_j = 0$ for every machine.
- The Expected Time to Compute (ETC) matrix ($nb\_tasks \times nb\_machines$) in which $ETC[i][j]$ is the expected execution time of task t_i on machine m_j.

Tasks must be processed by a single machine. Machines can process one task at a time, and are heterogeneous. The RSMP proposes allocating the tasks on the available machines minimizing the *makespan* of the schedule, while at the same time its *robustness* is maximized. These two objectives are defined below.

- Makespan is defined as the maximum completion time of all the resources used in the schedule. The completion time of a machine m_j in schedule S is defined in Eq. 1, $S(j)$ is the set of tasks assigned to m_j, and C is a matrix with the actual times to compute the tasks in every machine ($C = ETC$ when not considering errors in the estimation of the duration of tasks). The makespan function is defined in Eq. 2, where $\boldsymbol{x}$ represents an allocation.

$$F_j(C) = ready_j + \sum_{t \in S(j)} C_{t,j} \quad (1)$$

$$f_M(\boldsymbol{x}) = \max\{F_j(C)\} \quad (2)$$

– The robustness metric is defined as the minimum of the robustness radii of the machines [1]. The robustness radius is the smallest collective increase in the error that would cost the finishing time of a machine to be τ times the original, as defined by Eq. 3, where M_{orig} is the makespan of the schedule with the estimated ETC values and $F_j(C_{orig})$ is the ETC of tasks assigned to machine m_j. This robustness metric assures that if the c even when all the task execution times are increased in a percentage ρ, the makespan will not be increased beyond that percentage.

$$f_R(\boldsymbol{x}) = \min \frac{\tau \cdot M^{orig} - F_j(C^{orig})}{\sqrt{\text{number of tasks allocated to } m_j}} \tag{3}$$

The multi-objective RSMP propose finding schedules that minimizes the makespan (f_M) and maximizes the robustness (f_R). These objectives are in conflict, thus multi-objective optimization methods must be applied.

4 Algorithms for Robust Scheduling in Datacenters

This section presents the algorithms applied in this article to solve the RSMP.

4.1 Greedy Scheduling Heuristics

Greedy scheduling techniques are deterministic static scheduling methods that work by assigning priorities to tasks, based on a particular ad-hoc heuristic [18]. After that, the list of tasks is sorted in decreasing priority and each task is assigned to a processor, regarding the task priority and the processor availability.

Some of the most popular heuristics in this class include [4]:

- *MinMin*: starting from a set U of all unmapped tasks, determines the machine that provides the Minimum Completion Time (MCT) for each task in U, and assigns the task with the minimum overall MCT to its best machine. The mapped task is removed from U, and the process is repeated until all tasks are mapped.
- *MaxMin*: is similar to MinMin, but it assigns the task with the overall *maximum* MCT to its MCT machine. Therefore, larger tasks are allocated first in the most suitable machines and shorter tasks are mapped afterward, with the aim of balancing the load of all available machines.
- *Sufferage*: identifies the task that will *suffer* the most if it is not assigned to a certain host. The *sufferage value* is computed as the difference between the best MCT of the task and its second-best MCT. For a particular machine, Sufferage gives precedence to the task with highest sufferage value, assigning it to the machine that can complete it at the earliest time.

MinMin, MaxMin, and Sufferage follow a generic schema that applies two *phases* to perform the task-to-resource assignment: in the first phase, N pairs (task, machine) are selected considering a specific criterion, and in the second phase one of those N pairs is selected after an overall comparison. These heuristics were conceived for optimizing only one objective function (makespan), thus the criteria applied in the two phases are related to the makespan. This article extends the generic schema for the simultaneous optimization of the two objective functions considered in the RSMP (makespan and robustness).

4.2 Greedy Heuristics for Makespan and Robustness Optimization

The proposed heuristics work in two phases, considering makespan or robustness optimization criterion in each phase. Twenty-four heuristics were designed. They have been named using the following convention: the criteria that optimize the makespan metric are written in lower case (e.g.: Min, Max, Suff), and the criteria focusing on robustness are written in upper case (e.g.: MIN, MAX). The same convention applies to strategies that use a k-percent-best list: kpb in lower case indicates a list sorted by MCT, while KPB in upper case stands for a list sorted by robustness. The name starts with the kpb/KPB specification (if used), followed by the abbreviation of the second and first criteria, in that order:
`[kpb/KPB][.]<phase 2 criterion><phase 1 criterion>`

The studied heuristics for makespan and robustness ratio optimization are:

1. *MinMin*: the traditional MinMin heuristic, optimizing only makespan.
2. *MaxMin*: the traditional MaxMin heuristic, optimizing only makespan.
3. *Sufferage*: the traditional Sufferage heuristic, optimizing only makespan.
4. *MAXMin*: selects pairs (t_i, m_j) minimizing the MCT (phase 1), and then selects the pair that maximizes the robustness ratio (phase 2).
5. *MinMAX*: selects pairs (t_i, m_j) maximizing the robustness ratio (phase 1), and then selects the pair that that minimizes the MCT (phase 2).
6. *MAXMAX*: selects pairs (t_i, m_j) maximizing the robustness ratio (phase 1), and then selects the pair that maximizes the robustness ratio (phase 2).
7. *MINMAX*: selects pairs (t_i, m_j) maximizing the robustness ratio (phase 1), and then selects the pair that minimizes the robustness ratio (phase 2).
8. *SuffMAX*: selects pairs (t_i, m_j) maximizing the robustness ratio (phase 1), and then selects the pair that minimizes sufferage (phase 2).
9. *MAXSuff*: selects pairs (t_i, m_j) minimizing the sufferage (phase 1), and then selects the pair that maximizes the robustness ratio (phase 2).

The k-percent-best variants of the proposed heuristics consider the MCT or robustness to restrict the list of candidate machines for each task used in phase 1 to the best k machines, according to the selected criterion. We set k to 30% of the machines in each instance, after a preliminary sensitivity analysis. This way, kpb.MinMin, kpb.MaxMin, kpb.Sufferage, kpb.MAXMin, kpb.MinMAX, kpb.MAXMAX, kpb.MINMAX, kpb.SuffMAX, and kpb.MAXSuff only consider in phase 1 the 30% best machines, in terms of MCT for every task. Similarly, heuristics KPB.MinMin, KPB.MaxMin, KPB.Sufferage, KPB.MAXMin, KPB.SuffMAX, and KPB.MAXSuff only consider in phase 1 the top 30% machines with the largest robustness ratio for every task.

5 Experimental Analysis

This section describes the experimental evaluation of the proposed greedy scheduling heuristics for makespan and robustness optimization.

Table 1. Average makespan and robustness results for the greedy scheduling heuristics

Heuristic	HTRH		LTRH	
	Makespan	Robustness	Makespan	Robustness
MinMin	5797.4	303.3	18047.6	947.2
MaxMin	10235.0	505.1	28116.2	1452.7
Sufferage	**5045.1**	258.9	**17610.4**	914.1
MAXMin	6630.5	341.9	20971.3	1085.8
MinMAX	17472.6	1285.4	20404.4	1123.9
MAXMAX	22572.3	2880.7	24801.2	1635.9
MINMAX	26317.3	3664.2	26881.1	1915.9
MAXSuff	6630.5	341.9	20971.3	1085.8
SuffMAX	30023.0	4185.4	28014.2	2124.0
kpb.MinMin	5798.9	303.8	18152.7	949.9
kpb.MaxMin	8240.6	396.9	21526.3	1094.0
kpb.Sufferage	5155.9	263.5	17844.9	925.9
kpb.MAXMin	6471.3	334.2	19960.7	1036.5
kpb.MinMAX	6985.8	372.2	19478.5	1025.5
kpb.MAXMAX	10912.0	953.0	23562.8	1375.7
kpb.MINMAX	7919.1	473.5	20671.6	1134.1
kpb.SuffMAX	8932.2	738.7	21544.8	1260.1
kpb.MAXSuff	6462.9	333.4	19960.7	1036.5
KPB.MinMin	54771.3	2869.6	41964.9	2197.2
KPB.MaxMin	63537.5	3066.2	46450.6	2358.2
KPB.Sufferage	50758.7	2570.7	41463.9	2150.4
KPB.MAXMin	57419.3	2979.3	44165.9	2296.2
KPB.MAXSuff	57419.3	2979.3	44165.9	2296.2
KPB.SuffMAX	69956.5	**5180.4**	46972.4	**2705.8**

5.1 Methodology

Problem scenarios. Two different problem classes are considered: High Task and Resource Heterogeneity (HTRH) and Low Task and Resource Heterogeneity (LTRH). Task heterogeneity represents the variation of the tasks execution times for a given machine. Resource heterogeneity evaluates the variation of execution times for a given task across the available machines. **200** problem instances are solved (100 for each of the two studied problem classes, using 100 different ETC matrices). Instances were generated using the coefficient of variation method for ETC matrices. All instances are composed of 512 independent tasks that must be scheduled on a cluster of 16 machines, as proposed in related works [4,24]. The length of tasks in all instances are within the same range.

Development and execution platform. The proposed heuristics were implemented in C language. The experimental evaluation was performed on a Dell PowerEdge (QuadCore Xeon E5430 processor at 2.66 GHz, 8GB RAM and CentOS Linux), from Cluster FING, Universidad de la República, Uruguay [19].

Baseline for comparison. Four state-of-the-art MOEAs were selected for the results comparison: IBEA, MOCell, MOEA/D, and NSGA-II. These methods provide competitive results for the problem solved in this article [8].

5.2 Numerical Results and Discussion

Table 1 reports the average makespan and robustness results for the greedy scheduling heuristics when solving the HTRH and LTRH scenarios. The results indicate that the traditional Sufferage method is the best heuristic regarding the makespan optimization in the two scenarios. However, the combined heuristic *KPB.SuffMAX*, using a k-percent-best list of machines based on the maximum robustness ratio, is the best option for maximizing the robustness of solutions.

The KPB versions of the heuristics find more robust solutions than their original counterparts. For instance, SuffMAX is the second best heuristic for HTRH (after KPB.SuffMAX), with 19.2% worse robustness quality in average. For LTRH, SuffMAX is 21.5% worse than its KPB version in terms of robustness. Indeed, using KPB increases the robustness by up to 89.9% for Sufferage, but at the cost of 90.1% increase in makespan. Incorporating the kpb policy generally leads to makespan reductions, but it also worsens the robustness.

Figures 1 and 2 show how the heuristics find similar robustness and makespan values for all instances, with a few exceptions (mainly for HTRH). There are statistically significant differences when the notches of the boxes are not overlapped.

Several heuristics provide accurate trade-off values between makespan and robustness ratio, as shown in the sample trade-off analysis for HTRH scenario in Fig. 3. SuffMAX, MAXMAX, and MINMAX compute balanced schedules, providing the best compromise values for makespan and robustness. The KBP heuristics are located on the right-most part, and the rest are in the bottom left region. A similar situation holds for the LTRH scenario. The best heuristics for makespan have the worst robustness values, confirming the conflicting nature of the two objectives.

Table 2 reports the average ranks of each heuristic computed by the non-parametric Friedman statistical test for each objective and problem class, regarding makespan and robustness results. For example, a makespan rank value of 3 means that a given method occupied, on average, the third position when sorting the twenty heuristics regarding the makespan values (the average is computed

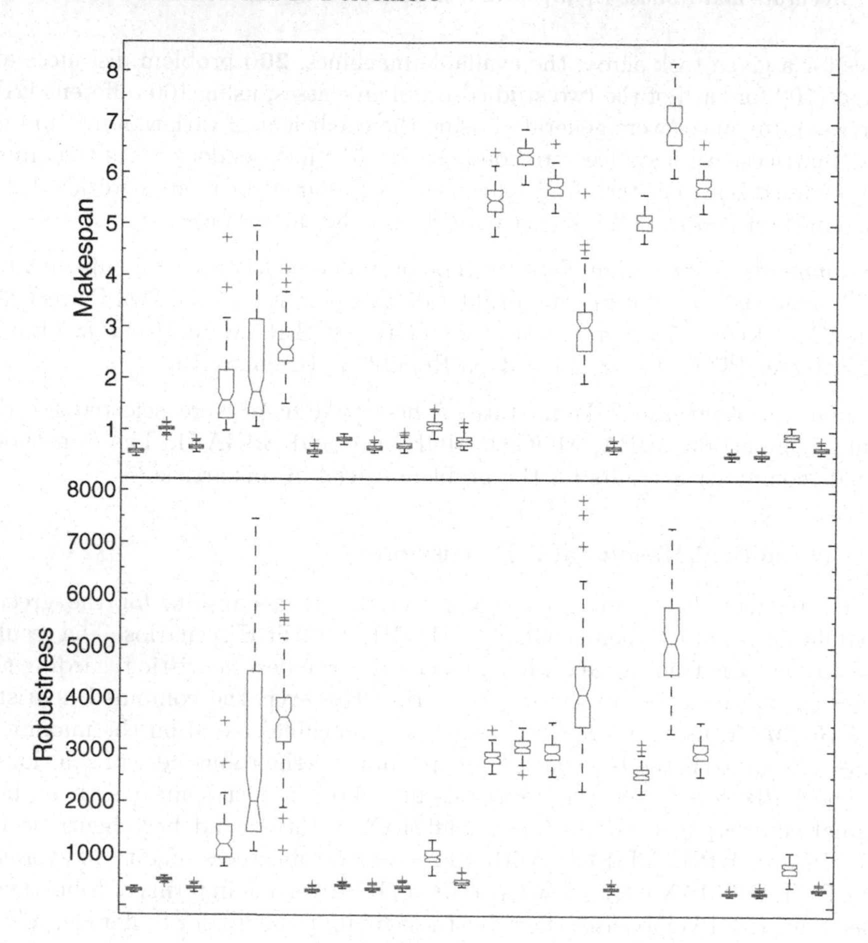

Fig. 1. Boxplot results for the 100 HTRH instances

over the 100 instances in each problem class). In all cases, the p-values of the Friedman test was lower than 0.095, therefore the differences between the algorithms are statistically significant with 95% confidence.

Table 2 confirms Sufferage as the best heuristic in terms of makespan, and KPB.SuffMAX in the case of robustness. The kpb and KPB versions of the heuristics usually lead to better quality solutions in terms of makespan and robustness, respectively. Additionally, the best heuristics for makespan are among the worst ones in terms of robustness, and vice-versa. The best trade-off algorithms are SuffMAX, MAXMAX, and MINMAX for HTRH, according to the analysis in Fig. 3. The same algorithms are the best ones for LTRH, together with kbp.MINMAX, which is slightly better positioned than they three.

Figure 4 shows representative cases of the comparison of the results of the proposed heuristics (plotted as '+') versus the Pareto fronts from [8], built with the best non-dominated solutions of four state-of-the-art MOEAs, hybridized

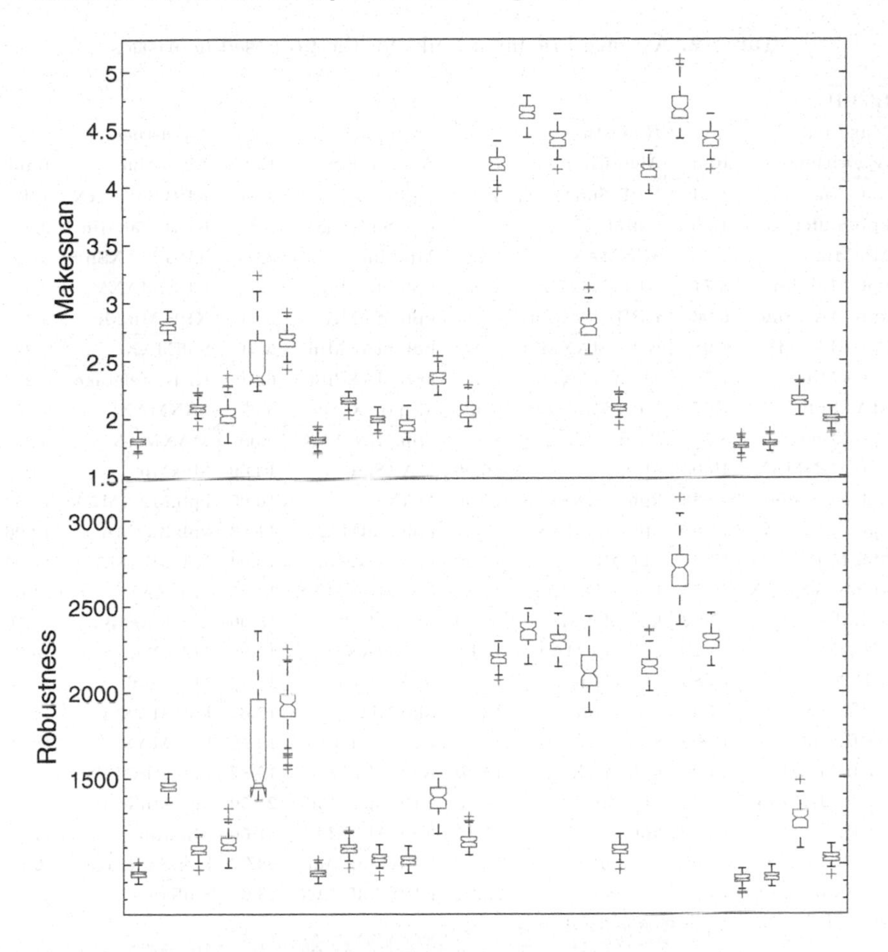

Fig. 2. Boxplot results for the 100 LTRH instances

with specific operators for the problem (plotted as '·'). Results are encouraging. In the region with the lowest makespan solutions, heuristics results are very close to the reference Pareto fronts, for all instances. In some cases, the best heuristics dominate solutions from the reference Pareto front, finding schedules with higher robustness at similar makespan values. This is the case of kpb.MINMAX for HTRH instance number 0 and KBP.SuffMax for LTRH instance number 35.

The heuristics find a large variety of results, covering a wide area of trade-off solutions and are much more efficient than state-of-the-art methods. Indeed, the runtime of the heuristics is about 0.01 seconds, while the reference Pareto front was built from the results of 100 independent runs of four MOEAs with runtime between 40 and 200 seconds per run, for every algorithm. Please note that offering a quick response is crucial for this problem.

Table 2. Average Friedman ranks for the proposed heuristics

HTRH				LTRH			
Makespan		Robustness		Makespan		Robustness	
Algorithm	Rank	Algorithm	Rank	Algorithm	Rank	Algorithm	Rank
Sufferage	1.24	KPB.SuffMAX	1.45	Sufferage	1.21	KPB.SuffMAX	1.01
kpb.Sufferage	1.85	SuffMAX	2.84	kpb.Sufferage	2.31	KPB.MaxMin	2.51
MinMin	3.65	MINMAX	3.65	MinMin	3.03	KPB.MAXSuff	3.52
kpb.MinMin	3.77	KPB.MaxMin	4.96	kpb.MinMin	3.45	KPB.MAXMin	3.52
kpb.MAXSuff	5.96	KPB.MAXSuff	5.66	kpb.MinMAX	5.71	KPB.MinMin	5.40
kpb.MAXMin	6.98	KPB.MAXMin	5.66	kpb.MAXMin	6.81	SuffMAX	6.22
MAXMin	7.57	MAXMAX	6.42	kpb.MAXSuff	6.81	KPB.Sufferage	6.27
MAXSuff	7.57	KPB.MinMin	6.49	MinMAX	8.37	MINMAX	8.06
kpb.MinMAX	8.15	KPB.Sufferage	8.12	kpb.MINMAX	9.38	MAXMAX	8.93
kpb.MINMAX	10.05	MinMAX	10.06	MAXSuff	10.09	MaxMin	9.92
kpb.MaxMin	10.88	kpb.MAXMAX	10.90	MAXMin	10.09	kpb.MAXMAX	10.84
kpb.SuffMAX	11.60	kpb.SuffMAX	11.97	kpb.SuffMAX	11.93	kpb.SuffMAX	12.06
MaxMin	13.21	MaxMin	13.20	kpb.MaxMin	12.04	kpb.MINMAX	14.00
kpb.MAXMAX	13.74	kpb.MINMAX	13.82	kpb.MAXMAX	14.45	MinMAX	14.66
MinMAX	15.43	kpb.MaxMin	15.49	MAXMAX	15.06	kpb.MaxMin	15.50
MAXMAX	16.29	kpb.MinMAX	16.63	MINMAX	15.98	MAXMin	15.76
MINMAX	16.85	MAXMin	18.34	SuffMAX	17.16	MAXSuff	15.76
SuffMAX	17.32	MAXSuff	18.34	MaxMin	17.32	kpb.MAXSuff	18.60
KPB.Sufferage	18.99	kpb.MAXMin	18.68	KPB.Sufferage	19.18	kpb.MAXMin	18.40
KPB.MinMin	20.20	kpb.MAXSuff	18.70	KPB.MinMin	19.82	kpb.MinMAX	18.97
KPB.MAXMin	21.40	MinMin	21.00	KPB.MAXSuff	21.50	kpb.MinMin	21.87
KPB.MAXSuff	21.40	kpb.MinMin	21.16	KPB.MAXMin	21.50	MinMin	21.88
KPB.MaxMin	23.08	kpb.Sufferage	23.12	KPB.MaxMin	23.37	kpb.Sufferage	22.90
KPB.SuffMAX	23.81	Sufferage	23.35	KPB.SuffMAX	23.63	Sufferage	23.30

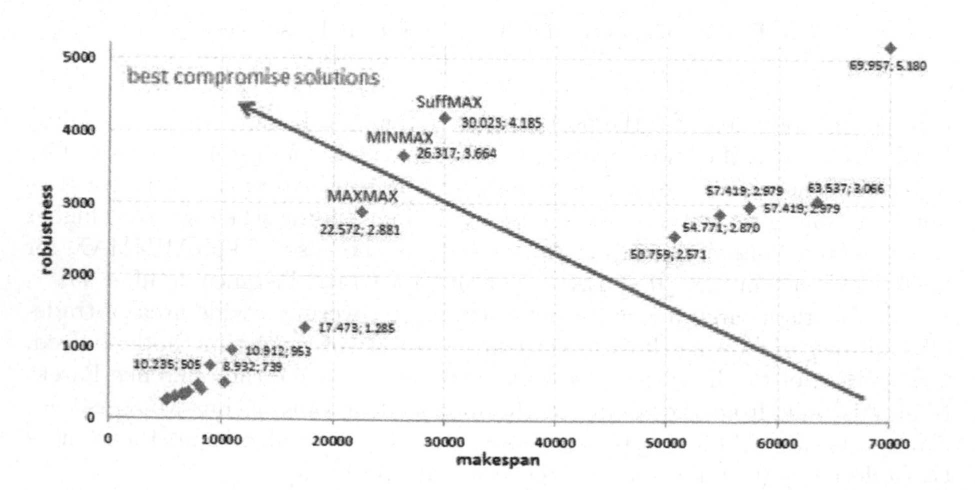

Fig. 3. Trade-off analysis for the HTRH scenario

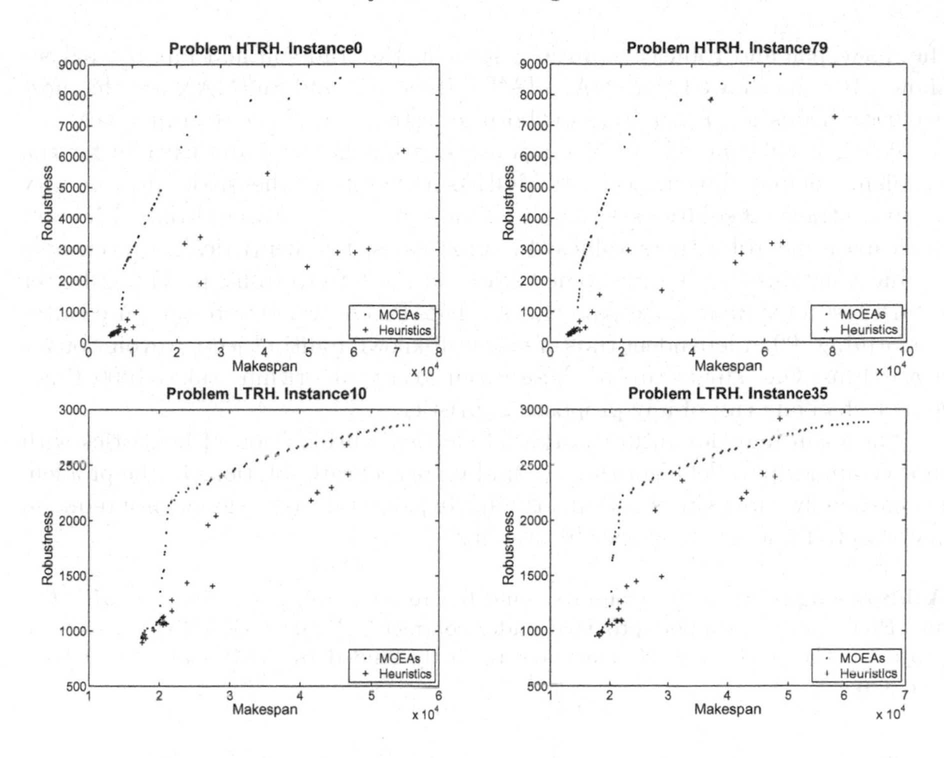

Fig. 4. Results comparison: heuristics ('+') and the reference Pareto front [8]

6 Conclusions and Future Work

This article addresses the robust scheduling problem in heterogeneous computing datacenters. This is a very important problem to address the uncertainties due to the lack of accurate estimations on the time required to perform tasks, trying to reduce the negative impact of such inaccuracies on the system performance.

The bi-objective problem of simultaneously optimizing the makespan and the robustness ratio of the computed schedules against inaccuracies in the execution time estimation of tasks, was formulated. These two objectives are in conflict, since reducing the makespan implies also reducing the robustness of the schedule.

Twenty-four greedy scheduling heuristics were proposed and evaluated. These heuristics are able to compute different trade-off solutions for the problem in efficient execution time, thus being more suitable for scheduling in real heterogeneous computing infrastructures. The proposed methods were designed following a two-phase approach considering the two objectives in the problem.

The experimental analysis demonstrated that accurate schedules are computed by the proposed heuristics. Results indicate that the traditional Sufferage method is the best heuristic for optimizing makespan. However, KPB.SuffMAX, applying the two-phase approach and a k-percent-best list of machines based on the maximum robustness ratio, is the best option for maximizing robustness. In addition, several other heuristics provide accurate trade-off values between the makespan and robustness ratio results, as the trade-off and Pareto analysis shows. Results showed that MAXMAX, MINMAX, and SuffMAX are the most accurate heuristics, accounting for both objectives on all problem instances.

When comparing the results against state-of-the-art Pareto fronts for the problem, computed using accurate MOEAs, the proposed heuristics provide very accurate trade-off solutions, specially in the region of the Pareto front with lower makespan and robustness values. In some cases, the heuristics are even able to find solutions dominating some regions of the Pareto front, providing better robustness for similar makespan values. These Pareto fronts were computed after performing 100 independent runs of four well-known multi-objective evolutionary algorithms. One single run of these evolutionary algorithms takes 1000 times longer than one run of any proposed heuristic.

The main lines for future work include the hybridization of heuristics with more complex parallel algorithms to find more accurate solutions to the problem in reasonably short times and identifying/exploring deeply regions of interest in the Pareto front for some specific systems.

Acknowledgment. B. Dorronsoro would like to acknowledge the Spanish MINECO and ERDF for the support provided under contract TIN2014-60844-R (the SAVANT project). The work of S. Nesmachnow is partly funded by ANII and PEDECIBA, Uruguay.

References

1. Ali, S., Maciejewski, A., Siegel, H., Kim, J.: Measuring the robustness of a resource allocation. IEEE Trans. Parall. Distrib. Syst. **51**(7), 630–641 (2004)
2. Artigues, C., Leus, R., Talla, F.: Robust optimization for resource-constrained project scheduling with uncertain activity durations. Flex. Serv. Manuf. J. **25**(1–2), 175–205 (2013)
3. Blazewicz, J., Ecker, K., Pesch, E., Schmidt, G., Weglarz, J.: Handbook on Scheduling-From Theory to Applications. Springer, Heidelberg (2007). https://doi.org/10.1007/978-3-540-32220-7
4. Braun, T., Siegel, H., Beck, N., Bölöni, L., Maheswaran, M., Reuther, A., Robertson, J., Theys, M., Yao, B.: A comparison of eleven static heuristics for mapping a class of independent tasks onto heterogeneous distributed computing systems. J. Parall. Distrib. Comput. **61**(6), 810–837 (2001)
5. Canon, L., Jeannot, E.: Evaluation and optimization of the robustness of DAG schedules in heterogeneous environments. IEEE Trans. Parall. Distrib. Systems **21**(4), 532–546 (2010)
6. Chiraphadhanakul, V., Barnhart, C.: Robust flight schedules through slack reallocation. EURO J. Transp. Logist. **2**(4), 277–306 (2013)
7. De Falco, I., Cioppa, A.D., Maisto, D., Scafuri, U., Tarantino, E.: A multiobjective extremal optimization algorithm for efficient mapping in grids. In: Mehnen, J., Köppen, M., Saad, A., Tiwari, A. (eds.) Applications of Soft Computing, vol. 58, pp. 367–377. Springer, Heidelberg (2009). https://doi.org/10.1007/978-3-540-89619-7_36

8. Dorronsoro, B., Bouvry, P., Cañero, J., Maciejewski, A., Siegel, H.: Multi-objective robust static mapping of independent tasks on grids. In: IEEE Congress on Evolutionary Computation, pp. 3389–3396 (2010)
9. Hart, E., Ross, P., Nelson, J.: Producing robust schedules via an artificial immune system. In: World Congress on Computational Intelligence, pp. 464–469 (1998)
10. Herrmann, J.: Handbook of Production Scheduling. International Series in Operations Research & Management. Springer, Heidelberg (2006). https://doi.org/10.1007/0-387-33117-4
11. Horowitz, E., Sahni, S.: Exact and approximate algorithms for scheduling nonidentical processors. J. ACM **23**, 317–327 (1976)
12. Iturriaga, S., García, S., Nesmachnow, S.: An empirical study of the robustness of energy-aware schedulers for high performance computing systems under uncertainty. In: Hernández, G., et al. (eds.) High Performance Computing, vol. 485, pp. 143–157. Springer, Heidelberg (2014). https://doi.org/10.1007/978-3-662-45483-1_11
13. Jensen, M.: Improving robustness and flexibility of tardiness and total flow-time job shops using robustness measures. Appl. Soft Comput. **1**, 35–52 (2001)
14. Kouvelis, P., Yu, G.: Robust Discrete Optimization and Its Applications. Kluwer Academic Publishers, Norwell (1997)
15. Leon, V., Wu, S., Storer, R.: Robustness measures and robust scheduling for job shops. IIE Trans. **26**(5), 32–43 (1994)
16. Leung, J.: Handbook of Scheduling-Algorithms, Models, and Performance Analysis. Chapman & Hall/CRC, New York (2004)
17. Luna, F., Chicano, F., Alba, E.: Robust solutions for the software project scheduling problem: a preliminary analysis. Int. J. Metaheuristics **2**(1), 59–79 (2012)
18. Luo, P., Lü, K., Shi, Z.: A revisit of fast greedy heuristics for mapping a class of independent tasks onto heterogeneous computing systems. J. Parall. Distrib. Comput. **67**(6), 695–714 (2007)
19. Nesmachnow, S.: Computación científica de alto desempeño en la Facultad de Ingeniería, Universidad de la República. Revista de la Asociación de Ingenieros del Uruguay **61**(1), 12–15 (2010). Text in Spanish
20. Nesmachnow, S., Cancela, H., Alba, E.: A parallel micro evolutionary algorithm for heterogeneous computing and grid scheduling. Appl. Soft Comput. **12**(2), 626–639 (2012)
21. Nesmachnow, S., Dorronsoro, B., Pecero, J., Bouvry, P.: Energy-aware scheduling on multicore heterogeneous grid computing systems. J. Grid Comput. **11**(4), 653–680 (2013)
22. Pinel, F., Dorronsoro, B., Bouvry, P.: Solving very large instances of the scheduling of independent tasks problem on the GPU. J. Parall. Distrib. Comput. **73**, 101–110 (2013)
23. Wu, S., Byeon, E., Storer, R.: A graph-theoretic decomposition of the job shop scheduling problem to achieve scheduling robustness. Oper. Res. **47**(1), 113–124 (1999)
24. Xhafa, F., Carretero, J., Dorronsoro, B., Alba, E.: A tabu search algorithm for scheduling independent jobs in computational grids. Comput. Inf. **28**, 1001–1014 (2009)

Applying Genetic Algorithms for the Improvement of an Autonomous Fuzzy Driver for Simulated Car Racing

Mohammed Salem[1(✉)], Antonio Miguel Mora[2], Juan Julian Merelo Guervós[3], and Pablo García-Sánchez[4]

[1] Department of Computer Sciences, University of Mascara, Mascara, Algeria
salem@univ-mascara.dz
[2] Department of Computer Sciences and Technology, ESIT, International University of La Rioja (UNIR), Logroño, Spain
antoniomiguel.mora@unir.net
[3] Department of Architecture and Computer Technology, University of Granada, Granada, Spain
jmerelo@ugr.es
[4] Department of Computer Science, ESI, University of Cádiz, Cádiz, Spain
pablo.garciasanchez@uca.es

Abstract. Games offer a suitable testbed where new methodologies and algorithms can be tested in a near-real life environment. For example, in a car driving game, using transfer learning or other techniques results can be generalized to autonomous driving environments. In this work, we use evolutionary algorithms to optimize a fuzzy autonomous driver for the open simulated car racing game TORCS. The Genetic Algorithm applied improves the fuzzy systems to set an optimal target speed as well as the instantaneous steering angle during the race. Thus, the approach offer an automatic way to define the membership functions, instead of a manual or hill-climbing descent method. However, the main issue with this kind of algorithms is to define a proper fitness function that best delivers the obtained result, which is eventually to win as many races as possible. In this paper we define two different evaluation functions, and prove that fine-tuning the controller via evolutionary algorithms robustly finds good results and that, in many cases, they are able to play very competitively against other published results, with a more relying approach that needs very few parameters to tune. The optimized fuzzy-controllers (one per fitness) yield a very good performance, mainly in tracks that have many turning points, which are, in turn, the most difficult for any autonomous agent. Experimental results show that the enhanced controllers are very competitive with respect to the embedded TORCS drivers, and much more efficient in driving than the original fuzzy-controller.

Keywords: Videogames · Fuzzy controller · TORCS
Steering control · Optimization · Genetic algorithms

J. Medina et al. (Eds.): IPMU 2018, CCIS 855, pp. 236–247, 2018.
https://doi.org/10.1007/978-3-319-91479-4_20

1 Introduction

Autonomous driving has become a hot research topic since many traditional and emerging companies have entered the arena of their design and manufacture. Automotive industry needs the creation of real self-driving cars that can travel in everyday roads and streets or, for that matter, in a desert or hostile environment. However, this is only an objective; autonomous vehicles should also optimize fuel consumption as well as car safety and, in some cases, occupant comfort [10].

Optimization in car racing simulators or games can be placed in that context, with solutions obtained there having utility beyond the game itself. For instance, The Open Racing Car Simulator (TORCS) [21] is a realistic racing simulator with a sophisticated engine which has been used for many standalone racing competition challenges. This fact, combined with the possibility to compare controllers, have made TORCS one of the most used simulator in the field of autonomous driving [4,6,9,16].

Among the different types of controllers that have been used in TORCS, fuzzy-based ones have proved to have one of the best performances, since they simulate in part the human reasoning when driving [12,17]. Thus, the authors presented previously an approach in which two specialized fuzzy controllers were combined to decide the car's steering angle and desired speed in every single point (or tick) during a race [19]. It yielded good results in several tracks, but the autonomous driver had some troubles in difficult circuits, such as those with many curves or competing against tough rivals.

Its main disadvantage is related with the parameters of the fuzzy systems, namely those which define the membership functions, which were set 'manually'. Thus, in this work we have considered this issue as an optimization problem, so we have applied a real-coded Genetic Algorithm (GA) [7] to obtain the best configuration.

This kind of algorithm mainly require the definition of a good evaluation or fitness function, which will determinate the final performance of the solution (an autonomous driver in this case). Thus, in order to evaluate an individual/solution of the algorithm (a potential driver), we set the parameters which define its Artificial Intelligence (AI) engine, and then we put it in a test race (with or without rivals). Then, the fitness function follows two different approaches for the evaluation: the first one is computed using the mean lap time and damage obtained during the race, while the second also considers the maximum speed reached.

Once optimized, the best genetic-fuzzy based controllers (one per fitness) have been evaluated in a practice race (without rivals) first, and then in a real race against different drivers in TORCS. According to the obtained results the enhanced controllers perform both much better than the original fuzzy controller, improving the lap time and reducing the received damage. Moreover they are much more competitive against tough rivals, reaching high ranks in the most difficult races.

The rest of the paper is organized as follows. Next we present the state of the art, to be followed by a description of the TORCS simulator in Sect. 3 and the method for optimizing fuzzy controllers will be presented in Section Results will be presented next in Sect. 4. Finally, conclusions and future lines of work will be presented in Sect. 5.

2 State of the Art

Evolutionary algorithms have targeted TORCS almost since its publication, for instance, for determining the optimal trajectory of a lap in a known circuit [18], but this approach suffers from the problem that the obtained trajectory in the evolving process strongly depends on the initial state of the car. In the same context, the authors in [14] tried to design a novel approach to compute the optimal racing line without any human intervention, using a GA to find the best trade-off between the minimization of two conflicting objectives: the length and the curvature of the racing line.

However, definitely, the most prolific area of application of EAs inside TORCS has been the optimization of autonomous controllers for car driving, i.e. conducting a meta-optimization process. Thus, EAs have been applied to 'refine' the parameters which define the driver's behavior [1,9], or to improve the structure/architecture of the models [11,20], working offline, or online (during the game) [2,22]. Our approach is focused in this line, proposing the application of an off-line genetic algorithm for the improvement of the parameters which determine the behavior of a controller for TORCS. We have focused on a Fuzzy-based model, as it is one of the best options for modeling human-like decisions and actions, as others authors have also used this kind of technique in the literature with good results [17]. For instance, in [6], a fuzzy rule-based car controller for a Car Racing Competition was built and tuned with co-evolutionary genetic algorithms. Two fuzzy sub-controllers were designed (acceleration and turning angle). But this approach was applied to a simpler simulator than TORCS which is a more realistic and time-constrained simulator.

Pérez et al. introduced an evolutionary fuzzy approach for TORCS in [12], where they applied EAs for improving fuzzy models to infer the acceleration and turning angle. However, the models were not so specialized as the proposed here, since their controller did not compute the target speed, which is a key factor for a competitive controller.

Onieva et al. [16] presented a parametrized modular architecture with a fuzzy system and a GA in the design of fuzzy logic controllers for steering wheel management that can reproduce human driver behavior, but it did not take the target speed into account, unlike our previous controller [19] which computed the target speed and the steer with two fuzzy sub-controllers and whose membership functions parameters were defined by trial/error process. In this paper, we propose to optimize these parameters using a real coded genetic algorithm aiming to improve the performance of the original fuzzy controller.

3 Experimental Setup

The Open Racing Car Simulator (TORCS) [21] is an open source, modern, multi-player, modular and portable racing simulator that allows users to race against computer-controlled opponents. There is a large set of sensors [13] which the car can consider during a race, such as distances to track borders, to rivals, current fuel, current gear, position in the race, speed, or damage, among others. A controller is a program, which runs inside TORCS, that automatically drives a car. It gets as input information about the current state of the car and its situation on the track (sensors). These collected data are used to decide actions to perform in the next simulation tick.

The initial proposed controller [19] has the same modular architecture as the simple TORCS driver, however, the target speed and steering angle are computed by means of two modular and specialized fuzzy sub-controllers, which consider five position sensors. This is the controller which will be improved by means of a GA in this work.

The *fuzzy target speed sub-controller* aims to estimate the optimal target speed of the car, both in straight parts and curves of the track, taking into account two criteria: move as fast as possible and be safe. This estimation is based on two general cases: if the car is in a straight line, the target speed will take a maximum value (*maxSpeed* km/h). However, if it is close to a curve, the controller will decrease the current speed to a value included in the interval *[minSpeed, maxSpeed]* km/h.

This fuzzy controller has an output, the speed, and three input values:

- Front = Track_9: front distance to the track border (angle 0°).
- M5 = max (Track_8, Track_10): max distance to the track border in an angle of +5° and -5° with respect to Front.
- M10 = max (Track_7, Track_11): max distance to track border in an angle of +10° and -10°.

It is a Mamdani-based fuzzy system [8] with three trapezoidal Membership Functions (MF) for every input variable. The description of these fuzzy inputs and output are represented in Table 1. In [19] we set the values by hand; previously we had made initial tests using an evolutionary algorithm. In this paper we will try to improve obtained results by fine-tuning this evolutionary algorithm.

Table 1. Fuzzy variables description.

Variable	Range	Name	MF	Low	Medium	High
Input	[0–100] m	Front	trapezoidal	[0–50]	[20–80]	[60–100]
Input	[0–100] m	M5	trapezoidal	[0–40]	[10–70]	[50–100]
Input	[0–100] m	M10	trapezoidal	[0–30]	[20–60]	[50–100]
Output	[0–200] m/s	TargetSpeed	singleton	/	/	/

The proposed optimization approach aims to find the optimal parameters of the membership functions of the two sub-controllers previously introduced. The GA starts by creating the initial population with random values for the parameters in the defined range [0, 100]. The fitness of each candidate solution is computed by injecting its gene values to the parameters of the membership functions of the two fuzzy sub-controllers. The defined autonomous controller is used to drive a car in a 20 laps race in a circuit without opponents, and the results (Top speed, Damage and Mean Lap time) are used to compute the fitness value.

As previously stated, the designed fuzzy controllers have trapezoidal membership functions. In such a controller, fuzzy rules are applied to linguistic terms. These terms, which qualify a linguistic variable, are defined through membership functions, which, in turn, depend on a set of parameters that 'describes' their shape (and operation). Using a GA we will optimize the parameters of the membership functions that constitute the fuzzy partition of the linguistic variable [16]. The input linguistic variables in our problem, *Front, Max5* and *Max10*, are represented by three trapezoidal membership functions (See Table 1).

And a fuzzy partition with n trapezoidal membership functions is defined by $2n$ variables ($a = x_1, x_2, ..., x_{2n} = b$) (Eq. 2). In this case, the representation is given by the Fig. 1 with:

$$a = x_1 \leq x_2 \leq ... \leq x_{2n-1} \leq x_{2n} = b \tag{1}$$

$$\begin{aligned} \mu_{A1}(x) &= \begin{cases} 1, & x_1 \leq x \leq x_2 \\ \frac{x_3 - x}{x_3 - x_2}, & x_2 \leq x \leq x_3 \\ 0, & x > x_3 \end{cases} \\ \mu_{Ai}(x) &= \begin{cases} 0, & x \leq x_{2i-2} \\ \frac{x - x_{2i-2}}{x_{2i-1} - x_{2i-2}}, & x_{2i-2} \leq x \leq x_{2i-1}, n = 2, ..., i-1 \\ 1, & x_{2i-1} \leq x \leq x_{2i} \\ \frac{x_{2i+1} - x}{x_{2i+1} - x_{2i}}, & x_{2i} \leq x \leq x_{2i+1} \\ 0, & x > x_{2i+1} \end{cases} \\ \mu_{An}(x) &= \begin{cases} 0, & x \leq x_{2n-2} \\ \frac{x - x_{2n-2}}{x_{2n-1} - x_{2n-2}}, & x_{2n-2} \leq x \leq x_{2n-1} \\ 1, & x > x_{2n-1} \end{cases} \end{aligned} \tag{2}$$

When the number of parameters is reduced and their ranges of variations are well defined, a GA with a binary coding is largely sufficient to find their optimal values. On the other hand, if the number of parameters becomes important, and their variation interval is not well known, the real coding is the most appropriate

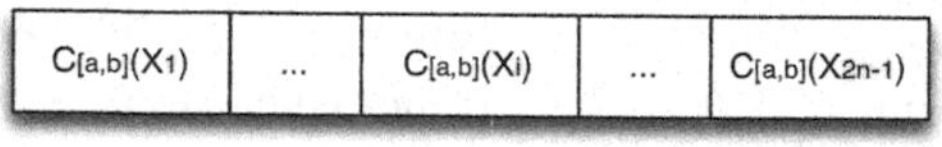

C[a,b](X1)	...	C[a,b](Xi)	...	C[a,b](X2n-1)

Fig. 1. Trapezoidal-shaped MFs coding

[3]. Since our work requires some precision and the variation interval of each parameter is not well known, we have considered a real coding implementation in a vector that includes all variables to optimize.

The initialization of the chromosomes (first population) is performed assigning random values inside a range of variation [5], in order to start from feasible values [19]. Tournament based selection has been used to elect chromosomes as parents for genetic operators, while simple arithmetic two point crossover [23] and non uniform mutation [15] have been chosen, as two of the most contrasted methods in the literature.

The objective of the car controller is to win as many races as possible. However, we have to optimize the most general case by carrying out solo *training races* in which we try to minimize the damage of the car (*damage*) and the average lap time *LapTime*, while maximizing *TopSpeed*. It is a multiobjective optimization problem, but since we want to obtain a single controller, we will have to use heuristics to derive two possible fitness functions:

GFC1:

$$f_1 = damage + \alpha \cdot LapTime \tag{3}$$

GFC2:

$$f_2 = damage + \alpha \cdot LapTime + \beta \cdot \frac{1}{TopSpeed} \tag{4}$$

α and β are two heuristic weights. The main difference is in the use of the *TopSpeed* in the fitness to enhance the controller performances in straight lines aiming to reduce lap time and the overall race time. To evaluate the candidate controllers during the evolutionary process, we will make each of them compete in a 20 laps practice race in a medium difficulty circuit without rivals. We have omitted the presence of opponents in order to avoid including additional uncertainty sources to the optimization process.

Then, the obtained output values *damage*, *LapTime* and *TopSpeed* are collected to compute the corresponding fitness value. As a clarification, *LapTime* is the average of the 20 laps time.

4 Experimental Results

We will first need to choose whose tracks and cars are going to be used in the experiment among the ones TORCS provides; in our case, we have selected the E-Track5 circuit as it is a quite complex one, with multiple turns. *car1-tbr1* has been selected as the driving car [19]. According to previous experiments, this is a fair choice due to its moderate performance. This will lead our controller to be prepared to drive in the most usual conditions.

We have evaluated the FGC with the two proposed fitness functions, comparing them for racing performance. Also, we have carried out two algorithm executions with two different population sizes: 20 and 50, respectively. The rest of the parameters are: Generations = 50, Crossover rate = 0.7, Mutation rate = 0.3, number of runs per configuration = 20.

- **GFC1**: GA-Fuzzy controller with fitness 1 (Eq. 3).
- **GFC2**: GA-Fuzzy controller with fitness 2 (Eq. 4).

The coefficients α and β are chosen to be 1 and $10 * MaxSpeed$ respectively, where $MaxSpeed$ is the maximum value of speed that *car1-tbr1* could take ($MaxSpeed = 300$) [19], this choice is motivated by the fact to normalize the Top speed values and make them in the same level as other fitness terms. The results of these runs are shown in Table 2. Wilcoxon rank sum non-parametric test is used to reject or accept the null hypothesis of equality of medians of the values of the two fitness functions for the 20 runs with 50 chromosomes. The obtained p-value was $p = 0.0011$, this result lead to the rejection of null hypothesis with a threshold $\alpha = 0.01$ which allows us to conclude that the two samples sets are different.

Table 2. Results of 10 runs of GA with the two fitness functions. Please bear in mind that fitness follow different formula, and thus cannot be compared; LapTime and Damage should be the quantities used for comparison.

	With population size 20							
	GFC1				GFC2			
	Min GFC1	*LapTime*	*Damage*	*TopSpeed*	Min fit. 2	*LapTime*	*Damage*	*TopSpeed*
Best	**29.44**	29.44	0	231	**39.74**	29.25	0	286
Mean	33.88	30.79	3.10	227.43	44.14	30.14	2.70	267.30
St. Dev.	5.61	1.18	4.58	32.55	6.73	0.78	5.81	22.03
	With population size 50							
	GFC1				GFC2			
	Min GFC2	*LapTime*	*Damage*	*TopSpeed*	Min fit. 2	*LapTime*	*Damage*	*TopSpeed*
Best	**28.78**	28.78	0	233	**28.11**	38.52	0	288
Mean	33.14	29.89	4.14	230.19	42.93	29.57	3.46	271.74
St. Dev.	4.98	1.32	4.22	31.14	5.63	0.93	5.19	23.90

Since the **GFC2** controller also optimizes the TopSpeed, we can notice that it is clearly superior to that of **GFC1** which surely influences the overall Lap time. This improvement in TopSpeed greatly increases the performance of the **GFC2**.

Increasing the population size to 50 has led to better values of the two controllers, yielding better values for lap time, speed and damage. This increase in the population size has allowed a better coverage of the research space, thus getting closer to the optimal solution. The best solution obtained with each fitness function from these runs will be used in races against selected opponents. The shapes of the obtained membership functions are completely different from those obtained by Trial/Error in the previous work [19] where the Medium linguistic variable of the new functions has bigger range. This makes the controller very sensitive to the middle distances of the inputs, like for a real driver who considers most of the cases the car distance from the borders in that range. The other remark from the obtained membership functions is the dimension of the common range between the LOW and MEDIUM, which provide a higher diversity in the output values.

The two best genetic based fuzzy controllers obtained in the previous experiments, one per fitness function, and thus named $GFC1$ and $GFC2$, have been tested in a practice race together. They have been run each one for 20 laps in E-Track5 circuit, which was the one used during the evolution; then, they will be tested also in a practice race in E-Road, a track not used previously. The obtained results are presented in Table 3.

Table 3. Results of the three controllers in a 20 laps practice race. Results of the *AD* controller [19], a hand-designed fuzzy controller, are included for comparison purposes where tested.

E-Track 5			
Results	**AD**	**GFC1**	**GFC2**
Best Lap Time	29:70	30:01	29:21
Top Speed	209	224	234
Min Speed	168	151	186
Damage	936	0	0

CG Track2		
Results	**GFC1**	**GFC2**
Best Lap Time	01:04:96	01:02:32
Top Speed	220	238
Min Speed	33	40
Damage	0	0

From the table, we can see that the fuzzy controllers optimized by the GA yield the best results, obtaining very good overall global race times and eliminating damage, which is reduced to 0. For the sake of comparison, we include the hand-designed AD controller [19], which finished the practice race where it was tested with a lot of damage, implying that it could, in some difficult cases, not finish the race. Testing the controllers in CG Track2, which is quite long and difficult as it can be seen by the time it takes to run a single lap, has proved their value in the adaptation to other tracks different from the one used for 'training', that is, the optimization of the fuzzy controllers.

The GFC2 controller has run with a higher speed (considering overall Top Speed and Min Speed) than GFC1 in the two tracks. This is a positive consequence of the inclusion of the *TopSpeed* variable in the fitness computation so the GA based fuzzy controller has optimized the speed of the car due to early braking and detection of turns and their curving angles. This ability of the GA-fuzzy controller collaborates to minimize the overall race time and thus the final ranking. According to these results, GFC2 seems to be the best controller.

Comparing average lap time gives us an overall idea of which controller performs the best; however, at the end of the day in a racing game the race has to be won. That is why we have tested every fuzzy separately from the others in a real race against five standard controllers from each team integrated with TORCS. Tables 4 and 5 illustrate their performance in two 5 laps real races.

$GFC1$ and $GFC2$ controllers are quite competitive in these races; GFC2 has got an excellent second position in the track used during optimization (E-Track), and it has also got a remarkable third rank in the unknown track (E-Road).

Table 4. Results of GFC1 in two real races (5 laps)

E-TRACK5	GFC1	berwin 10	bt 3	damned 2	inferno 5	tita 10
Ranking	3/6	4/6	1/6	5/6	2/6	6/6
Race Time	02:29:32 + 24:11	02:29:32 + 1 lap	02:29:32	02:29:32 + 1 lap	02:29:32 + 13:67	02:29:32 + 1 lap
Best Lap	33:79	35:39	28:09	36:73	31:49	34:12
Max Speed	199	206	233	198	229	219
Damages	0	0	0	599	7	566
CG Track2	GFC1	berwin 10	bt 3	damned 2	inferno 5	tita 10
Ranking	3/6	4/6	1/6	6/6	5/6	2/6
Race Time	05:10:66 + 25:43	05:10:66 + 55:65	05:10:66	05:10:66 + 1 lap	05:10:66 + 38:44	05:10:66 + 19:82
Best Time	1:03:65	1:04:21	1:00:57	1:04:26	1:03:19	1:03:98
Max Speed	233	236	288	200	238	229
Damage	112	376	433	988	541	890

Table 5. Results of GFC2 in two real races (5 laps)

E-TRACK5	GFC2	berwin 10	bt 3	damned 2	inferno 5	tita 10
Ranking	2/6	4/6	1/6	6/6	3/6	5/6
Race Time	02:30:83 + 03:99	02:30:83 + 1 lap	02:30:83	02:30:83 + 1 lap	02:30:83 + 08:35	02:30:83 + 1 lap
Best Time	29:82	36:38	28:35	37:04	30:53	36:00
Max Speed	214	202	230	188	226	204
Damage	0	0	343	1230	0	668
E-ROAD	GFC2	berwin 10	bt 3	damned 2	inferno 5	tita 10
Ranking	3/6	4/6	1/6	6/6	2/6	5/6
Race Time	05:38:23 + 17:72	05:38:23 + 1 lap	05:38:23	05:38:23 + 1 lap	05:38:23 + 10:73	05:38:23 + 1 lap
Best Time	1:17:34	1:16:29	1:14:97	1:20:80	1:13:98	1:15:29
Max Speed	221	206	228	178	228	206
Damage	120	356	753	2750	130	894

Both controllers have dealt very well for not being damaged, which even the winner, `bt 3`, could not avoid.

These results are a confirmation of the proper optimization done by the GA and mainly when the Top Speed was considered in the fitness. The obtained results in real races with opponents from tough teams of TORCS are encouraging even if the optimization process was in practice races. This good adaptation of the proposed controller in races with rivals is due to the fact the modular fuzzy controller takes into consideration the presence of opponents in the track [19]. The enhancement of that driver by the optimal values of the membership function values, allows it to detect and overtake the other cars with no damage or stuck.

In the last experiment, we tried to get the best of our controller by testing its limits in a disadvantageous track, so A-Speedway was selected. The results are shown in Table 6 where one could clearly notice the degradation of the performances of the optimized fuzzy controller. It was ranked in the fourth position just before `tita10` and `bt3` controllers. Its top speed is acceptable considering that the used car is not as fast as others but the fast lap time was higher and the GFC2 controller was not as competitive with its rivals in this race.

These results could be justified by the nature of the track. A-Speedway is an oval circuit with only four turns, this kind of tracks deprives our controller of its strongest point which is the late braking and the selection of an optimal trajectory in turns. The fact that the car has a worse average speed compared to the others and that our controller can not compensate for this loss of time in the turns will not work in favor of getting better best lap times.

Table 6. Results of GFC2 in a real race (5 laps)

A-Speedway	GFC2	berwin 10	bt 3	damned 2	inferno 5	tita 10
Ranking	4/6	2/6	1/6	6/6	3/6	5/6
Race Time	02:37:74 + 29:12	02:37:74 + 19:03	02:37:74	02:37:74 + 1 lap	02:37:74 + 28:89	02:37:74 + 1 lap
Best Time	35:78	32:83	29:40	39:49	34:86	40:92
Max Speed	239	251	266	223	259	238
Damage	119	615	1290	363	739	899

5 Conclusions and Future Work

In this work, we have presented an improved Genetic Algorithm implementation that optimizes and improves an autonomous driver using fuzzy systems for TORCS simulator [19]. It combines two sub-controllers, one to calculate the target speed and the other for the direction, that is, for driving the steering wheel.

After initial tests, that showed the promise of using evolutionary algorithms with two different fitness functions, one considering the average lap time and the car damage and another adding the top speed reached, we have fine-tuned some algorithm parameters to obtain better results.

The yielded results are very promising since the optimized controllers (one per fitness function) were ranked among the first ones in three different evaluation races with rivals, with the minimum of damage.

In the comparison with the original (before the optimization) fuzzy controller, the improvement can be clearly seen in the results. The new controllers are able to drive much faster than it, and moreover they manage to not receiving any damage, while the original controller even crashed the car in some races.

The results show that including the top speed in the calculus improves results, since the obtained drivers have proved to be able to run a 10 to 15% faster in the races. However the damage term must be also considered to 'compensate' somehow the influence of the top speed, otherwise the controller would be extremely aggressive and would not finish many of the races.

Thus, we can conclude from the results that the presented evolutionary algorithm with the proposed fitness functions are well suited for finding the best trade-off between the two objectives of any racing controller: damage and speed.

Nevertheless, these results can be improved by extending the evaluation of population controllers in the Genetic algorithm to other tracks and not just one, to allow the elected controller to adapt to many different situations during the races. The applied GA could be improved in different ways, for instance, reducing its computation time by means of the parallelization of the evaluation phase. Also, a multi-objective approach could be implemented, in which the main objectives to address by the controller could be optimized at once. Moreover, we could also try to generate, optimize and tune automatically the rule base of the fuzzy controller by means of a Genetic Programming algorithm.

Finally, the fuzzy controller could be evolved and adapted to be an efficient autonomous driver for a real car. This could be addressed by considering real-life traffic situations instead of races and, of course, redefining the fitness functions to accomplish other objectives, mainly related with security and comfort.

Acknowledgments. This work has been supported in part by: Ministerio español de Economía y Competitividad under project TIN2014-56494-C4-3-P (UGR-EPHEMECH), TIN2017-85727-C4-2-P (UGR-DeepBio) and TEC2015-68752 (also funded by FEDER).

References

1. Butz, M.V., Lönneker, T.D.: Optimized sensory-motor couplings plus strategy extensions for the TORCS car racing challenge. In: Lanzi, P.L. (ed.) Proceedings of the 2009 IEEE Symposium on Computational Intelligence and Games, CIG 2009, Milano, Italy, 7–10 September 2009, pp. 317–324. IEEE (2009)
2. Cardamone, L., Loiacono, D., Lanzi, P.L.: On-line neuroevolution applied to the open racing car simulator. In: Proceedings of the Eleventh Conference on Congress on Evolutionary Computation, CEC 2009, pp. 2622–2629. IEEE Press, Piscataway (2009)
3. Elsayed, S.M.M., Sarker, R., Essam, D.L.: A genetic algorithm for solving the CEC 2013 competition problems on real-parameter optimization. In: IEEE Congress on Evolutionary Computation, CEC 2013, Cancun, Mexico, 21–23 June 2013, pp. 356–360 (2013)
4. Floreano, D., Kato, T., Marocco, D., Sauser, E.: Coevolution of active vision and feature selection. Biol. Cybern. **90**, 218–228 (2004). https://doi.org/10.1007/s00422-004-0467-5
5. Goldberg, D.E.: Genetic Algorithms in search, optimization and machine learning. Addison Wesley, Reading (1989)
6. Guadarrama, S., Vazquez, R.: Tuning a fuzzy racing car by coevolution. In: Genetic and Evolving Systems, GEFS 2008. IEEE, March 2008. https://doi.org/10.1109/GEFS.2008.4484568
7. Herrera, F., Lozano, M., Verdegay, J.: Automatic track generation for high-end racing games using evolutionary computation. Artif. Intell. Rev. **12**(4), 265–319 (1998)
8. Iancu, I.: A Mamdani Type Fuzzy Logic Controller, pp. 325–352. InTech (2012)
9. Kim, T.S., Na, J.C., Kim, K.J.: Optimization of an autonomous car controller using a self-adaptive evolutionary strategy. Int. J. Adv. Robot. Syst. **9**(3), 73 (2012)

10. Kolski, S., Ferguson, D., Stacniss, C., Siegwart, R.: Autonomous driving in dynamic environments. In: Proceedings of the Workshop on Safe Navigation in Open and Dynamic Environments at the IEEE/RSJ International Conference on Intelligent Robots and Systems (IROS), Beijing, China (2006)
11. Koutnik, J., Cuccu, G., Schmidhuber, J., Gomez, F.: Evolving large scale neural networks for vision based TORCS. In: Foundations of Digital Games. J. Koutnik, Dipartimento tecnologie innovative Istituto Dalle Molle di studi sull'intelligenza artificiale, March 2013. http://repository.supsi.ch/id/eprint/4548
12. Liébana, D.P., Recio, G., Sáez, Y., Isasi, P.: Evolving a fuzzy controller for a car racing competition. In: Lanzi, P.L. (ed.) Proceedings of the 2009 IEEE Symposium on Computational Intelligence and Games, CIG 2009, Milano, Italy, 7–10 September 2009, pp. 263–270. IEEE (2009). https://doi.org/10.1109/CIG.2009.5286467
13. Loiacono, D., Cardamone, L., Butz, M., Lanzi, P.L.: The 2011 simulated car racing championship @ cig-2011. TORCS news (2011). http://cig.dei.polimi.it/wpcontent/
14. Loiacono, D., Lanzi, P.L., Bardelli, A.P.: Searching for the optimal racing line using genetic algorithms. In: 2010 IEEE Proceedings of the Symposium on Computational Intelligence and Games (CIG). IEEE Press, Copenhagen (2010). https://doi.org/10.1109/ITW.2010.5593330
15. Neubauer, A.: A theoretical analysis of the non-uniform mutation operator for the modified genetic algorithm. In: Proceedings of the IEEE International Conference on Evolutionary Computation. IEEE Press, Indianapolis (1997). https://doi.org/10.1109/ICEC.1997.592275
16. Onieva, E., Alonso, J., Perez, J., Milanés, V.: Autonomous car fuzzy control modeled by iterative genetic algorithms. In: Fuzzy Systems, pp. 1615–1620 (2009). https://doi.org/10.1109/FUZZY.2009.5277397
17. Onieva, E., Pelta, D., Godoy, J., Milanés, V., Rastelli, J.: An evolutionary tuned driving system for virtual car racing games: the autopia driver. Int. J. Intell. Syst. **27**, 217–241 (2012). https://doi.org/10.1002/int.21512
18. Saez, Y., Perez, D., Sanjuan, O., Isasi, P.: Driving cars by means of genetic algorithms. In: Rudolph, G., Jansen, T., Beume, N., Lucas, S., Poloni, C. (eds.) PPSN 2008. LNCS, vol. 5199, pp. 1101–1110. Springer, Heidelberg (2008). https://doi.org/10.1007/978-3-540-87700-4_109
19. Salem, M., Mora, A.M., Merelo, J.J., García-Sánchez, P.: Driving in TORCS using modular fuzzy controllers. In: Squillero, G., Sim, K. (eds.) EvoApplications 2017. LNCS, vol. 10199, pp. 361–376. Springer, Cham (2017). https://doi.org/10.1007/978-3-319-55849-3_24
20. SeongKim, T., Na, J.C., Kim, K.J.: Optimization of an autonomous car controller using a self-adaptive evolutionary strategy. Int. J. Adv. Robot. Syst. **9**(3), 73 (2012)
21. Sourceforge: Web TORCS. Web, November 2016. http://torcs.sourceforge.net/
22. Tan, C.H., Ang, J.H., Tan, K.C., Tay, A.: Online adaptive controller for simulated car racing. In: Proceedings of the IEEE Congress on Evolutionary Computation, CEC 2008, 1–6 June 2008, Hong Kong, China, pp. 2239–2245. IEEE (2008). https://doi.org/10.1109/CEC.2008.4631096
23. Varun Kumar, S.G., Panneerselvam, R.: A study of crossover operators for genetic algorithms to solve VRP and its variants and new sinusoidal motion crossover operator. Int. J. Comput. Intell. Res. **13**(7), 1717–1733 (2017)

A Self-Organizing Ensemble of Deep Neural Networks for the Classification of Data from Complex Processes

Niclas Ståhl(✉), Göran Falkman, Gunnar Mathiason, and Alexander Karlsson

School of Informatics, University of Skövde, Högskolevägen 28, SE 54145 Skövde, Sweden
niclas.stahl@his.se

Abstract. We present a new self-organizing algorithm for classification of a data that combines and extends the strengths of several common machine learning algorithms, such as algorithms in self-organizing neural networks, ensemble methods and deep neural networks. The increased expression power is combined with the explanation power of self-organizing networks. Our algorithm outperforms both deep neural networks and ensembles of deep neural networks. For our evaluation case, we use production monitoring data from a complex steel manufacturing process, where data is both high-dimensional and has many nonlinear interdependencies. In addition to the improved prediction score, the algorithm offers a new deep-learning based approach for how computational resources can be focused in data exploration, since the algorithm points out areas of the input space that are more challenging to learn.

Keywords: Self organisation · Ensemble methods
Complex processes · Artificial neural networks

1 Introduction

Our modern society generates tremendous amounts of data and the potential impact of data analysis is immense [16]. Using recent powerful data analysis techniques, data collected from *complex* processes containing multiple non-trivial dependences, can be analyzed to achieve new insights. Deep Learning has proven itself in several areas to be a powerful technique that is especially successful in the analysis of very complex and complicated systems, and it has therefore gained a lot of attention in the machine learning community [10]. Due to the large amounts of data in the manufacturing industry, combined with the potential of both process and product improvements for an increased competitiveness, new machine learning techniques start to gain attraction in that industry. In steel manufacturing, there are many complex and interdependent physical and chemical processes involved, and is it difficult to fully understand the entire process

J. Medina et al. (Eds.): IPMU 2018, CCIS 855, pp. 248–259, 2018.
https://doi.org/10.1007/978-3-319-91479-4_21

for full process control[1]. A transition into a more data-driven manufacturing process development has the potential to utilize intelligent data analysis for a more thorough process understanding and improved control. For the evaluation of our proposed algorithm, we have chosen one particular problem within the hot rolling steel industry. In hot rolling, coils of rolled steel sheets can be obliquely reeled, causing the final product to be askewed. This is called *telescoping*, and the underlying factors causing such defects, in this specific production setting, is not fully understood. The classification of finished coils, being telescoped of not, is conducted manually by an operator. In this paper, we present a novel approach for *self organizing ensemble of neural networks* (SOENN) to be used for such a classification tasks. Besides that this algorithm performs best in our experiments, it also allows for interpretation of what the algorithm has learned and decided to focus on in the input data.

The SOENN algorithm is inspired by the *supervised growing neural gas* (SGNG) algorithm [4], but in SOENN neural networks are dynamically added and removed from an ensemble, rather than individual neurons. Each network within this ensemble is assigned a position in the input space. These networks can then only focus on samples in their vicinity. Hence, the global classification problem is split into many smaller local problems. The self organizing property also allows SOENN to find good ways to distribute its networks in the input space and thus trying to find optimal sub problems that can be used to solve the global problem. The result is that in the ensemble more networks will be located in areas of the input space with samples that are more difficult to classify. The global solution consists of the combination of relative positions of the networks and the weights of the networks. For the data of this study, we believe that there are several regions of the input space that are independent of each other, since the data comes from different production stations in a complex production line where multiple types of products are produced. By modeling a complex mix of stations, using an expressive self-organizing network, we believe that SOENN can model both the overall structure of the input space, as well as the individual stations.

The hypothesis of this paper is that an algorithm that considers multiple local problems separately, such as the SOENN, would perform better than other algorithms, such as a Deep Neural Network (DNN). To this end, the SOENN algorithm is compared to a DNN, where the total number of neurons is the same in the ensemble as in the DNN. A common drawback of DNNs is that they are very difficult to interpret and thus, it is often intractable to understand and gain further knowledge about the data through these methods. The SOENN algorithm consists of several smaller artificial neural networks, easier to analyze and interpret that the full DNN. The smaller networks are distributed over the input space, and each (relative) position reveals where the ensemble needs to allocate more networks to model the data in that sub problem. Hence, these

[1] This work was supported by Vinnova and Jernkontoret under the project Dataflow. We would like to thank Andreas Persson at Outokumpu AB for the valuable collaboration.

positions can be used to find problematic sub problems, to target for deeper analysis. Further, we compare the self organizing ensemble with an ensemble without this property, to evaluate whether an improvement is due to the self organizing property itself. The SOENN algorithm is shown to significantly outperform the two other algorithms used for comparison. We conjecture that this is mainly due to the networks ability to assign networks to different subspaces of the input space and hence, split the problem into several smaller sub-problems. While there is still remaining work to be done in making the result of the SOENN algorithm interpretable, we argue for that the algorithm presents some insights about the data it is trained on and thus, is a step in the right direction in order to make deep learning methods interpretable in the context of complex processes and data.

2 Background

In this section, the three different methods and ideas which the SOENN algorithm is inspired by are presented. First, a description of self organizing networks is given. There will be a special focus on the supervised growing neural gas algorithm, which is the main inspiration to the presented algorithm. Secondly, a description of ensemble methods is presented. Finally, an overview to deep neural networks is given.

2.1 Self Organizing Neural Networks

While there exists a large variety of self organizing neural networks, they all relate to *Kohonens self-organizing feature map*, which was introduced by [9]. In a self-organizing feature map, neurons are arranged in a low dimensional discretized space. The most common approach is to use two or three dimensions so that the resulting configuration can be visualized. A mapping between the space of the map and the input space is learnt through competitive Hebbian learning and thus the self-organizing feature map finds a low dimensional representation of the data in the input space. This method has previously been applied to visualize and further understand steel manufacturing processes [2].

Another self organizing method is the *growing neural gas* algorithm that was first introduced by [4]. The main idea behind this algorithm was to combine the *neural gas* [13] and the *growing cell structure* algorithm [3]. Thereby, using the ability to learn general topologies without having to specify the number of neurons to use a priory. As with self-organizing feature maps, the GNG learns a mapping between each neuron and the input space. But, instead of organizing the neurons in a low dimensional space, the GNG organizes them into an arbitrary graph structure, which is generated by a successive addition and deletion of neurons. When new neurons are added to the network, they are added where the current configuration is least adapted. As with the growing cell structure, [4] points out that the GNG algorithm can be applied for both supervised and unsupervised learning. A major advantage of the GNG algorithm, and especially the SGNG algorithm, is that any arbitrary measure can be used to represent the goodness of a local configuration of the graph structure [5]. Hence, in the supervised case it is possible to take into account the difficulty of the classification in a certain area of the input space and therefore assign more neurons to areas where the classification is difficult. This makes it possible to effectively use each neuron for classification and [5] shows that a SGNG often achieves a much better classification accuracy than a radial basis network with the same number of neurons.

2.2 Ensemble Methods

Ensemble methods are learning algorithms that consist of a set of classifiers. When new predictions are made by the ensemble, each classifier presents its own prediction and these are then weighted together into a final prediction. It is generally believed that a good set of (even weak) classifiers can be combined into a strong classifier, assuming that each weak classifier are loosely correlated in making errors [6]. There are many different approaches to find such a set of classifiers. In for example bootstrap aggregating, abbreviated *bagging*, each classifier is trained on a random subset of the data and thus these classifiers will learn from different sets of data.

The most commonly used ensemble algorithm is the *random forest* algorithm, in which multiple decision trees are trained [1]. Besides random forests, there have been several works considering ensemble methods using neural networks as classifiers in the ensemble. An example of this is the empirical study by Opitz and Maclin [14], which studies the behaviour of ensembles of decision trees and neural networks on 25 different datasets. The general conclusion of this study is that the ensemble most often learnt to classify new examples better than any of its individual classifiers.

2.3 Deep Neural Networks

Much of the advances within AI in the last years can be attributed to deep learning and hence deep neural networks [10]. While there are multiple different architectures of such networks, this study will only consider deep feed forward neural networks. A feed forward neural network is neural network which has its neurons structured in multiple layers. There are no cyclical connections in such networks and all neurons in one layer is connected to all neurons in the next layer. It has been theoretically shown that such networks, using only one hidden layer, can approximate any given function to any degree. However, this is under the assumption that there are an infinite number of neurons in that layer [11]. Moreover, it has been shown, both with practical and theoretical results that it is often better to use a deep architecture for the network, hence networks with multiple layers. One finding that supports this is presented by Poole et al. [15], whom show that the expressibility of a network grows exponentially when more layers are added. As mentioned earlier, such networks have been successfully applied to many machine learning problems due to their flexibility and expressibility. However, one big disadvantage of such methods is that it is often intractable to interpret the final result and hence, it is intractable to gain any new knowledge about the data from the final configuration.

3 Self-Organizing Ensemble of Neural Networks

In this section, a novel algorithm using a self-organizing ensemble of neural networks (SOENN) is presented. This algorithm is mainly inspired by the different methods presented in the previous section. The main idea behind this algorithm is to start with just two networks in the ensemble. These two networks have the randomly initialized weights, $\mathbf{W_1}$ and $\mathbf{W_2}$, and are each assigned a position in the input space, annotated p_1 and p_2. As the training progresses more networks will be added to the ensemble and some will be removed. The conditions for this is described later on in this section. A basic visual description of the algorithm is shown in Fig. 1 and more specific pseudo code is given in Algorithm 1.

When calculating the prediction $\mathbf{\hat{y}_i}$ of the ensemble the prediction for each internal network is first calculated. These predictions are then weighed together to get the final prediction. The weight assigned to each prediction is dependent of the adjacency of the network and the sampled point. Hence, the prediction is calculated by:

$$\mathbf{\hat{y}_i} = \frac{1}{2} * \left(1 + \sum_{j=0}^{N} \left((2 * \hat{y}_{i,j} - 1) * \frac{d(p_j, x_i)}{\sum_{k=0}^{N} d(p_k, x_i)} \right) \right), \tag{1}$$

where $d(p_j, x_i)$ is the adjacency of sampled data-point x_i and the $j : th$ neural network in the ensemble. Thus, networks that are close to the sampled point will have a higher impact on the prediction than networks that are far away. The adjacency between a network and a sampled point is calculated as:

$$d(p_j, x_i) = e^{-\left(\frac{1}{M} \sum_{k=0}^{M} (x_{i,k} - p_j)^2 \right)}. \tag{2}$$

The binary cross entropy error is then minimized in order to train the SOENN and make it better at a given classification task. The binary cross entropy error is calculated by:

$$\Delta_i = -y_i \log \mathbf{\hat{y}_i} - (1 - y_i) \log(1 - \mathbf{\hat{y}_i}) \tag{3}$$

This error is then minimized using backpropagation. Hence, changing W_j and p_j in the opposite direction of the gradients $\frac{\partial \Delta_i}{\partial W_j}$ and $\frac{\partial \Delta_i}{\partial W_j}$. for all networks. Note that $\frac{\partial \Delta_i}{\partial W_j}$ depends on $d(p_j, x_i)$ and will therefore be much smaller if the sample x_i is far away from the position of the network. Hence, the internal networks will mostly be trained on samples in the vicinity of their position in the input space.

A new network is added to the ensemble after λ iterations. The position of this newly added network is assigned position γ in the input space. Here, γ is the same as the sample x_k in the last λ iterations with the maximal value of:

$$(1 - d(p_j, x_k)) * |y_k - \mathbf{\hat{y}_k}| \quad \text{where} \quad d(p_j, x_k) > d(p_j, x_k) \quad \forall j \neq l \tag{4}$$

Hence, γ is selected as a sampled data-point that is far away from any networks in the ensemble and that the ensemble is unable to predict well. All networks that have not been the closest or the second closest to any sampled point in the last λ iterations are then removed. The algorithm is furthermore described in greater detail in Algorithm 1. In this description, one additional variable, C is used to keep track of the networks that has been the closest or the next closest to any sampled data-point.

For intuition of the algorithm, we give and example (Fig. 1) of the progress of SOENN: In the initialization phase at (a), two neural networks are placed at the same location in the inputs as two randomly sampled points. At each iteration, a random data-point is sampled and the ensemble will provide a prediction,

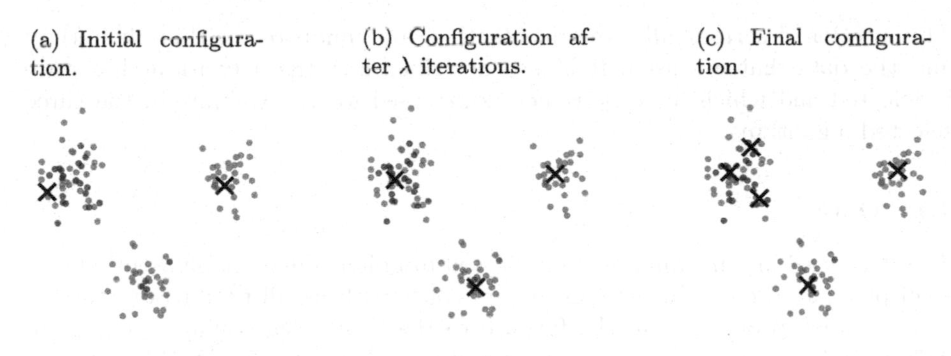

(a) Initial configuration.
(b) Configuration after λ iterations.
(c) Final configuration.

Fig. 1. Initial configuration.

Algorithm 1. The SOENN algorithm

1: Initialize the ensemble, containing only two networks with position p_1 and p_2.
2: $C = \{1, 2\}$
3: **for** η number of iterations **do**
4:     **for** λ number of iterations **do**
5:         $(x_i, y_i) \leftarrow$ random sample from dataset.
6:         Calculate $\hat{\mathbf{y}}_\mathbf{i}$, as defined in equation 1.
7:         Train all networks in the ensemble through one step of backpropagation in order to minimize Δ_i which is defined in equation 3.
8:         Find the two networks r and q which are closest to the sampled point x_i.
9:         Add r and q to C
10:         Calculate a temporal value for γ, annotated γ_{temp} using equation (4).
11:         $\gamma \leftarrow \gamma_{temp}$ if $\gamma_{temp} > \gamma$
12:     **end for**
13:     Remove network j from the ensemble if j is not in C.
14:     **if** the maximum networks has not been reached: **then**
15:         Add a new network to the ensemble.
16:         Set the position of the newly added network to γ.
17:     **end if**
18:     $\gamma \leftarrow 0$
19:     $C \leftarrow \{\}$
20: **end for**

---

according to Eq. (1), for that sample. The weights in the networks and the position of the networks will then be adjusted in order to minimize the difference between this prediction $\hat{\mathbf{y}}_\mathbf{i}$ and the real value of y_i. After λ iterations a new network will be added to the ensemble, this is shown in sub-figure (b). This new network is added at the position γ, how γ is calculated is described in Eq. (4). The final result, shown in sub-figure (c), is achieved after repeating this process multiple times. Here, the ensemble has organized itself to put more networks in the upper left cluster where the data distribution is very heterogeneous and hard to classify.

4 Evaluation

This section describes all experiments that are conducted within this study and the data that are used. It is also described how the network architecture is selected and which hyper-parameters are used for the training in the three selected algorithms.

4.1 Data

The data used in our evaluation is collected from hot rolling manufacturing at a steel plant in Sweden. Measurements are collected from all production stations of the rolling process, from the furnace to the final down coiling, resulting in 9276 samples where each sample consists of measurements of 1718 different variables. Each sample is manually labeled as either telescoped or not telescoped. In addition to the data complexity, there exist several non-linear dependencies between data. In addition, the dataset is unbalanced since the amount of telescoped samples are much fewer than the amount of non-telescoped samples. For several production stations, such as rolling, there are repeated operations, which produces multiples of the station data, and the number of repetitions differ for different product types. Data from repetitions are aggregated as mean and standard deviation. Further, each variable of the data is normalized to zero mean and unit variance.

4.2 Selecting Network Architecture

Theoretically, any arbitrary size and shape of network can be used within the algorithm presented in the previous section. But, in order to show that the concept of several weak classifiers can be used to create one strong classifier, several small, but deep, neural networks are selected. The networks consist of 3 hidden layers and 1 output layer. These networks have 4 neurons in the first hidden layer, 4 neurons in the second one and 2 neurons in the last hidden layer. Each hidden layer consists of leaky rectified linear units. The output layer consists of a single neuron, which uses the sigmoid function as its activation function.

One hypothesis of this work is that SOENN algorithm can outperform a single large DNN. Therefore, for a fair comparison, the same amount of neurons will be used within both algorithms. However, the SOENN contains an extra computational step where the results of all internal network are combined. Hence, to match the number of neurons, the large DNN has to contain one more layer than the networks in the SOENN. To not bias the results in favour of SOENN algorithm, experiments to find the best possible configuration for the large DNN is first conducted. To find such a configuration multiple experiments where multiple configurations of the large DNN are evaluated. This process is repeated 30 times and the configuration achieving the best result is selected for the comparison. The configurations that are evaluated correspond to between at most 10 and 75 networks in the SOENN algorithm. The maximal number of networks in the SOENN algorithm is then selected, to make sure that it cannot have more neurons than the best achieving DNN. A second aim of this work is to investigate the benefit of the self organizing property of the SOENN algorithm. Therefore, experiments using an ensemble with the same architecture and number of networks, but without the self organizing property are conducted.

4.3 Experiments

In the main experiment of this research, the SOENN algorithm, presented in Sect. 3, as well as a large DNN and a standard ensemble of neural networks, are applied to the presented problem of predicting if a rolled steel block, (referred to as a slab), should be classified as telescoped or not. The original data is divided into a training set of 70%, a validation set of 15% and a test set of 15%. The aim of this division is to be able to show how well each algorithm would perform on new unseen data. Each algorithm is trained using the training data and the validation data. The final performance of the algorithm is then evaluated on the test set. Since the dataset is unbalanced an evaluation metric that depends on both the precision and recall should be considered. Therefore, the area under the receiver operating characteristic curve (AUC-ROC) score is selected for the evaluation of the algorithm performance. To be able to give a fair comparison between the performance of each algorithm, this process is repeated over 30 random splits of the dataset and the average performance of each algorithm for all different test sets is reported. All these algorithms use backpropagation with the ADAM optimization algorithm [8] for the training. The values assigned to the hyper-parameters used when training the three different algorithms in the final experiment are listed in Table 1.

To determine if there is a significant difference between the average performance of the SOENN algorithm and the two other algorithms, t-tests are conducted. These are paired t-tests since the same division of the data into training, validation and test sets are used once for each algorithm. Since multiple tests are conducted, Holm–Bonferroni correction is used to adjust the achieved p-values.

Table 1. Algorithms evaluated, with their hyper-parameters.

Algorithm	Parameters
Simple ensemble of DNNs	Epochs = 150, learning rate = 0.003, total number of networks = 45, bootstrap samples = 1000, hidden layer sizes = $\{4, 4, 2\}$
Large DNN	Epochs = 150, learning rate = 0.003, hidden layer sizes = $\{180, 180, 90, 45\}$
SOENN	$\eta = 150$, $\lambda = 6500$, learning rate = 0.003, hidden layer sizes = $\{4, 4, 2\}$, maximum number of networks = 45

5 Result

The first experiment conducted aimed to find a good architecture for the large DNN that later could be used in the comparison with SOENN algorithm. The result from this experiment is shown in Fig. 2. Using these results, it was decided to use 45 networks in the two ensemble methods in the following experiments.

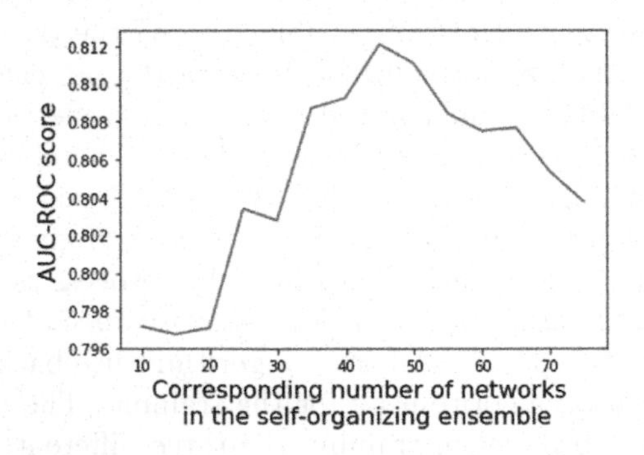

Fig. 2. The mean AUC-ROC score for the conducted experiments with different architectures for the large DNN. The x-axis shows the maximum number of networks in the SOENN algorithm that a given configuration would correspond to. Each measurement consist of 30 randomly selected instantiations of the DNN.

In the main experiment the three described methods are applied to the prediction problem of classifying slabs as either telescoped or not. The SOENN algorithm is the algorithm that achieves the highest AUC-ROC score on this problem.

This method achieves a mean AUC-ROC score of 0.822 while the large DNN and a standard ensemble of DNNs achive mean AUC-ROC scores of 0.812 and 0.804 respectively. The distribution of AUC-ROC scores for the performed analysis, using these three methods is shown in Fig. 3. The differences between the results are shown to be significant, using two separate paired t-tests which both yield a p-value that is smaller than 0.00001 after Holm–Bonferroni correction.

To show that the SOENN algorithm managed to exploit the internal structure of the data, the positions of the networks in relation to the data, are displayed. Since the input data are very high dimensional, the number of dimensions have to be reduced, using t-SNE [12], in order to plot it in two dimensions. The final result is shown in Fig. 4. This allows for the interpretation of where the network has decided to focus, in the input space, and thus where it is difficult to perform classifications.

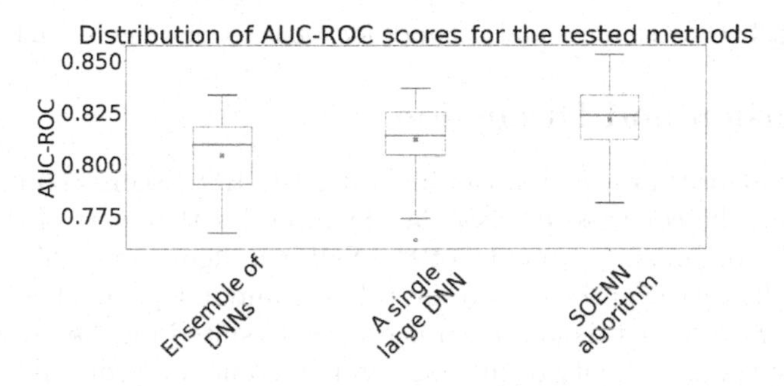

Fig. 3. The distribution of AUC-ROC scores for the three tested algorithms. The SOENN algorithm is positioned furthest to the right. The median AUC-ROC score is represented by an orange line, while the mean is represented by an orange x. The achieved mean value for the AUC-ROC scores are 0.804, 0.812 and 0.822. (Color figure online)

In Fig. 4 we show the distribution of data points in the training set, reduced to a two dimensional space using t-SNE. Telescoped samples are presented as red (darker) points and non-telescoped samples are presented as blue (lighter) points. Each network in the ensemble is presented as a black "x". The important concept that is shown by this figure is that the ensemble has managed to spread out the networks throughout the input space and hence, managed to exploit the distribution of samples in the input data. It can be seen that the networks in the ensemble has been distributed to areas where the distribution of classes is heterogeneous and thus, classification is difficult. A typical example of this can be observed in the bottom of the left bottom of the figure. A typical example of the opposite, i.e an area where it is trivial to classify samples, can be seen in the middle right area.

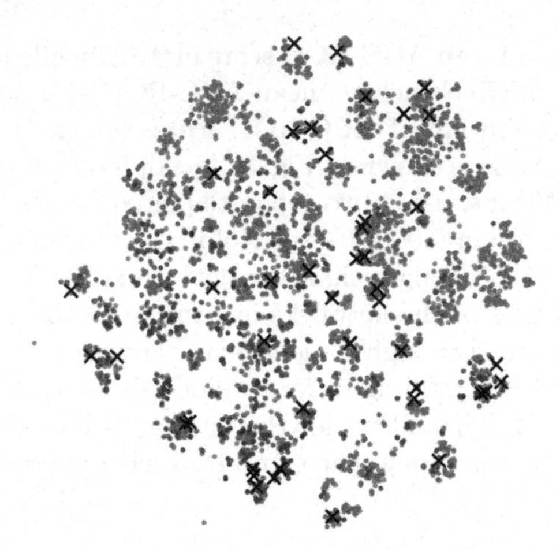

Fig. 4. Data samples and network locations. (Color figure online)

6 Conclusion and Discussion

The SOENN algorithm achieves the highest AUC-ROC score compared to the alternatives on the collected dataset. We hypothesize that part of this result is due to the self organizing property of the SOENN algorithm, which allows the algorithm to divide the problem into several local and independent sub-problems. We further believe that this approach is especially suited for the presented case, mainly due to the complexity of the data and different sub problems in the data. There is nothing suggesting that this result is only applicable to hot rolling. Instead, the same improvement is likely when applying SOENN to any data with similar complex structure and with several sub problems. However, more research is needed to further investigate this. Unbalanced data, typical in a production setting, can be difficult for a DNN to learn. Using SOENN for unbalanced data, the smaller networks of SOENN can learn individual sub-problems in spite of the imbalance, in contrast with a DNN that needs to learn the overall problem. It is easier to learn to classify local points with simpler classifiers in a ensemble. This is supported by He and Garcia [7] who shows that ensemble methods often perform better than other methods on unbalanced datasets.

At this point, SOENN is limited by the ad. hoc addition and removal of its internal networks, so there is a need to develop refined schemes for this. An improvement for network removal would be that removals would depend on the ensembles' ability to classify samples correctly without the given network.

In Fig. 4, it is shown that the SOENN algorithm manages to spread out the internal networks in the input space. Thus, SOENN is able to allocate more networks to areas where the classification is more difficult. This representation of what the SOENN algorithm has learned from the data is not optimal but it still offers some insights, and this is more than what can be expected from the other evaluated algorithms. One of the main utilities of the presented visualization is that it can be used to target further analyses, manual or not, in order to gain more insights about the process.

References

1. Breiman, L.: Random forests. Mach. Learn. **45**(1), 5–32 (2001). ISSN 1573–0565. https://doi.org/10.1023/A:1010933404324
2. Cuadrado, A.A., Diaz, I., Diez, A.B., Obeso, F., Gonzalez, J.A.: Visual data mining and monitoring in steel processes. In: Conference Record of the 2002 IEEE Industry Applications Conference, 37th IAS Annual Meeting (Cat. No. 02CH37344), vol. 1, pp. 493–500 (2002). https://doi.org/10.1109/IAS.2002.1044131
3. Fritzke, B.: Growing cell structures—a self-organizing network for unsupervised and supervised learning. Neural Netw. **7**(9), 1441–1460 (1994). https://doi.org/10.1016/0893-6080(94)90091-4
4. Fritzke, B.: A growing neural gas network learns topologies. In: Advances in Neural Information Processing Systems, pp. 625–632 (1995)
5. Fritzke, B.: Growing self-organizing networks-why? In: ESANN 1996, pp. 61–72 (1996)
6. Hansen, L.K., Salamon, P.: Neural network ensembles. IEEE Trans. Pattern Anal. Mach. Intell. **12**(10), 993–1001 (1990). https://doi.org/10.1109/34.58871
7. He, H., Garcia, E.A.: A learning from imbalanced data. IEEE Trans. Knowl. Data Eng. **21**(9), 1263–1284 (2009). https://doi.org/10.1109/TKDE.2008.239
8. Kingma, D., Ba, J.: Adam: a method for stochastic optimization. arXiv preprint arXiv:1412.6980 (2014)
9. Kohonen, T.: Self-organized formation of topologically correct feature maps. Biol. Cybern. **43**(1), 59–69 (1982). https://doi.org/10.1007/BF00337288
10. LeCun, Y., Bengio, Y., Hinton, G.: Deep learning. Nature **521**(7553), 436–444 (2015). https://doi.org/10.1038/nature14539
11. Leshno, M., Lin, V.Y., Pinkus, A., Schocken, S.: Multilayer feedforward networks with a nonpolynomial activation function can approximate any function. Neural Netw. **6**(6), 861–867 (1993). https://doi.org/10.1016/S0893-6080(05)80131-5
12. van der Maaten, L., Hinton, G.: Visualizing data using t-sne. J. Mach. Learn. Res. **9**, 2579–2605 (2008)
13. Martinetz, T.M., Berkovich, S.G., Schulten, K.J.: 'Neural-gas' network for vector quantization and its application to time-series prediction. IEEE Trans. Neural Netw. **4**(4), 558–569 (1993)
14. Opitz, D.W., Maclin, R.: Popular ensemble methods. J. Artif. Intell. Res. (JAIR) **11**, 169–198 (1999)
15. Poole, B., Lahiri, S., Raghu, M., Sohl-Dickstein, J., Ganguli, S.: Exponential expressivity in deep neural networks through transient chaos. In: Lee, D.D., Sugiyama, M., Luxburg, U.V., Guyon, I., Garnett, R. (eds.) Advances in Neural Information Processing Systems, vol. 29, pp. 3360–3368. Curran Associates Inc. (2016)
16. Sagiroglu, S., Sinanc, D.: Big data: a review. In: 2013 International Conference on Collaboration Technologies and Systems (CTS), pp. 42–47. IEEE (2013). https://doi.org/10.1109/CTS.2013.6567202

Dealing with Epistemic Uncertainty in Multi-objective Optimization: A Survey

Oumayma Bahri[1,2(✉)] and El-Ghazali Talbi[1]

[1] INRIA Laboratory, CRISTAL-CNRS, Lille, France
oumayma.b@gmail.com, el-ghazali.talbi@lifl.fr
[2] LARODEC Laboratory, ISG Tunis, Le Bardo, Tunisia

Abstract. Multi-objective optimization under epistemic uncertainty is today present as an active research area reflecting reality of many practical applications. In this paper, we try to present and discuss relevant state-of-the-art related to multi-objective optimisation with uncertain-valued objective. In fact, we give an overview of approaches that have already been proposed in this context and limitations of each one of them. We also present recent researches developed for taking into account uncertainty in the Pareto optimality aspect.

Keywords: Multi-objective optimization · Epistemic uncertainty
Uncertain-valued objectives · Pareto optimality

1 Introduction

Multi-objective optimization is a well-studied research field encountered in many academic and industrial applications such as in engineering, manufacturing and logistics. It rises as a salient paradigm of decision making in which the decision maker is always confronted with different conflicting objectives. For instance, a good purchase choice is associated with several factors like the price, the durability and the quality, etc. Hence, the most common purpose is to choose the best trade-off among all these factors. In that sense, it is practically impossible to find a single solution that optimize all predefined objectives at the same time but rather many efficient and incomparable solutions. Thus, the challenge of solving a combinatorial multi-objective problem lies in the difficulty to find a set of best compromise solutions between the different objectives. This set represents, in the objective space, the Pareto front from which the decision maker will subsequently choose one final alternative to realize. Then if the number of multiple objectives and/or decision variables grows, the problem becomes much more complex. A wide variety of resolution methods and techniques have been designed according to the complexity and way of solving such problems [19,23].

In addition, when dealing with real-life problems, the massive amounts of data are generally associated with unavoidable imperfections. In other words,

J. Medina et al. (Eds.): IPMU 2018, CCIS 855, pp. 260–271, 2018.
https://doi.org/10.1007/978-3-319-91479-4_22

practical applications are strongly connected to some uncertainties in inputs, parameters and environmental data. In fact, uncertain data may result from using unreliable information sources such as bad analysis or interpretation processes, faulty description, data incompleteness, ambiguity in perception and so on. Besides, it may be caused by poor decision maker opinions due to any lack of its background knowledge, absence of information or even difficulty of giving perfect qualification for some costly situations. Depending on the nature of imperfection and the problem context, two types of uncertainty can be distinguished: Aleatory (or objective) uncertainty which is characterized by natural randomness and variability and Epistemic (or subjective) uncertainty associated with ambiguity, fuzziness or any lack of information [29].

Indeed, the literature exposes over the years many modeling approaches for reasoning under uncertainty in single-objective optimization problems [10,22,25]. Nevertheless, this aspect is, until today, not well considered in the multi-objective setting that reflects more reality in every domain of our lives. Moreover, the few state-of-the-arts in this setting, such as [11,12,30], are primely focused on multi-objective problems under aleatory uncertainty (e.g. phenomenon of randomness and noise). Thereby, from our deep survey of existing approaches relative to epistemic uncertainty, we have identified that almost all of them have been limited to transform the problem into crisp or single-objective equivalents. Unfortunately, such transformation may affect the problem results and decision making process. Only some approaches have been proposed to address the uncertain multi-objective problem as-is, without ignoring any of its characteristics [3,18].

Hence, there is a significant need for examining and classifying the different approaches capable to deal with epistemic uncertainty in combinatorial multi-objective optimization. Our primary motivation in this paper is to give a global view of these approaches, while focusing on two major issues: *Where does the effects of uncertainty propagation occur?* and *How ranking uncertain valued-solutions in the sense of Pareto optimality?*

The remainder of the paper is organized as follows: Sect. 2 recalls some basic definitions of deterministic multi-objective optimization and highlights the main classes of resolution methods. Section 3 discusses the specific case of multi-objective problems with uncertain-valued objectives by defining the problem and reviewing some approaches in this field.

2 Deterministic Multi-objective Optimization

Multi-objective optimization is the process of optimizing systematically and simultaneously two or more conflicting objectives subject to certain constraints. Then, contrary to the single-objective case, multi-objective optimization does not restrict to find a unique global solution but it aims to find a set of efficient solutions.

Formally, a deterministic *multi-objective optimization problem (MOP)*, defined in the sense of minimization of all objectives, consists of solving the following mathematical program:

$$\min\ F(x) = (f_1(x), \ldots, f_n(x))\ s.t.\ \ x \in X \tag{1}$$

where $F(x)$ is the vector of n ($n \geq 2$) objective functions to be minimized and $x = (x_1, \ldots, x_k)$ is the vector of decision variables from the feasible decision space $X \subseteq \mathbb{R}^n$. In the objective space, F can be defined as a cost function by assigning an objective vector $y \in Y$ which represents the quality of solutions.

$$F : X \rightarrow Y \subseteq \mathbb{R}^n, \quad F(x) = y = \begin{pmatrix} y_1 \\ \vdots \\ y_n \end{pmatrix} \tag{2}$$

where $Y = F(S)$ represents the feasible points (solutions) in the objective space and $y_i = f_i(x)$ is a point of this space that represents the solution quality or fitness.

The objectives are often in conflict with each other (e.g., minimize cost and maximize profit), so that it is practically impossible to have a unique solution x^* optimal for all the objectives: $\forall i \in 1..n, \forall x^* \in S, f_i(x^*) \leq f_i(x))$. Otherwise, the main purpose of a MOP is to find the set of efficient solutions called as Pareto optimal (i.e. efficient, non-dominated and non-inferior solution). To this end, other concepts of optimality, namely Pareto dominance relations, should be applied to define an order relation between the optimal compromise solutions.

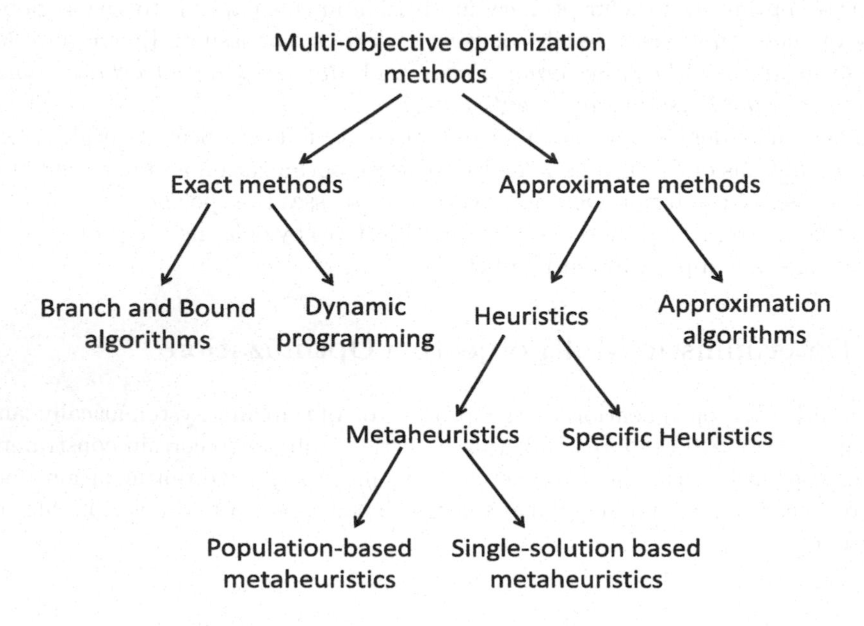

Fig. 1. Taxonomy of multi-objective methods

Various multi-objective resolution methods have been developed through years. They can be classified into different classes according to their effectiveness, applicability and problems complexity. Figure 1 describes a classification of the main multi-objective methods (for details see [19,23,26]).

All these methods have been initially developed to tackle deterministic MOPs, without taking into account any imperfect or inexact inputs data. Therefore, this may restrict their applicability to real-life problems and practical applications in which the big amount of data provides certainly some inevitable imperfections or uncertainties. However, the uncertainty aspect has been extensively discussed in the context of single-objective optimization, whereas its combination with the multi-objectivity aspect has not been deeply studied so far.

In the following, we first present a general description of uncertain MOPs and the major issues related to such problems. Then, we propose a classification of the existing approaches while discussing their advantages and limits.

3 Multi-objective Optimization Under Epistemic Uncertainty

In general, the aim of uncertain multi-objective optimization is to satisfy predefined objectives while considering that some information are uncertain and without knowing what their full effects will be. In other words, a MOP under uncertainty is characterized by the necessity of optimizing simultaneously several conflicting objectives in presence of some uncertain input data.

This field has attracted nowadays increasing attention since it appears in many real-life applications and poses very interesting challenges. One of these challenges is how to identify the type of inevitable uncertainties and their impacts on the results and optimal decision making. However, a great number of studies have been conducted to treat the case of aleatory uncertainty inherent to natural and stochastic behavior. A literature review can be found in [11,12,30]. Some studies exist today for dealing with the combination of multi-objectivity aspect and epistemic uncertainty i.e. resulting from imperfection and ambiguity. Their main purpose is to analyse the manner in which such uncertainty is modeled and propagated through the multi-objective optimization process. At this level, many important issues should be taken into account such as:

- *How propagating such uncertainty through the optimization process?*
- *What are the effects and consequences on the problem solutions?*
- *How to develop and perform a resolution method in this context?*

In that sense, the hardest issue is that disturbances in inputs data may be propagated through the model to the quantities of interest. Usually, the quantum of propagation depends on the problem context, the nature of uncertain inputs data, the form of uncertainty distributions and their transfer to the outputs through the functional relationship. Then, propagating uncertainties may affect the optimization process and even the key elements of decision making such as preference parameters, decision variables, constraints and/or objectives.

This may subsequently mislead the analyst into determining the optimal alternatives, leading to a final bad choice. All these factors contribute to an excessive increase in problem's complexity and difficulty in resolution stage. Thus, it is necessary to analyze the effects of uncertainty propagation before the optimization process.

Another important issue is the uncertainty inherent to the resolution stage. The question is how to explore an uncertain design space which often leads to very large-scales and complex optimization models. Evidently, as a deterministic MOP is already NP-hard and time-consuming, the consideration of uncertainty in the optimization process may lead to prohibitive computation. The increasing costs of such very complex problems motivates more and more the scientific researchers to develop efficient resolution methods. In this context, a resolution method may still contain some uncertainties due to its inability to provide exact results or to the lack of optimality proof.

In what follows, we focus on the specific case where epistemic uncertainty is assumed to occur in the objective functions. Indeed, uncertainty propagation to the objectives presents a critical and sensitive obstacle because it may affect the whole search process and consequently the optimality of solutions. Thus, we propose in the next section a classification of the existing approaches according to how uncertainty in objectives is treated.

3.1 Existing Approaches for MOPs with Uncertain Objectives

Formally, a minimization MOP with uncertain objectives may be defined as:

$$\min\ F(x,\xi) = \min[f_1(x,\xi), f_2(x,\xi), \ldots, f_n(x,\xi)]\ s.t.\ \ x \in X, \xi \in U \quad (3)$$

where F is the set of objective functions that may depend on some inputs uncertainty U. Each $f_i(x,\xi)$ represents an uncertain valued-objective, where x is a decision variable vector from its admissible region $X \subseteq \mathbb{R}^n$ and ξ represents a vector of uncertain distributions or quantities induced by U.

Once the nature of inputs uncertainties and their effects are identified, the second relevant challenge consists to find a suitable way for treating such uncertainty in the objective search space of any resolution method. Nonetheless, although uncertainty in the objective functions has gained attention in recent years, the efforts devoted to this problem are still limited. Figure 2 presents a taxonomy of the existing approaches.

The first attempts to cope with uncertainty in objectives belong to the category of aggregation-based approaches. The basic idea of these traditional approaches is to combine the multiple objectives into a single uncertain one. For example, in [8], authors convert the MOP into a one or a set of single-objective problems. Furthermore, the different objectives can be rewritten into an aggregate objective f_A by applying a weighted sum function as follows:

$$f_A(x,\xi) = \sum_{i=1}^{n} [f_1(x,\xi), f_2(x,\xi), \ldots, f_n(x,\xi)] \quad (4)$$

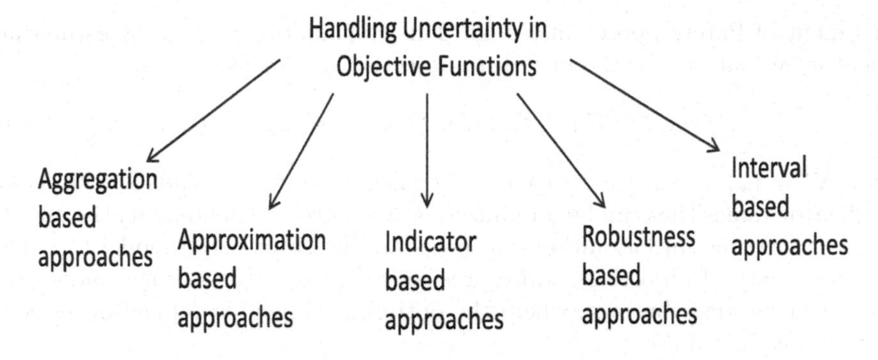

Fig. 2. Taxonomy of approaches for MOPs with uncertain objectives

In this case, the existing approaches designed for single-objective optimization problems under uncertainty can simply be applied. Clearly, aggregation-based approaches have the advantage of simplicity because they do not require a particular development for uncertain multi-objective optimization. Yet, they still not efficient since they limit the objective space, ignore the significant role of multi-objectivity and also relationship between the conflicting objectives. In consequence, the obtained results are very often useless and far from reality.

The second category encloses approximation-based approaches that use statistical functions to convert the uncertain objectives into their crisp equivalents [13,24]. Otherwise, these approaches still abide to the certainty of objectives and usually allow to carry out an approximation of observed uncertainty. In this case, a statistical function may be applied to approximate each objective function as follows:

$$\Phi(f_1(x,\xi)), \Phi(f_2(x,\xi)), \ldots, \Phi(f_n(x,\xi)) \tag{5}$$

where $\Phi(.)$ denotes the statistical operator which can be the expected function $E[.]$ with respect to ξ. This category includes also mean-value and mean-penalty approaches such as [21]. Commonly, each objective is approximated by estimating the mean value of each random sample. This allows to transform the uncertain MOP into a crisp problem that can be resolved using standard deterministic multi-objective optimizers. A major limit of approximation-based approaches is that the propagation and effects of uncertainty are neglected. Yet, ignoring the uncertainty propagation in the optimization process can lead to very poor decisions with often misleading simulation results. It is therefore necessary to account for the relationship between uncertain inputs and generated solutions, because if the input data or parameters are highly uncertain, how can the optimizer simply state that the outputs are exact values? It may be feasible only for simplicity or other practical reasons as long as the optimization performance will not be affected.

The third category includes different approaches [5,20] that combine uncertainty of objectives and quality indicators (i.e., real-valued functions which allow assessment of Pareto approximations). This combination is done by estimating indicator evaluations for the uncertain objective vectors as:

$$I(f_1(x,\xi), X^*), I(f_2(x,\xi), X^*), \ldots, I(f_n(x,\xi), X^*) \tag{6}$$

where $X^* = \{x_1^*, \ldots, x_r^*\}$ is a variable reference set and $I(.)$ stands for the vector of indicator values that can be minimized or maximized depending on the quality goal. For instance, in [5], authors proposed an indicator-based model to reflect the uncertainty of objectives. More precisely, the objective vector is associated with uncertain distributions, where the optimization goal is defined in terms of the $\epsilon-$indicator values.

Another category of approaches refers to the robustness aspect [1,9]. This aspect is connected to the idea that in presence of uncertain inputs, the outputs should be relatively insensitive (small uncertainty outputs). The robustness in objective functions can be modeled as:

$$(f_1(x,\xi), R_1), (f_2(x,\xi), R_2), \ldots, (f_n(x,\xi), R_n) \tag{7}$$

where R_i is the robustness criterion that should be maximized. It is defined in terms of the variation of $f_i(x)$ regarding the uncertainty associated with x. For instance, [9] proposed to estimate the expected uncertainty using Monte Carlo simulations based on effective objective function that takes into account robustness. In [1], a propagating approach based on the concept of robustness degrees of uncertain objectives is introduced.

However, the main drawback of robustness-based and indicator-based approaches is that they rely on the assumption of a priori knowledge about decisive information such as the reference set of solutions or the robustness confidence level. Evidently, if such information is inappropriate or incorrect, the outputs of theses approaches can be misleading.

Further studies assume to display uncertainty of objectives through intervals and thereby to perform the multi-objective optimization based on this uniform distribution. These studies fall under the category of interval-based approaches [7,17,18]. In this case, the cost of evaluating $f(x,\xi)$, namely Y is represented as intervals as:

$$F(x,\xi) = Y = ([\underline{y_1}, \overline{y_1}], \ldots, [\underline{y_n}, \overline{y_n}]) \tag{8}$$

where $\underline{y_i}$ and $\overline{y_i}$ are respectively lower and upper bounds of the corresponding interval-valued function i. For instance, in [7], authors expressed uncertainty via confidence intervals of fuzzy sets and then proposed a multi-objective genetic algorithm for handling such interval-valued functions.

Recently, in our previous work [3], we have suggested a novel different approach able to model uncertainty in objectives by means of an intuitive and natural shape of fuzzy sets, namely the triangular fuzzy numbers. Then, we have developed new multi-objective optimizers for solving such specific type of problems.

In this latter category of approaches, another challenging issue occurs which consists on identifying and/or ranking the uncertain solutions disrupted by the shape of objectives (i.e. modeled by crisp or non-crisp intervals). Then, as the classic Pareto concepts cannot be used in this case, a need for special dominance relations capable to handle uncertain solutions is evident In the following, we present an overview of some recent works related to the definition of Pareto optimality within an uncertain context.

3.2 Pareto Optimality with Uncertainty

Pareto dominance concepts are crucial when dealing with multi-objective optimization because we need to maintain a set of non-dominated solutions rather than only one. These solutions, called *Pareto optimal set*, correspond to a compromise between different objectives (i.e., achieving the optimal value for one objective requires some sacrifice of quality on at least one other objective). This notion of Pareto optimality is based on intuitive discrimination of what are the most good or desired alternatives among others. In the case of uncertain MOPs, the purpose becomes to explicitly consider uncertainty into account during the Pareto analysis. In fact, the solutions of such a problem are not modeled by single points in the objective space, but rather by a range or set of points. Then, in order to find out the best solutions in the sense of dominance, some contributions for exploring the Pareto front in uncertain context have been proposed in the literature such as [14,15,17,24,27].

On the one hand, most of existing studies are based on the idea of reducing the amount of uncertainty in the solution space. Indeed, they are often limited to quantify such uncertain solutions as crisp degrees or to estimate the whole Pareto set through some simulations. Thus, the visualization of solutions still absolutely equivalent as in classical Pareto front (where non-dominated solutions are exact points). For instance, in [4], a tool based on Gaussian conditional simulations is interpreted for quantifying how much uncertainty remains at every stage of Pareto exploration. Thereafter, an estimation of the total Pareto front is deduced based on meta-model notion. In [16,28], the solutions are firstly represented as fuzzy cost functions. Then, a ranking scheme is defined by assigning a dominance degree to each fuzzy solution. The advantage of these studies is their simplicity, computational efficiency and the possibility of application of classical concepts for any comparative analysis. Whereas, their principal drawback is that uncertainty of Pareto front is only visualised through approximations or simulations. Yet, the quality and efficiency of these latter depends entirely on the used tool or measures.

On the other hand, some other studies propose to extend the Pareto dominance relations for ranking interval-valued outcomes. For instance, in [14,24], a probabilistic dominance based on intervals is used to guide the selection process of Pareto-set. In [17,18], intervals of belief functions are used to represent the uncertain Pareto optimal solutions. Moreover, authors in [27] involve uncertainty as fuzzy coefficients in the objective functions. Then, an interval-based abstraction is introduced to generate the Pareto optimality of candidate solutions. In this case, a non-deterministic configuration or representation of Pareto front is provided. More precisely, the solutions are represented by finite bounding-boxes in the objective space, as shown in Fig. 3, where each bounding box represents a unique interval-valued solution. However, such a configuration can only express lower an upper bounds and it is often not enough to express uncertainty and its distribution.

Another important issue is *how to analyse the Pareto optimality when epistemic uncertainty is modeled by non-crisp intervals?* Unfortunately, there is a lack of research studies on this topic due to its increasing complexity. In other words, extending the Pareto dominance for ranking two or more non-crisp intervals like fuzzy ones is a very hard task. To the best of our knowledge, there is only our previous research work that addressed the Pareto optimality between

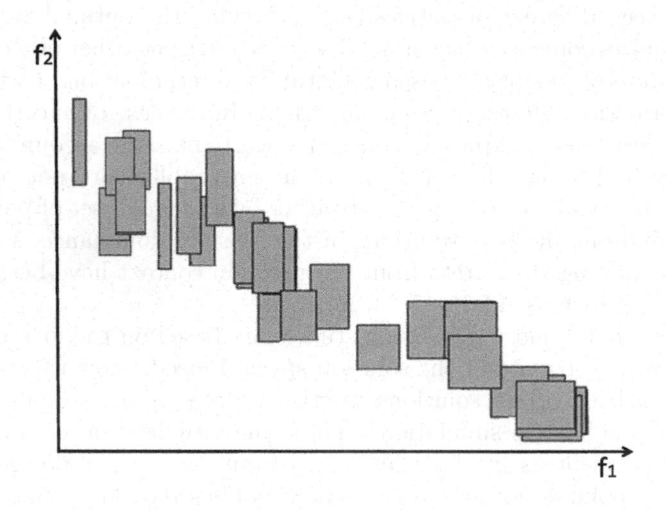

Fig. 3. Interval-based Pareto front

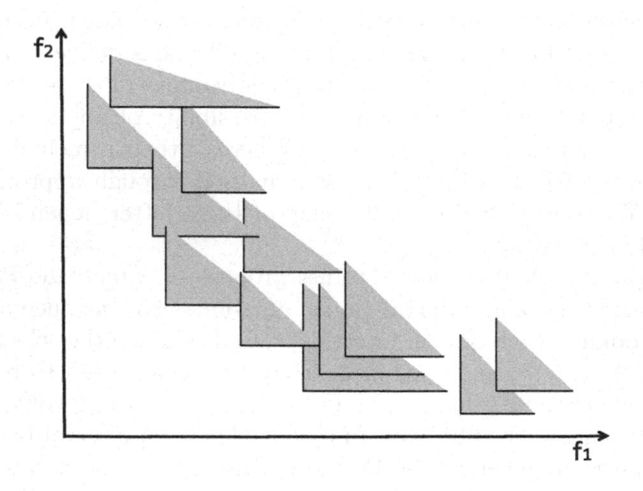

Fig. 4. Triangular-based Pareto front

triangular fuzzy-valued solutions [2]. Consequently, instead of the rectangular form in the case of crisp intervals, the Pareto front is composed by a set of triangular vectors, where each triangle represents one fuzzy solution as shown in Fig. 4. This alternative of representation offers much more flexibility depending on the nature of uncertainty.

4 Conclusion

In this paper, we have firstly presented the fundamental background of deterministic multi-objective optimization, starting with some classical definitions to a brief overview of resolution methods. Then, we have surveyed the state-of-the-art relative to uncertain multi-objective optimization. In particular, we have introduced a classification of the different existing approaches to handle epistemic uncertainty in objective functions. In the second part of the paper, we also gave a survey of some research works regarding Pareto-optimality notion in the uncertain context.

It will be interesting to study the consideration of uncertainty into the commonly used multi-objective quality indicators. This allows us not only evaluating the generated uncertain solutions, but also providing an accurate and better performance assessment. Finally, whatever the studied aspect, there are still many open questions and perspectives to investigate in this field.

References

1. Barrico, C., Antunes, C.H.: Robustness analysis in multi-objective optimization using a degree of robustness concept. In: IEEE CEC, pp. 1887–1892 (2006)
2. Bahri, O., Ben Amor, N., El-Ghazali, T.: New Pareto approach for ranking triangular fuzzy numbers. In: Laurent, A., Strauss, O., Bouchon-Meunier, B., Yager, R.R. (eds.) IPMU 2014, Part II. CCIS, vol. 443, pp. 264–273. Springer, Cham (2014). https://doi.org/10.1007/978-3-319-08855-6_27
3. Bahri, O., Ben Amor N., Talbi E.-G.: Optimization algorithms for multi-objective problems with fuzzy data. In: IEEE International Symposium on MCDM, pp. 194–201 (2014)
4. Binois, M., Ginsbourger, D., Roustant, O.: Quantifying uncertainty on Pareto fronts with Gaussian process conditional simulations. Eur. J. Oper. Res. **243**(2), 386–394 (2015)
5. Basseur, M., Zitzler, E.: Handling uncertainty in indicator-based multiobjective optimization. Int. J. Comput. Intell. Res. **2**(3), 255–272 (2006)
6. Coello Coello, C.A., Hernández Aguirre, A., Zitzler, E. (eds.): EMO 2005. LNCS, vol. 3410. Springer, Heidelberg (2005). https://doi.org/10.1007/b106458
7. Sánchez, L., Couso, I., Casillas, J.: A multiobjective genetic fuzzy system with imprecise probability fitness for vague data. In: IEEE International Symposium on Evolving Fuzzy Systems, pp. 131–136 (2006)
8. Goncalves, G., Hsu, T., Xu, J.: Vehicle routing problem with time windows and fuzzy demands: an approach based on the possibility theory. Int. J. Adv. Oper. Manage. **1**(4), 312–330 (2009)

9. Deb, K., Gupta, H.: Searching for robust Pareto-optimal solutions in multi-objective optimization. In: Coello Coello, C.A., Hernández Aguirre, A., Zitzler, E. (eds.) EMO 2005. LNCS, vol. 3410, pp. 150–164. Springer, Heidelberg (2005). https://doi.org/10.1007/978-3-540-31880-4_11
10. Diwekar, U.: Optimization under uncertainty. Introduction to Applied Optimization. SOIA, vol. 22, pp. 1–54. Springer, Boston (2008). https://doi.org/10.1007/978-0-387-76635-5_5
11. Fieldsend, J.E., Everson, R.M.: Multi-objective optimisation in the presence of uncertainty. In: IEEE CEC, vol. 1, pp. 243–250 (2005)
12. Goh, C.K., Tan, K.C.: Evolutionary multi-objective optimization in uncertain environments. J. Stud. Comput. Intell. **186**, 5–18 (2009)
13. Hughes, E.J.: Evolutionary multi-objective ranking with uncertainty and noise. In: Zitzler, E., Thiele, L., Deb, K., Coello Coello, C.A., Corne, D. (eds.) EMO 2001. LNCS, vol. 1993, pp. 329–343. Springer, Heidelberg (2001). https://doi.org/10.1007/3-540-44719-9_23
14. Haubelt, C., Teich, J.: Accelerating design space exploration using Pareto-front arithmetics. In: ACM Conference on Asia and South Pacific Design Automation, pp. 525–531 (2003)
15. Hendriks, M., Geile, M., Basten, T.: Pareto analysis with uncertainty. In: 9th International Conference on EUC, pp. 189–196 (2011)
16. Köppen, M., Vicente-Garcia, R., Nickolay, B.: Fuzzy-Pareto-dominance and its application in evolutionary multi-objective optimization. In: Coello Coello, C.A., Hernández Aguirre, A., Zitzler, E. (eds.) EMO 2005. LNCS, vol. 3410, pp. 399–412. Springer, Heidelberg (2005). https://doi.org/10.1007/978-3-540-31880-4_28
17. Limbourg, P.: Multi-objective optimization of problems with epistemic uncertainty. In: Coello Coello, C.A., Hernández Aguirre, A., Zitzler, E. (eds.) EMO 2005. LNCS, vol. 3410, pp. 413–427. Springer, Heidelberg (2005). https://doi.org/10.1007/978-3-540-31880-4_29
18. Limbourg, P., Aponte, D.E.S.: An optimization algorithm for imprecise multi-objective problem functions. In: IEEE CEC, vol. 1, pp. 459–466 (2005)
19. Liefooghe, A.: Methodes pour l'optimisation multiobjectif: Approche cooperative, prise en compte de l'incertitude et application logistique. PHD thesis, Universit de Lille 1, pp. 13–20 (2009)
20. Liefooghe, A., Jourdan, L., Talbi, E.G.: Indicator-based approaches for multiobjective optimization in uncertain environments. In: 25th Mini-EURO Conference URPDM (2010)
21. Meng, Z., Shen, R., Jiang, M.: An objective penalty functions algorithm for multiobjective optimization problem. J. Oper. Res. **1**(4), 229 (2011)
22. Petrone, G.: Optimization under Uncertainty: theory, algorithms and industrial applications. PHD thesis, Università degli Studi di Napoli Federico II, pp. 77–122 (2011)
23. Talbi, E.-G.: Metaheuristics: From design to implementation, vol. 74, pp. 309–373. John Wiley and Sons (2009)
24. Teich, J.: Pareto-front exploration with uncertain objectives. In: Zitzler, E., Thiele, L., Deb, K., Coello Coello, C.A., Corne, D. (eds.) EMO 2001. LNCS, vol. 1993, pp. 314–328. Springer, Heidelberg (2001). https://doi.org/10.1007/3-540-44719-9_22
25. Sahinidis, N.V.: Optimization under uncertainty: state-of-the-art and opportunities. J. Comput. Chem. Eng. **28**(6), 971–983 (2004)
26. Saka, M.P., Dogan, E.: Recent developments in metaheuristic algorithms: a review. J. Comput. Technol. Rev. **5**(4), 31–78 (2012)

27. Silva, R.C., Yamakami, A.: Definition of fuzzy Pareto-optimality by using possibility theory. In: IFSA/EUSFLAT Conference, pp. 1234–1239. Citeseer (2009)
28. Wang, G., Huawei, J.: Fuzzy-dominance and its application in evolutionary many objective optimization. In: IEEE International Conference on Computational Intelligence and Security Workshops CISW, pp. 195–198 (2007)
29. Zadeh, L.A.: Fuzzy sets. In: Fuzzy Sets, Fuzzy Logic and Fuzzy Systems: Selected Papers by Lotfi A. Zadeh, pp. 394–432 (1996)
30. Zhou, J., Yang, F., Wang, K.: Multi-objective optimization in uncertain random environments. J. Fuzzy Optim. Decis. Mak. **13**(4), 397–413 (2014)

Analyzing the Influence of LLVM Code Optimization Passes on Software Performance

Juan Carlos de la Torre[1], Patricia Ruiz[2], Bernabé Dorronsoro[1(✉)], and Pedro L. Galindo[1]

[1] Computer Sciences Engineering Department, University of Cadiz, Cádiz, Spain
{juan.detorre,bernabe.dorronsoro,pedro.galindo}@uca.es
[2] Mechanical Engineering and Industrial Design Department, University of Cadiz, Cádiz, Spain
patricia.ruiz@uca.es

Abstract. Sensitivity analysis is a mathematical tool that distributes the uncertainty of the output of a model among its different input variables. We use in this work the Extended Fourier Amplitude Sensitivity Test to carefully analyze the impact of 54 LLVM code optimization operators on the execution time of nine benchmark software programs. Experiments presented involve performing over 16 million executions. The results show that the different LLVM transformations have a low direct effect on the execution time, but it becomes meaningful when considering the transformation in combination with the others (almost 60% average impact by all passes on all considered benchmarks). These results provide slight indications on the transformations to apply for optimizing the software, revealing the extreme difficulty of the problem.

Keywords: Sensitivity analysis · Software optimization
LLVM transform passes

1 Introduction

LLVM [1] is a modern compilation framework, widely used nowadays, with very interesting features. The core of LLVM is the Intermediate Representation (IR), a novel programming language that LLVM defines, to which the software code is translated as an intermediate step during compilation. A salient characteristic of IR is that it is independent of both the language in which the software is coded and the target architecture. LLVM currently provides support for a number of well-known programming languages, as C, C++, Objective C, Java, or Python, among others.

LLVM implements a large number of tools for its IR, and programs implemented in any of the programming languages supported by LLVM can benefit from them. Because these tools work on the IR, they are generic for all supported programming languages. We are particularly interested in this work on

J. Medina et al. (Eds.): IPMU 2018, CCIS 855, pp. 272–283, 2018.
https://doi.org/10.1007/978-3-319-91479-4_23

LLVM *transform passes*. They are 54 IR-to-IR code transformations designed for software optimization. Designing software optimizers based on LLVM transform passes allows using exactly the same optimizer tool on software code implemented in numerous programming languages, with no changes. Additionally, working with IR prevents the introduction of any additional unpredictable optimization by the compiler.

Despite the popularity of LLVM and the usefulness of the transform passes it provides for software optimization, there is little known on how the different passes actually affect the performance of the program. Only a few works exist providing empirical experiments, and they do not allow extracting generic conclusions (they are reviewed in Sect. 4). Therefore, there is a need of thorough studies analyzing the impact of the different passes LLVM provides on the performance of software programs. Such a study could provide highly valuable information for compilers, and software optimization in general. This paper goes in that direction.

Sensitivity Analysis (SA) [2] is a mathematical method that allows quantifying the impact of the input parameter values of a model on the uncertainty on the output of the model. SA was used in the past, for instance, to analyze the impact of the values the parameters of a simulator can take on its output [3], or for parameter tunning [4,5].

The main contribution of this work is the systematic mathematical study of the influence of the different LLVM passes on the performance of several software benchmark programs. The impact on the software execution time of every transformation both itself and when combined with the others are analyzed. This will be done through the Extended Fourier Amplitude Sensitivity Test (EFAST) [2]. We analyze the impact of all LLVM transform passes on a selection of eight benchmark programs from the BEEBS benchmark suite [6], as well as on a new program we implemented as the sequential execution of all the benchmarks considered in this work, one after the other. Our main conclusion is that the influence of the different transformations on the performance of the software is highly unpredictable. Indeed, the direct influence of the transformations is almost negligible in most studied cases, but it becomes highly important when combined with the other passes. This result exposes the complexity of the problem of designing software optimizers.

The structure of this paper is as follows. We first give a short introduction to LLVM and SA in Sects. 2 and 3, respectively. Section 4 presents some relevant related works from the Literature. After that, Sects. 5 and 6 present the configuration we designed for our experiments, as well as an elucidation on the obtained results. Finally, the paper ends with our main conclusions and the most salient lines of future work we identify from this work.

2 LLVM Compiler Infrastructure

LLVM [1] is a modular compiler infrastructure that includes a collection of modular compiler and toolchain technologies. It is an open-source project implemented in C++, designed for the compilation and optimization of programs written in different programming languages. This is possible thanks to the Intermediate Representation (IR), which is the core of LLVM. IR is a back-end pseudo-assembly language, composed by a strongly typed RISC instruction set, which abstracts away details of the target architecture.

LLVM was originally written to be a replacement for the existing code generator in the GCC stack, and many of the existing GCC front-ends have been modified to work with it. Thanks to the widespread interest in LLVM, there are a large number of projects to develop new front ends for different languages. One of the most important examples is Clang, a new compiler for C, Objective-C and C++ supported by Apple.

LLVM implements a set of 54 different code optimizations, called *transform passes*. These IR-to-IR transformations do not alter the semantics of the program, and they are often used for code optimization. They are performed on the IR code, providing a generic solution to optimize software, whatever the programming language used and the target architecture, if they are supported in LLVM.

The application of the passes is handled by a class called PassManager, which allows applying the transformations in three different ways: Over the entire module, on individual functions or on individual basic blocks. The available passes target many different operations, as those focusing on memory (e.g., mem2reg, memcpyopt, reg2mem, etc.), loops (e.g., loop-extract, loop-reduce, or loop-unroll, among others), and many other optimizations, as the elimination of useless and/or redundant pieces of code (e.g., constprop. This pass looks for instructions involving only constant operands and replaces them with a constant value; *add i32 1,2* will be transformed into *i32 3*). Because LLVM is an open source project, it also offers the possibility of building custom passes.

3 Sensitivity Analysis

Sensitivity Analysis (SA) identifies how the uncertainty of the output of a given model is influenced by the uncertainty of in its input parameters [2]. SA performs a systematic study of the behavior of the model for a large set of different input values, and it offers multiple information about the model, such as: (i) verification of the initial hypothesis of the model designer by understanding its real behavior; (ii) identification of the most influential factors; (iii) simplification of the model by possibly setting fixed values to the least influential ones; (iv) ascertainment of some interaction effects within the model, or (v) help for designing experiments and setting parameters

There are basically two main approaches for SA, namely local and global ones. On the one hand, the local approach consists in calculating the partial derivatives of the model in a certain point of the space of variables. Therefore, it studies how little variations in a certain input value modify the output of the model (while holding the other inputs fixed), providing information about the space of the input parameters. On the other hand, the global approach does not need an initial set of values but considers the whole input domain, which is essential in non-linear systems. In this case, this global approach studies the model as a whole and not as a particular solution similar to another one previously defined.

We can find a number of techniques that allow performing SA, either visual (e.g., scatterplot) or quantitative [2]. A well-known representative of the latter group, is the Fourier Amplitude Sensitivity Test (FAST) [7], a tool that can be used to calculate the first order sensibility indexes, as well as superior ones. The SA method used in this work is the extended FAST (EFAST) [7]. It calculates the interrelation among parameters, where a global interaction happens when the influence of a set of parameter modifications over the output is not the sum of the individual effects. This method allows us analyzing not only the direct effect of an input parameter on the uncertainty of the model, but also its effect in combination with the values of the other input parameters. For a review on the different existing methods of global SA, please refer to [8].

4 Related Work

The problem of software optimization through the application of passes can be divided into two subproblems: (1) the selection of the best passes to apply, and (2) the order in which they should be applied. Each of them has an important influence on the other, meaning that the performance of a software program after being optimized with a given selection of passes strongly depends on the order in which they were applied, and vice versa. Covering both subproblems simultaneously is a complex task, as reflected in [9–11].

Different techniques are used in the Literature to predict the adequacy of a given sequence of code transformations. Thus, in [12], a two-level method is proposed to choose the most convenient selection of optimizations for an auto-tuning application. In [13] sub-sequences are determined by machine learning to accelerate the choice of LLVM optimizations. The same idea of grouping passes is followed in [14], but making use of data mining processes.

Evolutionary algorithms (EAs) and heuristic strategies have been used to select compiler flags for software optimization, focusing on different performance metrics as run time [15] or energy consumption [16]. In [17], statistical techniques are used together with EAs to help improving the solutions. EAs are also used to parallelise sequential programs by means of straight-line code parallelisation, loop transformations, or parallelisation of recursive routines, as reviewed in [18]. This approach led to the first auto-parallelising compiler, called Parafrase [19].

Another novel line of research is the use of statistical techniques to calculate the impact of the optimizations. In [20], a procedure based on Orthogonal Arrays is used to propose an iterative algorithm to activate or deactivate a subset of passes. In a similar approach to ours, Tuzov et al. [21] make use of SA to determine the list of flags offered by FPGA manufacturers to optimize a given design. There are 71 synthesis flags available to choose from, and the authors reduce that number to 9 factors.

Because of the difficulty of the considered problem, all commented related works focus their efforts on reducing the search space. This is done by grouping passes following different methods. However, to the very best of our knowledge, there is not any work that thoroughly analyzes the influence of the source code optimizations on the performance of the program. Such an study, although difficult and time consuming, could set a fundamental theoretical basis for future works on source code optimization. In this work, we perform a systematic mathematical study to analyze the influence of all available LLVM transform passes (up to 54), one of the most popular compilers nowadays, on nine benchmark problems.

5 Configuration of Experiments

The model we analyze with SA in this work is a simple program that first applies a number of transform passes to the target software and after that it compiles and executes it, measuring the run time. The decision on the passes to apply is made according to the input variables. There are 54 variables (one per considered transform pass available in LLVM), and its value is the probability to apply the corresponding transformation. SA requires the input variables of the model to be real values in a given (wide) interval. The interval of allowed values for all input variables is: $[0, 1]$ (because they are probabilities), and the distribution of the values variables can take is set to be uniform.

Table 1. Description of the benchmark programs used and the operations they contain.

Name	Branching	Memory	Integer	Floating point	License	Category
AES	High	Low	Medium	Low	GPL	Security
Blowfish	Low	Medium	High	Low	GPL	Security
CRC32	Medium	Low	High	Low	GPL	Network
Dijkstra	Medium	Low	High	Low	GPL	Network
FDCT	High	High	Low	High	None	Consumer
FIR 2D	High	Medium	Low	High	None	Automotive/Consumer
Quicksort	High	Medium	High	Low	None	Automotive
SHA	High	Medium	Medium	Low	GPL	Network/Security

Among the multiple SA tools available, we used an implementation of EFAST called *Fast99*, which is available in package *sensitivity* of the R language [22]. In the studies made in this work, we considered eight programs from the BEEBS benchmark suite [6] with different features, as presented in Table 1. They are chosen because they expose processor and memory's performance. The selected benchmark programs are Advanced Encryption Standard (AES), Blowfish, CRC32, the Dijkstra shortest path algorithm, Finite Discrete Cosine Transform (FDCT), 2D FIR filters, Quicksort, and Secure Hashing Algorithm (SHA).

They contain algorithms typically found in sorting, security, networking, and telecommunications applications. In addition, we implemented a ninth program that executes all the chosen benchmarks, one after the other. The idea behind this, is to analyze how the influence of the passes applied to an individual program can change when it becomes part of a larger program.

The SA method used generates 54,000 parameter combinations, *numberOfSamples*numberOfParamaters* (the number of samples is typically 1,000). Every benchmark program is executed with these 54,000 different parameterizations, and every parameterization is executed for 32 independent runs, since the program that we evaluate is non-deterministic. The same 32 seeds were used in the independent runs, and the average execution time was selected to evaluate the performance of the algorithm. This means that the results presented in this paper are obtained after performing 1,728,000 executions for every benchmark program, making **15,552,000 experiments in total**.

The execution platform is an Intel Core i7-7700K 4.20 GHz with 16 GB 2400 MHz RAM memory. The operative system is a clean installation of Ubuntu 14.04.5 LTS and we worked with LLVM version 3.8.1.

6 Results

We proceed to summarize in this section our main findings. Figures 1 and 2 present the results of our SA test on the different benchmarks considered. In the plots, the white box means the direct influence of the corresponding transformation on the performance of the benchmark program evaluated. Value 0.0 means that applying the transformation has no effect on the program execution time, while value 1.0 indicates that the execution time of the program only depends on that transformation. The grey box indicates the effect of the transformation in combination with the others. Large grey boxes indicate strong dependencies among all variables.

With the purpose of clarity, we do not plot the results for all the 54 transform passes but just for those with a direct effect higher than 1%. on average, the number of transform passes exceeding this threshold is 9.6, supposing around 18% of the total number of transformations analyzed. The benchmark programs with the highest number of passes over 1% direct effect are FDCT and SHA, with 13 passes each. The one with less passes over 1% effect is CRC32, with 5.

We analyze in Fig. 1a the influence of the LLVM transform passes on SHA. SHA is a hashing algorithm used for verification in data streams. It has low memory requirements, and mostly integer operations. The average direct effect of all passes is 0.76% for this program, and 74,21% in the case of the effect with interactions. For SHA, `loop.rotate` and `loop.extract` are the transformations with the strongest effect on the run time of the program, with 2.3% and 1.9% direct effect, respectively. It calls our attention that, despite these two passes present the highest direct effect, their effect in interaction with the others does not stand out: it is roughly 74% in both cases, matching the average value for all passes in this benchmark. The transformation with the highest effect in

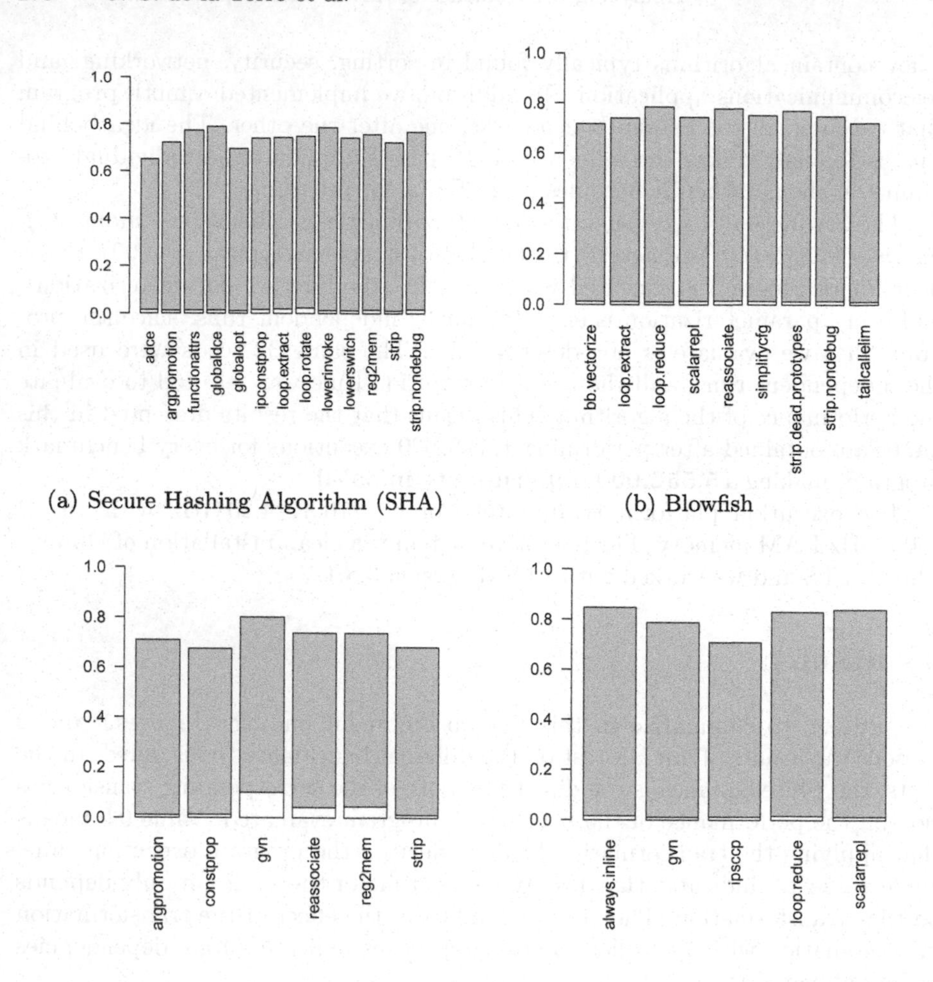

(a) Secure Hashing Algorithm (SHA) (b) Blowfish

(c) Finite Impulse Response 2D (FIR 2D) (d) Cyclic Redundancy Check 32 (CRC32)

Fig. 1. Impact of LLVM transform passes on the uncertainty of the performance of AES, Blowfish, CRC32, and Dijkstra shortest-path. Only those passes with over 1% main effect are displayed.

interaction with the others is `sink`, with 80.1% effect (but only 0.16% direct effect, which is actually the minimum one). This transformation tries to move instructions into successor blocks, so that they are executed just when they are needed, and not before. It makes sense that this transformation can be very useful after applying a number of automatic transformations to the code.

We can find similar behaviors for Blowfish (Fig. 1b), FIR 2D (Fig. 1c), Dijkstra (Fig. 2a), and FDCT (Fig. 2b). The main effect of all passes is almost negligible in all of them (below 2%). In contrast, the impact of the passes on the performance of the program when interacting with the others is meaningful

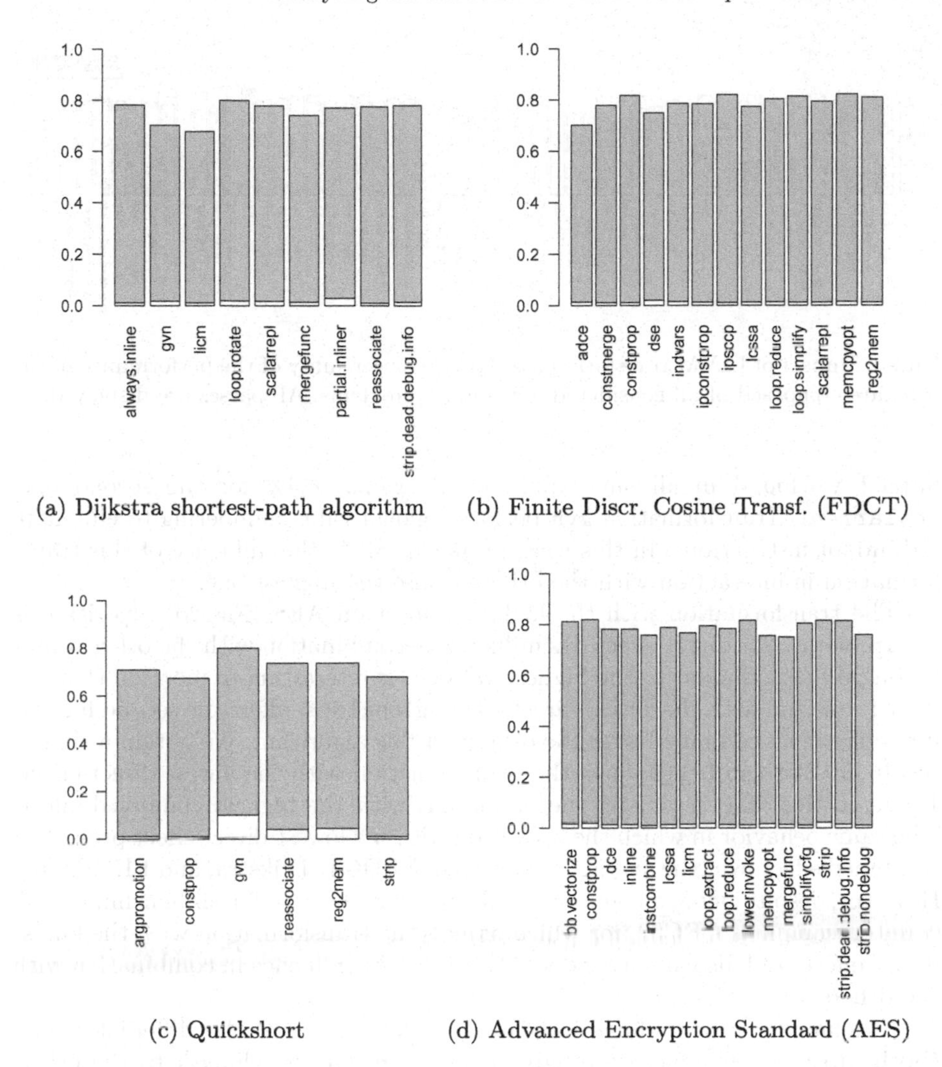

(a) Dijkstra shortest-path algorithm

(b) Finite Discr. Cosine Transf. (FDCT)

(c) Quickshort

(d) Advanced Encryption Standard (AES)

Fig. 2. Impact of LLVM transform passes on the uncertainty of the performance of FIR 2D, FDCT, Quicksort, and SHA. Only those passes with over 1% main effect are displayed.

(over roughly 70%, and up to 82%). Similar results are also observed for CRC32 (Fig. 1d), where the highest direct effect is 1.6%, obtained for `always.inline` operator. However, we can appreciate in this case a different behavior of the passes in combination with the others: although most of them provide an effect over 70%, as before, there are a number of passes with very low effect (14 passes below 30%), and this is not observed in any other studied benchmark. This issue cannot be appreciated in Fig. 1d because most passes are not displayed.

In the case of Quicksort algorithm (Fig. 2c), we can see how `gvn` is clearly more important than the other passes, with a direct effect of 10% (the highest

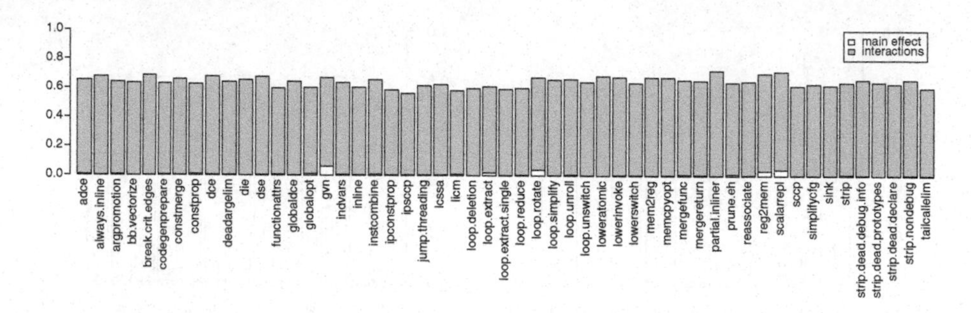

Fig. 3. Impact of LLVM transform passes on the uncertainty of the performance of the software composed of all considered benchmark problems. All passes are displayed.

impact we found in all our experiments) against 4.4% for the second one, `scalarrepl`. Transformation `gvn` performs global value numbering to eliminate redundant instructions. In this case, unlike for SHA, the influence of this transformation in interaction with the others is also the highest one.

The transformation with the highest impact on AES (Fig. 2d) algorithm is `strip`, with 2.2% direct effect. Its influence in combination with the other passes is roughly 82%, similar to the highest value. It is interesting to mention that the transformation with the lowest direct effect, `globaldce`, offers the second highest influence when combined with the others for this algorithm. We obtain a similar result for Blowfish benchmark: the transformation with the lowest direct effect, `lcssa` in this case, is among the operations with the highest combined effect. The same behavior in which the operation with the lowest direct effect presents a combined effect over the average is observed in FDCT, Dijkstra, and FIR 2D too. However, despite being a common result, it is not the case for all benchmarks. A counterexample is CRC32, for which `sink` is the transformation with the lowest direct effect, and also among those with the lowest influence in combination with the others.

In summary, our findings indicate that the effects of the different passes (both direct or combined) strongly change from one benchmark to the other. Indeed, the transformation with the strongest effect for some software might be the less important one in another, as it is the case of `deadargelim`, providing the highest combined effect for CRC32 and FDCT, and the weakest one on Dijkstra. In order to extract some conclusions, we computed the average effect of every transformation in all studied benchmark programs. We obtained that the transformation with the highest average effect was `gvn`, with 2.01%, followed by `reg2mem` (1.19%), `prune.eh` (1.04%), `strip` (1.02%), and `loop.rotate` and `mergefunc`, both with 0.98% effect. On the contrary, the passes with the lowest average effect were found to be `break.crit.edges` (0.38%), `sink` (0.44%), `mem2reg` (0.48%), and `partial.inliner` and `die` (0.49%).

Finally, we also evaluated the effects of all passes on a new benchmark that is composed by the sequential run of all studied benchmarks. The results are shown in Fig. 3 (please, note that here we are not filtering those passes with less than 1% effect). In this case, the average main effect of all passes is 8.7%, while it is 64.3% when considering interactions. These values are slightly higher than the average effect of all passes when running the benchmarks independently, which is 7.5%, while the average effect of passes in interaction with the others is 58, 3%. We can see how the best (and worst) passes identified before are generally the ones with the strongest (and weakest) effect in Fig. 3. It is also interesting to see how some passes have strong influence (in comparison with the others) for some benchmarks, but not for the aggregation of all of them. For instance, `globalopt` has 1.3% influence in the latter, while it is below 1% for all benchmarks but SHA (1.17%).

7 Conclusions and Future Work

We perform in this work a thorough study on the influence of LLVM source code optimization transformations (called passes) through sensitivity analysis. We considered for the study eight benchmark programs, taken from BEEBS benchmark suite [6], as well as another one we implemented as the sequential run of all considered benchmarks, in order to analyze the influence of the different passes in a larger software.

After our studies, involving over fifteen million experiments, we can identify `gvn`, `reg2mem`, `prune.eh`, and `strip` as the LLVM passes showing the highest effect on the performance of the considered benchmark programs, on average. However, we obtained very low direct effect values for all passes in all studied benchmarks (7.5% on average). The effect of passes when combined with the others is roughly one order of magnitude higher. This fact shows the difficulty of software optimization problem.

As future work, we are currently considering extending the study to a larger benchmark suite. Additionally, we consider of high interest studying the impact of the passes on the energy consumption of the execution platform, in addition to the run time.

Acknowledgment. The authors would like to acknowledge the Spanish MINECO and ERDF for the support provided under contracts TIN2014-60844-R (the SAVANT project) and RYC-2013-13355.

References

1. The LLVM Compiler Infrastructure. http://llvm.org. Accessed 2018
2. Saltelli, A., Tarantola, S., Campolongo, F., Ratto, M.: Sensitivity Analysis in Practice: A Guide to Assessing Scientific Models. Wiley, New York (2004)
3. Etxeberria, L., Trubiani, C., Cortellessa, V., Sagardui, G.: Performance-based selection of software and hardware features under parameter uncertainty. In: Proceedings of the 10th International ACM SIGSOFT Conference on Quality of Software Architectures (Part of CompArch 2014), QoSA 2014, pp. 23–32 (2014)

4. Iturriaga, S., Ruiz, P., Nesmachnow, S., Dorronsoro, B., Bouvry, P.: A parallel multi-objective local search for AEDB protocol tuning. In: Proceedings - IEEE 27th International Parallel and Distributed Processing Symposium Workshops and PhD Forum, IPDPSW 2013 (Sect. VI), pp. 415–424 (2013)
5. Srinivas, C., Reddy, B., Ramji, K., Naveen, R.: Sensitivity analysis to determine the parameters of genetic algorithm for machine layout. Procedia Mater. Sci. **6**(Icmpc), 866–876 (2014)
6. Pallister, J., Hollis, S., Bennett, J.: BEEBS: Open benchmarks for energy measurements on embedded platforms. Technical report 1308.5174, arXiv (2013)
7. Saltelli, A., Tarantola, S., Chan, K.S.: A quantitative model-independent method for global sensitivity analysis of model output. Technometrics **41**(1), 39–56 (1999)
8. Iooss, B., Lemaître, P.: A review on global sensitivity analysis methods. In: Dellino, G., Meloni, C. (eds.) Uncertainty Management in Simulation-Optimization of Complex Systems. ORSIS, vol. 59, pp. 101–122. Springer, Boston, MA (2015). https://doi.org/10.1007/978-1-4899-7547-8_5
9. Fitzgerald, T., Sullivan, B.: Analysing the effect of candidate selection and instance ordering in a realtime algorithm configuration system. In: Proceedings of the Symposium on Applied Computing, pp. 1003–1008 (2017)
10. Chebolu, N., Wankar, R.: Tuning compilations landscape. In: Proceedings of the 2016 2nd International Conference on Contemporary Computing and Informatics, IC3I 2016, pp. 575–583 (2016)
11. Ashouri, A., Palermo, G., Cavazos, J., Silvano, C.: Automatic Tuning of Compilers using Machine Learning. Springer, Heidelberg (2017)
12. Ding, Y., Ansel, J., Veeramachaneni, K., Shen, X., O'Reilly, U.M., Amarasinghe, S.: Autotuning algorithmic choice for input sensitivity. In: Proceedings of the 36th ACM SIGPLAN Conference on Programming Language Design and Implementation (PLDI), pp. 379–390 (2015)
13. Ashouri, A., Bignoli, A., Palermo, G., Silvano, C., Kulkarni, S., Cavazos, J., Bignoli, A., Palermo, G., Silvano, C.: MiCOMP: mitigating the compiler phase-ordering problem using optimization sub-sequences and machine learning. ACM Trans. Archit. Code Optim. **14** (2017). Article No. 29
14. Martins, L., Nobre, R., Cardoso, J., Delbem, A., Marques, E.: Clustering-based selection for the exploration of compiler optimization sequences. ACM Trans. Archit. Code Optim. **13**(1), 1–28 (2016)
15. Sandran, T., Zakaria, N., Pal, A.: An optimized tuning of genetic algorithm parameters in compiler flag selection based on compilation and execution duration. In: Deep, K., Nagar, A., Pant, M., Bansal, J. (eds.) SocProS 2011. AINSC, vol. 131, pp. 599–610. Springer, New Delhi (2012)
16. Varrette, S., Dorronsoro, B., Bouvry, P.: An LLVM-based approach to generate energy aware code by means of MOEAs. In: 7th European Symposium on Computational Intelligence and Mathematics (ESCIM), pp. 198–204 (2015)
17. Garciarena, U., Santana, R.: Evolutionary optimization of compiler flag selection by learning and exploiting flags interactions. In: Proceedings of the 2016 on Genetic and Evolutionary Computation Conference (GECCO), pp. 1159–1166. ACM (2016)
18. Banerjee, U., Eigenmann, R., Nicolau, A., Padua, D.A.: Automatic program parallelization. Proc. IEEE **82**, 211–243 (1993)
19. Polychronopoulos, C.: Parallel Programming and Compilers. Kluwer Academic Publishers, Norwell (1988)

20. Pinkers, R., Knijnenburg, P., Haneda, M., Wijshoff, H.: Statistical selection of compiler options. In: Proceedings - IEEE Computer Society's Annual International Symposium on Modeling, Analysis, and Simulation of Computer and Telecommunications Systems, MASCOTS, pp. 494–501 (2004)
21. Tuzov, I., Andrés, D., Ruiz, J.C.: Tuning synthesis flags to optimize implementation goals: performance and robustness of the LEON3 processor as a case study. J. Parallel Distrib. Comput. **112**, 84–96 (2018)
22. R: A Language and Environment for Statistical Computing: R foundation for statistical computing. https://www.R-project.org. Accessed 2018

Finding the Most Influential Parameters of Coalitions in a PSO-CO Algorithm

Patricia Ruiz[1(✉)], Bernabé Dorronsoro[2], Juan Carlos de la Torre[2], and Juan Carlos Burguillo[3]

[1] Mechanical Engineering and Industrial Design Department, University of Cadiz, Cadiz, Spain
patricia.ruiz@uca.es

[2] Computer Sciences Engineering Department, University of Cadiz, Cadiz, Spain
{bernabe.dorronsoro,juan.detorre}@uca.es

[3] Telecommunications Engineering School, Universidad de Vigo, Vigo, Spain
j.c.burguillo@det.uvigo.es

Abstract. Literature reveals that optimization algorithms are generally composed of a large number of parameters that highly influence on its performance. In the early stages of the definition of a new algorithm, it is crucial to know how the uncertainty in the input parameters affects the behavior of the algorithm, influencing on its final output, so that it is possible to set up the most efficient configuration.

In this work, we are making a sensitivity analysis using the Extended Fourier Amplitude Sensitivity Test to compute the first order effects and interactions for each parameter on a recently proposed particle swarm optimization algorithm that implements a dynamic structured swarm, based on coalitions. This technique, inherited from game theory, includes four new parameters that are analyzed and tested on a well-known benchmark for continuous optimization. Results give interesting insights of the importance of one of the parameters over the rest.

Keywords: Sensitivity analysis · Particle Swarm Optimization Optimization · Coalitions

1 Introduction

It is well known that decentralizing the population helps Evolutionary Algorithms (EAs) keeping diversity of solutions for longer and therefore, contributing to mitigate premature convergence in the population [1,2]. This is done at the cost of reducing the convergence speed, so algorithms implementing such decentralized populations might take longer to find accurate solutions with respect to panmictic ones. However, they can find accurate solutions to problems for which panmictic EAs get stuck in low quality local optimal solutions.

Not all problems require the same convergence speed. Panmictic populations are, generally, more effective optimization tools for problems displaying simple

J. Medina et al. (Eds.): IPMU 2018, CCIS 855, pp. 284–296, 2018.
https://doi.org/10.1007/978-3-319-91479-4_24

landscapes with a low number of local optimal solutions, while structured populations are a more appropriate configuration when dealing with multi-modal and complex functions. Algorithms implementing dynamic population topologies [3,4] aim at efficiently solving problems of different features.

We recently introduced in [5] a novel Particle Swarm Optimization algorithm [6,7] that implements a dynamic topology in the swarm. The algorithm is called PSO-CO, and it implements coalitions of particles in the swarm. The different coalitions can be understood as independent swarms so that particles can only interact with other particles in the same swarm. Particles can leave their swarm and join another, or build their own, at any moment, based on their own expected benefit. The resulting algorithm clearly outperformed the equivalent PSO implementation with a panmictic population, but at the cost of introducing four new parameters in the algorithm, required for implementing the coalitions.

The main contribution of this work is the performance of a rigorous systematic study to mathematically evaluate the impact of the already mentioned four new parameters of PSO-CO on the uncertainty of the performance of the algorithm, in terms of both, the quality of the solutions found and execution time. This is done through the extended Fourier Amplitude Sensitivity Test [8] (EFAST for short). We carefully analyze the output of EFAST method and results show the importance of one of the parameters over the rest.

This paper is organized as follows. Next Section introduces the concept of sensitivity analysis and its purpose. Section 3 presents some relevant work on the topic. The optimization algorithm we are tackling, PSO-CO is explained in Sect. 4. Both, the experiments and the results obtained are shown in Sects. 5 and 6, respectively. Finally, Sect. 7, concludes the papers and presents our main future research lines.

2 Sensitivity Analysis

The objective of performing a sensitivity analysis (SA) of a model is to determine the dependency of its output on its input factors [8]. SA allows quantifying the effects of the values these parameters can take on the uncertainty of the output of the model [9].

SA helps to understand the behavior of the model and how different parts of the model interplay. It also identifies the most and the least influencing parameters, and measures how variations of a specific input parameter affects on the uncertainty of the output of the model. This study allows to fix non-influential inputs to nominal values and to fine tune the influencing ones using available information. Moreover, in the design of an algorithm, the developer usually implements an algorithm considering his initial idea of its behavior. SA provides verification of the designer's hypothesis and allows him/her to modify the model, if necessary.

We can find in the Literature two different techniques related to sensitivity analysis: local and global SA. The former is the first approach to SA and modifies one parameter at a time around a nominal value (keeping the others fixed) and analyses the local variation of the output of the model. In order to overcome its limitations, global SA methods appear, considering the whole input domain.

SA is often applied to analyze simulators. Simulations allow to accomplish experiments that otherwise can be very costly in terms of time or resources in multiple areas (physics, engineering, etc.). They generally rely on mathematical models that require the estimation of different parameters to determine the behavior of the experiment. Therefore, parameter analysis is essential in the development of the final model. SA allows measuring and quantifying how input parameters can influence output values. It is a basic tool for extracting knowledge from the system, as well as for modeling and development.

There are many techniques to perform SA, either visual (e.g., scatterplot) or quantitative [9] (e.g., the Sobal method or the extended Fourier's amplitude sensibility test –EFAST–). For a more detailed review on the different existing global analysis methods, please refer to [10]. The Sobol and EFAST methods can be used to calculate the first order sensibility indexes or even superior ones. The implementation used in this work is Fast99, and can be found in the R project [11]. It is an implementation of the –EFAST– algorithm originally proposed in [8,9], that calculates the interrelation among parameters, where a global interaction happens when the influence of a set of parameter modifications over the output is not the sum of the individual effects.

3 The Influence of Parameters on Software Performance, A Review

Methods for Sensitivity analysis are inestimable tools in any field. We can find in the Literature many works facing the challenging problem of trying to understand how the different input factors interact and influence on the behavior of the algorithm (the uncertainty of the output). In this section, we point out some the most relevant works that use these tools in many different domains.

In software configurable systems, performance is very dependent on the selected characteristics of the system. In [12], the authors perform two SA analysis levels (uniform binding and statistical sampling), to determine their influence over the software and hardware contexts.

Concerning Science of Materials, the work described in [13] performs a SA to optimize crossover and mutation parameters over a basic genetic algorithm (GA) used to manage machinery configurations.

Considering search and optimization methods, [14] proposes a SA algorithm (Local-Global SA algorithm) for its Artificial Bee Colony algorithm (ABC). Also in [15], authors performed an exhaustive SA on an evolutionary algorithm (EA) and pointed out that the metaheuristic parameters have little influence on the accuracy of the algorithm but some local search operators had strong impact.

In the Telecommunications area, the work presented in [16] uses *Fast99* to determine the influence of the parameters used by the algorithm AEDB-MLS, and this analysis is used to design local search operators.

In nuclear engineering, uncertainties in the pressurized thermal shock of a nuclear pressurized reactor are determined in [17], simulating an accidental scenario. A very complete review of different SA methods tested on environmental models is presented in [18]. In the field of stealth technology, Lefebvre uses a SA for finding the most influential parameters in an aircraft infrared signature simulation model with more than 30 input factors [19].

Finally, in Biomedical Informatics, [20] proposes a framework to speed up the SA and the parameter tuning processes in a framework for performing tissue image analysis.

4 PSO with Coalitions, PSO-CO

Particle Swarm Optimization algorithms (PSO hereinafter) are well-known metaheuristics that emulate the social behavior of bird flocks or school of fish for solving complex continuous optimization problems. The population of the algorithm is called swarm. In this swarm, the best particle found so far is considered the leader, and the rest of the swarm follows it during the search. In every iteration, each particle updates its position in terms of the speed and the current position as shown in Eq. 1 for particle i:

$$\vec{x_i}(t) = \vec{x_i}(t-1) + \vec{v_i}(t). \tag{1}$$

Particles are in pursuit of the leader by updating their speed in terms of the leader's position (the global best position in the swarm), and the best position the particle ever visited (i.e. historical local best position). The speed is defined as follows:

$$\vec{v_i}(t) = w \cdot \vec{v_i}(t-1) + C_1 \cdot r_1 \cdot (\vec{x_{li}} - \vec{x_i}) + C_2 \cdot r_2 \cdot (\vec{x_{gi}} - \vec{x_i}), \tag{2}$$

where $\vec{x_{li}}$ represents the best solution visited by particle i, $\vec{x_{gi}}$ refers to the best particle of the swarm (the leader), w is the weight of the particle's inertia (controls the balance between local and global experience), r_1 and r_2 are two random numbers, and C_1 and C_2 are two specific parameters that control the effect of the position of the best particle and the leader.

PSO provides a simple and effective implementation for solving continuous optimization problems. However, its centralized population and movement strategy allow a fast convergence, that generally gets stuck in local optima. Decentralizing the population prevents that fast convergence, preserving at the same time the diversity of the population for longer [1].

In [5], we proposed PSO-CO, a PSO where we applied cooperative game theory techniques (formation of coalitions) for decentralizing the swarm. These coalitions are known to help the agents achieving their goals and even accomplishing others they might not be able to fulfill independently [21].

By applying the coalitions into the swarm of the PSO-CO we are benefiting from two of the most well-known population structures: island and cellular. Inside the coalition every particle interacts with any other particle of the same coalition, behaving like a subpopulation of an algorithm structured in islands. However, isolated or frontier particles that just left a coalition, analyze neighboring ones (like a cellular algorithm) before deciding whether to join one or not.

Particles in the swarm can only belong to one coalition, and they do not follow the leader of the swarm anymore, but the leader of the coalition they belong to. Coalitions are measured in terms of their quality (explained later) and it is desirable they join the most profitable coalition. In the same way, particles that consider the benefit of belonging to a coalition is low, can leave it and analyze neighboring ones. If the expected reward of joining an adjacent coalition is low, particles can remain isolated.

The pseudocode of PSO-CO is detailed, in Algorithm 1.

Algorithm 1. Pseudocode of PSO-CO

1: *InitialiseSwarm*();
2: *EvaluateSwarm*();
3: *BelongCoalition*();
4: $numEvals = 0$;
5: **while** $numEvals < maxNumEvals$ **do**
6:    **for** $i = 1 : NumCoalitions$ **do**
7:       *CalculateSpeeds*();
8:       *UpdatePositions*();
9:       *EvaluateCoalition*();
10:    **end for**
11:    *DecideBelongCoalition*();
12: **end while**
13: *ReturnBestSolutionFound*();

Particles are able to evaluate the potential reward (in case of merger) and compare different coalitions by using a quality value associated to them. It is known that, in optimization algorithms, keeping the diversity of the population is key for spacing from local optima and obtaining good results. Therefore, it is desirable to have large coalitions with diverse solutions. In PSO-CO, the quality of the coalition is determined by its size, as well as the quality and diversity of the particles composing it. This quality value is defined as follows:

$$Quality(C_i) = \alpha \cdot Size(C_i) + \beta \cdot Var(C_i) + \gamma \cdot Avg(C_i) \tag{3}$$

where $\alpha + \beta + \gamma = 1$. These coefficients model the importance given to the size, diversity and average of the fitness value of all the solutions composing the coalition, respectively.

Additionally, in order to promote the exchange of information between coalitions, a parameter called Independent Coefficient, Ind_c is also introduced and models the desire of a particle of remaining independent (probabilistic value).

In [5], the values of these four parameters were experimentally chosen, and the performance of the original PSO was improved. However, a deeper study on the influence of each the parameters over the results of the algorithm is needed to better understand the impact of the coalitions on the solutions. These interaction and influence between the parameters and the performance of the algorithm is analyzed here through a sensitivity analysis.

5 Experiments

We now proceed to present and explain the experiments done in order to understand the uncertainty of the output that is derived from the coalition parameters. These four new parameters required by PSO-CO for implementing the coalitions are: Ind_c, α, β, andγ. Ind_c represents the probability of individuals to leave the coalition and remain independent; α, β, and γ, are used to evaluate the expected benefit of joining a given coalition (as described in Eq. 3). The rest of the parameters of PSO-CO are the same of the classical PSO algorithm (introduced in Sect. 4), and they are described in Table 1. These values have also been experimentally chosen after an exhaustive preliminary analysis. The size of the swarm is larger than the typically used values in PSO, in order to allow the simultaneous formation of a number of coalitions in the swarm. The inertia (W) has a fixed value of 0.1, and the other four parameters take random values for each particle in each iteration. Parameters r_1 and r_2 are uniformly distributed in the [0, 1] interval, whereas C_1 and C_2 are in the [1.5, 2.5] interval. The maximum allowed number of generations if the optimum is not found is 2500 (i.e. 1,000,000 of evaluations).

The SA method requires to set up a wide range for the four targeted variables. As already mentioned in Eq. 3, the sum of α, β and γ must be equal to 1, and at the same time, Ind_c is a probabilistic value. Therefore, we use the range $[0, 1]$ for the four parameters, as presented in Table 2.

In our SA study, we use an implementation of the Extended Fourier Amplitude Sensitivity Test proposed in [8] by Saltelli et al., known as *fast99*, that is included in the *R* project [11]. All variables must be independent for this test. As we previously said in Sect. 4, there exists a restriction with three of the studied variables: $\alpha + \beta + \gamma = 1$. However, this restriction implies there are just two degrees of freedom, but we still have three independent variables. Therefore, we can apply the *fast99* test.

Table 1. Parameters of PSO-CO

Parameter	Value
Swarm size	400
W	0.1
C_1	random value in [1.5, 2.5]
C_2	random value in [1.5, 2.5]
r_1	random value in [0, 1]
r_2	random value in [0, 1]
Max evaluations	1,000,000

Table 2. Range of parameters of coalitions

Parameter	Value
α	[0, 1]
β	[0, 1]
γ	[0, 1]
Ind_c	[0, 1]

The method generates a large number of parameter combinations. Then, the algorithm is run with the prepared parameter combinations. The number of combinations is $N_{samples} \times N_{parameters}$ (typically, $N_{samples} = 1000$). Considering we have to cope with the already mentioned restriction of the parameters, we normalize the values before using these combinations in PSO-CO. For that, we just divide each of the three parameters α, β and γ by the sum of the three of them $(\alpha + \beta + \gamma)$.

The benchmark used for analyzing our algorithm was first proposed in *CEC 2015 Competition on Learning-based Real-Parameter Single Objective Optimization* [22]. This benchmark is composed of 15 minimization problems with dimensions 10, 30, 50, 100 (we consider in this study the two largest ones). Each function has different features and they all have been shifted and rotated. There are two *unimodal* functions (F1 and F2), three simple *multimodal* (F3–F5), three *hybrid* (F6–F8) and seven *composition* functions (F9–F15). The performance of the algorithm is compared in terms of both, the quality of the solution found and the execution time.

6 Results

In this section, we are presenting the results obtained after performing SA to PSO-CO. As explained in Sect. 5, we are performing the sensitivity analysis over the well-known competition for optimization algorithms of CEC 2015. We present here only representative results, so we show the results obtained for one problem of each the four different types proposed (*unimodal*, *multimodal*, *hybrid* and *composition*).

Results are shown in stacked bar charts presenting for each factor, their linear and non-linear (or interaction) effects on the output. The gray color represents the interaction that the parameter has along with the others on the algorithm, whilst the white color represents the direct effect this parameter exerts.

Figures 1 and 2 present the results obtained in terms of the accuracy of the algorithm for F1, F3 and F6, and F9 respectively, for the two studied dimensions (50 and 100). The benefits of the sensitivity analysis are immediately visible.

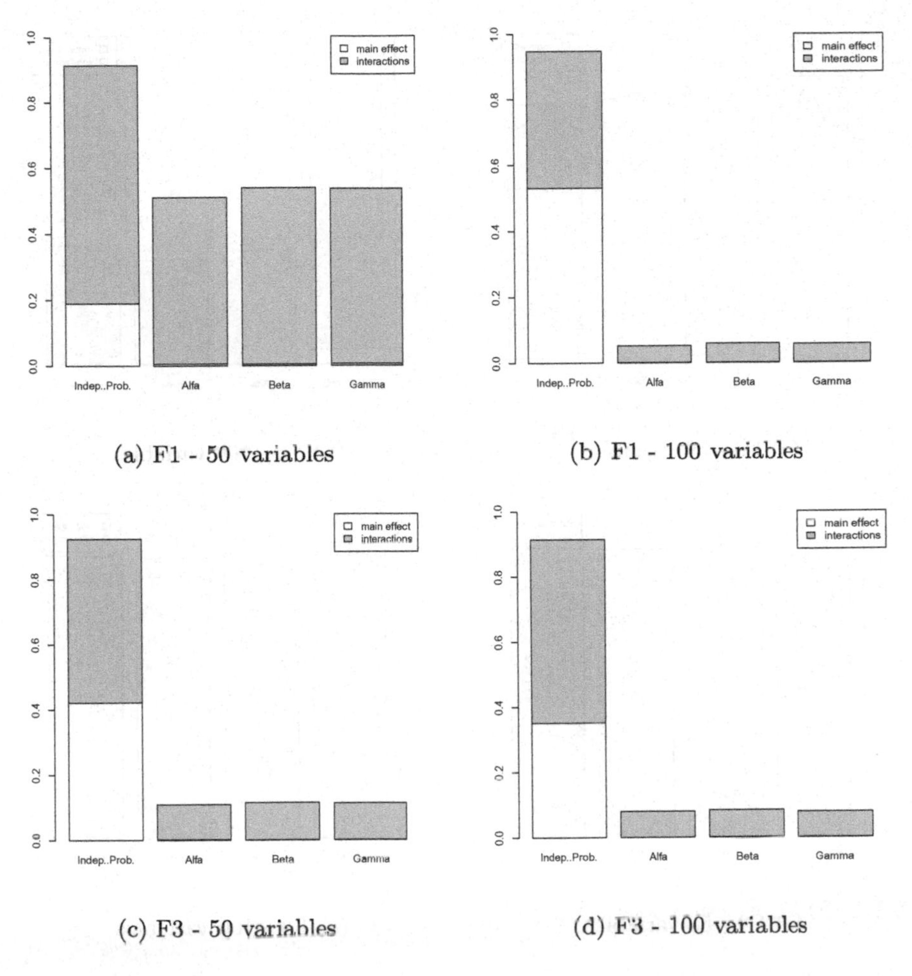

(a) F1 - 50 variables

(b) F1 - 100 variables

(c) F3 - 50 variables

(d) F3 - 100 variables

Fig. 1. Impact of the variables used for coalitions on the accuracy of the algorithm.

Results show that the only parameter that directly influences, to some extent, in the accuracy of the algorithm is Ind_c (the independent coefficient), regardless of the dimension considered. In Figs. 1a, b, c and d, we can see that for both *unimodal* and *multimodal* problems, Ind_c influences the most the output, while the influence of the three parameters α, β, and γ is negligible. However, for the *hybrid* and *composition* problems, F6 and F9, the direct influence is reduced almost to 0 for all parameters, but the interaction of all parameters is high, which means that all parameters combined, influence the output more than individually (see Figs. 2a, b, c and d).

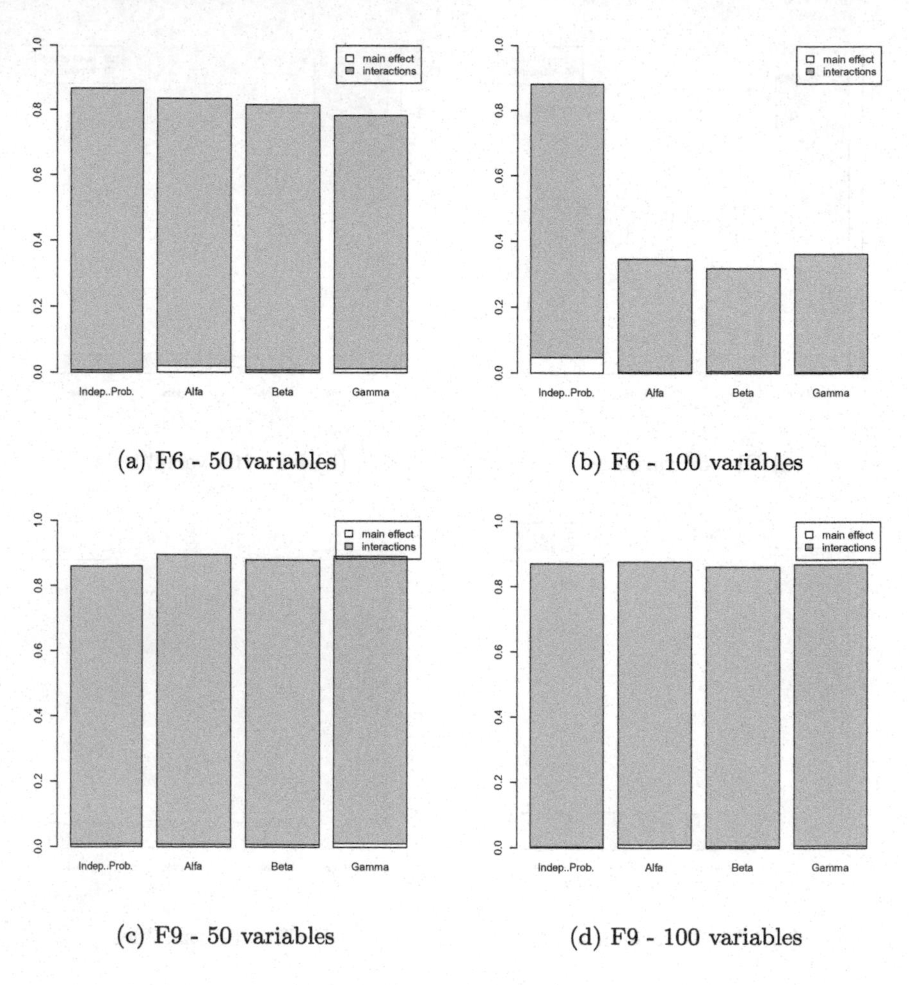

(a) F6 - 50 variables

(b) F6 - 100 variables

(c) F9 - 50 variables

(d) F9 - 100 variables

Fig. 2. Impact of the variables used for coalitions on the accuracy of the algorithm.

In terms of the execution time, Figs. 3 and 4 reveal that regardless the type of problem faced, the influence of α, β, and γ is insignificant. Nevertheless, the Ind_c highly and directly influences the execution time. This result is consistent with the expected behavior of PSO-CO because the creation of the coalitions delays the execution time (as it was shown in [5], PSO-CO takes longer than classical PSO). Additionally, while α, β, and γ determine the quality of the coalition, Ind_c is the only parameter involved in the actual creation of the coalition. High values of Ind_c would lead to classical PSO, where coalitions do not exist.

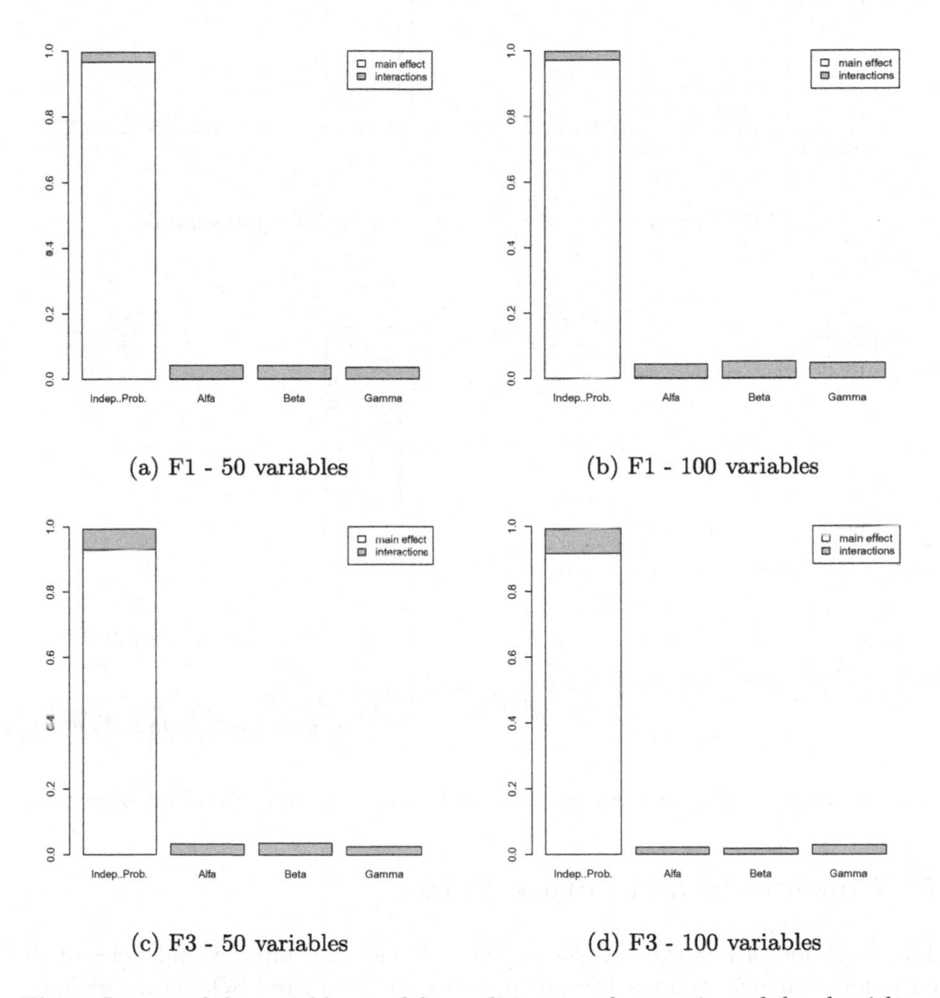

(a) F1 - 50 variables

(b) F1 - 100 variables

(c) F3 - 50 variables

(d) F3 - 100 variables

Fig. 3. Impact of the variables used for coalitions on the run time of the algorithm.

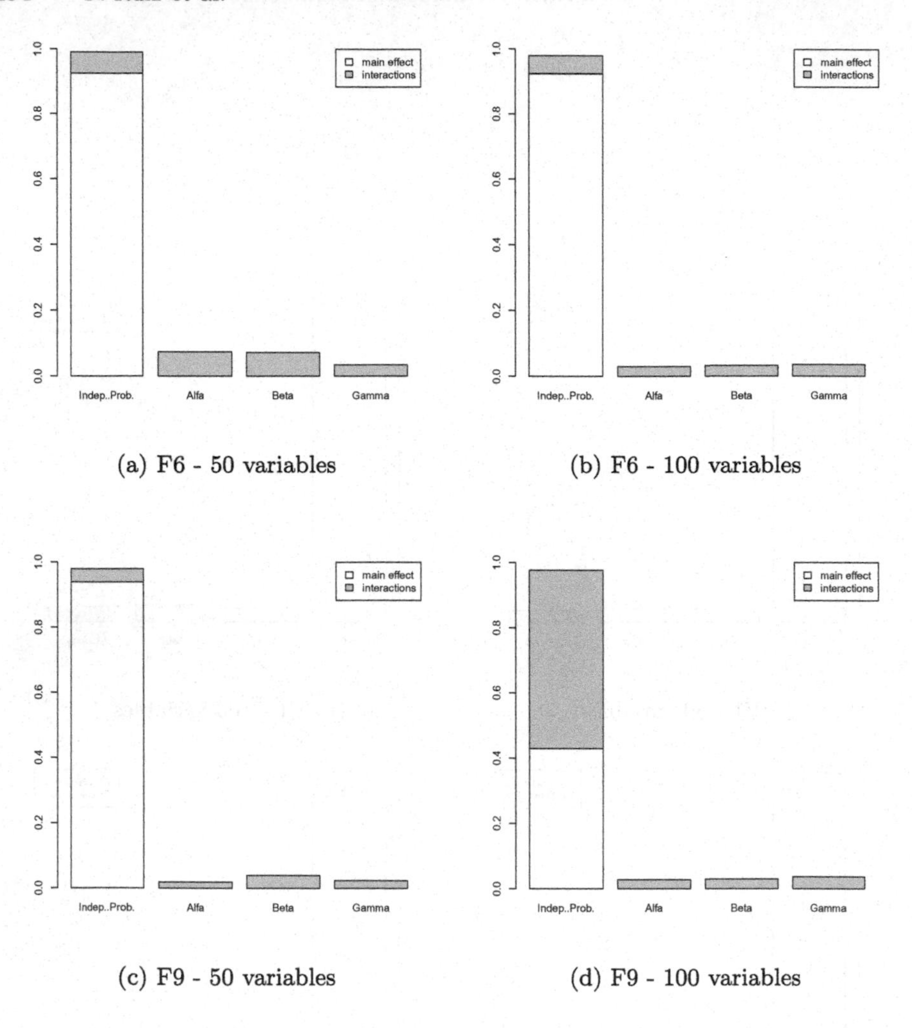

Fig. 4. Impact of the variables used for coalitions on the run time of the algorithm.

7 Conclusions and Future Work

This work focuses on the analysis of the influence and interdependencies of the parameters arisen from the formation of coalitions of the PSO-CO algorithm.

This influence is measured, through a sensitivity analysis, that helps to understand how the different variables of the algorithm interact, in terms of both the accuracy and the execution time.

Results show that the most impacting parameter is the Ind_c, which has the highest influence on the accuracy as well as on the execution time regardless the problem or the dimension studied.

In terms of the execution time, SA shows that the influence of the other three parameters (α, β, and γ) is negligible. That was expected because Ind_c models the desire of particles to remain independent. If this value is 1, PSO-CO behaves as a classical PSO (there are no coalitions in the swarm).

Regarding the accuracy, Ind_c also generally shows the highest impact. For the *unimodal* and *multimodal* problems there exists direct influence whilst for the hybrid and composite there is not. At the same time, the interaction of α, β, and γ for these problems is not relevant. For the two last problems, F6 and F9, the sensitivity analysis does not show any significant direct influence for any of the parameters. However, in these cases, the common interaction of α, β, and γ is the highest.

As future work, we are considering redesigning the algorithm so that the number of parameters is reduced by grouping the three parameters whose direct influence is less relevant, α, β, and γ, or even assigning them a fixed value.

Acknowledgment. The authors would like to acknowledge the Spanish MINECO and ERDF for the support provided under contracts TIN2014-60844-R (the SAVANT project) and RYC-2013-13355.

References

1. Alba, E., Dorronsoro, B.: Cellular Genetic Algorithms. Operations Research/Compuer Science Interfaces. Springer, Heidelberg (2008). https://doi.org/10.1007/978-0-387-77610-1
2. Dorronsoro, B., Bouvry, P.: Improving classical and decentralized differential evolution with new mutation operator and population topologies. IEEE Trans. Evol. Comput. **15**(1), 67–98 (2011)
3. Dorronsoro, B., Bouvry, P.: Cellular genetic algorithms without additional parameters. J. Supercomputing **63**(3), 816–835 (2013)
4. Dorronsoro, B., Burguillo, J.C., Peleteiro, A., Bouvry, P.: Evolutionary algorithms based on game theory and cellular automata with coalitions. In: Zelinka, I., Snášel, V., Abraham, A. (eds.) Handbook of Optimization. Intelligent Systems Reference Library, vol. 38, pp. 481–503. Springer, Heidelberg (2013). https://doi.org/10.1007/978-3-642-30504-7_19
5. Ruiz, P., Dorronsoro, B., Torre, J., Burguillo, J.: Including dynamic adaptative topology to particle swarm optimization algorithms. In: Proceedings of the 21 Congreso Int. de Dirección e Ingeniería de Proyectos. Lecture Notes in Management and Industrial Engineering. Springer (2018, in press)
6. Clerc, M., Kennedy, J.: The particle swarm - explosion, stability, and convergence in a multidimensional complex space. IEEE Trans. Evol. Comput. **6**(1), 58–73 (2002)
7. Clerc, M.: Particle Swarm Optimization. ISTE (International Scientific and Technical Encyclopedia) (2006)
8. Saltelli, A., Tarantola, S., Chan, K.P.S.: A quantitative model-independent method for global sensitivity analysis of model output. Technometrics **41**(1), 39–56 (1999)
9. Saltelli, A., Tarantola, S., Campolongo, F., Ratto, M.: Sensitivity Analysis in Practice: A Guide to Assessing Scientific Models. Wiley (2004)
10. Iooss, B., Lemaître, P.: A review on global sensitivity analysis methods. In: Dellino, G., Meloni, C. (eds.) Uncertainty Management in Simulation-Optimization of Complex Systems. ORSIS, vol. 59, pp. 101–122. Springer, Boston, MA (2015). https://doi.org/10.1007/978-1-4899-7547-8_5

11. R Core Team: R: a language and environment for statistical computing. https://www.R-project.org. Accessed 2017
12. Etxeberria, L., Trubiani, C., Cortellessa, V., Sagardui, G.: Performance-based selection of software and hardware features under parameter uncertainty. In: Proceedings of the 10th International ACM SIGSOFT Conference on Quality of Software Architectures, pp. 23–32 (2014)
13. Srinivas, C., Reddy, B.R., Ramji, K., Naveen, R.: Sensitivity analysis to determine the parameters of genetic algorithm for machine layout. Procedia Mater. Sci. **6**(Icmpc), 866–876 (2014)
14. Loubière, P., Jourdan, A., Siarry, P., Chelouah, R.: A modified sensitivity analysis method for driving a multidimensional search in the artificial bee colony algorithm. In: IEEE Congress on Evolutionary Computation, pp. 1453–1460 (2016)
15. Pinel, F., Danoy, G., Bouvry, P.: Evolutionary algorithm parameter tuning with sensitivity analysis. In: Bouvry, P., Kłopotek, M.A., Leprévost, F., Marciniak, M., Mykowiecka, A., Rybiński, H. (eds.) SIIS 2011. LNCS, vol. 7053, pp. 204–216. Springer, Heidelberg (2012). https://doi.org/10.1007/978-3-642-25261-7_16
16. Iturriaga, S., Ruiz, P., Nesmachnow, S., Dorronsoro, B., Bouvry, P.: A parallel multi-objective local search for AEDB protocol tuning. Proceedings of the IEEE 27th International Parallel and Distributed Processing Symposium Workshops and Ph.D. Forum, IPDPSW 2013 (Section VI), pp. 415–424 (2013)
17. Auder, B., Crécy, A., Iooss, B., Marqués, M.: Screening and metamodeling of computer experiments with functional outputs. Application to thermal-hydraulic computations. Reliab. Eng. Syst. Safety **107**, 122–131 (2012)
18. Hamby, D.M.: A review of techniques for parameter sensitivity analysis of environmental models. Environ. Monit. Assess. **32**, 135–154 (1994)
19. Lefebvre, S., Roblin, A., Varet, S., Durand, G.: A methodological approach for statistical evaluation of aircraft infrared signature. Reliab. Eng. Syst. Safety **95**, 484–493 (2010)
20. Teodoro, G., Kurç, T., Taveira, L., Melo, A., Gao, Y., Kong, J., Saltz, J.: Algorithm sensitivity analysis and parameter tuning for tissue image segmentation pipelines. Bioinformatics **33**(7), 1064–1072 (2017)
21. Li, X.: Improving multi-agent coalition formation in complex environments. Ph.D. thesis, University of Nebraska (2007)
22. Liang, J.J., Qu, B.Y., Suganthan, P.N., Chen, Q.: Learning-based real-parameter single objective optimization. In: IEEE Congress on Evolutionary Computation, Nanyang Technological University, Singapore (2015)

Optimization Models for Modern Analytics

Implementing Data Envelopment Analysis in an Uncertain Perception-Based Online Evaluation Environment

Debora Di Caprio[1,2] and Francisco Javier Santos-Arteaga[3(✉)]

[1] Department of Mathematics and Statistics, York University, Toronto M3J 1P3, Canada
dicaper@mathstat.yorku.ca

[2] Polo Tecnologico IISS G. Galilei, Via Cadorna 14, 39100 Bolzano, Italy

[3] Faculty of Economics and Management, Free University of Bolzano, Bolzano, Italy
fsantosarteaga@unibz.it

Abstract. Consider a decision maker (DM) who must select an alternative to evaluate when using an online recommender engine that displays multiple evaluations from unknown raters regarding the different characteristics of the available alternatives. The evaluations of the raters do not necessarily coincide with those that would be provided by the DM, who must consider the differences existing between the ratings observed and his subjective perception and subsequent potential evaluations. We formalize the incentives of the DM to observe and evaluate an alternative through a function that accounts for these differences in a multi-criteria decision making setting. The resulting perception-based framework is implemented in a data envelopment analysis (DEA) scenario to analyze the effects of perception differentials on the evaluation and ranking behavior of DMs.

Keywords: Uncertainty · Subjective perception · Data envelopment analysis Online evaluations · Multi-criteria decision making

1 Introduction

Consider a decision maker (DM) who must select an alternative to evaluate when using an online recommender engine that provides him with multiple evaluations from unknown raters regarding the different characteristics that compose the available alternatives (Li et al. 2014; Cyr 2014).

Psychologists have consistently illustrated the dependence of the evaluations provided on the differences in the beliefs and experience of the DMs (Kimmel 2012). Consequently, the marketing literature has shown how the attitude of DMs towards an alternative is determined by the subjective importance assigned to its different characteristics by the DM together with his beliefs (Blackwell et al. 2006). Cognitive sciences have focused on the substantial importance given to the subjective perception of DMs in the evaluation of the characteristics composing an alternative (Bartels and Johnson 2015; Chater 2015). This branch of the literature has illustrated that the

J. Medina et al. (Eds.): IPMU 2018, CCIS 855, pp. 299–309, 2018.
https://doi.org/10.1007/978-3-319-91479-4_25

perceived value of an alternative is affected by the attention and memory capacities of the DMs (Stewart et al. 2006), with differences in perception arising between genders (Bae and Lee 2011). As a result, information scientists have highlighted the importance of accounting for the differences existing between the perception of the DM and that of the raters when formalizing different evaluation scenarios (Tavana et al. 2015a, 2015b).

The intuition on which the current formal structure is built follows from the differences in the perception and evaluation of the characteristics composing a given set of alternatives. In other words, when presented with a set of alternatives rated by a group of unknown users, the DM must consider the differences existing between these ratings and his subjective perception and subsequent potential evaluations. The DM has to account for two main sources of uncertainty

- First, the distribution of the realizations of each characteristic is unknown to the DM. Thus, independently of the evaluations provided by the raters, if the DM were to observe and evaluate an alternative, the potential realizations assigned to each characteristic should follow a uniform probability distribution.
- Second, the potential evaluations provided by the DM after observing the characteristics of a given alternative are subjective and must therefore be contained within an uncertain interval determined by the inaccuracy inherent to his perception.

Thus, the value assigned by the DM to a characteristic must account for the uncertainty regarding the distribution of its realizations and the subjective quality of the perception determining his own evaluation. We formalize the incentives of the DM to observe and evaluate an alternative given the realizations received from the raters and the width of the uncertain intervals determining his subjective evaluations. The resulting perception-based framework is implemented within a data envelopment analysis (DEA) scenario to study numerically the effects that differences in perception have on the evaluation and ranking behavior of DMs.

2 Basic Assumptions

Assume that the DM must evaluate different alternatives composed by finite sets of characteristics and let X_1 be the set of potential values that may be taken by a given characteristic. We will actually identify each alternative with its numerical evaluation, $x_1 \in X_1$, such as those that can be retrieved from online recommender engines. In other words, the initial part of the paper focuses on one characteristic per alternative, while multiple uncertain ones defining the evaluations of different alternatives will be considered when defining the perception-based online-evaluation DEA framework.

Assume that $X_1 = [x_1^m, x_1^M]$ corresponds to the set of potential evaluations, where x_1^m and x_1^M are two real numbers such that $x_1^m \neq x_1^M$. Moreover, we will use D to denote a generic DM. The traditional approach to decision-making under uncertainty (Mas-Collel et al. 1995), requires D to define a

- strictly increasing continuous utility function $u_1 : X_1 \rightarrow \mathbb{R}$ that represents his preferences on X_1;

- continuous probability density function $\mu : X_1 \longrightarrow [0,\ 1]$ that expresses his subjective beliefs regarding the potential realizations that could be observed from a randomly selected alternative within X_1. Without loss of generality, it will be assumed that $Support(\mu) = X_1$.

2.1 Uncertain Evaluations

It will also be assumed that after observing the realization x_1^r of a characteristic, D expects its potential evaluation to be defined within $[x_1^r - \varepsilon_1, x_1^r + \varepsilon_1]$, with $\varepsilon_1 > 0$ determining the spread of his perception. That is, given the subjectivity inherent to his evaluations, D defines an interval of viable evaluations around the potential realizations that he could be retrieving from each characteristic if he were to observe the alternative in detail.

Therefore, D must consider the set of evaluations associated to each x_1^r potential realization together with the probability density value assigned to both each realization and its associated evaluations. It will be assumed that the evaluations are uniformly distributed over $[x_1^r - \varepsilon_1, x_1^r + \varepsilon_1]$, maximizing information entropy and uncertainty on the side of D

$$\vartheta(x_1|\varepsilon_1) = \begin{cases} \frac{1}{2\varepsilon_1} & \text{if } x_1 \in [x_1^r - \varepsilon_1, x_1^r + \varepsilon_1] \\ 0 & \text{otherwise} \end{cases} \tag{1}$$

It should be emphasized that the distribution of potential evaluations defined within $[x_1^r - \varepsilon_1, x_1^r + \varepsilon_1]$ can be adapted to the degree of optimism or pessimism with which D is assumed to be endowed.

3 Perception-Based Evaluation Intervals

Let $x_o^r \in X_1$ be the rating received by D regarding a given characteristic of an alternative. We define below the sets $I^+(x_o^r)$ and $I^-(x_o^r)$ contained in X_1 that account for all the potential evaluations $x_1 \in [x_1^r - \varepsilon_1, x_1^r + \varepsilon_1]$ that may be defined by D for each x_1^r realization and deliver a utility higher or lower than $x_o^r \in X_1$, respectively. That is, any potential evaluation derived from x_1^r must be defined with respect to x_o^r, while considering the uncertainty inherent to the new realization observed by D, i.e. $[x_1^r - \varepsilon_1, x_1^r + \varepsilon_1]$.

The subset $I^+(x_o^r)$, which defines the potential improvements relative to $x_o^r \in X_1$, is given by

$$I^+(x_o^r) \overset{def}{=} \{x_1 \in [x_1^r - \varepsilon_1, x_1^r + \varepsilon_1] \cap \mathrm{Support}(\mu) : \ x_1^r \in X_1 \ \wedge \ u_1(x_1) \geq u_1(x_o^r)\} \tag{2}$$

Similarly, the subset $I^-(x_o^r)$, which defines the potential worsenings relative to $x_o^r \in X_1$, is given by

$$I^-(x_o^r) \overset{def}{=} \{x_1 \in [x_1^r - \varepsilon_1, x_1^r + \varepsilon_1] \cap \mathrm{Support}(\mu) : \ x_1^r \in X_1 \ \wedge \ u_1(x_1) < u_1(x_o^r)\} \tag{3}$$

Finally, note that both improvements and worsenings relative to x_o^r must be considered whenever $x_o^r \in [x_1^r - \varepsilon_1, x_1^r + \varepsilon_1]$.

4 Perception-Based Value Function

We analyze the evaluation behavior of D if he were to observe a given alternative, while acknowledging that his behavior is determined by the subjective perception sets $I^+(x_o^r)$ and $I^-(x_o^r)$ defined in Eqs. (2) and (3). We introduce the notion of perception-based value function, determined by the initial observation x_o^r and the potential evaluations composing the sets $I^+(x_o^r)$ and $I^-(x_o^r)$ defined $\forall x_1^r \in X_1$.

Letting x_1^r vary above x_o^r but below $x_o^r + \varepsilon_1$, or below x_o^r but above $x_o^r - \varepsilon_1$, the perception-based value function $V : X_1 \times R^+ \to R$ is given by

$$V(x_o^r, \varepsilon_1) = \int\limits_{I^+(x_0^r) \cup I^-(x_0^r)} \mu(x_1^r) \left[\begin{array}{l} \int\limits_{x_1^r - \varepsilon_1}^{x_o^r} \vartheta(x_1|\varepsilon_1)(u(x_1))\,dx_1 + \\ \int\limits_{x_o^r}^{x_1^r + \varepsilon_1} \vartheta(x_1|\varepsilon_1)\left(u(x_1) + \left[\frac{x_1 - x_o^r}{x_1^M - x_o^r}\right]\right) dx_1 \end{array} \right] dx_1^r \quad (4)$$

The potential improvements and worsenings relative to x_o^r that may be realized by D together with their corresponding subjective evaluation spreads condition the value of the function defined in Eq. (4). In particular, the second right hand side (RHS) term defined for $x_1 \in [x_o^r, x_1^r + \varepsilon_1]$ constitutes an improvement over x_o^r, leading D to increment his utility proportionally by $(x_1 - x_o^r)/(x_1^M - x_o^r)$.

Given the fact that D has no information about the distribution of potential realizations of the characteristics, he must account for all the values of $x_1^r \in X_1$ when computing $V(x_o^r, \varepsilon_1)$ for each and every $x_o^r \in X_1$. The corresponding value of $V(x_o^r, \varepsilon_1)$ is obtained via $I^+(x_o^r)$ and $I^-(x_o^r)$, $\forall x_1^r \in X_1$. Figure 1 illustrates the behavior of several x_1^r potential realizations and their respective uncertain evaluation intervals $[x_1^r - \varepsilon_1, x_1^r + \varepsilon_1]$ considered by D when computing $V(x_o^r, \varepsilon_1)$ for different values of $x_o^r \in X_1$.

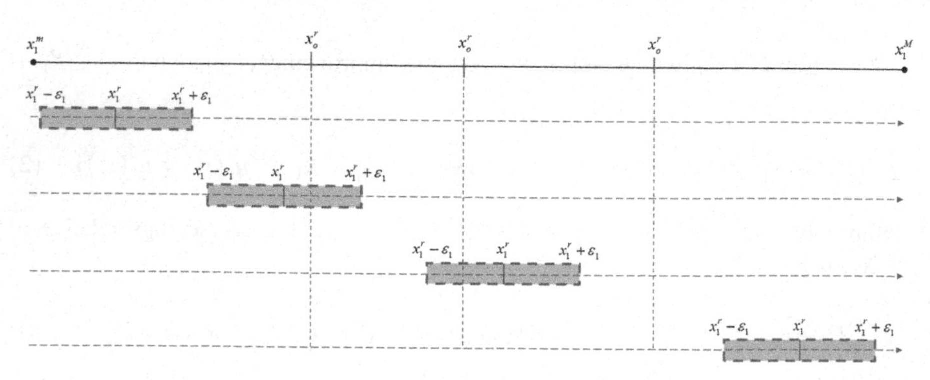

Fig. 1. Computation of $V(x_o^r, \varepsilon_1)$ by D through $[x_1^r - \varepsilon_1, x_1^r + \varepsilon_1]$ for different x_o^r realizations.

We describe the different cases defining the function $V(x_o^r, \varepsilon_1)$ in the next sections. Five different cases determined by the relative position of x_o^r within X_1 and the spread of the subjective uncertain evaluation intervals, ε_1, will be analyzed.

4.1 Lower Interval Case: $x_o^r \in [x_1^m, x_1^m + \varepsilon_1]$

$$
\begin{aligned}
& V(x_0^r, \varepsilon_1 | x_o^r < x_1^m + \varepsilon_1) \\
&= \int_{x_1^m}^{x_1^m + \varepsilon_1} \mu(x_1^r) \left[\begin{array}{l} \int_{x_1^m}^{x_o^r} \frac{1}{(x_1^r + \varepsilon_1) - x_1^m} (u(x_1)) dx_1 + \\ \int_{x_o^r}^{x_1^r + \varepsilon_1} \frac{1}{(x_1^r + \varepsilon_1) - x_1^m} \left(u(x_1) + \left[\frac{x_1 - x_o^r}{x_1^M - x_o^r} \right] \right) dx_1 \end{array} \right] dx_1^r \\
&+ \int_{x_1^m + \varepsilon_1}^{x_o^r + \varepsilon_1} \mu(x_1^r) \left[\int_{x_1^r - \varepsilon_1}^{x_o^r} \frac{1}{2\varepsilon_1} (u(x_1)) dx_1 + \int_{x_o^r}^{x_1^r + \varepsilon_1} \frac{1}{2\varepsilon_1} \left(u(x_1) + \left[\frac{x_1 - x_o^r}{x_1^M - x_o^r} \right] \right) dx_1 \right] dx_1^r \\
&+ \int_{x_o^r + \varepsilon_1}^{x_1^M - \varepsilon_1} \mu(x_1^r) \left[\int_{x_1^r - \varepsilon_1}^{x_1^r + \varepsilon_1} \frac{1}{2\varepsilon_1} \left[u(x_1) + \left[\frac{x_1 - x_o^r}{x_1^M - x_o^r} \right] \right] dx_1 \right] dx_1^r \\
&+ \int_{x_1^M - \varepsilon_1}^{x_1^M} \mu(x_1^r) \left[\int_{x_1^r - \varepsilon_1}^{x_1^M} \frac{1}{x_1^M - (x_1^r - \varepsilon_1)} \left[u(x_1) + \left[\frac{x_1 - x_o^r}{x_1^M - x_o^r} \right] \right] dx_1 \right] dx_1^r
\end{aligned} \tag{5}
$$

$V(x_o^r, \varepsilon_1 | \cdot)$ has been defined so as to consider all the values of x_1^r that may be realized when observing an alternative. Note that when $x_1^r < x_1^m + \varepsilon_1$, part of the interval $[x_1^r - \varepsilon_1, x_1^r + \varepsilon_1]$ surpasses the lower limit of X_1, i.e. x_1^m, while x_1^r improves potentially upon x_o^r. This possibility is described by the first RHS term within Eq. (5), where the $\vartheta(x_1 | \varepsilon_1)$ expressions have been adapted to account for the corresponding support limits of the density function.

As already stated, the weight $(x_1 - x_o^r)/(x_1^M - x_o^r)$ has been included to represent the increments in utility obtained by D as he observes alternatives considered to be potentially better than x_o^r, and is determined by the relative location of the x_o^r and x_1^r realizations within X_1.

Note that the value function defined by D must account for two distinct levels of uncertainty:

- $\mu(x_1^r)$, reflecting the uncertainty of the $x_1^r \in X_1$ potential realizations;
- $\vartheta(x_1 | \varepsilon_1)$, reflecting the uncertainty inherent to x_1^r while adapting to the limits of the domain of X_1.

4.2 Lower-Middle Interval Case: $x_o^r \in [x_1^m + \varepsilon_1, x_1^m + 2\varepsilon_1]$

$$
\begin{aligned}
&V(x_0^r, \varepsilon_1 | x_1^m + \varepsilon_1 \le x_o^r \le x_1^m + 2\varepsilon_1) \\
&= \int_{x_1^m}^{x_o^r - \varepsilon_1} \mu(x_1^r) \left[\int_{x_1^m}^{x_1^r + \varepsilon_1} \frac{1}{(x_1^r + \varepsilon_1) - x_1^m} (u(x_1)) dx_1 \right] dx_1^r \\
&+ \int_{x_o^r - \varepsilon_1}^{x_1^m + \varepsilon_1} \mu(x_1^r) \left[\begin{array}{l} \int_{x_1^m}^{x_o^r} \frac{1}{(x_1^r + \varepsilon_1) - x_1^m} (u(x_1)) dx_1 \\ + \int_{x_o^r}^{x_1^r + \varepsilon_1} \frac{1}{(x_1^r + \varepsilon_1) - x_1^m} \left(u(x_1) + \left[\frac{x_1 - x_o^r}{x_1^M - x_o^r} \right] \right) dx_1 \end{array} \right] dx_1^r \\
&+ \int_{x_1^m + \varepsilon_1}^{x_o^r + \varepsilon_1} \mu(x_1^r) \left[\int_{x_1^r - \varepsilon_1}^{x_o^r} \frac{1}{2\varepsilon_1} (u(x_1)) dx_1 + \int_{x_o^r}^{x_1^r + \varepsilon_1} \frac{1}{2\varepsilon_1} \left(u(x_1) + \left[\frac{x_1 - x_o^r}{x_1^M - x_o^r} \right] \right) dx_1 \right] dx_1^r \\
&+ \int_{x_o^r + \varepsilon_1}^{x_1^M - \varepsilon_1} \mu(x_1^r) \left[\int_{x_1^r - \varepsilon_1}^{x_1^r + \varepsilon_1} \frac{1}{2\varepsilon_1} \left[u(x_1) + \left[\frac{x_1 - x_o^r}{x_1^M - x_o^r} \right] \right] dx_1 \right] dx_1^r \\
&+ \int_{x_1^M - \varepsilon_1}^{x_1^M} \mu(x_1^r) \left[\int_{x_1^r - \varepsilon_1}^{x_1^M} \frac{1}{x_1^M - (x_1^r - \varepsilon_1)} \left[u(x_1) + \left[\frac{x_1 - x_o^r}{x_1^M - x_o^r} \right] \right] dx_1 \right] dx_1^r
\end{aligned} \tag{6}
$$

Equation (6) accounts for the progressive overtaking of x_o^r by x_1^r until reaching x_1^M. Note that the first RHS term represents potential evaluations contained exclusively within $I^-(x_o^r)$, the second and third terms those contained within both $I^-(x_o^r)$ and $I^+(x_o^r)$, while the last two terms consider evaluations contained exclusively within $I^+(x_o^r)$. The remaining expressions for the $V(x_o^r, \varepsilon_1)$ function complement these previous ones through the different realizations of x_o^r and x_1^r composing the domain of X_1 until the upper interval case $x_o^r \in [x_1^M - \varepsilon_1, x_1^M]$ is reached.

4.3 Middle Interval Case: $x_o^r \in [x_1^m + 2\varepsilon_1, x_1^M - 2\varepsilon_1]$

$$
\begin{aligned}
&V(x_0^r, \varepsilon | x_1^m + 2\varepsilon_1 \le x_o^r \le x_1^M - 2\varepsilon_1) \\
&= \int_{x_1^m}^{x_1^m + \varepsilon_1} \mu(x_1^r) \left[\int_{x_1^m}^{x_1^r + \varepsilon_1} \frac{1}{(x_1^r + \varepsilon_1) - x_1^m} (u(x_1)) dx_1 \right] dx_1^r \\
&+ \int_{x_1^m + \varepsilon_1}^{x_o^r - \varepsilon_1} \mu(x_1^r) \left[\int_{x_1^r - \varepsilon_1}^{x_1^r + \varepsilon_1} \frac{1}{2\varepsilon_1} (u(x_1)) dx_1 \right] dx_1^r \\
&+ \int_{x_o^r - \varepsilon_1}^{x_o^r + \varepsilon_1} \mu(x_1^r) \left[\int_{x_1^r - \varepsilon_1}^{x_o^r} \frac{1}{2\varepsilon_1} (u(x_1)) dx_1 + \int_{x_o^r}^{x_1^r + \varepsilon_1} \frac{1}{2\varepsilon_1} \left(u(x_1) + \left[\frac{x_1 - x_o^r}{x_1^M - x_o^r} \right] \right) dx_1 \right] dx_1^r \\
&+ \int_{x_o^r + \varepsilon_1}^{x_1^M - \varepsilon_1} \mu(x_1^r) \left[\int_{x_1^r - \varepsilon_1}^{x_1^r + \varepsilon_1} \frac{1}{2\varepsilon_1} \left[u(x_1) + \left[\frac{x_1 - x_o^r}{x_1^M - x_o^r} \right] \right] dx_1 \right] dx_1^r \\
&+ \int_{x_1^M - \varepsilon_1}^{x_1^M} \mu(x_1^r) \left[\int_{x_1^r - \varepsilon_1}^{x_1^M} \frac{1}{x_1^M - (x_1^r - \varepsilon_1)} \left[u(x_1) + \left[\frac{x_1 - x_o^r}{x_1^M - x_o^r} \right] \right] dx_1 \right] dx_1^r
\end{aligned} \tag{7}
$$

4.4 Upper-Middle Interval Case: $x_o^r \in [x_1^M - 2\varepsilon_1, x_1^M - \varepsilon_1]$

$$\begin{aligned}
&V(x_0^r, \varepsilon_1 | x_1^M - 2\varepsilon_1 \le x_o^r \le x_1^M - \varepsilon_1) \\
&= \int_{x_1^m}^{x_1^m+\varepsilon_1} \mu(x_1^r)\left[\int_{x_1^m}^{x_1^r+\varepsilon_1} \tfrac{1}{(x_1^r+\varepsilon_1)-x_1^m}(u(x_1))dx_1\right]dx_1^r \\
&+ \int_{x_1^m+\varepsilon_1}^{x_o^r-\varepsilon_1} \mu(x_1^r)\left[\int_{x_1^r-\varepsilon_1}^{x_1^r+\varepsilon_1} \tfrac{1}{2\varepsilon_1}(u(x_1))dx_1\right]dx_1^r \\
&+ \int_{x_o^r-\varepsilon_1}^{x_1^M-\varepsilon_1} \mu(x_1^r)\left[\int_{x_1^r-\varepsilon_1}^{x_o^r} \tfrac{1}{2\varepsilon_1}(u(x_1))dx_1 + \int_{x_o^r}^{x_1^r+\varepsilon_1} \tfrac{1}{2\varepsilon_1}\left(u(x_1)+\left[\tfrac{x_1-x_o^r}{x_1^M-x_o^r}\right]\right)dx_1\right]dx_1^r \\
&+ \int_{x_1^M-\varepsilon_1}^{x_o^r+\varepsilon_1} \mu(x_1^r)\left[\begin{array}{l}\int_{x_1^r-\varepsilon_1}^{x_o^r} \tfrac{1}{x_1^M-(x_1^r-\varepsilon_1)}(u(x_1))dx_1 \\ + \int_{x_o^r}^{x_1^M} \tfrac{1}{x_1^M-(x_1^r-\varepsilon_1)}\left(u(x_1)+\left[\tfrac{x_1-x_o^r}{x_1^M-x_o^r}\right]\right)dx_1\end{array}\right]dx_1^r \\
&+ \int_{x_o^r+\varepsilon_1}^{x_1^M} \mu(x_1^r)\left[\int_{x_1^r-\varepsilon_1}^{x_1^M} \tfrac{1}{x_1^M-(x_1^r-\varepsilon_1)}\left(u(x_1)+\left[\tfrac{x_1-x_o^r}{x_1^M-x_o^r}\right]\right)dx_1\right]dx_1^r
\end{aligned} \tag{8}$$

4.5 Upper Interval Case: $x_o^r \in [x_1^M - \varepsilon_1, x_1^M]$

$$\begin{aligned}
&V(x_0^r, \varepsilon_1 | x_o^r + \varepsilon_1 > x_1^M) \\
&= \int_{x_1^m}^{x_1^m+\varepsilon_1} \mu(x_1^r)\left[\int_{x_1^m}^{x_1^r+\varepsilon_1} \tfrac{1}{(x_1^r+\varepsilon_1)-x_1^m}(u(x_1))dx_1\right]dx_1^r + \\
&+ \int_{x_1^m+\varepsilon_1}^{x_o^r-\varepsilon_1} \mu(x_1^r)\left[\int_{x_1^r-\varepsilon_1}^{x_1^r+\varepsilon_1} \tfrac{1}{2\varepsilon_1}(u(x_1))dx_1\right]dx_1^r \\
&+ \int_{x_o^r-\varepsilon_1}^{x_1^M-\varepsilon_1} \mu(x_1^r)\left[\int_{x_1^r-\varepsilon_1}^{x_o^r} \tfrac{1}{2\varepsilon_1}(u(x_1))dx_1 + \int_{x_o^r}^{x_1^r+\varepsilon_1} \tfrac{1}{2\varepsilon_1}\left(u(x_1)+\left[\tfrac{x_1-x_o^r}{x_1^M-x_o^r}\right]\right)dx_1\right]dx_1^r \\
&+ \int_{x_1^M-\varepsilon_1}^{x_1^M} \mu(x_1^r)\left[\begin{array}{l}\int_{x_1^r-\varepsilon_1}^{x_o^r} \tfrac{1}{x_1^M-(x_1^r-\varepsilon_1)}(u(x_1))dx_1 \\ + \int_{x_o^r}^{x_1^M} \tfrac{1}{x_1^M-(x_1^r-\varepsilon_1)}\left(u(x_1)+\left[\tfrac{x_1-x_o^r}{x_1^M-x_o^r}\right]\right)dx_1\end{array}\right]dx_1^r
\end{aligned} \tag{9}$$

As was the case in Eq. (5), the density $\vartheta(x_1|\varepsilon_1)$ must be adapted to the realizations of x_1^r as the upper domain limit of X_1, x_1^M, is exceeded.

5 DEA with Subjective Perception-Based Intervals

After observing x_o^r, D computes the expected utility from evaluating an alternative using $V(x_o^r, \varepsilon_1|\cdot)$, which, at the same time, is based on the ε_1 width defining his subjective uncertain evaluation intervals. We implement this perception-based value function $V(x_o^r, \varepsilon_1|\cdot)$ within an output-oriented DEA framework determined by objective inputs and subjectively evaluated outputs.

A basic DEA framework consists of a set of n decision making units (DMUs) or alternatives using m inputs, x_i, $x_i \neq 0$, to produce s outputs, y_r, $y_r \neq 0$. Given the ε_1-based subjectivity inherent to the evaluation of the outputs, D transforms y_{rj} into $V(y_{rj}, \varepsilon_1)$, $j = 1, \ldots, n$, $r = 1, \ldots, s$. An output-oriented DEA model determining the efficiency of the DMU_o under variable returns to scale is given by

$$\begin{aligned}
&\max \quad \theta_o \\
&\text{subject to} \\
&\sum_{j=1}^{n} \lambda_j x_{ij} \le x_{ip} \qquad i = 1, 2, \ldots, m, \\
&\sum_{j=1}^{n} \lambda_j V(y_{rj}, \varepsilon_1) \ge \theta_o V(y_{ro}, \varepsilon_1) \qquad r = 1, 2, \ldots, s, \\
&\sum_{j=1}^{n} \lambda_j = 1, \\
&\lambda_j \ge 0 \qquad j = 1, 2, \ldots, n.
\end{aligned} \tag{10}$$

Model (10) assigns an efficiency score of $1/\theta_o$ to DMU_o, with $\theta_o \ge 1$. A value of $\theta_o > 1$ implies that DMU_o is inefficient and θ_o defines the output increment required for DMU_o to become efficient. After deriving the optimal values of θ_o^* from Model (10), the following linear programming problem is solved to obtain the reference set associated with DMU_o

$$\begin{aligned}
&\max \quad \sum_{i=1}^{m} s_i^- + \sum_{r=1}^{s} s_r^+ \\
&\text{subject to} \\
&\sum_{j=1}^{n} \lambda_j x_{ij} + s_i^- = x_{io} \qquad i = 1, 2, \ldots, m, \\
&\sum_{j=1}^{n} \lambda_j V(y_{rj}, \varepsilon_1) - s_r^+ = \theta_o^* V(y_{ro}, \varepsilon_1) \qquad r = 1, 2, \ldots, s, \\
&\sum_{j=1}^{n} \lambda_j = 1, \\
&\lambda_j \ge 0,\ j = 1, 2, \ldots, n; \quad s_i^- \ge 0,\ i = 1, 2, \ldots, m; \quad s_r^+ \ge 0,\ r = 1, 2, \ldots, s.
\end{aligned} \tag{11}$$

Definition 1. Let θ_o^* be the optimal solution of Model (10). Let $s_i^{-*}(i = 1, 2, \ldots, m)$ and $s_r^{+*}(r = 1, 2, \ldots, s)$ be the optimal solutions of Model (11). DMU_o is

- efficient if $\theta_o^* = 1$ and $\sum_{r=1}^{s} s_r^{+*} + \sum_{i=1}^{m} s_i^{-*} = 0$;

- weakly efficient if $\theta_o^* = 1$ and $\sum_{r=1}^{s} s_r^{+*} + \sum_{i=1}^{m} s_i^{-*} \neq 0$;
- inefficient if $\theta_o^* > 1$.

The next section illustrates numerically the effect that different values of ε_1 have on the ranking obtained by D when applying Models (10) and (11) to a set of objective and subjective evaluation categories such as those commonly observed in recommender engines.

6 Numerical Evaluation

We provide now a numerical example describing the ranking obtained by D as the values of ε_1 defining the $[x_1^r - \varepsilon_1, x_1^r + \varepsilon_1]$ intervals are modified. Consider a set of alternatives whose online ratings are defined within [0, 10] and account for both objective characteristic as well as subjective evaluation categories. For illustrative purposes, assume that all the alternatives are endowed with the same input and output evaluations, given by $(x_{1j}, x_{2j}, x_{3j}) = (5, 5, 5), j = 1, \ldots, n$, and $(y_{1j}, y_{2j}, y_{3j}) = (4, 5, 6), j = 1, \ldots, n$.

Consider also a basic setting with a risk neutral D, i.e. $u(x_1) = x_1$, and $\mu(x_1^r) = 1/10$, $\forall x_1^r \subset X_1$, with $X_1 = [0, 10]$. The $V(\cdot, \varepsilon_1)$ functions generated by D for four different ε_1 values are presented in Fig. 2. The horizontal axis corresponds to the set of realizations within X_1 that may be received from the raters as well as those potentially observable by D. The vertical axis defines the subjective evaluations derived from the corresponding value functions. We have applied the following simplification to deal with the non-linear expressions of the value function: the x_1^r terms have been removed from the $\vartheta(x_1|\varepsilon_1)$ expressions dealing with the domain limits of X_1. Consequently, $1/[(x_1^r + \varepsilon_1) - x_1^m]$ and $1/[x_1^M - (x_1^r - \varepsilon_1)]$ have both been transformed into functions of ε_1, i.e. $1/\varepsilon_1$, with $\varepsilon_1 > 0$.

An important result can be directly derived from comparing the different $V(\cdot, \varepsilon_1)$ obtained. Increasing ε_1 leads to an increase in the $V(x_o^r, \varepsilon_1)$ obtained by D for identical x_o^r realizations. Thus, the incentives of D to evaluate an alternative differ substantially depending on the value of ε_1 assumed. Note also the decreasing trend exhibited by the $V(x_o^r, \varepsilon_1)$ function as the value of x_o^r increases. This is the case since the uncertainty faced by D regarding the realizations of an alternative implies that potential improvements become more plausible for relatively lower realizations of x_o^r. At the same time, these improvements are less probable and lead to lower expected $V(x_o^r, \varepsilon_1)$ as the value of x_o^r increases.

The subjective evaluations of $(y_{1j}, y_{2j}, y_{3j}) = (4, 5, 6)$, $j = 1, 2, \ldots, n$, computed by D for eight different values of ε_1 and the corresponding efficiency results derived from implementing the output-oriented perception-based DEA model described in the previous section are reported in Table 1. Figure 3 illustrates the efficiencies obtained from the DEA model, which are clearly increasing in the value of ε_1. That is, higher spreads lead to a higher efficiency for the set of x_o^r realizations selected. It can be inferred from Fig. 2 that when considering the previous and lower realizations of x_o^r, the efficiency of

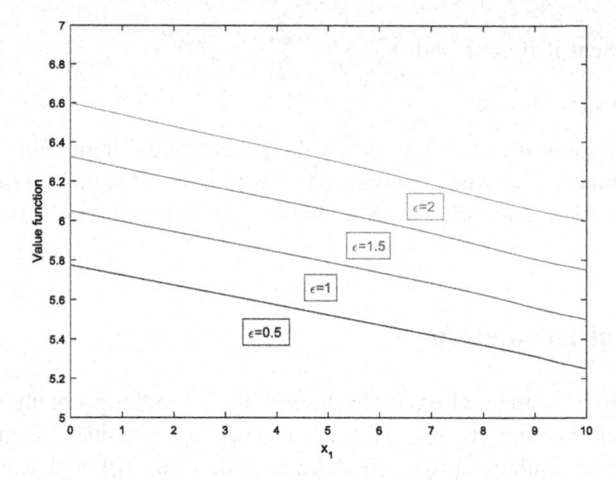

Fig. 2. Perception-based $V(x_o^r, \varepsilon_1)$ corresponding to different evaluation spreads.

Table 1. Inputs, evaluation spreads and the relative efficiency of the DMUs/alternatives

DMU	x_1	x_2	x_3	y_1	y_2	y_3	ε	θ
1	5	5	5	4	5	6	–	**1.0417**
2	5	5	5	5.4370	5.3869	5.3367	0.25	**1.1710**
3	5	5	5	5.5729	5.5225	5.4719	0.5	**1.1422**
4	5	5	5	5.7078	5.6569	5.6055	0.75	**1.1150**
5	5	5	5	5.8417	5.7900	5.7375	1	**1.0893**
6	5	5	5	5.9745	5.9219	5.8680	1.25	**1.0651**
7	5	5	5	6.1062	6.0525	5.9969	1.5	**1.0422**
8	5	5	5	6.2370	6.1819	6.1242	1.75	**1.0205**
9	5	5	5	6.3667	6.3100	6.2500	2	**1.0000**

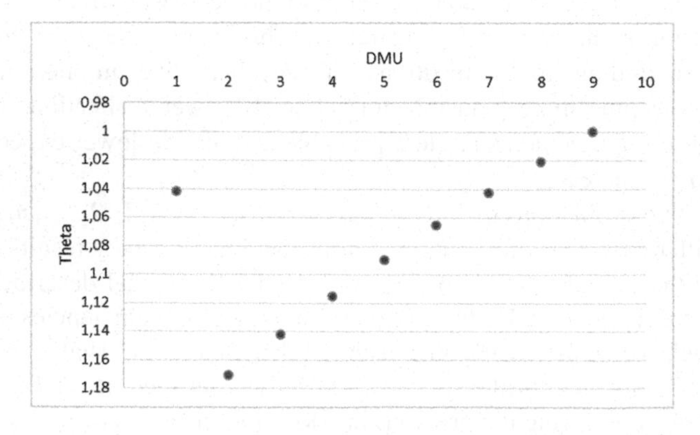

Fig. 3. θ values per DMU/alternative with different evaluation spreads.

the corresponding reference alternative will be lower than that of those alternatives facing relatively higher spreads. On the other hand, such a feature does not prevail for relatively higher realizations of x_o^r, opening the way for strategic considerations within the current evaluation environment.

7 Conclusion

We have studied a formal setting determined by the subjective uncertainty inherent to the potential realizations expected to be observed by D when evaluating a set of alternatives. A function has been introduced to determine the value that D expects to assign to the characteristics composing an alternative when his evaluations are contained within subjective uncertain intervals defined with respect to all potentially observable realizations. Leaving aside the possibility of reporting strategically (Di Caprio and Santos Arteaga 2011; Tavana et al. 2017), we have analyzed the effect that the width of these subjective evaluation intervals has on the rankings derived from an output-oriented DEA model.

We conclude by emphasizing that the current framework can be applied to any multi-criteria decision making technique such as TOPSIS in order to account for differences in perception and their effect on the evaluations and rankings derived by D.

References

Bae, S., Lee, T.: Gender differences in consumers' perception of online consumer reviews. Electron. Commer. Res. **11**, 201–214 (2011)

Bartels, D.M., Johnson, E.J.: Connecting cognition and consumer choice. Cognition **135**, 47–51 (2015)

Blackwell, R.D., Miniard, P.W., Engel, J.F.: Consumer Behavior. Thomson/South-Western, Mason (2006)

Chater, N.: Can cognitive science create a cognitive economics? Cognition **135**, 52–55 (2015)

Cyr, D.: Return visits: a review of how web site design can engender visitor loyalty. J. Inf. Technol. **29**, 1–26 (2014)

Di Caprio, D., Santos Arteaga, F.J.: Strategic diffusion of information and preference manipulation. Int. J. Strateg. Decis. Sci. **2**, 1–19 (2011)

Kimmel, A.J.: Psychological Foundations of Marketing. Routledge, East Sussex (2012)

Li, Y.M., Chou, C.L., Lin, L.F.: A social recommender mechanism for location-based group commerce. Inf. Sci. **274**, 125–142 (2014)

Mas-Colell, A., Whinston, M.D., Green, J.R.: Microeconomic Theory. Oxford University Press, New York (1995)

Stewart, N., Chater, N., Brown, G.D.: Decision by sampling. Cogn. Psychol. **53**, 1–26 (2006)

Tavana, M., Di Caprio, D., Santos-Arteaga, F.J.: A multi-criteria perception-based strict-ordering algorithm for identifying the most-preferred choice among equally-evaluated alternatives. Inf. Sci. **381**, 322–340 (2017)

Tavana, M., Di Caprio, D., Santos-Arteaga, F.J.: A bilateral exchange model: The paradox of quantifying the linguistic values of qualitative characteristics. Inf. Sci. **296**, 201–218 (2015a)

Tavana, M., Di Caprio, D., Santos-Arteaga, F.J.: An ordinal ranking criterion for the subjective evaluation of alternatives and exchange reliability. Inf. Sci. **317**, 295–314 (2015b)

Constraint Shortest Path Problem in a Network with Intuitionistic Fuzzy Arc Weights

Homayun Motameni[1](✉) and Ali Ebrahimnejad[2]

[1] Department of Computer Engineering, Sari Branch, Islamic Azad University, Sari, Iran
motameni@iausari.ac.ir
[2] Department of Mathematics, Qaemshahr Branch, Islamic Azad University, Qaemshahr, Iran
a.ebrahimnejad@qaemiau.ac.ir

Abstract. The Shortest Path (SP) problem is one of the most widely used problems in network optimization which has a wide range of applications in various fields of science and engineering such as communication, transportation, routing and scheduling. The aim of this problem is to find a minimum cost path between two specified nodes. In the present communication, we consider a modified version of the SP known as constraint SP (CSP) problem with an additional constraint that establishes an upper limit on the travel time for the path. The objective of the CSP problem is to determine a minimum cost path between two specified nodes that the traversal time of the path does not exceed from a specified time. Traditional CSP problems assume the arc weights represented by time and cost are specified precisely. However, these weights can fluctuate with traffic conditions, weather, or payload. For this reason, being able to deal with vague and imprecise data may greatly contribute to the application of CSP problems. Here, we first formulate a CSP problem in a directed network where the arc weights represented by cost and time are intuitionistic trapezoidal fuzzy numbers. We then develop an approach for solving the intuitionistic fuzzy CSP problem under consideration. Finally, we present a small numerical example to illustrate the proposed approach.

Keywords: Constraint shortest path · Intuitionistic fuzzy numbers
Intuitionistic fuzzy ranking

1 Introduction

The shortest path (SP) problem is an important network optimization. The aim of the SP problem is to find a path between two nodes and optimizing the weight of the path. In this paper, we consider a generalized version of the SP problem known as constraint SP (CSP) problem with an additional constraint that establishes an upper limit on the travel time for the path [16]. In the network considered here, the arc weights represent transportation cost and time. As time and cost fluctuate with traffic conditions, weather and payload, fuzzy numbers based on fuzzy set theory or intuitionistic fuzzy numbers

J. Medina et al. (Eds.): IPMU 2018, CCIS 855, pp. 310–318, 2018.
https://doi.org/10.1007/978-3-319-91479-4_26

(IFNs) based on intuitionistic fuzzy theory can be utilized to represent such arc weights. This type of the problem is called fuzzy CSP (FCSP) problem or intuitionistic FCSP (IFCSP) problem, respectively.

Many researchers have focused on fuzzy (SP) and intuitionistic FSP (IFSP) formulations and solution approaches. Okada and Soper [14] developed an algorithm for solving the FSP problem on the basis of the multiple labeling methods for a multi–criteria shortest path. Okada [15] proposed an algorithm for solving FSP problem. Based on possibility theory, the degree of possibility for each arc is determined by this algorithm. Chuang and Kung [2] proposed a heuristic procedure to find the length of FSP among all possible paths. Moazeni [12] developed a new algorithm for finding the set of non–dominated paths with respect to the extension principle. Hernandes et al. [9] considered a generic algorithm for solving the FSP problem. Mahdavi et al. [11] proposed a dynamic programming approach to solve the fuzzy shortest chain problem using a suitable ranking method. The other approaches for solving the FSP problem can be found in [3–5, 8]. Mukherjee [13] considered the SP problem in an intuitionistic fuzzy environment. Geetharamani and Jayagowri [7] proposed a new algorithm to deal with the IFSP problem using intuitionistic fuzzy shortest path length procedure and similarity measure. Biswas et al. [1] developed a method to search for an intuitionistic fuzzy shortest path from a source node to a destination node. Kumar et al. [10] proposed an algorithm to find the shortest path and shortest distance in a network with nodes and arcs being crisp but the arcs weights will be interval–valued IFNs. Sujatha and Hyacinta [17] proposed two different approaches for solving the IFSP problem. To the best of our knowledge, there is no method in the literature to find the optimal solution of IFCSP problems. In this study, we formulate IFCSP problem and propose a solution technique to find the optimal solution of the problem.

The rest of the paper is organized as follows: In Sect. 2, some basic concepts of IFNs are reviewed. In Sect. 3, the mathematical formulation of the IFCSP problem is given. In Sect. 4, a new method is proposed for solving the same problem. In Sect. 5, the application of the proposed method is illustrated by using a numerical example. Finally, we conclude the paper in Sect. 6.

2 Preliminaries

This section is devoted to review some necessary background and notions of the intuitionistic fuzzy numbers which are applied throughout this paper [6].

Definition 1: Let X denote the universe set. An intuitionistic fuzzy set (IFS) $\tilde{A}^I$ in X is defined by a set of ordered triple $\tilde{A}^I = \{\langle x, \mu_{\tilde{A}^I}(x), \upsilon_{\tilde{A}^I}(x)\rangle; x \in X\}$ where the functions $\mu_{\tilde{A}^I}(x) : X \to [0, 1]$ and $\upsilon_{\tilde{A}^I}(x) : X \to [0, 1]$, respectively represent the membership degree and non-membership degree of x in $\tilde{A}$ such that for each element $x \in X$, $0 \le \mu_{\tilde{A}^I}(x) + \upsilon_{\tilde{A}^I}(x) \le 1$.

Definition 2: An intuitionistic fuzzy set $\tilde{A}^I = \{\langle x, \mu_{\tilde{A}^I}(x), \upsilon_{\tilde{A}^I}(x)\rangle; x \in X\}$ is called normal if there is any $x_\circ \in X$ such that $\mu_{\tilde{A}^I}(x_\circ) = 1$ (so $\upsilon_{\tilde{A}^I}(x_\circ) = 0$).

Definition 3: An intuitionistic fuzzy set $\tilde{A}^I = \{\langle x, \mu_{\tilde{A}^I}(x), \upsilon_{\tilde{A}^I}(x)\rangle; x \in X\}$ is called intuitionistic fuzzy convex if its membership function is fuzzy convex, i.e. $\forall x_1, x_2 \in X, \forall \lambda \in [0, 1], \mu_{\tilde{A}^I}(\lambda x_1 + (1-\lambda)x_2) \geq \min\{\mu_{\tilde{A}^I}(x_1), \mu_{\tilde{A}^I}(x_2)\}$ and its non-membership function is concave, i.e. $\upsilon_{\tilde{A}^I}(\lambda x_1 + (1-\lambda)x_2) \leq \max\{\upsilon_{\tilde{A}^I}(x_1), \upsilon_{\tilde{A}^I}(x_2)\}$.

Definition 4: A normal and convex intuitionistic fuzzy set $\tilde{A}^I = \{\langle x, \mu_{\tilde{A}^I}(x), \upsilon_{\tilde{A}^I}(x)\rangle; x \in R\}$ defined on the set of real numbers R is called an intuitionistic fuzzy number (IFN) if $\mu_{\tilde{A}^I}$ is upper semi-continuous and $\upsilon_{\tilde{A}^I}$ is lower semi-continuous.

Definition 5: A Trapezoidal Intuitionistic Fuzzy Number (TrIFN) $\tilde{A}^I$, denoted by $\tilde{A}^I = (a_1, a_2, a_3, a_4;\ a'_1, a'_2, a'_3, a'_4)$, is an especial IFN with the membership function non-membership function defined as follows:

$$\mu_{\tilde{A}^I}(x) = \begin{cases} \frac{x-a_1}{a_2-a_1}, & a_1 < x \leq a_2, \\ 1, & a_2 < x \leq a_3, \\ \frac{a_4-x}{a_4-a_3}, & a_3 \leq x < a_4 \\ 0, & \text{Otherwise.} \end{cases} \quad \text{and } \upsilon_{\tilde{A}^I}(x) = \begin{cases} \frac{a'_2-x}{a'_2-a'_1}, & a'_1 < x \leq a'_2, \\ 0, & a'_1 < x \leq a'_3, \\ \frac{x-a'_3}{a'_4-a'_3}, & a'_3 \leq x < a'_4, \\ 1, & \text{Otherwise.} \end{cases}$$

where $a'_1 \leq a_1 \leq a'_2 \leq a_2 \leq a_3 \leq a'_3 \leq a_4 \leq a'_4$.

Definition 6: The summation operation between two TrIFNs $\tilde{A}^I = (a_1, a_2, a_3, a_4;\ a'_1, a'_2, a'_3, a'_4)$ and $\tilde{B}^I = (b_1, b_2, b_3, b_4;\ b'_1, b'_2, b'_3, b'_4)$ is defined as $\tilde{A}^I \oplus \tilde{B}^I = (a_1 + b_1, a_2 + b_2, a_3 + b_3, a_4 + b_4;\ a'_1 + b'_1, a'_2 + b'_2, a'_3 + b'_3, a'_4 + b'_4)$.

Definition 7: Given TrIFNs $\tilde{A}^I = (a_1, a_2, a_3, a_4;\ a'_1, a'_2, a'_3, a'_4)$ and real number $k \geq 0$, the scaler multiplication operation is defined as $k\tilde{A}^I = (ka_1, ka_2, ka_3, ka_4;\ ka'_1, ka'_2, ka'_3, ka'_4)$.

Definition 8: A TrIFN $\tilde{A}^I = (a_1, a_2, a_3, a_4;\ a'_1, a'_2, a'_3, a'_4)$ is said to be a non–negative TrIFN if and only if $a'_1 \geq 0$.

Definition 9: For TrIFN $\tilde{A}^I = (a_1, a_2, a_3, a_4;\ a'_1, a'_2, a'_3, a'_4)$, its accuracy function is defined as follows:

$$H(\tilde{A}^I) = \frac{(a_1 + a_2 + a_3 + a_4) + (a'_1 + a'_2 + a'_3 + a'_4)}{8} \tag{1}$$

Definition 10: Let $\tilde{A}^I = (a_1, a_2, a_3, a_4;\ a'_1, a'_2, a'_3, a'_4)$ and $\tilde{B}^I = (b_1, b_2, b_3, b_4;\ b'_1, b'_2, b'_3, b'_4)$ be two TrIFNs. Then $\tilde{A}^I \preceq \tilde{B}^I$ if $a_1 \leq b_1, a_2 \leq b_2, a_3 \leq b_3, a_4 \leq b_4;\ a'_1 \leq b'_1, a_2 \leq b'_2, a'_3 \leq b'_3, a'_4 \leq b'_4$.

3 Mathematical Formulation of the IFCSP Problem

In this section, the mathematical formulation of the constraint shortest path problem with intuitionistic fuzzy numbers is presented.

We consider a directed network $G = (V, E)$, where $V = \{1, 2, \ldots, m\}$ is the set of nodes and $E = \{(i,j) : i,j \in V,\ i \neq j\}$ is the set of arcs. Each arc is denoted by an ordered pair (i,j), where $i,j \in E$. The network has two distinguished nodes s and t, called the source node and the destination node, respectively. It is supposed that there is only one directed arc (i,j) from node i to node j. A path p_{ij} from node i to node j is a sequence of arcs $p_{ij} = \{(i, i_1), (i_1, i_2), \ldots, (i_k, j)\}$ in which the initial node of each arc is same as the terminal node of preceding arc in the sequence. We define the length (weight) of a directed path as the sum of the lengths (weights) of arcs in the path. It is supposed that the network contains a directed path from the source node to every other node in the network.

Two non–negative weights c_{ij} and t_{ij} are associated with each arc (i,j) representing the length (or cost) and the travel time associated with the respective arc, respectively. The objective of the CSP problem is to determine a minimum cost path between the source node s and the destination node t such that the traversal time of the path does not exceed from the maximum allowable time to transverse the path. Conventional CSP problems assume precise values for the cost and time weights. But, this may not be suitable for situations where one has to deal with uncertainty as well as with hesitation. In such situations, intuitionistic fuzzy numbers are used to represent the imprecise parameters of the CSP problem under consideration. The resulting problem is therefore referred to as an Intuitionistic Fuzzy CSP Problem (IFCSP).

An IFCSP problem having uncertainty and hesitation for the cost and time weights can be formulated as follows:

$$
\begin{aligned}
\min\ & \tilde{Z}^I = \sum_{i=1}^{m}\sum_{j=1}^{m} \tilde{c}^I_{ij} x_{ij} \\
s.t.\ & \\
& \sum_{j=1}^{n} x_{ij} - \sum_{k=1}^{m} x_{ki} = \begin{cases} 1, & i = s, \\ 0, & i \neq s, t, \\ -1, & i = t, \end{cases} \\
& \sum_{i=1}^{m}\sum_{j=1}^{m} \tilde{t}^I_{ij} x_{ij} \preceq \tilde{T}^I, \\
& x_{ij} \geq 0,\ i,j = 1, 2, \ldots, m.
\end{aligned}
\tag{2}
$$

In model (2), x_{ij} are binary variables associated with each arc (i,j). If arc (i,j) is included in the optimal path, then $x_{ij} = 1$; otherwise $x_{ij} = 0$. The intuitionistic fuzzy parameter, $\tilde{T}^I$, represents the maximum value allowed for the sum of the $\tilde{t}^I_{ij}$ arc weights.

Let P_{st} denotes the set of all paths from node s to node t. Define $\tilde{C}^I(p) = \sum_{(i,j)\in p} \tilde{c}_{ij}^I$ and $\tilde{D}^I(p) = \sum_{(i,j)\in p} \tilde{t}_{ij}^I$. Given $\tilde{T}^I \succsim 0^I$, let $P_{st}(\tilde{T}^I)$ be the set of all paths p_{st} from node s to node t such that $\tilde{D}(p_{st}) \preceq \tilde{T}^I$, i.e. $P_{st}(\tilde{T}^I) = \{p_{st} : \tilde{D}(p_{st}) \preceq \tilde{T}^I\}$. Each path belonging to the set $P_{st}(\tilde{T}^I)$ is called an intuitionistic fuzzy feasible path. In this case, the IFCSP problem is to find a minimum cost intuitionistic fuzzy feasible path.

Definition 11: The intuitionistic fuzzy optimal path of IFCSP problem (2) is an intuitionistic fuzzy feasible path p_{st}^* such that $H(\tilde{C}^I(p_{st}^*)) < H(\tilde{C}^I(p_{st}))$ for any intuitionistic fuzzy feasible path p_{st}.

4 Solution Approach

In this section, a solution approach is proposed for solving the IFCSP problem (2).

Assume that intuitionistic fuzzy parameters of model (2) are all trapezoidal. Therefore, $\tilde{c}_{ij}^I$, $\tilde{t}_{ij}^I$ and $\tilde{T}^I$ are all represented by intuitionistic trapezoidal fuzzy numbers $(c_{ij,1}, c_{ij,2}, c_{ij,3}, c_{ij,4};\ c'_{ij,1}, c'_{ij,2}, c'_{ij,3}, c'_{ij,4})$, $(t_{ij,1}, t_{ij,2}, t_{ij,3}, t_{ij,4};\ t'_{ij,1}, t'_{ij,2}, t'_{ij,3}, t'_{ij,4})$ and $(T_1, T_2, T_3, T_4;\ T'_1, T'_2, T'_3, T'_4)$, respectively. Thus, the IFCSP problem (2) can be rewritten as follows:

$$
\begin{aligned}
&\min\ \tilde{Z}^I = \sum_{i=1}^{m}\sum_{j=1}^{m} (c_{ij,1}, c_{ij,2}, c_{ij,3}, c_{ij,4};\ c'_{ij,1}, c'_{ij,2}, c'_{ij,3}, c'_{ij,4}) x_{ij} \\
&\quad s.t. \\
&\quad \sum_{j=1}^{n} x_{ij} - \sum_{k=1}^{m} x_{ki} = \begin{cases} 1, & i = s, \\ 0, & i \neq s, t, \\ -1, & i = t, \end{cases} \\
&\quad \sum_{i=1}^{m}\sum_{j=1}^{m} (t_{ij,1}, t_{ij,2}, t_{ij,3}, t_{ij,4};\ t'_{ij,1}, t'_{ij,2}, t'_{ij,3}, t'_{ij,4}) x_{ij} \preceq (T_1, T_2, T_3, T_4;\ T'_1, T'_2, T'_3, T'_4), \\
&\quad x_{ij} \geq 0,\ i, j = 1, 2, \ldots, m.
\end{aligned} \tag{3}
$$

Regarding Definitions 6 and 7, the intuitionistic trapezoidal fuzzy objective function of model (3), can be rewritten as follows:

$$
\begin{aligned}
\min\ \tilde{Z}^I = \Bigg(&\sum_{i=1}^{m}\sum_{j=1}^{m} c_{ij,1}x_{ij}, \sum_{i=1}^{m}\sum_{j=1}^{m} c_{ij,2}x_{ij}, \sum_{i=1}^{m}\sum_{j=1}^{m} c_{ij,3}x_{ij}, \sum_{i=1}^{m}\sum_{j=1}^{m} c_{ij,4}x_{ij}, \\
&\sum_{i=1}^{m}\sum_{j=1}^{m} c'_{ij,1}x_{ij}, \sum_{i=1}^{m}\sum_{j=1}^{m} c'_{ij,2}x_{ij}, \sum_{i=1}^{m}\sum_{j=1}^{m} c'_{ij,3}x_{ij}, \sum_{i=1}^{m}\sum_{j=1}^{m} c'_{ij,4}x_{ij} \Bigg)
\end{aligned} \tag{4}
$$

Similarly, with regard to Definitions 6–9, the inequality constraint of model (3), can be rewritten as follows:

$$\begin{aligned}&\sum_{i=1}^{m}\sum_{j=1}^{m} t_{ij,1}x_{ij} \le T_1,\ \sum_{i=1}^{m}\sum_{j=1}^{m} t_{ij,2}x_{ij} \le T_2,\ \sum_{i=1}^{m}\sum_{j=1}^{m} t_{ij,3}x_{ij} \le T_3,\ \sum_{i=1}^{m}\sum_{j=1}^{m} t_{ij,4}x_{ij} \le T_4,\\ &\sum_{i=1}^{m}\sum_{j=1}^{m} t'_{ij,1}x_{ij} \le T'_1,\ \sum_{i=1}^{m}\sum_{j=1}^{m} t'_{ij,2}x_{ij} \le T'_2,\ \sum_{i=1}^{m}\sum_{j=1}^{m} t'_{ij,3}x_{ij} \le T'_3,\ \sum_{i=1}^{m}\sum_{j=1}^{m} t'_{ij,4}x_{ij} \le T'_4.\end{aligned} \tag{5}$$

Now, in order to obtain the intuitionistic fuzzy optimal path of IFCSP problem (2) which satisfies the properties of Definition 11, we solve the following problem with regard to Eqs. (4) and (5):

$$\begin{aligned}\min\ H(\tilde{Z}^I) = H\Bigg(&\sum_{i=1}^{m}\sum_{j=1}^{m} c_{ij,1}x_{ij},\ \sum_{i=1}^{m}\sum_{j=1}^{m} c_{ij,2}x_{ij},\ \sum_{i=1}^{m}\sum_{j=1}^{m} c_{ij,3}x_{ij},\ \sum_{i=1}^{m}\sum_{j=1}^{m} c_{ij,4}x_{ij},\\ &\sum_{i=1}^{m}\sum_{j=1}^{m} c'_{ij,1}x_{ij},\ \sum_{i=1}^{m}\sum_{j=1}^{m} c'_{ij,2}x_{ij},\ \sum_{i=1}^{m}\sum_{j=1}^{m} c'_{ij,3}x_{ij},\ \sum_{i=1}^{m}\sum_{j=1}^{m} c'_{ij,4}x_{ij}\Bigg)\\ s.t.\quad &\sum_{j=1}^{n} x_{ij} - \sum_{k=1}^{m} x_{ki} = \begin{cases}1, & i = s,\\ 0, & i \ne s,t,\\ -1, & i = t,\end{cases}\\ &\sum_{i=1}^{m}\sum_{j=1}^{m} t_{ij,1}x_{ij} \le T_1,\ \sum_{i=1}^{m}\sum_{j=1}^{m} t_{ij,2}x_{ij} \le T_2,\ \sum_{i=1}^{m}\sum_{j=1}^{m} t_{ij,3}x_{ij} \le T_3,\ \sum_{i=1}^{m}\sum_{j=1}^{m} t_{ij,4}x_{ij} \le T_4,\\ &\sum_{i=1}^{m}\sum_{j=1}^{m} t'_{ij,1}x_{ij} \le T'_1,\ \sum_{i=1}^{m}\sum_{j=1}^{m} t'_{ij,2}x_{ij} \le T'_2,\ \sum_{i=1}^{m}\sum_{j=1}^{m} t'_{ij,3}x_{ij} \le T'_3,\ \sum_{i=1}^{m}\sum_{j=1}^{m} t'_{ij,4}x_{ij} \le T'_4.\\ &x_{ij} \ge 0,\ i,j = 1,2,\ldots,m.\end{aligned} \tag{6}$$

The above model is obviously a linear program and can be solved using the standard LP algorithms.

5 Numerical Example

In this section, for the illustration of the proposed approach, a simple IFCSP problem is solved.

Let us consider the network in Fig. 1 with arc intuitionistic fuzzy weights associated with cost and time as given in Table 1.

The maximum intuitionistic fuzzy value allowed for the sum of the $\tilde{t}^I_{ij}$ arc weights is $\tilde{T}^I = (11, 13, 14, 16; 10, 12, 15, 17)$. It is desired to find minimum cost intuitionistic fuzzy feasible path from node 1 to node 5.

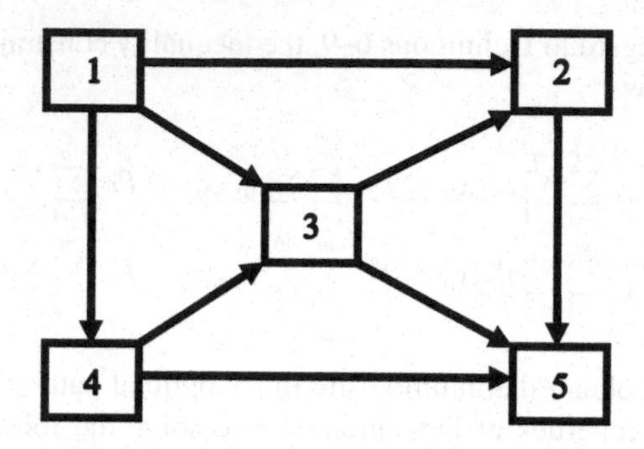

Fig. 1. An example of IFCSP network

Table 1. Arc information in terms of interval numbers

Arc	Intuitionistic fuzzy cost	Intuitionistic fuzzy time
(1, 2)	(12, 14, 15, 17; 11, 13, 16, 18)	(2, 4, 5, 7; 1, 3, 6, 8)
(1, 3)	(8, 10, 11, 13; 7, 9, 12, 14)	(2, 4, 5, 7; 1, 3, 6, 8)
(1, 4)	(10, 12, 13, 15; 9, 11, 14, 16)	(3, 5, 6, 8; 2, 4, 7, 9)
(2, 5)	(5, 7, 8, 10; 4, 6, 9, 11)	(5, 7, 8, 10; 4, 6, 9, 11)
(3, 2)	(3, 5, 6, 8; 2, 4, 7, 9)	(4, 6, 7, 9; 3, 5, 8, 10)
(3, 5)	(11, 13, 14, 16; 10, 12, 15, 17)	(3, 5, 6, 8; 2, 4, 7, 9)
(4, 3)	(5, 7, 8, 10; 4, 6, 9, 11)	(1, 3, 4, 6; 0, 2, 5, 7)
(4, 5)	(8, 10, 11, 13; 7, 9, 12, 14)	(3, 5, 6, 8; 2, 4, 7, 9)

Based on model (2), we should solve the following IFCSP problem:

$$
\begin{aligned}
\min\ \tilde{Z}^I = {} & (12,14,15,17;11,13,16,18)x_{12} + (8,10,11,13;7,9,12,14)x_{13} \\
& + (10,12,13,15;9,11,14,16)x_{14} + (5,7,8,10;4,6,9,11)x_{25} \\
& + (3,5,6,8;2,4,7,9)x_{32} + (11,13,14,16;10,12,15,17)x_{35} \\
& + (5,7,8,10;4,6,9,11)x_{43} + (8,10,11,13;7,9,12,14)x_{45} \\
s.t.\quad & \\
& x_{12} + x_{13} + x_{14} = 1, \\
& x_{25} - x_{12} - x_{32} = 0, \\
& x_{32} + x_{35} - x_{13} - x_{43} = 0, \\
& x_{45} + x_{43} - x_{14} = 0, \\
& -x_{25} - x_{35} - x_{45} = -1, \\
& (2,4,5,7;1,3,6,8)x_{12} + (2,4,5,7;1,3,6,8)x_{13} \\
& + (3,5,6,8;2,4,7,9)x_{14} + (5,7,8,10;4,6,9,11)x_{25} \\
& + (4,6,7,9;3,5,8,10)x_{32} + (3,5,6,8;2,4,7,9)x_{35} \\
& + (1,3,4,6;0,2,5,7)x_{43} + (3,5,6,8;2,4,7,9)x_{45} \preceq (11,13,14,16;10,12,15,17), \\
& x_{12}, x_{13}, x_{14}, x_{25}, x_{32}, x_{35}, x_{43}, x_{45} \geq 0.
\end{aligned}
\tag{7}
$$

This model is transformed to the following linear program with regard to model (6):

$$\min\ H(\tilde{Z}^{I}) = \frac{1}{8}[116x_{12} + 84x_{13} + 100x_{14} + 60x_{25} + 44x_{32} + 108x_{35} + 60x_{43} + 84x_{45}]$$

$$\begin{aligned}
& s.t. \\
& x_{12} + x_{13} + x_{14} = 1, \\
& x_{25} - x_{12} - x_{32} = 0, \\
& x_{32} + x_{35} - x_{13} - x_{43} = 0, \\
& x_{45} + x_{43} - x_{14} = 0, \\
& -x_{25} - x_{35} - x_{45} = -1, \\
& x_{12} + x_{13} + 2x_{14} + 4x_{25} + 3x_{32} + 2x_{35} + 2x_{45} \le 10, \\
& 2x_{12} + 2x_{13} + 3x_{14} + 5x_{25} + 4x_{32} + 3x_{35} + x_{43} + 3x_{45} \le 11, \\
& 3x_{12} + 3x_{13} + 4x_{14} + 6x_{25} + 5x_{32} + 4x_{35} + 2x_{43} + 4x_{45} \le 12, \\
& 4x_{12} + 4x_{13} + 5x_{14} + 7x_{25} + 6x_{32} + 5x_{35} + 3x_{43} + 5x_{45} \le 13, \\
& 5x_{12} + 5x_{13} + 6x_{14} + 8x_{25} + 7x_{32} + 6x_{35} + 4x_{43} + 6x_{45} \le 14, \\
& 6x_{12} + 6x_{13} + 7x_{14} + 9x_{25} + 8x_{32} + 7x_{35} + 5x_{43} + 7x_{45} \le 15, \\
& 7x_{12} + 7x_{13} + 8x_{14} + 10x_{25} + 9x_{32} + 8x_{35} + 6x_{43} + 8x_{45} \le 16, \\
& 8x_{12} + 8x_{13} + 9x_{14} + 11x_{25} + 10x_{32} + 9x_{35} + 7x_{43} + 9x_{45} \le 17, \\
& x_{12}, x_{13}, x_{14}, x_{25}, x_{32}, x_{35}, x_{43}, x_{45} \ge 0.
\end{aligned} \tag{8}$$

The optimal solution of the crisp linear model (8) is as follows:

$$x^{*}_{12} = 0,\ x^{*}_{13} = 1,\ x^{*}_{14} = 0,\ x^{*}_{25} = 0, x^{*}_{32} = 0, x^{*}_{35} = 1, x^{*}_{43} = 0,\ x^{*}_{145} = 0. \tag{9}$$

This means that the intuitionistic fuzzy optimal path is $p^{*}_{15} : 1 \rightarrow 3 \rightarrow 5$. By substituting the optimal solution (9) in the intuitionistic fuzzy objective function of model (7), we obtain:

$$\begin{aligned}
\tilde{C}^{I}(p^{*}_{15}) &= (8, 10, 11, 13; 7, 9, 12, 14) + (11, 13, 14, 16; 10, 12, 15, 17) \\
&= (19, 23, 25, 29; 17, 21, 27, 31)
\end{aligned}$$

Moreover, we have:

$$\begin{aligned}
\tilde{D}^{I}(p^{*}_{15}) &= \sum_{(i,j)\in p^{*}_{15}} \tilde{t}^{I}_{ij} = \tilde{t}^{I}_{13} + \tilde{t}^{I}_{35} \\
&= (2, 4, 5, 7; 1, 3, 6, 8) + (3, 5, 6, 8; 2, 4, 7, 9) = (5, 9, 11, 15; 3, 7, 13, 17).
\end{aligned}$$

6 Conclusions

On the basis of the presented study, it can be concluded that there is no method in the literature for solving the constraint shortest path problems in an intuitionistic environment. In this paper, a CSP problem having uncertainty as well as hesitation in prediction of the arc weights has been investigated. In the CSP problem considered in this study, the arc weights represented by cost and time were intuitionistic triangular fuzzy numbers. Here, we proposed a new solution approach for solving intuitionistic fuzzy CSP problem. We converted the IFCSP problem under consideration into a linear programming problem which can be solved using the standard linear programing algorithms.

References

1. Biswas, S.S., Bashir, A., Doja, M.N.: An algorithm for extracting intuitionistic fuzzy shortest path in a graph. Appl. Comput. Intell. Soft Comput. **2013**, 1–5 (2013)
2. Chuang, T.N., Kung, J.Y.: The fuzzy shortest path length and the corresponding shortest path in a network. Comput. Oper. Res. **32**(6), 1409–1428 (2005)
3. Dou, Y., Zhu, L., Wang, H.S.: Solving the fuzzy shortest path problem using multi-criteria decision method based on vague similarity measure. Appl. Soft Comput. **12**(6), 1621–1631 (2012)
4. Ebrahimnejad, A., Karimnejad, Z., Alrezaamiri, H.: Particle swarm optimization algorithm for solving shortest path problems with mixed fuzzy arc weights. Int. J. Appl. Decision Sci. **8**(2), 203–222 (2015)
5. Ebrahimnejad, A., Tavana, M., Alrezaamiri, H.: A novel artificial bee colony algorithm for shortest path problems with fuzzy arc weights. Measurement **93**, 48–56 (2016)
6. Ebrahimnejad, A., Verdegay, J.L.: An efficient computational approach for solving type–2 intuitionistic fuzzy numbers based transportation problems. Int. J. Comput. Intell. Syst. **9**(6), 1154–1173 (2016)
7. Geetharamani, G., Jayagowri, P.: Using similarity degree approach for shortest path in Intuitionistic fuzzy network. In: International Conference on Computing, Communication and Applications (ICCCA), pp. 1–6, IEEE, Dindigul (2012)
8. Hassanzadeh, R., Mahdavi, I., Mahdavi-Amiri, N., Tajdin, A.: A genetic algorithm for solving fuzzy shortest path problems with mixed fuzzy arc lengths. Math. Comput. Model. **57**(1–2), 84–99 (2013)
9. Hernandes, F., Lamata, M.T., Verdegay, J.L., Yamakami, A.: The shortest path problem on networks with fuzzy parameters. Fuzzy Sets Syst. **158**(14), 1561–1570 (2007)
10. Kumar, G., Bajaj, R.K., Gandotra, N.: Algorithm for shortest path problem in a network with interval–valued intuitionistic trapezoidal fuzzy number. Procedia Comput. Sci. **70**, 123–129 (2015)
11. Mahdavi, I., Nourifar, R., Heidarzade, A., Mahdavi Amiri, N.: A dynamic programming approach for finding shortest chains in a fuzzy network. Appl. Soft Comput. **9**, 503–511 (2009)
12. Moazeni, M.: Fuzzy shortest path problem with finite fuzzy quantities. Appl. Math. Comput. **183**, 160–169 (2006)
13. Mukherjee, S.: Dijkstra's algorithm for solving the shortest path problem on networks under intuitionistic fuzzy environment. J. Math. Model. Algorithm **11**(4), 345–359 (2012)
14. Okada, S., Super, T.: A shortest path problem on a network with fuzzy arc lengths. Fuzzy Sets Syst. **109**, 129–140 (2000)
15. Olada, T.: Fuzzy shortest path problems incorporating interactivity among paths. Fuzzy Sets Syst. **142**, 335–357 (2004)
16. Santos, L., Coutinho-Rodrigues, J., Current, J.R.: An improved solution algorithm for the constrained shortest path problem. Transp. Res. Part B **41**, 756–771 (2007)
17. Sujatha, L., Hyacinta, J.D.: The shortest path problem on networks with intuitionistic fuzzy edge weights. Global J. Pure Appl. Math. **13**(7), 3285–3300 (2017)

MOLP Approach for Solving Transportation Problems with Intuitionistic Fuzzy Costs

Ali Ebrahimnejad[1(✉)] and José L. Verdegay[2]

[1] Department of Mathematics, Qaemshahr Branch, Islamic Azad University, Qaemshahr, Iran
a.ebrahimnejad@qaemiau.ac.ir

[2] Department of Computer Science and A.I., Universidad de Granada, Granada, Spain
verdegay@decsai.ugr.es

Abstract. Many researchers have focused on a Transportation Problem (TP) in uncertain environment because of its importance to various applications. This paper is concerned with the solution procedure of a TP in which transportation costs are represented in terms of intuitionistic triangular fuzzy numbers and supplies and demands are real numbers. We first formulate the intuitionistic fuzzy TP (IFTP) and then propose a new solution technique to solve the problem. Based on the proposed approach, the IFTP is converted into a Multi Objective Linear Programming (MOLP) problem with five objective functions. Then, a lexicographic approach is used to obtain the efficient solution of the resulting MOLP problem. The optimization process confirms that the optimum intuitionistic fuzzy transportation cost preserves the form of an intuitionistic triangular fuzzy number. A simple numerical example is included to illustrate of the proposed technique. The obtained results confirm the reliability and applicability of the proposed approach.

Keywords: Transportation problem · Intuitionistic fuzzy number
MOLP

1 Introduction

The transportation problem (TP) is a special class of linear programming (LP) problems, and widely used in the areas of inventory control, communication network, aggregate planning, logistic, supply chains, personal management and so on. The central concept in this problem is to determine the minimum total transportation cost of a commodity for satisfying the demand at destinations using the available supply at the origins. In classical TP it is assumed that the transportation costs are exactly known. In real-life transportation cases, decision makers may face with many uncertainties on the cost of transportation because of changing weather, social, or economic conditions. In other words, the decision makers cannot exactly know the transportation costs of a TP in reality. Additionally, they are not stable since this imprecision may follow from the lack of exact information or data, uncertainty in judgment and high information cost. This imprecision embedded into the transportation costs can be handled via fuzzy

J. Medina et al. (Eds.): IPMU 2018, CCIS 855, pp. 319–329, 2018.
https://doi.org/10.1007/978-3-319-91479-4_27

parameters and a Fuzzy Transportation Problem (FTP) appears in a natural way. From this point of view, numerous researchers have devoted their efforts to using fuzzy numbers in real life TP [3–7, 11, 12].

In spite of this, although fuzzy numbers are commonly used for modeling imprecise data when one has to cope with real TP, this may not be suitable for situations where one has to deal with uncertainty as well as with hesitation. In such situations, Intuitionistic Fuzzy Numbers (IFNs) [2] are used to represent the imprecise transportation costs of the TP under consideration. The resulting problem is therefore referred to as an Intuitionistic Fuzzy Transportation Problem (IFTP). Thus, determining solutions for the IFTPs is a relatively new and active research topic. There are just a few research papers in this subject.

As there is a hesitation in the parameters of TP many authors have solved this problem under intuitionistic fuzzy environment successfully. Hussain and Kumar [10] proposed an intuitionistic fuzzy zero point method to solve a TP in which supply and demand were intuitionistic fuzzy numbers. Singh and Yadav [15] developed a new ordering procedure using accuracy function of triangular IFN (TIFN) and used this ordering to develop an algorithm for finding optimal solution of the same IFTP. But, these methods [10, 15] cannot be used for solving fully IFTP, where transportation costs are IFNs as well. To overcome this shortcoming, Kumar and Hussain [13] transformed the fully IFTP into a crisp one and applied the conventional method to solve the problem. But, all of these proposed methods [10, 13, 15] cannot provide non-negative intuitionistic fuzzy optimal solution and optimal cost for the IFTP under consideration. To overcome this shortcoming, Ebrahimnejad and Verdegay [9] proposed a novel solution approach for solving fully IFTP based on classical LP algorithms.

Although the fully IFTP problem is the general case of the IFTP, it may not be suitable for all IFTP problems with different assumptions and sources of imprecision. The IFTP model proposed in this study belongs to those categories in which the transportation costs are represented in terms of intuitionistic triangular fuzzy numbers (ITFNs) and supplies and demands are real numbers. Antony et al. [1] have used Vogel's approximation method for solving TP with TIFNs. They claimed that the solution obtained by this method is optimal without providing any optimality conditions. Singh and Yadav [14] formulated a same IFTP. They first used an accuracy function defined on score functions for membership and non-membership functions of TIFNs for ordering of IFNs. Then, they used this ordering to develop methods for finding an initial basic feasible solution and optimal solution of IFTP in terms TIFNs. However their method [14], in spite of its merits, requires a lot of intuitionistic fuzzy arithmetic operations and a lot of comparisons on TIFNs. For this reason, Ebrahimnejad and Verdegay [8] proposed an efficient computational solution approach for solving the same problem based on classical transportation algorithms. However, such solution methods [1, 8, 14] neglect the valuable uncertain information in the optimization process because of using linear accuracy function for defuzzification the IFTP in the intuitionistic fuzzy sense. In this paper, we propose a new solution technique to solve the same IFTP without neglecting valuable uncertain information. In the proposed approach, the IFTP is converted into a Multi Objective Linear Programming (MOLP) problem with five objective functions. Then, a lexicographic approach is used to obtain the efficient solution of the resulting MOLP problem.

The rest of the paper is organized as follows: In Sect. 2, some basic concepts of intuitionistic fuzzy sets theory are reviewed. In Sect. 3, the TP with intuitionistic fuzzy transportation costs is formulated. In Sect. 4, a new method is proposed for obtaining the optimal solution of the IFTP. In Sect. 5, the application of the proposed method is illustrated by using a simple numerical example. Section 6, including the main conclusions as well as some interesting future research lines, ends the paper.

2 Preliminaries

This section is devoted to review some necessary background and notions of the intuitionistic fuzzy numbers which are applied throughout this paper [8, 10, 14].

Definition 1. Let X denote the universe set. A fuzzy set $\tilde{A}$ in X is defined by a set of ordered pairs $\tilde{A} = \{(x, \mu_{\tilde{A}}(x)); x \in X\}$ where $\mu_{\tilde{A}}(x) \in [0, 1]$ represents the membership degree of x in $\tilde{A}$, and is called the membership function of $\tilde{A}$.

Definition 2. Let X denote the universe set. An intuitionistic fuzzy set (IFS) $\tilde{A}^I$ in X is defined by a set of ordered triple $\tilde{A}^I = \{\langle x, \mu_{\tilde{A}^I}(x), \upsilon_{\tilde{A}^I}(x)\rangle; x \in X\}$ where the functions $\mu_{\tilde{A}^I}(x) : X \to [0, 1]$ and $\upsilon_{\tilde{A}^I}(x) : X \to [0, 1]$, respectively represent the membership degree and non-membership degree of x in $\tilde{A}$ such that for each element $x \in X$, $0 \leq \mu_{\tilde{A}^I}(x) + \upsilon_{\tilde{A}^I}(x) \leq 1$.

Definition 3. For each intuitionistic fuzzy set $\tilde{A}^I = \{\langle x, \mu_{\tilde{A}^I}(x), \upsilon_{\tilde{A}^I}(x)\rangle; x \in X\}$ in X, the value $h_{\tilde{A}^I}(x) = 1 - \mu_{\tilde{A}^I}(x) - \upsilon_{\tilde{A}^I}(x)$ is called degree of hesitancy of x to $\tilde{A}^I$.

Definition 4. An intuitionistic fuzzy set $\tilde{A}^I = \{\langle x, \mu_{\tilde{A}^I}(x), \upsilon_{\tilde{A}^I}(x)\rangle; x \in X\}$ is called normal if there is any $x_\circ \in X$ such that $\mu_{\tilde{A}^I}(x_\circ) = 1$(so $\upsilon_{\tilde{A}^I}(x_\circ) = 0$).

Definition 5. An intuitionistic fuzzy set $\tilde{A}^I = \{\langle x, \mu_{\tilde{A}^I}(x), \upsilon_{\tilde{A}^I}(x)\rangle; x \in X\}$ is called intuitionistic fuzzy convex if its membership function is fuzzy convex, i.e. $\forall x_1, x_2 \in X, \forall \lambda \in [0, 1], \quad \mu_{\tilde{A}^I}(\lambda x_1 + (1 - \lambda)x_2) \geq \min\{\mu_{\tilde{A}^I}(x_1), \mu_{\tilde{A}^I}(x_2)\}$ and its non-membership function is concave, i.e. $\upsilon_{\tilde{A}^I}(\lambda x_1 + (1 - \lambda)x_2) \leq \max\{\upsilon_{\tilde{A}^I}(x_1), \upsilon_{\tilde{A}^I}(x_2)\}$.

Definition 6. An intuitionistic fuzzy set $\tilde{A}^I = \{\langle x, \mu_{\tilde{A}^I}(x), \upsilon_{\tilde{A}^I}(x)\rangle; x \in R\}$ of the real number R is called an intuitionistic fuzzy number if

- $\tilde{A}^I$ is intuitionistic fuzzy normal and intuitionistic fuzzy convex.
- $\mu_{\tilde{A}^I}$ is upper semi continuous and $\upsilon_{\tilde{A}^I}$ is semi lower continuous.
- $Supp\,\tilde{A}^I = \{x \in R;\ \upsilon_{\tilde{A}^I}(x) < 1\}$ is bounded.

Definition 7. A Triangular Intuitionistic Fuzzy Number (TIFN) $\tilde{A}^I$ is an especial IFN with the membership function non-membership function defined as follows:

$$\mu_{\tilde{A}^I}(x) = \begin{cases} \frac{x-a_1}{a_2-a_1}, & a_1 < x \leq a_2, \\ \frac{a_3-x}{a_3-a_2}, & a_2 \leq x < a_3 \\ 0, & \text{Otherwise.} \end{cases} \quad \text{and} \quad \upsilon_{\tilde{A}^I}(x) = \begin{cases} \frac{a_2-x}{a_2-a'_1}, & a'_1 < x \leq a_2, \\ \frac{x-a_2}{a'_3-a_3}, & a_2 \leq x < a'_3, \\ 1, & \text{Otherwise.} \end{cases}$$

where $a'_1 \leq a_1 \leq a_2 \leq a_3 \leq a'_3$. This TIFN is denoted by $\tilde{A}^I = (a_1, a_2, a_3;\ a'_1,\ a_2, a'_3)$.

Definition 8. The arithmetic operations between two TIFNs $\tilde{A}^I = (a_1, a_2, a_3;\ a'_1,\ a_2, a'_3)$ and $\tilde{B}^I = (b_1, b_2, b_3;\ b'_1,\ b_2, b'_3)$ are defined as follows:

(i) $\tilde{A}^I \oplus \tilde{B}^I = (a_1 + b_1, a_2 + b_2, a_3 + b_3;\ a'_1 + b'_1,\ a_2 + b_2, a'_3 + b'_3)$
(ii) $\tilde{A}^I \ominus \tilde{B}^I = (a_1 - b_3, a_2 - b_2, a_3 - b_1;\ a'_1 - b'_3, a_2 - b_2, a'_3 - b'_1)$,
(iii) $k\tilde{A}^I = (ka_1, ka_2, ka_3;\ ka'_1,\ ka_2, ka'_3)$, $k > 0$,
(iv) $k\tilde{A}^I = (ka_3, ka_2, ka_1;\ ka'_3,\ ka_2, ka'_1)$, $k < 0$.

Definition 9. Two TIFNs $\tilde{A}^I = (a_1, a_2, a_3;\ a'_1,\ a_2, a'_3)$ and $\tilde{B}^I = (b_1, b_2, b_3;\ b'_1,\ b_2, b'_3)$ are said to be equal, i.e. $\tilde{A}^I = \tilde{B}^I$ if and only if $a_1 = b_1, a_2 = b_2, a_3 = b_3, a'_1 = b'_1, a'_3 = b'_3$.

Definition 10. A TIFN $\tilde{A}^I = (a_1, a_2, a_3;\ a'_1,\ a_2, a'_3)$ is said to be a non-negative TIFN if and only if $a'_1 \geq 0$.

Definition 11. For TIFN $\tilde{A}^I = (a_1, a_2, a_3;\ a'_1,\ a_2, a'_3)$, its accuracy function is defined as follows:

$$H(\tilde{A}^I) = \frac{(a_1 + 2a_2 + a_3) + (a'_1 + 2a_2 + a'_3)}{8} \tag{1}$$

Definition 12. Let $\tilde{A}^I = (a_1, a_2, a_3;\ a'_1,\ a_2, a'_3)$ and $\tilde{B}^I = (b_1, b_2, b_3;\ b'_1,\ b_2, b'_3)$ be two TIFNs. Then $\tilde{A}^I \geq \tilde{B}^I$ if $H(\tilde{A}^I) \geq H(\tilde{B}^I)$, and $\tilde{A}^I \leq \tilde{B}^I$ if $H(\tilde{A}^I) \leq H(\tilde{B}^I)$.

3 Intuitionistic Fuzzy Transportation Problems

In this section, the linear programming formulation of TP in intuitionistic fuzzy environment is presented.

The IFTP in which a decision maker considers the cost as TIFN to deal efficiently with the uncertainty as well as hesitation arising in prediction of transportation cost, but (s)he is sure about the availability and demand of the product, can be formulated as follows [8, 9, 13]:

$$\min \ \tilde{Z}^I = \sum_{i=1}^{m}\sum_{j=1}^{n} \tilde{c}_{ij}^I x_{ij}$$
$$s.t. \sum_{j=1}^{n} x_{ij} = a_i, \quad i = 1,2,\ldots,m,$$
$$\sum_{i=1}^{m} x_{ij} = b_j, \quad j = 1,2,\ldots,n, \tag{2}$$
$$x_{ij} \geq 0, i = 1,2,\ldots,m, j = 1,2,\ldots,n.$$

where a_i is: the total availability of the product at i^{th} source; b_j: the total demand of the product at j^{th} destination; $\tilde{c}_{ij}^I = (c_1^{ij}, c_2^{ij}, c_3^{ij}; c_1^{ij'}, c_2^{ij}, c_3^{ij'})$: the intuitionistic cost for transporting one unit quantity of the product from the ith source to the jth destination; x_{ij}: the quantity transported from the ith source to the jth destination or decision variables; $\sum_{i=1}^{m}\sum_{j=1}^{n} \tilde{c}_{ij}^I x_{ij}$: total intuitionistic fuzzy transportation cost.

Because negative transportation costs have no physical meaning, it is assumed that intuitionistic fuzzy transportation costs of the IFTP (2) are non-negative TIFNs. Hence, with regard to Definition 8, IFTP (2) can be reformulated as follows:

$$\min \ \tilde{Z}^I = \sum_{i=1}^{m}\sum_{j=1}^{n} (c_1^{ij} x_{ij}, c_2^{ij} x_{ij}, c_3^{ij} x_{ij}; c_1^{ij'} x_{ij}, c_2^{ij} x_{ij}, c_3^{ij'} x_{ij})$$
$$s.t. \ \sum_{j=1}^{n} x_{ij} = a_i, \qquad i = 1,2,\ldots,m,$$
$$\sum_{i=1}^{m} x_{ij} = b_j, \qquad j = 1,2,\ldots,n, \tag{3}$$
$$x_{ij} \geq 0, \ i = 1,2,\ldots,m, \ j = 1,2,\ldots,n.$$

Equivalently, with regard to Definitions 8, IFTP (3) may be rewritten as follows:

$$\min \ \tilde{Z}^I = \left(\sum_{i=1}^{m}\sum_{j=1}^{n} c_1^{ij} x_{ij}, \sum_{i=1}^{m}\sum_{j=1}^{n} c_2^{ij} x_{ij}, \sum_{i=1}^{m}\sum_{j=1}^{n} c_3^{ij} x_{ij}; \sum_{i=1}^{m}\sum_{j=1}^{n} c_1^{ij'} x_{ij}, \sum_{i=1}^{m}\sum_{j=1}^{n} c_2^{ij} x_{ij}, \sum_{i=1}^{m}\sum_{j=1}^{n} c_3^{ij'} x_{ij} \right)$$
$$s.t. \ \sum_{j=1}^{n} x_{ij} = a_i, \qquad i = 1,2,\ldots,m,$$
$$\sum_{i=1}^{m} x_{ij} = b_j, \qquad j = 1,2,\ldots,n, \tag{4}$$
$$x_{ij} \geq 0, \ i = 1,2,\ldots,m, \ j = 1,2,\ldots,n.$$

4 New Method

Model (4) is an IFTP with one intuitionistic fuzzy variable in the objective function. This study proposes a new method for solving model (4) by converting this model into an MOLP problem with five objective functions.

The resulting intuitionistic fuzzy objective function of IFTP (4) has different five components that can be considered as a multi objective function. Here, we present a lexicography approach for solving IFTP (5). The steps of the proposed method are as follows:

Step 1: Solve the following crisp LP problem using the standard LP algorithms:

$$\begin{aligned} Z_1^{*I'} &= \min\ Z_1^{I'} = \sum_{i=1}^{m}\sum_{j=1}^{n} c_1^{ij'} x_{ij} \\ s.t.\ &\sum_{j=1}^{n} x_{ij} = a_i, \qquad i = 1,2,\ldots,m, \\ &\sum_{i=1}^{m} x_{ij} = b_j, \qquad j = 1,2,\ldots,n, \\ &x_{ij} \geq 0,\ i = 1,2,\ldots,m,\ j = 1,2,\ldots,n. \end{aligned} \tag{5}$$

The optimal value of the objective function of model (5), $Z_1^{*I'}$, is the fourth component of the intuitionistic fuzzy optimal total transportation cost.

Step 2: Solve the following crisp LP problem using the standard LP algorithms:

$$\begin{aligned} Z_1^{*I} &= \min\ Z_1^{I} = \sum_{i=1}^{m}\sum_{j=1}^{n} c_1^{ij} x_{ij} \\ s.t.\ &\sum_{i=1}^{m}\sum_{j=1}^{n} c_1^{ij'} x_{ij} = Z_1^{*I'} \\ &\text{Constraints of Model (5).} \end{aligned} \tag{6}$$

The optimal value of the objective function of model (6), Z_1^{*I}, is the first component of the intuitionistic fuzzy optimal total transportation cost.

Proposition 1. The optimal value of the objective function of model (5) is less than or equal to that of the objective function of model (6).

Step 3: Solve the following crisp LP problem using the standard LP algorithms:

$$\begin{aligned} Z_2^{*I} &= \min\ Z_2^{I} = \sum_{i=1}^{m}\sum_{j=1}^{n} c_2^{ij} x_{ij} \\ s.t.\ &\sum_{i=1}^{m}\sum_{j=1}^{n} c_1^{ij} x_{ij} = Z_1^{*I} \\ &\text{Constraints of Model (6).} \end{aligned} \tag{7}$$

The optimal value of the objective function of model (7), Z_2^{*I}, is the second and fifth components of the intuitionistic fuzzy optimal total transportation cost.

Proposition 2. The optimal value of the objective function of model (6) is less than or equal to that of the objective function of model (7).

Step 4: Solve the following crisp LP problem using the standard LP algorithms:

$$\begin{aligned} Z_3^{*I} &= \min Z_3^I = \sum_{i=1}^{m}\sum_{j=1}^{n} c_3^{ij} x_{ij} \\ s.t.\ & \sum_{i=1}^{m}\sum_{j=1}^{n} c_2^{ij} x_{ij} = Z_2^{*I} \\ & \text{Constraints of Model (7).} \end{aligned} \tag{8}$$

The optimal value of the objective function of model (8), Z_3^{*I}, is the third component of the intuitionistic fuzzy optimal total transportation cost.

Proposition 3. The optimal value of the objective function of model (7) is less than or equal to that of the objective function of model (8).

Step 5: Solve the following crisp LP problem using the standard LP algorithms:

$$\begin{aligned} Z_3^{*I'} &= \min Z_3^I = \sum_{i=1}^{m}\sum_{j=1}^{n} c_3^{ij'} x_{ij} \\ s.t.\ & \sum_{i=1}^{m}\sum_{j=1}^{n} c_3^{ij} x_{ij} = Z_3^{*I} \\ & \text{Constraints of Model (8).} \end{aligned} \tag{9}$$

The optimal value of the objective function of model (9), $Z_3^{*I'}$, is the sixth component of the intuitionistic fuzzy optimal total transportation cost.

Proposition 4. The optimal value of the objective function of model (8) is less than or equal to that of the objective function of model (9).

Theorem 1. The intuitionistic fuzzy optimal total transportation cost, $\tilde{Z}^{*I} = \left(Z_1^{*I}, Z_2^{*I}, Z_3^{*I'}; Z_1^{*I'}, Z_2^{*I}, Z_3^{*I'}\right)$, maintains the form of a non-negative intuitionistic triangular fuzzy number.

5 Numerical Example

In this section, in order to demonstrate the effectiveness of the proposed method, an intuitionistic fuzzy TP taken from Singh and Yadav [14] is considered.

Table 1 gives the crisp supply (a_i) of the product available at four origins S_i ($i = 1, 2, 3, 4$) and the crisp demand (b_j) at four destinations D_j($j = 1, 2, 3, 4$). The transportation costs from origins to destinations are represented by ITFNs. The aim is to find

Table 1. Summary of the intuitionistic fuzzy transportation problem

	D_1	D_2	D_3	D_4	a_i
S_1	(2, 4, 5; 1, 4, 6)	(2, 5, 7; 1, 5, 8)	(4, 6, 8; 3, 6, 9)	(4, 7, 8; 3, 7, 9)	11
S_2	(4, 6, 8; 3, 6, 9)	(3, 7, 12; 2, 7, 13)	(10, 15, 20; 8, 15, 22)	(11, 12, 13; 10, 12, 14)	11
S_3	(3, 4, 6; 1, 4, 8)	(8, 10, 13; 5, 10, 16)	(2, 3, 5; 1, 3, 6)	(6, 10, 14; 5, 10, 15)	11
S_4	(2, 4, 6; 1, 4, 7)	(3, 9, 10; 2, 9, 12)	(3, 6, 10; 2, 6, 12)	(3, 4, 5; 2, 4, 8)	12
b_j	16	10	8	11	45

the least total intuitionistic fuzzy transportation cost of the commodity in order to satisfy demands at destinations using available availabilities at origins.

Mathematically, IFTP is formulated as follows:

$$\begin{aligned}
\min\ & (2,4,5;1,4,6)x_{11} + (2,5,7;1,5,8)x_{12} + (4,6,8;3,6,9)x_{13} \\
& + (4,7,8;3,7,9)x_{14} \\
& + (4,6,8;3,6,9)x_{21} + (3,7,12;2,7,13)x_{22} + (10,15,20;8,15,22)x_{23} \\
& + (11,12,13;10,12,14)x_{24} \\
& + (3,4,6;1,4,8)x_{31} + (8,10,13;5,10,16)x_{32} + (2,3,5;1,3,6)x_{33} \\
& + (6,10,14;5,10,15)x_{34} \\
& + (2,4,6;1,4,7)x_{41} + (3,9,10;2,9,12)x_{42} + (3,6,10;2,6,12)x_{43} \\
& + (3,4,5;2,4,8)x_{44} \\
s.t.\ & x_{11} + x_{12} + x_{13} + x_{14} = 11, \\
& x_{21} + x_{22} + x_{23} + x_{24} = 11, \\
& x_{31} + x_{32} + x_{33} + x_{34} = 11, \\
& x_{41} + x_{42} + x_{43} + x_{44} = 12, \\
& x_{11} + x_{21} + x_{31} + x_{41} = 16, \\
& x_{12} + x_{22} + x_{32} + x_{42} = 10, \\
& x_{13} + x_{23} + x_{33} + x_{43} = 8, \\
& x_{14} + x_{24} + x_{34} + x_{44} = 11, \\
& x_{ij} \geq 0,\ i,j = 1,2,3,4.
\end{aligned} \tag{10}$$

The fuzzy optimal solution of IFTP (10) can be obtained using the proposed method in Sect. 4, as follows:

Step 1: We solve the following classical transportation problem using the standard LP algorithms:

$$\begin{aligned}
Z_1^{*I'} = \min Z_1^{I'} = & x_{11} + x_{12} + 3x_{13} + 3x_{14} + 3x_{21} + 2x_{22} + 8x_{23} + 10x_{24} \\
& + x_{31} + 5x_{32} + x_{33} + 5x_{34} + x_{41} + 2x_{42} + 2x_{43} + 2x_{44} \\
s.t.\ & \text{Constraints of Model (10)}.
\end{aligned} \tag{11}$$

The optimal solution of Model (11) is as follows:

$$\begin{aligned}&x_{11}=11,\ x_{12}=0,\ x_{13}=0,\ x_{14}=0,\\&x_{21}=1,\ \ x_{22}=10,\ x_{23}=0,\ x_{24}=0,\\&x_{31}=3,\ \ x_{32}=0,\ \ x_{33}=8,\ x_{34}=0,\\&x_{41}=1,\ \ x_{42}=0,\ \ x_{43}=0, x_{44}=11.\\&Z_1^{*I'}=68\end{aligned} \tag{12}$$

Step 2: We solve the following classical transportation problem using the standard LP algorithms:

$$\begin{aligned}Z_1^{*I}=\min\ Z_1^I&=2x_{11}+2x_{12}+4x_{13}+4x_{14}+4x_{21}+3x_{22}+10x_{23}+11x_{24}\\&\quad+3x_{31}+8x_{32}+2x_{33}+6x_{34}+2x_{41}+3x_{42}+3x_{43}+3x_{44}\\ s.t.\ &x_{11}+x_{12}+3x_{13}+3x_{14}+3x_{21}+2x_{22}+8x_{23}+10x_{24}\\&\quad+x_{31}+5x_{32}+x_{33}+5x_{34}+x_{41}+2x_{42}+2x_{43}+2x_{44}=68\\&\text{Constraints of Model (11).}\end{aligned} \tag{13}$$

The optimal solution of Model (13) is as follows:

$$\begin{aligned}&x_{11}=11,\ x_{12}=0,\ \ x_{13}=0,\ x_{14}=0,\\&x_{21}=1,\ \ x_{22}=10,\ x_{23}=0,\ x_{24}=0,\\&x_{31}=3,\ \ x_{32}=0,\ \ x_{33}=8, x_{34}=0,\\&x_{41}=1,\ \ x_{42}=0,\ \ x_{43}=0, x_{44}=11.\\&Z_1^{*I}=116\end{aligned} \tag{14}$$

Similarly, using the Steps 3–5 of the proposed method, the optimal solution and the minimum total intuitionistic fuzzy transportation cost of IFTP (10) are obtained as follows:

$$\begin{aligned}&x_{11}=11,\ x_{12}=0,\ \ x_{13}=0,\ x_{14}=0,\\&x_{21}=1,\ \ x_{22}=10,\ x_{23}=0,\ x_{24}=0,\\&x_{31}=3,\ \ x_{32}=0,\ \ x_{33}=8,\ x_{34}=0,\\&x_{41}=1,\ \ x_{42}=0,\ \ x_{43}=0, x_{44}=11.\end{aligned} \tag{15}$$

By substituting the optimal solution (15) in the objective function of the IFTP (10), the total intuitionistic fuzzy transportation cost is determined as follows:

$$\tilde{Z}^{*I}=(116,204,302;68,204,372). \tag{16}$$

The degree of acceptance of the transportation cost for the decision maker (DM) increases if the cost increases from 116 to 204; while it decreases if the cost increases from 204 to 302. Beyond (126, 282), the level of acceptance or the level of satisfaction for the DM is zero. The DM is totally satisfied or the transportation cost is totally acceptable if transportation cost is 204. The degree of non-acceptance of the transportation cost for the DM decreases if the cost increases from 68 to 204 while it increases if the cost increases from 204 to 372. Beyond (68, 272), the cost is totally un-acceptable.

6 Conclusions

In this paper, a TP having uncertainty as well as hesitation in prediction of the transportation cost has been investigated. In the TP considered in this study, the values of transportation costs are represented by triangular intuitionistic fuzzy numbers and the values of supply and demand of the products are represented by real numbers. Here, we proposed a new solution approach for solving IFTP. We converted the IFTP under consideration into a MOLP problem with five objective functions and used a lexicographic approach to obtain the efficient solution of the resulting MOLP problem. In contrast to the existing methods [1, 8, 10, 14], the proposed algorithm in this study kept the valuable uncertain information in the optimization process. Here, we shall point out that the IFTP studied in this paper is not in the form of a problem whose demands and supplies are as triangular intuitionistic fuzzy numbers too. Therefore, further research on extending the proposed method to overcome these shortcomings is an interesting stream of future research.

References

1. Antony, R.J.P., Savarimuthu, S.J., Pathinathan, T.: Method for solving the transportation problem using triangular intuitionistic fuzzy number. Int. J. Comput. Algorithm **03**, 590–605 (2014)
2. Atanassov, K.T.: Intuitionistic fuzzy sets. Fuzzy Sets Syst. **20**, 87–96 (1986)
3. Ebrahimnejad, A.: A simplified new approach for solving fuzzy transportation problems with generalized trapezoidal fuzzy numbers. Appl. Soft Comput. **19**, 171–176 (2014)
4. Ebrahimnejad, A.: Note on "A fuzzy approach to transport optimization problem". Optim. Eng. **17**(4), 981–998 (2015)
5. Ebrahimnejad, A.: An improved approach for solving fuzzy transportation problem with triangular fuzzy numbers. J. Intell. Fuzzy Syst. **29**(2), 963–974 (2015)
6. Ebrahimnejad, A.: New method for solving fuzzy transportation problems with LR flat fuzzy numbers. Inf. Sci. **357**, 108–124 (2016)
7. Ebrahimnejad, A.: Fuzzy linear programming approach for solving transportation problems with interval-valued trapezoidal fuzzy numbers. Sadhana **41**(3), 299–316 (2016)
8. Ebrahimnejad, A., Verdegay, J.L.: An efficient computational approach for solving type-2 intuitionistic fuzzy numbers based transportation problems. Int. J. Comput. Intell. Syst. **9**(6), 1154–1173 (2016)
9. Ebrahimnejad, A., Verdegay, J.L.: A new approach for solving fully intuitionistic fuzzy transportation problems. Fuzzy Optim. Decis. Making (2017). https://doi.org/10.1007/s10700-017-9280-1
10. Hussain, R.J., Kumar, P.S.: Algorithmic approach for solving intuitionistic fuzzy transportation problem. Appl. Math. Sci. **6**(80), 3981–3989 (2012)

11. Jimenez, F., Verdegay, J.L.: Uncertain solid transportation problem. Fuzzy Sets Syst. **100**, 45–57 (1998)
12. Jimenez, F., Verdegay, J.L.: Solving fuzzy solid transportation problems by an evolutionary algorithm based parametric approach. Eur. J. Oper. Res. **117**, 485–510 (1999)
13. Kumar, P.S., Hussain, R.J.: Computationally simple approach for solving fully intuitionistic fuzzy real life transportation problems. Int. J. Syst. Assur. Eng. Manag. **17**(1), 90–101 (2016)
14. Singh, S.K., Yadav, S.P.: A new approach for solving intuitionistic fuzzy transportation problem of type-2. Ann. Oper. Res. **243**(1), 349–363 (2014)
15. Singh, S.K., Yadav, S.P.: Efficient approach for solving type-1 intuitionistic fuzzy transportation problem. Int. J. Syst. Assur. Eng. Manag. **6**(3), 259–267 (2015)

Context-Based Decision and Optimization: The Case of the Maximal Coverage Location Problem

María T. Lamata[1], David A. Pelta[1](✉), Alejandro Rosete[2], and José L. Verdegay[1]

[1] Department of Computer Science and Artificial Intelligence, Universidad de Granada, Granada, Spain
{mtl,dpelta,verdegay}@decsai.ugr.es

[2] Universidad Tecnológica de La Habana "José Antonio Echeverría" (Cujae), Havana, Cuba
rosete@ceis.cujae.edu.cu

Abstract. Every decision problem, understood as the need to take the best decision in some sense, leads to an optimization problem. There is a need to consider the "context" where each decision is made because it directly affects the underlying decision/optimization model with obvious implications in the change of the optimal solutions.

In this contribution this topic is further explored using the problem of locating emergency services (ambulances) in a set of available locations. A number of different contexts are considered and how they can be defined from an operational point of view is shown. The results obtained allowed to show how the best solutions of the problem may change.

Even using this simple example, we can conclude that the role of the context in decision/optimization problems and the need to properly define it should not be underestimated.

Keywords: Decision · Optimization · Contexts
Maximal coverage location problem

1 Introduction

Decision making and optimization are ubiquitous tasks in our modern societies. From which clothes to wear, to the selection of the cheapest flight or to find the best route to work, decisions must be taken with the aim of optimizing the results associated with the final decision.

M. T. Lamata—This work is partially supported by projects TIN2014-55024-P from the Spanish Ministry of Economy and Competitiveness and P11-TIC-8001 from Junta de Andalucía (both including European Regional Funds).

J. Medina et al. (Eds.): IPMU 2018, CCIS 855, pp. 330–341, 2018.
https://doi.org/10.1007/978-3-319-91479-4_28

Decision Theory has been traditionally linked with Economics, Statistics and Operations Research. Also the connections between Decision Theory and Artificial Intelligence, aiming at providing more or less autonomous behaviour to different devices or software artifacts (like non-player characters in videogames) are gaining increasing attention.

In many cases, most of the studies are focused on how to take the "best decision", which in turn, means that under every decision problem, there is an underlying optimization problem.

Another key aspect is that decisions are not taken in a vacuum space. There is always a *context* where a decision should be made. Recently Lamata, Pelta and Verdegay [7] explored the connections between decision problems, optimization and contexts. They showed the close relation between decision problems and optimization problems and they also illustrated how the so called "framework of behaviour" (the context, in what follows) where each decision is made directly affects the underlying decision/optimization model with obvious implications in the change of the optimal solutions. They develop the idea of this decision/optimization synergy in the context of fuzzy optimization problems.

The aim of this contribution is to further explore this topic using the maximal covering location problem (MCLP) as an example. MCLP aims at selecting a number of location in which a number of service units should be located in such a way that the highest coverage is attained. By service units, one may understand a variety of possibilities ranging from taxis stops or WiFi hotspots to supermarkets or hospitals. We will show how a number of different contexts can be defined from an operational point of view and, in turn, how the best solutions of the problem may change.

The paper is organized as follows: Sect. 2 briefly describe the main aspects about the connections among decisions, optimization and context. Then we present the standard mathematical formulation of the Maximal Coverage Location Problem (MCLP) [3] and we pose it later under the terms of the decision-making process approach presented in [7]. We will focus on the problem of allocating emergency services (ambulances) in a city. We consider different contexts, their corresponding mathematical formulations and the optimal solution for each case. Finally, Sect. 6 is devoted to conclusions and further research.

2 Preliminaries

A one step unipersonal decision problem can be represented by a sextet $(X, E, f, \leq, H, K)$ where X is the set of available actions to take, E is the environment, f measures the consequences or results produced by the actions, $\leq$ is the relation that sorts the results, H is the available information and K is the context in which the decision-maker makes decisions [6,7].

Departing from this formalization, the optimization problem associated with a given decision problem can be represented as a tuple $(X^{H_K}, E^{H_K}, f^{H_K}, \leq^{H_K})$, where H_k stands for a specific type of information (H) and a context (K).

In this optimization problem (we assume a maximization one), the objective is to find the action/alternative $x \in X^{H_K}$ such that

$$f^{H_K}(x^*) = Max^{\leq^{H_K}} \{f^{H_K}(x) : x \in X^{H_K}, f^{H_K} : X^{H_K} \times E^{H_K} \rightarrow U^{H_K}\} \quad (1)$$

Now, when different definitions for the elements of the problem are considered, several variations are obtained.

In order to properly describe our ideas, in what follows we rely on two assumptions: (1) we will focus in the particular case where the information H of the problem is complete and precise; and (2) as a consequence, there is only one possible state or environment E to consider.

Both elements H and E are removed from the formulation in order to simplify the notation. Then, the problem may be represented in a shorter way as $(X^K, f^K, \leq^K)$, where K, as before, is the context. In this optimization problem, the objective is to find that alternative $x \in X^K$ such that

$$f^K(x^*) = Max^K \{f^K(x) : x \in X^K; f^K : X^K \rightarrow U^K\} \quad (2)$$

In the following sections, we will illustrate how different contexts (the definition of K) can be represented in an optimization problem and how the corresponding solutions vary.

3 The Maximal Coverage Location Problem

In this contribution we will focus on the problem of locating emergency service units (e.g., ambulances). Departing from a set of J potential locations, we need to decide where p available ambulances should be located in order to maximize the population coverage. We say that a location is open if it is used to place an ambulance and it is closed if not.

The demand of an area (a node) w_j represents the number of emergency calls issued from that area in the past. A demand node is covered if there is an ambulance located at a distance or travel time closer/shorter than certain reference value S (usually called the service time or coverage radius).

This problem can be modeled as a Maximal Covering Location Problem (MCLP) [3]. MCLP has been widely studied [1,4,5,10] and applied to solve problems in different domains [2,8–12]. The mathematical formulation is as follows:

Sets

i, I - index and set of the demand nodes.
j, J - index and set of potential locations for the ambulances.
N_i - $\{j \in J | d_{ij} \leqslant S\}$ the set of potential ambulance locations that can cover the node i within the time or distance S, d_{ij} is the distance between the node i and the potential location for the facility j.

Input Parameters

p - the number of ambulances to be located
S - the maximum allowed time or distance to respond to a request.
w_i - value that represent the demand associated to node i.

Decision Variables

x_j - 1 if an ambulance is located at the node j, 0 otherwise.
y_i - represents the coverage of node i, 1 if node is covered $(\exists_j \mid x_j = 1 \wedge j \in N_i)$, 0 otherwise.

Mathematical Model

$$\max Z = \sum_{i \in I} w_i y_i \tag{3}$$

subject to

$$\sum_{j \in N_i} x_j \geqslant y_i \quad \forall i \in I \tag{4}$$

$$\sum_{j \in J} x_j = p \tag{5}$$

$$x_j = \{0, 1\} \quad \forall j \in J \tag{6}$$

$$y_i = \{0, 1\} \quad \forall i \in I \tag{7}$$

The objective function (3) maximizes the demand covered by the set of established facilities (ambulances). Constraints (4) state that one or more facilities will be located within the distance or travel time pre-defined S from the demand node i. Constraint (5) ensures that the number of ambulances to be located is p. Finally, the constraints (6) and (7) indicate binary restrictions on the decision variables x_j and y_i.

4 Possible Context: Examples

Before trying to solve the problem in a given context K, it is necessary to provide a proper definition of the elements of the tuple $(X^K, f^K, \leq^K)$. Such definitions depart from a "context-independent" $(X, f, \leq)$ where the elements are:

- X: a binary vector $(x_1, x_2, \ldots, x_n)$ of size n that represents the set of decisions x_i with respect to use (or not) a particular location point i.
- $\sum_{i=1}^{n} x_i = p$,
- f: $\sum_{j=1}^{m} y_j w_j$, where w_j is the demand in the demand point j, and y_j is a binary variable that reflects that the demand j is covered by an open location i, i.e. $Y_j = 1 \iff \exists_{i \in N} x_i = 1 \wedge D_{ij} \leq S$, $y_j = 0$ otherwise.
- $\leq$: As the objective is to maximize the demand, the order relation $\leq$ between two decisions A and B is as follows, $A \leq B \iff f(A) \geq f(B)$

The decisions and the corresponding demands covered in the "context independent" MCLP can be used as reference values for what can be attained.

The key question here is how the value K is defined in every context considered. In the following subsections we describe an instance of the MCLP problem and the particular contexts considered: an ethical context, an emergency context, and a sustainability context.

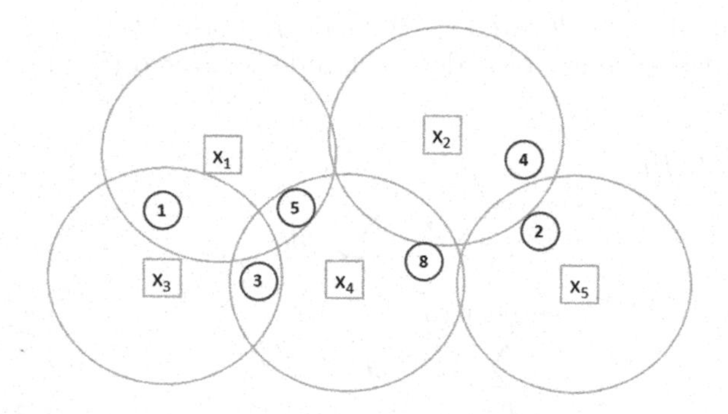

Fig. 1. A simple example of the maximal covering location problem.

4.1 Context Independent

This case represents a reference situation, where no specific context is considered. Figure 1 presents an instance of the MCLP. In this example there are five locations where the ambulances may be located ($n = 5$) which are represented by rectangles. Each big circle (with radius S) represents the area covered by such locations. There are six demand points ($m = 6$) represented by small circles with a number inside representing the demands (w_j): $w_1 = 1$, $w_2 = 5$, $w_3 = 4$, $w_4 = 3$, $w_5 = 8$, and $w_6 = 2$.

Figure 2 shows an example where $p = 2$ ambulances are located. The selected locations $x_2 = 1, x_4 = 1$ are marked with filled rectangles and the demand covered is highlighted with a darker circle.

The number of potential decisions for this "context-independent" example is $2^n = 2^5 = 32$ binary vectors indicating if each one of the 5 locations points is used or not. In this case, we consider that only two ambulances are available ($p = 2$). Thus, some of these 32 vectors do not represent feasible solutions (they violate the constraint $\sum_{i=1}^{n} x_i = p$ either due to the use of a higher or lower number of ambulances). Consequently, the number of feasible solutions is $\binom{n}{p} = \binom{5}{2} = 10$. The space of feasible solutions contains 00011, 00101, 00110, 01001, 01010, 01100, 10001, 10010, 10100 and 11000. Table 1 shows the coverage attained by each feasible solution. The optimal solution 01010 with a coverage of 20 units is the one displayed in Fig. 2.

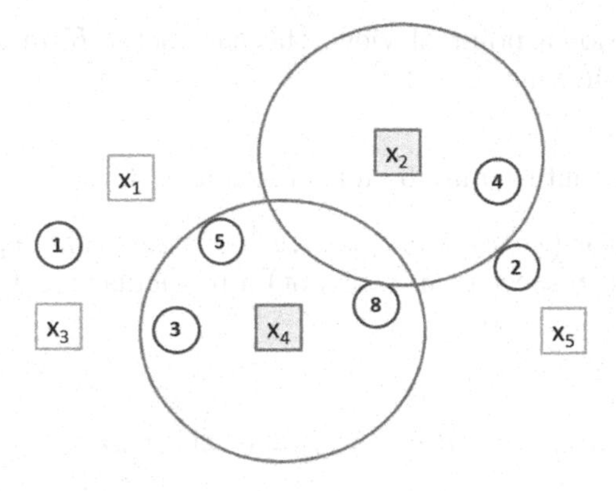

Fig. 2. A simple example of maximal covering location problem with $p = 2$. Ambulances are located at x_2, x_4.

Table 1. Set of feasible solutions for the MCLP instance with $p = 2$. The optimum is marked in bold.

Decision	Covered demand	Uncovered demand	f
00011	3+5+8+2	1+4	18
00101	1+3+2	4+5+8	6
00110	1+3+5+8	4+2	17
01001	4+2	1+3+5+8	6
01010	**4+3+5+8**	**1+2**	**20**
01100	4+1+3	5+8	8
10001	1+5+2	3+8+4	8
10010	1+3+5+8	4+2	17
10100	1+3+5	4+8+2	9
11000	1+5+4	3+8+2	10

4.2 Ethical Context

Motivation
The ethical context arise when the decision is not taken just from a maximization point of view. An example could be the need to provide an ambulance in a neighborhood where most of the population is old.

As a consequence a new location point appears, and an ambulance should be located there. In the simplest case, only one specific neighborhood should be covered but this may be extended to considered that a set C of new locations ($|C| = c$, $c \leq p$) is now available to place ambulances on them. Thus, the set of available locations is $R = N \cup C$ and the number of available locations $r = n + c$. The ethical context imposes that one ambulance must be placed in the new locations.

From an operational point of view, the parameter K in an ethical context gives raise to the following model.

Model
The model for this context may be formalized as follows:

- X: a binary vector $(x_1, x_2, \ldots, x_n, \ldots, x_r)$ of size r that represents the set of decisions x_i with respect to use (or not) a particular location point i.
- $\forall_{i \in C}\, x_i = 1$
- $f = \sum_{j=1}^{m} y_j w_j$
- $\sum_{i=1}^{n} x_i = p - c$
- $y_j = 1 \iff \exists_{i \in R}\, x_i = 1 \wedge d_{ij} \leq S$, $y_j = 0$ otherwise
- $\preceq^k = \leq$

Example
For illustrating the ethical context, let's suppose that the local government identified that a new location x_6 should be considered and, in this ethical context, an ambulance should be located there, so $x_6 = 1$. This is shown in Fig. 3.

This implies that $r = n + c = 5 + 1 = 6$. In this simple case, the real decision is where to place the remaining ambulance, i.e. originally $p = 2$ and $c = 1$, then we have $p - c = 1$ ambulances to place. The optimal solution for this modified instance is 000101 presented in Fig. 3(a), with a coverage of $1+3+5+8=17$.

In this case the coverage is lower than in the context-independent problem but this is not always the case. For example, if the location of the neighborhood to cover varies, as shown in Fig. 3(b) (now, in the right-most filled rectangle), the optimal solution 000101 has a coverage of $3+5+8+4+2=22$ which is greater than the coverage attained in the context independent problem.

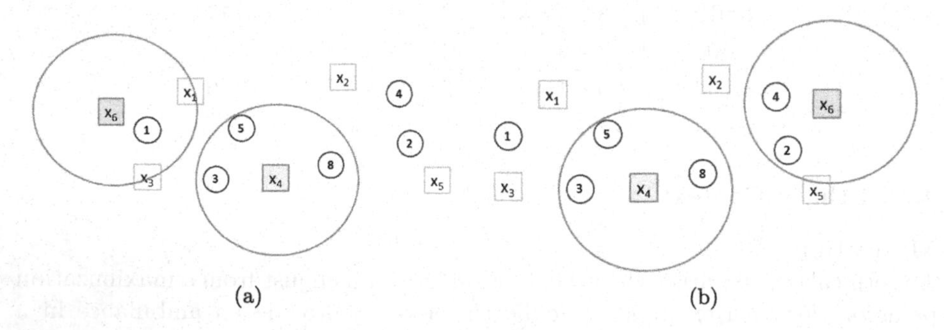

Fig. 3. Two examples of the ethical context where a new location appears and needs to be covered $x_6 = 1$. In (a), the total coverage is lower than in the context independent case. While in (b) the total coverage is higher.

4.3 Emergency Context

Motivation

In an emergency context, for example a heavy snow, it may happen that a subset $U \subset N$ of location points becomes unavailable. This implies that $u = |U|$ locations must be closed

From an operational point of view, the parameter K in an emergency context gives raise to the following model.

Model

The model is addressed as follows.

- $X^k = X$
- $\forall_{i \in U} \; x_i = 0$
- $\sum_{i=1}^{n} x_i = p$
- $f^k = f$ and $\leq^k = \leq$

Example

Let's suppose now that due to a contingency (for example a heavy snow), the locations points $U = \{1, 2\}$ are unavailable. This implies that ambulances will not be located there ($x_1 = 0$ and $x_2 = 0$). Under the emergency context, the optimal solution is 00011 presented in Fig. 4, attaining a coverage of 18 units.

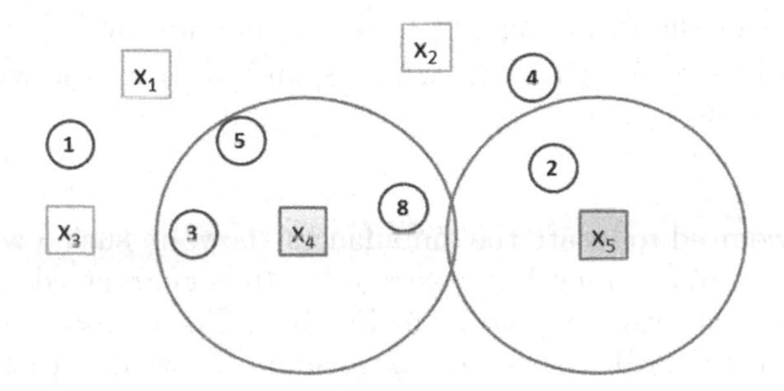

Fig. 4. An instance of the emergency context where two location points are unavailable ($x_1 = 0$ and $x_2 = 0$).

An emergency context may have much more implications than the one considered here. Different combinations of several conditions (less ambulances, modified distances, unavailable locations, reduced radius of coverage) may occur in real emergency situations. It is out of the scope of this contribution the consideration of such combinations.

4.4 Sustainability Context

Motivation
In this context, a decision maker should take now the best decision without compromising (in some sense) the future. This context implies that the decision maker has certain expectations about how certain conditions/parameters will vary in the future. For example, if snowfall happens frequently, it is very predictable its occurrence in the future.

In the problem of locating ambulances, the sustainability context may imply that the current location of ambulances should guarantee a minimum level of coverage (l) when a snowfall happens. As the reader may notice, information about the change of conditions should be available. For example, we may know how the distances d_{ij} between each possible location and each demand node will change (denoted as d^T_{ij}) due to snow.

From an operational point of view, the parameter K in a sustainability context gives raise to the following model where an additional constraint appears.

Model
This situation may be formalized as follows.

- $X^k = X$
- $\sum_{i=1}^{n} x_i = p$
- d^T_{ij} are the distances in the future (when the snowfall)
- the coverage in the future should have a minimum level: $\sum_{j=1}^{m} y^T_j w_j \geq l$, where $y^T_j = 1 \iff \exists_{i \in N}\, x_i = 1 \wedge d^T_{ij} \leq S$, and $y^T_j = 0$, otherwise
- $f^k = f$ and $\preceq^k = \preceq$

Example
Suppose that we need to locate the ambulances "now" in such a way that, after a heavy snow, a minimum level of coverage $l = 16$ is guaranteed.

Let's assume the situation shown in Fig. 5(a). The original solution having $x_2 = 1, x_4 = 1$ attained a coverage of 20 units. Now, due to the snow, the path (marked with a line) connecting demand node 2 (with $w_2 = 5$) with the potential location x_4 is not available anymore. In other words, $d^t_{42} = \infty$. Under this circumstance, the coverage will degrade down to 15 units.

However, if we had considered the solution shown in Fig. 5(b), we observe that if demand node 2 is now served by location x_4, the solution attains a coverage of 17 units. When we consider the new conditions ($d^t_{42} = \infty$) the coverage is also 17 (thus condition $17 \geq 16$ holds), because demand node 2 is served by x_1.

Table 2 shows, for every potential solution, the current and the future coverage attained. In this example, just one feasible solution exists: 10010. Other solutions do not guarantee the expected coverage required in the sustainability context.

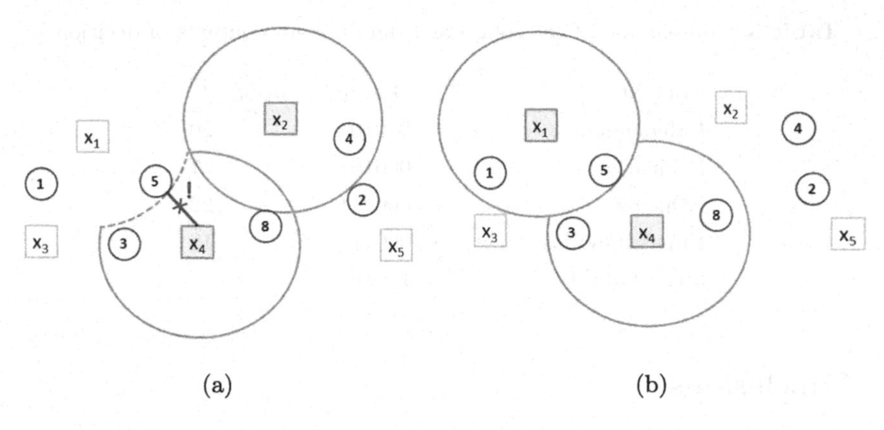

Fig. 5. The original solution (a) having $x_2 = 1, x_4 = 1$ cannot satisfy the level of coverage desired when the conditions changed (the path marked in red is not available anymore). However, the solution in (b), although suboptimal in "normal conditions" fits the requirements in the sustainability context. (Color figure online)

Table 2. Solutions in the sustainability context with $l = 16$. Just the first solution is feasible (the future coverage is higher than l).

Solutions	f	Future coverage
10010	**17**	**17**
00011	18	13
00101	6	6
00110	17	12
10010	6	6

Solutions	f	Future coverage
01010	20	15
01100	8	8
10001	8	8
10100	9	9
11000	10	10

5 Summary of the Contexts

Table 3 show the optimal solution and the coverage attained in the different contexts exemplified.

It may be observed that even in this illustrative example, the optimal solution changes as a function of the context considered.

If we analyze the coverage, the optimal solution in the second example of the ethical context is better than the one from the context independent case. However, in the remaining cases, the coverage value attained with the context independent solution represents an upper bound for the attainable coverage in the other contexts.

If we analyze the solutions, we observe that $x_4 = 1$ is part of the optimal decision in the five examples considered. In other words, we can consider x_4 as a critical location. In turn $x_3 = 0$ in all the cases, meaning that it is not an "attractive" location. No other clear pattern is detected for the other potential locations.

Table 3. Comparison of the solutions from different contexts of decision

Context	Optimal solution	*f*
Independent	01010	20
Ethical1	000101	17
Ethical2	000101	22
Emergency: less locations	00011	18
Sustainability	10010	17

6 Conclusions

In [7], the authors explored the close relation between decision problems and optimization problems and they also illustrate how the context, where each decision is made directly affects the underlying decision/optimization model with obvious implications in the change of the optimal solutions. They developed the idea of this decision/optimization synergy in the context of fuzzy optimization problems.

In order to emphasize the role of the context in decision/optimization, in this contribution we further explored the topic using the problem of locating p emergency services (ambulances) in a set of available locations. The problem was modelled as a maximal covering location problem (MCLP). We considered a number of different contexts and we showed how they can be defined from an operational point of view. The results obtained allowed us to show how the best solutions of the problem may change.

Even using this simple example, we can conclude that the role of the context in decision/optimization problems and the need to properly define it should not be underestimated.

References

1. Bagherinejad, J., Shoeib, M.: Dynamic capacitated maximal covering location problem by considering dynamic capacity. Int. J. Ind. Eng. Comput. **9**(2), 249–264 (2018)
2. Basu, S., Sharma, M., Ghosh, P.S.: Metaheuristic applications on discrete facility location problems: a survey. Opsearch **52**(3), 530–561 (2015)
3. Church, R., ReVelle, C.: The maximal covering location problem. Pap. Reg. Sci. Assoc. **32**, 101–118 (1974). https://doi.org/10.1111/j.1435-5597.1974.tb00902.x
4. Colombo, F., Cordone, R., Lulli, G.: The multimode covering location problem. Comput. Oper. Res. **67**(C), 25–33 (2016). https://doi.org/10.1016/j.cor.2015.09.003
5. Guzmán, V.C., Pelta, D.A., Verdegay, J.L.: An approach for solving maximal covering location problems with fuzzy constraints. Int. J. Comput. Intell. Syst. **9**(4), 734–744 (2016). https://doi.org/10.1080/18756891.2016.1204121

6. Lamata, M.T., Verdegay, J.L.: On new frameworks for decision making and optimization. In: Gil, E., Gil, E., Gil, J., Gil, M.Á. (eds.) The Mathematics of the Uncertain: A Tribute to Pedro Gil, pp. 629–641. Springer, Cham (2018). https://doi.org/10.1007/978-3-319-73848-2_58
7. Lamata, M., Pelta, D., Verdegay, J.: Optimisation problems as decision problems: the case of fuzzy optimisation problems. Inf. Sci. (2017). https://doi.org/10.1016/j.ins.2017.07.035
8. Li, X., Zhao, Z., Zhu, X., Wyatt, T.: Covering models and optimization techniques for emergency response facility location and planning: a review. Math. Methods Oper. Res. **74**(3), 281–310 (2011). https://doi.org/10.1007/s00186-011-0363-4
9. Peng, H., Qin, Y., Yang, Y.: Relationship between set covering location and maximal covering location problems in facility location application. In: Qin, Y., Jia, L., Feng, J., An, M., Diao, L. (eds.) Proceedings of the 2015 International Conference on Electrical and Information Technologies for Rail Transportation. LNEE, vol. 378, pp. 711–720. Springer, Heidelberg (2016). https://doi.org/10.1007/978-3-662-49370-0_74
10. Shi, Y., Tian, Y., Kou, G., Peng, Y., Li, J.: MCLP extensions. In: Optimization Based Data Mining: Theory and Applications, pp. 133–156. Springer, London (2011). https://doi.org/10.1007/978-0-85729-504-0_8
11. Sorensen, P., Church, R.: Integrating expected coverage and local reliability for emergency medical serviceslocation problems. Soc. Econ. Plann. Sci. **44**(1), 8–18 (2010)
12. Zhang, B., Peng, J., Li, S.: Covering location problem of emergency service facilities in an uncertain environment. Appl. Math. Model. **51**(Suppl. C), 429–447 (2017). https://doi.org/10.1016/j.apm.2017.06.043

Increasing Performance via Gamification in a Volunteer-Based Evolutionary Computation System

Mario García-Valdez[1(✉)], Juan Julian Merelo Guervós[2], Lucero Lara[1], and Pablo García-Sánchez[3]

[1] Instituto Tecnológico de Tijuana, Tijuana, BC, Mexico
mario@tectijuana.edu.mx
[2] Universidad de Granada, Granada, Spain
jmerelo@geneura.ugr.es
[3] Universidad de Cádiz, Cádiz, Spain
pablo.garciasanchez@uca.es

Abstract. Distributed computing systems can be created using volunteers, users who spontaneously, after receiving an invitation, decide to provide their own resources or storage to contribute to a common effort. They can, for instance, run a script embedded in a web page; thus, collaboration is straightforward, but also ephemeral, with resources depending on the amount of time the user decides to lend. This implies that the user has to be kept engaged so as to obtain as many computing cycles as possible. In this paper, we analyze a volunteer-based evolutionary computing system called NodIO with the objective of discovering design decisions that encourage volunteer participation, thus increasing the overall computing power. We present the results of an experiment in which a gamification technique is applied by adding a leader-board showing the top scores achieved by registered contributors. In NodIO, volunteers can participate without creating an account, so one of the questions we wanted to address was if the need to register would have a negative impact on user participation. The experiment results show that even if only a small percentage of users created an account, those participating in the competition provided around 90% of the work, thus effectively increasing the performance of the overall system.

Keywords: Distributed evolutionary algorithms
Volunteer computing · Socio-technical systems

1 Introduction

The World Wide Web is an increasingly reliable and high-performance operating system for running distributed applications. Besides the maturity of HTTP, the underlying protocol itself, there are other factors: the JavaScript virtual machine embedded in every browser and the REST interface, providing a *de facto* standard interface for communication. A distributed system based on the browser can

J. Medina et al. (Eds.): IPMU 2018, CCIS 855, pp. 342–353, 2018.
https://doi.org/10.1007/978-3-319-91479-4_29

be easily created using these mechanisms, which lately have been enhanced by WebSockets for lower latency; you can start to gather users just by announcing the URL. This approach for creating distributed experiments is called *volunteer*, *cycle-scavenging*, or *opportunistic* computing [24] and it dates back, in different shapes and underlying mechanisms, to the origin of the Internet [1].

In this context, we are mainly interested in evolutionary algorithms [19, 22, 26]. The fact that they are population-based makes them suitable for a straightforward distribution of work to different clients, and since they have an asynchronous nature, there is no considerable impact on their performance, and might even be beneficial for a number of reasons, including an increased diversity [4]. All these properties make them ideal candidates for volunteer computing setups such as the one presented in this paper.

But even if adaptation is straightforward, the application of volunteer computing techniques to the development of distributed evolutionary algorithms still has several open issues. An important research aspect is approaching the problem as a socio-technical system [18, 27], which integrates user decisions and behavioral patterns in the model; this includes trying to optimize the number of users or the overall time they contribute in a particular experiment. The challenge is to design a system that, whatever the number of users available and willing to perform the experiment, it is able to maximize their contribution to the evolutionary algorithm. One obvious way of achieving that is trying to find as many users as possible. But another approach, and the one that we will be following in this paper, is to delve into the socio-technical nature of the system, using gamification techniques to try and improve the amount of time every user will lend to the system. However, there is a trade-off with gamification, since users have to be identified in some way to participate, and thus it has to include some way of user authentication, which is a hurdle that might take users away. However, the open design of the system we use means that we can actually mix anonymous (with no gamification) and authenticated (gamified) users, with only the small overhead of adding the gamification mechanism.

This trade-off between ease of use, allowing users to participate by just visiting a web page, and customization, making registration a requirement in order to participate, is what we will analyze in this paper. In a socio-technical system such as this one, every decision the user has to take has an influence on the overall performance, and registration could be seen as a hurdle to volunteers trying to participate in an ephemeral experiment, so it is a challenge to analyze what actually happens in this case; quite clearly, the inclusion of gamification techniques has an impact on page loading and complexity of server-side programming; if the number of users or the amount of time they contribute to the experiment is not enough, it would be better to revert NodIO to a purely anonymous system. The rest of the paper is organized as follows: The state of the art in opportunistic/volunteer distributed evolutionary computation (EC) is presented next. The proposed Gamification Technique is described in Sect. 3. Section 4 will describe the framework and the problem used in the experiments, which are publicly available under a free license. The results of each of the steps in the incremental design are presented in Sect. 5, to finally wrap up with the conclusions.

2 Related Work

Volunteer computing involves users who decide to run a program that acts as a client or as a peer in a distributed computing experiment and, as such, has been deployed in many different ways since the beginning of the Internet, starting with the SETI@home framework for processing extraterrestrial signals [1], which used a client that acted as an screensaver. However, the dual facts of the introduction of JavaScript as a universal language for the browser and the browser itself as both an ubiquitous web and Internet client has made this combination the most popular for volunteer computing frameworks in what is generally described as BBVC or Browser-Based Volunteer Computing [8].

In order to attract volunteers, many systems rely on downloadable clients; Fabisiak [8] analyzes the different capabilities of Internet and browser-based systems, including their ability to handle large sets of input data and time-consuming problems, concluding that BBVC is mostly apt for short, data-light tasks. As indicated in the introduction, the human is an integral part of the system, which we can consider a *human computation* [23]. Giving the user more control, and adding visibility to their participation might have positive effects. In fact, the BOINC project applies this technique by having a web page presenting the "Top 100 multi-project BOINC participants" in which the name of the volunteer, the number of projects, GFLOPS, country and team are displayed.

Several authors have already described systems using JavaScript either for unwitting [2,3,12] or volunteer [13,15] distributed EC and it has been used by several authors ever since. In fact, [21] performs an analysis of what is called *Gray computing* covering aspects of feasibility, cost-effectiveness, change in the userś experience and the architectural optimization needed, concluding that the computing power available is vast and it can be cost-effective to use it.

Many of the systems described above do not go any further than trying to find out how many users join the effort, and how many of them the system can support. In fact, systems such as the one described in [15] had severe scaling issues; some of them also tried to find out how much time was needed to find the solution or, alternatively, how many users would be needed to be competitive against single-user single-computer implementations of the same algorithm. Lately, researchers have tried to integrate volunteer computing techniques as a part of a larger distributed evolutionary computing effort [14]. In fact, systems using volunteers exclusively, exhibit a certain amount of unpredictability, and they might be better used in combination with ready-available computing power.

3 Proposed Gamification Technique

A definition of Gamification given by Huotari [11] is "Gamification is the process by which gaming concepts are brought to real world tasks associated with real people". Gamification uses game design elements out of the domain of games with the objective of enhancing the user's experience, engagement, productivity, learning, among others. Deterding et al. proposes the following definition: "Gamification is the use of game design elements in non-game contexts" [6]. Gamification techniques in a volunteer context seeks to persuade users to use their natural desire to compete, learn and socialize in given non-game context application [5,9]. By Making the rewards for tasks achievements visible to other players or providing leader boards are ways of encouraging players to compete [10].

In this work, a leader board was implemented in order to promote competition, presenting only the all-time top five participating users.

Another gamification technique employed in this work is based on a rewarding mechanism [7]. In general rewards consist of a reputation system with score points, levels and leader boards. Points are awarded to users in response of the accomplishment of certain activities that need to be encouraged. In these case a point is awarded for each HTTP PUT Request sent to the server. Levels are a long term achievement, in this case the level depends on the score:

$$\text{level}(\text{score}; a, b) = \begin{cases} 0, & \text{score} \le a \\ 2(\frac{\text{score}-a}{b-a})^2, & a \le \text{score} \le \frac{a+b}{2} \\ 1 - 2(\frac{\text{score}-b}{b-a})^2, & \frac{a+b}{2} \le \text{score} \le b \\ 1, & \text{score} \ge b \end{cases}$$

The function returns a normalized value that is multiplied by the maximum level in this case 100. The variables $a = 100$ and $b = 7360$ are set to give users a rapid increase of levels at the beginning. The effect a and b have on the results was not determined, and other techniques could be used.

4 Experimental Setup and Execution

For this experiment the gamification technique is applied to the NodIO volunteer computing framework, using the particular version described in detail in [16,17]. The objective of the experiment is to test the kind of impact applying a gamification technique has on user engagement in this particular application. In order to apply a rewarding system user authentication had to be developed first. In earlier designs this functionality was not desired because it was seen as a barrier for participation. After all, the advantage of a browser based volunteer system is precisely the minimum amount of user intervention needed to start. The last thing a user wants to see is yet another registration form. Then, the first design decision for this version is that registration is optional and simple.

After login to the NodIO web application, registered users are welcomed by their name, and they can see their current score and level. All users can see the leader board in a modal dialog. While the modal is opened, the scores are refreshed every second. All other functionality is available to all users and is the same as the previous NodIO version, showing the state of the current algorithm. In the front page a link to the open source code of the experiment was also available in https://github.com/lucero21/login-master.

4.1 Problem

In the browser, each page visiting the experiment loads an HTTP Web Worker that runs a local island of an evolutionary algorithm to solve a multi-modal problem called *l-trap* which has been used extensively as a benchmark for evolutionary algorithms [20]. This function counts the number of bits in a sequence l and assigns the local maximum a if it has 0 bits and the global maximum b if it has l bits, this makes the fitness fall into a *trap* as the number of bits increases, and decreasing linearly until a change in slope is reached at point z, adding a deceptive component for evolutionary algorithms. In order to increase difficulty, trap functions can be concatenated, in our case we have used 40 concatenated traps. The trap function is defined as:

$$f(u) = \begin{cases} \frac{a}{z}(z-u) & \text{if } u \leq z \\ \frac{b}{l-z}(u-z) & \text{otherwise} \end{cases}$$

Each local GA had the following parameters, the initial population was randomly generated with a size between 128 and 256 individuals, the period to send individuals to the server was set at 100 local generations, the parameters for the trap function are fixed at $l = 4$, $a = 1$, $b = 2$, $z = 3$ with a chromosome length of 160 bits.

4.2 Experiment Execution

We made a call to participation on November 23th, 2016 through the authors' social networks: "Asking again for your help, we are conducting a computational experiment that requires computer power. Can we borrow some of your CPU? just visit the web page (link) and leave the tab open. Be part of the TOP TEN, register so we can track your participation. The experiment will run until November 27th, Thanks!" As you can see, the message mentioned a deadline; a link that could be used to register was also included. Users visiting the page after the deadline were presented with a thank you message and a final leader board. Considering the number of friends, followers and the organic sharing of the posts, we estimate the post was visited by around 2000 users. Only some of them registered, but in fact relatively small samples of users (10±) are sufficient for discovering 80% of usability problems [25]. In fact, we were not so much interested in making a precise model but on how the behavior of the registered users was different from non-registered ones; in non-parametric comparisons, using 15 sample units is in general good enough, as seems to be the case in this particular experiment, where we were also interested in computing the number of users who actually registered, proving that, in fact, for most users the fact that they have to register represents a bigger effort they seem to be able to muster for this kind of experiments.

5 Results

At the end of the experiment the resulting log file contained 933,513 contributions considering both registered and anonymous volunteers. Registered users were responsible for about 90% of the contributions, while being only 16% of the total number of unique users (assuming each IP is from a single user).

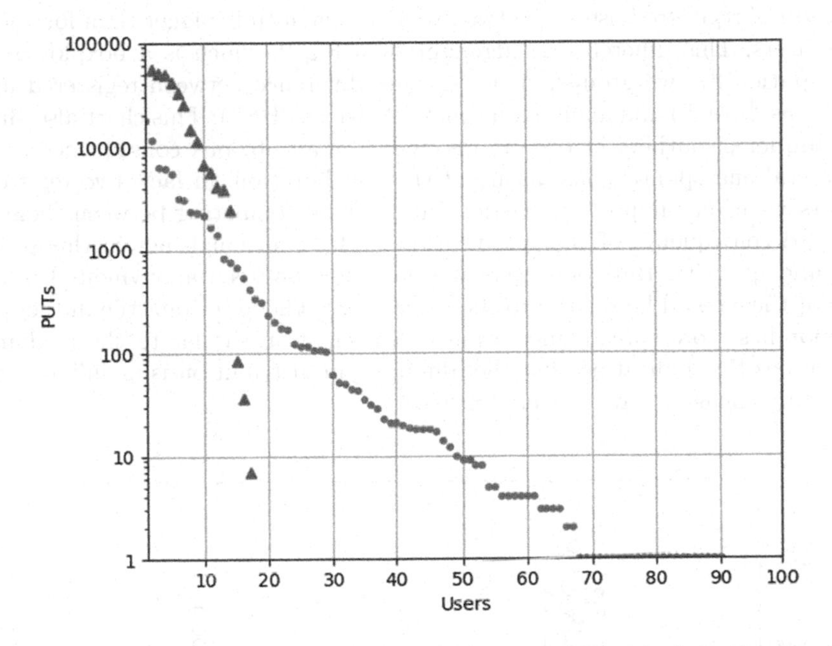

Fig. 1. Usage by registered (green triangles) and anonymous (red dots) users. Users are ranked by the number of requests sent, in the x axis is the rank, and in the y axis number of PUTs in a logarithmic scale. (Color figure online)

Contributions are counted by the number of HTTP PUT requests sent to the server. In Fig. 1 a comparison between the contribution of registered vs anonymous users is presented.

The first observation arises from the length of the x axis. There were only 18 registered users vs 91 distinct IPs; the exact number of anonymous users is upper-bounded by that amount but is not known precisely as we only recorded the IP of the request, we did not use cookies or other means of identification since it was not actually needed for the experiment. Besides, a registered user could sometimes be anonymous too, which could be caused by the fact that the first time they visited the application they automatically begin to work as anonymous, so the number of non-anonymous users, 91, is actually the amount of unique IPs that had 'anonymous' as the user. In the plot of anonymous users, some of the lasts in the rank could in fact be registered users, momentarily participating as anonymous. Even if this is not considered, only about 16% of users attending the call decided to participate as named users and thus enter the game dynamics. The second observation is apparent in the y axis of Fig. 1. There is a notable difference in the amount of participation between the two groups, and also in the slope of the plot. An independent-samples t-test was conducted to compare participation, and there was a significant difference in the scores for anonymous users ($M = 1057.97$, $SD = 3669.86$) and non-anonymous users ($M = 46513.55$, $SD = 93537.20$); t-test = 4.70, $p = 7.5e{-}06$; that is, the contribution of registered users was two orders of magnitude bigger than for anonymous users. This difference is highlighted in Fig. 2, which is a boxplot of the participation by two groups, shows the big difference between registered users (shown as "users") and anonymous ones (shown as "IPs"). This chart also shows the number of outliers in the groups, which is an obvious consequence of the ephemeral and spontaneous nature of the collaboration; in fact, two registered outliers were very important participants, maybe competing between them, an expected consequence of the gamification, in fact, although not in this precise form and quantity. But there were more outliers between anonymous IPs, and some of them could be related to the same users. The lower quartile in the users box-plot has more spread, indicating a participation similar to the median of anonymous IPs, indicating that the difference in distribution is mainly due to a few highly engaged and registered users.

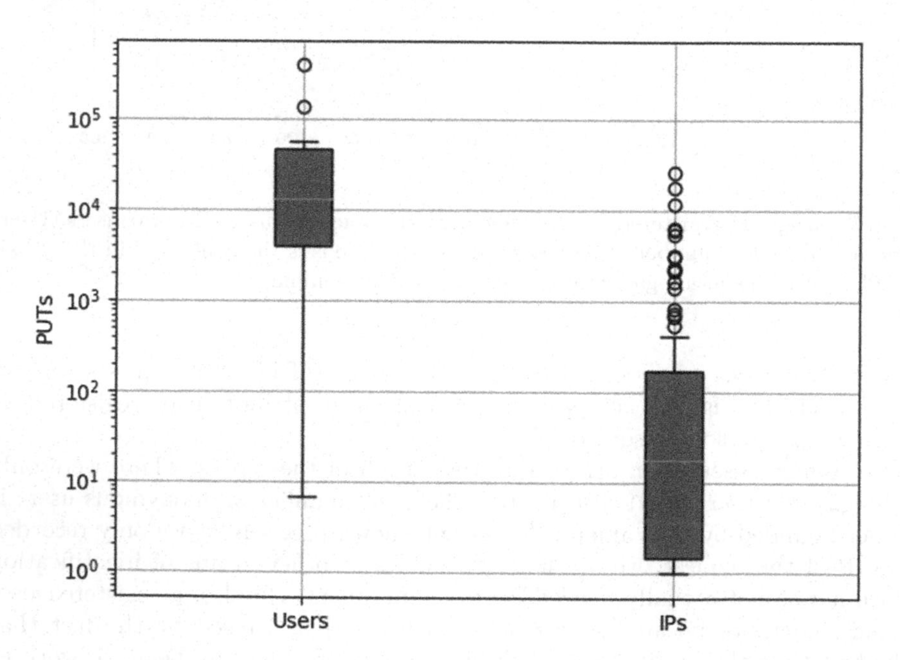

Fig. 2. Box-plot of the number of requests sent by registered users and anonymous IPs. In the y axis are number or PUTs in a logarithmic scale.

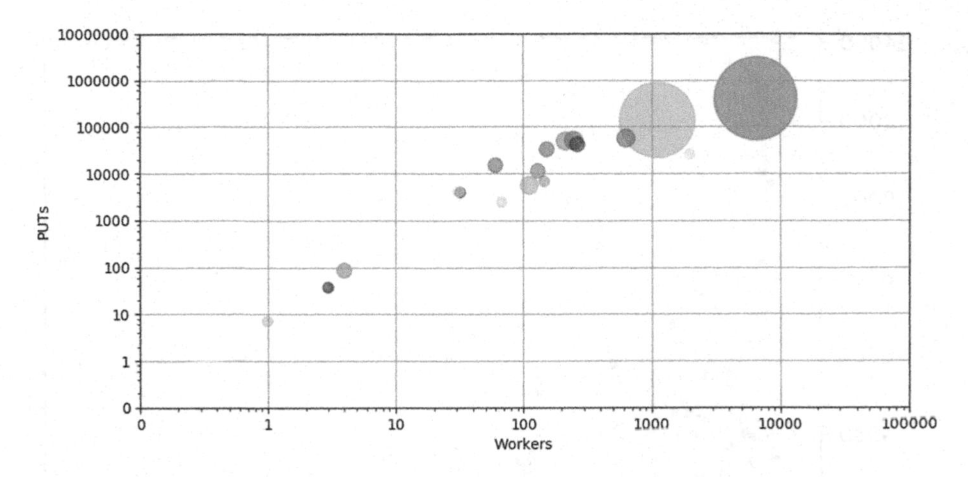

Fig. 3. Participation of registered users. Each circle represents a user relating in the x axis the number of workers used and in the y axis number or PUTs in a logarithmic scale. The area of the circle is the number of unique IPs associated to every user.

Figure 3 gives an interesting view on the amount of resources shared by registered users. The area of the circle is proportional to the number of unique IPs used by each user; the user with the most participation used a total of 66 different IPs. This number could be related with the number of different devices used during the participation. There are some users that using less devices participated more, this could mean a more powerful device, faster Internet connection or simply more time spent participating. At any rate, the chart seems to indicate that the main mechanism registered users employ to increase their rank seems to be using all devices available, and maxing out their CPUs by increasing the number of workers. This indicates that, in fact, registered users who have incurred in additional time expenses by performing registration and login are ready to work even more by manually deploying more workers and using more computers, which is another proof of the socio-technical nature of volunteer computing systems.

User engagement is related to the amount of time a user spends in the application. Figure 4 shows two important aspects of engagement, the overall time spent and the amount of resources shared in that time. Registered users were more engaged during the experiment. They contribute more resources during longer periods of time.

The amount of participation of the top ranked user is presented in 15 s time slots in Fig. 5. In the beginning this user was using more than 30 workers, this means that more than 30 tabs of the page were open at the same time, perhaps using several computing devices. Some users reported that they wanted to test the limits of their own systems, checking the percentage of CPU they were using.

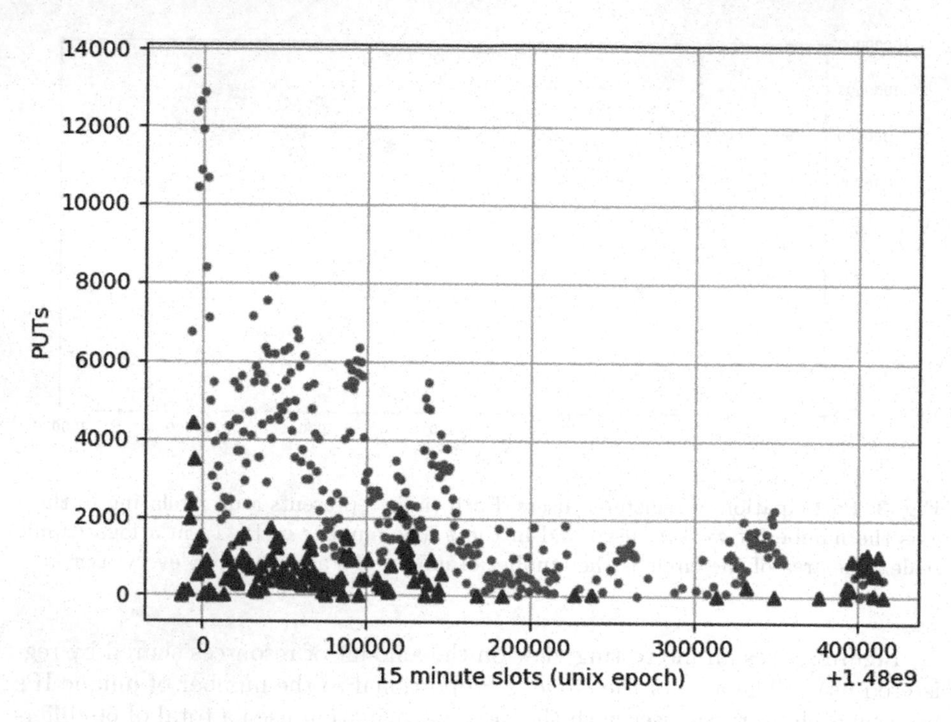

Fig. 4. Number of PUTs by User (green dots) or IP (blue triangles) in 15 min slots. (Color figure online)

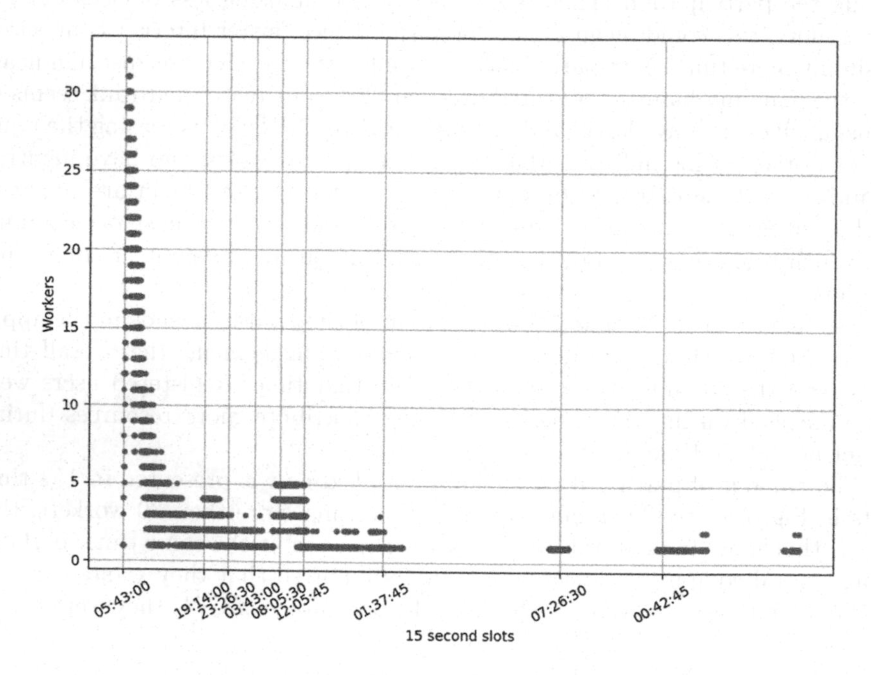

Fig. 5. Number of workers used by the top ranked user in 15 min slots.

6 Conclusions, Discussion and Future Work

In this paper we have added gamification to a volunteer computing system via a leaderboard of contributions of registered users. The objective of this work was to test the kind of impact applying a gamification technique has on user engagement in the NodIO volunteer evolutionary computing framework; in concordance with the results obtained in other browser-based volunteer systems, after applying the gamification techniques, registered users participated much more intensely than anonymous ones.

In fact, we have proved that, although only a minority of the users opt for registration, their individual contributions are on average much higher than for non-authenticated/non-gamified users. The fact that less than one fifth of users actually registering would discard the exclusive use of this feature; however, it is an interesting addition to the volunteer system that in fact enhances its computing power.

The fact that applying a gamification technique improved user participation highlights the social nature of volunteer computing systems, piling on the fact that successful *social clouds* must embed games and other social mechanisms to enhance participation. That is why future lines of work will look for intermediate ways of including gamification without registration, as well as some ways of making the user increase control over the algorithm that would go beyond simply staying for longer or leaving the page. Since it is a sociotechnical system, including social features in it and studying emerging social networks is an interesting line of work too. Other factors, such as the language used, the way the experiment is announced, are also very important in this social context, which is why they will be included in future lines of work too, along with the study of the behavior of users as they are participating in the web system.

Another line of work would be to study the possible negative effects of using gamification techniques to improve engagement, like cheating or literally *gaming* the system to defeat competition. We already found some hints of this behavior, but more subtle effect could be taking place. Finally, the refinement of the proposed framework will need more case studies and further multi-disciplinary research.

Acknowledgments. This work has been supported in part by: Ministerio español de Economía y Competitividad under projects TIN2014-56494-C4-3-P (UGR-EPHEMECH) and TIN2017-85727-C4-2-P (UGR-DeepBio).

References

1. Anderson, D.P., Cobb, J., Korpela, E., Lebofsky, M., Werthimer, D.: SETI@home: an experiment in public-resource computing. Commun. ACM **45**(11), 56–61 (2002)
2. Apolónia, N., Ferreira, P., Veiga, L.: Enhancing online communities with cycle-sharing for social networks. In: Abraham, A., Hassanien, A.E. (eds.) Computational Social Networks. Springer, London (2012). https://doi.org/10.1007/978-1-4471-4048-1_7

3. Boldrin, F., Taddia, C., Mazzini, G.: Distributed computing through web browser. In: 2007 IEEE 66th Vehicular Technology Conference (VTC-2007) Fall, pp. 2020–2024. IEEE (2007)
4. Cantú-Paz, E.: Migration policies, selection pressure, and parallel evolutionary algorithms. J. Heuristics **7**(4), 311–334 (2001)
5. Deterding, S., Dixon, D., Khaled, R., Nacke, L.E.: From game design elements to gamefulness: defining gamification. In: Proceedings of MindTrek 2011, pp. 9–15. ACM, Tampere (2011). https://doi.org/10.1145/2181037.2181040
6. Deterding, S., Sicart, M., Nacke, L., O'Hara, K., Dixon, D.: Gamification: using game-design elements in non-gaming contexts. In: CHI'11 Extended Abstracts on Human Factors in Computing Systems, pp. 2425–2428. ACM (2011)
7. Dubois, D.J., Tamburrelli, G.: Understanding gamification mechanisms for software development. In: Proceedings of the 2013 9th Joint Meeting on Foundations of Software Engineering, pp. 659–662. ACM, NY (2013)
8. Fabisiak, T., Danilecki, A.: Browser-based harnessing of voluntary computational power. Found. Comput. Decis. Sci. **42**(1), 3–42 (2017)
9. Hamari, J., Koivisto, J., Sarsa, H.: Does gamification work? A literature review of empirical studies on gamification. In: 2014 47th Hawaii International Conference on System Sciences, pp. 3025–3034. IEEE (2014)
10. Hickman, T.: Total engagement: using games and virtual worlds to change the way people work and businesses (2010)
11. Huotari, K., Hamari, J.: Defining gamification: a service marketing perspective. In: Proceeding of the 16th International Academic MindTrek Conference, pp. 17–22. ACM (2012)
12. Klein, J., Spector, L.: Unwitting distributed genetic programming via asynchronous JavaScript and XML. In: Proceedings of the 9th Annual Conference on Genetic and Evolutionary Computation (GECCO 2007), pp. 1628–1635. ACM, New York (2007)
13. Langdon, W.B.: Pfeiffer - a distributed open-ended evolutionary system. In: Edmonds, B., Gilbert, N., Gustafson, S., Hales, D., Krasnogor, N. (eds.) AISB 2005: Proceedings of the Joint Symposium on Socially Inspired Computing (METAS 2005), pp. 7–13, sSAISB 2005 Convention. University of Hertfordshire, Hatfield (2005). http://www.cs.ucl.ac.uk/staff/W.Langdon/ftp/papers/wbl_metas2005.pdf
14. Leclerc, G., Auerbach, J.E., Iacca, G., Floreano, D.: The seamless peer and cloud evolution framework. In: Proceedings of the 2016 on Genetic and Evolutionary Computation Conference, pp. 821–828. ACM (2016)
15. Merelo, J.J., García, A.M., Laredo, J.L.J., Lupión, J., Tricas, F.: Browser-based distributed evolutionary computation: performance and scaling behavior. In: GECCO 2007: Proceedings of the 2007 GECCO Conference Companion on Genetic and Evolutionary Computation, pp. 2851–2858. ACM Press, New York (2007)
16. Merelo, J.J., García-Valdez, M., Castillo, P.A., García-Sánchez, P., de las Cuevas, P., Rico, N.: NodIO, a JavaScript framework for volunteer-based evolutionary algorithms: first results. Technical report, GeNeura/UGR/CITIC (2016). http://arxiv.org/abs/1601.01607
17. Merelo, J.J., Castillo, P.A., García-Sánchez, P., de las Cuevas, P., Rico, N., Valdez, M.G.: Performance for the masses: experiments with a web based architecture to harness volunteer resources for low cost distributed evolutionary computation. In: Friedrich, T., Neumann, F., Sutton, A.M. (eds.) Proceedings of the 2016 on Genetic and Evolutionary Computation Conference, Denver, CO, USA, 20–24 July 2016, pp. 837–844. ACM, USA (2016). https://doi.org/10.1145/2908812.2908849

18. Merelo-Guervós, J.J., García-Sánchez, P.: Designing and modeling a browser-based distributed evolutionary computation system. In: Proceedings of the Companion Publication of the 2015 Annual Conference on Genetic and Evolutionary Computation, pp. 1117–1124. ACM (2015)
19. Milani, A.: Online genetic algorithms. Technical report, Institute of Information Theories and Applications FOI ITHEA (2004). http://hdl.handle.net/10525/838
20. Nijssen, S., Back, T.: An analysis of the behavior of simplified evolutionary algorithms on trap functions. IEEE Trans. Evol. Comput. **7**(1), 11–22 (2003)
21. Pan, Y., White, J., Sun, Y., Gray, J.: Gray computing: an analysis of computing with background Javascript tasks. In: Proceedings of the 37th International Conference on Software Engineering, vol. 1, pp. 167–177. IEEE Press (2015)
22. Peñalver, J.G., Merelo, J.J.: Optimizing web page layout using an annealed genetic algorithm as client-side script. In: Eiben, A.E., Bäck, T., Schoenauer, M., Schwefel, H.-P. (eds.) PPSN 1998. LNCS, vol. 1498, pp. 1018–1027. Springer, Heidelberg (1998). https://doi.org/10.1007/BFb0056943
23. Quinn, A.J., Bederson, B.B.: Human computation: a survey and taxonomy of a growing field. In: Proceedings of the SIGCHI Conference on Human Factors in Computing Systems, pp. 1403–1412. ACM, New York(2011)
24. Sarmenta, L.F.: Volunteer computing. Ph.D. thesis, Massachusetts Institute of Technology (2001)
25. Schmettow, M.: Sample size in usability studies. Commun. ACM **55**(4), 64–70 (2012). https://doi.org/10.1145/2133806.2133824
26. Sherry, D., Veeramachaneni, K., McDermott, J., O'Reilly, U.-M.: Flex-GP: genetic programming on the cloud. In: Chio, C., et al. (eds.) EvoApplications 2012. LNCS, vol. 7248, pp. 477–486. Springer, Heidelberg (2012). https://doi.org/10.1007/978-3-642-29178-4_48
27. Vespignani, A., et al.: Predicting the behavior of techno-social systems. Science **325**(5939), 425 (2009)

A Decision Support System Based on a Hybrid Genetic Local Search Heuristic for Solving the Dynamic Vehicle Routing Problem: Tunisian Case

Ines Sbai[1(✉)], Olfa Limam[2], and Saoussen Krichen[1]

[1] LARODEC Laboratory, ISG Tunis, Université de Tunis, Le Bardo, Tunis, Tunisia
sbai_ines@yahoo.fr, Krichen_s@yahoo.fr
[2] ISI Tunis, Université de Tunis El Manar, Tunis, Tunisia
limemolfa@yahoo.fr

Abstract. Vehicle Routing Problem is the most common and simplest routing problems. One of its important variants is the Dynamic Vehicle Routing Problem in which a new customer orders and order cancellations continually happen over time and thus perturb the optimal routing schedule that was originally invented. The Dynamic Vehicle Routing Problem is an *NP*-*H*ard problem aims to design the route set of minimum cost for a homogenous feet of vehicles, starting and terminating at the depot, to serve all the customers. In this paper, we propose a prototype of a Decision Support System that integrates a hybrid of Genetic Algorithm and Local Search to solve the Dynamic Vehicle Routing Problem. The performance of the proposed algorithm is highlighted through the implementation of the Decision Support System. Some benchmark problems are selected to test the performance of the proposed hybrid method. Our approach is better than the performance of compared algorithms in most cases in terms of solution quality and robustness. In order to demonstrate the performance of the proposed Decision Support System in term of solution quality, we apply it for a real case of the Regional Post Office of the city of Kef in the north west of Tunisia. The results are then highlighted in a cartographic format using Google Maps.

Keywords: Dynamic Vehicle Routing Problem · Genetic algorithm
Local search · Decision Support System

1 Introduction

The delivery of goods to customers is considered to be one of the most challenging activities in logistic sectors. One of the most studied problems in supply chain and logistics-related areas is the Vehicle Routing Problem (VRP). VRP was first introduced by Dantzig and Ramser [1] as a generalization of the Traveling Salesman Problem (TSP). The objective consists in designing a set of

J. Medina et al. (Eds.): IPMU 2018, CCIS 855, pp. 354–365, 2018.
https://doi.org/10.1007/978-3-319-91479-4_30

trips, starting and terminating at a central depot, minimizing the total transportation cost with a homogenous fleet of vehicles based on a depot node. In some cases, not all customers are known in advance, but are revealed as the system progresses. One of those formulations is VRP with Dynamic Requests, referred to a Dynamic Vehicle Routing Problem (DVRP). Although, in the literature, several types of metaheuristic methods have been applied to solve DVRP. The first one is due to Wilson and Colvin [2], where a single vehicle DARP is studied and customer requests are trips from an origin to a destination that appear dynamically. The approach uses insertion heuristics. Pillac et al. [20] presented a review of the DVRP and introduce the notion of degree of dynamism. Elhassania et al. [3] presented an Ant Colony Optimization (ACO) with a Large Neighbourhood Search (LNS) algorithm for depot visits order requests. Okulewicz and Mandziuk [4] used a standard continuous Particle Swarm Optimization (PSO) and a cluster-based heuristic generating initial solutions for priorities separate requests and clusters centers multi requests. Mandziuk and Zychowski [5] presented a Memetic Algorithm (MA) consisting of GA with a local search based on adaptive heuristic operators sequences for order requests. Okulewicz and Mandziuk [6] solved requests-to-vehicles assignment by the PSO algorithm, route optimization by a separate instance of the PSO algorithm. Recently, [21] proposed an ant colony based meta-heuristic for solving the DVRPP with time windows and also, AbdAllah et al. [19] solve DVRP using GA. They propose a weighted fitness evaluation approach as an alternative for the biased time-based approach.

In this paper, In order to enhance the exploitation ability of GA, we propose a hybrid Genetic algorithm (GA) with a local search (LS) algorithm for solving a DVRP for order depot visits requests. In DVRP, new customer demands are received along the day. Hence, they must be serviced at their locations by a set of vehicles in real time. The effectiveness of our approach is demonstrated through experiments on widely used benchmark instances. Numerical experiments show that the proposed method outperforms other local searches and metaheuristics. We also, present a Decision Support System (DSS) that integrates a Geographical Information System (GIS) for solving the addressed problem. The proposed DSS is applied for a real case of the Regional Posts Office of Kef (RPOK) on the city of kef in the north west of Tunisia in order to demonstrate the performance of the solution quality. The results are then highlighted in a cartographic format using Google Maps.

The remainder of this paper is structured as follows. The DVRP is stated mathematically in Sect. 2. In Sect. 3 the main steps of the proposed DSS are outlined. Section 4 provides a description of the resolution methodology HGLS. Section 5 describes the computational results. Section 6 details the case study. Finally, Sect. 7 concludes the paper.

2 Problem Description

The standard VRP is defined on a connected graph $G = (V,A)$, where V defines a set of $n+1$ vertices and $E = ((i,j)|i,j \in V)$ enumerates the availability of direct routes between vertices i and j of V. Vertices v_0 and v_{n+1} correspond to the depot at which K homogeneous vehicles are based, and the remaining vertices denote the customers. Each arc (v_i,v_j) is associated with a non-negative weight $c_{v_i v_j}$, which represents the travel distance from v_i to v_j. Each customer i has a delivery demand q_i. The CVRP consists of determining a set of least cost vehicle routes such that:

- Each route starts and ends at the depot,
- Each customer is visited exactly once by exactly one vehicle,
- The total demand of the customers assigned to any vehicle must not exceed the vehicle capacity Q.

To ensure the dynamic aspect to the VRP, we additionally two parameters:

t_{r_i} is a point in time when their is a new request and T is the duration (time) of a service.

The decision variables of the problem are, x_{ij} defined as:

x_{ij}^k takes 1 if the vehicle k travels from customer i to j (0 otherwise).

Where n presents the total number of customers, k is the total number of vehicles, Q is the capacity of each vehicles and c_{ij} is the cost of traveling from customers i, j.

Figure 1 presents a simple example of a dynamic vehicle routing situation is shown. The advance request customers are represented by black nodes, while those that are immediate requests are depicted by white nodes.

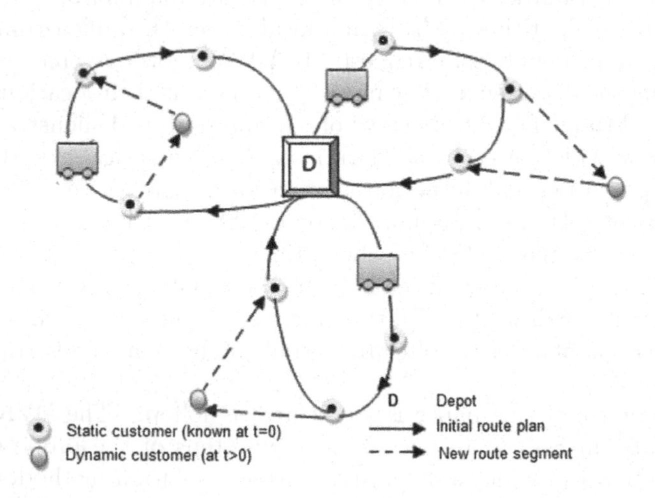

Fig. 1. An example of a DVRP solution

3 Problem Formulation

We state in what follows the mathematical model of the DVRP:

$$MinZ(x) = \sum_{k=1}^{k}\sum_{i=0}^{n}\sum_{j=0,j\neq i}^{n} c_{ij}x_{ij}^{k} \quad (1)$$

S.t.

$$\sum_{j=1}^{n} x_{0j}^{k} = \sum_{i=1}^{n} x_{i0}^{k} = 1, k \in \{1,\dots,n\} \quad (2)$$

$$\sum_{j=0,j\neq i}^{n}\sum_{k=1}^{k} x_{ij}^{k} = 1, i \in \{1,\dots,n\} \quad (3)$$

$$\sum_{i\in S_t}\sum_{j\in S_t} x_{ij}^{k} \leq |S_t| - 1, k \in \{1,\dots,n\}, S_t \subseteq V \quad (4)$$

$$tr_i \leq T \quad (5)$$

The objective function (1) consists of minimising the total cost of a fleet of vehicles, Constraints (2) express that each travel should begin and end at the depot. In addition, Constraints (3) provide that a single vehicle leaves each client i. Constraints (4) eliminate the sub tour ($S_t \subseteq V$) with $|S_t| = i$ is $C_n^i \times K$. Finally, constraints (5) ensure that the time of the dynamic request is less or equal to the total time T of service.

4 Decision Support System Architecture

We develop DSS based on an HGLS that satisfies all customer requests in order to generate the optimum vehicle paths (see Fig. 3). The first step of our DSS is to input problem parameters, namely the number of customers to be served, vehicles capacity and the number of available vehicles. The second step after providing data, geographical coordinates and customer demands are to be set. The Genetic algorithm proceeds iteratively by an alternative use of the Local Search in order to diversify the search. Once the numerical solution is generated, the DSS moves to design the cartographical solution that well illustrates the real itinerary. Finally, Vehicles pathways are then highlighted.

5 The HGLS Approach for DVRP

The DVRP is obviously *NP-H*ard problem [7]. In this study we propose a hybridization based on an Genetic Algorithm and Local Search algorithm to solve the DVRP presented in Algorithm 1.

Algorithm 1. The proposed HGLS approach for the DVRP

```
1: Begin
2: t=0
3: Create an Initial Population
4: Evaluate each chromosome in P(t)
5: while stopping criterion is not satisfied do
6:    Initialize a temporary population p'
7:    for i=1 to —P(t)— do
8:       Select two parents from P(t)
9:       Apply Crossover (offspring)
10:      if the offspring and parents are identical then
11:         Improve each offspring by the Mutation operator
12:      end if
13:   end for
14:   t=t+1;
15:   Replace the old Population P(t) by the new P';
16: end while
17: Apply the Local search procedure (Algorithm2)
18: The best solution founded is returned
19: END
```

5.1 Genetic Algorithm

Since their introduction by Holland [8], based on the concept of natural selections and genetics, Genetic algorithms have become popular in a wide variety of *NP-Hard* combinatorial optimization problems such as the Traveling Salesman Problem(TSP) [9], the Vehicle Routing Problem (VRP)[14], the Job-shop Scheduling Problem [11] and the Quadratic Assignment Problem (QAP) [10], because of ease of operation (Selection, Crossover, Mutation), its simplicity, and global perspective. The different steps of the GA are stated as follows:

GA starts by generating some feasible solutions (individuals) after initialized the parameters used in the algorithm, and then using the objective (fitness) function each of them is evaluated. The GA is iterated until the terminal condition is satisfied. In each iteration (generation). The second step is to select probabilistically individuals from the population according to selection methods. The third one is to generate offspring by applying the crossover and mutation operators. Then, the worst individual in population is removed to keep the size of population constant if the generated offspring is not the same as any individual in population. Finally, the pheromone trails are updated. In the reminder of this section, we introduce the main processes of the proposed algorithm.

Chromosome Representation and Initial Population Creation
In our approach, we use a variable length chromosome representation that includes some information on previously visited customers [12]. There exists awaiting customers that have recently been added to the days planning but

not yet allocated to any vehicle and engaged customers that have already been visited by a given vehicle, since any given chromosome represents a number of partial tours at any time step. Figure 2 represents an example of a DVRP chromosome representation where the positive nodes represent the static customers and the negative one represent the dynamic customers (when a new customer is newly added). In our contribution, the initial population of DVRP is generated randomly.

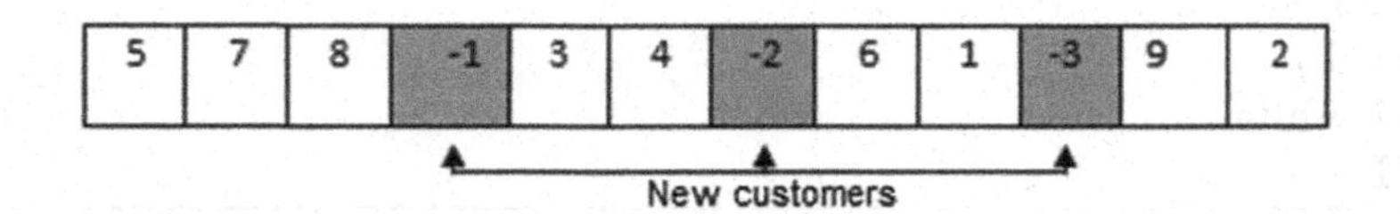

Fig. 2. An example of a DVRP chromosome representation

Fitness Evaluation
Each individual is evaluated using the fitness value $F_{DVRP}(x)$. according to $F_{DVRP}(x) = \sum_{k=1}^{k} cost(E_i)$ where E_i represents a set of routes.

Crossover Operator
In the crossover operator, two parents P_1 and P_2 are selected from the population. A route is randomly selected from each parent chromosome and the customer orders present in each route are eliminated from the other parent. In this paper we use the Ordered Crossover (OX) [13] like a crossover operator, where two points are randomly selected. The substring between the two sections points on the first parent is copied to the offspring. Then, the remaining positions are replaced by following the customer order on the second parent, starting at the position just after the second cross point.

Mutation Operator
we use Exchange Mutation to select two random positions in new chromosome and those positions are interchanged.

Our Proposed algorithm is iterated until the best solution doesn't change to a better value for a predefined value of generations during the evolution process.

5.2 Local Search procedure

The role of local search is to encourage better convergence and to discover any missing trade-off regions in evolutionary optimization. The general scheme of LS procedure is presented in Algorithm 2. An initial solution S_0 is generated and improved. The local optimum that is obtained is indicated by S. The following steps are repeated, until predetermined termination criteria are not met. The solution S is perturbed and a new current solution S' is obtained. The LS is applied to S' and a solution S" is obtained. If S" is accepted, it becomes the new current local optimum.

Algorithm 2. Local Search procedure

1: Begin
2: Create a solution S_0
3: Apply LS to S_0 and obtain S
4: **while** Termination criteria are not met do **do**
5: Perturb S to obtain S'
6: Apply LS to S' and obtain S"
7: **if** S" is accepted **then**
8: $S' = S$"
9: **end if**
10: **end while**
11: END

6 Computational Experiments

The proposed solution approach is implemented using Java Language version 7. All experiments were performed on a PC equipped Intel (R) Core (TM) i3-4005U CPU with 4 Go RAM under Microsoft Windows 7.

Table 1. Comparison between the published systems and the HGLS proposed system.

Problem	M-VRPDR		GA-based DVRP		Proposed HGLS	
	Bst	*Avg*	*Bst*	*Avg*	*Bst*	*Avg*
c50D	524.61	548.10	566.01	597.34	**520.11**	550.30
c75D	852.95	885.00	944.46	990.78	853.5	794.66
c100D	860.56	913.81	943.89	988.15	**855.4**	911.23
c100bD	820.92	864.03	869.41	904.03	**820.50**	866.52
c120D	1189.06	1295.31	1288.66	1399.40	**1189.01**	1290.63
c150D	1083.79	1142.24	1273.50	1359.25	1200.99	1141.25
c199D	1377.26	1466.30	1646.36	1700.54	**1083.77**	1466.25
tai75aD	1656.12	1705.76	1744.78	1823.71	**1656.09**	1704.77
tai100aD	2093.63	2168.87	2181.31	2290.95	2093.63	2166.99
tai150bD	2847.08	2980.33	2885.94	3073.58	**2844.89**	2999.40
tai75cD	1355.41	1427.74	1433.73	1502.56	**1344.90**	1425.88
tai100bD	1990.99	2097.86	2119.03	2212.58	**1998.50**	2096.58
tai150dD	2737.37	2903.68	2911.47	3010.34	2740.23	2901.30
f71D	260.17	297.61	288.30	309.49	260.17	296.65
f134D	11815.95	12299.59	14871.40	15789.80	11815.80	12278.55

The dynamic problems adopted in this paper have been originally proposed in Kilby et al. [15]. They are derived from some very popular static VRP benchmark datasets, namely 6 problems are taken from Taillard [16] 7 problems are from [17] and 2 problems are from [18]. These problems range from 50 to 199 customers. The number of customers can be inferred from the name of each instance. Table 1 illustrates the comparison between the optimal/best objective values of our best objective values in the literature for these benchmark instances. Average performances are computed over all problem instances of the corresponding data set. In Table 1, best results reported in two recently published heuristic which are the Memetic Algorithm [5] and GA-based DVRP [19] papers for comparison. The proposed HGLS DVRP finds 9 new best solutions and 12 better averages out of the 15 problems in this comparison

7 Real Case Study

In order to test the proposed approach, we experiment it for a Regional Post Office of the city of Kef in the north west of Tunisia. To ensure the distribution of letters and postal products in different countries offices and postal cells that cover the governorate of Kef, the RPOK has three paths: Tajerouine-Kalaat Senan, Sakia-Nebeur and Dahmani-El Kssour (Table 2). In order to achieve this task, the DRPK has awarded three vehicles. The choice of the vehicle depends on the driver. Table 2 explains the existing paths.

Table 2. The RDPK existing paths

Path		Traveled distance
Sakia-Nebeur	Distribution centre $\longrightarrow$ $Quickpostoffice$ $\longrightarrow$ Ain Karma $\longrightarrow$ $Sakia$ $\longrightarrow$ $Touiref$ $\longrightarrow$ Nebeur $\longrightarrow$ $Maleg$ $\longrightarrow$ $SidiKhiar$ $\longrightarrow$ Tal el Ghoelene $\longrightarrow$ $Nebeur$ $\longrightarrow$ $BorjelAifa$ $\longrightarrow$ Sakia $\longrightarrow$ $AinKarma$ $\longrightarrow$ $Distributioncentre$ $\longrightarrow$ $Quickpostoffice$	273 Km
Dahmani-ElKssou	Distribution $\longrightarrow$ $CenterCEF$ $\longrightarrow$ $WestKef$ $\longrightarrow$ Dahmani $\longrightarrow$ $ElKssour$ $\longrightarrow$ $Ezzaitouna$ $\longrightarrow$ $LeSers$ $\longrightarrow$ $WedSouani$ $\longrightarrow$ $LeSers$ $\longrightarrow$ $Elles$ $\longrightarrow$ $ElKssour$ $\longrightarrow$ $Dahmani$ $\longrightarrow$ CEF $\longrightarrow$ $WestKef$ $\longrightarrow$ $DistributionCenter$ $\longrightarrow$ $Quickpostoffice$	250 Km
Tajrouine -Kaat Senan	Distribution Center $\longrightarrow$ $Quickpostoffice$ $\longrightarrow$ Sidi Mtir $\longrightarrow$ $Jezza$ $\longrightarrow$ $Tajerouine$ $\longrightarrow$ $Jerissa$ $\longrightarrow$ $ManzelSalem$ $\longrightarrow$ $KalaatSenan$ $\longrightarrow$ $AinSenan$ $\longrightarrow$ $Mahjouba$ $\longrightarrow$ $Boujaber$ $\longrightarrow$ $KalaatSenan$ $\longrightarrow$ $SidiAhmedSaleh$ $\longrightarrow$ $KalaaKhasba$ $\longrightarrow$ $SidiAhmedSaleh$ $\longrightarrow$ $Jerissa$ $\longrightarrow$ $Tajerouine$ $\longrightarrow$ $Jezza$ $\longrightarrow$ $SidiMtir$ $\longrightarrow$ $DistributionCenter$ $\longrightarrow$ $Quickpostoffice$	270 Km

Study of the Transportation Cost within the Regional Posts Direction of Kef
The goal of our work is the resolution of the DVRP within the DRPK. Our problem lies in the planning of the distribution rounds some mail; in other words, the planner has no idea about the appearance of the dynamic node, therefore the information necessary for the planning tours is not known entirely by the planner when the process planning begins and can change after the initial tours were built. Our goal is to study all the dynamic cases that may appear once the vehicle starts its tour and try to find a better solution while minimizing the total cost of transportation as well as the distance traveled to get the shortest way. Table 3 details the cost of transportation or the distribution of the mail and postal assets and shows the total cost of each route before minimization.

Table 3. Paths characteristic

Path	Distance	Cost
Sakia-Nebeur	273	155.610
Dahmani-El Kssour	250	142.500
Tajerouine-Kallat Senan	270	153.900

Minimization of the Distance
In order to minimize the cost of transportation, we will try to minimize the total distance traveled for each route without touching the progress of the mail distribution operation. According to Tables 2, 3 and 4 we notice at the level of each path a sub tour. A sub tour designs that the vehicle of mail visits the majority of post offices twice a day during of his distribution work. That's why we're going to eliminate the sub-tour, we will obtain new short paths. Table 5 shows the new routes:

Table 4. New obtained paths

Path		New distance
Sakia-Nebeur	Distribution Center $\longrightarrow$ *Quickpostoffice* $\longrightarrow$ AinKarma $\longrightarrow$ *Sakia* $\longrightarrow$ *Touiref* $\longrightarrow$ Sidi Khiar $\longrightarrow$ *Talelghozlne* $\longrightarrow$ *Malg* $\longrightarrow$ *Nebeur* $\longrightarrow$ *BorjelAifa* $\longrightarrow$ *DistributionCenter*	185 Km
Dahmani-ElKssou	Distribution Center $\longrightarrow$ *Quickpostoffice* $\longrightarrow$ Exploitation center $\longrightarrow$ *KefOuest* $\longrightarrow$ *Dahmani* $\longrightarrow$ *ElKssour* $\longrightarrow$ *Ezzaitouna* $\longrightarrow$ *Sers* $\longrightarrow$ *WedSouani* $\longrightarrow$ *DistributionCenter*	169 Km
Tajrouine -Kaat Senan	Distribution Center $\longrightarrow$ *Quickpostoffice* $\longrightarrow$ Sidi Mtir $\longrightarrow$ *Jezza* $\longrightarrow$ *Tajerouine* $\longrightarrow$ *Jerissa* $\longrightarrow$ *Mahjoub* $\longrightarrow$ *ManzelSalem* $\longrightarrow$ *KalaatSenan* $\longrightarrow$ *AinSenan* $\longrightarrow$ *Boujaber* $\longrightarrow$ *SidiAhmedSaleh* $\longrightarrow$ *KalanKhasba* $\longrightarrow$ *Distributioncenter*	245 Km

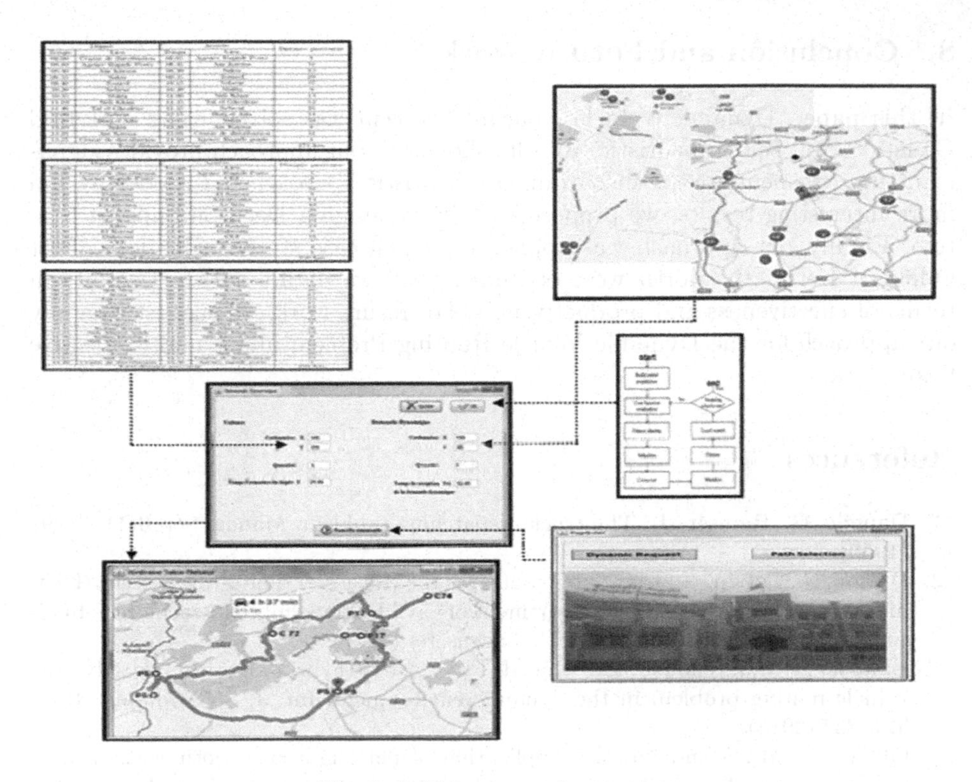

Fig. 3. DSS inputs and outputs

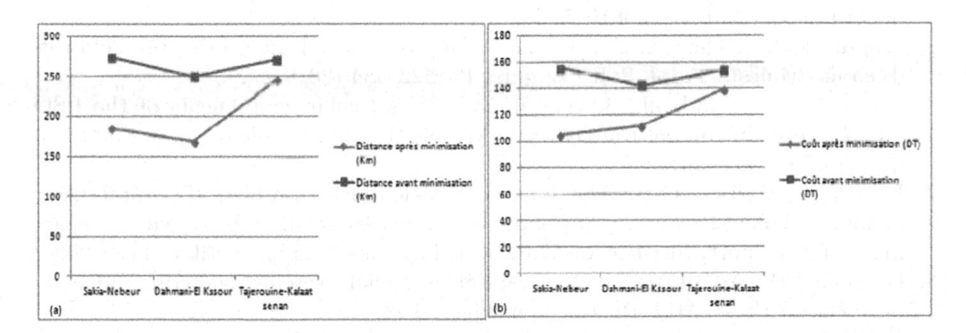

Fig. 4. (a) Illustrates the distance before and after minimization (b) Illustrates the cost before and after minimization

The minimization of distance leads to a minimization of the cost, which shows Table 6.

As a conclusion, Fig. 4 Illustrates the distance and the cost before and after minimization.

8 Conclusion and Future Work

In this paper, Dynamic Vehicle Routing Problem was solved using a Hybrid Genetic Local Search heuristic. Results show that our approach provides significant improvements over other examined heuristic approaches. In order to find more interesting results, we proposed a GIS to design a Decision Support System. Finally, our approach was applied on a real case study in Regional Post Office of Kef in the north west of Tunisia and we obtain satisfied results in terms of effectiveness and productiveness. For future work, we suggest to apply our approach for the Dynamic Vehicle Routing Problem under multi objective framework.

References

1. Dantzig, G., Ramser, J.: The truck dispatching problem. Manag. Sci. **6**(1), 80–91 (1959)
2. Wilson, N., Colvin, N.: Computer control of the Rochester dial-a-ride system. Technical report Report R77–31, Department of Civil Engineering, Massachusetts Institute of Technology, Cambridge, Massachusetts (1977)
3. Elhassania, M.J., Jaouad, B., Ahmed, E.A.: A new hybrid algorithm to solve the vehicle routing problem in the dynamic environment. Int. J. Soft Comput. **8**(5), 327–334 (2013)
4. Okulewicz, M., Mańdziuk, J.: Application of particle swarm optimization algorithm to dynamic vehicle routing problem. In: Rutkowski, L., Korytkowski, M., Scherer, R., Tadeusiewicz, R., Zadeh, L.A., Zurada, J.M. (eds.) ICAISC 2013. LNCS (LNAI), vol. 7895, pp. 547–558. Springer, Heidelberg (2013). https://doi.org/10.1007/978-3-642-38610-7_50
5. Mandziuk, J., Zychowski, A.: A memetic approach to vehicle routing problem with dynamic requests. Appl. Soft Comput. **48**, 522–534 (2016)
6. Okulewicz, M., Madziuk, J.: The impact of particular components of the PSO-based algorithm solving the Dynamic Vehicle Routing Problem. Appl. Soft Comput. **58**, 586–604 (2017)
7. Larsen, A., Madsen, O.B.: The dynamic vehicle routing problem (Doctoral dissertation, Technical University of Denmark, Danmarks Tekniske Universitet, Department of Transport, Institut for Transport, Logistics & ITSLogistik & ITS) (2000)
8. Holland, J.H.: Adaptation in Natural and Artificial Systems, 2nd edn. University of Michigan Press, MIT Press, Cambridge (1992)
9. Nguyen, H.D., Yoshihara, I., Yamamori, K., Yasunaga, M.: Implementation of an effective hybrid GA for large-scale traveling salesman problems. IEEE Trans. Syst. Man Cybern. Part B Cybern. **37**(1), 92–99 (2007)
10. Misevicius, A.: An improved hybrid genetic algorithm: new results for the quadratic assignment problem. Knowl. Based Syst. **17**(2–4), 65–73 (2004)
11. Park, B.J., Choi, H.R., Kim, H.S.: A hybrid genetic algorithm for the job shop scheduling problems. Comput. Ind. Eng. **45**(4), 597–613 (2003)
12. Hanshar, F.T., Ombuki-Berman, B.M.: Dynamic vehicle routing using genetic algorithms. Appl. Intell. **27**(1), 89–99 (2007)
13. Karakatic, S., Podgorelec, V.: A suvey of genetic algorithms for solving multi depot vehicle routing problem. Appl. Soft Comput. **27**, 519–532 (2015)

14. Prins, C.: A simple and effective evolutionary algorithm for the vehicle routing problem. Comput. Oper. Res. **31**(2), 1985–2002 (2004)
15. Kilby, P., Prosser, P., Shaw, P.: Dynamic VRPs: A study of scenarios, Technical report APES-06-1998. University of Strathclyde, UK (1998)
16. Taillard, E.: Parallel iterative search methods for vehicle-routing problems. Networks **23**(8), 661–673 (1994)
17. Christofides, N., Beasley, J.: The period routing problem. Networks **14**, 237–256 (1984)
18. Fisher, M., Jakumar, R., van Wassenhove, L.: A generalized assignment heuristic for vehicle routing. Networks **11**, 109–124 (1981)
19. AbdAllah, A.M.F., Essam, D.L., Sarker, R.A.: On solving periodic re-optimization dynamic vehicle routing problems. Appl. Soft Comput. **55**, 1–12 (2017)
20. Pillac, V., Gendreau, M., Guret, C., Medaglia, A.L.: A review of dynamic vehicle routing problems. Eur. J. Oper. Res. **225**(1), 1–11 (2013)
21. Yang, Z., van Osta, J.P., van Veen, B., van Krevelen, R., van Klaveren, R., Stam, A., Kok, J., Bäck, T., Emmerich, M.: Dynamic vehicle routing with time windows in theory and practice. Nat. Comput. **16**(1), 119–134 (2017)

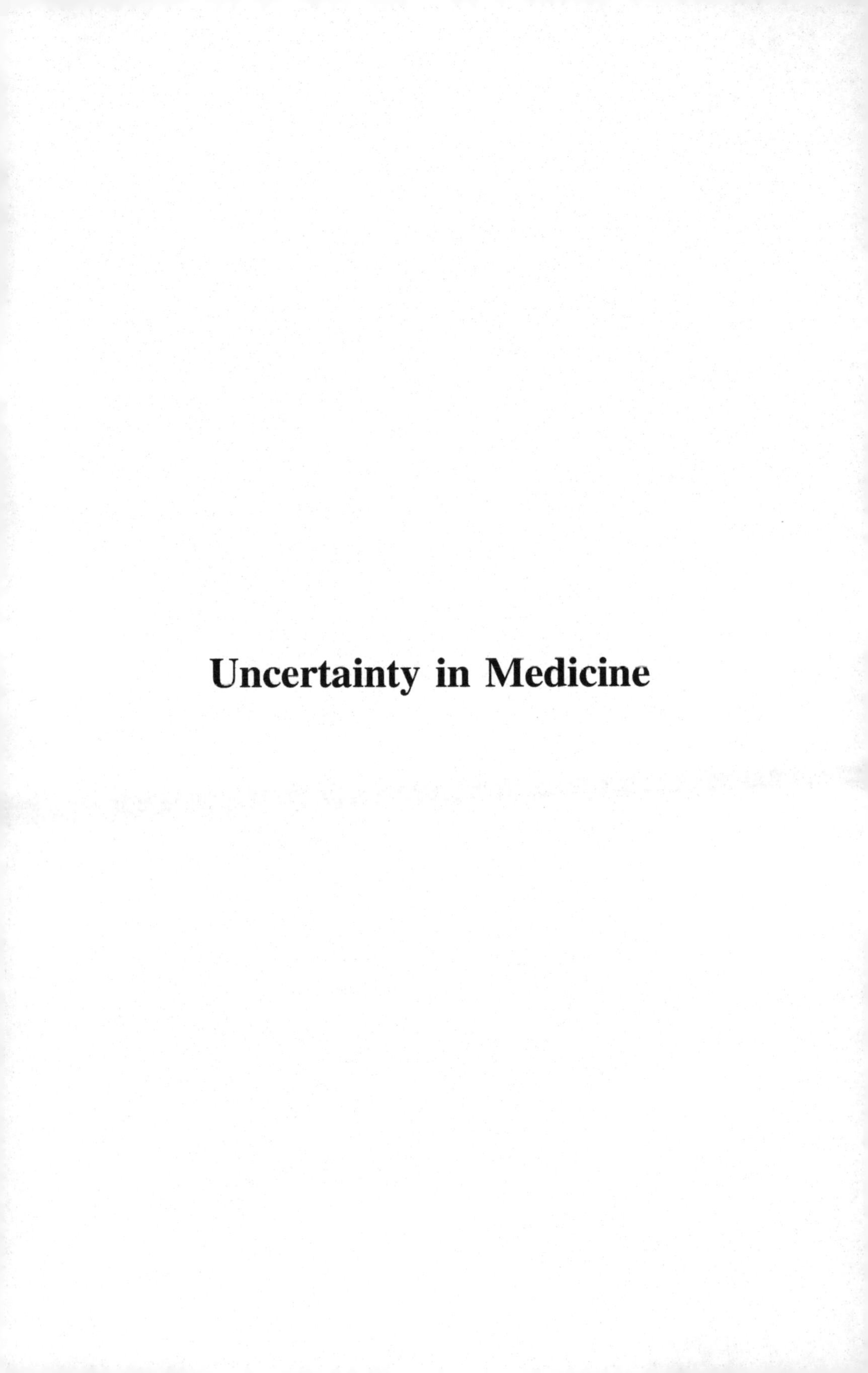

Uncertainty in Medicine

Modelling Medical Uncertainties with Use of Fuzzy Sets and Their Extensions

Patryk Żywica(✉)

Faculty of Mathematics and Computer Science,
Adam Mickiewicz University in Poznań, Poznań, Poland
bikol@amu.edu.pl
http://min.wmi.amu.edu.pl/en/staff/patryk-zywica/

Abstract. This work presents an approach to deal with uncertainty in patient's medical record. After giving a brief characterisation of possible sources of uncertainty in medical records, the paper introduces fuzzy set based approach that allows modelling of such information. First, heterogeneous data is converted to homogeneous model with the use of Feature Set structure. With such model uncertainty may be represented directly as Fuzzy Membership Function Families (FMFFs). Some theoretical results connecting FMFFs with Hesitant Fuzzy Sets and Type-2 Fuzzy Sets are also given.

Keywords: Medical data · Hesitant Fuzzy Sets
Imperfect information

1 Introduction

Over the years, many models have been developed to extend and generalise the fuzzy sets theory. Motivation for a large part of them was the modelling of broadly understood data uncertainty. Most commonly used in practice are: IVFS [1], AIFS [2,3], HFS [4], and T2FS [1,5,6]. All these extensions enrich the set of values that membership function can take, replacing a single number with an interval, set, and even a fuzzy number in $[0,1]bib0, bib1$. Depending on the interpretation, new values of membership can be considered as separate "logical" values (ontic interpretation) or as an approximation of our imperfect knowledge of a certain value (epistemic interpretation) [7,8].

This work is devoted to uncertainty in an epistemic sense, i.e. one that results from limited, incomplete or imperfect knowledge. The simplest case is the complete lack of knowledge – the case of incomplete data. However, even here one should be careful. Missing values are often marked with a very ambiguous abbreviation *NA*. *NA* – not available – the value is not available, no one conducted an examination or test. This is a perfect example of total epistemic uncertainty. On the other hand, another interpretation of NA – not applicable – has nothing to do with the incompleteness of the data. After all, this is completely certain

J. Medina et al. (Eds.): IPMU 2018, CCIS 855, pp. 369–380, 2018.
https://doi.org/10.1007/978-3-319-91479-4_31

information that a given parameter is not suitable for describing a particular case. This fact further highlights how important it is to understand the context of modelled data.

The primary objective of this paper is to describe how one can use fuzzy sets for patient record modelling. The two main problems will be addressed. The first involves the process of transforming heterogeneous medical data into a fuzzy set with clearly defined semantics. The second concerns the modelling of epistemic uncertainty.

The rest of the paper is organised as follows. Section 2 gives some background information about fuzzy sets and their extensions. The third section covers the issue of medical data normalisation. Section 4 deals with uncertainty model for medical data. The concept of Fuzzy Membership Function Family (FMFF) is defined and some theoretical results connecting it with Hesitant Fuzzy Sets and Type-2 Fuzzy Sets are given. Finally, in Sect. 5 some conclusions and areas for further research are given.

2 Definitions

Let $U = \{x_1, x_2, \dots\}$ be a crisp, at most countable universal set. A mapping $A : U \to [0,1]$ is called a fuzzy set in U. For each i, the value $A(x_i)$ (a_i for short) represents the membership grade of x_i in A. Let $\mathcal{FS}(U)$ be the family of all fuzzy sets in U.

Interval-valued fuzzy set theory, which is a special case of Type–2 Fuzzy Set (T2FS) theory, was introduced by Zadeh [1]. Let $\mathcal{I}$ be the set of all closed subintervals of $[0,1]$. A mapping $\hat{A} : X \to \mathcal{I}$ is called an Interval-Valued Fuzzy Set. For each $1 \leq i \leq n$, the value $\hat{A}(x_i) = [\underline{A}(x_i), \overline{A}(x_i)] \in \mathcal{I}$ represents the membership of an element x_i in $\hat{A}$. Usually $\underline{A}$ and $\overline{A}$ are called the lower and upper membership functions of $\hat{A}$ respectively. In epistemic approach, interval $\hat{A}(x_i)$ is understood to contain the true membership degree of x_i in some incompletely known fuzzy set A represented by $\hat{A}$. We denote the set of all interval-valued fuzzy sets in U by $\mathcal{IVFS}(U)$.

In 2010, Torra defined Hesitant Fuzzy Sets (HFS [4]). They perfectly combine the simplicity of IVFS and the ability to model very complex data provided by Type–2 Fuzzy Sets [9]. The important fact about Hesitant Fuzzy Sets is that they were created with the aim of representing the uncertain, epistemic data. A Hesitant Fuzzy Set is a mapping $A^H : X \to 2^{[0,1]}$. It is worth noting that this is an equivalent concept to the notion of Set-Valued Fuzzy Sets [9,10]. Type–2 Fuzzy Sets were proposed by Zadeh in 1971 [11]. They generalise most of the known extensions of fuzzy sets. A T2FS is defined as a mapping $\tilde{A} : X \to \mathcal{FS}([0,1])$.

3 Modelling Medical Data as Fuzzy Membership Function

In applications such as classification or decision support, it is often required that individual instances (patients) should be represented as homogeneous real vectors. Given the wide variety of data types encountered in the patient record,

Table 1. Sample medical data represented with original values.

No	Age	Gender	Tumor in family	WBC	Qualitative assessment	Quantitative assessment
1	38	F	Yes	2817.3	1/brown	2/medium
2	18	F	No	2181.8	3/blue	4/very high
3	64	F	No	8611.3	4/yellow	1/low
4	40	F	Yes	3017.1	2/grey	2/medium

Table 2. Data from Table 1 after min/max normalisation.

No	Age	Gender	Tumor in family	WBC	Qualitative assessment	Quantitative assessment
1	0.38	1	1	0.282	0.0	0.333
2	0.18	1	0	0.218	0.667	1
3	0.64	1	0	0.861	1.0	0
4	0.40	1	1	0.302	0.333	0.333

direct conversion/normalisation of data can lead to some anomalies. In this section we will present a procedure for converting heterogeneous medical data into a homogeneous model that allow to preserve full semantics of data.

3.1 Data Normalisation and Fuzzy Sets

Let start with the example data presented in Table 1. The simplest methods of transforming these data are min/max normalisation and standardisation. Table 2 shows the results of the min/max normalisation.

Many researchers (including the author of this work [12,13]) have tried to treat these vectors as fuzzy sets. This is a convenient approach because there are many useful tools in the field of Fuzzy Logic such as similarity measures or rule based models. This approach leads to the following fuzzy sets in the universal set $U = \{\text{age}, \text{gender}, \text{in family}, \text{WBC}, \text{qualitative}, \text{quantitative}\}$:

$$A_1 = {}^{0.38}/_{\text{age}} + {}^{1}/_{\text{gender}} + {}^{1}/_{\text{in family}} + {}^{0.282}/_{\text{WBC}} + {}^{0}/_{\text{qualitative}} + {}^{0.333}/_{\text{quantitative}} . \quad (1)$$

This design is correct (at least formally). In addition, through the use of appropriate methods, it is possible to achieve very good results, for example, in supporting medical diagnosis [12,14].

But can A_1 be called a (fuzzy) set? What would it contain? It turns out that the interpretation of such fuzzy set is difficult to determine. Of course, according to the characteristics given in the classic work of Dubois and Prade [15], we are dealing here with the “degree of similarity” semantics. However, to be able to interpret the degrees of membership as the similarity, a reference point

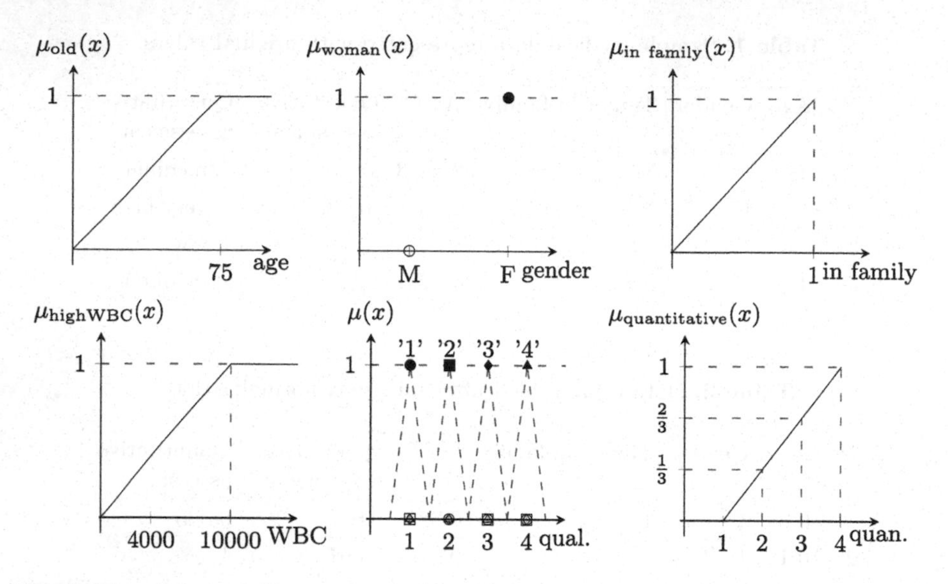

Fig. 1. Linguistic variables used to define feature set for data from previous subsection.

(prototype) is required. Was it determined in this case? Certainly not explicitly. For some attributes, definition of the "prototype" is a simple task: age – old person; WBC – high serum level; quantitative assessment – very high. The most difficult part is to determine prototype for the qualitative assessment, and it is an attribute type that most often cause problems. Choosing the highest value (4/yellow) as a prototype is as unjustified as any other. There is no naturally defined order between the values. It can not be said that 2/green is somehow similar to 4/yellow, since they are not related to each other and cannot be compared.

In this section a method of data normalisation will be given, which in particular:

- allows to specify a clear semantics of a (fuzzy) set for a patient representation,
- describes how to deal with different types of data in order to preserve their interpretation,
- can be easily extended to model the uncertainty.

3.2 Proposed Approach to Normalisation

The basic idea is to represent each instance (the patient) as a fuzzy set. For this purpose, the concept of *feature set* will be introduced. Each instance is described by a fuzzy set of features that it has. The *feature set* itself is defined in terms of the mappings from any number (k) of attributes to a single feature $f_i : \mathrm{R}^k \rightarrow [0, 1]$. Each such mapping should also give clear description of the new feature meaning. In this paper we will present simple approach where each term of linguistic variable defines new mapping. Interpretations of terms for linguistic variables for the data from the previous example are shown on Fig. 1.

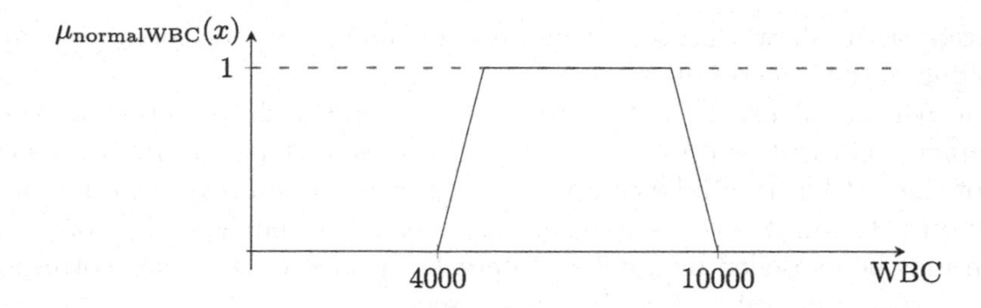

Fig. 2. Alternative definition of linguistic variable for WBC attribute.

Particular attention should be paid to the attribute "qualitative assessment". The corresponding linguistic variable has 4 terms, one for each value. In this situation, the fuzzy set representing the patient will contain 4 four features for one attribute. Moreover, each of these features can have a different level of membership, symbolically indicated by the dashed line in the graph. Whether a given attribute, and later a feature allows partial membership depends only on the semantics of the attribute (e.g. gender where a partial membership to "women" feature is rather difficult to interpret).

Also for numerical attributes, conversion to features using the linguistic variable brings some certain advantages. In addition to clearly defining the meaning of membership degree, we also get the possibility of a more advanced, but still easy to interpret, transformation of numerical values. For example, for a WBC attribute the medically recognised norm is 4000–10000. Hence, instead of defining a feature as "high", you can specify the "normal" feature, as shown in the Fig. 2. Although it's just a small change, experience shows that it can significantly affect model quality, especially when we are limited to simple models such as regression [16,17]. In addition, the medical norm, rather than embedded into the decision model, is explicitly included in the data model, which in some applications can be very useful.

What will be the fuzzy representation of patients from the previous subsection? First, the universal set is now a set of all features in feature set. Membership values are now calculated with respect to each term interpretation. The patient record is represented by the following fuzzy set:

$$P_1 = {}^{0.5}/_{\text{old}} + {}^{1}/_{\text{woman}} + {}^{1}/_{\text{in family}} + {}^{0.28}/_{\text{high WBC}} + {}^{1}/_{1/\text{brown}} + {}^{0.33}/_{\text{quantitative}} \,. \quad (2)$$

Apparently little has changed compared to (1). Referring to the three semantics of a fuzzy set, we are still in the "degree of similarity" semantics. However, now P_1 can certainly be called a fuzzy set. It is a set of features possessed by the first patient. Answer to the question whether $a \in P_1$ has simple interpretation and tells whether and to what extent a feature a describes the specific medical case. All membership degrees, regardless of the data type of the corresponding attribute, now have a uniform and clearly defined interpretation.

However, the greatest advantage of this data model, is the ability to directly model incomplete and, more generally, uncertain data. The proposed data uncertainty model will be formalised in the next section. The aim of the following discussion is to show that an adequate data model can considerably simplify modelling of the uncertainty.

The benefits of using the feature set based model for the representation of incomplete data will be presented on the examples of attributes from the beginning of the section. It will be compared to the classical min/max normalisation. Singleton notation will be used, where membership value depending on the situation may be modelled by number, interval or arbitrary set which corresponds to the normal, interval and hesitant fuzzy sets.

Lack of knowledge of the patient's age may occur for elderly patients whose documentation is missing, and getting information from them is not possible. For example, if a patient was born before 1939, then in the new model, thanks to the use of non-linear normalisation with the use of the linguistic variable we have ${}^{1}/_{\text{old}}$. With min/max normalisation, it would be necessary to use the interval membership degree ${}^{[0.79,1]}/_{\text{age}}$. The uncertainty for the gender attribute may result from the refusal to provide information. In our model, this means that the "woman" feature can simultaneously belong and not belong to the feature set. Hence, we get ${}^{\{0,1\}}/_{\text{woman}}$. Because for this feature there is no sense of partial membership, we use a two-element set instead of interval. In the classic approach, you can use the same model ${}^{\{0,1\}}/_{\text{gender}}$. In the case of cancer in the family, the patient can know only the immediate family, or part of it. For example, if it is known that the cancer did not occurred in parents and the distant family is unknown, then such situation can be modelled in both approaches as ${}^{[0,0.8]}/_{\text{in family}}$.

The most interesting, however, is the case of qualitative assessment (in the example of eye colour), where modelling of some uncertainty variants posed problems. Suppose that the only thing that is known about eye colour is that they are not 3/blue. In the original model we get ${}^{\{0,0.33,1\}}/_{\text{qualitative}}$. However, such a degree of membership can not be easily converted to interval, which makes it impossible to use simple modelling using for example IVFS. Thanks to new normalisation scheme we have 4 features which allows us to use following representation ${}^{[0,1]}/_{1/\text{brown}} + {}^{[0,1]}/_{2/\text{grey}} + {}^{[0,1]}/_{4/\text{yellow}}$. Therefore, the same knowledge was successfully modelled using only intervals. What if the doctor evaluating a given parameter is leaning towards option 1/brown but does not exclude 4/yellow? Unfortunately, this information can not be reproduced in the classic model, where at most we can write ${}^{\{0,1\}}/_{\text{qualitative}}$ (solution using T2FS is possible, although is not feasible here). Thanks to the introduction of feature set, we have ${}^{[0,0.75]}/_{1/\text{brown}} + {}^{[0,0.25]}/_{4/\text{yellow}}$.

3.3 Evaluation

Evaluation was based on test dataset from recent research on application of aggregation operators to incomplete data classification [18]. Original study group consists of 388 patients diagnosed and treated for ovarian tumor in the Division of Gynecological Surgery, Poznan University of Medical Sciences, between 2005 and 2015. Among them, 61% were diagnosed with a benign tumor and 39% with a malignant one. Moreover, 56% of the patients had no missing values in the

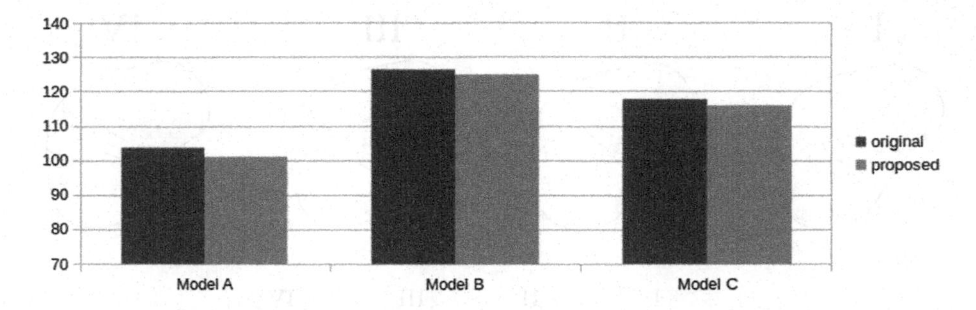

Fig. 3. Total cost of classification based on data normalised using some variant of min/max (original research) and using proposed approach. The lower the value the better classification is. Models A, B and C are selected best models form original research.

attributes required by diagnostic models, 40% had a percentage of missing values in the range (0%, 50%], and the remainder had more than 50% missing values. The test set consists of patients with real missing data and some proportion of patients with a complete set of features. As a result, the test set consisted of 175 patients. Patients with more than 50% missing values were excluded from the study. The results from [19] are reproduced and compared with the same classification scheme applied to data normalised using proposed approach. For more information regarding dataset we refer the reader to original papers [18–20].

Performance of classification based on data normalised using some variant of min/max (original research) and using proposed approach are presented in Fig. 3. Total cost method was used to measure model performance since accuracy does not fit the medical diagnosis problem. They show that change of normalisation scheme may give slight improvement in overall classification cost. The main reason of this improvement is proper handling of qualitative attributes.

4 Families of Fuzzy Membership Functions

The previous section shows that the patient's medical record can be reliably and unambiguously modelled while still preserving the semantics of the fuzzy set. It was also presented that it may have positive impact on modelling of incompleteness and uncertainty. Although the examples presented here will relate to medical issues, the model and results of this section are generic to any fuzzy set, regardless of interpretation.

4.1 Motivation

In this subsection, two example problems will be presented. They aim is to show that classic approaches to uncertainty modelling based on fuzzy sets may not be sufficient to fully take into account all available knowledge. First example shows that interval representation of uncertainty, though effective, is not sufficient to

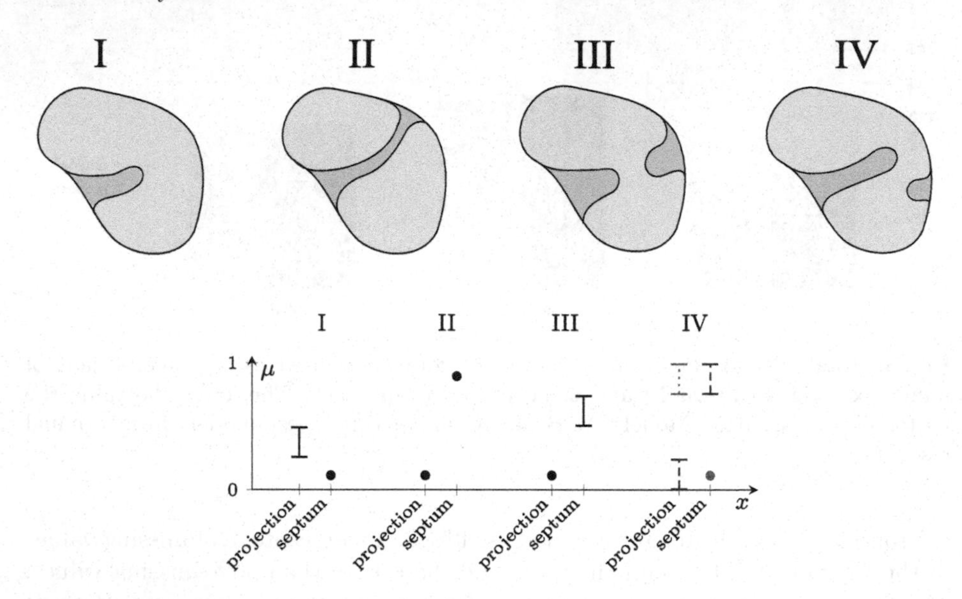

Fig. 4. Drawing of ovaries from Example 1 along with their interval representations. (Color figure online)

fully reflect the real data. Second one presents situation where some information is lost during conversion to fuzzy based representation.

Example 1. Figure 4 presents drawings of ovaries, which in actual medical practise are obtained with the help of ultrasound. Among the many features evaluated by a physician, very interesting is the case of papillary projections and septum. Drawings I and II show an average size papillary projection and total septum, respectively (along with interval representation below). Sometimes the septum is not yet completely developed which has been depicted in the third drawing. The fourth drawing shows a situation where you can not explicitly qualify whether we are dealing with a large papillary projection and the lack of septum (red, dotted line), or maybe it is a small papillary projection and almost full septum (blue, dashed line). The inability to distinguish between these two situations arises only from lack of knowledge (the impossibility of observing the ovary from the other side). As can be seen interval representation is not sufficient to cover both situations in one data description (it is necessary to assign two disjoint intervals). This example shows the limitations of interval representation of incomplete data, indicating a real need for more general formalism with greater power of expression such as HFS.

Example 2. Let assume we have information on the volume of tumor calculated as $V = \gamma abc$. Unfortunately, some diagnostic models use a combination of the two largest dimensions of the tumor $M = \alpha a + \beta b$. For simplicity, we can assume that $\alpha = \beta = \gamma = 1$ and that $a, b, c \geq 1$.

Let for a certain patient V=200 mm^2. Representing the three dimensions of the tumor as intervals we get $a = [1, 200]$, $b = [1, 200]$ and $c = [1, 200]$. Calculating the model using interval arithmetic we get $M = [2, 400]$.

The same data, however, can be better represented by the use of appropriate set-based approach $\mathcal{A} = \{(a, b, c) : abc = 200, a \geq b \geq c\}$. The M model with such a knowledge representation gives:

$$M' = M(\mathcal{A}) = \{M(a, b) : (a, b, c) \in \mathcal{A}\} \approx [11.696, 201] . \tag{3}$$

It is easy to see that the resulting model values are much more specific and $M' \subset M$.

Using interval representation and arithmetic we lose some important information. This knowledge is given in the form of dependencies between several membership degrees, and therefore can not be represented using three independent intervals.

4.2 Proposed Approach to Uncertainty Modelling

This subsection will present a new approach to uncertainty modelling. During its design, it was assumed that it should be as simple as possible and based on fuzzy sets, and most importantly allow for modelling the situations presented in Examples 1 and 2. For the purposes of modelling medical data, we can narrow the discussion to the situation in which the universal set U is at most countable. Then, $\mathcal{FS}(U)$ can be treated as a subset of $\mathbb{R}^n$ or l^∞.

Any closed subset $\mathcal{A}$ of $\mathcal{FS}(U)$ will be called Fuzzy Membership Function Family (FMFF). Most known extensions of fuzzy sets can be fully accurately represented using FMFF. For example, interval–valued fuzzy set $\hat{A}$ corresponds to the following FMFF

$$\mathcal{A} = \left\{A \in \mathcal{FS}(U) : \forall_{x \in U} \quad \underline{A}(x) \leq A(x) < \overline{A}(x)\right\} . \tag{4}$$

This approach is largely inspired by the Mendel representation theorem and his Wavy-Slice representation [21, 22]. Referring to this theory for the above IVFS we have $\mathcal{A} = \mathrm{FOU}(\hat{A})$.

Returning to the examples from the previous subsection, one can see that both situations can be directly modelled using FMFF.

Example 3 (Solution to Example 1). The situation shown in Fig. 4 IV can be represented by the following FMFF

$$\mathcal{A}_{\mathrm{IV}} = \Big\{{}^{\alpha}/_{\mathrm{pap}} + {}^{\beta}/_{\mathrm{septum}} \in \mathcal{FS}(\{\mathrm{pap}, \mathrm{septum}\}) : \\ (\alpha, \beta) \in [0, 0.25] \times \{1\} \cup [0, 0.25] \times [0.75, 1]\Big\} . \tag{5}$$

Example 4 (Solution to Example 2). The solution to this problem has already been implicitly given. Now it will be shown in the formalism of fuzzy sets. The first step is normalisation of input data. Because they are numeric attributes, simple approach is sufficient. For all three tumor dimensions, we define the same mapping shown in Fig. 5. Thus, we obtain following FMFF representing patient condition

$$\mathcal{A} = \left\{{}^{\mu_{\mathrm{big}}(a)}/_{\mathrm{first}} + {}^{\mu_{\mathrm{big}}(b)}/_{\mathrm{second}} + {}^{\mu_{\mathrm{big}}(b)}/_{\mathrm{third}} : \ abc = 200\right\} . \tag{6}$$

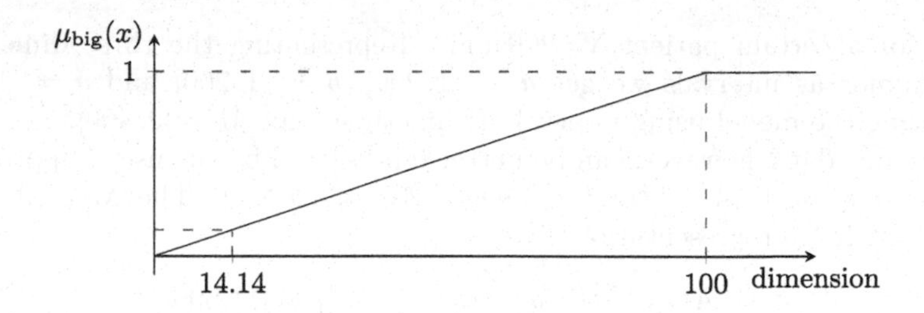

Fig. 5. Linguistic variable used to convert tumor dimensions from Example 4.

Diagnostic model needs to be adapted to handle normalised data

$$M_{\text{norm}}\left({}^{\mu_{\text{big}}(a)}/_{\text{first}} + {}^{\mu_{\text{big}}(b)}/_{\text{second}} + {}^{\mu_{\text{big}}(b)}/_{\text{third}}\right) = a + b\,. \tag{7}$$

With this model we obtain $M_{\text{norm}}(\mathcal{A}) = [11.696, 201]$ which is an optimal solution.

Definition 1. *Let $\sim$ be an equivalence relation between FMFF such that*

$$\mathcal{A} \sim \mathcal{B} \iff \forall_{x \in U}\,\{\mu_A(x) : A \in \mathcal{A}\} = \{\mu_B(x) : B \in \mathcal{B}\}\,. \tag{8}$$

Two FMFF are $\sim$–equivalent if they have exactly the same membership values for the same elements of the universal set. In accordance with this observation it should not be surprising that the equivalence classes $[\mathcal{A}]_\sim$ can be identified with Hesitant Fuzzy Sets.

Remark 1. Hesitant Fuzzy Sets are the smallest extension of fuzzy sets containing all FMFF in which there are no dependencies between the membership degrees of different elements of U.

Since HFS accurately describes the values of membership, ignoring the dependencies, they can be extended with the description of the dependency, gaining much wider data modelling capabilities.

Consider now the situation in which individual membership functions belonging to the FMFF have the weights assigned to them. Weighted FMFF (WFMFF) is defined as

$$\mathcal{A}^* \subseteq \mathcal{FS}(U) \times [0,1]\,. \tag{9}$$

Similarly as before, equivalence relation can be defined.

Definition 2. *Let $\sim^*$ be an equivalence relation between WFMFF such that*

$$\mathcal{A}^* \sim^* \mathcal{B}^* \iff \forall_{w \in [0,1]}\forall_{x \in U}\,\{\mu_A(x) : (A, w) \in \mathcal{A}^*\} = \{\mu_B(x) : (B, w) \in \mathcal{B}^*\}\,. \tag{10}$$

In this case, the equivalence classes $[\mathcal{A}^*]_{\sim^*}$ can be identified with Type–2 Fuzzy Sets.

Remark 2. Type–2 Fuzzy Sets are the smallest extension of fuzzy sets containing all WFMFF in which there are no dependencies between the membership degrees of different elements of U.

4.3 Computational Issues

In order for the proposed model to be used in practice, it is necessary to solve computational problems. The model in the proposed form is computationally inefficient at least using classical computation methods. Of course, by applying appropriate restrictions to HFS and representation of dependencies, this task can be reduced to various known computational problems, for example to linear programming, for which there is an effective calculation method. The second approach is to use highly parallel and increasingly accessible computing platforms such as GPGPU (general-purpose computing on graphics processing units) to effectively find satisfactory solutions. The author of this work believes that the combination of these two approaches will allow the construction of computationally efficient models based on FMFF.

5 Conclusions and Further Research

The presented results are the starting point for a comprehensive approach to modelling medical data. By defining an unambiguous and uniform interpretation of the patient record as a fuzzy set, further efforts can be focused on the problems of modelling data uncertainty and developing methods of supporting medical diagnostics.

The most important direction of further research, is the research on the effective use of this model in practical computations. However, for this to be possible, at least a few theoretical issues must be solved. It is necessary to examine the theoretical properties of the FMFFs which may be required to simplify computations. Moreover, it is important to solve the problem of additional elements introduced in feature set, which in some cases can increase the computational complexity and even reduce the efficiency [23]. This can be done thought introduction of hierarchical data model. It still needs to be developed formally. There is proposition to use lattice theory and L–Fuzzy Sets to solve this problem. The last step is to reconstruct some operations on FMFFs, important from the practical point of view, such as similarity measures.

Acknowledgments. This work was supported by the Polish National Science Centre grant number 2016/21/N/ST6/00316.

References

1. Zadeh, L.A.: The concept of a linguistic variable and its application to approximate reasoning—I. Inf. Sci. **8**(3), 199–249 (1975)
2. Atanassov, K.: Intuitionistic fuzzy sets. Fuzzy Sets Syst. **20**, 87–96 (1986)
3. Atanassov, K.: Intuitionistic Fuzzy Sets: Theory and Applications. Studies in Fuzziness and Soft Computing, vol. 35. Physica-Verlag, Heidelberg (1999)
4. Torra, V.: Hesitant fuzzy sets. Int. J. Intell. Syst. **25**(6), 529–539 (2010)
5. John, R.: Type 2 fuzzy sets: an appraisal of theory and applications. Int. J. Uncertain. Fuzziness Knowl. Based Syst. **6**(6), 563–576 (1998)

6. Mendel, J.M., John, R.B.: Type-2 fuzzy sets made simple. IEEE Trans. Fuzzy Syst. **10**(2), 117–127 (2002)
7. Dubois, D., Prade, H.: Gradualness, uncertainty and bipolarity: making sense of fuzzy sets. Fuzzy Sets Syst. **192**, 3–24 (2012)
8. Couso, I., Sánchez, L.: Machine learning models, epistemic set-valued data and generalized loss functions: an encompassing approach. Inf. Sci. **358**, 129–150 (2016)
9. Bustince, H., Barrenechea, E., Pagola, M., Fernandez, J., Xu, Z., Bedregal, B., Montero, J., Hagras, H., Herrera, F., De Baets, B.: A historical account of types of fuzzy sets and their relationships. IEEE Trans. Fuzzy Syst. **24**(1), 179–194 (2016)
10. Grattan-Guinness, I.: Fuzzy membership mapped onto intervals and many-valued quantities. Math. Log. Q. **22**(1), 149–160 (1976)
11. Zadeh, L.A.: Quantitative fuzzy semantics. Inf. Sci. **3**(2), 159–176 (1971)
12. Żywica, P., Wójtowicz, A., et al.: Improving medical decisions under incomplete data using interval-valued fuzzy aggregation. In: Proceedings of 9th European Society for Fuzzy Logic and Technology (EUSFLAT), Gijón, Spain, pp. 577–584 (2015)
13. Stachowiak, A., Żywica, P., Dyczkowski, K., Wójtowicz, A.: An interval-valued fuzzy classifier based on an uncertainty-aware similarity measure. In: Angelov, P., et al. (eds.) Intelligent Systems 2014. AISC, vol. 322, pp. 741–751. Springer, Cham (2015). https://doi.org/10.1007/978-3-319-11313-5_65
14. Żywica, P., Dyczkowski, K., Wójtowicz, A., Stachowiak, A., Szubert, S., Moszyński, R.: Development of a fuzzy-driven system for ovarian tumor diagnosis. Biocybern. Biomed. Eng. **36**(4), 632–643 (2016)
15. Dubois, D., Prade, H.: The three semantics of fuzzy sets. Fuzzy Sets Syst. **90**(2), 141–150 (1997)
16. Moszyński, R., Żywica, P., et al.: Menopausal status strongly influences the utility of predictive models in differential diagnosis of ovarian tumors: an external validation of selected diagnostic tools. Ginekol. Pol. **85**(12), 892–899 (2014)
17. Wójtowicz, A., Żywica, P., et al.: Dealing with uncertainty in ovarian tumor diagnosis. In: Atanassov, K., Homenda, W., et al. (eds.) Modern Approaches in Fuzzy Sets, Intuitionistic Fuzzy Sets, Generalized Nets and Related Topics, Volume II: Applications, SRI PAS, Warsaw, pp. 151–158 (2014)
18. Wójtowicz, A., Żywica, P., et al.: Solving the problem of incomplete data in medical diagnosis via interval modeling. Appl. Soft Comput. **47**, 424–437 (2016)
19. Żywica, P.: Similarity measures of interval–valued fuzzy sets in classification of uncertain data. Applications in Ovarian Tumor Diagnosis, Ph.D. thesis, Faculty of Mathematics and Computer Science of Adam Mickiewicz University, in Polish, June 2016
20. Dyczkowski, K.: Intelligent Medical Decision Support System Based on Imperfect Information. SCI, vol. 735. Springer, Cham (2018). https://doi.org/10.1007/978-3-319-67005-8
21. Mendel, J.M.: Tutorial on the uses of the interval type-2 fuzzy set's Wavy Slice Representation Theorem. In: Proceedings of Annual Meeting of the North American Fuzzy Information Processing Society (NAFIPS), New York City, USA, pp. 1–6 (2008)
22. Mendel, J.M., John, R.I., Liu, F.: Interval type-2 fuzzy logic systems made simple. IEEE Trans. Fuzzy Syst. **14**(6), 808–821 (2006)
23. Hughes, G.: On the mean accuracy of statistical pattern recognizers. IEEE Trans. Inf. Theor. **14**(1), 55–63 (1968)

An Uncertainty Aware Medical Diagnosis Support System

Krzysztof Dyczkowski(✉), Anna Stachowiak, Andrzej Wójtowicz, and Patryk Żywica

Department of Imprecise Information Processing Methods, Faculty of Mathematics and Computer Science, Adam Mickiewicz University, Umultowska 87, 61-614 Poznań, Poland
chris@amu.edu.pl

Abstract. In the paper we describe a computer system that store and process uncertain data in such a way as to be able to obtain information essential to make an effective diagnosis while also indicating the uncertainty level of that diagnosis. We consider the problem of incompleteness and imprecision of medical data and discuss some issues connected with such kind of information - like modeling, making decision that is aware of the imperfection of data, evaluating results in the context of uncertain medical data. As an example we describe a method of supporting medical decision implemented in the OvaExpert system that is based on interval-valued fuzzy sets cardinality.

Keywords: Diagnosis support · Medical data · Uncertainty Imperfect information

1 Introduction

Computer decision-making systems are highly effective in terms of prognosis when solving many diagnostic problems in medical context. This is true especially for common diseases for which there is access to large number of cases. The situation is less satisfactory for diseases which are less common and thus the access to large number of well-depicted cases is limited. Lack of centralized system for gathering uniform data from many medical institutions is also a problem. If such databases exist they are gathered in a specific medical center and are not accessible to others. Another problem is lack of access to full required diagnostics (e.g. due to unavailability of proper diagnostic equipment or high cost of diagnostic examinations), which contributes to ambiguities and omissions in patient's record. In addition, by their very nature, medical descriptions are often imprecise and ambiguous. In most cases, they are descriptive and terminology used in them is not standardized. Their quality often depends on the education of the doctor (including the center where he or she was educated) as well as the doctor's experience. The existing situation calls for the use of unconventional data modeling and reasoning methods. It requires methods factoring in both the

J. Medina et al. (Eds.): IPMU 2018, CCIS 855, pp. 381–390, 2018.
https://doi.org/10.1007/978-3-319-91479-4_32

imprecision and incompleteness of the data. Those methods must ensure high efficacy for disease entities for which there are no sufficiently large databases available.

In this paper we demonstrate some part of a bigger concept - the OvaExpert system - that was meant to deal with a forementioned situation. A set of concepts and methods cover the problem at every stage - collecting, modeling and processing of uncertain data. They combine theoretical knowledge with the capabilities of a computer system. We propose how to maximize the use of such a system and of computing power to solve efficiently the problem of uncertain data.

In Sect. 2 we give a brief view on the OvaExpert system and two research path that we have taken. Section 3 is devoted to one of the implemented methods, among many others, that supports gynaecologists in a diagnosis of ovarian tumors. In Sect. 4 we present the results of the analysis of methods based on counting. Section 5 gives some conclusions and areas for further research.

2 OvaExpert System - Two Research Tracks

OvaExpert, the intelligent system for ovarian tumor diagnosis, introduces a completely novel approach to the imprecision connected with data imperfection (see [1–4]), The aim of the system is to store and process uncertain data in such a way as to be able to obtain information essential to make an effective diagnosis while also indicating the uncertainty level with which the information is suggested.

Traditionally, gynaecologists are assisted by many prognostic models, ultrasonographic morphological scales, and other risk of malignancy calculators that are used for differential diagnosis of ovarian tumors. The most common diagnostic models are based on scoring systems [5,6] and logistic regressions [7]. Another predictive models were proposed by IOTA group: the most recent one is ADNEX [8].

The starting point for presented research was finding out that some of those models in some specific cases are complementing each other, i.e. applying them simultaneously yields better diagnostic efficacy as opposed to applying them separately (see [9]).

Consequently, there were two research tracks. The first one concerned the design of a decision model while the other involved using the synergy of the existing diagnostic models. Both tracks used interval-valued fuzzy sets in an epistemic sense which allowed us to include imperfect input data.

The first research track resulted in the concept of interval valued classifier based on similarity measures allowing imperfect input data. The results of this part of the research have been published in [10–13]. The method based on this algorithm will be marked as IVFC (method in the Ovaexpert system based on similarity measures).

The other research track involved using the method of aggregation/synergy of imperfect knowledge from several decision models. Our previous research has shown that fuzzy aggregation methods prove to be very effective in improving the quality of diagnosis and minimizing the impact of lack of data and imprecision. This is due to the variety of models and their different levels of efficacy across different patient groups. Many models, when used simultaneously, considerably improve the quality of the decision. As a part of this research path we applied the theory of interval-value fuzzy set cardinality, that will be described in more details in the next section. This approach allows to make a decision supported by majority of data sources (models) preserving the information about the level of uncertainty about this decision. The other research centers are also currently developing this approach using intuitionistic fuzzy preference relations (cf. [14–17]).

3 Algorithm for Decision Support Based on Interval-Valued Fuzzy Set Cardinality

Interval-valued fuzzy sets (IVFS) are a special variant of type-2 fuzzy sets, also introduced by Zadeh (see [18]). The notion of interval-valued fuzzy sets is a generalization of the notion of a usual fuzzy set. Its significant role is to introduce uncertainty as an actual value of membership function (epistemic interpretation of interval-valued fuzzy set (see [19])) that can be anywhere between the given interval values. Two approaches to interval-value fuzzy set cardinality were used: scalar (sigma f-Count) and fuzzy (f-FECount). Both make use of the cardinality patterns - functions that help determine the influence of single elements of an interval-valued fuzzy set on the value of its cardinality. In the case of interval-valued fuzzy set its cardinality is an interval or an interval-valued fuzzy set, and a notion of interval representative was introduced to compare cardinalities. It is a single real number belonging to this interval. The most obvious interval representatives include: interval center, right limit (minimum value) and left limit (maximum value).

The idea behind the algorithm conforms with a usual method of making decisions by counting crisp sets. We make a decision supported by majority of data sources, on condition that they are more numerous that the reverse option by a specified value. If both options have the same support of decision sources (or the difference is minimal), then we do not decide. The idea behind decision algorithm is to use bipolar perspective on IVFS. Because an IVFS contains information about uncertainty level, it carries both information supporting and rejecting the decision. This property of IVFS is used in decision algorithm.
The basic idea behind this algorithm consists of a couple of steps:

1. On the basis of input data, we define two IVFSs: P ("Pro") modeling support level for a positive decision and C ("Contra") mirroring support level for a negative decision.
2. We calculate cardinalities of these IVFSs with the selected calculation method.
3. We compare cardinalities to find out whether we can make a decision i.e. whether one of them significantly outweighs the other, and if so we select the decision supported by greater cardinality.

In order to make a decision we need to determine a method for comparing cardinality intervals. For this purpose, we defined two approaches (modes):

- interval approach consisting in comparing overlap of intervals of respective distances between their endpoints,
- numerical approach consisting in determining numerical interval representatives.

Depending on selection of calculation methods and comparison methods we obtain various decision algorithms based on sigma f-count from specific groups:

- SC-cen – based on interval center representatives
- SC-int – based on interval comparison method
- SC-max – based on left limit representatives.

and based on $f - FEcount$:

- FE-cen – based on interval center representatives
- FE-int – based on interval comparison method
- FE-max – based on left limit representatives.

An outline of the solution is presented in Fig. 1.

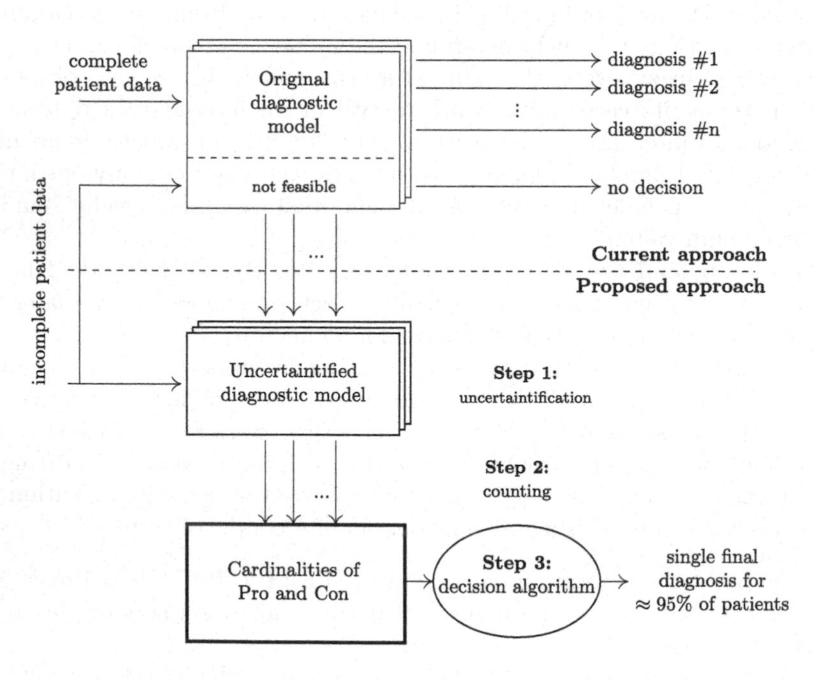

Fig. 1. A solution based on the cardinality of IVFSs.

The interval mode is much more restrictive and only efficient in situations with small amount of missing information (size of ignorance intervals). Definition of cardinality pattern is also of key importance. If using identity function as cardinality pattern (using sigma-count for calculating cardinality) cardinalities of both IVFSs P and C are symmetrical and decision is only made if both IVFSs are sufficiently similar to crisp sets. It is also important that thresholds (parameters defining cardinality pattern) are selected in such a way as to reduce the significance of input decisions close to 0.5.

4 Evaluation of Efficiency - Results

A very important aspect of construction and application of tumor malignancy classification (prediction) methods is to evaluate their efficiency (prediction quality). In a binary classification, we divide the decisions into two classes: positive (malignant tumors, which also include borderline malignant tumors which require the same treatment as ovarian cancer), and negative (benign tumors and non-neoplastic changes). In addition, in our research we allow a situation in which a classifier may not make a decision due to data being of too low quality. In medicine, numerous quality classification measures are applied, i.e. sensitivity, specificity, accuracy, f-measure etc. For classifiers that operate on data of poor quality (e.g. incomplete data), in some applications, it is necessary to consider a situation in which the classifier has insufficient information to make a sufficiently certain decision. This is often the case in medical applications when insufficiently certain decision can have serious consequences for the patient. This is why an additional measure has been introduced – decisiveness – which determines in how many cases the classifier was able to make a decision.

In many applications (often including medical ones) the above measures do not reflect the actual required quality of the classifier. This is the case when the significance of the individual classes of errors (actual effects of wrong decisions) are different. For example, in the medical diagnosis of ovarian tumors the situation when the system diagnoses a tumor as benign and, in fact, it was malignant causes much more significant effects for the patient as opposed to the situation when the benign tumor is diagnosed as cancer. In such models, the concept of cost matrix (cost function) is used where for each error type a weight (penalty) is assigned for a wrong decision. The quality value is the sum of costs (penalties) assigned to the classifier for making wrong decisions. Such a cost matrix will be used to evaluate the quality of classifications in our system.

The presented algorithms have been tested on real medical data. These data described 388 cases of patients diagnosed and treated in the Division of Gynecological Surgery, Poznan University of Medical Sciences, between 2005 and 2015. Out of them 61% have been diagnosed as suffering from benign tumors and 39% as suffering from malign tumors. Moreover, 56% of patients had full diagnostic (no test required by diagnostic scales was missing), 40% had significant amounts of missing data varying from 0% to 50%, and for the remaining ones 50% of data was missing. Detailed description of data used for evaluation can be found in [9]. More information on the data format and technical details can be found in [20].

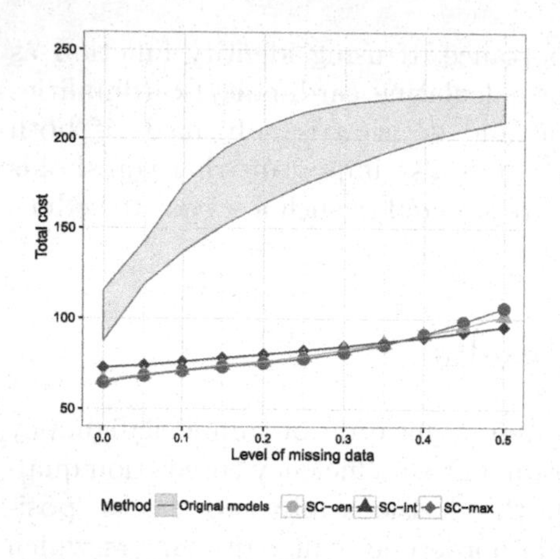

Fig. 2. Decision making efficacy of algorithms based on sigma f-count

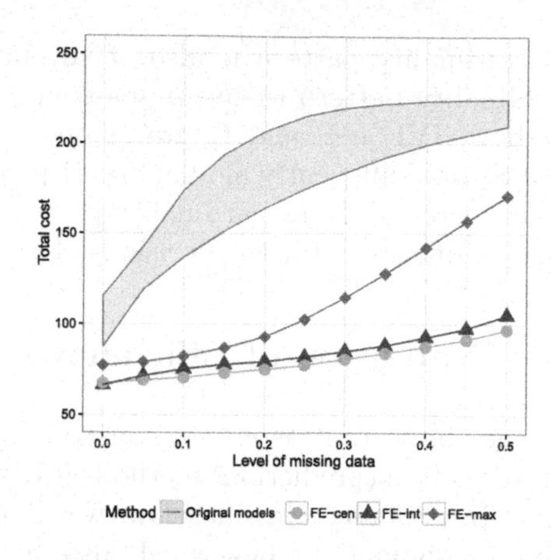

Fig. 3. Decision making efficacy of algorithm based on f-FEcount

Figures 2 and 3 present classification results based on the proposed algorithms with the best versions obtained from optimization in specific groups. Efficiency area of original models has been marked in grey. The graphs show the total cost (the higher the cost the lower the classification quality) in relation to the level of missing data. The graph in Fig. 2 presents the best three algorithms based on sigma f-count: SC-cen, SC-int, SC-max, whereas, the graph presented in Fig. 3 presents three best algorithms based on $f-FEcount$: FE-cen, FE-int, FE-max.

As a result of analysis of the obtained decision efficiency, the algorithm FE-cen has been selected as the best for application in the OvaExpert system from amount the counting methods. A method based on this algorithm with the use of cardinality pattern is designated as FSC (the OvaExpert system method based on counting). The prognostic results of all three decision modules implemented in the OvaExpert system - OEA, IVFC and FSC - are presented in Table 1 and for the purpose of comparison the results for the original diagnostic models are also presented.

The original diagnostic models differ in their classification properties: some of them tend to make more conservative decisions (i.e. LR1, LR2, SM), and some of them are more liberal (i.e. RMI, Tim.). This can be observed in discernible differences in values between sensitivity and specificity. Only one of these models ensures the balance of both factors (Alc.). It should be noted that all original models have very low decisiveness (due to deficiencies in diagnostic data), which results in high total cost.

The new models implemented in the OvaExpert system have high sensitivity and specificity values. Two of them tend to be more conservative (OEA and IVFC), while FSC is more balanced. All three models provide a high level of decisiveness because they are able to deal with deficiencies in data. This is why their total cost is much lower than the original models.

Table 1. The results of the decision-making quality of the original models compared to the OvaExpert methods

		Total cost	Dec.	Sen.	Spec.	Acc.
Original models	Alc. [5]	189.0	20.6 %	88.2 %	89.5 %	88.9 %
	LR1 [21]	184.0	27.4 %	92.6 %	57.1 %	77.1 %
	LR2 [21]	164.0	33.1 %	94.3 %	65.2 %	82.8 %
	RMI [22]	156.0	56.6 %	75.9 %	87.1 %	83.8 %
	SM [23]	142.0	62.9 %	94.6 %	71.2 %	79.1 %
	Tim. [24]	159.0	47.4 %	66.7 %	97.1 %	91.6 %
New diag.modules	OEA	72.0	96.6 %	90.2 %	86.4 %	87.6 %
	IVFC	72.5	100.0 %	90.4 %	84.6 %	86.3 %
	FSC	67.0	93.7 %	90.0 %	90.2 %	89.4 %

It can be noted that the diagnostic models of the OvaExpert system differ significantly from the original models in terms of classification. Although diagnostic modules differ in classification quality indicators from one another, the differences in classification are not statistically significant.

In the light of these results, the OvaExpert system based on the presented modules is a promising tool for supporting the prognosis of ovarian cancers, especially in the case of partial gaps in diagnostic data that are common in the everyday medical practice.

5 Conclusions and Further Research

At the moment, the OvaExpert system is tested in several medical centers offering the diagnosis and treatment of gynecological tumors. Its further implementation depends on overcoming legal and organizational obstacles concerning medical systems in Poland.

The demo version of the system is available on the project website http://ovaexpert.pl where one can get acquainted with the functions and possibilities offered by the system.

Statistical evaluation and implementation of the proposed methods have been performed with R, version 3.1.2. Scripts, documentation and non-sensitive data are available at GitHub (see http://ovaexpert.github.io/ovarian-tumor-aggregation). Because of large amount of calculations needed to do the research, we decided to use Microsoft Azure cloud service available to our team under Microsoft Azure Research Grant "Azure Machine Learning – Development of an Intelligent System for Ovarian Tumor Diagnosis".

The OvaExpert system was designed to take advantage of the synergy of many classic diagnostic models and those newly created based on the knowledge derived from the data. The system has implemented all well-known prediction models since medical specialists trust their results. Additionally, new diagnostic methods have been implemented - among others a method based on IVFSs cardinalities, described in this paper. This method, based on a solid theoretical foundation, is relatively easy to implement and interpret, and, most importantly, achieves very good effectiveness in real-life applications such as medical diagnostics. Our approach is meant to be adapted also to non-medical problems where data quality is a matter of concern. It could be applied when the information that comes from independent experts is imperfect and it is important to preserve information about this imperfection in the final result. By returning bipolar information – concerning the quantities of positive and negative premises – we are able to evaluate that imperfection and the quality of the information.

All of our effort may be summarized with the following achievements:

- Development of computational intelligence methods that help make decisions based on low quality data, in particular:
  - Development of representation and processing methods for low quality data using interval-valued fuzzy sets.
  - Development of selection and optimization methods for decision making algorithms based on interval-valued fuzzy sets.
  - Development of methods calculating the cardinalities of interval-valued fuzzy sets.
  - Development of decision making algorithms based on the cardinalities of interval-valued fuzzy sets.
- Application of the above-mentioned methods in designing the intelligent system OvaExpert supporting medical diagnosis.
- Pilot implementation of the OvaExpert system that supports gynecologists and helps gathering data for further research and development.

References

1. Dyczkowski, K.: Intelligent Medical Decision Support System Based on Imperfect Information: The Case of Ovarian Tumor Diagnosis. SCI, vol. 735. Springer, Cham (2018). https://doi.org/10.1007/978-3-319-67005-8
2. Dyczkowski, K., Wójtowicz, A., Żywica, P., Stachowiak, A., Moszyński, R., Szubert, S.: An intelligent system for computer-aided ovarian tumor diagnosis. In: Filev, D., Jabłkowski, J., Kacprzyk, J., Krawczak, M., Popchev, I., Rutkowski, L., Sgurev, V., Sotirova, E., Szynkarczyk, P., Zadrozny, S. (eds.) Intelligent Systems'2014. AISC, vol. 323, pp. 335–343. Springer, Cham (2015). https://doi.org/10.1007/978-3-319-11310-4_29
3. Wójtowicz, A., Żywica, P., et al.: Dealing with uncertainty in ovarian tumor diagnosis. In: Atanassov, K., Homenda, W., et al. (eds.) Modern Approaches in Fuzzy Sets, Intuitionistic Fuzzy Sets, Generalized Nets and Related Topics, Volume II: Applications, SRI PAS, Warsaw, pp. 151–158 (2014)
4. Żywica, P., Dyczkowski, K., Wójtowicz, A., Stachowiak, A., Szubert, S., Moszyński, R.: Development of a fuzzy-driven system for ovarian tumor diagnosis. Biocybernetics Biomed. Eng. **36**(4), 632–643 (2016)

5. Alcazar, J.L., Merce, L.T., et al.: A new scoring system to differentiate benign from malignant adnexal masses. Obstet. Gynecol. Surv. **58**(7), 462–463 (2003)
6. Szpurek, D., Moszyński, R., et al.: An ultrasonographic morphological index for prediction of ovarian tumor malignancy. Eur. J. Gynaecol. Oncol. **26**(1), 51–54 (2005)
7. Timmerman, D., Testa, A.C., et al.: Logistic regression model to distinguish between the benign and malignant adnexal mass before surgery: a multicenter study by the international ovarian tumor analysis group. J. Clin. Oncol. **23**(34), 8794–8801 (2005)
8. Van Calster, B., Van Hoorde, K., et al.: Evaluating the risk of ovarian cancer before surgery using the adnex model to differentiate between benign, borderline, early and advanced stage invasive, and secondary metastatic tumours: prospective multicentre diagnostic study. BMJ **349**, 5920 (2014)
9. Moszyński, R., Żywica, P., et al.: Menopausal status strongly influences the utility of predictive models in differential diagnosis of ovarian tumors: an external validation of selected diagnostic tools. Ginekol. Pol. **85**(12), 892–899 (2014)
10. Stachowiak, A., Żywica, P., Dyczkowski, K., Wójtowicz, A.: An interval-valued fuzzy classifier based on an uncertainty-aware similarity measure. In: Angelov, P., Atanassov, K.T., Doukovska, L., Hadjiski, M., Jotsov, V., Kacprzyk, J., Kasabov, N., Sotirov, S., Szmidt, E., Zadrożny, S. (eds.) Intelligent Systems 2014. AISC, vol. 322, pp. 741–751. Springer, Cham (2015). https://doi.org/10.1007/978-3-319-11313-5_65
11. Żywica, P.: Similarity measures of interval-valued fuzzy sets in classification of uncertain data. Applications in Ovarian Tumor Diagnosis, Ph.D. thesis, Faculty of Mathematics and Computer Science of Adam Mickiewicz University, in Polish, June 2016
12. Żywica, P., Stachowiak, A., Wygralak, M.: An algorithmic study of relative cardinalities for interval-valued fuzzy sets. Fuzzy Sets Syst. **294**, 105–124 (2016)
13. Żywica, P., Stachowiak, A.: A new algorithm for computing relative cardinality of intuitionistic fuzzy sets. In: Atanassov, K., et al. (eds.) Modern Approaches in Fuzzy Sets, Intuitionistic Fuzzy Sets, Generalized Nets and Related Topics, Volume I: Foundations, IBS PAN - SRI PAS, Warsaw, pp. 181–189 (2014)
14. Bentkowska, U., Król, A.: Preservation of fuzzy relation properties based on fuzzy conjunctions and disjunctions during aggregation process. Fuzzy Sets Syst. **291**, 98–113 (2016)
15. Dudziak, U., Pękala, B.: Intuitionistic fuzzy preference relations. In: Proceedings of the 7th Conference of the European Society for Fuzzy Logic and Technology, pp. 529–536. Atlantis Press (2011)
16. Pękala, B.: Properties of interval-valued fuzzy relations, Atanassov's operators and decomposable operations. In: Hüllermeier, E., Kruse, R., Hoffmann, F. (eds.) IPMU 2010. CCIS, vol. 80, pp. 647–655. Springer, Heidelberg (2010). https://doi.org/10.1007/978-3-642-14055-6_68
17. Pękala, B.: Operations on interval matrices. In: Rough Sets and Intelligent Systems Paradigms, pp. 613–621 (2007)
18. Zadeh, L.A.: The concept of a linguistic variable and its application to approximate reasoning–I. Inf. Sci. **8**(3), 199–249 (1975)
19. Dubois, D., Liu, W., Ma, J., Prade, H.: The basic principles of uncertain information fusion. an organised review of merging rules in different representation frameworks. Inf. Fusion **32**, 12–39 (2016)
20. Wójtowicz, A., Żywica, P., et al.: Solving the problem of incomplete data in medical diagnosis via interval modeling. Appl. Soft Comput. **47**, 424–437 (2016)

21. Timmerman, D., Van Calster, B., Testa, A.C., Guerriero, S., Fischerova, D., Lissoni, A., Van Holsbeke, C., Fruscio, R., Czekierdowski, A., Jurkovic, D., et al.: Ovarian cancer prediction in adnexal masses using ultrasound-based logistic regression models: a temporal and external validation study by the iota group. Ultrasound Obstet. Gynecol. **36**(2), 226–234 (2010)
22. Jacobs, I., Oram, D., et al.: A risk of malignancy index incorporating CA 125, ultrasound and menopausal status for the accurate preoperative diagnosis of ovarian cancer. BJOG: Int. J. Obstet. Gynaecol. **97**(10), 922–929 (1990)
23. Szpurek, D., Moszynski, R., Zietkowiak, W., Spaczynski, M., Sajdak, S.: An ultrasonographic morphological index for prediction of ovarian tumor malignancy. Eur. J. Gynaecol. Oncol. **26**(1), 51–54 (2005)
24. Timmerman, D., Bourne, T.H., Tailor, A., Collins, W.P., Verrelst, H., Vandenberghe, K., Vergote, I.: A comparison of methods for preoperative discrimination between malignant and benign adnexal masses: the development of a new logistic regression model. Am. J. Obstet. Gynecol. **181**(1), 57–65 (1999)

Diverse Classes of Interval-Valued Aggregation Functions in Medical Diagnosis Support

Urszula Bentkowska and Barbara Pękala(✉)

Interdisciplinary Centre for Computational Modelling, University of Rzeszów, Pigonia 1, 35-310 Rzeszów, Poland
{ududziak,bpekala}@ur.edu.pl

Abstract. In this contribution results connected with using new types of aggregation functions in medical diagnosis support are presented. These aggregation functions belong to the recently introduced families of possible and necessary aggregation functions as well as aggregation functions with respect to admissible linear orders. Examples of the mentioned families of aggregation functions proved to be comparably effective (if it comes to statistical measures and lower cost of prediction) to the previously used aggregation functions in medical diagnosis support systems. The considered classes of aggregation functions differ from the ones previously applied by the comparability relations between intervals involved in the monotonicity conditions.

Keywords: Aggregation functions with respect to admissible linear orders · Possible aggregation functions
Necessary aggregation functions · Uncertainty · Medical diagnosis
Decision making

1 Introduction

Interval-Valued Fuzzy Sets (IVFSs) [24,34], which are extensions of fuzzy sets [33], are applied in many areas like databases, pattern recognition, neural networks, fuzzy modeling, economy, medicine or multicriteria decision making. Also in classification the use of IVFSs has led to improvements of the performance of some of the state-of-the-art algorithms for fuzzy rule based classification systems. More precisely, the use of intervals for the creation of the rules gives more flexibility to the algorithms, which leads to better results [25].

One of the real-life diagnosis support systems for ovarian tumor diagnosis is OvaExpert [15,16,19,26,28,31,32,36,37]. It enables to make accurate and high-quality decisions under incomplete information and uncertainty. For the evaluation process in [32] there were selected six diagnostic models: two scoring systems SM and Alc [1,27] and four regression models LR1, LR2, Tim and RMI [17,29,30]. OvaExpert uses interval modeling of incomplete data and thanks to this uncertaintification of the classical mentioned models is possible. The classical models were created by individual research units, such as the Alcazar model

J. Medina et al. (Eds.): IPMU 2018, CCIS 855, pp. 391–403, 2018.
https://doi.org/10.1007/978-3-319-91479-4_33

and SM, other methods (like LR1) by organizations (incorporating a number of research centers), such as IOTA (The International Ovarian Tumor Analysis group which was founded in 1999 by Dirk Timmerman, Lil Valentin and Tom Bourne). The majority are scoring models and models based on logistic regression. In decision module of diagnostic models aggregation of the existing models may be applied in order to take advantage of the synergy of data. Moreover, aggregation methods applied to diverse structures and problems proved to be effective (cf. [2,3,8,13,14,20,21]).

In OEA (cf. [16,32]) the main approach deployed in the system is based on Ordered Weighted Averaging operation (OWA). In ovarian tumor diagnosis the problem of missing data is commonly encountered. The results presented in [37] confirmed that methods based on interval modeling and aggregation make it possible to reduce the negative impact of lack of data and lead to meaningful and accurate decisions. A diagnostic model developed in this way proved better than classical diagnostic models for ovarian tumor. OEA is based on the binary classifier (malignant or benign). In OvaExpert [16] there are applied diverse modules where multi-class classification is possible. In OEA the aim of the training phase was to optimize the parameters of the aggregation operators and thresholding strategies on different simulated percentages of missing features. In the testing phase, the optimized aggregation operators and thresholding strategies were examined on the test set. This step checked the performance of these aggregation operators on data with the actual missing values. Although there were considered interval-valued functions representing models, there were applied two possible modes of aggregation. The first, called numerical, uses a single value that represents the whole interval (e.g. the interval's center, lower bound or upper bound). The interval mode, on the other hand, utilizes the whole of the interval information. The study group consisted of 388 patients diagnosed and treated for ovarian tumor in the Division of Gynecological Surgery, Poznan University of Medical Sciences, between 2005 and 2015.

The aim of this contribution is to show the performance of diverse classes of aggregation functions in medical decision support systems using aggregation methods. One of such systems is OvaExpert with its module OEA [32]. Here we present the results on aggregation methods connected with recently introduced possible and necessary aggregation functions [4] and also aggregation functions with respect to admissible linear orders [35]. These new concepts of aggregation functions follow from diverse concepts concerning comparability of intervals. Namely, these are possible and necessary comparability relations connected with epistemic and ontic setting of interval-valued calculus [11,12] and the concept of linear orders introduced in [5]. It turned out that on OEA dataset the considered examples of aggregation functions are statistically comparable with the previously applied operators and in is some cases they obtained lower cost of prediction. However, the presented results were obtained with less number of repetitions in the training phase what may result in lower stability of the obtained results.

This work is composed of the following parts. Firstly, some concepts and results connected with aggregation functions are presented (Sect. 2). Next, basic information about OvaExpert diagnosis support system are recalled and information about the performance of the new types of aggregation functions in OEA in connection with decision support are provided (Sect. 3).

2 Aggregation Functions for Interval-Valued Fuzzy Setting

Firstly, we recall definition of an interval-valued fuzzy set and the classically applied order for this setting.

Definition 1 (cf. [24,34]). *An interval-valued fuzzy set F in X is a mapping $F : X \to L^I$ such that $F(x) = [\underline{F}(x), \overline{F}(x)] \in L^I$ for $x \in X$, where*

$$L^I = \{[x_1, x_2] : x_1, x_2 \in [0,1], \; x_1 \leqslant x_2\}.$$

The well-known classical monotonicity (partial order) for intervals is of the form

$$[x_1, y_1] \preceq [x_2, y_2] \Leftrightarrow x_1 \leqslant x_2, \; y_1 \leqslant y_2. \tag{1}$$

The family of all fuzzy sets on a given universe X with $\preceq$ is partially ordered and moreover it is a lattice. We may also consider the following comparability relations on L^I (cf. [22]):

$$[x_1, y_1] \preceq_\pi [x_2, y_2] \Leftrightarrow x_1 \leqslant y_2, \tag{2}$$

$$[x_1, y_1] \preceq_\nu [x_2, y_2] \Leftrightarrow y_1 \leqslant x_2. \tag{3}$$

These relations, including classical order, follow from the epistemic setting of interval-valued fuzzy sets and form the full possible set of interpretations of comparability relations on intervals. Relation $\preceq_\pi$ is an interval order and the relation $\preceq_\nu$ is antisymmetric and transitive on L^I. Detailed discussion on this subject will be presented in [23].

Classical order (1) is not complete, which is important for application reasons for example in decision making problems. In the paper [5] the general method to build different linear orders for L^I, covering some of the known linear orders for intervals, such as lexicographical orders, the Xu and Yager order, was presented. An order $\leq_{L^I}$ on L^I is called admissible if it is linear and for all $x, y \in L^I$, such that if $x \preceq y$, then $x \leq_{L^I} y$. For example in [6], this class of linear orders on L^I was used to extend the definition of OWA operators to interval-valued fuzzy setting. We recall here only the notion of Xu and Yager $\leq_{XY}$ linear order which will be applied in this paper.

Let $x = [\underline{x}, \overline{x}], y = [\underline{y}, \overline{y}]$. Xu and Yager linear order is defined as follows $[\underline{x}, \overline{x}] \leq_{XY} [\underline{y}, \overline{y}]$ if and only if $\underline{x}+\overline{x} < \underline{y}+\overline{y}$ or ($\underline{x}+\overline{x} = \underline{y}+\overline{y}$ and $\overline{x}-\underline{x} \leqslant \overline{y}-\underline{y}$).

In this paper we will present the behavior of some operations for interval-valued fuzzy setting, namely possible and necessary aggregation functions and aggregation functions with respect to admissible linear orders. Before giving these definitions we recall the notion of interval-valued aggregation functions, already applied on OEA dataset [16], and an aggregation function on $[0, 1]$.

Definition 2 (cf. [7], p. 6**).** *An increasing function* $A : [0,1]^n \to [0,1]$, $n \in \mathbb{N}$, $n \geqslant 2$, *is called an aggregation function if* $A(0,\ldots,0) = 0$, $A(1,\ldots,1) = 1$.

Definition 3 (cf. [18]**).** *An operation* $\mathcal{A} : (L^I)^n \to L^I$ *is called an aggregation function on* L^I *if it is increasing, i.e.*

$$\forall_{\boldsymbol{x}_i, \boldsymbol{y}_i \in L^I} \quad \boldsymbol{x}_i \preceq \boldsymbol{y}_i \Rightarrow \mathcal{A}(\boldsymbol{x}_1, \ldots, \boldsymbol{x}_n) \preceq \mathcal{A}(\boldsymbol{y}_1, \ldots, \boldsymbol{y}_n) \tag{4}$$

and $\mathcal{A}(\underbrace{\boldsymbol{0}, ..., \boldsymbol{0}}_{n\times}) = \boldsymbol{0}$, $\mathcal{A}(\underbrace{\boldsymbol{1}, ..., \boldsymbol{1}}_{n\times}) = \boldsymbol{1}$.

For the simplicity of notations all examples and representations of aggregation functions will be presented for two–argument cases (n-argument versions may be obtained recursively).

Definition 4 (cf. [9]**).** *Let* $\mathcal{A} : (L^I)^2 \to L^I$ *be an aggregation function.* $\mathcal{A}$ *is said to be a representable aggregation function on* L^I *if there exist two aggregation functions* $A_1, A_2 : [0,1]^2 \to [0,1]$, $A_1 \leqslant A_2$ *such that, for every* $[x_1, x_2], [y_1, y_2] \in L^I$ *it holds that*

$$\mathcal{A}([x_1, x_2], [y_1, y_2]) = [A_1(x_1, y_1), A_2(x_2, y_2)].$$

Representability is not the only possible way to build interval-valued aggregation functions.

Definition 5 (cf. [10]**).** *Let* $\boldsymbol{x} = [x_1, x_2], \boldsymbol{y} = [y_1, y_2] \in L^I$ *and let* $A_1, A_2 : [0,1]^2 \to [0,1]$, $A_1 \leqslant A_2$ *be aggregation functions. The aggregation function* $\mathcal{A}$ *on* L^I *is called pseudomax* A_1A_2 *- representable if*

$$\mathcal{A}(\boldsymbol{x}, \boldsymbol{y}) = [A_1(x_1, y_1), \max(A_2(x_1, y_2), A_2(x_2, y_1))], \tag{5}$$

and pseudomin A_1A_2*-representable if*

$$\mathcal{A}(\boldsymbol{x}, \boldsymbol{y}) = [\min(A_1(x_1, y_2), A_1(x_2, y_1)), A_2(x_2, y_2)]. \tag{6}$$

By replacing in the monotonicity condition (4) the natural order $\preceq$ with the admissible linear orders or relations $\preceq_\pi$ and $\preceq_\nu$, new types of aggregation functions are obtained [4,35].

Definition 6. *$\mathcal{A} : (L^I)^n \to L^I$ is called an aggregation function with respect to $\leq_{L^I}$, if*

$$\forall_{\boldsymbol{x}_i, \boldsymbol{y}_i \in L^I} \ \boldsymbol{x}_i \leq_{L^I} \boldsymbol{y}_i \Rightarrow \mathcal{A}(\boldsymbol{x}_1, \ldots, \boldsymbol{x}_n) \leq_{L^I} \mathcal{A}(\boldsymbol{y}_1, \ldots, \boldsymbol{y}_n)$$

and $\mathcal{A}(\underbrace{\boldsymbol{0}, ..., \boldsymbol{0}}_{n\times}) = \boldsymbol{0}, \ \mathcal{A}(\underbrace{\boldsymbol{1}, ..., \boldsymbol{1}}_{n\times}) = \boldsymbol{1}.$

Definition 7. *An operation $\mathcal{A} : (L^I)^n \to L^I$ is called a possible aggregation function (for short pos-aggregation function) if*

$$\forall_{\boldsymbol{x}_i, \boldsymbol{y}_i \in L^I} \ \boldsymbol{x}_i \preceq_{\pi} \boldsymbol{y}_i \Rightarrow \mathcal{A}(\boldsymbol{x}_1, \ldots, \boldsymbol{x}_n) \preceq_{\pi} \mathcal{A}(\boldsymbol{y}_1, \ldots, \boldsymbol{y}_n) \tag{7}$$

and $\mathcal{A}(\underbrace{\boldsymbol{0}, ..., \boldsymbol{0}}_{n\times}) = \boldsymbol{0}, \ \mathcal{A}(\underbrace{\boldsymbol{1}, ..., \boldsymbol{1}}_{n\times}) = \boldsymbol{1}.$

Definition 8. *An operation $\mathcal{A} : (L^I)^n \to L^I$ is called a necessary aggregation function (for short nec-aggregation function) if*

$$\forall_{\boldsymbol{x}_i, \boldsymbol{y}_i \in L^I} \ \boldsymbol{x}_i \preceq_{\nu} \boldsymbol{y}_i \Rightarrow \mathcal{A}(\boldsymbol{x}_1, \ldots, \boldsymbol{x}_n) \preceq_{\nu} \mathcal{A}(\boldsymbol{y}_1, \ldots, \boldsymbol{y}_n) \tag{8}$$

and $\mathcal{A}(\underbrace{\boldsymbol{0}, ..., \boldsymbol{0}}_{n\times}) = \boldsymbol{0}, \ \mathcal{A}(\underbrace{\boldsymbol{1}, ..., \boldsymbol{1}}_{n\times}) = \boldsymbol{1}.$

Let $\mathbf{x} = [x_1, x_2]$, $\mathbf{y} = [y_1, y_2]$. The following functions are aggregation functions on L^I (non-representable) and they are also pos-aggregation functions (but they are not nec-aggregation functions):

$$\mathcal{A}_{pi1}(\mathbf{x}, \mathbf{y}) = \begin{cases} [1, 1], & (\mathbf{x}, \mathbf{y}) = ([1, 1], [1, 1]) \\ [\frac{y_1 \frac{x_1+x_2}{2}}{2}, \frac{x_2+y_2}{2}], & \text{otherwise} \end{cases}$$

$$\mathcal{A}_{pi2}(\mathbf{x}, \mathbf{y}) = \begin{cases} [1, 1], & (\mathbf{x}, \mathbf{y}) = ([1, 1], [1, 1]) \\ [\frac{x_1 \frac{y_1+y_2}{2}}{2}, \frac{x_2+y_2}{2}], & \text{otherwise} \end{cases}$$

The following function is a nec-aggregation function (it is a pseudomax $A_1 A_2$ - representable aggregation function) but it is not a pos-aggregation function:

$$\mathcal{A}_{nu}(\mathbf{x}, \mathbf{y}) = [\frac{x_1 + y_1}{2}, \max(\frac{x_1 + y_2}{2}, \frac{x_2 + y_1}{2})].$$

As a useful example of aggregation function with respect to an admissible linear order we may put OWA operator. Firstly, we recall definition of OWA operator in a numeric case.

Definition 9. *Let $w = (w_1, ..., w_n) \in [0, 1]^n$ be a weighted vector (i.e., $w_i \in [0, 1]$ and $\sum_{i=1}^{n} w_i = 1$). An OWA operator of dimension n associated with the weighted vector w is a function $OWA : [0, 1]^n \to [0, 1]$ defined by*

$$OWA(x_1, ..., x_n) = \sum_{i=1}^{n} w_i x_{(i)},$$

where (.) denotes a permutation of $\{1, ..., n\}$ such that $x_{(1)} \geqslant x_{(2)} \geqslant ... \geqslant x_{(n)}$.

And for an interval case we have

Definition 10 [6]. *Let $\leq$ be an admissible order on L^I, and let $w = (w_1, \ldots, w_n) \in [0,1]^n$, with $w_1 + \cdots + w_n = 1$. The Interval-Valued OWA operator (IVOWA) associated with $\leq$ and w is a mapping $IVOWA_{\leq,w} : (L^I)^n \to L^I$, given by*

$$IVOWA_{\leq,w}([a_1, b_1], \ldots, [a_n, b_n]) = \sum_{i=1}^{n} w_i \cdot [a_{(i)}, b_{(i)}],$$

where $[a_{(i)}, b_{(i)}]$, $i = 1, \ldots, n$, denotes the i-th greatest of the inputs with respect to the order $\leq$ and $w \cdot [a,b] = [wa, wb]$, $[a_1, b_1] + [a_2, b_2] = [a_1 + a_2, b_1 + b_2]$.

Let us note that $IVOWA_{\leq,w}$ is not an aggregation function with respect to $\preceq$ (see [6]). This is why it is necessary to use linear orders in its notion.

3 Performance of New Types of Aggregation Functions in Ovarian Tumor Diagnosis

As we have already mentioned in the Introduction, application of aggregation methods in medical diagnosis support proved to be fruitful and improved the ability to obtain the final diagnosis (cf. [31]). In this contribution we would like to present the results connected with potential application of the possible and necessary aggregation functions as well as aggregation functions with respect to admissible linear orders. In fact we would also like to check the performance of diverse comparability relations for interval comparing, i.e. possible, necessary and linear orders. We use analogous methods to the ones presented in [29] and apply our operators on OEA dataset from [16] designed for ovarian tumor diagnosis. However, in the training phase (due to the ability of the applied equipment) we performed less repetitions of the evaluation. For some of the examples of aggregation functions we obtained comparable results and for the others even better results, i.e. these examples of aggregation functions proved to be comparable statistically but with lower cost of prediction. We have tested several examples of each considered class of aggregations and diverse tresholding strategies (cf. [37]). We have considered three classes of aggregation functions. We have chosen the best representatives of each class. Moreover, diverse methods of creating n-argument versions of binary aggregation functions [4] were considered. If it comes to the class of possible and necessary aggregation functions we had to take into account the properties of comparability relations involved in the notions of these aggregation functions. Since, $\preceq_\pi$ is connected but it is not antisymmetric, we also took into account the width of intervals and the position of endpoints while creating the n-argument versions of aggregation functions. The inputs were sorted increasingly and decreasingly. Since relation $\preceq_\nu$ is not connected, we also had to perform analogous procedure as for creating pos-aggregation functions. For both types of aggregation functions we considered methods of sorting involving $\preceq_\pi$ and $\preceq_\nu$. In the class of aggregation functions with respect to linear orders we have considered OWA operator (cf. [6]) with

Xu and Yager order and we compared the obtained results to the ones for OEA from [32].

Concerning the classifier we used the same assumptions as in original approach. To select the best aggregation operator from these returned in the training and testing phases, it was required that the following conditions be satisfied:

- sensitivity $\geqslant 90\%$
- specificity $\geqslant 80\%$
- sensitivity $>$ specificity
- decisiveness $< 100\%$.

The first two rules choose aggregation operators with high sensitivity and specificity values. The third rule reflects the fact that in a medical context sensitivity is more important than specificity. Since these two measures are correlated there may be some models (aggregation operators) that trade off sensitivity for specificity. Such models were rejected. Finally, models with 100% decisiveness were excluded in order not to impose diagnoses that lack sufficient justification. No decision, leading to further examinations, is better than a wrong decision.

Moreover, in OvaExpert interval representation of data was applied. This approach enabled effective decision-making, in spite of missing data. The novel approach applied in OvaExpert was to describe the value of each attribute of a patient by an interval, regardless of whether or not the description of the attribute was given. If the value was not provided, then the proposed representation had the form of a set containing all possible values for the attribute. If the value was given, it was represented by an interval reduced to a point. The main advantage of this approach is that all patients can be described in the same, uniform way and can be processed with the same diagnostic model. There are many different diagnostic models for ovarian tumor, which use different attributes describing the patient, and are therefore subject to different levels of uncertainty. The main idea was to improve the final diagnosis by taking advantage of the models? diversity. Given n models $m_1, m_2, ..., m_n$, an aggregation function is used and as result a new diagnosis is obtained that gathers information from the input models. There are two possible modes of such aggregation. The first, called numerical, uses a single value that represents the whole interval (the most common choices are the interval's center, lower bound and upper bound). The interval mode utilizes the whole of the interval information. At the end some tresholding strategies are applied. They have the aim of converting a numerical or interval decision into a final diagnosis. For the numerical case there was only one class of thresholding strategies, i.e. thresholding with margin $\epsilon \in [-0.5, 0.5]$ given by

$$\tau_\epsilon(a) = \begin{cases} B, & a < 0.5 - \epsilon \\ M, & a \geqslant 0.5 + \epsilon\,, \\ NA, & otherwise \end{cases} \quad a \in [0,1].$$

For interval mode there were evaluated three thresholding strategies. The first approach was to apply a numerical threshold to the interval representative. The second was to calculate the common part between intervals (cf. [37]). Finally,

Actual	Predicted		
	Benign	Malignant	NA
Benign	0	2.5	1
Malignant	5	0	2

Fig. 1. Cost matrix

the third one was the interval version of thresholding with a margin given for each $\epsilon \in [-0.5, 0.5]$ by

$$\tau_\epsilon([a,b]) = \begin{cases} B, & b < 0.5 + \epsilon \\ M, & a \geqslant 0.5 - \epsilon, \\ NA, & otherwise \end{cases} \quad [a,b] \in L^I.$$

We used the notation B for benign tumor, M for the malignant one, and NA for the case, where there was no diagnosis.

In the medical diagnosis of ovarian tumors the situation when the system diagnoses a tumor as benign and, in fact, it was malignant causes much more significant effects for the patient, as opposed to the situation when the benign tumor is diagnosed as cancer. In such models the concept of cost matrix (cost function) is used where for each error type a weight (penalty) is assigned for a wrong decision. The quality value is the sum of costs (penalties) assigned to the classifier for making wrong decisions. If it comes to the cost matrix the costs have been selected in cooperation with experts in ovarian cancer diagnosis. Figure 1 presents costs (penalties) attributed to classifiers for incorrect decisions. Correct decisions did not receive a penalty. A classifier receives top penalty in the case if a patient with malign tumor is classified as a benign case. Penalty for the case if a patient with benign tumor was classified as malignant was half of it, as unjustified operation is still dangerous for a patient but death risk is much lower. Additionally, there were also penalties for the classifier for failure to make a decision (NA). The penalty is lower, as in such a case the patient needs additional diagnostics and will probably be directed to a more experienced specialist who would make a correct diagnosis. However, penalties for lack of decision in positive (malignant) was twice as high as in the negative (benign) case. For more details we refer the readers to [32,37].

If it comes to the presented in this paper study the best result for possible aggregation functions were obtained for $\mathcal{A}_{pi1}$, $\mathcal{A}_{pi2}$, and for the necessary aggregation functions for $\mathcal{A}_{nu}$. We present in Fig. 2 the measures of performance and cost matrix for the mentioned above the best representatives of the families of aggregation functions. We analyze the results regarding accuracy, sensitivity, specificity and decisiveness. Where accuracy is a measure of the ratio of the number of correctly classified objects to all evaluated objects. This is the most intuitive measure, but despite its simplicity it is not always the best measure. Especially in the case of unbalanced number of positive and negative cases in test sample. Additionally, accuracy does not work in situations where it is more important for us to have no errors of a given type. While, sensitivity

Method	Cost matrix	Accuracy	Decisiveness	Sensitivity	Specificity
orig.Alc	186.5	0.889	0.206	0.941	0.842
orig.LR1	182.0	0.771	0.274	0.963	0.524
orig.LR2	161.5	0.828	0.331	0.971	0.609
orig.RMI	161.5	0.818	0.566	0.759	0.843
orig.SM	140.0	0.791	0.629	0.973	0.699
orig.Tim	161.5	0.904	0.474	0.667	0.956
unc.Alc	146.5	0.851	0.537	0.897	0.831
unc.LR1	152.0	0.721	0.594	0.978	0.517
unc.LR2	154.5	0.712	0.594	0.978	0.500
unc.RMI	128.0	0.858	0.766	0.767	0.885
unc.SM	111.5	0.790	0.926	0.935	0.733
unc.Tim	129.5	0.893	0.691	0.724	0.946
A_pi	74.0	0.851	0.960	0.980	0.797
A_nu	66.5	0.871	0.977	0.942	0.840
OEA	72.0	0.876	0.966	0.902	0.864
FSC	67.0	0.894	0.937	0.900	0.902
OWA_XY	65.5	0.887	0.960	0.922	0.872

Fig. 2. Comparison of performance for diverse aggregation functions

specifies how many of the positive class objects are properly categorized. It can be interpreted as the probability that the classification will be correct, provided that the case is positive, i.e. the probability that the test performed for a cancer patient will show that the tumor is malignant. Specificity shows how often the model correctly classifies objects from the negative class. In other words, it is the probability that the classification will be correct, provided that the case was negative, i.e. the probability that a person with a benign tumor will show that the tumor is benign. Moreover, decisiveness measure determines in how many cases the classifier was able to make a decision.

In Fig. 2 we present the results for the original models [1,17,27,29,30] (denoted *orig.* for short) and the uncertaintified ones (denoted *unc.* for short). The uncertaintified models are the original models adjusted to uncertainty data, i.e. data presented with the use of intervals. We also recall the results for the operator with best results in OEA. It is an OWA operator with central element of interval as a representative selector and $\tau_{0.025}$ as threshold. We also give the results for FSC module (involving methods based on counting) which was presented in detail in [16]. Considered here aggregation operators (not included in the original system) were tested with diverse comparability relations if needed (for creating n-argument versions of operators or for ordering inputs). In Fig. 2 there are given results for the best obtained options. Since $\mathcal{A}_{pi1}$ and $\mathcal{A}_{pi2}$ gave similar results, we put only the data for $\mathcal{A}_{pi1}$ and denote it by $\mathcal{A}_{pi}$ which were obtained in interval mode with $\tau_{0.025}$, sorting with respect to $\preceq_\pi$ and width of intervals, the recurrence for obtaining n–argument versions of operations beginning from the right end. We see that these operators have weaker results than assumed concerning specificity. For $\mathcal{A}_{nu}$ the results were obtained in interval mode with $\tau_{0.025}$, sorting with respect to $\preceq_\pi$ and width of intervals (what is interesting for the relation $\preceq_\nu$ there were worse results), the recurrence for obtaining n–argument versions of operations beginning from the right end. Moreover, to obtain better results the inputs were also taken in the reverse form (i.e., $1 - x$) and the final result was again reversed. The results for OWA operator with Xu and Yager order were equally good both for the numerical and the interval mode. The treshold $\tau_{0.025}$ there was used in both cases and in the numerical case the central element of interval was taken as a representative selector.

We may conclude that the presented representatives of newly tested aggregation operators obtained comparable values in comparison to the operators and methods of ordering for inputs from [32]. Furthermore, in our experiment OWA operator with the Xu and Yager order applied for inputs ordering, obtained the lowest cost matrix among all the considered methods. However, the results were obtained for less number of repetitions than in [32].

4 Conclusions

We showed that application of aggregation functions defined with respect to possible and necessary relations as well as admissible linear orders may result in good results of cost prediction of the classifier in systems applying aggregation methods (cf. [32]). However, due to the equipment ability we performed the evaluation with less repetitions which may result in lower stability of the obtained results. For the future work we would like to study what are the properties of the presented here aggregation functions that allowed to obtain better results of accuracy, sensitivity, specificity, decisiveness and cost matrix than other examples of aggregation operators from [4].

Acknowledgements. The authors would like to express gratitude to the OvaExpert authors for their valuable comments concerning the presented material.

This contribution was supported by the Centre for Innovation and Transfer of Natural Sciences and Engineering Knowledge of University of Rzeszów, Poland, the project RPPK.01.03.00-18-001/10.

References

1. Alcázar, J.L., Mercé, L.T., et al.: A new scoring system to differentiate benign from malignant adnexal masses. Obstet. Gynecol. Surv. **58**(7), 462–463 (2003)
2. Bentkowska, U., Król, A.: Preservation of fuzzy relation properties based on fuzzy conjunctions and disjunctions during aggregation process. Fuzzy Sets Syst. **291**, 98–113 (2016)

3. Bentkowska, U.: Aggregation of diverse types of fuzzy orders for decision making problems. Inf. Sci. **424**, 317–336 (2018)
4. Bentkowska, U.: New types of aggregation functions for interval-valued fuzzy setting and preservation of pos-B and nec-B-transitivity in decision making problems. Inform. Sci. **424**, 385–399 (2018)
5. Bustince, H., Fernandez, J., Kolesárová, A., Mesiar, R.: Generation of linear orders for intervals by means of aggregation functions. Fuzzy Sets Syst. **220**, 69–77 (2013)
6. Bustince, H., Galar, M., Bedregal, B., Kolesárová, A., Mesiar, R.: A new approach to interval-valued Choquet integrals and the problem of ordering in interval-valued fuzzy sets applications. IEEE Trans. Fuzzy Syst. **21**(6), 1150–1162 (2013)
7. Calvo, T., Kolesárová, A., Komorníková, M., Mesiar, R.: Aggregation operators: properties, classes and construction methods. In: Calvo, T., et al. (eds.) Aggregation Operators, pp. 3–104. Physica-Verlag, Heidelberg (2002)
8. De Miguel, L., Bustince, H., Pękala, B., Bentkowska, U., Da Silva, I., Bedregal, B., Mesiar, R., Ochoa, G.: Interval-valued atanassov intuitionistic owa aggregations using admissible linear orders and their application to decision making. IEEE Trans. Fuzzy Syst. **24**(6), 1586–1597 (2016)
9. Deschrijver, G.: Arithmetic operators in interval-valued fuzzy set theory. Inform. Sci. **177**, 2906–2924 (2007)
10. Deschrijver, G.: Quasi-arithmetic means and OWA functions in interval-valued and Atanassov intuitionistic fuzzy set theory. In: Galichet, S., et al. (eds.) Proceedings of EUSFLAT-LFA 2011, 18–22 July 2011, Aix-les-Bains, France, pp. 506–513 (2011)
11. Dubois, D., Prade, H.: Possibility Theory. Plenum Press, New York (1988)
12. Dubois, D., Prade, H.: Gradualness, uncertainty and bipolarity: making sense of fuzzy sets. Fuzzy Sets Syst. **192**, 3–24 (2012)
13. Dudziak, U.: Weak and graded properties of fuzzy relations in the context of aggregation process. Fuzzy Sets Syst. **161**, 216–233 (2010)
14. Dudziak, U., Pękala, B.: Intuitionistic fuzzy preference relations. In: Galichet, S., et al. (eds.) Proceedings of the 7th Conference of the European Society for Fuzzy Logic and Technology (EUSFLAT-2011) and LFA-2011, pp. 529–536. Atlantis Press (2011)
15. Dyczkowski, K., Wójtowicz, A., Żywica, P., Stachowiak, A., Moszyński, R., Szubert, S.: An intelligent system for computer-aided ovarian tumor diagnosis. In: Filev, D., et al. (eds.) Intelligent Systems'2014. AISC, vol. 323, pp. 335–343. Springer, Cham (2015). https://doi.org/10.1007/978-3-319-11310-4_29
16. Dyczkowski, K.: Intelligent Medical Decision Support System Based on Imperfect Information. SCI, vol. 735. Springer, Cham (2018). https://doi.org/10.1007/978-3-319-67005-8
17. Jacobs, I., Oram, D., et al.: A risk of malignancy index incorporating CA 125, ultrasound and menopausal status for the accurate preoperative diagnosis ofovarian cancer. BJOG **97**(10), 922–929 (1990)
18. Komorníková, M., Mesiar, R.: Aggregation functions on bounded partially ordered sets and their classification. Fuzzy Sets Syst. **175**, 48–56 (2011)
19. Moszyński, R., Żywica, P., Wójtowicz, A., Szubert, S., Sajdak, S., Stachowiak, A., Dyczkowski, K., Wygralak, M., Szpurek, D.: Menopausal status strongly influences the utility of predictive models in differential diagnosis of ovarian tumors: an external validation of selected diagnostic tools. Ginekol. Pol. **85**(12), 892–899 (2014)

20. Pękala, B.: Operations on interval matrices. In: Kryszkiewicz, M., Peters, J.F., Rybinski, H., Skowron, A. (eds.) RSEISP 2007. LNCS (LNAI), vol. 4585, pp. 613–621. Springer, Heidelberg (2007). https://doi.org/10.1007/978-3-540-73451-2_64
21. Pękala, B.: Properties of interval-valued fuzzy relations, atanassov's operators and decomposable operations. In: Hüllermeier, E., Kruse, R., Hoffmann, F. (eds.) IPMU 2010. CCIS, vol. 80, pp. 647–655. Springer, Heidelberg (2010). https://doi.org/10.1007/978-3-642-14055-6_68
22. Pękala, B., Bentkowska, U., De Baets, B.: On comparability relations in the class of interval-valued fuzzy relations. Tatra Mountains Math. Publ. **66**, 91–101 (2016)
23. Pękala B., De Baets B.: Structures of the class of interval-valued fuzzy relations created by different comparability relations. Inf. Sci. (submitted)
24. Sambuc, R.: Fonctions ϕ-floues: Application á l'aide au diagnostic en pathologie thyroidienne. Ph.D. Thesis, Universitê de Marseille, France (1975) (in French)
25. Sanz, J., Fernandez, A., Bustince, H., Herrera, F.: A genetic tuning to improve the performance of fuzzy rule-based classification systems with intervalvalued fuzzy sets: degree of ignorance and lateral position. Int. J. Approx. Reason. **52**(6), 751–766 (2011)
26. Stachowiak, A., Dyczkowski, K., Wójtowicz, A., Żywica, P., Wygralak, M.: A bipolar view on medical diagnosis in OvaExpert system. Flexible Query Answering Systems 2015. AISC, vol. 400, pp. 483–492. Springer, Cham (2016). https://doi.org/10.1007/978-3-319-26154-6_37
27. Szpurek, D., Moszyński, R., et al.: An ultrasonographic morphological index forprediction of ovarian tumor malignancy. Eur. J. Gynaecol. Oncol. **26**(1), 51–54 (2005)
28. Szubert, S., Wójtowicz, A., Moszyński, R., Żywica, P., Dyczkowski, K., Stachowiak, A., Sajdak, S., Szpurek, D., Alcázar, J.L.: External validation of the IOTA ADNEX model performed by two independent gynecologic centers. Gynecol. Oncol. **142**(3), 490–495 (2016)
29. Timmerman, D., Bourne, T.H., et al.: comparison of methods for preoperative discrimination between malignant and benign adnexal masses: the development of a new logistic regression model. Am. J. Obstet. Gynecol. **181**(1), 57–65 (1999)
30. Timmerman, D., Testa, A.C., et al.: Logistic regression model to distinguish between the benign and malignant adnexal mass before surgery: amulticenter study by the international Ovarian tumor analysis group. J. Clin. Oncol. **23**(34), 8794–8801 (2005)
31. Wójtowicz, A., Żywica, P., Szarzyński, K., Moszyński, R., Szubert, S., Dyczkowski, K., Stachowiak, A., Szpurek, D., Wygralak, M.: Dealing with uncertinity in ovarian tumor diagnosis. In: Modern Approaches in Fuzzy Sets, Intuitionistic Fuzzy Sets, Generalized Nets and Related Topics, Volume II: Applications, pp. 151–158. IBS PAN-SRI PAS (2014)
32. Wójtowicz, A., Żywica, P., Stachowiak, A., Dyczkowski, K.: Solving the problem of incomplete data in medical diagnosis via interval modeling. Appl. Soft Comput. **47**, 424–437 (2016)
33. Zadeh, L.A.: Fuzzy sets. Inform. Control **8**, 338–353 (1965)
34. Zadeh, L.A.: The concept of a linguistic variable and its application to approximate reasoning-I. Inf. Sci. **8**, 199–249 (1975)
35. Zapata, H., Bustince, H., Montes, S., Bedregal, B., Dimuro, G.P., Takáč, Z., Baczyński, M., Fernandez, J.: Interval-valued implications and interval-valued strong equality index with admissible orders. Internat. J. Approx. Reason. **88**, 91–109 (2017)

36. Żywica, P., Wójtowicz, A., Stachowiak, A., Dyczkowski, K.: Improving medical decisions under incomplete data using intervalvalued fuzzy aggregation. In: Proceedings of IFSA-EUSFLAT 2015, pp. 577–584. Atlantis Press (2015)
37. Żywica, P., Dyczkowski, K., Wójtowicz, A., Stachowiak, A., Szubert, S., Moszyński, R.: Development of a fuzzy-driven system for ovarian tumor diagnosis. Biocybern. Biomed. Eng. **36**(4), 632–643 (2016)

On Fuzzy Compliance for Clinical Protocols

Anna Wilbik[1(✉)], Ivo Kuiper[1], Walther van Mook[2], Dennis Bergmans[2], Serge Heines[2], and Irene Vanderfeesten[1]

[1] Information Systems, School of Industrial Engineering, Eindhoven University of Technology, Eindhoven, The Netherlands
a.m.wilbik@tue.nl

[2] Department of Intensive Care, Maastricht University Medical Centre+, Maastricht, The Netherlands

Abstract. Clinical protocols are introduced in hospitals to standardize the care delivery process. Compliance is a measure used to determine whether the protocol has been followed. However, so far an activity in the protocol could be either compliant or non-compliant. In this paper we consider the compliance of a single activity as a fuzzy term. We propose to define the rules which can assess the compliance degree of an activity. We proposed the fuzzy compliance measure of clinical protocol that aggregates those compliance degrees. We demonstrate a case of glucose management protocol at Intensive Care Unit (ICU). Initial results are promising.

Keywords: Clinical protocols · Compliance · Fuzzy evaluation

1 Introduction

Contemporary advances in medicine result in better diagnosis, more and better treatment options, but also increased complexity and costs of healthcare. In order to deal with this complexity and to increase the efficiency of the care, many hospitals and healthcare providers introduce standardization of care, e.g. by using clinical guidelines and protocols. There is compelling evidence of the positive effects of such approaches, e.g., reduced hospital complications without increased length of stay and cost [1], reduced treatment cost [2].

Clinical practice guidelines [3] are defined as "systematically developed statements to assist practitioners and patient decisions about appropriate health care for specific circumstances". They are designed based on best available evidence. Clinical protocols can be seen as more specific than guidelines, defined in greater detail. Protocols provide "a comprehensive set of rigid criteria outlining the management steps for a single clinical condition or aspects of organization" [4]. Examples of clinical protocols used in practice are glucose management protocol which defines how much insulin or glucose should be administered to the

J. Medina et al. (Eds.): IPMU 2018, CCIS 855, pp. 404–413, 2018.
https://doi.org/10.1007/978-3-319-91479-4_34

patients, or weaning protocol which defines step-by-step the process of switching the patient from the artificial breathing.

In many intervention case studies, it was shown, that by changing the way of working (by implementing a protocol), better results could be obtained, such as improved patient outcomes or lower costs [5,6]. These papers often measure the compliance level to the new protocol, i.e. whether the protocol has been followed. However each paper employs a different metric, since compliance is not standardized nor well defined.

In this paper we address the issue of compliance measurement and propose a new measure for (fuzzy) compliance of a clinical protocol. We define compliance as "the degree to which the behavior of the executors of the clinical protocol corresponds to the behavior described in the clinical protocol". Please note that in the literature and also here, the terms adherence and compliance are used interchangeably. Another aspect is patients' compliance or adherence to the doctors' recommendations, e.g. regarding drug intake [7]. We believe that our metric could be also used for this purpose, although patients' compliance is out of scope for this study.

This paper is structured as follows. The next section provides the overview of several compliance measures found in the literature. Section 3 describes the proposed method, which is followed by an illustrative case study of compliance of a glucose management protocol in Sect. 4. The paper is finished with concluding remarks.

2 Background

In this section we describe the compliance measures found in literature. Generally, it can be seen that there is no consensus on what compliance/adherence is. In order to measure compliance many metrics has been proposed, but none of them has been evaluated.

2.1 Definitions of Clinical Protocol Compliance

In many of those examples compliance for a patient is treated as a binary (crisp) notion. For instance, [6] defines overall adherence to a prescribed distress screening protocol were calculated based on documentation in the EHR that screening adherence and an appropriate clinical response had occurred. Only when both the screening and response took place, overall adherence per patient was considered as "yes". The authors calculated the overall adherence rate as number of overall adherent patients per number of all patients.

A different metric was adopted by Lobach et al. [8]. The authors were evaluating compliance of the recommendations given. First the authors individually evaluated all recommendations as complaint, if the physicians followed the guideline's recommendation within prescribed time interval. Moreover the clinicians are also considered as compliant with a guideline, if they state why the guideline was not being followed. The authors calculate two rates, compliance rate, calculated as percentage of compliant recommendations during one encounter. Adherence rate is defined as the rate of all complaint recommendations to all recommendations made.

Some authors like Stewart et al. [9] introduced a scoring system. Each protocol execution could receive a score between 0 and 5, where score 4 or 5 allowed one deviation from the protocol and meant excellent protocol execution. The adherence was reflected by the percentages of the protocol executions with each of the six scores. Similarly, Drews et al. [10] calculated the number of steps that were compliant to report the percentages of executions with zero, one, two non-compliant activities.

Another way of reporting of compliance is to report the compliance per activity [5,11], presenting the percentages of compliant activities per treatment path. Such analysis can go even further, to distinguish whether an activity was done within certain time frame or in general [12]. Moreover one more measure is reported, that is the fraction of cases in which all activities were performed in due time.

In a few cases the authors used the questionnaires to evaluate the compliance to the protocol [13,14].

Lauzier et al. [15] introduces another measure for medication compliance, i.e. evaluation to what extent patient receives the correct amount of medication in the correct times. However the case, the authors consider, concerns continuous drug administration, namely vasopressors. As a simple solution, the authors defined potential protocol deviations when patient had pressure outside predefined target range for more than four hours without adjustment of the vasopressor dose. Those deviations were next divided into two groups as clinically-justified and non-clinically-justified. Beside the number of deviation events, the number of days and number of patients with at least one deviation event were also reported, as well as their percentages.

2.2 Conformance Checking in Process Mining

Clinical protocol can be considered as the process model. In the process mining field [16], conformance (compliance) checking is one of the tasks, next to model discovery and model enhancement. We can consider a protocol as a process model. The observed behavior is recorded in an execution log. It is assumed that such log consists of a list of the tasks that were performed, the time at which each task was finished and an identifier of the (patient) case for which the task was performed. Additional information, such as the time at which the task started, the healthcare provider who performed the task, and information about the patient can enable further analysis. In [17] two different types of metric were proposed for compliance checking:

- fitness, defined as the degree to which log traces can be linked with valid execution paths in the process model, and
- appropriateness, defined as the degree of accuracy to which the process model describes the observed behavior.

In our context, appropriateness can be seen as the validity measure of a clinical protocol, that is beyond scope of this paper.

3 Fuzzy Compliance

In all above examples of the compliance assessment, a single activity within the protocol could be either compliant to the clinical protocol or not compliant. Based on this information, two metrics can be easily derived: percentage of protocol executions with all activities compliant and percentage of compliant activities.

However, can the activity be only compliant or non-compliant? Consider a case when a patient receives 2.5 units of a drug instead of 2 units, or a protocol in which activity B was done 35 min after activity A, 5 min too late according to the protocol. Are the following examples truly non-compliant? There may be good reasons for slight deviations, as doctors see every patient as a unique and complex case.

We believe that compliance of a single activity is not a binary state, but a matter of degree. This was also reflected in the conversations with many healthcare providers. Therefore each activity should have defined a compliance level. This compliance level in case of drug administration can be defined simply, using a fuzzy set with trapezoidal membership function, e.g. as shown in Fig. 1, with clearly defined four threshold points, where dose deviation that is perfectly ok (compliance degree of 1), and dose deviation which is completely not OK (compliance degree of 0). For the values in between we use linear interpolation. The use of such membership function was already advocated by Zadeh [18], since it is easy to understand by domain experts, and provides a good compromise between a so-called cointension and computational complexity.

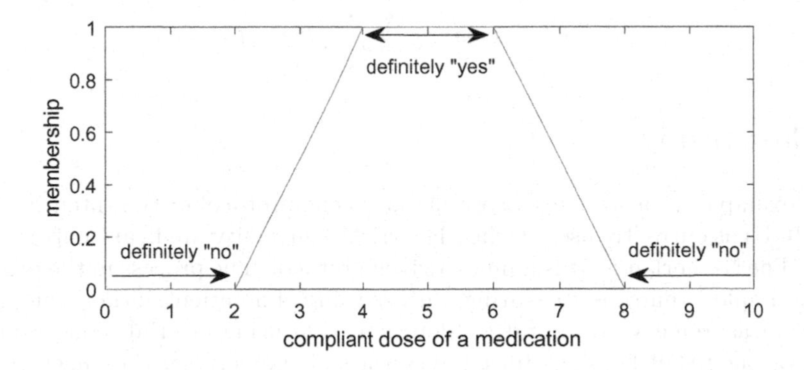

Fig. 1. Membership of a fuzzy set defining compliant dose of a medication

In the remainder of this paper we will use the following notation. Let us consider a protocol P, which consists of n unique activities p_i, $i = 1, ..., n$.

The protocol P may include decision points, as well as loops in which certain activities can be repeated. For instance, weaning protocol starts with doing the check of the patients. Next assisted spontaneous breathing (ASB) is reduced in steps and after each step the patient status is checked. Last activity is extubation of the patient.

Behavior of healthcare providers is captured in forms of traces T. Let us assume that, there are m executions, traces of this protocol t_j, $j = 1, ..., m$. For instance, it can mean that for m patients the weaning protocol was followed. A trace t_j contains information about r_j activities performed related to this protocol a_{jk}, $k = 1, ..., r_j$. In case of a weaning protocol, a trace t_j may contain information, that first activity, a_{j1}, was the check. Next ASB was reduced followed by a patient check. This happened three times. At the end patient was extubated.

Therefore for each activity p_i, the healthcare providers may define rules defining the degree of compliance. For instance, a compliance degree $\mu_{p_i}(\cdot)$ of an activity p_i can be described as the membership value of a fuzzy set of the difference between the drug dose administered and described in the protocol.

In order to calculate the compliance of a trace t_j, we align the trace with the protocol execution. There are many algorithms for this purpose [16, 19]. For each activity a_{jk} from the trace t_j that is in the protocol P we calculate its compliance degree $\mu_P(a_{jk})$. It may happen that some activities were not performed. During the alignment procedure, such missing activities are added to the execution log and marked with special symbol. For such activities the compliance degree is naturally 0. Please note that this also implies that length of the trace r_j is longer.

Once we have the compliance degree of each activity $\mu_P(a_{jk})$ for $j = 1, ..., m$ and $k = 1, ..., r_k$, we can aggregate them. In this paper we are using the average. Hence the compliance degree of the protocol is calculated as

$$C(P) = \frac{1}{\sum_{j=1}^{m} r_j} \sum_{j=1}^{m} \sum_{k=1}^{r_j} \mu_P(a_{jk}). \tag{1}$$

4 Case Study

As an example we used the glucose management protocol of the Intensive Care Unit (ICU) as currently used in the Maastricht University Medical Centre, Maastricht, The Netherlands. It is a nurse-driven protocol. The process in the protocol is very simple, a nurse is measuring glucose value of a patient, either using point of care measurement or laboratory determination using arterial blood samples. Based on the result he/she adjust the settings of the perfusor (i.e. machine that administers drug to the patient on continuous basis).

The ICU consists of three different wards. Two of them, lets denote them A and B, treat mixed medical/surgical patients, where most patients (90%) are acute patients (requiring immediate assessment or treatment). On the wards A and B the same glucose management protocol was implemented around 7 years ago, while on third ward a different protocol was implemented. Therefore in this paper we will analyze only wards A and B. The protocol under consideration concerns intravenous administration of short-acting insulin and is considered for each patient till the moment when she/he can receive oral nutrition. The protocol assumes that the target values of blood glucose are between 4.5 and 7.0 mmol/l. The protocol consists of four tables, that define perfusor settings as well as additional actions. In Table 1 we show the starting scheme, so perfusor settings for a new patient. Table 2 shows perfusor settings adaptation schemes.

Table 1. Starting scheme of the glucose management protocol used in the study (translation from Dutch).

Starting schema		
Blood glucose (mmol/l)	**Perfusor setting** (50 units/50 ml NaCl 0.9%)	**Insulin bolus**
<7.0	-	
7.0–8.0	1 unit/hour	
8.0–10.0	2 unit/hour	
10.0–15.0	4 unit/hour	4 units insulin
15.0–20.0	6 unit/hour	6 units insulin
>20.0	6 unit/hour	8 units insulin

In our study we first interviewed the wards' coordinators and had a survey among nurses regarding the compliance of this protocol. Only afterwards we analyzed the data, and draw conclusions combining also the baseline information.

We interviewed the wards' coordinators and nurses working on those two wards regarding the awareness, use and reasons for (non)adherence to this protocol. In the ward A the protocol was used for about six months after implementation, but it is not used anymore. Controlling the blood glucose level is done based on the experience and insight of nurses. The staff believes that everyone is roughly aware of the protocol, but stated that it is less of a protocol act, but more about own insight. This is confirmed by the survey results, where 57% of respondents were aware of the protocol, but 42% of them responded that they don't use the protocol at all.

In the ward B the protocol is still in use, and its physical version is available behind the desk. It was confirmed also by the survey among the nurses, where all of them confirmed that they are aware of the protocol and most of them are following the protocol to a huge extend. Hence we can conclude that according to the nurses the protocol is still being adhered to, however, slightly less than the ward's coordinator believes.

We obtained data from those two wards, A and B, regarding the glucose management from year 2014 and 2015. Those data included the perfusor settings and glucose results of 297 patients in Ward A and 274 patients in Ward B.

Table 2. Perfusor adaptation scheme for glucose management protocol used in the study (translation from Dutch)

Blood glucose decreased >30% or increased		
Blood glucose (mmol/l)	**Perfusor setting** (50units/50ml NaCl 0.9%)	**Action**
<3.5	stop	act according hypoglycemia protocol
3.5- 4.5	-1.0 unit/hour	check glucose after 30 minutes
4.5 - 7.0	no changes	
7.0 - 8.8	+0.5 unit/hour	
8.0-9.0	+1.0 unit/hour	
9.0-10	+1.5 unit/hour	
10-15	+2.0 unit/hour	2 units insulin as bolus
>15	+3.0 unit/hour	4 units insulin as bolus

Blood glucose decreased <30%		
Blood glucose (mmol/l)	**Perfusor setting** (50units/50ml NaCl 0.9%)	**Action**
<3.5	stop	act according hypoglycemia protocol
3.5 - 4.5	stop, if glucose > 5 mmol/l then start with half of last dose	check glucose every 15 minutes till glucose > 5 mmol/l
4.5-7	half the dosage	
7-10	-1 unit/hour	
>10	no changes	

hypoglycemia (blood glucose <3.5 mmol/l)		
Blood glucose (mmol/l)	**Perfusor setting** (50units/50ml NaCl 0.9%)	**Action**
<3.5	stop	50 ml glucose 50% in 10 minutes and check glucose level
> 4.5 (after 1st glucose bolus)	start with last setting -1 unit/hour	check glucose after 30 minutes
<4.5 (after 1st glucose bolus)	keep perfusor stopped	30ml glucise 50% in 10 minutes and check glucose level
> 4.5 (after 2st glucose bolus)	start with last setting -2 unit/hour	
< 4.5 (after 2st glucose bolus)	keep perfusor stopped	50ml glucose 50% in 10 minutes and consult the doctor

Since the protocol concentrates on the medication dosages, and the glucose measurement activity is implicit, in this case study we assess the compliance only with respect to the aspect of medication dosage on those three aspects: insulin perfusor settings, insulin bolus (injection) and glucose bolus (injection).

For comparison purposes, we first calculated the strict compliance of the protocol. In Ward *A* only 5 executions (1.7%) were completely according to the protocol. In total 37.8% of separate activities were according to the protocol. For Ward *B* only 2 executions (0.7%) were completely according to the protocol, and 34.8% of separate activities were compliant. Those very low values for

Ward B are surprising in comparison to the results of the interview, as both the coordinator and nurses believed that they are following the protocol.

However, in the interviews it was indicated that the protocol should be treated more like a guideline, and some deviations are possible. Therefore in consultation with healthcare providers we defined clinically accepted and non-accepted deviations. We decided to use the trapezoidal membership function, because it was comprehensive for the medical experts. We asked them to provide two threshold values (completely acceptable deviation and completely unacceptable deviation) for each of the actions (perfusor settings, insulin bolus and glucose bolus). Those values are presented in Table 3. The degree of compliance of an activity was defined as the minimal value of the compliance with respect to the three medication types.

Table 3. Thresholds for accepted and non-accepted deviations

Medication	Totally acceptable deviation	Totally unacceptable deviation
Perfusor setting	1 unit/hour	2 units/hour
Insulin bolus	2 units	3 units
Glucose bolus	10 mL	15 mL

The fuzzy compliance values (on activity level) for Wards A and B are both around 82%. All values are shown in Table 4. Of course those values are much higher. But does it mean that this protocol is executed very well? Not necessarily. Because the fuzzy compliance is taking into consideration acceptable deviations from the protocol, a score below 100% still indicates noncompliance to a certain extent. Further analysis of the protocol is required to pinpoint the reasons the non-acceptable deviations.

Table 4. Compliance values for Wards A and B according to different metrics.

Method	Ward A	Ward B
% of compliant executions	1.7%	0.7%
% of compliant activities	37.8%	34.8%
Fuzzy compliance	82.6%	82.8%

Furthermore, it would be beneficial to introduce the concept of "comply or explain", in which the healthcare providers are given a choice either to comply with the protocol or explain, why they think they should deviate. Such information, combined with the outcome measures would be extremely beneficial for the clinical protocol improvement.

5 Concluding Remarks

Clinical protocols are introduced in hospitals to standardize the care delivery process. Compliance is a measure used to determine whether the protocol has been followed or not and to what extent. However, an activity in the protocol could be either compliant or non-compliant. In this paper we proposed to treat the compliance of a single activity as a fuzzy term described by the compliance degree. We proposed the fuzzy compliance measure of clinical protocol that aggregates those compliance degrees.

Further work will focus on using different aggregation functions instead of averaging and incorporating the importance of different activities. We will also investigate in more detailed non-compliance in the clinical protocol executions, since a single number is not enough to understand the situation.

References

1. Rotter, T.: Clinical pathways in hospitals: evaluating effects and costs. Ph.D. thesis, Erasmus University Rotterdam (2013)
2. Haddock, B., Goulart, L.: Cancer care pathways: hopes, facts, and concerns. AJMC Evid. Based Oncol. **5** (2016)
3. Field, M.J., Lohr, K.N. (eds.): Clinical Practice Guidelines: Directions for a New Program. The National Academies Press, Institute of Medicine, Washington, DC (1990)
4. Grune, F., Ottens, T., Klimek, M.: Standards of care in operating theatres. In: Gullo, A. (ed.) Anaesthesia, Pain, Intensive Care and Emergency A.P.I.C.E.: Proceedings of the 21st Postgraduate Course in Critical Care Medicine Venice-Mestre, Italy, 10–13 November 2006, pp. 281–290. Springer, Milan (2007)
5. Bruce, H.R., Maiden, J., Fedullo, P.F., Kim, S.C.: Impact of nurse-initiated ed sepsis protocol on compliance with sepsis bundles, time to initial antibiotic administration, and in-hospital mortality. J. Emergency Nurs. **41**(2), 130–137 (2015)
6. Zebrack, B., Kayser, K., Bybee, D., Padgett, L., Sundstrom, L., Jobin, C., Oktay, J.: A practice-based evaluation of distress screening protocol adherence and medical service utilization. JNCCN J. Nat. Compr. Cancer Netw. **15**(7), 903–912 (2017)
7. World Health Organization: Adherence to long-term therapies: evidence for action. Technical report (2012)
8. Lobach, D., Hammond, W.: Computerized decision support based on a clinical practice guideline improves compliance with care standards. Am. J. Med. **102**(1), 89–98 (1997)
9. Stewart, S., Stocks, N.P., Burrell, L.M., de Looze, F.J., Esterman, A., Harris, M., Hung, J., Swemmer, C.H., Kurstjens, N.P., Jennings, G.L., Carrington, M.J.: More rigorous protocol adherence to intensive structured management improves blood pressure control in primary care: results from the valsartan intensified primary care reduction of blood pressure study. J. Hypertens. **32**(6), 1342–1350 (2014)
10. Drews, F., Wallace, J., Benuzillo, J., Markewitz, B., Samore, M.: Protocol adherence in the intensive care unit. Hum. Factors Ergon. Manuf. **22**(1), 21–31 (2012)
11. Mehrara, M., Tavakoli, N., Fathi, M., Mahshidfar, B., Zare, M.A., Asadi, A., Hosseinzadeh, S., Safdarian, M.: Protocol adherence in prehospital medical care provided for patients with chest pain and loss of consciousness; a brief report. Emergency **5**(1), e40 (2017)

12. UK National Surgical Research Collaborative: Multicentre observational study of adherence to sepsis six guidelines in emergency general surgery. BJS **107**, e165–e171 (2017)
13. Healy, J.M.: How hospital leaders implemented a safe surgery protocol in Australian hospitals. Int. J. Qual. Health Care **24**(1), 88–94 (2012)
14. Sitaresmi, M.N., Mostert, S., Gundy, C.M., Sutaryo, Veerman, A.J.: Health-care providers' compliance with childhood acute lymphoblastic leukemia protocol in Indonesia. Pediatr. Blood Cancer **51**(6), 732–736 (2008)
15. Lauzier, F., Adhikari, N.K., Seely, A., Koo, K.K.Y., Belley-Côté, E.P., Burns, K.E.A., Cook, D.J., D'Aragon, F., Rochwerg, B., Kho, M.E., Oczkowksi, S.J.W., Duan, E.H., Meade, M.O., Day, A.G., Lamontagne, F.: Protocol adherence for continuously titrated interventions in randomized trials: an overview of the current methodology and case study. BMC Med. Res. Methodol. **17**(1), 106 (2017)
16. van der Aalst, W.M.: Process Mining - Discovery. Conformance and Enhancement of Business Processes. Springer, Heidelberg (2011). https://doi.org/10.1007/978-3-642-19345-3
17. Rozinat, A., van der Aalst, W.: Conformance checking of processes based on monitoring real behavior. Inf. Syst. **33**(1), 64–95 (2008)
18. Zadeh, L.A.: Computation with imprecise probabilities. In: Proceedings of the 12th International Conference Information Processing and Management of Uncertainty in Knowledge-based Systems (2008)
19. Yan, H., Gorp, P.V., Kaymak, U., Lu, X., Vdovjak, R., Korsten, H.H.M., Duan, H.: Analyzing conformance to clinical protocols involving advanced synchronizations. In: Li, G., Kim, S., Hughes, M., McLachlan, G.J., Sun, H., Hu, X., Ressom, H.W., Liu, B., Liebman, M.N. (eds.) 2013 IEEE International Conference on Bioinformatics and Biomedicine, Shanghai, China, 18–21 December 2013, pp. 61–68. IEEE Computer Society (2013)

Uncertainty in Video/Image Processing (UVIP)

Foreground Detection Enhancement Using Pearson Correlation Filtering

Rafael Marcos Luque-Baena, Miguel A. Molina-Cabello(✉), Ezequiel López-Rubio, and Enrique Domínguez

Department of Computer Languages and Computer Science, University of Málaga, Bulevar Louis Pasteur, 35, 29071 Málaga, Spain
{rmluque,miguelangel,ezeqlr,enriqued}@lcc.uma.es

Abstract. Foreground detection algorithms are commonly employed as an initial module in video processing pipelines for automated surveillance. The resulting masks produced by these algorithms are usually postprocessed in order to improve their quality. In this work, a post-processing filter based on the Pearson correlation among the pixels in a neighborhood of the pixel at hand is proposed. The flow of information among pixels is controlled by the correlation that exists among them. This way, the filtering performance is enhanced with respect to some state of the art proposals, as demonstrated with a selection of benchmark videos.

Keywords: Foreground detection · Postprocessing
Pearson correlation · Morphological operators · Background modeling

1 Introduction

Extracting objects of interest from a video or image is a very important task in computer vision applications. Numerous features can be extracted from the foreground to develop classifying and recognizing processes in subsequent steps. For instance, autonomous visual systems must be able to recognize relevant objects and its movements in order to maintain an internal representation of the environment and understand the scene. Most of surveillance systems are only interested on the moving objects, so the aim of these segmentation algorithms [6,10] consists in separating the foreground pixels from the background pixels.

In the literature, many foreground detection algorithms [2,7–9] have been proposed, whose internal parameters (thresholds, sizes of regions...) and post-processing techniques are fixed to obtain meaningful results depending on the application.

The result produced by foreground object detection algorithms usually contains noise and it is not suitable to carry out high-level processing like object tracking or the analysis of the behavior of these objects. So that, it is necessary to execute a filtering of the binary mask obtained in the segmentation step. Some of the possible causes that can produce this kind of noise are listed below:

J. Medina et al. (Eds.): IPMU 2018, CCIS 855, pp. 417–428, 2018.
https://doi.org/10.1007/978-3-319-91479-4_35

- Camera noise. It is produced by the quality of the acquisition of the images from the camera. Sometimes, a pixel from an image presents a color tone, and the same pixel in the next frame (without any movement in the scenario) exhibits a different one. In this kind of noise, the noise produced by the video downsampling applied in the hardware device can be included.
- Reflection noise. The movement of a spotlight, for example, the sun, produces that some background parts reflect the light and the result of the foreground object detection algorithm is affected by this effect, considering those zones as foreground.
- Noise in the background objects. Several parts of the objects have the same color (or tone in the case of grayscale images) as the background behind them. This similarity produces that some algorithms do not detect these pixels as foreground objects, so they are not correctly detected.
- Shadows and abrupt illumination changes. Most algorithms detect the projected shadows of the objects as foreground. The illumination changes (for example, turn on a light in a room) also produce that the algorithms fail in the detection of the foreground objects.

All of these commented failures can not be solved by the objects in motion segmentation pixel-level algorithms. Thus, it is necessary to develop post-processing techniques in order to improve the quality of the final segmentation.

2 Model Approach

In this section, the foreground object detection algorithm (2.1), considered as our baseline method, is described. After that, the morphological operators are depicted (Subsect. 2.2), since they are the most applied technique in order to remove the noise. And finally we also propose the employ of the Pearson correlation to reduce the noise (Subsect. 2.3).

2.1 Foreground Object Detection Algorithm

A foreground object detection algorithm provides a mask for each frame of a sequence where it presents the same size than the input frame and each pixel has a value in the range $[0, 1]$ that represents the likelihood (in order to manage the uncertainty) of belonging to the foreground. Thus, pixels from this mask with a value close to 1 (white pixels) represent the foreground objects, while pixels with a value close to 0 (black pixels) are considered as background.

In order to quantify the improvement of the application of a postprocessing method to the result produced by the algorithms of this kind, we need to compare the output of the algorithm and the result provided by the postprocessing method.

In this case, we have considered as our baseline method the algorithm described in [5], and noted as *AE*. This proposal uses mixtures of uniform distributions and multivariate Gaussians with full covariance matrices, and it is indicated to detect foreground objects in complex context, like videos which exhibit

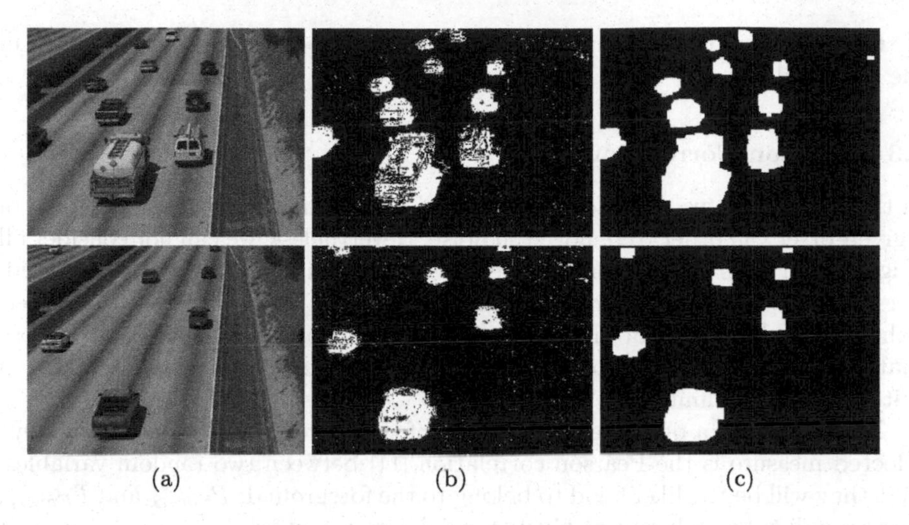

Fig. 1. Result after the noise removal. (a) Frames corresponding to a traffic sequence. (b) Segmentation by Gaussian distributions. (c) Noise removal by morphological operators.

dynamic backgrounds or shadow appearances. In addition, due to the employ of a stochastic approximation, the computational complexity of this algorithm is low, so it is a suitable method for real time applications.

2.2 Morphological Operators

Morphological operators like erosion or dilation are applied to the segmentation of the foreground objects in order to remove spurious pixels [1,3], that cause the three first items referred in Sect. 1. The aim of the application of this kind of process is to remove those pixels that do not belong to the foreground (NFN, *non-foreground noise*) and to delete those that they are detected as background (NBN, *non-background noise*) in the closest zones and the interior of the objects that actually belong to the foreground.

The erosion process erodes a unit over the external limits of the objects. The dilation is the opposite process, expanding the limits of the foreground objects. The decision about the order and the quantity of filters to be applied is quite significant. The order of the operators affects the quality and the quantity affects the quality and time complexity of the algorithm.

For example, if we apply the dilation and then the erosion process, we can not remove isolated unit pixels (NFN), since the dilation operator increases its limits with one pixel and the erosion will remove the added pixels, thereby keeping the spurious original pixels. On the other hand, this order will remove some NBN noise from the inside of the objects (gaps). In the case of the application of this operations in inverse order, so that, dilation after erosion, it will remove isolated unit pixels that do not belong to the foreground (NFN), but the filling of the existing gaps inside the objects (NBN) can not be carried out. Thus, the sequence of the application of the morphological operators depends on the peculiarities of the analyzed scene.

2.3 Pearson Correlation

In this subsection we propose the employ of information from the 8 neighbors of a given pixel $\mathbf{x}$ in order to remove the noise. Nevertheless, we can not consider all neighbors in the same proportion due to several adjoining pixels are not related. For example, this can be produced in the border of a road: the outside pixels (where the vehicles do not drive along it) are practically independent from those that belong to the interior of the road (where the vehicles usually circulate) in spite of their proximity.

Thus, we need a quantitative measure of the correlation of pixels pairs. Our selected measure is the Pearson correlation [11] between two random variables, that they will be the likelihood to belong to the foreground, $P_{Fore,\mathbf{x}}$ and $P_{Fore,\mathbf{y}}$, corresponding to each pair of pixels 8 neighbors $\mathbf{x}$ and $\mathbf{y}$:

$$\rho_{\mathbf{x},\mathbf{y}} = \frac{\phi_{\mathbf{x},\mathbf{y}}}{\sqrt{\nu_{\mathbf{x}}}\sqrt{\nu_{\mathbf{y}}}} \tag{1}$$

$$\begin{aligned} \phi_{\mathbf{x},\mathbf{y}} &= \operatorname{cov}\left(P_{Fore,\mathbf{x}}, P_{Fore,\mathbf{y}}\right) \\ &= E\left[\left(P_{Fore,\mathbf{x}} - E\left[P_{Fore,\mathbf{x}}\right]\right)\left(P_{Fore,\mathbf{y}} - E\left[P_{Fore,\mathbf{y}}\right]\right)\right] \end{aligned} \tag{2}$$

$$\nu_{\mathbf{x}} = \operatorname{var}\left(P_{Fore,\mathbf{x}}\right) = E\left[\left(P_{Fore,\mathbf{x}} - E\left[P_{Fore,\mathbf{x}}\right]\right)^2\right] \tag{3}$$

$$\nu_{\mathbf{y}} = \operatorname{var}\left(P_{Fore,\mathbf{y}}\right) = E\left[\left(P_{Fore,\mathbf{y}} - E\left[P_{Fore,\mathbf{y}}\right]\right)^2\right] \tag{4}$$

where we have that:

$$E\left[P_{Fore,\mathbf{x}}\right] = \pi_{Fore,\mathbf{x}} \tag{5}$$

$$E\left[P_{Fore,\mathbf{y}}\right] = \pi_{Fore,\mathbf{y}} \tag{6}$$

and π_{Fore} is the likelihood to belong to the foreground that is updating throughout the time.

Please, note that the properties of the Pearson correlation imply that:

$$\rho_{\mathbf{x},\mathbf{y}} \in [-1, 1] \tag{7}$$

$$\rho_{\mathbf{x},\mathbf{y}} = \rho_{\mathbf{y},\mathbf{x}} \tag{8}$$

where the last equation saves half of the required calculations.

The values of $\rho_{\mathbf{x},\mathbf{y}}$ are high and positive if and only if the pixels $\mathbf{x}$ and $\mathbf{y}$ are usually assigned to the same class, i.e. both pixels belong to the background or the foreground. On the other hand, if the pixels are quite independent, then we will have $\rho_{\mathbf{x},\mathbf{y}} = 0$; we remember that independence implies no correlation, but the reverse implication is not true, i.e. if it does not exist a correlation between variables then this does not imply that these variables are independent.

In relation to the negative correlations, it is expected that they will be infrequent, and theoretically they are related to pairs of pixels that are usually assigned to opposite classes. Nevertheless, in practice, the negative correlations are obtained due to the noise in the estimations of the input data.

The correlations $\rho_{\mathbf{x},\mathbf{y}}$ allow us to obtain a free-noise version of $P_{Fore,\mathbf{x}}(\mathbf{t})$, combining it with the information of the 8 neighbors of $\mathbf{x}$:

$$\widetilde{P}_{Fore,\mathbf{x}}(\mathbf{t}) = \text{ramp}\left(\frac{1}{9}\sum_{\mathbf{y}\in Neigh(\mathbf{x})} \rho_{\mathbf{x},\mathbf{y}} P_{Fore,\mathbf{y}}(\mathbf{t})\right) \quad (9)$$

where $Neigh(\mathbf{x})$ contains the pixel $\mathbf{x}$ and its 8 neighbors, and

$$\rho_{\mathbf{x},\mathbf{x}} = 1 \quad (10)$$

$$\text{ramp}(z) = \begin{cases} z \text{ iff } z \geq 0 \\ 0 \text{ in other case} \end{cases} \quad (11)$$

ramp function is used in the Eq. (9) in order to fix the variable in the range $[0, 1]$ of belonging likelihood to the foreground, when a excess of negative correlations related to the noise is presented. Empirically, it is observed that the argument of the trunc function in the Eq. (9) is negative in less than 0.5% of the cases, since $\rho_{\mathbf{x},\mathbf{y}}$ are nearly always positive.

The $\widetilde{P}_{Fore,\mathbf{x}}(\mathbf{t})$ values correspond to a probability of belonging to the foreground class, which can be considered a measure of uncertainty. Thus, a pixel whose $\widetilde{P}_{Fore,\mathbf{x}}(\mathbf{t})$ value is close to one will be part of a foreground object with great certainty, while if its value is close to 0.5 it is not possible to deduce anything (it has the same probability of belonging to the foreground as to the background).

3 Experimental Results

3.1 Parameter Selection

In order to appreciate what kind of filtering is the most convenient, we are going to make a comparison between the two proposed alternatives, the application of morphological operators and the Pearson correlation. For this last case, we have to consider that the the baseline algorithm (*AE*) has to provide the likelihood that a pixel belongs to the foreground class. We will note it as *Basic* and different learning rate options are considered: $\varepsilon_0 = \{1e^{-4}, 5e^{-4}, 0.001, 0.005, 0.01, 0.05, 0.1\}$. In the case of the morphological operators, we will distinguish in the order application. We note as $OMDE$ the application of the morphological operators in the *dilation - erosion* order, and $OMED$ is related to the *erosion - dilation* order, due to the highly significant influence of a change in the order. We will have the number of applied iterations as a parameter, where $\#\mathcal{ED} = \#\mathcal{DE} = \{1, 2, 3\}$. For example, a value of $\#\mathcal{DE} = 2$ indicates that two iterations of the dilation operator are carried out, followed by two iterations of the erosion operator.

3.2 Sequences

We have used several sequences in order to make the comparison. This repository of indoor and outdoor videos provides a wide range of the existing variability in real scenes. We have selected the sequences named Video2 (V2) and Video4 (V4) from a synthetic dataset (the sequences are composed by motion objects generated by software with real background scenes) designed for the algorithm competition of foreground objects detection methods created by the International Conference VSSN'06[1].

Several sequences have been chosen from the dataset[2] developed by Li et al. [4], namely: water surface (WS), moving escalators in a subway station (SS), campus with plentiful vegetation moving continuously (CAM), meeting room with moving curtain (MR) and public fountain throwing water (FT).

We also have taken three videos from the IPPR (Image Processing and Pattern Recognition)[3] contest held in Taiwan in 2006, whose names are IPPRData1 (IP1), IPPRData2 (IP2) and IPPRData3 (IP3). The two first sequences are indoor videos where a corridor is observed from different points of view, and the third video is a highway recorded from an elevated position.

Finally, we have included in our test dataset two sequences from CAVIAR dataset[4]: a sequence where people are walking on the corridor (OC) and an outdoor level crossing scene (LC).

3.3 Qualitative and Quantitative Results

From a qualitative point of view, a comparison of the four commented options can be observed in Figs. 2 and 3, one of them without the application of any postprocessing method (*Basic AE*) and the remaining options by applying the different alternatives described previously (*Pearson*, *OMED* and *OMDE*). The results are obtained with the best tuned configuration that we have tested. It is obvious that the postprocessing method improves and removes the existing noise (false positives) in scenes with a high variability in the background. It can be observed in the sequence *CAM*, in the two first columns of the Fig. 2, and *V4* and *WS*, two last columns of the Fig. 3. In these videos, the *dilation - erosion* strategy in morphological operators (*OMDE*) is not the best appropriate because the application of the dilation operator in the first step maintains the noise. On the other hand, *Pearson* and *OMED* largely remove the spurious pixels, without any considerable penalization in the real objects in motion. Nevertheless, note that *Pearson* maintains the shape of the object better than *OMED* in spite of filtering the noise of the background in a lower grade.

Nevertheless, in scenes where the appearance of false negatives is prevailing, the employ of *OMED* is worse than the basic output of the segmentation algorithm, as it can be observed in the third and forth columns in the Fig. 2. In this case, the postprocessing method with *OMED* produces more false negatives than

---

[1] http://mmc36.informatik.uni-augsburg.de/VSSN06_OSAC.

[2] http://perception.i2r.a-star.edu.sg/bk_model/bk_index.html.

[3] http://media.ee.ntu.edu.tw/Archer_contest/.

[4] http://homepages.inf.ed.ac.uk/rbf/CAVIARDATA1/.

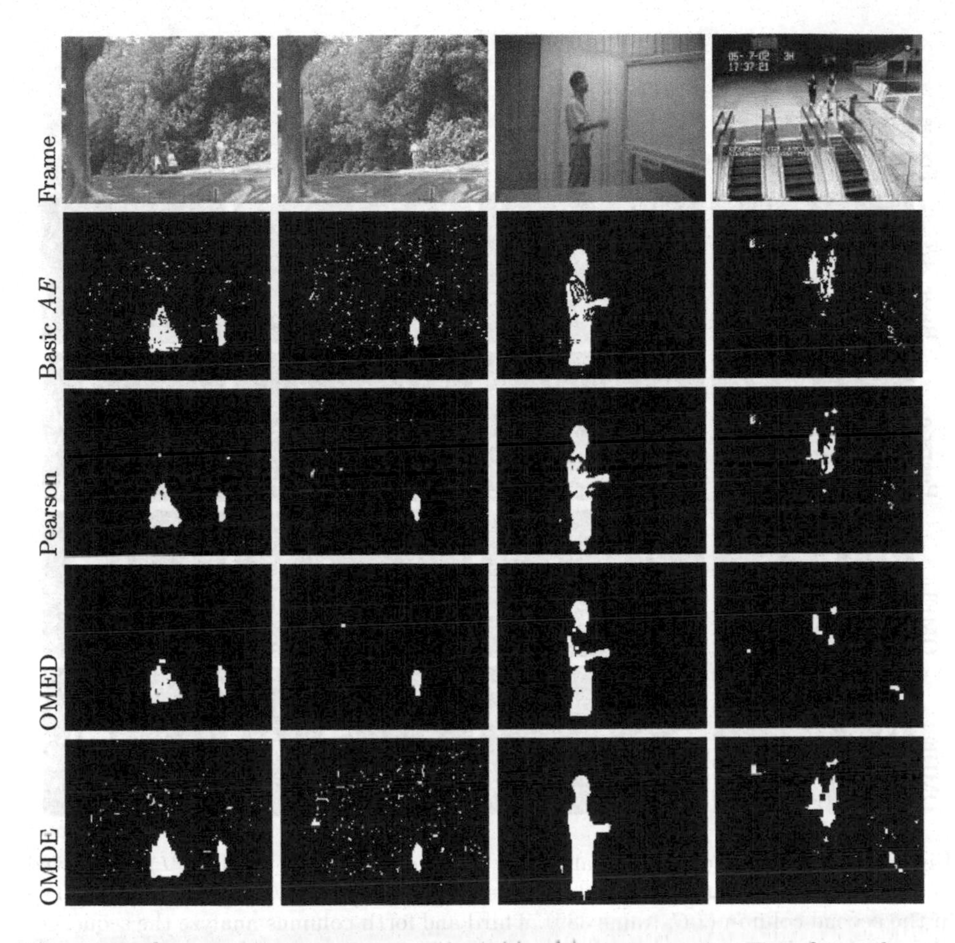

Fig. 2. Results over complex scenes with stable objects in motion. Two first columns exhibit the exterior sequence *CAM* with vegetation continuously moving on it (frames 1372 and 1392). Third and forth columns show the analysis of the sequences *MR*, meeting room, and *SS*, escalators, over the frames 3242 and 4558, respectively.

the original ones, while the output with *OMDE* is the optimal in comparison with the remaining alternatives.

In order to compare the performance of each proposed method from a quantitative point of view, we have chosen several well-known measures [4]. The spatial accuracy (AC), the precision (PR) and the recall (RC) are considered in this work in make the comparison. A value in the range $[0, 1]$, where higher is better, is provided by each measure. We also consider True positives (TP), True negatives (TN), False negatives (FN) and false positives (FP) rates. A good overall evaluation of the performance of a given method is offered by AC, while PR must be considered against RC. Each measure can be defined as follow:

$$AC = \frac{TP}{TP + FN + FP} \tag{12}$$

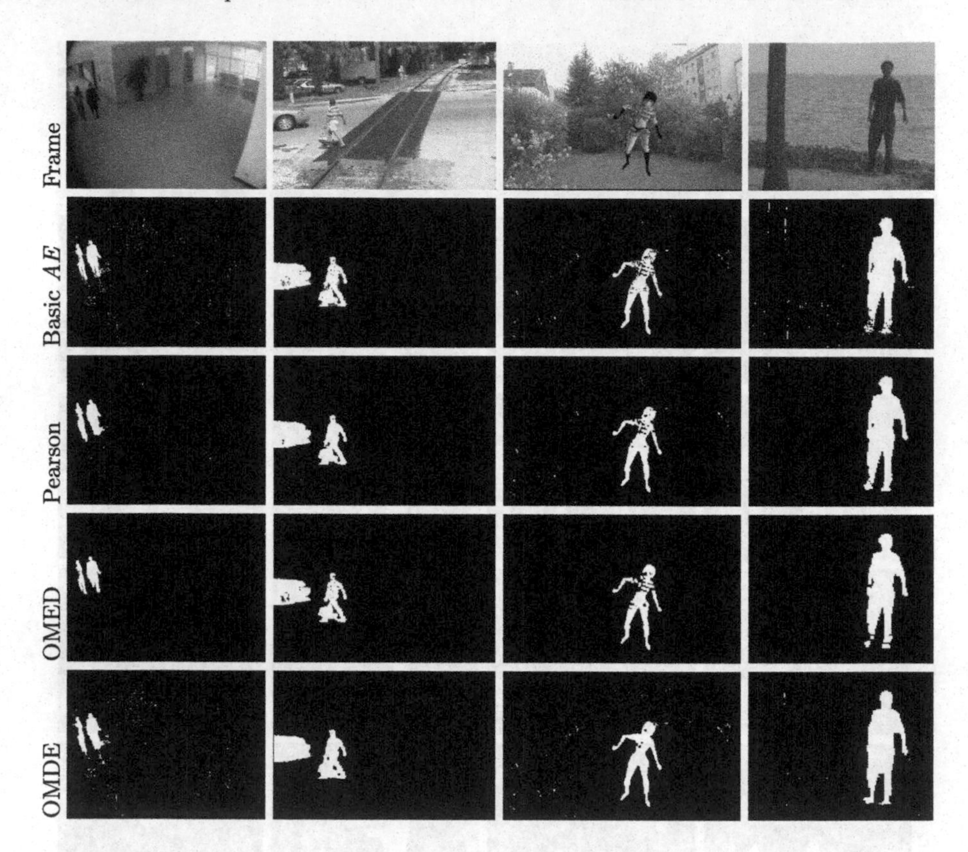

Fig. 3. Another set of experimental results. An interior scene of a hall (*IP1*) is observed in the first column (frame 116) while a scene of a level crossing without barriers is shown in the second column (*LC*, frame 389). Third and forth columns analyze the sequences *V4*, synthetically generated, and *WS*, where the movement of the waves of the sea, with frames 815 and 1624, respectively.

$$RC = \frac{TP}{TP + FN} \tag{13}$$

$$PR = \frac{TP}{TP + FP} \tag{14}$$

In general, and for all the tested sequences, we can indicate that *Pearson* is the most suitable method together with *OMED* for scenes with false positives, while those with false negatives is not as effective as *OMDE*, but it is quite competitive. This information is shown in Tables 1, 2 and 3.

In the first of them (Table 1) we can assume the comments from a visual point of view of the qualitative comparison from Figs. 2 and 3. We could consider *OMED* as the best suitable method because it is the best in five of the twelve analyzed sequences, while *Pearson* and *OMDE* are better in three videos each one. But, because of the fact that *OMED* is the worst in quality terms in

Table 1. Quantitative assessment using the measure AC. The mean and the standard deviation are shown after the analysis of each set of data, where best result is highlighted in **bold**. The last row indicates the average performance of each method, where the best method obtains one point, the second two points, and so on. The fewer score the better the postprocessing on average. Rows with a star indicate that the results are statistically significant.

	Basic	Pearson	OMED	OMDE
Campus (*CAM*)	0.594 ± 0.168	0.725 ± 0.093	$\mathbf{0.735 \pm 0.075}$	0.607 ± 0.182
*Meeting Room (*MR*)	0.775 ± 0.060	0.803 ± 0.058	0.752 ± 0.072	$\mathbf{0.866 \pm 0.036}$
Subway Station (*SS*)	0.458 ± 0.124	0.462 ± 0.128	0.424 ± 0.140	$\mathbf{0.511 \pm 0.144}$
Fountain (*FT*)	0.407 ± 0.150	0.435 ± 0.143	0.463 ± 0.086	$\mathbf{0.480 \pm 0.194}$
IPPRData1 (*IP1*)	0.536 ± 0.169	0.620 ± 0.170	$\mathbf{0.632 \pm 0.183}$	0.539 ± 0.181
*IPPRData2 (*IP2*)	0.451 ± 0.134	0.469 ± 0.140	$\mathbf{0.628 \pm 0.133}$	0.459 ± 0.137
IPPRData3 (*IP3*)	0.606 ± 0.174	$\mathbf{0.620 \pm 0.172}$	0.603 ± 0.194	0.619 ± 0.171
*Level Crossing (*LC*)	0.878 ± 0.035	$\mathbf{0.897 \pm 0.030}$	0.879 ± 0.035	0.883 ± 0.033
Corridor (*OC*)	0.709 ± 0.037	0.712 ± 0.038	$\mathbf{0.723 \pm 0.039}$	0.706 ± 0.038
Video2 (*V2*)	$\mathbf{0.920 \pm 0.019}$	0.919 ± 0.023	0.910 ± 0.024	0.916 ± 0.027
Video4 (*V4*)	0.682 ± 0.111	0.722 ± 0.112	$\mathbf{0.723 \pm 0.115}$	0.714 ± 0.128
WaterSurface (*WS*)	0.870 ± 0.023	$\mathbf{0.902 \pm 0.025}$	0.890 ± 0.026	0.899 ± 0.022
Puntuación	41	**22**	29	28

several sequences (see the performance for the *MR*, *IP3* and *V2* scenes) we have incorporated a ranking index which scores each method in ascending order of its performance. Thus, the winner method for a sequence will obtain one point, the second two points, the third three points and the forth four points. The last row of the Table 1 represents the sum of these scores for all the sequences. From these results, we can conclude that *Pearson* is the best method on average because, although it is not always the best, it is the second best method in most of the sequences (see the ranking of the methods in Table 3). Therefore, this approach could be viewed as the most stable method regardless of the type of sequence. It is interesting to observe that, except one case and with low margin (sequence *V2*), the *Pearson* correlation filtering always improves the result produced by the *Basic* segmentation algorithm.

Another non-trivial advantage of the postprocessing with *Pearson* is that it is not necessary to adjust any kind of parameters, except those required by the segmentation algorithm (ε_0 for AE in this experiments). As it is observed in Table 2, the number of iterations of *erosion - dilation* or *dilation - erosion* for the morphological operators is not always the same, so it is needed to estimate the optimal value to avoid performing the search exhaustively.

Finally, in Table 3 a statistical significance test employing the *Student's t test* is carried out. A two-tailed test between the best method and the remaining alternatives is computed, where the result is considered statistically significant if the *p-value* is lower than 0.05 for all the comparisons in the same sequence.

Table 2. Best configuration for each method and each sequence.

Seq.	Metd.	Accuracy	*Precision*	*Recall*	Parameters
CAM	Basic	0.594 ± 0.168	0.65 ± 0.19	0.88 ± 0.06	ε_0=0.01
	Pearson	0.725 ± 0.093	0.79 ± 0.11	0.90 ± 0.05	ε_0=0.005
	OMED	$\mathbf{0.735 \pm 0.076}$	0.85 ± 0.09	0.85 ± 0.08	ε_0=0.001, $\#\mathcal{ED}$=1
	OMDE	0.607 ± 0.182	0.64 ± 0.20	0.93 ± 0.05	ε_0=0.01, $\#\mathcal{DE}$=1
MR	Basic	0.775 ± 0.060	0.96 ± 0.03	0.80 ± 0.06	ε_0=0.0005
	Pearson	0.803 ± 0.058	0.96 ± 0.03	0.83 ± 0.06	ε_0=0.0005
	OMED	0.752 ± 0.072	0.98 ± 0.02	0.76 ± 0.08	ε_0=0.0005, $\#\mathcal{ED}$=1
	OMDE	$\mathbf{0.866 \pm 0.036}$	0.95 ± 0.03	0.91 ± 0.02	ε_0=0.001, $\#\mathcal{DE}$=3
SS	Basic	0.458 ± 0.124	0.70 ± 0.18	0.59 ± 0.16	ε_0=0.01
	Pearson	0.462 ± 0.128	0.72 ± 0.19	0.59 ± 0.17	ε_0=0.01
	OMED	0.424 ± 0.140	0.68 ± 0.17	0.54 ± 0.17	ε_0=0.1, $\#\mathcal{ED}$=1
	OMDE	$\mathbf{0.511 \pm 0.144}$	0.65 ± 0.19	0.72 ± 0.14	ε_0=0.01, $\#\mathcal{DE}$=2
FT	Basic	0.407 ± 0.150	0.52 ± 0.18	0.64 ± 0.11	ε_0=0.005
	Pearson	0.435 ± 0.143	0.52 ± 0.17	0.72 ± 0.10	ε_0=0.001
	OMED	0.463 ± 0.086	0.71 ± 0.08	0.57 ± 0.10	ε_0=0.0001, $\#\mathcal{ED}$=1
	OMDE	$\mathbf{0.480 \pm 0.194}$	0.51 ± 0.20	0.87 ± 0.10	ε_0=0.01, $\#\mathcal{DE}$=2
IP1	Basic	0.536 ± 0.169	0.66 ± 0.18	0.73 ± 0.17	ε_0=0.01
	Pearson	0.620 ± 0.170	0.81 ± 0.11	0.72 ± 0.19	ε_0=0.005
	OMED	$\mathbf{0.632 \pm 0.183}$	0.93 ± 0.06	0.67 ± 0.20	ε_0=0.01, $\#\mathcal{ED}$=1
	OMDE	0.539 ± 0.181	0.63 ± 0.19	0.78 ± 0.17	ε_0=0.01, $\#\mathcal{DE}$=1
IP2	Basic	0.451 ± 0.134	0.53 ± 0.18	0.76 ± 0.08	ε_0=0.01
	Pearson	0.469 ± 0.140	0.56 ± 0.19	0.77 ± 0.08	ε_0=0.005
	OMED	$\mathbf{0.628 \pm 0.133}$	0.77 ± 0.15	0.78 ± 0.12	ε_0=0.01, $\#\mathcal{ED}$=2
	OMDE	0.459 ± 0.137	0.52 ± 0.17	0.81 ± 0.08	ε_0=0.01, $\#\mathcal{DE}$=1
IP3	Basic	0.606 ± 0.174	0.70 ± 0.17	0.81 ± 0.14	ε_0=0.05
	Pearson	$\mathbf{0.619 \pm 0.172}$	0.71 ± 0.17	0.83 ± 0.14	ε_0=0.05
	OMED	0.603 ± 0.194	0.79 ± 0.14	0.72 ± 0.20	ε_0=0.05, $\#\mathcal{ED}$=1
	OMDE	0.619 ± 0.171	0.68 ± 0.17	0.87 ± 0.11	ε_0=0.05, $\#\mathcal{DE}$=1
LC	Basic	0.879 ± 0.035	0.91 ± 0.03	0.96 ± 0.01	ε_0=0.01
	Pearson	$\mathbf{0.897 \pm 0.030}$	0.91 ± 0.03	0.98 ± 0.01	ε_0=0.01
	OMED	0.879 ± 0.034	0.93 ± 0.02	0.94 ± 0.02	ε_0=0.01, $\#\mathcal{ED}$=1
	OMDE	0.883 ± 0.033	0.89 ± 0.03	0.99 ± 0.01	ε_0=0.01, $\#\mathcal{DE}$=1
OC	Basic	0.709 ± 0.037	0.75 ± 0.02	0.92 ± 0.04	ε_0=0.01
	Pearson	0.712 ± 0.038	0.75 ± 0.02	0.92 ± 0.04	ε_0=0.01
	OMED	$\mathbf{0.723 \pm 0.039}$	0.81 ± 0.02	0.87 ± 0.04	ε_0=0.01, $\#\mathcal{ED}$=3
	OMDE	0.706 ± 0.038	0.74 ± 0.02	0.94 ± 0.04	ε_0=0.01, $\#\mathcal{DE}$=1
V2	Basic	$\mathbf{0.920 \pm 0.019}$	0.94 ± 0.01	0.97 ± 0.02	ε_0=0.001
	Pearson	0.919 ± 0.023	0.95 ± 0.01	0.97 ± 0.02	ε_0=0.0005
	OMED	0.910 ± 0.024	0.94 ± 0.01	0.96 ± 0.02	ε_0=0.0001, $\#\mathcal{ED}$=1
	OMDE	0.916 ± 0.027	0.94 ± 0.01	0.97 ± 0.03	ε_0=0.005, $\#\mathcal{DE}$=1
V4	Basic	0.682 ± 0.111	0.75 ± 0.13	0.88 ± 0.03	ε_0=0.005
	Pearson	0.722 ± 0.112	0.79 ± 0.13	0.90 ± 0.03	ε_0=0.005
	OMED	$\mathbf{0.723 \pm 0.115}$	0.82 ± 0.15	0.87 ± 0.03	ε_0=0.001, $\#\mathcal{ED}$=1
	OMDE	0.714 ± 0.128	0.74 ± 0.14	0.96 ± 0.02	ε_0=0.005, $\#\mathcal{DE}$=1
WS	Basic	0.870 ± 0.023	0.95 ± 0.02	0.91 ± 0.02	ε_0=0.001
	Pearson	$\mathbf{0.902 \pm 0.025}$	0.97 ± 0.01	0.93 ± 0.02	ε_0=0.0005
	OMED	0.890 ± 0.026	0.98 ± 0.01	0.91 ± 0.02	ε_0=0.0001, $\#\mathcal{ED}$=1
	OMDE	0.899 ± 0.022	0.96 ± 0.02	0.93 ± 0.02	ε_0=0.005, $\#\mathcal{DE}$=2

Table 3. *P-values* after the application of the *Student's t test* between the best method and the remaining competitors. If all the *p-values* of the same sequence are lower than 0.05, the result of the best technique is regarded as significant, whose name in the table is formatted in **bold** style.

Seq.	Best metd.	Competitor 1		Competitor 2		Competitor 3	
CAM	OMED	Pearson	0.7079	OMDE	6.10e − 003	Basic	1.43e − 003
MR	**OMDE**	Pearson	**2.06e − 004**	Basic	**1.10e − 006**	OMED	**2.05e − 007**
SS	OMDE	Pearson	0.2647	Basic	0.2172	OMED	0.0606
FT	OMDE	OMED	0.7237	Pearson	0.4092	Basic	0.1884
IP1	OMED	Pearson	0.5246	OMDE	2.93e − 006	Basic	5.38e − 007
IP2	**OMED**	Pearson	**0.00e + 000**	OMDE	**0.00e + 000**	Basic	**0.00e + 000**
IP3	Pearson	OMDE	0.9538	Basic	0.4230	OMED	0.3761
LC	**Pearson**	OMDE	**6.25e − 005**	OMED	**1.00e − 006**	Basic	**3.75e − 007**
OC	OMED	Pearson	0.4916	Basic	0.3821	OMDE	0.2957
V2	Basic	Pearson	0.3784	OMDE	4.96e − 003	OMED	8.45e − 013
V4	OMED	Pearson	0.9188	OMDE	0.3522	Basic	5.65e − 006
WS	Pearson	OMDE	0.6302	OMED	0.1377	Basic	1.37e − 004

Consequently, the obtained results are quite similar, since the improvement of the method is significant in only three of the tested sequences. Nevertheless, if we observe, for example, the *p-values* of the *CAM* sequence, we can see that *OMED* is not significantly better than *Pearson*, but, in fact, it is regarding *OMDE* and the proposal without any postprocessing method (*Basic*). In addition, it can be observed that *Pearson* is the second method (*Competitor 1* in Table 3) in practically all the sequences where it is not the winner.

4 Conclusion

We can conclude that the postprocessing method, in general, is necessary because it substantially improves the result obtained by the segmentation algorithms. In addition, it has been observed that *Pearson* is the best method on average for all the tested sequences, although with no statistically significant evidence.

Acknowledgments. This work is partially supported by the Ministry of Economy and Competitiveness of Spain under grants TIN2014-53465-R, project name Video surveillance by active search of anomalous events, and TIN2016-75097-P. It is also partially supported by the Autonomous Government of Andalusia (Spain) under projects TIC-6213, project name Development of Self-Organizing Neural Networks for Information Technologies; and TIC-657, project name Self-organizing systems and robust estimators for video surveillance. All of them include funds from the European Regional Development Fund (ERDF). The authors thankfully acknowledge the computer resources, technical expertise and assistance provided by the SCBI (Supercomputing and Bioinformatics) center of the University of Málaga.

References

1. Burdick, H.E.: Digital Imaging - Theory and Applications. McGraw-Hill, New York (1997)
2. Gamarra, M., Zurek, E., San-juan, H., Eng, S., Norte, U.: A study of image analysis algorithms for segmentation, feature extraction and classification of cells. J. Inf. Syst. Eng. Manag. **2**(4), 20 (2017)
3. Kastrinaki, V., Zervakis, M., Kalaitzakis, K.: A survey of video processing techniques for traffic applications. Image Vis. Comput. **21**(4), 359–381 (2003)
4. Li, L., Huang, W., Gu, I.Y.H., Tian, Q.: Statistical modeling of complex backgrounds for foreground object detection. IEEE Trans. Image Process. **13**(11), 1459–1472 (2004)
5. López-Rubio, E., Luque-Baena, R.M.: Stochastic approximation for background modelling. Comput. Vis. Image Underst. **115**(6), 735–749 (2011)
6. Lopez-Rubio, E., Luque-Baena, R.M., Dominguez, E.: Foreground detection in video sequences with probabilistic self-organizing maps. Int. J. Neural Syst. **21**(3), 225–246 (2011)
7. Masood, S., Sharif, M., Masood, A., Yasmin, M., Raza, M.: A survey on medical image segmentation. Curr. Med. Imaging Rev. **11**(1), 3–14 (2015)
8. Nilakant, R., Menon, H.P., Vikram, K.: A survey on advanced segmentation techniques for brain MRI image segmentation. Int. J. Adv. Sci. Eng. Inf. Technol. **7**(4), 1448–1456 (2017)
9. Palomo, E.J., Domínguez, E., Luque-Baena, R.M., Muñoz, J.: Image compression and video segmentation using hierarchical self-organization. Neural Process. Lett. **37**(1), 69–87 (2013)
10. Luque, R.M., Domínguez, E., Palomo, E.J., Muñoz, J.: A neural network approach for video object segmentation in traffic surveillance. In: Campilho, A., Kamel, M. (eds.) ICIAR 2008. LNCS, vol. 5112, pp. 151–158. Springer, Heidelberg (2008). https://doi.org/10.1007/978-3-540-69812-8_15
11. Rodgers, J.L., Nicewander, W.A.: Thirteen ways to look at the correlation coefficient. Am. Stat. **42**(1), 59–66 (1988)

Identifying Pixels Classified Uncertainties ckMeansImage Algorithm

Rogério R. de Vargas[1], Ricardo Freddo[1], Cristiano Galafassi[1(✉)], Sidnei L. B. Gass[1], Alexandre Russini[1], and Benjamín Bedregal[2]

[1] Laboratório de Sistemas Inteligentes e Modelagem (LabSIM), Federal University of Pampa, Itaqui, RS, Brazil
cristianogalafassi@unipampa.edu.br

[2] Logic, Language, Information, Theory and Applications (LoLITA), Federal University of Rio Grande do Norte, Natal, RN, Brazil
bedregal@dimap.ufrn.br

http://www.labsim.unipampa.edu.br

Abstract. Floods may occur in rivers when the flow rate exceeds the capacity of the river channel, particularly at bends or meanders in the waterway. Floods often cause damage to homes and businesses becoming the most prevalent type of disaster in the world and the one with the highest number of events, causing the greatest economic losses, affecting a large number of people. This paper has the objective of mapping and identifying the flooding areas of a selected region in the municipality of Itaqui-RS using remote sensing. In order to do it, we used the Fuzzy ckMeansImage Algorithm to group and to classify the image into similarity clusters. The methodology consists in processing satellite images before and after the flooding occurs. Finally, we discuss the processed images and present the flooded area.

Keywords: ckMeans · Clustering · Flood · Fuzzy · Sensing remote

1 Introduction

Several of the human senses gather their awareness of the external world almost entirely by perceiving a variety of signals, either emitted or reflected, actively or passively, from objects that transmit these information in waves or pulses. Thus, one hears disturbances in the atmosphere carried as sound waves, experiences sensations such as heat, reacts to chemical signal form food through taste and smell, is knowledge of certain material properties such as roughness through touch, and recognizes shapes, colors, and relative positions of exterior objects and classes of materials by means of seeing visible light and issuing from them. In the previous sentence, all sensations that are not received through direct contact, they are remotely sensed [1].

Remote sensing, according to [2], can be defined as a science and art of obtaining information about an object, area or phenomenon through the analysis

J. Medina et al. (Eds.): IPMU 2018, CCIS 855, pp. 429–440, 2018.
https://doi.org/10.1007/978-3-319-91479-4_36

of data acquired by instruments that are not in direct contact with object, area or phenomenon. Still, the analysis and interpretation of such data, the representations of images, tables or graphs, also integrate a science of remote sensing.

One of the areas in which the application of remote sensing has a vast field of work is flood monitoring. As presented by [3], flooding is the most prevalent type of disaster in the world and the one with the highest number of events, causing the greatest economic losses, affecting a large number of people. Considering all sorts of natural disasters, floods are probably the most devastating and which occurs more frequently. It is a natural and recurrent event for a river and, statistically, the average annual flood is matched or exceeded every 2.33 years. In order to be able to have a flood coverage dimension, it is worth mentioning, as recorded by the authors, that they are responsible for almost 55% of all recorded disasters and for approximately 72.5% of total economic losses around the world.

In the conception work, the analysis and interpretation of such data, the representations of images, the tables or the graphs, also integrate a science of remote sensing. The extraction of remote sensing information demands clear and logical methods and analyzes so that they can be applied to any product [4]. One of the tools of analysis is the classification, which can be understood as a technique of recognizing patterns represented in a multispectral image, by grouping pixels so that similar pixels belong to the same cluster [5].

In [6], the authors concluded that fuzzy logic becomes more appropriate when it is necessary to work with information that addresses possible ambiguities or inaccuracies. Thus, it is possible to note the potential of the methodology to be applied in the field of remote sensing.

In this context, this work applies the fuzzy ckMeansImage algorithm [7] to a multispectral image with the objective of detecting and measuring the extent of the flood occurred on the municipality of Itaqui-RS. For this, two images of the selected area, one with the river in its normal level and other in its flood period. The two images are processed by the algorithm and the total area of the river in the selected region is calculated to estimate the extension of the flood.

2 Background

In this section we present the Fuzzy ckMeans algorithm, proposed by [8] and the definition of the α-cut [9].

2.1 ckMeans Algorithm

The idea is basically to share the fuzzy set $X = \{x_1, x_2, \ldots, x_n\}$ in p clusters where μ_{ij} is the membership degree of the sample x_i that belongs to the j-th cluster and the result of clustering is expressed by membership degrees on matrix μ.

The ckMeans use techniques of FCM [10] and K-Means [11] algorithms attempts to partitionate sets of data by minimizing an objective function shown in Eq. (1):

$$J = \sum_{i=1}^{n} \sum_{j=1}^{p} \mu_{ij}^{m} d_{ij} \left(x_i; c_j \right)^2 \tag{1}$$

where:

- n is the number of data;
- p is the number of clusters considered in the algorithm, which must be decided before execution;
- m is a fuzzification parameter in the range $(1; w)$, indicating the width of n dimensional cluster perimeter. Usually, m is the range [1.25; 2] [12] - we only consider rational values to simplify the calculation of Eqs. (1), (2) and (4). Actually, it is used rational m's;
- x_i a vector of training data, where $i = 1, 2, \ldots, n$. These are the cluster attributes selected from the source data elements (such as columns in a database table);
- c_j the centroid (or centrer) of a fuzzy cluster ($j = 1, 2, \ldots, p$);
- $d_{ij} \left(x_i; c_j \right)$ is the distance between x_i and c_j - when the values are numbers, it is usually used the Euclidean distance;

The input of the algorithm are n data, the number of clusters p and value m. Its steps are:

1. Starts μ (membership degree) with a continuous random value between zero (no relevance) and 1 (total relevance) where the sum of pertinence must be one.
2. Calculate the centroid of the cluster j as follows: We stabilize the centroid of each cluster as in the K-Means algorithm [11]. However, in our algorithm we first create a new matrix, which is called μCrisp, containing values 1 or zero. Each line of this new matrix has 1 in the positions with the greatest value of this line in the μ and zero in the other positions of the line. When a column of the matrix μCrisp, after this step, has only zeros in it, it is assigned the value 1 in the position that corresponds to the largest value of the same column in the matrix μ.
   The ckMeans algorithm returns a matrix μCrisp, which the values of the elements belong to the set {0, 1} as shown in Eq. (2). Thus, μCrisp_{ij} the content of matrix at the position (ij), is defined in Eq. (2).

$$\mu Crisp_{ij} = \max \left(\left\lfloor \frac{\mu_{ij}}{\max_{l=1}^{p} \mu_{il}} \right\rfloor, \left\lfloor \frac{\mu_{ij}}{\max_{l=1}^{n} \mu_{lj}} \right\rfloor \right) \tag{2}$$

   The first argument of the right side of Eq. (2) is for any datum whose value is 1 for the cluster it belongs with the greatest degree, and 0 for the others.

The second argument is for the greatest degree of each column (cluster) is 1, so as to ensure that all clusters have at least one element. Thus, on rare occasions may happen that a line has more than one value 1 (which does not occur in the algorithm K-Means), but as this matrix is only auxiliary, this does not bring any inconvenience.
The steps of the algorithm to calculate μCrisp_{ij} are performed as follows (there may be a situation where the result of μCrisp_{ij} is not completely accurate in Eq. (2). In this case, the greatest value of the column μ_{ij} have 1 in μCrisp_{ij}):
(a) Read μ;
(b) Find the larger value at the first line of the matrix μ. After that, assign, on μCrisp matrix, the value 1 to the position corresponding to the larger value position on matrix μ and 0 to the others. To complete the process, repeat the same procedure to the other lines;
(c) Store in a vector the number of 1's that each column μCrisp has.
If a column in μCrisp has no 1's, assign 1 in the position of the largest value of that column of the matrix μ.
After calculating the matrix μCrisp calculate the new centroids of clusters as in Eq. (3):

$$c_j = \frac{\Sigma_{i=1}^{n} x_i \mu Crisp_{ij}}{\Sigma_{i=1}^{n} \mu Crisp_{ij}} \tag{3}$$

c_j is calculated by adding the data belonging to cluster (in crisp form) and dividing it by the number of classified objects as 1 in the matrix μCrisp for this cluster.

3. Calculate an initial value (a data) for J using the Eq. (1);
4. Calculate the table of the fuzzy membership function as shown in Eq. (4):

$$\mu_{ij} = \frac{\left(\frac{1}{d_{ij}(x_i;c_j)}\right)^{\frac{2}{m-1}}}{\sum_{k=1}^{p}\left(\frac{1}{d_{ik}\left(x_i;c_k\right)}\right)^{\frac{2}{m-1}}} \tag{4}$$

5. Return to step 2 until a convergence condition is reached.

Some possible stopping conditions are:

- A fixed number of iterations is executed;
- The user reports a value $\epsilon > 0$ of convergence, and if

$$d_{ij}\left(J_U; J_A\right) \leq \epsilon$$

then the algorithm stops, where J_A is the objective function (Eq. (1)) calculated in the previous iteration and J_U is the objective function of the last iteration.

2.2 α-cut

A fuzzy set is a collection of objects with various membership degree. Often it is useful to consider those elements that have at least some minimal membership degree α. This is liking asking who has a passing grade in a class, or a minimum height to ride on a roller coaster [13]. We call this process an αcut.

For every $\alpha \in [0; 1]$, a given fuzzy set A yields a crisp set A^α which contains those elements of the universe X who have membership grade in A of at least α (Eq. 5):

$$A^\alpha = \{x \in X | A(x) \geq \alpha\} \tag{5}$$

where α is in the range of $0 < \alpha \leq 1$ and "|" stands for "such that" [14].

We can not emphasize enough that an α-cut of a fuzzy set is not a fuzzy set, it is a crisp set.

In other works, several authors use the idea of the α-cut in the clusterization process. In [15], the authors proposed the alpha-cut implemented in fuzzy clustering algorithms, called FCMalpha. It allows the data points for being able to completely belong to one cluster. The proposed FCMalpha algorithms may form a cluster core for each cluster, where data points inside a cluster core will have a membership value of 1 so that it can resolve the drawbacks of FCM.

A study in [16] shows an analyses of FCM using a literature review based on various articles ranging from 1987 to 2015 with the keywords to find out how FCM along with alpha-cuts and various similarity or dissimilarity measures have advanced in this period. On the basis of 75 articles, this work has classified the previous FCM classification works using the four categories such as: Land Cover Classification method; Fuzzy c-Classification; Measures of similarity and dissimilarity; and Fuzzy alpha-cuts in accordance with various research problems and domains.

Similarly to the work of [15] where as defined a value α cut a priori and as discussed in several other articles, as shown in [16], we also chose to use the α-cut concept to create and highlight an uncertainty data cluster. Thus, it allows the knowledge specialist to visualize the uncertainty pixels that the algorithm was not so certain in its classification.

3 Methodology

The images selected for the experiment are a fragment of the T21JWH tile of the Sentinel-2 satellite, dated 06/02/2017 (Fig. 5(a)), when the Uruguay River was 10.96 m above the normal level, and 08/17/2017 (Fig. 3(a)), when the river quota was normal. The spatial resolution of the images was 10 m and the following spectral bands with their respective central wavelengths were used: band 3 (green, 560 nm), band 4 (red, 665 nm) and band 8 (infrared, 842 nm).

Using TerrSet's remote sensing tools [17], the RGB color composition 483 was run, which was exported to the JPG format for further processing by the Fuzzy ckMeans algorithm. The choice of the spectral bands and the color composition in question was determined by the possibility of a good distinction between the targets, especially the water slides, which are the subject of the present study. The image used has the following characteristics: 660 rows and 629 columns, with a total of 415140 pixels, making a covered area of 41,51 km^2.

The area covered by the images selected for the present experiment, represent the city of itaqui and adjacencies, located in the western border of Rio Grande do Sul, Brazil (Fig. 1). The region is characterized by a flat relief, and the city being situated between the Uruguay River to the north, the Cambai stream to the east, and the Olarias stream to the west. The morphological characteristics of the region, associated with a large area drained by the Uruguay River basin, cause successive floods in Itaqui, which cause disorders to the population and must be a constant object of planning processes by the municipal authority. The center point of the used images are defined with the following geographical coordinates: 29° 08' 22.81"S and 56° 32' 32.04"W.

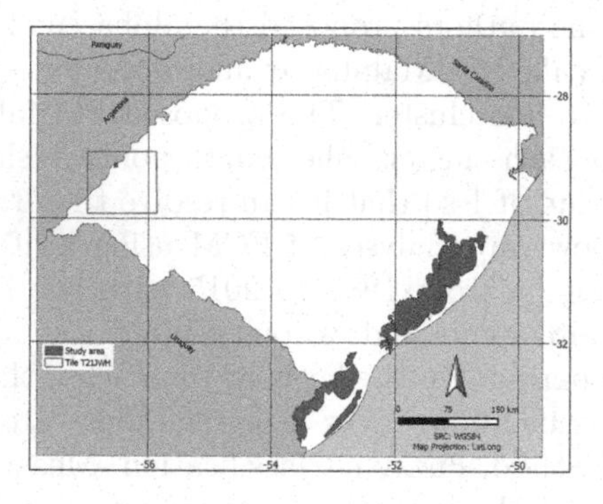

Fig. 1. Study area

The procedure to obtain an image pass trough a five-steps process, describes how we obtained the river images. From steps 2 to 5, the TerrSet software was used.

1. Data acquisition - Acquisition from LandViewer (Avaliable in: https://lv.eosda.com);
2. Image importation - Image casting from *tiff* to *img*;
3. Area extraction - Image cropping (window);
4. False-color band combination - Image composition;
5. Image exportation - Image casting from *img* to *jpg*.

4 Proposal

The process, since the image reading until the characterized image, follows the steps:

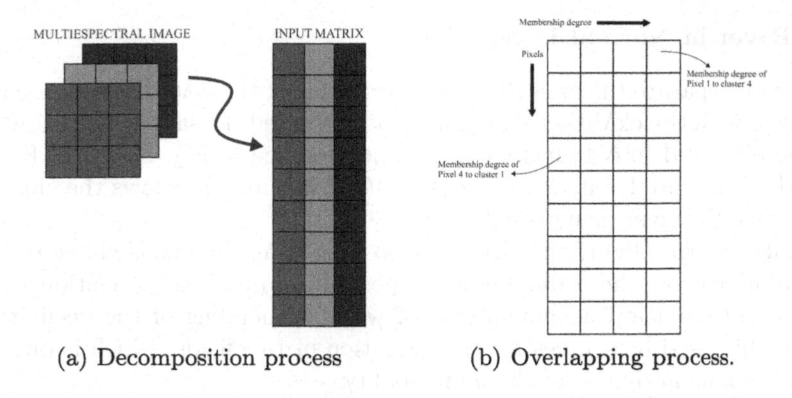

(a) Decomposition process (b) Overlapping process.

Fig. 2. Decomposition and overlapping process (Color figure online)

1. Image Read - extracts the image digital number for RGB;
2. Input Matrix - converts the digital numbers in each pixel into a three column matrix (Input Matrix), where the columns represent RED, GREEN and BLUE digital numbers. The process is shown in Fig. 2(a). It is important to mention that the ckMeans outputs a membership degree for each line in the Input Matrix, it is necessary to convert the pixel matrix into a decomposed input matrix. In this way, it is necessary to recreate the pixel matrix to be able to interpret the patterns in the characterized image;
3. Fuzzy ckMeans - runs the ckMeans with the Input Matrix, as describe in Subsect. 2.1;
4. Define a priori the α-cut parameter;
5. Membership Degree Matrix - resulting matrix with the membership degree for each line in the Input Matrix (Fig. 2(b));
6. Characterized Image - the membership degree matrix generates three pixels matrix, one for each RGB spectrum, and the overlay generates the Characterized Image. Also, when the algorithm sets a pixel to a cluster, in the overlapping process, the cluster size is updated. In the end of the process, the Fuzzy ckMeansImage returns, besides the clustered image, the pixel size (*i.e.*, the amount of pixels in each cluster).

   This process was namely ckMeansImage algorithm for adapting ckMeans to the image segmentation process and by using alpha-cut conceptualization.

5 Results

The experiment consist in processing two images, river in normal level (Fig. 3(a)) and river in flood period (Fig. 5(a)) of the same area: before and after the flooding occurrence using the K-Means, ckMeans and ckMeansImage algorithms. The images were processed with 3, 4 and 5 clusters and the best result was chose. The cluster number may be different due to the image characteristics. This image are evaluated qualitatively, by an specialist, and quantitatively by the measurement of its areas (the cluster size). The calculated area is a simple product between the number os pixels in each cluster and its size (10 m).

5.1 River in Normal Level

In order to separate the area effectively occupied by the water bodies, the resulting image with the ckMeans algorithm that presented the satisfactory results was the one clustered into 4 clusters, as can be observed in Fig. 4(b). The K-Means algorithm presented similar results (Fig. 4(a)). Figure 3(b) shows the clusterized image with the river using α-cut = 0.6.

Analyzing this figure, it is possible to infer that the pixels clustered in the blue color refer to the water bodies, especially those already mentioned previously, besides natural accumulations of water, depending of the declivity that are possibly used as sources for the irrigation of rice fields, which is one of the main economic activities of the municipality.

The pixels clustered in dark green color represent areas with denser vegetation such as riparian forests, urban afforestation, grazing areas or grasses with a lower accumulation of moisture during the image capture period. On the other hand, the pixels clustered in the light green color are associated to the areas with less dense or senescent vegetation, or, for the most part, to the areas traditionally occupied by rice fields, and that, during the recording period of the image by the satellite, maintained a higher concentration of moisture in the soil.

All the other elements, which had their pixels clustered in yellow, represent the great majority of the urban structure of Itaqui and other elements of human interference as irrigation, compacted soil and road structures.

Analysing the Fig. 3(a), comparing with Figs. 4(a) and (b), there is a region that has been clustered the wrong way, for both K-Means and ckMeans. The ckMeansImage algorithm, using the α-cut, shows exactly that region on the circled area (Fig. 3(b)).

5.2 River in Flood Period

In the image processing of the flood season, the image generated with the Fuzzy ckMeans algorithm presented satisfactory results with 3 clusters, as can be observed in Fig. 6(b). Still, it obtained better clusters compared to the result generated with the K-Means algorithm (Fig. 6(a)), where the ckMeansImage clustered the urban structure more accurately.

Analysing these figures, it is possible to infer that the pixels that represent the water slides, were clustered in the blue color. The pixels of the image representing vegetative elements and exposed soil were clustered in green. In turn, the other elements, especially the urban structure and clouds were represented in the yellow cluster.

The ckMeansImage algorithm with the α-cut highlights an area (presented in Fig. 5(b)) where the pixels do not have a acceptable membership degree. For this region, the ckMeansImage algorithm clustered some area as vegetative elements and some area as urban structure. The K-Means algorithm clustered some area as vegetative elements too, but the other area as water bodies.

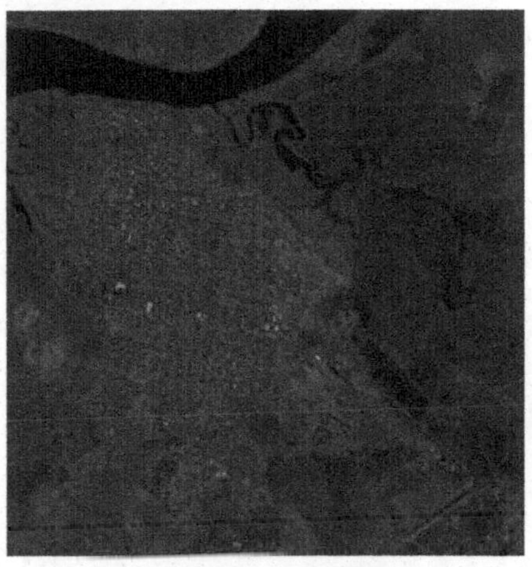

(a) River in normal level

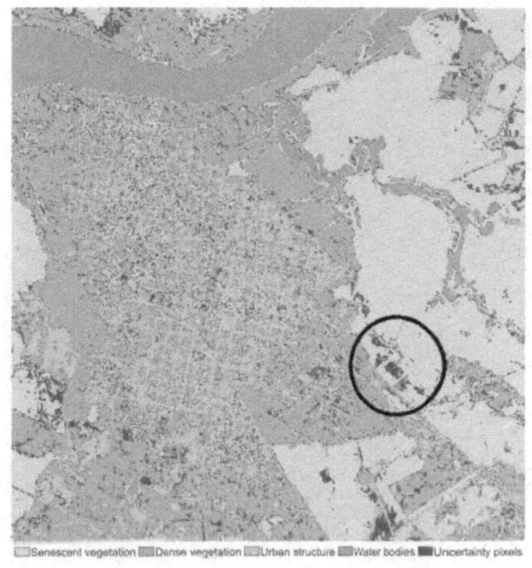

(b) Clusterized image with ckMeansImage using α-cut in 0.6

Fig. 3. Selected area with the river in normal level (Color figure online)

(a) Clusterized image with K-Means

(b) Clusterized image with ckMeans

Fig. 4. Selected area with the river in normal level (Color figure online)

5.3 Floodplain Analysis

From the counting of pixels per cluster that was made by the algorithm, it was possible to estimate the area effectively flooded in the image used in the experiments. Considering that the area of each pixel is 0.0001 km^2 (considering the spatial resolution of 10 m of the image), the results presented in Table 1 were obtained.

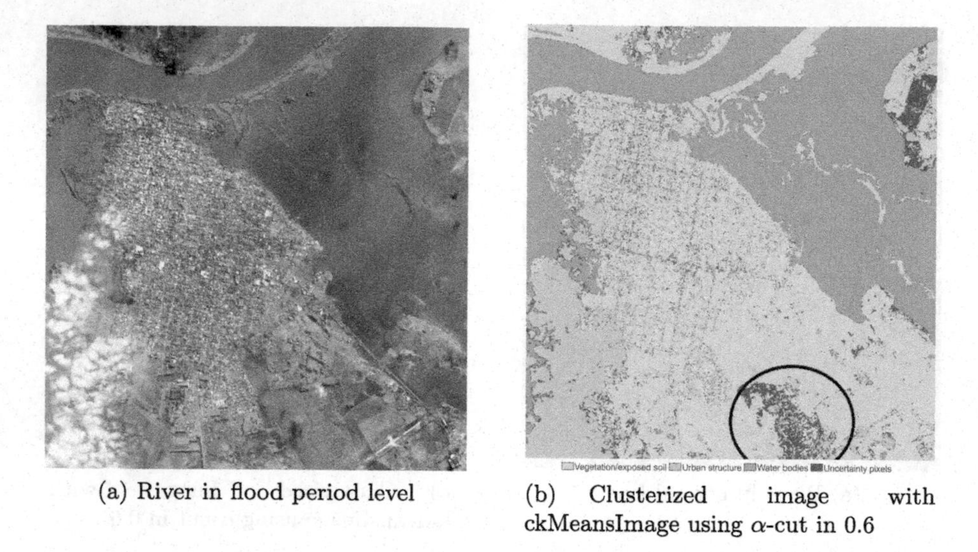

(a) River in flood period level

(b) Clusterized image with ckMeansImage using α-cut in 0.6

Fig. 5. Selected area with the river in flood period (Color figure online)

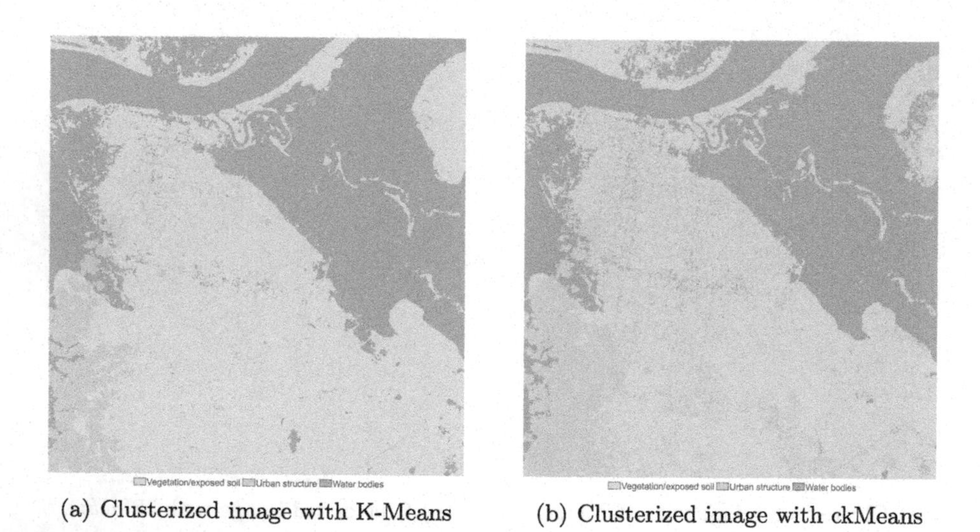

(a) Clusterized image with K-Means

(b) Clusterized image with ckMeans

Fig. 6. Selected area with the river in flood period (Color figure online)

Through observation of the image after processing, it becomes possible to clearly define the existence of a transition area located between the urban perimeter of the municipality and the rice fields, which presents a significant probability of flooding. However, using ckMeansImage algorithm with the α-cut equal to 0.6 shows that it is necessary the attention from the expert, showing that there are

Table 1. Analysis of the floodplain area for K-Means and ckMeans

	N° of pixels		Area (km^2)		%	
	K-Means	ckMeans	K-Means	ckMeans	K-Means	ckMeans
Normal level	32.267	34.534	3.23	3.45	7.77	8.32
Flood period	160.145	154.105	16.01	15.41	38.57	37.12

areas where the membership degree is lower, being able to group in the imprecise form. The total number of imprecise pixels for the river in normal level was 39.759, using the α-cut equals to 0.6. For the river in the flood period, the number of imprecise pixels was 25.979.

In this way, it is possible to infer beforehand the areas with potential flood risk, serving as subsidies for the implementation of public policies aimed at planning the expansion of the urban area. A non-observation of the areas considered at risk may cause serious social problems, resulting from the relocation of families, sanitation and infrastructure, as well as economic losses arising from the adversities provided in routine, regarding the activities of commerce and industry located in the central part, as well as in the primary production within the municipality.

6 Conclusions

We compared the proposed method to others robusts clustering algorithms such as K-Means and ckMeans. Note that Fuzzy ckMeansImage is a robust version of ckMeans by implementing the α-cut concept. It will be equivalent to ckMeans when α-cut = 0.

The present work shows a comparison between a normal and flooded stage in the municipality of Itaqui, western border of Rio Grande do Sul. Two images were used to analyse and infer the floodplains allowing us to estimate the area size and the region where it occurs. It is also possible to infer that the results presented by the K-Means, ckMeans and Fuzzy ckMeansImage algorithms are accurate in cases where the original images present a more homogeneous set of pixels, thus allowing the extraction of core areas, as it was possible to verify with the water masses.

Given the characteristic of the Fuzzy ckMeansImage algorithm in dealing with membership degree, such as future work, we intend to map areas of uncertainty. Since each pixel is classified in every clusters, it is intended to create a threshold, highlighting the pixels with low membership degree, relying on the expert's knowledge for decision making.

References

1. Kumar, S.: Basics of Remote Sensing and GIS. Laxmi Publications (P) Ltd., New Delhi (2005)
2. Jensen, J.: Remote Sensing of the Environment: An Earth Resource Perspective, 2nd edn. Pearson Eduaction, South Asia (2007)
3. Sausen, T., Narvaes, S.: Sensoriamento remoto para inundações e enxurradas. In: Sausen, T.M., Lacruz, M.S.P. (eds.) Sensoriamento remoto para desastres. Oficina de textos, São Paulo (2015)
4. Meneses, R., Sano, E.: Classificação pixel a pixel de imagens. In: Meneses, P.R., Almeida, T. (eds.) Introdução ao processamento de imagens de sensoriamento remoto. UnB, Brasília (2012)
5. Melgani, F., Al Hashemy, B.A.R., Taha, S.: An explicit fuzzy supervised classification method for multispectral remote sensing images. IEEE Trans. Geosci. Remote Sens. **38**, 287–295 (2000)
6. Zadeh, L.: Fuzzy sets. Inf. Control **8**(3), 338–353 (1965)
7. Freddo, R., de Vargas, R., Galafassi, C., Russini, A., Pinto, T.: Desenvolvimento de uma ferramenta web para a execução do algoritmo Fuzzy ckMeans no processamento de imagens. Revista Jr de Iniciação Científica em Ciências Exatas e Engenharia (ICCEEg) **1**(15), 1–10 (2017)
8. de Vargas, R., Bedregal, B.: A comparative study between fuzzy C-means and ckMeans algorithms. In: Proceedings of the Conference on North American Fuzzy Information Processing Society (NAFIPS 2010), Toronto, Canada (2010)
9. Zadeh, L.: The concept of a linguistic variable and its application to approximate reasoning. Inf. Sci. **8**(3), 199–249 (1975)
10. Bezdek, J.: Pattern Recognition with Fuzzy Objective Function Algorithms. Kluwer Academic Publishers, Norwell (1981)
11. MacQueen, J.: Some methods for classification and analysis of multivariate observations. In: Proceedings of the Fifth Berkeley Symposium on Mathematical Statistics and Probability, pp. 281–297. University of California Press, Berkeley (1967)
12. Cox, E. (ed.): Fuzzy Modeling and Genetic Algorithms for Data Mining and Exploration. Elsevier/Morgan Kaufmann, San Francisco (2005)
13. Wierman, M.: An Introduction to the Mathematics of Uncertainty, Center for the Mathematics of Uncertainty. Creighton University (2010)
14. Ponce-Cruz, P., Ramírez-Figueroa, F.: Intelligent Control Systems with LabVIEW^{TM}. Springer, London (2009). https://doi.org/10.1007/978-1-84882-684-7
15. Yang, M.S., Wu, K.L., Hsieh, J.N., Yu, J.: Alpha-cut implemented fuzzy clustering algorithms and switching regressions. Trans. Sys. Man Cyber. Part B **38**(3), 588–603 (2008)
16. Mukhopadhaya, S.: Fuzzy c-means classifier with alpha-cuts in application for similarity and dissimilarity measures: a literature survey. J. Basic Appl. Eng. Res. **3**, 2350–77 (2016)
17. Eastman, R.: TerrSet Geospatial Monitoring and Modeling Software. ClarkLabs, Worcester (2016). http://clarklabs.org/wp-content/uploads/2016/03/TerrSet18-2_Brochure_WEB.pdf

Automatic Detection of Thistle-Weeds in Cereal Crops from Aerial RGB Images

Camilo Franco[1(✉)], Carely Guada[2], J. Tinguaro Rodríguez[2], Jon Nielsen[3], Jesper Rasmussen[3], Daniel Gómez[4], and Javier Montero[2,5]

[1] Department of Industrial Engineering, Andes University, Bogotá, Colombia
c.franco31@uniandes.edu.co

[2] Department of Statistics and OR, Complutense University, Madrid, Spain
cguada@ucm.es, {jtrodrig,monty}@mat.ucm.es

[3] Department of Plant and Environmental Sciences, University of Copenhagen, Copenhagen, Denmark
{jon,jer}@plen.ku.dk

[4] Department of Statistics and OR III, Complutense University, Madrid, Spain
dagomez@estad.ucm.es

[5] Geosciences Institute (CSIC-UCM), Complutense University, Madrid, Spain

Abstract. Capturing aerial images by Unmanned Aerial Vehicles (UAV) allows gathering a general view of an agricultural site together with a detailed exploration of its relevant aspects for operational actions. Here we explore the challenging task of detecting *cirsium arvense*, a thistle-weed species, from aerial images of barley-cereal crops taken from 50 m above the ground, with the purpose of applying herbicide for site-specific weed treatment. The methods for automatic detection are based on object-based annotations, pointing out the RGB attributes of the Weed or Cereal classes for an entire group of pixels, referring to a crop area which will have to be treated if it is classified as being of the Weed class. In this way, an annotation belongs to the Weed class if more than half of its area is known to be covered by thistle weeds. Hence, based on object and pixel-level analysis, we compare the use of k-Nearest Neighbours (k-NN) and (feed-forward, one-hidden layer) neural networks, obtaining the best results for weed detection based on pixel-level analysis, based on a soft measure given by the proportion of predicted weed pixels per object, with a global accuracy of over 98%.

Keywords: Image analysis · k-Nearest Neighbours · Neural networks
Soft measures · Weed detection · Precision agriculture

1 Introduction

Unmanned Aerial Vehicles (UAV) allow gathering images with high potential for knowledge discovery, having added attractiveness due to their low operational costs and flexible driving capabilities [15]. Focusing on images of agricultural sites, UAVs stand as ideal tools for monitoring seasonally variable crop/soil

J. Medina et al. (Eds.): IPMU 2018, CCIS 855, pp. 441–452, 2018.
https://doi.org/10.1007/978-3-319-91479-4_37

conditions in time-specific and time-critical crop management [13]. In this way, images can be stitched together into a detailed map with the position and distribution of the objects of interest (see e.g. [15,19]), or can be directly interpreted for real-time implementation of the required actions (see e.g. [6]). Either way, images are the raw visual input on the characteristics of the site, holding rich information that can be understood by (knowledge-based) data-driven models for its efficient use (see [10], but also [9] for a review on different image processing techniques). By means of these models, automatic or semi-automatic implementation of certain operational actions can be undertaken in the field.

Detection and identification of weeds under a wide range of conditions is in fact an important challenge for weed control systems [12,18,19] and remote sensing (i.e., unmanned, air and space borne sensing) in precision agriculture (see e.g. [1,4,17,20]). Here, we explore the use of the k-Nearest Neighbours (k-NN) and (feed-forward, one-hidden layer) neural network methodologies for inducing *relevant knowledge* from aerial images, examining standard (RGB) aerial images captured at 50 m above ground, and running (supervised) learning-based simulations on late-stage barley fields for thistle weed (*cirsium arvense*) detection.

In order to train the models, we used a set of aerial images with two types of object-based annotations. These annotations indicated the RGB attributes of the Weed and the Cereal classes, grouping under a unique label an entire group of pixels to the Weed or the Cereal class. In this way, if more than half of the annotated area was known to be covered by thistle or cereal, then it would be marked accordingly. Hence, examining these images at pixel-level, there is an *inherent uncertainty* regarding the membership of the pixel to either the Weed or the Cereal class.

Thereby, weed detection can be carried out at two different levels: *pixel* and *object*. In the first approach, the focus is placed on analysing and being able to effectively separate the weed and crop classes at a pixel basis. In the second one, the interest is rather placed on compact groups of pixels, usually referred to as objects, somehow identifiable with relatively small plots of land, that are in this way analysed in order to assign each object as a whole to either decision class. In the set of images we analyse in this paper, the supervision was made at the object level, that is, a priori information about the decision classes is only available for certain predefined objects (a set of square plots of land). Therefore, at pixel-level, there is an inherent uncertainty of the membership of pixels to either class, since the pixels belonging to a given object may not necessarily all belong to the object class.

In the existing literature, the k-NN has been previously applied to remote sensing data (see e.g. [5]). On this line of research, such a methodology has been examined for pixel-based classification on multi-spectral ASTER data [21], comparing the results with an object-based approach relying on maximum likelihood. Although the latter (the object-based approach) seems to obtain better results (see again [21]), it can also be that pixel-level analysis allows arriving at results with similar significance under less complex procedures [16].

Reflecting on the *object-based paradigm* (see [2]), its popularity can be justified by the particular resolution of the remote sensing images that are commonly studied, mainly coming from space borne platforms, together with the nature of the objects under study. In this sense, with increasing spatial resolution, objects are made up of several pixels, which implicitly suggests the use of object-oriented methods (see [3] for an extensive discussion). Different researchers have focused on comparison studies between the performance of object and pixel based methods (see e.g. [4,14,21]), trying to find evidence supporting the use of one or another. But it remains clear that the suitability of either approach depends on the specific classification problem and its level of difficulty, as well as the image-space resolution. At the same time, addressing the value of information, it is noticed here that pixel level analysis allows a very precise application of decision making actions (such as fumigation), which require the appropriate instruments for implementing those actions (see e.g. [6]).

Taking into account these ideas, in the present study a mixed pixel-object approach is developed, in which objects are classified attending to a soft measure obtained from a previous classification of the pixels belonging to these objects. This mixed approach allows combining some of the advantages of both the pixel-based and object-based paradigms. Particularly, it allows for pixel-based recognition of areas of interest (i.e. plots to be fumigated) from the distribution of predicted weed pixels in an image, as well as to treat these whole areas as objects upon which a decision has to be made. As we shall see, an added value of the proposed approach is that it seems to enable a more robust and accurate classification of objects.

This paper is organized as follows. In the first section we present the experimental setting for collecting and capturing the input (raw data) images, briefly reviewing the k-NN and the neural network procedures and their use of the object-based annotations. In Sect. 3 results are discussed together with the statistical performance of the algorithms, also addressing their potential impact for decision making. Lastly, in Sect. 4, some final comments are given along with open lines for future research.

2 Methodology

In this section, the setting for the site experiments is introduced, also explaining the theoretical framework for the k-NN and the neural networks.

2.1 Instrumentation and Data Sets

Images were captured with a camera mounted on an UAV, consisting in a six-rotor hexacopter equipped with a (Canon G15) camera (see [15,19] for the full details). The UAV flew over the agricultural sites located at Taastrup, Denmark, focusing on barley fields infested with thistle. The camera had a 10 megapixel CCD sensor, with a Field-Of-View of roughly 60 m by 45 m, flying at 50 m from the ground. In this way, an image size is 4000 pixel wide and 3000 pixel high, thus corresponding to a Ground-Sampling-Distance of 15 mm and an area of 2700 m^2. The computational procedures were developed in Matlab (R2014), treating images according to their RGB pixel-color composition.

The available set of images in this study consists of 28 images (as the one shown in Fig. 1a), coming from 5 different days at an advanced growth stage of the crops. All images were taken under sunny conditions during day light. There are respectively 3, 2, 11, 2 and 10 images in the different days. Through an expert knowledge, coming e.g. from farmers or specialists, a set of reference ground in the field is classified in weed and thistle and they are represented in the set of images. That is, as shown in Fig. 1, each RGB image is accompanied of two annotated templates, one identifying plots of land (i.e. objects) associated to the crop class (see Fig. 1b), and the other identifying objects associated to the weed class (see Fig. 1c). There are fewer crop annotations than weed ones, but the former usually contain far more pixels than the latter. In addition, the annotations have different colours to distinguish between them, but there is no other meaning associated to these colours.

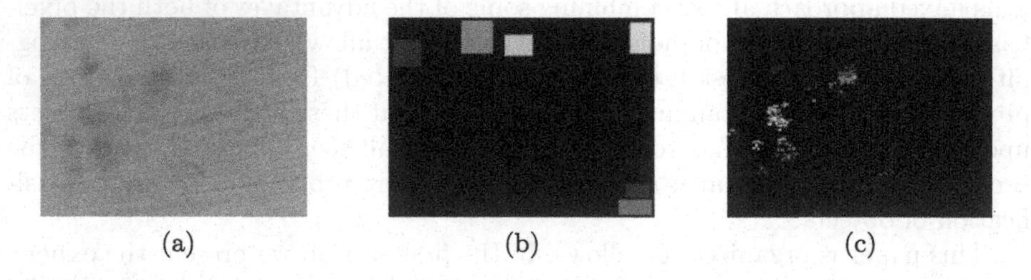

Fig. 1. An (a) aerial image and its annotations, (b) crop and (c) weed.

Such an identification can be developed directly on the field, demanding a very expensive practice if the whole area covered by the UAVs is to be covered, or can be inferred directly from the images, carefully examining a representative set of sample images. Here, the available annotations come from gathered knowledge on the field, where a given area of the crop is annotated as belonging to the Weed class if more than half of the area is covered by the thistle weed. Then, taking a set of annotated samples, the algorithms can be trained on the basis of the knowledge captured for a given group of pixels, not for pixels themselves, making it difficult to develop robust pixel-based procedures.

From these images, we have obtained two kinds of datasets: one at object level, in which each annotated object is stored as a single observation or row, measuring its class and the average R, G and B intensities of the pixels it contains. The second type of dataset, at pixel level, contains an observation for each pixel in an annotated object, storing the RGB intensity of each pixel together with the class and the ID of the object it belongs to. We also record the ID of each image for the objects and pixels it contains, as well as the date at which the image was taken. Particularly, we organize this information in pairs of object-level and pixel-level datasets for each of the 5 days at which the different available RGB images were taken.

2.2 Experimental Setting

At an experimental level, we focus on measuring the performance of our approach when the information of 4 of the 5 different days is used to train the classifiers, and the information of the remaining day is used as test data to validate the trained models. The idea is thus to analyse the suitability of effectively detecting weed pixels and objects in an RGB image from supervised information coming from images taken in potentially different conditions (e.g. differences in luminosity or wind intensity and direction among different days). This would allow in practice to perform weed detection when desired or necessary, without needing to invest funds and time in acquiring annotated information with the particular conditions of the specific moment at which the task is performed.

This 4-days-vs.-1-day cross-validation setting is applied at both pixel and object level, thus feeding the classifiers with a training sample composed of either pixels or objects from images of the 4 of the 5 available days, and then predicting all the annotated pixels or objects of the images of the remaining day. Because of the huge number of pixels available at the training stage, the pixel level experiments proceed by selecting as training sample a relatively small random sample of pixels from the selected 4 days, from which all the test sample pixels are predicted. This process is then replicated a number N of times (it will be $N = 15$) in order to obtain more robust performance estimations. At object level, this randomization and replication scheme is not necessary as the number of objects is easily manageable, and thus the training sample for each 4 vs. 1 combination is composed of all the annotated objects in all the images of the 4 days used for learning.

Finally, once pixel level predictions are available, it is possible to exploit its distribution in order to derive an alternative prediction for the test objects. This is carried out by means of a soft measure given by the proportion of predicted weed pixels in each object, and a threshold $\tau \in [0, 1]$ controlling the proportion required to assign each object to the weed class. Formally,

$$p(object\ i) = \frac{\#\{pixels\ assigned\ to\ the\ weed\ class\ in\ object\ i\}}{\#\{pixels\ in\ object\ i\}}, \quad (1)$$

and thus object i is assigned to the weed class whenever $p(object\ i) > \tau$. Notice that different values of this threshold τ, allows setting up different sensitivity levels in order to assign objects to the weed class, as different decision-requirements may be imposed. Also, the soft measure defined in (1) allows considering *soft accuracy measures*, as the ones proposed in [7], avoiding the assumption that all errors in classification should be taken as equally important (see also [8]).

2.3 Classification Methodologies

In the present study, we have mainly focused on the k-NN supervised classification methodology. Let us recall that k-NN classifiers proceed by assigning a query or item to be classified to the most frequent class in the set of the k training instances which are closer to such query. That is, given a query x, it is necessary to measure the distance $d(x,t)$ from the query to each instance t in the training sample, then find the k nearest training instances, so finally x is assigned to the majority class of this set of k nearest-neighbours of x. Thus, the distance d is a key ingredient of the k-NN methodology, as it is in charge of defining the notion of *proximity* being applied to find the nearest neighbours.

Despite its conceptual simplicity, k-NN classifiers may achieve a rather good performance, and its low computational complexity and suitability for parallelization make its usage appealing in contexts with massive datasets. These reasons have often motivated its application in the context of image processing, in which relatively big datasets are easily obtained when combining information at a pixel level from several images. We adhere to these motives in the present study, and as we describe below, the results obtained through this classification methodology are quite competitive even for $k = 1$.

Regarding the applied metric $d(x,t)$, we have used a weighted Euclidean distance scheme, in such a way that the distance between two instances $x = (R(x), G(x), B(x))$ and $y = (R(y), G(y), B(y))$ in the RGB space is obtained as

$$d(x,y) = w_R(R(x) - R(y))^2 + w_G(G(x) - G(y))^2 + w_B(B(x) - B(y))^2,$$

with $w = (w_R, w_G, w_B)$ being a vector of weights. Particularly, we experimentally found that assigning nearly three times more weight to the R-coordinate than to the other two, provided the best results in our application context.

In order to compare and extend the results obtained through the previous k-NN methodology, we have also conducted some similar experiments using artificial neural networks. Here we use a simple, single-layer, learning network with error back propagation [11]. The activation function of the neurons in the single hidden layer was the hyperbolic tangent-sigmoid function, the initialization of weights and bias was randomly repeated 100 times, and the training for the updating of the synaptic weights and the bias values was done using the Levenberg-Marquardt optimization algorithm. Under this framework, all the possible combinations from 5 to 15 neurons were tested in the hidden layer, applying the early stopping technique for setting the optimal number of iterations of the network model. Thus, the best neural network corresponded with the model specification on number of neurons and the required iterations, which accomplished the minimum mean squared error (MSE).

3 Experimental Results and Discussion

In this section we present and discuss the results obtained by applying the classification methodologies described in the experimental setting exposed above in Sect. 2.2. Basically, we describe here the results of three different experiments: (i) an object level experiment, in which classifiers are fed with the object-level dataset containing the average RGB of each object in order to predict test objects; (ii) a pixel level experiment, using the pixel level datasets to predict test pixels; and (iii) a mixed approach experiment, using the test pixels predictions obtained in (iv) to provide an alternative prediction of the corresponding test objects.

In the experiment at the pixel level, all 28 images are used as test image, one at a time, gathering the training data from all the images taken in different days than the test image. Particularly, the training data is composed of a randomly selected 1% of each image pixels. This process is repeated 15 times for each test image, thus replicating the random sampling step in order to obtain a more robust estimation of the performance indexes.

We just show the results of the k-NN classifiers for $k = 1$. We experimentally found that greater values of k do not provide improvements of the results as significant as to make up for the increased computational costs.

3.1 Object Level Experiment

We start by describing the results obtained on the object-level data. Here we are measuring the ability of the classifiers to accurately detect weed objects in the test images (all taken the same day) from the object-level information of the remaining 4 days. Therefore, a total of 5 experiments were conducted, each taking the objects in all the images of a single, different day as test sample, while the objects in all the images of the remaining 4 days are used as training sample. Let us recall that objects are described through the average R, G and B intensities of the pixels that compose them.

There is a total of 1955 annotated objects in all images, of which 738 belong to the crop class (i.e., the negative class), while the remaining 1216 belong to the weed class (the positive class). The combined confusion matrix of the 5 experiments is given in Table 1, achieving a sensitivity of 0.9762 and a specificity of 0.7249, for a global accuracy of 88.13%.

Table 1. Confusion matrix of the object level experiment

		Predicted condition	
		Negative	Positive
True condition	Negative	535	203
	Positive	29	1188

These results point to the 1-NN classifier being quite effective at detecting positive, weed objects, although it does not perform that well regarding negative, crop objects. In other words, although almost all weed plots will be detected, there is a significant risk that many actual crop plots are identified as weed objects, leading to unnecessarily expenses for fumigating or treating (possibly) weed-free crop areas.

Nevertheless, it is important to remark that most of the false positives and negatives were obtained in 2 of the 5 experiments. Table 2 shows the pairs of sensitivity-specificity indexes attained when using the images of each day to form the test sample, together with the number of test objects in each case. Clearly, most false positives are committed in the first day, while the majority of false negatives occur in the second day.

Table 2. Sensitivity, specificity and number of test objects of the 4 vs. 1 experiments

	Day 1	Day 2	Day 3	Day 4	Day 5
Sensitivity	1	0.4782	0.9914	0.9941	0.9881
Specificity	0.6808	1	1	1	1
Number of objects	1267	53	146	207	282

3.2 Pixel Level Experiment

Now we describe the results obtained when using the pixel-level data. Here the aim is to predict the class of the object each pixel belongs to, since pixels are not directly annotated, but only indirectly, through the objects they make part of. Let us recall that, in this case, the experimental setting proceeds by randomly drawing 1% of the pixels in the images taken in a different day than the test image. This process is replicated 15 times for each test image, and thus the confusion matrix in Table 3 counts 15 times the actual number of pixels (i.e. replication results are summed rather than averaged). For this confusion matrix, specificity is 0.9315 and sensitivity is 0.9016, for a total accuracy rate of 92.69%.

Table 3. Confusion matrix of the experiments using the pixel-level data

		Predicted condition	
		Negative	Positive
True condition	Negative	270481707	19879713
	Positive	5201477	47667883

A few (2 or at most 3) specific test images provide somehow worse results than the other, with a similar pattern to the one detected in the previous object-level experiment, i.e. 2 of these images were taken at days 1 and 2, which obtained a poorer object-level performance. However, other images also taken during these first 2 days obtained much better results, so actually at the pixel level evidence is not clear regarding the potential worse behaviour of the classifiers for some of the days. Anyway, these are quite competitive results for a classification methodology as simple as the 1-NN.

3.3 Mixed Approach Experiment

Now we use the pixel predictions obtained in the previous experiment to compute the soft index $p(o_i)$ introduced in Sect. 2.2 for each object o_i, $i = 1, \ldots, 1955$. That is, we now try to predict test objects (i.e. plots of land) attending only to the available pixel-level class information of those pixels that compose each test object. Here we provide results for $\tau = 0.5$, as a neutral value not specifically biasing the results towards a greater specificity or sensitivity. Again, the confusion matrix in Table 4 combines the results of the 15 replications for each test image, leading to a total of $15 \times 1955 = 29325$ test objects classified. In this case, specificity is 0.9949 and sensitivity is 0.9854, for a global accuracy of 98.90%.

Table 4. Confusion matrix of the mixed approach experiment

		Predicted condition	
		Negative	Positive
True condition	Negative	11013	57
	Positive	267	17988

Interestingly, the results at the object-level are significantly better when objects are not directly predicted through their average RGB (see Table 1 above), but through the predicted class proportions of the pixels they contain. In this sense, it is better to assume the uncertain, indirectly obtained class annotations of pixels in the training objects, than to discard it and proceed by averaging the RGB intensities of the pixels in each object.

Furthermore, now all days' objects are predicted with an almost uniform effectivity, as shown in Table 5. Particularly, days 1 and 2 now obtain a similar performance to the other days (compare with the results in Table 2).

Table 5. Sensitivity, specificity and number of test objects of the 4 vs. 1 experiments

	Day 1	Day 2	Day 3	Day 4	Day 5
Sensitivity	0.9978	0.9246	0.9675	0.9831	0.9749
Specificity	0.9984	1	1	1	0.9034
Number of objects	1267	53	146	207	282

3.4 Experiments with Artificial Neural Networks

In an equivalent way as previous subsections, we have also used the artificial neural networks using the object and pixel level data for the detection of weed.

In the first place, a one-day image was used to train the model, testing it on the remaining set of images. The best model was found consisting of six neurons

Table 6. Confusion matrix of pixel-level experiment using neural networks

		Predicted condition	
		Negative	Positive
True condition	Negative	16286621	449076
	Positive	490187	2585985

(see Table 6), obtaining a sensitivity of 0.8407 and a specificity of 0.9732, for a total accuracy rate of 95.26%.

Then, we tested the previous best model over the remaining images (from all the other days) to detect weed objects (see Table 7). The results obtained a sensitivity of 0.9819 and a specificity of 1.0000, for a total accuracy rate of 98.87%.

Table 7. Confusion matrix of object level experiment using neural networks

		Predicted condition	
		Negative	Positive
True condition	Negative	738	0
	Positive	22	1195

An excellent performance is obtained with this model compared with the 1-NN method, where the results for the neural network obtains a 100% specificity. Such results imply that there is a minimum (≈ 0) risk in predicting the crop areas as weed areas, entailing higher savings for the farmer regarding the site-specific application of herbicide. Nonetheless, there is still an error of less than 2% in detecting the thistle weeds, which should be carefully analyzed to understand its implications, from an economic/agricultural viewpoint, according to its effects over the yield and quality of the barley-agricultural product. Also, the accuracy of 98% obtained at object level is higher than the one obtained in [19], where the convolutional networks obtained a 96–97% accuracy with other images of crops taken at 10 and 50 m in early and late growth stages.

4 Conclusions

In this paper, k-NN and neural network methodologies have been examined for thistle weed detection from object-based annotations. The relevance of this approach focuses on the treatment of such (pixel-level) imprecise annotations, due to the high cost of gathering precise annotations on the occurrence of thistle weed on the field. Both methods achieve a satisfactory performance on images taken from 50 m above ground by UAV, with respect to both their statistical reliability and expected implications for undertaking the required actions.

It is relevant to point out that the best results were obtained from the pixel-level analysis, based on the computation of a soft index $p(o_i)$ introduced in Sect. 2.2. The suggested measure allows taking into account in a basic and flexible way, the uncertainty in detecting the decision object from the aerial image of the crop, allowing further interaction through the fine-tuning of the free parameter τ. As it was mentioned earlier, this soft index was fixed at 0.5, but a robust methodology should be implemented for optimizing its value. On the other hand, different metrics could be explored, such as the *intersection-over-union*, which is commonly used for evaluating *semantic segmentation* approaches.

For future research, soft accuracy statistics could be further explored together with testing the models under broader, general conditions, regarding the resolution of the images, ranging e.g. from 10 m to 100 m above ground, and the different growth stages of the crop. Besides, neural networks could be further examined paying special attention to the random sampling in their training, and the balance/imbalance ratio between the classes. Finally, building on the present proposal, it is suggested that the neural networks and the stored knowledge gained while solving the particular problem of this study can be exploited by means of semi-supervised learning and transfer learning, developing efficient algorithms for thistle-weed detection under different and changing crop conditions.

Acknowledgement. This research has been partially supported by the Government of Spain (grant TIN2015-66471-P), the Government of Madrid (grant S2013/ICE-2845, CASICAM-CM), Complutense University (UCM research group 910149), and the Danish Environmental Protection Agency.

References

1. Andrew, M.E., Ustin, S.L.: The role of environmental context in mapping invasive plants with hyperspectral image data. Remote Sens. Environ. **112**, 4301–4317 (2008)
2. Blaschke, T.: Object based image analysis for remote sensing. ISPRS J. Photogrammetry Remote Sens. **65**, 2–16 (2010)
3. Blaschke, T., Burnett, C., Pekkarinen, A.: New contextual approaches using image segmentation for object-based classification. In: De Meer, F., de Jong, S. (eds.) Remote Sensing Image Analysis: Including the Spatial Domain, pp. 211–236. Kluver Academic Publishers, Dordrecht (2004)
4. Duro, D.C., Franklin, S.E., Dubé, M.G.: A comparison of pixel-based and object-based image analysis with selected machine learning algorithms for the classification of agricultural landscapes using SPOT-5 HRG imagery. Remote Sens. Environ. **118**, 259–272 (2012)
5. Egorov, A.V., Hansen, M.C., Roya, D.P., Kommareddy, A., Potapov, P.V.: Image interpretation-guided supervised classification using nested segmentation. Remote Sens. Environ. **165**, 135–147 (2015)
6. Franco, C., Pedersen, S.M., Papaharalampos, H., Ørum, J.E.: The value of precision for image-based decision support in weed management. Precis. Agric. **18**, 366–382 (2017)

7. Gómez, D., Biging, G., Montero, J.: Accuracy statistics for judging soft classification. Int. J. Remote Sens. **29**, 693–709 (2008)
8. Gómez, D., Biging, G.S., Montero, J.: Accuracy assessment for soft classification maps. In: Wang, G., Weng, Q. (eds.) Remote Sensing of Natural Resources, pp. 57–86. CRC Press, Boca Raton (2014)
9. Guada, C., Gómez, D., Rodríguez, J.T., Yáñez, J., Montero, J.: Classifying image analysis techniques from their output. Int. J. Comput. Intell. Syst. **9**, 43–68 (2016)
10. Guada, C., Zarrazola, E., Yáñez, J., Rodríguez, J.T., Gómez, D., Montero, J.: A novel edge detection algorithm based on a hierarchical graph-partition approach. J. Intell. Fuzzy Syst. **34**(3), 1875–1892 (2018, in press)
11. Haykin, S.: Neural Networks. A Comprehensive Foundation. Prentice Hall International, Upper Saddle River (1999)
12. Lamb, D.W., Brown, R.B.: Remote-sensing and mapping of weeds in crops. Agri. Eng. Res. **78**, 117–125 (2001)
13. Moran, M.S., Inoue, Y., Barnes, E.M.: Opportunities and limitations for image-based remote sensing in precision crop management. Remote Sens. Environ. **61**, 319–346 (1997)
14. Myint, S.W., Gober, P., Brazel, A., Grossman-Clarke, S., Weng, Q.: Per-pixel vs. object-based classification of urban land cover extraction using high spatial resolution imagery. Remote Sens. Environ. **115**, 1145–1161 (2011)
15. Rasmussen, J., Nielsen, J., Garcia-Ruiz, F., Christensen, S., Streibig, J.C.: Potential uses of small unmanned aircraft systems (UAS) in weed research. Weed Res. **53**, 242–248 (2013)
16. Robertson, L.D., King, D.J.: Comparison of pixel and object based classification in land cover mapping. Int. J. Remote Sens. **32**, 1505–1529 (2011)
17. Seelan, S.K., Laguette, S., Casady, G.M., Seielstad, G.A.: Remote sensing applications for precision agriculture: a learning community approach. Remote Sens. Environ. **88**, 157–169 (2003)
18. Slaughter, D.C., Giles, D.K., Downey, D.: Autonomous robotic weed control systems: a review. Comput. Electron. Agric. **61**, 63–78 (2008)
19. Sørensen, R., Rasmussen, J., Nielsen, J., Jørgensen, R.N.: Thistle detection using convolutional neural networks. In: Proceedings EFITA-WCCA 2017 Conference, Montpellier, France, paper 75, 2–6 July 2017
20. Tellaeche, A., Pajares, G., Burgos-Artizzu, X.P., Ribeiro, A.: A computer vision approach for weeds identification through support vector machines. Appl. Soft Comput. **11**, 908–915 (2011)
21. Whiteside, T.G., Boggs, G.S., Maier, S.W.: Comparing object-based and pixel-based classifications for mapping savannas. Appl. Earth Obs. Geoinf. **13**, 884–893 (2011)

Meaning and Uncertainty Inherent in Understanding Images, Spatial-Taxon Hierarchy, Word Annotation and Relevant Context

Lauren Barghout(✉)

Berkeley Institute for Soft Computing (BISC), Visiting Scholar 2014 - 2017,
U.C. Berkeley Electrical Engineering and Computer Sciences, Berkeley, CA, USA
lbarghout@eecs.berkeley.edu
http://www.laurenbarghout.org

Abstract. This paper explores the meaning and uncertainty inherent in (a) understanding image hierarchies; (b) describing them with words; and (c) navigating the abstraction context of the viewer. A spatial-taxon hierarchy, a standardized scene architecture, partitions an image into a foreground, subject and salient objects and/or sub-objects. The introduction starts with a thought experiment (Thought experiments, borrowed from the model-theoretic isomorphism standard of structure-mapping theory, enable readers to compare two systems thought to be similar. It's a form of inductive reasoning that expands knowledge in the face of uncertainty (Holland et al. 1986 [13]) by providing an explicit representation of how two systems are similar. Though the conclusion that the two systems do share an isomorphic structure can only be supported via various degrees of truth (fuzzy membership), it establishes its plausibility. Analogical reasoning is natural to human thought and communication making it useful for scientific papers.) based on a poem & an image landscape. The thought experiment is intended to provide analogical inference as scaffolding for the rest of the paper. The results of experimental data of human annotated spatial-taxon and corresponding word descriptions of two images are presented. The experimental results are analyzed in terms of spatial-taxon designation and the meaning & uncertainty presented by the human annotations. The results support the fuzzy spatial-taxon hierarchy of human scene perception described by other works, show that word descriptions depend on spatial-taxon designation and that long tail word distributions require unbounded possibility with semantic uncertainty (type 2 fuzzy sets) for the word counts in the probability distribution. Deep learning image recognition, Zadeh information restriction principal, Shannon's distinction between information content and semantics, customized image descriptions and fuzzy inference techniques are explored.

Keywords: Segmentation · Recognition · Classification
Image labeling · Semantic · Spatial-taxon · Fuzzy inference
Possibility theory · Deep learning · Information theory
Visual-taxometrics · Machine vision · Human vision · Attention
Visual & Image cognition

J. Medina et al. (Eds.): IPMU 2018, CCIS 855, pp. 453–465, 2018.
https://doi.org/10.1007/978-3-319-91479-4_38

1 Introduction

As homage to Lotfi Zadeh, I describe a novel & preliminary method for examining two entwined classes of uncertainty described in Professor Zadeh's 2013 paper titled "Toward a Restriction-centered Theory of Truth and Meaning." [37] In his paper he distinguishes two classes of uncertainty: *perceptual and other-mind.* Both are invoked when two or more people use language and pictures to reach mutual understanding. This paper uses experimental data that collected language and spatial annotation of two pictures to explore these two classes of uncertainty.

In day-to-day speech, the truism "a picture is worth a thousand words" shows the usefulness of pictures for bridging communication when language falls short. When people share an image to augment language they invoke what cognitive psychologist refer to as the 'theory of other-minds'[1] and 'directed gaze' [24]. Both rest on an implicit assumption that though individual minds are separate, they are similar enough to perceive roughly the same thing when they direct their gaze upon the same point in an image. Descartes subject-object distinction underlies much of the scientific method[2]. The distinction between self and others underlies human understanding. If telepathy, humans directly experiencing the phenomenology of other humans, existed it could minimize perceptual and other-mind uncertainty. The prevalence of telepathy in myth and fantasy draws on the frustration caused by other-mind uncertainty.

1.1 Thought Experiment

Consider the frustration on the limitations of language as expressed in lyrics of the song "Language" by Suzanne Vega shown on the left of Fig. 1. Since this paper presents data combining image and language descriptions of images, the lyrics of this song are useful for analogical reasoning. As a thought experiment, let's walk through each idea expressed in the lyrics and link them to their corresponding analogue.

The first stanza analogizes between liquefied words and visual attention as it flows within contextual abstraction of an image. For the purpose of this thought experiment, let's assume that "eloquent silence" refers to a picture's ability to "speak a thousand words." Yet as observed by John Berger in the 1972 BBC television series Ways of Seeing, "images are still in a sense that information never is" [12]. In other words, visual attention and the thoughts (words) associated with them change quickly. [18] As described by Barghout (2016) [8]

[1] I chose the term "other mind" used in cognitive psychology, after personal communication with Professor Zadeh regarding an example of unprecisiated restriction, which he described as "a perception evoked in one's mind." In this paper "other-mind uncertainty" refers to the unknown differences between each human's individual mental construct. It does not refer to uncertainty in the word semantics, context or particular definitions.

[2] Excluding Quantum Mechanics.

"Language" By *Suzanne Vega*

A If language were liquid
It would be rushing in
Instead here we are
In a silence more eloquent
Than any word could ever be

B These words are too solid
They don't move fast enough
To catch the blur in the brain
That flies by and is gone

[...]

C I won't use words again
They don't mean what I meant
They don't say what I said

They're just the crust of the meaning
D With realms underneath
Never touched
Never stirred
Never even moved through

Fig. 1. (*A*) "Language" by Suzanne Vega and Michael Visceglia. Each Stanza labelled A through D. (*B*) An ambiguous figure [23] with liquefied words (analogous to the first line of lyrics "if language were liquid") poured over a landscape an image comprised of a hierarchy of nested spatial-taxons. The least abstract taxon is at the highest point in landscape. In this 3 dimensional version of the image the third dimension (height) designates the abstraction level of the spatial taxon.

> "Images convey multiple meanings that depend on the context in which the viewer perceptually organizes the scene. Spatial-taxon granularity is dynamically linked to this context and the viewer's evolving understanding as he/she navigates physically, intellectually or emotionally through the scene. The phenomenology of information granule, defined by Bargiela & Pedrycz (2003) as conceptual entities that compactly encapsulate information at specific level of abstraction giving rise to cognitive hierarchies (such as the nested spatial-taxon hierarchy)."–L. Barghout

Stanza B (as shown in Fig. 1b) captures the phenomena of words not coming fast enough to keep up with the "blur in the brain", where blur in the brain analogizes to information granules changing at the speed of thought. Visual attention [24,31], a cognitive process that filters the granularity of detailed perception to a limited region, shifts quickly which is why blur in the brain works as an analogy.

Stanza C (as shown in Fig. 1c) works to describe the semantic uncertainty of words and the perception of words. Stanza D (as shown in Fig. 1d) alludes to semantic cognitive hierarchies. The phrase "crust of the meaning" analogizes to a primary word definition, but invokes the nuance that underlying the common definition is the context each human brings to a conversation. Though statistical analysis can capture common and co-correlated word descriptions, it can't get at context without a proxy. The phrase "never even stirred" analogizes to: not statistically examined.

Consider the image abstraction landscape on the right side of Fig. 1. The pitcher pouring liquefied words over the landscape corresponds to Stanza A.

Liquid flows down hill. The liquefied words fill in at the most abstract spatial-taxon of the image. As explained in Barghout (2014) and shown again in Fig. 2, the most abstract spatial-taxon is the whole image.

1.2 Spatial-Taxon Hierarchy

Definition: Spatial-Taxon. Let X be the universe of discourse consisting of all pixels within the rectangular (or square) pixel array of an image, such that $X_{1,1}$ is located at the upper left corner, and pixel $X_{I,J}$ at the lower left corner. Let ST_0 be a non-empty set that contains all pixels in the universe of discourse (the image). ST_0 has two mutually exclusive children ST_1 and ST_0 - ST_1 such that $ST_1 \wedge (ST_0 - ST_1) = \emptyset$ and $ST_1 \vee (ST_0 - ST_1) = ST_0$ (the parent). We have now defined abstraction level 0 and level 1. The most abstract information granule is the whole image and the second most abstract level contains two mutually exclusive children subsets.

Let's next define the set ST_1 as having two children subsets: ST_2, (ST_1 - ST_2). As before, these children are mutually exclusive, such that $ST_2 \wedge (ST_1 - ST_2) = \emptyset$ and $ST_1 \vee (ST_0 - ST_1) = ST_0$ (the parent). This is the third most abstract level in nested spatial-taxon hierarchy.[3]

Using this definition of spatial-taxons lets return to the analogy liquefied words to predict how people would annotate images. By analogy liquefied words should fill in the most abstract spatial-taxon ST_0 (shown as the checkerboard pattern in Fig. 1(B) before filling in the child spatial-taxon ST_1) shown as the embracing figures that compose the foreground of the image.

Foreshadowing the experimental results, the pitcher pours liquefied words into a channel hugging the crevice along the edge boundary between the child spatial-taxon of the left figure and the right figure. Interesting things happen at spatial-taxon boundaries. Research on human perceptual organization show people remember outlines of figures as opposed to ground [22]. Tracing the outline of an object in the dark often provides enough information to identify the object. Therefore Fig. 1 provides a deep channel in the crevice in which liquefied words may pool.

The different abstraction heights of the spatial-taxon as illustrated, leads us to expect that different volumes of liquefied words would pool within each spatial-taxon.

1.3 Zadeh External and Internal Restriction on Truth and Meaning

Zadeh observes that the external truth value, which in our thought experiment refers to a spatial-taxon and the words used to describe it, relates the degree of agreement with factual information such that it induces a possibilistic restriction

[3] I could have used subset (ST_1 - ST_2) as a root for a new child subset. However, to make this readable, I limit the definition to spatial-taxon children stemming from a single initial image root.

on instantiated facts [35,37]. It follows then that experimentally we expect linguistic descriptions, like the liquified words in the thought experiment, to be constrained by perceptual organization of the spatial-taxon hierarchy. Specifically, a human annotation word set $w_i =$ [words of a descriptive phrase] restricts the external numerical truth value T_i of the spatial-taxon ST_i within an explanatory database induced by the instantiated words. The explanatory database refers to the semantics induced by the specific words and constrains the possibilistic interpretation of that spatial-taxon. This class of perceptual uncertainty lends itself to be precisiated via sampling of natural language descriptions. Other-mind uncertainty lends itself to indirect modeling of an internal truth value that can modeled by its agreement to factual external information. Other-mind uncertainty is an unbounded set as alluded to by Stanza C in the thought experiment.

2 Methods

Human subjects filled out paper surveys containing a bird (as shown in the bottom layer of Fig. 2.a) or the ambiguous ghost-woman (as shown in the bottom layer Fig. 3.b). Participants were asked to mark the center of the subject of the image and label it. Spatial-taxons were determined via k-means clustering of location measurements (as shown by each layer in Fig. 2). Word and word phrases for each spatial taxon were grouped by the ambiguous figure to which they referred, rank & corresponding word are reported in the table under each image. Words were counted and ranked by frequency, such that the most commonly occurring word was given a rank of one. Words with the same count were given the same rank. Surveys were conducted at the Burningman Art Festival in NV, U.S.A. the Macworld conference in CA U.S.A. and at department of motor vehicles in Raleigh, NC, U.S.A. [4–6,10].

The simple image of a still bird was chosen to contrast with the complex ambiguous figure. Because ambiguous figures are not perceived simultaneously, the data collected can be grouped according to which ambiguous figure the subject identifies. In addition, the dichotomous interpretations do not belong to the same superordinate spatial-taxon, enabling us to avoid "word overlap" due to subject choosing word labels at various abstraction levels. If as our paradigm assumes, the center of the subject serves as a proxy of figural status then the words used to label the spatial-taxon should correspond to the name of the figure perceived.

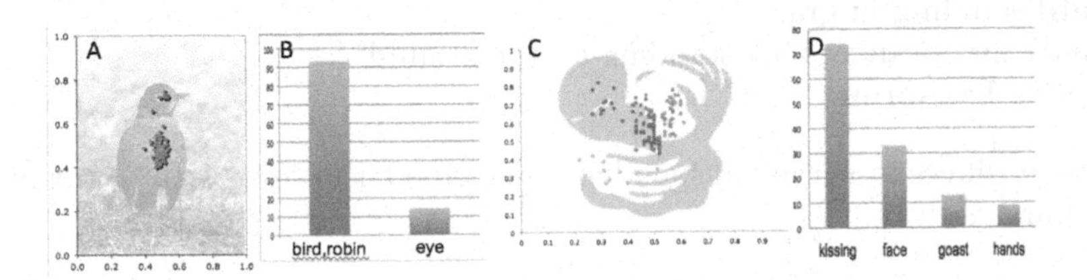

Fig. 2. A & B Scatter plot human chosen centers.C & D. Most frequent words for each spatial taxon or boundary

3 Results

For the bird image an ANOVA of the positions selected by human subjects yields two spatial-taxons ($F = 742.55$, $df = 1/106$, $p < .01$)) with a long tail of single occurring words. Statistics enable us to quantify the probability distribution of words, but reveal little of the semantic uncertainty due to the unbounded event set of the meaning of the words used by individual. Each spatial-taxon and associated word frequency vs word rank layers are shown in Fig. 2. For brevity, scatter plots of the raw data are not included in this paper (see [4,5] www.burningeyedeas.com for raw data). Below are three example descriptions provided by human subjects for the whole bird spatial-taxon.

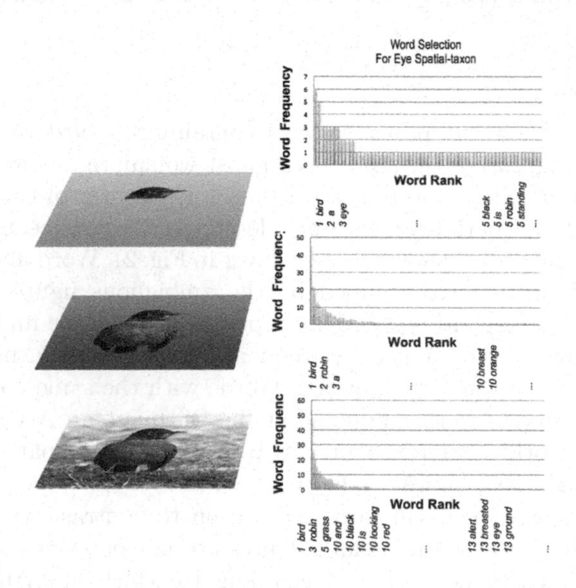

Fig. 3. A. Spatial-Taxon hierarchy of the bird image as derived from statistical analysis of human annotated surveys. B. Corresponding word count verses word rank for human subject word descriptions of the spatial-taxon layer on the left. An additional x-axis shows some example word ranks and words to provide context. Bottom chart shows words for the full image.

1. "bird standing in grass"
2. "black and orange, bird standing on the ground"
3. "spring has sprung"

Three different survey respondents described the 'eye' spatial-taxon with these phrases:

1. "black eye"
2. "pure noble basic glorious"
3. "bird small orange chest, eye is actual center"

For the ambiguous Ghost image, K-means cluster analysis was performed on the normalized horizontal and vertical location data to classify spatial-taxons. An ANOVA of these positions yielded $F(2,58) = 146$, $p < .01$ and vertical: $F = (2,58) = 16.1$, $p < .01$)). To help the reader visualize these spatial-taxons, they were segmented, artistically modified to illustrate the figure and organized into layers (as shown in Fig. 3). For each spatial-taxon, its associated word frequency vs word rank layer is shown on the right Fig. 3.

The word choices for each spatial taxon are strongly correlated with the figure descriptions. As foreshadowed in the thought-experiment, ghost-woman spatial-taxon boundaries had a super-ordinate interpretation - ie 'two beings kissing'. Both the ghost and face spatial-taxons contained descriptive present participles 'kissing'.

As Professor Zadeh identified as other-mind uncertainty, the meaning as it occurred for individuals, is unbound. The perception uncertainty, may be precisiated as shown in the three sample phrases for the spatial-taxons (below). The motif similarity to the yin-yang icon, increase the statistical co-occurrence of the terms "yin & yang" but the possibility space needs to include uncertainty as to whether objects are human, aliens or an octopus. In other words, we have unbounded semantic universe of discourse and fuzzy memberships (type 2 fuzzy sets) for each word.

Three different survey respondents described the "Kissing" boundary between the ghost and "Face" spatial-taxons with these phrases:

1. "It is whatever you can get away with"
2. "Its like a love yin-yang or a woman kissing an octopus"
3. "Cliche, really looks like a woman kissing an alien who is trying to escape"

Three different survey respondents described the "Face" spatial-taxon with these phrases:

1. "Surrounded by shadow in love"
2. "white face of woman and black ghost"
3. "unity"

And to the "Ghost" spatial-taxon with:

1. "it looks like a black ghost that is getting smothered with kisses"
2. "Ghost"
3. "dark side shadow self foreboding"

Thus stanza D from the thought experiment "never even stirred through" analogizes to unbounded sets unavailable to statistics[4]. Bounding the event space by calculating only the statistical occurrence of a particular word ignores the semantic uncertainty of the perception and other-minds.

[4] Since deep learning neural nets learn only possible classification within bound of composite training data, they are also limited by statistics.

4 Discussion

The thought experiment in Sect. 1.1 established the plausibility of an isomorphic structure between two types of uncertainty inherent in both images and language. The data collected by the experiment presented provide concrete examples of these uncertainties. The word frequency data anchor with associated spatai-taxon in manner consistent with hierarchical spatial-taxon scene architecture [4,5,10]. The last step in this paper is to discuss this isomorphic structure in terms of Professor Zadeh's information theory proposal [37].

Discussions on information theory are generally dominated by metrics introduced by Shannon, specifically that the Information of an outcome is $h(x = a_i) = log_2 \frac{1}{P(x=a_i)}$ and that Entropy is expected information for a series of outcomes $H(X) = \sum_x P(x) log_2 \frac{1}{P(x=a_i)}$. Where x is a random variable in the ensemble[5], P the probability function, and a_i an instance of the ensemble. Yet as shown plausible by both the thought experiment and image annotation data collected by human subjects, the meaning transmitted through human communication transfers more semantic information then the simple sum of pixels and words. The Gestalt truism that the message is more then the sum of its parts holds for image and language mediated communication. Other-mind uncertainty, perceptual uncertainty and attentional shifts between abstraction levels within a spatial-taxon hierarchy yield an unbounded possibility fuzzy event space. When these issues are considered, Zadeh's assertion that restriction equals information makes intuitive sense [37]. In his 2016 Cognitive Informatics talk, Rodolfo Fiorini [15] used an ambiguous figure from Douglas Hofstader (as shown in Fig. 5) to illustrate his point on computational information conservation theory. This same figure captures the conundrum quoted below by Shannon in his 1948 paper considered a founding paper of information theory.

> "The fundamental problem of communication is that of reproducing at one point either exactly or approximately a message selected at another point. Frequently, the messages have meaning: that is they refer to or are correlated according to some system with certain physical or conceptual entities. The semantic aspects of communication are irrelevant to the engineering problem."–C.E. Shannon [28]

Specifically, according to Shannon a perfect reproduction of the data, event set of pixels delineating the figure, reproduces the signal but uncertainty as to the meaning of the signal. As pointed out by Fiorini [15] the Hofstader ambiguous figure can mean either a 'wave' or a 'particle'. Its meaning is uncertain. Zadeh

[5] An ensemble, often referred to as an alphabet, is the set of possible outcomes of word annotations and pixel designations. In this paper, allowed pixel designations are spatial-taxons or the edges that outline spatial-taxons. The word annotations collected and shown in the results are outcomes. Note that since nested spatial-taxons are not mutually exclusive, the results count word annotations for each child spatial-taxon [5], which by the definition of spatial-taxons are included in both the parent taxon and child taxons [8,9].

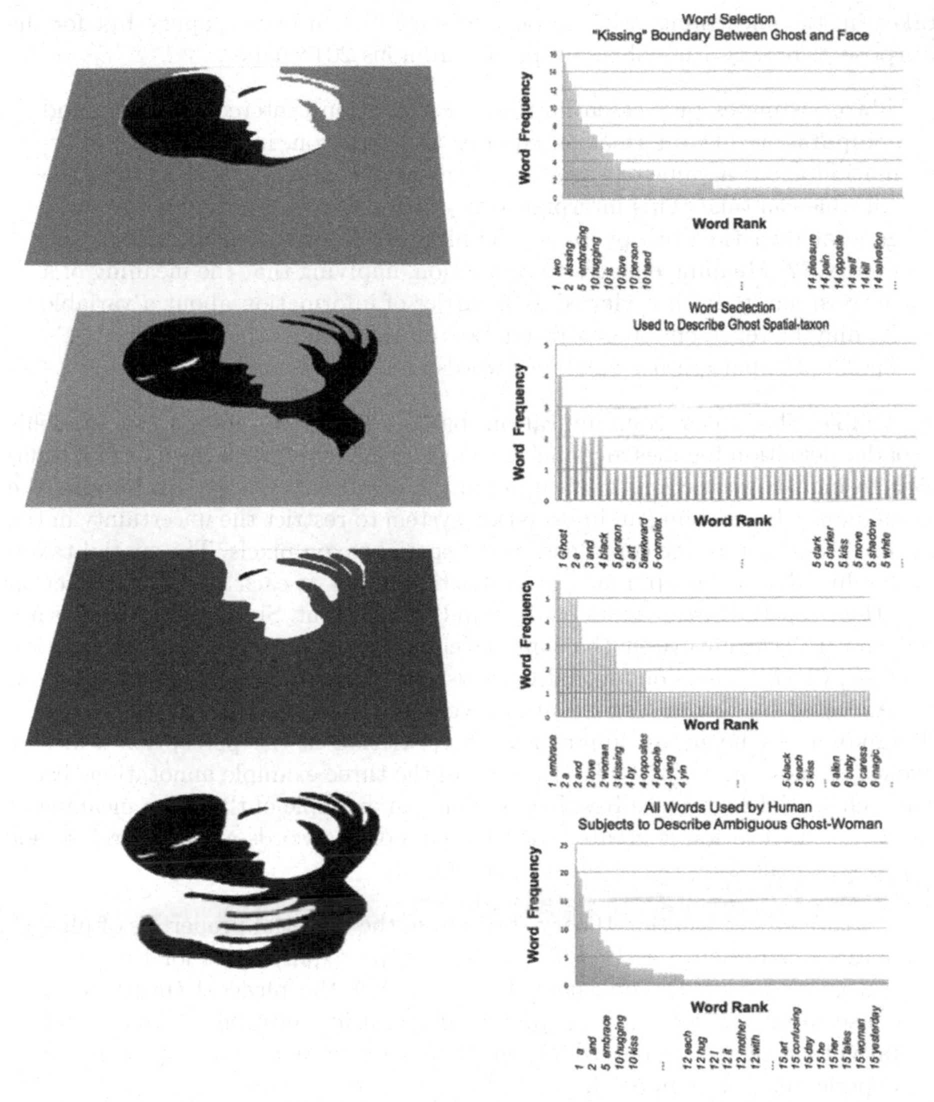

Fig. 4. A. Spatial-Taxon hierarchy of images used in thought experiment as derived from statistical analysis of human annotated surveys. B. Corresponding word count verses word rank for human subject word descriptions of the spatial-taxon layer on the left. An additional x-axis shows some example word ranks and words to provide context. Bottom chart shows words for the full image. The top chart word descriptions of boundary between the ghost and face spatial-taxon.

LIGHT IS A

Fig. 5. Douglas Hofstader ambiguous wave-particle figure.

takes up the uncertainty with respect to semantics in many papers, but for the purpose of brevity I use this paraphrase from his 2013 paper.

> 'Two postulates play essential roles in restriction centered reasoning and computation. (1) Information equates to restriction, implying that information about a value or variable is conveyed by restricting the values the variable can take. This interpretation of information is considerably more general then the entropy based definition of information in information theory. (2) Meaning equates to restriction, implying that the meaning of a proposition, p, with p viewed as a carrier of information about a variable X, may be represented as a restriction on the values which X can take.'– Zadeh [37] (paraphrase combines words and equations)

Unlike Shannon's communication based theory of information, Zadeh's broader definition handles meaning & semantics by requiring a theory of information to include restriction on meaning in the possibility space. This handles the ambiguouity by enabling an information system to restrict the uncertainty in the meaning of the message contained in the spatial-taxon pixels. The spatial-taxon hierarchy follows this approach by restricting pixels at each level of abstraction to either a spatial-taxon or its background complement. Since spatial-taxons are only mutually exclusive at the same level of abstraction, Shannon information content, which relieves on mutually exclusive events, does represent information.

As noted by Zadeh restriction is precisiated if information granules can be designated. As he notes "unprecisiated restriction is the perception which it evokes in ones mind" [37]. The diversity of the three example annotations listed for each spatial taxon (see Results) provides an example of this. The meaning of the whole annotation is greater than the sum of the words and is therefore not captured by the long tail of precisiated words.

> "Integrated information theory starts from the essential properties of phenomenal experience, from which it derives the requirements for the physical substrate of consciousness. It argues that the physical substrate of consciousness must be a maximum of intrinsic cause effect power and provides a means to determine, in principle, the quality and quantity of experience."– Tononi et al. [30]

Integrated information theory provides a means to restrict a perceptual unprecisiated value to is most basic phenomenology (also known as qualia [7–9, 19, 24, 30]) described as other-mind uncertainty [37]. This method was used in another paper presented in these proceedings in 2014 [8]. Future work on information metrics needs to Zadeh information restriction for full image hierarchies to the depth of qualia and handle uncertainty does to noise (Shannon), other-mind and unprecisiated perception.

To conclude, this paper explored methods for understanding the meaning and uncertainty within images and their linguistic descriptions. Uncertainty caused by image hierarchy was analyzed using spatial-taxons. Uncertainty due to ambiguity in language annotation was analyzed by word frequency subdivided by spatial-taxon abstraction. Perceptual and other-mind uncertainty was explored through using a thought-experiment that invoked analogical inference by assuming the plausibility of model-theoretic isomorphism between language (poem) and visual perception. Finally, the implications were explored in terms of Zadeh information restriction principal, integrated information theory, and Shannon's distinction between information content and semantics.

Acknowledgments. I thank Dr. Lotfi Zadeh for mentoring me on my graduate work at U.C. Berkeley and for inviting as a visiting scholar in the Electrical Engineering and Computer Science Department at University of California at Berkeley (2014 2017) to work on my upcoming book on Fuzzy, Bayesian, Hybrid methods for Computer Vision. I thank my collaborators Haley Winter, Analucia DaSivla, Yurik Riegal, Colin Rhodes, Eric Rabinowitz and Shawn Silverman. I thank Ralph Schmidt-Dunker for help with editing. BurningEyeDeas LLC, an organization that does research at the Burningman Art Festival. Data posted at www.burningeyedeas.com and/or laurenbargout.com.

References

1. Barghout, L., Lee, L.: Perceptual information processing system. USPTO patent application number: 20040059754 (2003)
2. Barghout, L.: System and Method for Edge Detection in Image Processing and Recognition. WIPO Patent Application. WO/2007/044828 (2006)
3. Barghout, L.: Linguistic Image Label Incorporating Decision Relevant Perceptual, Semantic, and Relationships Data. USPTO Patent Application 20080015843 (2007)
4. Barghout, L.: Empirical data on the configural architecture of human scene perception using natural images. J. Vis. **9**(8), 964 (2009). https://doi.org/10.1167/9.8.964
5. Barghout, L., Winter, H., Riegel, Y.: Empirical data on the configural architecture of human scene perception and linguistic labels using natural images and ambiguous figures. J. Vis. **11**, 1077 (2011)
6. Barghout, L., Sheynin, J.: Real-world scene perception and perceptual organization: lessons from computer vision. J. Vis. **13**(9), 709 (2013)
7. Barghout, L.: Image segmentation using fuzzy spatial-taxon cut: comparison of two different stage one perception based input models of color (Bayesian Classifier and Fuzzy Constraint). Electron. Imaging **2016**(16), 1–6 (2016)
8. Barghout, L.: Visual taxometric approach to image segmentation using fuzzy-spatial taxon cut yields contextually relevant regions. In: Laurent, A., Strauss, O., Bouchon-Meunier, B., Yager, R.R. (eds.) IPMU 2014. CCIS, vol. 443, pp. 163–173. Springer, Cham (2014). https://doi.org/10.1007/978-3-319-08855-6_17
9. Barghout, L.: Spatial-taxon information granules as used in iterative fuzzy-decision-making for image segmentation. In: Pedrycz, W., Chen, S.-M. (eds.) Granular Computing and Decision-Making: Interactive and Iterative Approaches. SBD, vol. 10, pp. 285–318. Springer, Cham (2015). https://doi.org/10.1007/978-3-319-16829-6_12
10. Barghout, L.: Image Segmentation Using Fuzzy-Spatial Taxon Cut (2015). https://docs.lib.purdue.edu/modvis/2015/session05/6/
11. Barghout, L.: Using the 5th dimensions of human visual perception to inspire automated edge and texture segmentation: a fuzzy spatial-taxon approach. In: 2016 IEEE 15th International Conference on Cognitive Informatics and Cognitive Computing (ICCI*CC). IEEE (2016)

12. Berger, J.: Ways of Seeing - Based on the BBC Television Series with John Berger. Transcribed by Pelican Books, p. 31 (1985)
13. Cancho, R.F., Sole, R.V.: Zipf 's law and random texts. Adv. Complex Syst. **5**(1), 1–6 (2002)
14. Deng, Y., Manjunath, B., Shin, H.: Color image segmentation. In: IEEE Computer Society Conference on Computer Vision and Pattern Recognition, vol. 2 (1999)
15. Fiorini, R.A.: Deep learning and deep thinking: new application framework by CICT. In: 2016 IEEE 15th International Conference on Cognitive Informatics and Cognitive Computing (ICCI*CC), pp. 117–128 (2016)
16. Hawkins, J., Blakeslee, S.: On Intelligence: How a New Understanding of the Brain Will Lead to the Creation of Truly Intelligent Machines. Macmillan, New York (2007)
17. James, W.: Principles of Psychology, 403 p. Holt, New York (1890)
18. Jolicoeur, P., Gluck, M.A., Kosslyn, S.M.: Pictures and names: making the connection. Cogn. Psychol. **16**, 243–275 (1984)
19. Klein, S.A.: Will robots see? In: Spatial Vision in Humans and Robots: The Proceedings of the 1991 York Conference on Spatial Vision in Humans and Robots. Cambridge University Press (1993)
20. Marcus, G.: Deep learning: a critical appraisal. arXiv preprint arXiv:1801.00631 (2018)
21. Miyajima, K., Ralescu, A.: Spatial organization in 2D segmented images: representation and recognition of primitive spatial relations. Fuzzy Sets Syst. **65**(2-3), 225–23 (1994)
22. Nelson, R., Palmer, S.E.: Of holes and wholes: the perception of surrounded regions. Perception **30**(10), 1213–1226 (2001)
23. O'Regan, K.: Experience is not something we feel but something we do: a principled way of explaining sensory phenomenology, with Change Blindness and other empirical consequences. Talk given at Bressanone on 24 January 2001. Source for Ghost-woman image
24. Palmer, S.: Vision Science: Photons to Phenomenology, MIT Press, Cambridge (1999)
25. Rosch, E.H.: Cognitive representation of semantic categories. J. Exp. Psychol. **104**(3), 192–233 (1975)
26. Rosch, E.H., Mervis, C.B., Gray, W.D., Johnson, D.M., Boyes-Braem, P.: Basic objects in natural categories. Cogn. Psychol. **8**(3), 382–439 (1976)
27. Ruscio, J., Haslam, N., Ruscio, A.: Introduction to Taxometric Method. Lawrence Erlbaum Associates, Mahwah (2006)
28. Shannon, C.E.: A mathematical theory of communication. Bell Syst. Tech. J. **27**, 623–656 (1948)
29. Simon, H.: The architecture of complexity. Proc. Am. Philos. Soc. **106**(6), 467–482 (1962)
30. Tononi, G., Boly, M., Massimini, M., Koch, C.: Integrated information theory: from consciousness to its physical substrate. Nat. Rev. Neurosci. **17**(7), 450–461 (2016)
31. Treisman, A.M.: Strategies and models of selective attention. Psychol. Rev. **76**(3), 282–299 (1969)
32. Wertheimer, M.: Laws of Organization in Perceptual Forms (partial translation) A Sourcebook of Gestalt Psychology. In: Ellis, W.B. (ed.), pp. 71–88, Harcourt Brace (1938)
33. Zadeh, L.: Outline of a new approach to the analysis of complex systems and decision processes. IEEE Trans. Syst. Man Cybern. **1**, 28–44 (1973)

34. Zadeh, L.A.: Toward a theory of fuzzy information granulation and its centrality in human reasoning and fuzzy logic. Fuzzy Sets Syst. **90**, 111–127 (1997)
35. Zadeh, L.: Generalized theory of uncertainty (GTU): principal concepts and ideas. Comput. Stat. Data Anal. **51**, 15–26 (2006)
36. Zadeh, L.: Toward a restriction-centered theory of truth and meaning (RCT). Inf. Sci. **248**, 1–14 (2013)
37. Zipf, G.K.: Human Behavior and the Principle of Least Effort: An introduction to human Ecology. Addison-Wesley, Cambridge (1972)

General Track

An IoT Control System for Wind Power Generators

Marouane Salhaoui[1(✉)], Mounir Arioua[1],
Antonio Guerrero-González[2], and María Socorro García-Cascales[3]

[1] Team of New Technology Trends, National School of Applied Sciences of Tetuan, Tétouan, Morocco
{marouane.salhaoui,m.arioua}@ieee.org

[2] Department of Automation and Systems Engineering, Universidad Politécnica De Cartagena, Cartagena, Spain
antonio.guerrero@upct.es

[3] Department of Electronics, Computer Science and Engineering Projects, Universidad Politécnica De Cartagena, Cartagena, Spain
socorro.garcia@upct.es

Abstract. New technology deployment for facilitating the control and managing huge amount of data and its uncertainty is very important challenge in the industry field. Energy sector as important part of the industry knows nowadays a high transformation towards renewable energy, and one of important solution is the wind energy. The wind control system must guarantee safe and reliable operation, monitor components and variables, and check that these variables are in an admissible range and must perform the detection and prediction of faults. We propose in this paper a new Internet of Things solution to control and monitor a wind energy system. The IoT gateway is used as a bridge between the different devices in the wind turbine control system and Internet. We adopted OPC Unified Architecture, as a protocol of communication, and we implemented the new IoT tool Node-RED in the gateway, in order to facilitate the link between OPC UA client and IBM cloud. The obtained results are evaluated in real-time in the cloud platform which eventually provides a consistent analysis and interpretation, and making better decision.

Keywords: Wind energy · Industry 4.0 · IoT gateway · OPC UA
Node-RED · IBM cloud

1 Introduction

As an important source of energy of different countries, renewable energy is widely used nowadays, it accounts for around 16% of global power generation as reported in IRENA (2017) [1]. This number is expected to double in the next 15 years and 65% of energy use could be provided from renewable resources by 2050. This is due to the rapid growth of different renewable energy supplied by sources like wind and solar photovoltaic. Wind power penetration has increased significantly. Across the global market, over 54 GW of clean renewable wind power was installed in 2016. Which currently comprises more than 90 countries, and allows building new industries,

J. Medina et al. (Eds.): IPMU 2018, CCIS 855, pp. 469–479, 2018.
https://doi.org/10.1007/978-3-319-91479-4_39

creating new jobs and more importantly leading the way towards a clean energy future [2]. The renewable energy industry benefits from the use of digital technologies. The existing machines used for manufacturing already support analog or digital sensing that is reported to a central control station for monitoring over a wired Ethernet systems [3]. However these systems are typically unconnected to the internet [4]. This is the era which meets the important evolution of the industry and the internet. To follow up this important evolution of wind energy, it is mandatory to apply the internet capacity to assess every collected data from different industrial components, motors, sensors, actuators, etc. This can be performed by building applications able to predict, prescribe, and order maintenance in order to improve significantly the efficiency of the control system [5], and to have a clear vision of the entire system, in real-time, without need to be physically in the area of the installation. Consequently, this will cut off waiting times and diminish unnecessary costs. Digitization is setting off a radical transformation of the manufacturing environment. With the advent of Internet of Things, data and services, stand on the verge of the fourth industrial revolution (Industry 4.0), where subjects and objects alike can communicate in real-time, as well as the convergence of the real and virtual worlds [6]. Each renewable energy resource is considered an object and it is assigned a unique IP address. Using bidirectional communication, it becomes possible to monitor and control each object [7]. All data obtained by sensors and actuators can be acquired, analyzed and managed, through the cloud-based platform. The communications between devices and objects can be performed through wired and wireless networks, using different technologies such as RS485, PLC, I2C, Z-Wave, WiFi, and ZigBee [8]. IoT is considered of one of the complex systems, and this complexity hails from the interaction of the environment, the inter-connectivity of the components of the IoT, and the number of the networks involved. The IoT gateway is the component that allows those different networks to communicate [9].

In order to acquire the significance of IoT in renewable energy, we perform in this paper, the different steps to forward data from sensors used in wind energy, to the cloud, using the new IoT gateway from Siemens. We provide a clear presentation of the state of each sensor in real-time in the cloud. We describe the different protocols and application used in the IoT gateway so as to maintain a reliable connection between the objects and the cloud. Moreover, we examine the possibilities that can be done using the transmitted data in real-time. The remainder of this paper is organized as follows. Section 2 discusses the IoT solution for the wind control system. The concept of an IoT gateway is described in Sect. 3. Section 4 talks over the IoT and the cloud service. Section 5 depicts the proposed IoT control system for power generation and the simulation results in the Cloud. Section 6 concludes the paper.

2 IoT as a Solution for the Wind Control System

The concept of Industry 4.0 was initially proposed for developing German economy in 2011 [10], as a next industrial revolution. The industrial internet affords a way to get better visibility and insight into the company's operations and assets through integration of machine sensors, middle-ware, software, and back-end cloud compute and storage systems [11]. One of the significant reasons of this revolution is that the complexity of traditional industrial has outpaced the human operator's capability to recognize and manage the inefficiencies. A new term of the industry evolution appeared (CPS: Cyber Physical Systems) that connect virtual and physical worlds, it is perceived to be a core ingredient in the so-called 4th Industrial Revolution [12]. Computers elements collaborate for the control of physical entities together to build a networked world and act as intelligent agents in the IoT and represent the basic framework of a smart factory [13]. The renewable energy as a type of industrialization face also important challenges, and the harnessing of the green energy is not always stable, thus the key of that is the optimization. This is where new technologies such as IoT, machine learning, cloud, big data, come into the picture. They can facilitate better usage of resources and help to harness clean power along with optimization. The IoT has important benefits for energy sector, especially wind energy, often this technology is applied to inaccessible environments, and remote areas, such as mountains, seas and volcanoes [14]. In our case, inside a wind energy system, a wind turbine converts wind energy into electrical energy. It consists firstly of a rotor, which transforms the aero-dynamic thrust in rotation movement, then a Multiplier, which adapts the rotation speed to the speed of the generator, also an Alternator, which transforms the energy of rotation into electrical energy, and finally a dump to the network, which injects energy into the electrical network.

The wind turbine control system must guarantee safe and reliable operation, monitor components and variables, check that the variables are in an admissible range and must perform the detection and prediction of faults. In a wind turbine, a yaw-drive motor turns the nacelle to face the wind, and the motor movement is based on the data from wind-direction sensors. Indeed, a predictive analytic will alert operators in advance if a component needs repairs or inspection. In addition, the sophisticated units that are embedded in equipment require frequent maintenance [15]. Real-time control and maintenance are the main purposes for the proposed sensing intelligence in the renewable energy industry [16].

3 Concept of an IoT Gateway

3.1 Definition

IoT gateway is one of the most important components of IoT, and is considered as bridge which connects traditional network and sensors network. It is the master device in charge of protocol conversion and data fusion of different sensor data [17]. The IoT gateway can also transmit data to application platform, not only receives sensed data from sensor node and commands from application, it is the middle layer between sensor node and application platform [18]. In other words, IoT gateway acts as a proxy for the sensing domain and network domain towards the things that are connected to it [19].

Actually, the IoT gateways become smarter, they are now physical devices with software programs and protocols that act as intermediaries between the sensors, con-trollers, intelligent devices, and the cloud. And provide the needed connectivity, security, and manageability; while some existing devices cannot share any data with the cloud [20] (Fig. 1). They are in charge of transfer data between the device network and the LAN (Local Area Network), reaching the Cloud service [21].

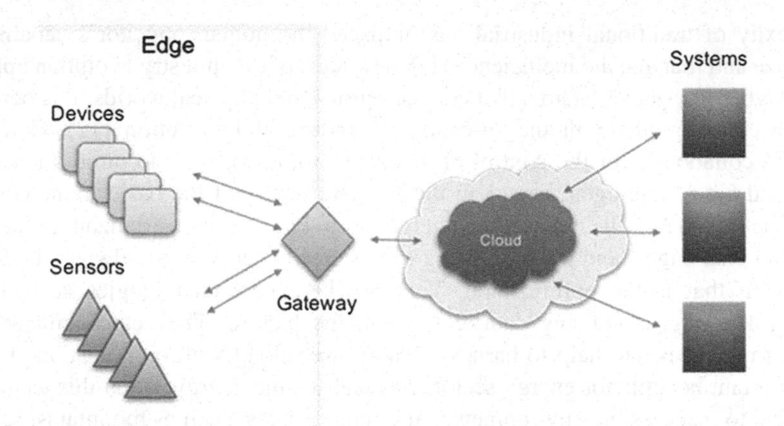

Fig. 1. General graph of an IoT communication system

In the industry field, there are different objects that have to be controlled and connected such as sensors, actuators, motors, programmable logic controllers (PLCs), SCADA [22]. In order to be part of the new revolution industry, Industry 4.0, it is necessary to cover the lack of connection protocols between industrial nodes and the IoT architecture. The IoT gateway links different objects to the internet in the industrial environment and able to operate as a joint interface among different networks and support different communication protocols.

3.2 Siemens IoT Gateway for Industry 4.0

One of the major challenges in IoT, is that different technologies and machines do not speak the same data language. In industry, every system consists of many subsystems, and each subsystem consumes and produces data. It is important to solve the cooperation problematic in industry and to make these subsystems working together [23]. One of the consistent solution to deal with this challenge is the SIMATIC IOT2040 (Table 1). It is a reliable open platform for collecting, treatment and transferring data in the production environment. It is used as gateway between the company's production and the cloud. It is a salient interface to use in both directions, to transfer also the data from the cloud to the production control [24].

Table 1. Overview of enhancement possibilities of IOT2040

Hardware extensions	Communication interfaces	I/Os	Sensors/specific functionality	Software extensions	Communication protocols
Arduino shields	Ethernet / CAN/RS485 NFC/RFID ZigBee	Digital, analog, relays	GPS (position), DCF77 (radio clock)	Arduino lib	MODBUS RTU, MODBUS TCP, (Basic) CAN
mPCIe	WLAN, Bluetooth, UMTS, LTE, ZigBee			Yocto Linux	PROFINET RT planned, AMQP, MQTP, OPC UA, MODBUS RTU, MODBUS TCP

4 IoT and the Cloud Service

One of the best solution to allow the real-time computation and delivery of high value information, is to combine optimally the IoT with cloud services, which brings important values to manufacturing, so to gain unprecedented operation effectiveness, increase profits, and reduce costs [25]. The cloud service is responsible to receive data from industrial devices and monitor industrial machines on a larger scale using data analytic algorithms, and also produce commands to be sent back to the device network [21]. The cloud has two principal functions, it affords a compatible environment and a configurable services, by integrating a dynamic flow of resources, and all kinds of heterogeneous software and hardware [26].

Companies such as Google, IBM, Microsoft, Oracle, SAP and Amazon have developed their own cloud OS [27–30]. For example, IBM Watson IoT Platform, allows for any device to publish data to a back-end message broker and also to receive control messages from other devices or IoT applications. Furthermore, this solution gives the possibility to create applications that communicate to the IBM Platform, so that applications could use data provided from devices or send control messages. Microsoft has developed a solution called Azure, which provides software as a service (SaaS), platform as a service and infrastructure as a service. With AWS IoT Core from Amazon, it can be directly filter, transform, and leverage data from designated devices based on the business rules you have defined.

5 Proposed IoT Control System for Power Generation

5.1 Hardware Description

According to this diagram (Fig. 2), we employed a solution that uses the Siemens technology, using two types of PLCs: the PLC 1214 and the PLC 1512 for control, and the new industrial IoT gateway IOT2040, which is the most important device for forwarding the data from devices to the cloud. In this proposition, we present different sensors used in a wind energy system; the most important are the wind direction sensor (Wind Vane) and the wind power direction sensor (Anemometer). They are directly connected to the 1214 Siemens PLC to control the state of the wind and send an order to the motor generator to change the direction of the blades to maximize the use of the system. Moreover, it allows switching off the operation in case of a strong flue of wind. Additionally, the quality of the energy can be monitored and visualized using the SENTRON PAC (3200), which provides the important data to assess the quality of an electrical network.

Figure 3 shows the connection between the sensors, two PLCs and the IoT gateway. The two devices are connected to the PLC 1214 and all information for this sensors are sent from the PLC 1214 to the PLC 1512 using the industrial communication standard PROFINET over Ethernet.

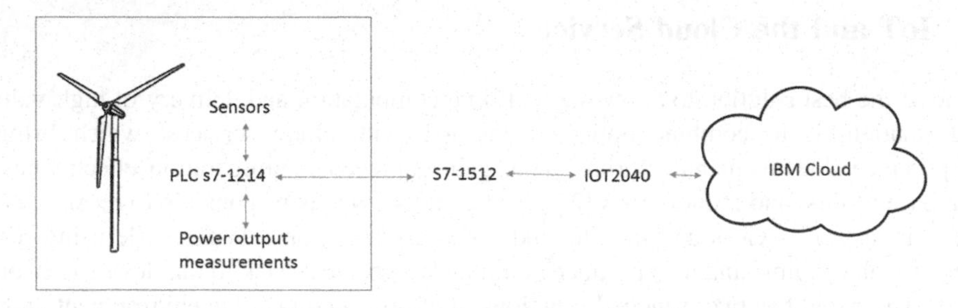

Fig. 2. General Block diagram

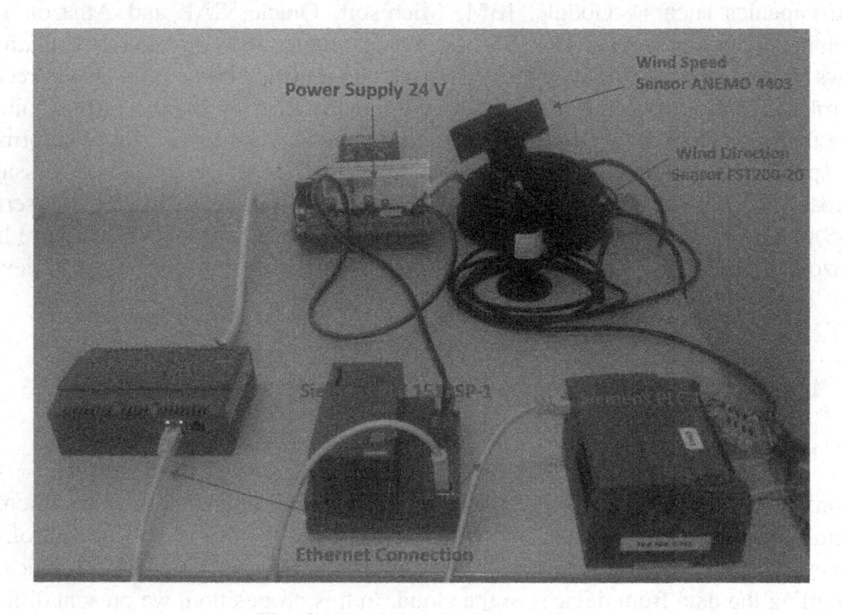

Fig. 3. Hardware implementation

5.2 Software Description

The communication between the devices and the cloud, is performed through the IOT2040 gateway. In order to send data from the SIMATIC IOT2040 to the cloud, it is required an OPC UA protocol to be implemented for industry and IoT. The OPC Unified Architecture (UA) is an independent service-oriented architecture that integrates all the functionality of the individual OPC Classic specifications into one extensible framework [31]. OPC UA is also an M2M communication protocol, developed to create an inter-operable, secure and reliable communication protocol. Based on these properties, OPC UA increasingly predominates as standard in the industrial plant communication and environment [32].

OPC-UA uses client-server architecture with a clear assigned roles. Servers are applications that present information following the OPC-UA information model, and clients are applications getting back information from servers by querying and browsing the information model. In each server, an address space is defined containing nodes of the OPC-UA model. These nodes represent software or real physical objects [33].

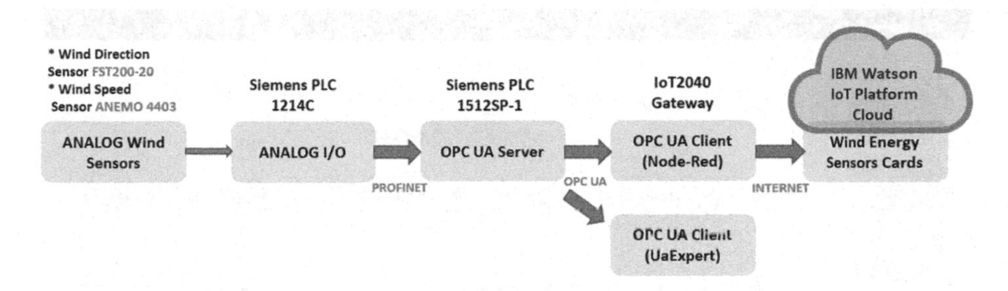

Fig. 4. Evaluation setup

Node-RED is a programming tool for wiring together hardware devices, APIs and online services. It is a solution to control flows to be designed and managed graphically. Node-RED has a sample set of nodes that we can use for the communication between different protocols and platforms.

Before using OPC Client in Node-RED in the IOT2040 gateway, we used the UaExpert tool in our local machine to test the communication between our OPC UA server, which is the PLC 1512, and the OPC UA client (Fig. 4). UaExpert can be used as server and Client as well, in this case we use it as a general purpose test client, connecting with UA Server (PLC1512) in order to show the UA Server information model, like tags, blocks, etc.

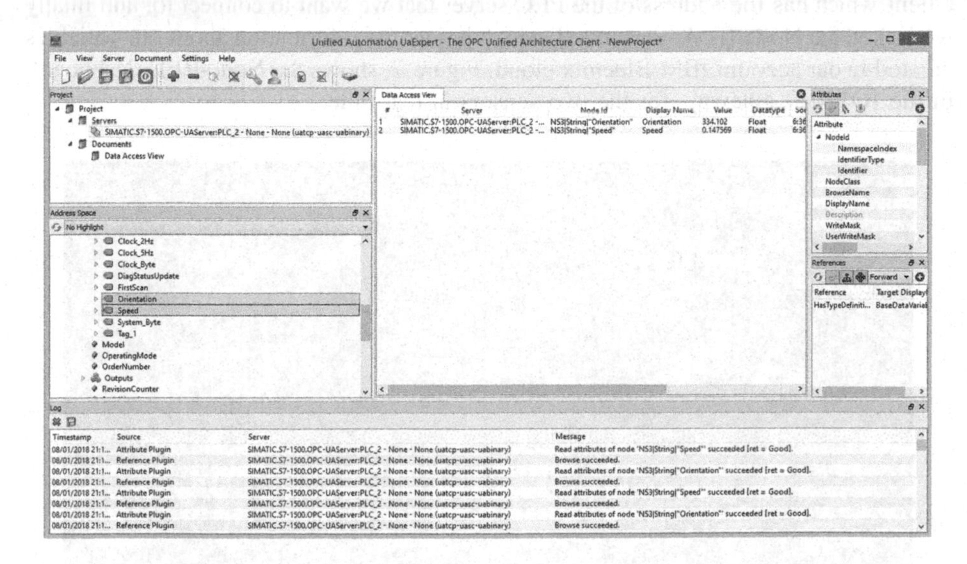

Fig. 5. Checking OPC UA connection using UaExpert Software

After connecting to UaExpert, we can test the connection to the OPC UA server, and display every information from the PLC. In our case, we need to control the two variables of the orientation sensor and the speed sensor (Fig. 5). Besides, we check the Node-ID of each variable in UaExpert, which is the most important ID used in Node-RED, in order to be connected to the PLC server.

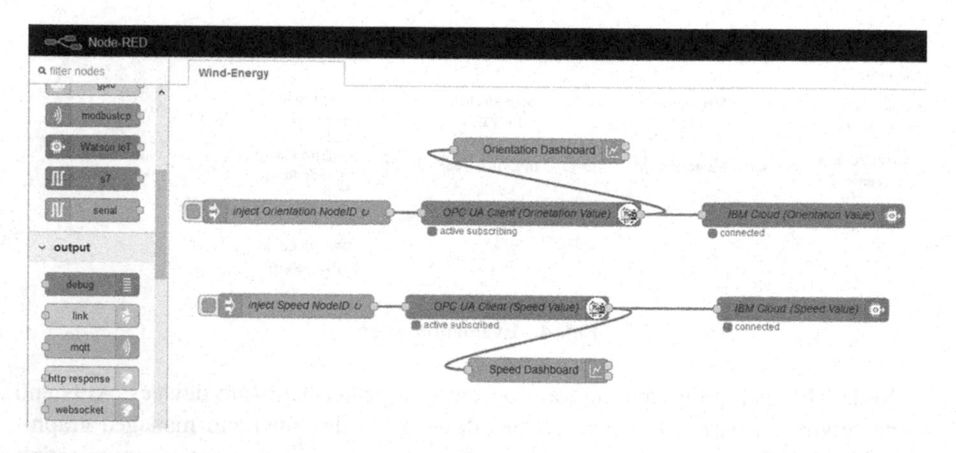

Fig. 6. Communication between the PLC 1512 and IBM Cloud through OPC UA protocol using Node-RED in the industrial Gateway IOT2040

Afterward, we used Node-RED the Internet of Thing tool to connect the different protocols and hardware, using just nodes, and a link between them (Fig. 6). In the first blue Node (Inject Node) we have introduced the topic, used to connect to the variable Orientation in the PLC, this topic is called also Node-ID that can be taken from the software UaExpert, after that, we connected the Inject Node to the Node OPC UA Client which has the address of the PLC server that we want to connect to, and finally we linked the Node IBM Watson IoT which has all the information about our variables created in our account IBM Bluemix cloud. Figure 7, shows the Node-RED dashboard, in the IOT2040 gateway, for the two sensors in real-time.

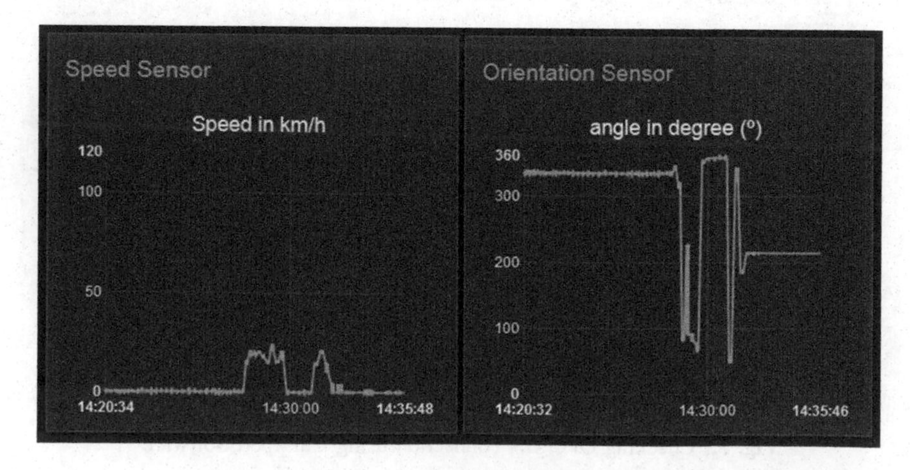

Fig. 7. Dashboard Data of wind Sensors in the IoT2040 Gateway

After having all the information about our sensors in the IOT2040 gateway, we have created an account in IBM Bluemix, then we created a device in this account, which is the IOT2040 gateway in order to connect it to the IBM Watson node in Node-RED. IBM allows to create different boards, and for each board it is possible to create cards that present your data and each data is a representation for your devices, sensors, actuators, or other. In the IBM Watson IoT Platform, we created the board Wind-Energy, in order to present in a real-time the two wind sensors (Fig. 8).

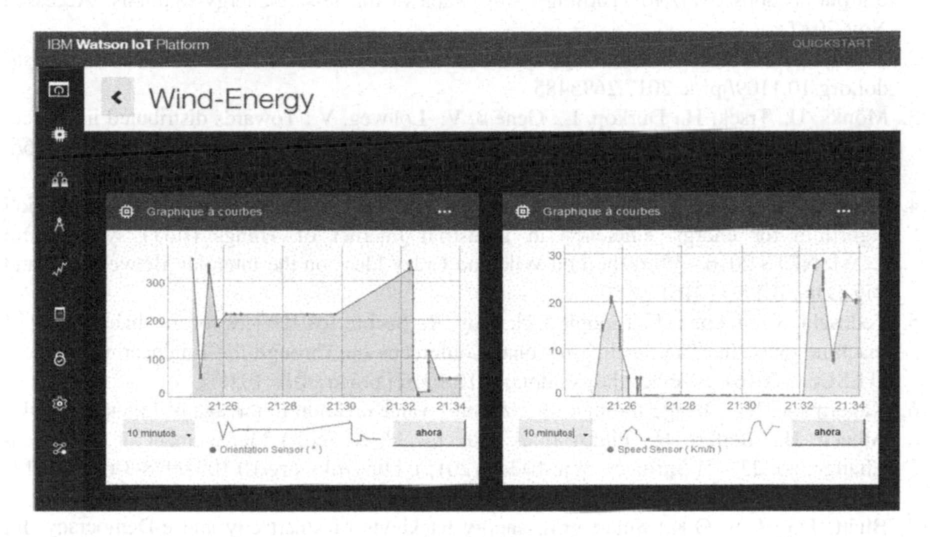

Fig. 8. Data sensors in IBM Watson IoT Platform

6 Conclusion

This paper presented a control system using a smart IoT gateway to create a connection between an industrial case and the cloud. We have provided in this paper, a solution for a wind energy system in order to visualize in a real-time and remotely the different components and devices inside a wind turbine control system. In this work, we proposed, the IOT2040 gateway from Siemens, and we have installed, several tools that helped us connect our device's information. It is simple to connect each sensor information of the wind turbine to the cloud by using the tool Node-RED, and through different communication protocols like OPC UA. This solution can really ease the control system of wind energy, by collecting, saving and communicating relevant data in real-time.

With an IoT gateway, is possible to transfer analyzed data from the cloud to the control system and devices. In the future research directions, we will focus more on how to treat an important amount of data and the uncertainty in the cloud, coming from different types of devices and protocols, and using an IoT Bot application that execute automated tasks over the cloud and send control messages to the devices.

Acknowledgments. This work was partially supported by projects TIN2014-55024-P from the Spanish Ministry of Science and Innovation and P11-TIC-8001 from Junta de Andalucía (including FEDER funds), SENECA Foundation 19882-GERM-15.

References

1. IRENA: Turning to Renewables: Climate-safe Energy Solutions (2017). http://www.irena.org/publications/2017/Nov/Turning-to-renewables-Climate-safe-energy-solutions. Accessed Nov 2017
2. Blaabjerg, F., Ma, K.: Wind energy systems. Proc. IEEE **105**(11), 2116–2131 (2017). https://doi.org/10.1109/jproc.2017.2695485
3. Mönks, U., Trsek, H., Dürkop, L., Geneiß, V., Lohweg, V.: Towards distributed intelligent sensor and information fusion. Mechatronics **34**, 63–71 (2016). https://doi.org/10.1016/j.mechatronics.2015.05.005
4. Dhondge, K., Shorey, R., Tew, J.: HOLA: heuristic and opportunistic link selection algorithm for energy efficiency in Industrial Internet of Things (IIoT) systems. In: COMSNETS 2016 - Workshop on Wild and Crazy Ideas on the Interplay Between IoT and Big Data. IEEE (2016)
5. Tedeschi, S., Mehnen, J., Tapoglou, N., Roy, R.: Secure IoT devices for the maintenance of machine tools. In: The 5th International Conference on Through-life Engineering Services (TESConf 2016). Elsevier. https://doi.org/10.1016/j.procir.2016.10.002
6. Kagermann, H.: Change through digitization—value creation in the age of Industry 4.0. In: Albach, H., Meffert, H., Pinkwart, A., Reichwald, R. (eds.) Management of Permanent Change, pp. 23–45. Springer, Wiesbaden (2015). https://doi.org/10.1007/978-3-658-05014-6_2
7. Bhatt, J.G., Jani, O.K.: Smart grid: energy backbone of smart city and e-Democracy. In: Vinod Kumar, T.M. (ed.) E-Democracy for Smart Cities. ACHS, pp. 319–366. Springer, Singapore (2017). https://doi.org/10.1007/978-981-10-4035-1_11
8. Hafeez, A., Kandil, N.H., Al-Omar, B., Landolsi, T., Al-Ali, A.R.: Smart home area networks protocols within the smart grid context. J. Commun. **9**(9), 665–671 (2014)
9. Altamimi, A.B., Ramadan, R.A.: Towards internet of things modeling: a gateway approach. Complex Adapt. Syst. Model **4**, 25 (2016). https://doi.org/10.1186/s40294-016-0038-3
10. Vogel-Heuser, B., Hess, D.: Guest editorial Industry 4.0–prerequisites and visions. IEEE Trans. Autom. Sci. Eng. **13**(2), 411–413 (2016)
11. Gilchrist, A.: Introducing Industry 4.0. In: Industry 4.0. Apress, Berkeley (2016). https://doi.org/10.1007/978-1-4842-2047-4_13
12. German National Academy of Science and Engineering (ACATECH): Cyber-physical systems: Driving force for innovation immobility, health, energy and production. Technical report, December 2011
13. Möller, D.P.F.: Digital Manufacturing/Industry 4.0. In: Guide to Computing Fundamentals in Cyber-Physical Systems. CCN, pp. 307–375. Springer, Cham (2016). https://doi.org/10.1007/978-3-319-25178-3_7
14. Chen, Y., Lee, G.M., Shu, L., Crespi, N.: Industrial Internet of Things-based collaborative sensing intelligence: framework and research challenges. Sensors **16**, 215 (2016). https://doi.org/10.3390/s16020215
15. Tavner, P., Xiang, J., Spinato, F.: Reliability analysis for wind turbines. Wind Energy **10**, 1–18 (2007)

16. Cecati, C., Guinjoan, F., Siano, P., Spagnuolo, G.: Introduction to the special section on smart devices for renewable energy systems. IEEE Trans. Ind. Electron. **60**, 1119–1121 (2013)
17. Shang, G., Chen, Y., Zuo, C., Zhu, Y.: Design and implementation of a smart IoT gateway. In: 2013 IEEE International Conference on Green Computing and Communications and IEEE Internet of Things and IEEE Cyber, Physical and Social Computing. https://doi.org/10.1109/greencom-ithings-cpscom.2013.130
18. Zhu, Q., Wang, R., Chen, Q., Liu, Y., Qin, W.: IOT gateway: bridging wireless sensor networks into Internet of Things. IEEE (2010). https://doi.org/10.1109/euc.2010.58
19. Chen, H., Jia, X., Li, H.: A brief introduction to IoT gateway. In: Communication Technology and Application (ICCTA 2011), IET International Conference, 14–16 October 2011. https://doi.org/10.1049/cp.2011.0740
20. Ferrández-Pastor, F.J., García-Chamizo, J.M., Nieto-Hidalgo, M., Mora-Pascual, J., Mora-Martínez, J.: Developing ubiquitous sensor network platform using Internet of Things: application in precision agriculture. Sensors **16**(7), 1141 (2016). https://doi.org/10.3390/s16071141
21. da Silva, F., Ohta, R.L., dos Santos, M.N., dos Binotto, A.P.D.: A cloud-based architecture for the Internet of Things targeting industrial devices remote monitoring and control. IFAC-PapersOnLine **49**(30), 108–113 (2016)
22. Hemmatpour, M., Ghazivakili, M., Montrucchio, B., Rebaudengo, M.: DIIG: a distributed industrial IoT gateway. In: 2017 IEEE 41st Annual Computer Software and Applications Conference. https://doi.org/10.1109/compsac.2017.110
23. Lojka, T., Miškuf, M., Zolotová, I.: Industrial IoT gateway with machine learning for smart manufacturing. In: Nääs, I., Vendrametto, O., Reis, J.M., Gonçalves, R.F., Silva, M.T., von Cieminski, G., Kiritsis, D. (eds.) APMS 2016. IAICT, vol. 488, pp. 759–766. Springer, Cham (2016). https://doi.org/10.1007/978-3-319-51133-7_89
24. The Intelligent Gateway for Industrial IoT Solutions. http://w3.siemens.com/mcms/pc-based-automation/en/industrial-iot/pages/default.aspx
25. Georgakopoulos, D., Jayaraman, P.P., Fazia, M., Villari, M., Ranjan, R.: Internet of Things and edge cloud computing roadmap for manufacturing. IEEE Cloud Comput. **3**(4), 66–73 (2016). https://doi.org/10.1109/mcc.2016.91
26. Xiong, G., Ji, T., Zhang, X., Zhu, F., Liu, W.: Cloud operating system for industrial application. In: 2015 IEEE International Conference on Service Operations and Logistics, and Informatics (SOLI)
27. Oracle Cloud, Complete, Integrated Cloud. https://cloud.oracle.com/en_US/home
28. IBM Cloud Computing is Designed for Business. https://www.ibm.com/cloud/
29. Microsoft Cloud. https://azure.microsoft.com/en-us/
30. Cloud Computing at SAP. https://www.sap.com/trends/cloud.html
31. The OPC Unified Architecture (UA). https://opcfoundation.org/about/opc-technologies/opc-ua/
32. Forsström, S., Jennehag, U.: A performance and cost evaluation of combining OPC-UA and Microsoft Azure IoT Hub into an industrial Internet-of-Things System. In: Global Internet of Things Summit (GIoTS). IEEE (2017). https://doi.org/10.1109/giots.2017.8016265
33. Bangemann, T., et al.: State of the art in industrial automation. In: Colombo, A.W., Bangemann, T., Karnouskos, S., Delsing, J., Stluka, P., Harrison, R., Jammes, F., Lastra, J.L. (eds.) Industrial Cloud-Based Cyber-Physical Systems, pp. 23–47. Springer, Cham (2014). https://doi.org/10.1007/978-3-319-05624-1_2

A Novel Uncertainty Quantification Method for Efficient Global Optimization

Bas van Stein(✉), Hao Wang, Wojtek Kowalczyk, and Thomas Bäck

Leiden Institute of Advanced Computer Science, Leiden University,
Niels Bohrweg 1, Leiden, The Netherlands
{b.van.stein,h.wang,w.j.kowalczyk,t.h.w.baeck}@liacs.leidenuniv.nl

Abstract. For most regression models, their overall accuracy can be estimated with help of various error measures. However, in some applications it is important to provide not only point predictions, but also to estimate the "uncertainty" of the prediction, e.g., in terms of confidence intervals, variances, or interquartile ranges. There are very few statistical modeling techniques able to achieve this. For instance, the Kriging/Gaussian Process method is equipped with a theoretical mean squared error. In this paper we address this problem by introducing a heuristic method to estimate the uncertainty of the prediction, based on the error information from the k-nearest neighbours. This heuristic, called the *k-NN uncertainty measure*, is computationally much cheaper than other approaches (e.g., bootstrapping) and can be applied regardless of the underlying regression model. To validate and demonstrate the usefulness of the proposed heuristic, it is combined with various models and plugged into the well-known Efficient Global Optimization algorithm (EGO). Results demonstrate that using different models with the proposed heuristic can improve the convergence of EGO significantly.

Keywords: Efficient global optimization · Uncertainty quantification
Expected error

1 Introduction

Statistical models, and more specifically, regression models are widely used in a large variety of fields. From estimating the chance of a certain disease given a set of symptoms to estimating the next best move in a game of chess. The correctness of a prediction is in many cases of extreme importance. For example: in the aviation industry, regression models can be used to estimate the drag and lift of a specific wing shape. Big differences between these estimates and the actual behaviour of the designed prototypes might have significant impact on the overall design process. The traditional way of estimating model accuracy is by using cross-validation. In general, the prediction error of a statistical model can be decomposed according to the well-known *Bias-Variance decomposition* [3].

J. Medina et al. (Eds.): IPMU 2018, CCIS 855, pp. 480–491, 2018.
https://doi.org/10.1007/978-3-319-91479-4_40

On one hand, a model can perfectly fit the training data (zero bias) and fail to generalize the prediction on a test data set (high variance), which is considered as over-fitting on the training data. On the other hand, a model can generalize the global trend of the test data very well (low variance) while having a larger error on the known data points (high bias). Besides the bias and variance of the predictor, there is also the irreducible error or noise, that is caused by the noise in the function value.

The prediction error plays a vital role in many algorithms and application, e.g., the field of efficient global optimization (EGO) [8]. In EGO, the goal is to optimize an expensive black-box function under a very small evaluation budget. This is achieved by heavily exploiting a regression model on the objective function. In this algorithm, the prediction error measures the expected risk when optimizing only the response surface of the model [14].

Gaussian Process Regression [11], also known as Kriging [9] is the ideal choice for EGO, as it models the prediction error as a normal distribution and thus a theoretical prediction error follows in a straightforward manner. However, many regression models lack such a theoretical uncertainty measure and cannot be easily adopted to the EGO algorithm. This paper proposes a novel heuristic to estimate the prediction error, regardless of the knowledge of the regression model. The proposed heuristic is beneficial for the EGO algorithm because it allows for using regression models other than Kriging, e.g., Random Forests. Since Kriging is computationally very expensive (its time complexity is $O(n^3)$) and known to work less effective in higher dimensions, using alternative models would boost the performance of EGO. Next to being useful in an EGO context, the heuristic can be used as general uncertainty measure for any predictive model, aiding decision makers to trust or reject predictions.

2 Efficient Global Optimization

The efficient global optimization algorithm was proposed to optimize expensive objective functions by sequentially choosing new candidate solutions from an underlying regression model. In principle, there are two key features exploited here: the model prediction and the uncertainty of the prediction. These two features balance the exploration - exploitation trade-off of the global search. Although the Kriging model (or Gaussian process regression) is the main model for EGO, we would like to relax such a model dependence and extend the EGO algorithm to other regression methods. To achieve this, we first re-visit the EGO algorithm from the perspective of uncertainty quantification and propose a novel empirical uncertainty quantification method in the next section.

2.1 Uncertainty Quantification in EGO

In the EGO algorithm, to optimize a computational expensive function, a regression model is used to approximate the (noisy) objective function based on a training data set $\mathcal{D} = \{\mathcal{X}, \mathbf{y}\}$: input data points $\mathcal{X} = \{\mathbf{x}^{(1)}, \mathbf{x}^{(2)}, \ldots, \mathbf{x}^{(n)}\} \subset \mathbb{R}^d$ with their corresponding observations $\mathbf{y} = \{y^{(1)}, y^{(2)}, \ldots, y^{(n)}\} \subset \mathbb{R}$. The standard assumption in regression is adopted here, in which the noisy objective function is linked to the regression function $f(\mathbf{x})$ (to be estimated) via an additive Gaussian white noise process ε:

$$y(\mathbf{x}) = f(\mathbf{x}) + \varepsilon, \quad \varepsilon \sim \mathcal{N}(0, \sigma_\varepsilon^2) \tag{1}$$

The noise variance σ_ε^2 is either estimated from the data or specified by the user. Regardless of the regression model assumed in Eq. 1, the model gives an estimate of the regression function $\hat{f}(\cdot)$ after fitting on the data $\mathcal{D}$. As the estimate $\hat{f}$ is stochastic (a function of the data), it is possible to formulate the expected error of the model prediction:

$$s^2(\mathbf{x}) = \mathbb{E}\left[\left(f(\mathbf{x}) - \hat{f}(\mathbf{x})\right)^2\right] \tag{2}$$

Note that, $\hat{f}$ is an unbiased estimate of f (meaning that $\mathbb{E}[\hat{f}] = f$) according to the construction of this method. As a result, the expected error above is equivalent to the variance of the prediction. Given such a model uncertainty measurement, it is not plausible to optimize the objective function f by simply searching for the optimality of $\hat{f}$. In addition, the uncertainty associated with each unseen point needs to be considered. In EGO, the so-called *infill-criterion* is designed to integrate the models response surface and the uncertainty. One of the most frequently used criteria, the expected improvement (EI) [8], calculates the expected amount of the improvement provided by an unseen point over the current best solution $\min(\mathbf{y})$:

$$\mathrm{EI}(\mathbf{x}) = (\min(\mathbf{y}) - \hat{f}(\mathbf{x}))\Phi\left(\frac{\min(\mathbf{y}) - \hat{f}(\mathbf{x})}{s(\mathbf{x})}\right) + s(\mathbf{x})\phi\left(\frac{\min(\mathbf{y}) - \hat{f}(\mathbf{x})}{s(\mathbf{x})}\right), \tag{3}$$

where $s(\mathbf{x}) = \sqrt{s^2(\mathbf{x})}$ and $\Phi(\cdot), \phi(\cdot)$ denote the cumulative distribution function and the probability density function of the standard normal distribution, respectively. It takes into account the quantity of the improvement and rewards high uncertainty: it monotonically increases with increasing uncertainty measure $s(\mathbf{x})$ and decreases with increasing prediction $\hat{f}$. Because of this nice property of EI, it is sensible to adopt it for any model on f as long as such a model provides good uncertainty quantification, although EI is originally proposed under the assumption that the regression function is modeled by Kriging.

The new candidate solution $\mathbf{x}^*$ is obtained by maximizing the EI function over the optimization domain. Then the new solution and its fitness value $y(\mathbf{x}^*)$ are appended to the data set $\mathcal{D}$ and the regression model $\hat{f}$ is re-trained on the extended data set. In this manner, the fitness function is optimized iteratively until a budget of function evaluations is reached. The EGO algorithm is summarized in Algorithm 1.

Algorithm 1. Efficient Global Optimization

1 Fit a regression model $\hat{f}$ on the initial data set $\mathcal{X}, \mathbf{y}$.
2 **while** the stop criteria are not fulfilled **do**
3 Find global optimum of the infill criterion:
$$\mathbf{x}^* = argmax_{\mathbf{x}} \, \mathrm{EI}(\mathbf{x})$$
4 Evaluate $\mathbf{x}^*$: $y^* = y(\mathbf{x}^*)$ and append $\mathbf{x}^*, y^*$ to $\mathcal{X}, \mathbf{y}$.
5 Re-estimate the model $\hat{f}$
6 **end while**

2.2 Kriging

In the original EGO algorithm, Kriging is used as the regression model because in addition to predicting a value of a function, it also provides an estimation of the variance of such a prediction. It is a stochastic interpolation method where the unseen value of a stochastic process (random field) is estimated as a linear function of the observed values. Kriging models the distribution of an unknown function by placing a prior Gaussian process on it. The data set $\mathcal{D}$ is used to compute the posterior process [6]:

$$\begin{aligned} f(\mathbf{x}) \mid \mathcal{X}, \mathbf{y} &\sim \mathcal{N}\left(\hat{f}(\mathbf{x}), s^2(\mathbf{x})\right) \\ \hat{f}(\mathbf{x}) &= \hat{\mu} + \mathbf{c}^\top \boldsymbol{\Sigma}^{-1} (\mathbf{y} - \hat{\mu}\mathbf{1}_n) \\ s^2(\mathbf{x}) &= \sigma_\varepsilon^2 - \mathbf{c}^\top \boldsymbol{\Sigma}^{-1} \mathbf{c} + \frac{(1 - \mathbf{c}^\top \boldsymbol{\Sigma}^{-1} \mathbf{1}_n)^2}{\mathbf{1}_n^\top \boldsymbol{\Sigma}^{-1} \mathbf{1}_n} \end{aligned}$$

Note that, originally the EI function (Eq. 3) is derived from the posterior Gaussian distribution at an unseen point $\mathbf{x}$. For the explanation on the terms above (e.g., $\hat{\mu}$), see [10].

Despite the elegance of the Kriging model, it also suffers from some disadvantages. One of the main bottlenecks of Kriging is the computational complexity of fitting the model which is $O(n^3)$. Another downside of Kriging, and more specifically, the Kriging variance, is that it is known that the Kriging variance can be over-optimistic as shown in [2]. Moreover, Kriging might not be the optimal regression model for some data set. For example, when the data set doesn't come from a Gaussian process or when the dimensionality of data is high, Kriging can be outperformed by other regression methods such as tree-based models. In other words, it is important to adopt the EGO algorithm for other regression models instead of Kriging.

2.3 Alternative Prediction Variances

Next to the prediction variance given by the Kriging model, there are a few proposed alternatives. One specific alternative for Kriging is an interpolation variance proposed in [15]. The interpolation variance is defined as follows

$$s_o^2 = \sum_{i=1}^{n} \lambda_i [\mathbf{z}(x_i) - \mathbf{z}^*(x_o)]^2 \tag{4}$$

where λ_i's are the ordinary Kriging weights. The properties of this interpolation variance are very similar to the properties of the proposed heuristic, but the downside of the Kriging interpolation variance is that it can only be used together with a Kriging model.

An alternative that seems to be most promising and model independent is using Bootstrap [4,13]. Bootstrapping is a popular technique to reduce the bias of a (simple) predictor by using several predictors on different samples of the data set and averaging their predictions. The set of predictions gained from the simple predictors can then also be used to give an indication of the prediction variance, as in the variance of the prediction set. This procedure is for example used in Sequential Model-Based Algorithm Configuration (SMAC) [7], by using a Random Forest [1] model where the variance is calculated using the predictions of the trees of the Random Forest. This method however, might not work well when the modeling assumptions of the individual predictors are wrong. Another downside of using bootstrapping is that depending on the model it may require a lot of additional computational resources, especially in EGO, since each iteration the regression models need to be refitted.

3 *k*-NN Uncertainty Measure

In this section, we propose a novel empirical measure of the prediction uncertainty. Such a measure aims at the following objectives: (1) It should operate independently of the modeling assumptions. (2) It should be exploitable by the Efficient Global Optimization algorithm, making it possible to use any regression model in the EGO framework.

In the nonparametric settings, when estimating the mean squared error of the predictor, the available information are the data set $\mathcal{D} := \{\mathcal{X}, \mathbf{y}\}$ and the prediction $\hat{f}(\mathbf{x})$ at $\mathbf{x}$. Intuitively, this empirical uncertainty measure should be zero at correctly predicted known observations and increase for the data points that are far from the observations. Given these preferred properties, a distance-weighted measure $\widehat{U}_{k\text{-NN}}$ is proposed as follows:

$$\widehat{U}_{k\text{-NN}} = \underbrace{\frac{\sum_{i \in N(\mathbf{x})} w_i^k \left| \hat{f}(\mathbf{x}) - y_i \right|}{\sum_{i \in N(\mathbf{x})} w_i^k}}_{\text{empirical prediction error}} + \underbrace{\frac{\min_{i \in N(\mathbf{x})} d(\mathbf{x}_i, \mathbf{x})}{\max_{\mathbf{x}_i, \mathbf{x}_j \in \mathcal{X}} d(\mathbf{x}_i, \mathbf{x}_j)} \widehat{\sigma}}_{\text{variability of the observation}} . \tag{5}$$

where

$$w_i = 1 - \frac{d(\mathbf{x}_i, \mathbf{x})}{\sum_{i \in N(\mathbf{x})} d(\mathbf{x}_i, \mathbf{x})}, \quad \widehat{\sigma} = \sqrt{\mathrm{Var}\left[\{y_i\}_{i \in N(\mathbf{x})} \cup \{\hat{f}(\mathbf{x})\}\right]}.$$

Note that $N(\mathbf{x})$ collects the indices of k nearest neighbours to $\mathbf{x}$ and $d(\cdot,\cdot)$ denotes the Euclidean distance metric. $\widehat{\sigma}$ is computed as the standard deviation of the observations in the neighbourhood with the prediction $\hat{f}(\mathbf{x})$.

The proposed uncertainty quantification consists of two components: *(1) the empirical error of the prediction and (2) the variability of the observed outputs y.* Intuitively, less empirical prediction error leads to higher certainty of the prediction. Moreover, when comparing two different regression tasks, a large variability of the observations y could contribute to the high uncertainty of the prediction, even if the predictor $\hat{f}$ were making the same empirical error on both tasks. The empirical error is computed from the difference between the prediction $\hat{f}(\mathbf{x})$ and the observations at the k-nearest neighbours. Such differences are linearly scaled where the weights are inversely proportional to Euclidean distances to the neighbours. This heuristic is based on the intuition that the closer neighbours have more influence than neighbours that are further away. To quickly diminish the effect of far-away neighbours, the exponent k (the number of neighbours) is applied to the weights. The variability of the observation is estimated by calculating the standard deviation of the observations at the nearest neighbours and the predicted point. The resulting value is then rescaled by the distance to the nearest neighbour. Using the distances to scale the heuristic error prediction, we make sure that the uncertainty goes to zero at correctly predicted known points and that it increases when predicting points further away from the known observations.

A good number of neighbours depends on the number of known points and the data dimensionality. For most of the experiments in this paper k is set to 20, more neighbours will provide a more smooth but also slightly more pessimistic prediction error while less neighbours make the expected prediction error more optimistic and less smooth. To illustrate the behaviour of k-NN uncertainty, a 1-D function $f(x) = x\sin(x)$ is used in Fig. 1. Note that k-NN uncertainty (green area) progresses very similarly to the Kriging uncertainty quantification (blue area). It can also be observed that at the known observations the prediction error given by the k-NN uncertainty algorithm is exactly the error between the prediction and the known observation. Note that the SVR model is badly fitted, and different hyperparameters would result in a much better fit. This is on purpose to illustrate how the k-NN uncertainty would look like using less fitted models. Lastly, it can be observed that when the to be predicted point is far away from the known data points the uncertainty of the prediction increases.

4 Experimental Setup and Results

Two different experimental setups are used to demonstrate the properties and effectiveness of k-NN uncertainty in Efficient Global Optimization. First, we validate k-NN uncertainty by visual inspection of plotted two and five dimensional benchmark functions that are often used in the field of optimization (in this case the *Ackley* and *Schaffer* function). In Fig. 2 it can be observed that k-NN uncertainty is quite similar to the Kriging variance as shown in the lower subplots of

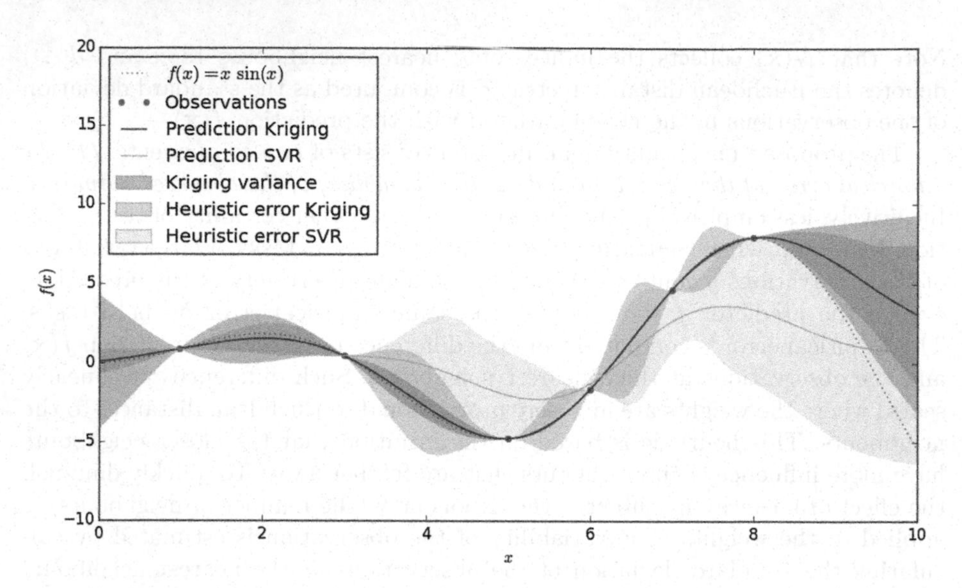

Fig. 1. Best viewed in color. Visualization of k-NN uncertainty heuristic. The dotted red line is the real function $f(x) = x \sin x$, the red dots are the observed points. The blue line is the predicted mean of the Kriging model with the shaded blue area showing the standard Kriging variance and the shaded green area is k-NN uncertainty on the same Kriging model. The yellow line shows the predictions of a Support Vector Regression (SVR) model with default hyper-parameters ($C = 1$, RBF kernel, $\epsilon = 0.1$) with the shaded yellow area denoting k-NN uncertainty using the SVR model. The number of neighbours for k-NN uncertainty is set to 4 in this case. (Color figure online)

Fig. 2a. k-NN uncertainty is a bit less optimistic than the Kriging variance but shows roughly the same areas with higher variance. When looking at the Random Forest bootstrapping variance and k-NN uncertainty it can be observed that the bootstrapping variance is very blocky, due to the Random Forest model assumptions. k-NN uncertainty however does not use the individual tree predictions of the Random Forest model and because of the interpolation effect using the distances to the known observations, it creates a much more smooth surface. In Fig. 3 the models are trained on samples of the *Schaffer* function in the space of -50 to 50 for both dimensions while tested on the complete range of -100 to 100. It can be observed that the k-NN uncertainty gradually increases when moving away from the known observations, while the Kriging variance almost immediately explodes to a flat high value. When looking at the same benchmark function but now in five dimensions, we can plot a one-dimensional slice of the function (using the first dimension) to show the local behaviour of k-NN uncertainty versus the Kriging variance in Fig. 4. Here it can be observed that the Kriging variance is over-optimistic and actually wrong, while k-NN uncertainty is much more pessimistic and actually captures very well the shape of the underlying function.

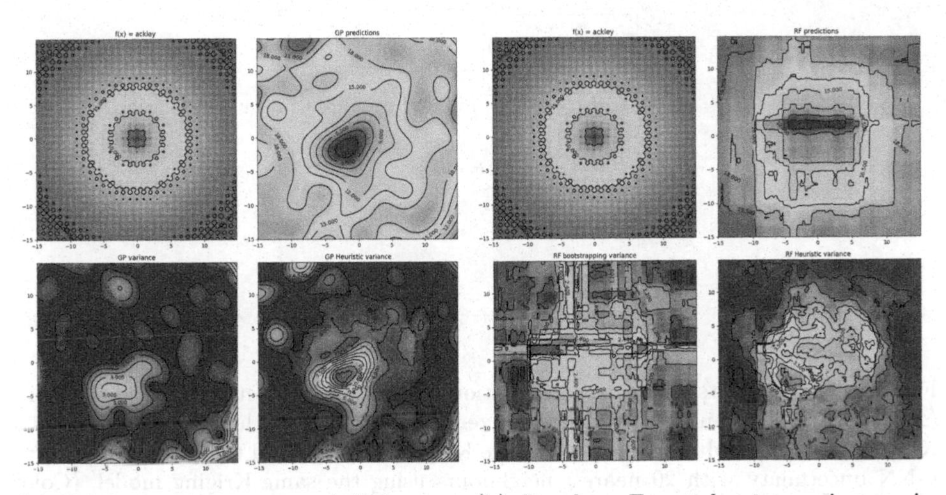

(a) Kriging variance versus k-NN uncertainty

(b) Random Forest bootstrapping variance versus k-NN uncertainty

Fig. 2. (a) Upper-left plot is the *Ackley* function in $2D$, up-right the Kriging prediction of this function using 100 data points for training. Lower-left plot shows the Kriging variance and bottom-right shows k-NN uncertainty using the same Kriging model. (b) Upper-left plot is the same as (a), up-right shows a Random Forest predictor with 50 trees using 100 data points for training. Lower-left plot shows the variance given by the Random Forest and bottom-right shows k-NN uncertainty using the same Random Forest model. The number of nearest neighbours for k-NN uncertainty is set to 20.

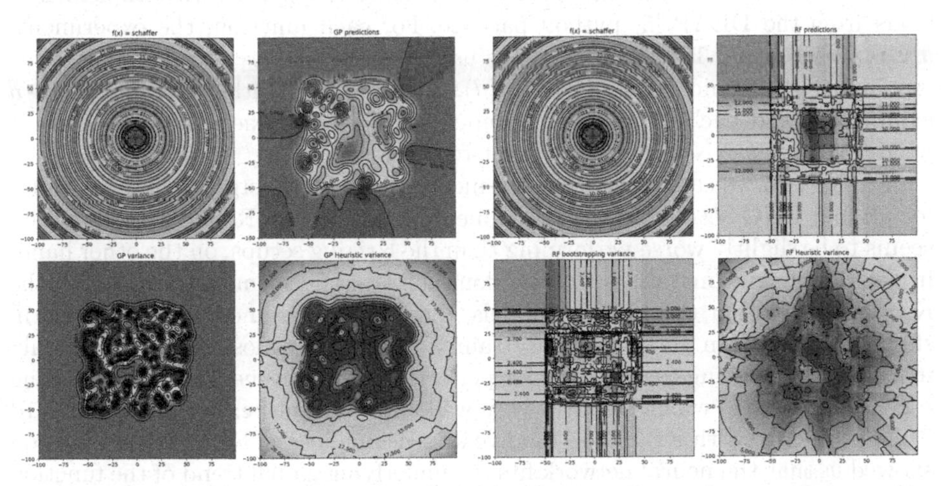

(a) Kriging variance versus k-NN uncertainty

(b) Random Forest bootstrapping variance versus k-NN uncertainty

Fig. 3. Same as in Fig. 2 but now using the *Schaffer* benchmark function.

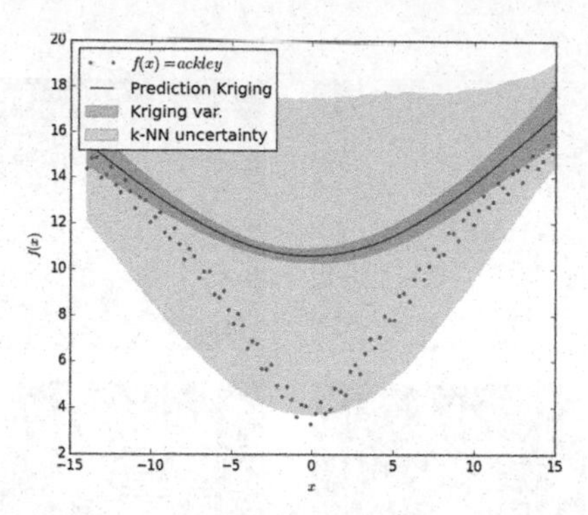

Fig. 4. The green dots are unseen observations of the *Ackley* function in five dimensions (slice with the last four dimensions set at zero), the blue line is the predicted mean of a Kriging model, the blue shaded area is the Kriging variance and the green shaded area k-NN uncertainty with 20 nearest neighbours using the same Kriging model. (Color figure online)

The second experiment is more quantitative as we compare the performance of k-NN uncertainty in the setting of Efficient Global Optimization. We compare the convergence speed of the original EGO with Kriging, EGO with Kriging using k-NN uncertainty instead of the Kriging variance, EGO with a Random Forest model using k-NN uncertainty as the prediction variance and finally EGO with a multi-layer perceptron using k-NN uncertainty (using two hidden layers of size 100 and 50 nodes respectively). The experiment is carried out using three different benchmark functions *Ackley*, *Rastrigin* and *Schaffer* with implementations from the DEAP [5] python package. For each function, the experiments are repeated using 100 initial samples using a Latin hypercube sampling strategy, and in 2, 5 and 10 dimensions (d). The EGO algorithm is run for $10 \cdot d$ evaluations and each experiment is repeated 40 times. The number of nearest neighbours for k-NN uncertainty is set to 20.

From Fig. 5 we can observe that in most cases the convergence is very similar for all four EGO setups. In the two dimensional cases the Random Forest setup seems to be slightly worse performing than the Kriging setups, on the other hand, in the ten dimensional cases the Random Forest setup seems to outperform the Kriging setups. For the Kriging models, the Kriging variance seems to perform slightly better than the k-NN uncertainty, however in most cases this is only marginal. Interesting is to note the performance of the neural network using k-NN uncertainty, which perform very well and even outperforms the standard EGO procedure with Kriging in the five dimensional cases. Further investigation showed us that the neural network fits the underlying global trend of the function much more accurate than the Kriging or Random Forest model, allowing the EGO procedure to quickly convert to the global optimum.

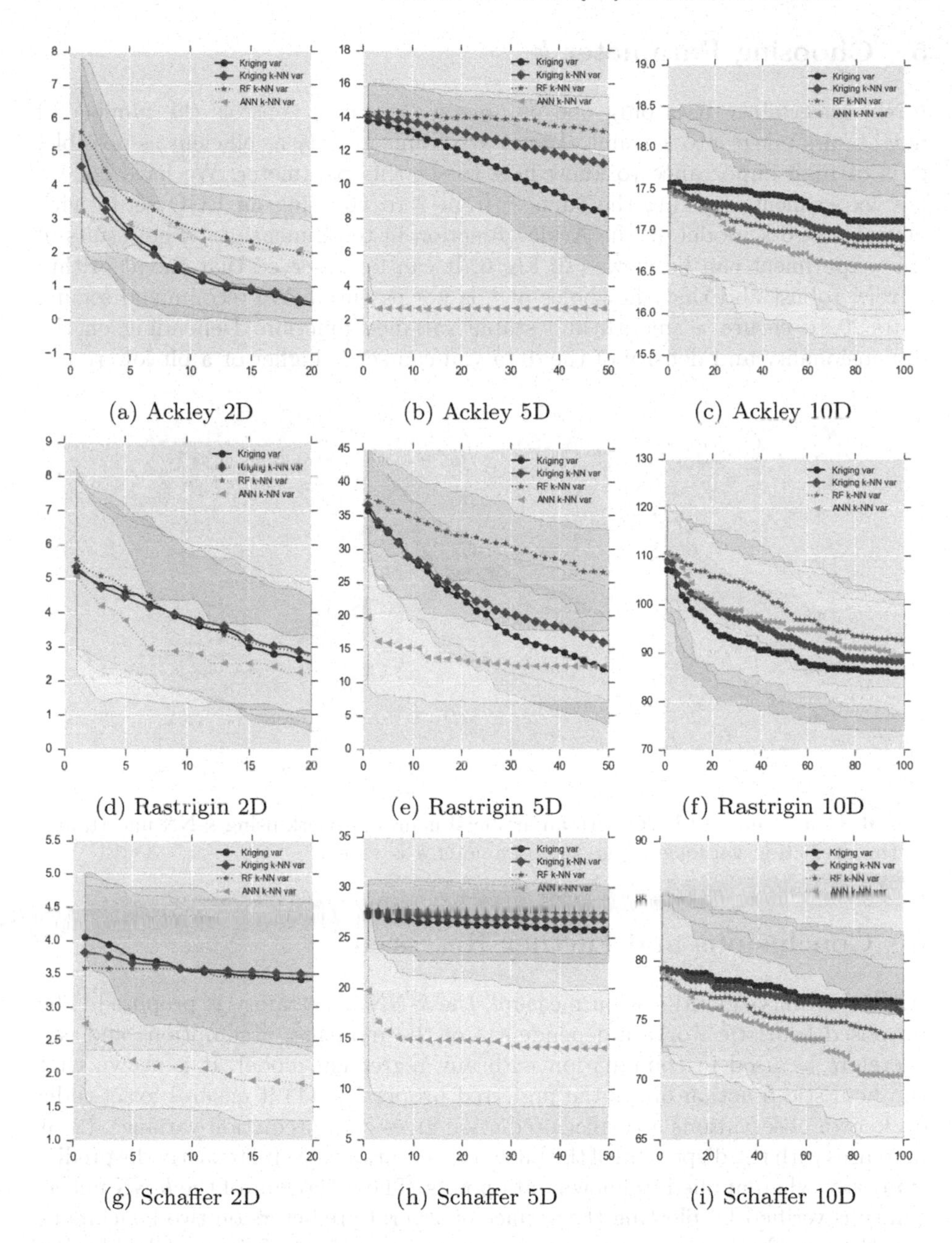

(a) Ackley 2D (b) Ackley 5D (c) Ackley 10D

(d) Rastrigin 2D (e) Rastrigin 5D (f) Rastrigin 10D

(g) Schaffer 2D (h) Schaffer 5D (i) Schaffer 10D

Fig. 5. Efficient Global Optimization convergence on the three benchmark functions. As blue line with dots the original EGO procedure with Kriging variance, the red line with stars shows the convergence of EGO with a Random Forest model using k-NN uncertainty, the green diamonds illustrate the convergence of EGO with Kriging using k-NN uncertainty and the yellow triangles illustrate te convergence of EGO with a multi-layer perceptron using k-NN uncertainty. The shaded areas show the 95% confidence interval over the 40 runs. (Color figure online)

5 Choosing Parameter k

k-NN uncertainty uses only one parameter, k, which controls the number of neighbours taken into account. To use k-NN uncertainty as efficient as possible, it is of high importance to know how to set this parameter. We have done a few experiments ranging the value k from 2 to 50 using an EGO setup with neural network model on the Ackley function in ten dimensions. The results of this experiment can be viewed in Fig. 6. It can be observed that the algorithm is very robust and that the choice of k is not essential. We recommend setting k to 20 to ensure a smooth and stable variance function. Depending on the dimensionality and density of the data you can set it higher or a bit lower.

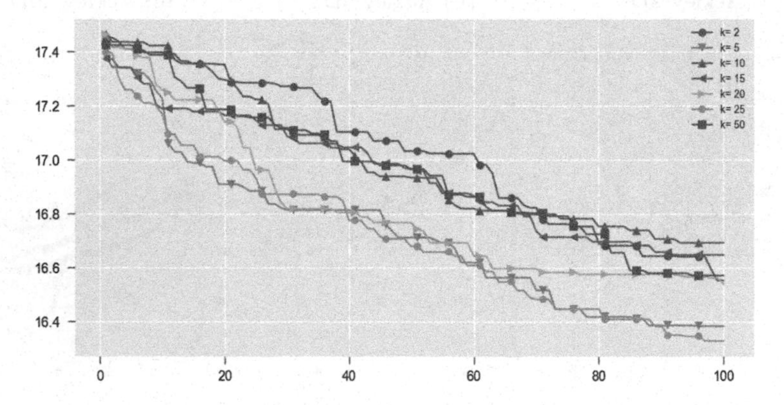

Fig. 6. Convergence of EGO with an artificial neural network using k-NN uncertainty as the prediction variance where the parameter k is varied.

6 Conclusions and Further Research

An uncertainty quantification measure, the k-NN uncertainty is proposed. The proposed heuristic works independently of the modeling assumptions and can therefore be used in combination with any regression model. It is shown that the heuristic function obeys the preferred properties: (1) it ensures exactitude; on known observations a correct prediction gives zero prediction variance. (2) it increases with the dispersion of the known observations. (3) It is exactly the prediction error when applied to known data points. The behaviour of the k-NN uncertainty is verified by plotting the surface of several predictors on two benchmark functions and by running a wide set of experiments using the Efficient Global Optimization framework. Results of the EGO experiments show that the heuristic can be used in such optimization settings and that the performance in both high and low dimensions, using different statistical models, can even outperform the original EGO concept that uses Kriging. It also shows that different regression models can be used in Efficient Global Optimization using such a heuristic as prediction variance, making EGO more widely applicable. It is shown that the proposed heuristic is robust with respect of its parameter k, the number of neighbours, and a recommendation of $k = 20$ is given.

For future research, one interesting direction would be to replace the nearest neighbour approach with Locality-sensitive hashing [12], to make the heuristic faster.

Acknowledgment. The authors acknowledge support by NWO (Netherlands Organisation for Scientific Research) PROMIMOOC project (project number: 650.002.001).

References

1. Breiman, L.: Random Forests. Mach. Learn. **45**(1), 5–32 (2001)
2. Den Hertog, D., Kleijnen, J.P., Siem, A.: The correct kriging variance estimated by bootstrapping. J. Oper. Res. Soc. **57**(4), 400–409 (2006)
3. Domingos, P.: A unified bias-variance decomposition. In: Proceedings of 17th International Conference on Machine Learning, pp. 231–238. Morgan Kaufmann, Stanford CA (2000)
4. England, P., Verrall, R.: Analytic and bootstrap estimates of prediction errors in claims reserving. Insur. Math. Econ. **25**(3), 281–293 (1999)
5. Fortin, F., Michel, F., Gardner, M.A., Parizeau, M., Gagné, C.: DEAP: evolutionary algorithms made easy. J. Mach. Learn. Res. **13**, 2171–2175 (2012)
6. Ginsbourger, D., Le Riche, R., Carraro, L.: Kriging is well-suited to parallelize optimization. In: Tenne, Y., Goh, C.-K. (eds.) Computational Intelligence in Expensive Optimization Problems. ALO, vol. 2, pp. 131–162. Springer, Heidelberg (2010). https://doi.org/10.1007/978-3-642-10701-6_6
7. Hutter, F., Hoos, H.H., Leyton-Brown, K.: Sequential model-based optimization for general algorithm configuration. In: Coello, C.A.C. (ed.) LION 2011. LNCS, vol. 6683, pp. 507–523. Springer, Heidelberg (2011). https://doi.org/10.1007/978-3-642-25566-3_40
8. Jones, D.R., Schonlau, M., Welch, W.J.: Efficient global optimization of expensive black-box functions. J. Global Optim. **13**(4), 455–492 (1998)
9. Krige, D.G.: A statistical approach to some basic mine valuation problems on the witwatersrand. J. Chem. Metall. Mining Soc. S. Afr. **52**(6), 119–139 (1951)
10. Rasmussen, C.E.: Gaussian Processes in Machine Learning. In: Bousquet, O., von Luxburg, U., Rätsch, G. (eds.) ML -2003. LNCS (LNAI), vol. 3176, pp. 63–71. Springer, Heidelberg (2004). https://doi.org/10.1007/978-3-540-28650-9_4
11. Rasmussen, C., Williams, C.: Gaussian Processes for Machine Learning. Adaptative computation and machine learning series, University Press Group Limited (2006). http://books.google.nl/books?id=vWtwQgAACAAJ
12. Slaney, M., Casey, M.: Locality-sensitive hashing for finding nearest neighbors [lecture notes]. IEEE Signal Process. Mag. **25**(2), 128–131 (2008)
13. Stine, R.A.: Bootstrap prediction intervals for regression. J. Am. Stat. Assoc. **80**(392), 1026–1031 (1985)
14. Wang, H., Emmerich, M., Back, T.: Balancing risk and expected gain in kriging-based global optimization. In: 2016 IEEE Congress on Evolutionary Computation (CEC), pp. 719–727. IEEE (2016)
15. Yamamoto, J.K.: An alternative measure of the reliability of ordinary kriging estimates. Math. Geol. **32**(4), 489–509 (2000)

Singular Outliers: Finding Common Observations with an Uncommon Feature

Mark Pijnenburg[1,2] and Wojtek Kowalczyk[1(✉)]

[1] Leiden Institute of Advanced Computer Science, Leiden, The Netherlands
{m.g.f.pijnenburg,w.j.kowalczyk}@liacs.leidenuniv.nl
[2] Netherlands Tax and Customs Administration, Utrecht, The Netherlands

Abstract. In this paper we introduce the concept of *singular outliers* and provide an algorithm (SODA) for detecting these outliers. Singular outliers are multivariate outliers that differ from conventional outliers by the fact that the anomalous values occur for only one feature (or a relatively small number of features). Singular outliers occur naturally in the fields of fraud detection and data quality, but can be observed in other application fields as well. The SODA algorithm is based on the local Euclidean Manhattan Ratio (LEMR). The algorithm is applied to five real-world data sets and the outliers found by it are qualitatively and quantitatively compared to outliers found by three conventional outlier detection algorithms, showing the different nature of singular outliers.

Keywords: Anomaly detection · Outliers · LOF · Tax administration

1 Singular Outliers

Currently, many outlier detection algorithms exist [5]. However, hardly any of these is able to detect a particular type of outliers, which we will call *singular outliers*. These outliers are met frequently in practice and often represent observations with an interesting characteristic that is mostly ignored by conventional outlier detection algorithms.

Roughly speaking, singular outliers are observations that show an anomalous value for one feature (or a relatively small set of features), while displaying common behavior on all other features. The feature with the anomalous value will be called the *discriminating feature*. The discriminating feature may be different for each outlier and finding the discriminating feature is part of the learning process. Singular outliers are typically overlooked by current outlier detection algorithms, since these algorithms prefer observations with anomalous values on as many features as possible.

To explain the concept of singular outliers more clearly, suppose we have a set of points $\mathbf{X}$ in a 5-dimensional space and a point $\mathbf{x} \in \mathbf{X}$, surrounded by its 20 nearest neighbors. Denote the (local) mean vector of the 20 neighbors with $\mathbf{m}$. This allows calculating the absolute distances between $\mathbf{x}$ and $\mathbf{m}$ for each dimension. When visualized in a needle plot, the result may look as one of

J. Medina et al. (Eds.): IPMU 2018, CCIS 855, pp. 492–503, 2018.
https://doi.org/10.1007/978-3-319-91479-4_41

subplots of Fig. 1. The subplot in the left shows large deviation from the local mean in all dimensions. This observation can be labeled a conventional outlier. However, the right subplot shows an observation with small deviations to the local mean on all but one dimension (dimension 4). This is what we will call a singular outlier.

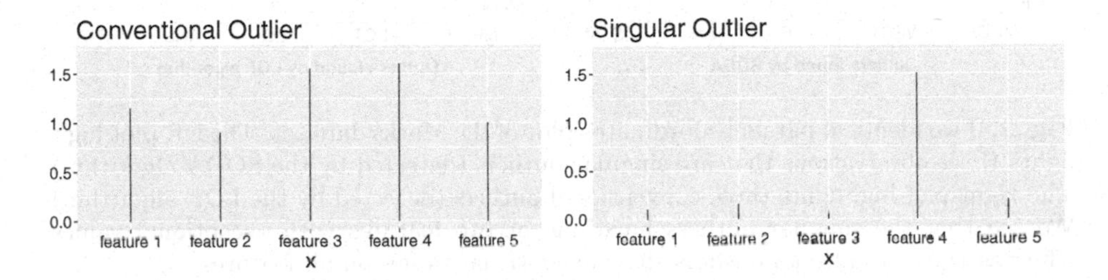

Fig. 1. Needle plots of the absolute differences $|x_i - m_i|$ $(i \in 1, \ldots, 5)$ of a point $\mathbf{x}$ and the (local) center $\mathbf{m}$ of its 20 nearest neighbors.

Singular outliers can provide interesting insights in the field of fraud detection or data quality. We will illustrate this by our experiences at the Netherlands Tax and Customs Authority (NTCA), where erroneous VAT tax returns had to be detected. It turns out that tax returns containing many unusual values, are often less risky than tax returns containing only one (or two) anomalous fields. This apparent contradiction can be explained partly by efforts of taxpayers to conceal tax evasion. But also by the existence of unconventional businesses with non-standard business models, that produce uncommon values on many fields of a tax declaration, without any increased risk of tax evasion. Moreover, it was noted at the NTCA that taxpayers sometimes unintentionally change two adjacent fields, leading to an example of singular outliers in the field of data quality. The examples in Sect. 4 point to other application areas.

Singular outliers usually differ from those found by conventional outlier detection algorithms. Let us consider a real life example from Sect. 4, where two algorithms, our SODA and the popular LOF algorithm, [3], are used to find outliers in a data set containing the marks on 5 courses of 88 students. Figure 2 shows two identical parallel coordinates plots (for more information on parallel coordinates plots see e.g., [4]) of this data set, but each plot highlights different outliers. The left plot highlights three singular outliers. We see that the singular outliers are students that have common marks on at least three subjects, but one or two exceptional marks. In contrast, the right subplot shows the same parallel coordinates plot but with three *conventional* outliers highlighted. These students have exceptional marks on all subjects. The singular outliers are interesting, since valuable insight might be gained in exploring the cause of the exceptional mark of these students.

We can now give a formal definition of a singular outlier.

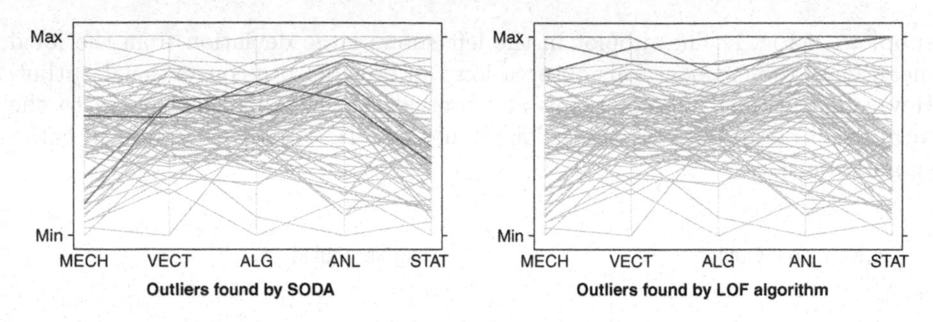

Fig. 2. Two identical parallel coordinates plot of the Marks data set. The left plot highlights three observations that are singular outliers (detected by the SODA algorithm). The right plot highlights three conventional outliers (detected by the LOF algorithm). We see that the singular outliers have one or two features with exceptional values, whereas the conventional outliers show exceptional values on all features.

Definition 1. *An observation is called a* singular outlier *if there exits one feature (or a relatively small number of features), called the* discriminating feature(s) *such that the observation is an outlier when the discriminating feature is taken into account, but no outlier when the features are restricted to the non-discriminating features.*

The purpose of this paper is to point to the class of singular outliers and present an algorithm for finding them. The paper is organized as follows. Section 2 gives a short overview of classes of outlier detection algorithms. Subsequently Sect. 3 describes the SODA algorithm for finding singular outliers and explains the experimental setup to test the algorithm on five public data sets. Section 4 contains the results of the experiments. Finally, Sect. 5 contains the conclusions of the paper and a discussion of the results.

2 Related Work

Chandola et al. [5] present an overview of commonly used outlier detection techniques. They distinguish five classes of unsupervised outlier detection algorithms: nearest neighbor-based algorithms (including density based approaches), clustering-based algorithms, statistical algorithms, information theoretic algorithms, and spectral algorithms. Goldstein and Uchida [6] present a similar categorization.

Nearest neighbor and clustering-based outlier detection are by far the most used categories in practice, according to Goldstein and Uchida. Of these two categories, nearest-neighbor based algorithms perform better in most cases [6]. Moreover it is useful to split the class of nearest neighbor algorithms in two subclasses: *local* algorithms and *global* algorithms. In the remaining categories outlier detection algorithms, the statistical algorithm HBOS (Histogram-based Outlier Score) performs remarkably well in experiments [6].

We will briefly describe three algorithms that we used in our comparisons to SODA: Local Outllier Factor, k^{th} Nearest Neighbor and HBOS. The Local Outlier Factor (LOF) is introduced by Breunig et al. [3] The basic idea is to compare the density of an observation $\mathbf{x}$ with the density of its k closest neighbors $N^k(\mathbf{x})$. As a density estimator is taken the mean *reachability distance* from an observation to its neighbors. To compute this distance, first the Euclidean distance is calculated between an observation $\mathbf{x}$ and its kth nearest neighbor $\mathbf{x}_{(k)}$. Subsequently, the distance is calculated of this k^{th} nearest neighbor $\mathbf{x}_{(k)}$ to *its* k^{th} nearest neighbor $\mathbf{x}_{(k)_{(k)}}$. Finally the maximum of these two is taken. In equations,

$$LOF(\mathbf{x}) = \frac{\text{density}(\mathbf{x})}{\underset{y \in N^k(\mathbf{x})}{\text{mean}}(\text{density}(\mathbf{y}))}, \text{ where} \tag{1}$$

$$\text{density}(\mathbf{x}) = \frac{1}{\max\{d_{Eucl}(\mathbf{x}, \mathbf{x}_{(k)}), d_{Eucl}(\mathbf{x}_{(\mathbf{k})}, \mathbf{x}_{(k)_{(k)}})\}} \tag{2}$$

The kth nearest neighbors outlier detection algorithm is straightforward: the distance to the kth neighbor is used as an anomaly score [9]. By taking the distance to the kth neighbor instead of the average distance to the kth nearest neighbors, the algorithm is better able to detect a small cluster of outliers.

The Histogram Based Outlier Score [6] starts with making a histogram for each feature in the data set. Then, for each observation, the height of the bins it resides are multiplied. This will result in a positive number. Subsequently the negative of this number is taken as an outlier score. In the experiments, the frequently used Sturges' formula is applied to compute the number of bins.

3 Singular Outlier Detection Algorithm

3.1 The Algorithm

We propose an algorithm called SODA (Singular Outlier Detection Algorithm) to find singular outliers. The algorithm involves several steps which are specified in Algorithm 1. The input of the algorithm is a data set with n observations and p numeric features as well as a parameter k that specifies the number of nearest neighbors. The output of the algorithm is an outlier score for each observation; large scores represent outliers. Additionally, the discriminating feature for each observation is given as output. The latter is a convenient starting point when manual inspection of outliers is required.

In the first step, for each observation $\mathbf{x_i}$, $i = 1, \ldots, n$, its k neighbors N_i^k are found. Subsequently, the center of these neighbors, $\mathbf{m_i}$, is determined as the trimmed mean, i.e., the mean calculated after removing the largest value and the smallest value along each dimension. These extreme values are ignored to limit their impact on $\mathbf{m_i}$.

As a next step, the Euclidean distance $d_{Eucl}(\mathbf{x_i}, \mathbf{m_i})$ as well as the Manhattan distance $d_{Manh}(\mathbf{x_i}, \mathbf{m_i})$ are calculated. We will call the ratio of these two distances LEMR (*Local Euclidean Manhattan Ratio*), i.e.:

$$\text{LEMR}(\mathbf{x_i}) = \frac{d_{Eucl}(\mathbf{x_i}, \mathbf{m_i})}{d_{Manh}(\mathbf{x_i}, \mathbf{m_i})}. \tag{3}$$

This ratio is the singular outlier score for observation $\mathbf{x_i}$. It is an indicator of the 'spread' of the values of the vector $\mathbf{v_i} = |\mathbf{x_i} - \mathbf{m_i}|$. The left subplot of Fig. 3 displays a 2-dimensional example with the Euclidean unit sphere (i.e., the circle) as well as the Manhattan unit sphere (the square: all points with a Manhattan distance 1 to the origin). It is clear from the figure that points on the coordinate axes (i.e., singular outliers) are in both spheres, so the ratio of the Euclidean to the Manhattan distance is 1. On the other hand, points that are on the diagonals (the opposite of singular outliers) have the largest difference in Euclidean and Manhattan distance to the origin and consequently have a low LEMR value. The right subplot of Fig. 3 shows the LEMR value in this 2-dimensional example.

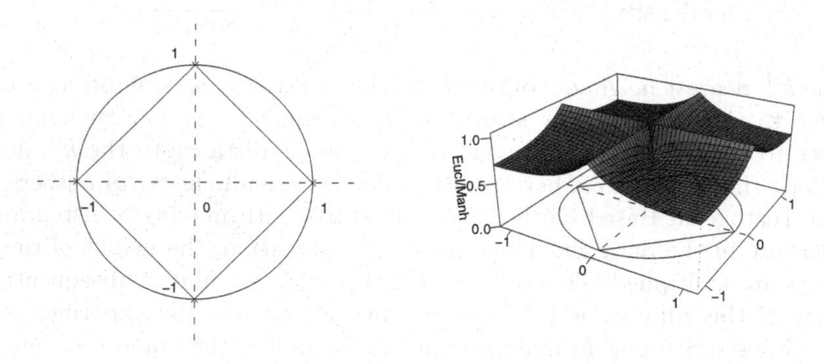

Fig. 3. Left: the Euclidean unit sphere and the Manhattan unit sphere in 2 dimensions. Points on coordinate axes (singular outliers) have a maximal value of the Local Euclidean Manhattan Ratio (LEMR = 1). Points on the diagonals (i.e., conventional outliers) have a minimal LEMR value (LEMR = $1/\sqrt{2}$), see subplot right. In more dimensions the difference between the Euclidean and the Manhattan distance is more pronounced.

The value of LEMR cannot exceed 1 since the Euclidean length of a vector $||\mathbf{v}||$ is always smaller or equal to the L_1-norm $||\mathbf{v}||_1 = \sum |v_j|$. The LERM value of 1 occurs when all but one components of $\mathbf{v}$ equal 0. The value of LEMR is a minimum in case the elements of $\mathbf{v}$ are all of equal length (and unequal 0). The minimal value depends on the number of features p and is given by $1/\sqrt{p}$.

The calculation of the nearest neighbors in the first step of the algorithm is performed using the Manhattan distance in contrast to the more frequently used Euclidean distance. The reason is that the discriminating feature of a singular outlier ideally deviates strongly from other observations. This deviation may however be so large that it can have too big influence on the determination of the nearest neighbors. In the worst case, this would lead to neighbors that only share a similar value for the discriminating feature. To diminish the effect, we prefer the Manhattan distance that is less influenced by extreme values in one feature.

Algorithm 1. Singular Outlier Detection Algorithm (SODA)

Input : A data set $\mathbf{X} = \{\mathbf{x_1}, \mathbf{x_2}, \ldots\}$ with p (numeric) features, k the number of nearest neighbors

Output: 1) outlier score. The observations with the largest scores represent the singular outliers,
2) discriminating feature for each $\mathbf{x}_i$. This indicates the feature that contains the anomalous value.

1 Find the k nearest neighbors N_i^k for each observation $\mathbf{x_i}$ in *df* based on the Manhattan distance d_{Manh}.
2 **for** *each observation (row)* $\mathbf{x_i}$ *in df* **do**
3 Compute the vector of trimmed means $\mathbf{m_i}$ of the neighbors N_i^k: $\mathbf{m_i} = \underset{\mathbf{x_j} \in N_i^k}{\text{mean}_{trimmed}} \{\mathbf{x_j}\}$
4 Compute
5 $d_{Eucl}(\mathbf{x_i}, \mathbf{m_i}) = \sqrt{\sum_{s=1}^{p} (x^s - m^s)^2}$
6 $d_{Manh}(\mathbf{x_i}, \mathbf{m_i}) = \sum_{s=1}^{p} |x^s - m^s|$
7 outlier score$(\mathbf{x_i}) = LEMR(\mathbf{x_i}) = \frac{d_{Eucl}(\mathbf{x_i}, \mathbf{m_i})}{d_{Manh}(\mathbf{x_i}, \mathbf{m_i})}$
8 discriminating feature$(\mathbf{x_i}) = \underset{s \in \{1, \ldots, p\}}{\text{argmax}} |x^s - m^s|$
9 **end**

The algorithm has one parameter k. A (too) small value of k will lead to observations $\mathbf{x}$ with very few neighbors. Consequently, it might result in outliers whose non-discriminating features have a risk of being not so common. A (too) large value of k usually has less severe consequences. However, in theory it can lead to bad results if the data points are gathered in many small clusters and k exceeds the cluster size. For instance this may happen if the data contains binary features. In our experiments we used $k = n/5$.

The time complexity of the singular outlier detection algorithm is $\mathcal{O}(n^2pk)$.

3.2 Measurement of Characteristics of Singular Outliers

To determine whether an outlier can be called *singular*, two characteristics must hold. These two characteristics follow from Definition 1 and are listed below.

No outlier with respect to non-discriminating features When the discriminating feature is removed, the observation stops to be an outlier.

Outlier with the discriminating feature The observation must become an outlier when the discriminating feature(s) is taken into account.

To quantify these two qualitative characteristics, we will formulate two measures. These measures will be applied in Sect. 4, Table 2 to see whether the outliers found by SODA can be called singular outliers.

The first characteristic is measured for an outlier by removing the discriminating feature from the data set and then calculate an outlier score. This outlier score must be low for the characteristic to be valid. In principle any outlier score can be used. In this paper the LOF score is chosen, since it is a well established outlier score that performs well on a broad variety of data sets [6]. A LOF score around 1 will be accepted as a sign that the first characteristic is applicable.

The second characteristic is measured for an outlier by computing an outlier score with and without the discriminating feature. Subsequently the ratio of the outlier scores is taken as a measure. Again, in principle any outlier score can be taken, but in this paper we choose the LOF score. Hence, the measure of the second characteristic becomes $\frac{\mathrm{LOF}_{all}(\mathbf{x})}{\mathrm{LOF}_{w.o.discr}(\mathbf{x})}$. A large ratio signifies that the addition of the discriminating feature has substantially increased the outlier score (LOF) and is taken as a sign that the second characteristic is applicable.

In Sect. 4 the outliers found by SODA will be compared with outliers of the three conventional outlier algorithms mentioned in Sect. 2. In order to apply the two measurements above, we have to determine the discriminating feature for outliers found by these algorithms as well. For the algorithms based on nearest neighbors (LOF and $k^{th}NN$) a natural choice is to take:

$$\text{discriminating feature}(\mathbf{x}) = \underset{s \in \{1,\ldots,p\}}{\operatorname{argmax}} |x^s - x^s_{(k)}|, \tag{4}$$

where $x^s_{(k)}$ is the sth feature of the k-nearest neighbor of $\mathbf{x}$. For the HBOS algorithm we take the feature with the lowest bin height as the most discriminating feature.

4 Comparison

This section compares the outliers found by SODA with outliers detected by three conventional outlier detection algorithms (mentioned in Sect. 2) to five real world data sets. The comparison consists of parallel coordinates plots and the two measurements described in Sect. 3.2.

The selected data sets are listed in Table 1, along with some key properties. One of these properties is the value k that is used in the SODA algorithm and the two conventional outlier detection algorithms based on nearest neighbors (LOF and k^{th}NN). In the experiments, k is set to (approximately) one-fifth of the number of observations, see Sect. 3.1.

4.1 Marks Data

The first data set, *Marks*, comes from Mardia et al. [8] and consists of the examination marks of 88 students in the five subjects mechanics, vectors, algebra, analysis and statistics. We used these unstandardized marks.

The parallel coordinates plot of this data set is displayed in the introduction, see Fig. 2. Clearly, the SODA algorithm picks up students that perform commonly on most subjects, except for one subject. In contrast, the LOF algorithm

Table 1. Data sets used in the experiments with key indicators.

Data set name	# rows	# columns	k
Marks	88	5	20
Istanbul stock exchange	536	9	100
Wholesale customers	440	6	80
Polish bankruptcy	5907	5	1000
Algae	306	8	60

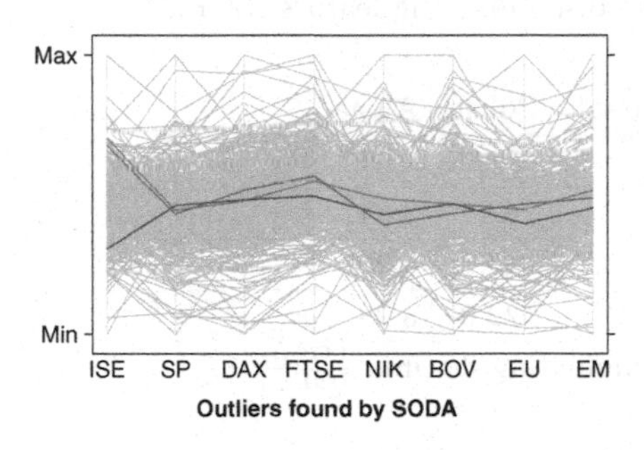

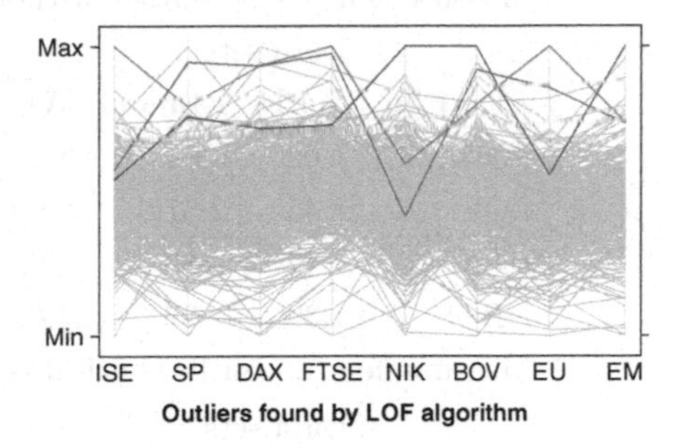

Fig. 4. Parallel coordinates plot for the Istanbul Stock Exchange data set with the three largest outliers highlighted.

selects students that perform exceptionally well on all subjects. The measurements of Table 2 confirm that the SODA outliers can be called singular outliers: the average value for the SODA outliers on measurement 1 equals 1.02, which is close to 1 and much lower compared to the values of other algorithms. The average value on the second measurement (1.09) is not very large, but at least larger than the values for the outliers found by the other algorithms.

4.2 Istanbul Stock Exchange Data

The Istanbul Stock Exchange data set [2] displays the daily increase in eight major stock exchange indices for the period January 5 2009 to February 22, 2011. The data are already standardized.

The parallel coordinates plots of Fig. 4 show that the SODA algorithm has picked three days where the average stock return is common (0.3 %), but the Istanbul stock exchange performed deviantly. In contrast, the LOF algorithm has selected three days on which most stock indices got exceptional good returns (average return of these days was 4.1 %, average of all days is less than 0.1 %). Table 2 confirms the differences of the outliers found by SODA and the three conventional algorithms.

Table 2. Measurements of the two characteristics of singular outliers (see Sect. 3.2) for the outliers selected by SODA, LOF, k^{th}NN and HBOS on five data sets. Each algorithm is allowed to pick 3 outliers and the values reported are averages over these three outliers. The first measurement computes an outlier score (LOF) without the discriminating feature. The second measurement computes the ratio of the LOF score with all features and the LOF score without the discriminating feature. Singular outliers are characterized by a value close to 1 for measurement 1 and a large value for measurement 2. We see that the outliers selected by SODA can be termed 'singular outliers' and differ on these characteristics from the outliers found by conventional outlier detection algorithms.

Measurement 1 - no outlier without discriminating feature: LOF_{-d}					
	Data sets				
Algorithm	Marks	Istanbul	Wholesale	Polish	Algae
SODA	1.02	0.99	1.16	1.01	1.01
LOF	1.45	2.57	2.98	147.7	3.84
k^{th}NN	1.42	2.56	3.69	147.7	4.20
HBOS	1.39	2.57	4.02	122.0	3.28
Measurement 2 - outlier with discriminating feature: $\frac{LOF_{all}}{LOF_{-d}}$					
	Data sets				
Algorithm	Marks	Istanbul	Wholesale	Polish	Algae
SODA	1.09	1.28	2.31	5.75	1.99
LOF	1.04	1.08	2.11	1.46	1.73
k^{th}NN	1.02	1.04	1.60	1.46	1.27
HBOS	1.06	1.06	1.06	1.13	1.08

4.3 Wholesale Customers Data

The Wholesale Customers data set [1] contains the annual spending on six product categories for 440 customers of a wholesaler in Portugal. These not standardized spending amounts were input for the algorithms. It is clear from the parallel coordinates plots in Fig. 5 that the SODA algorithm selects customers with average sales on all product categories, but one. The LOF algorithm selects customers with exceptional high purchases on at least three product categories.

The statement that SODA picks out singular outliers is confirmed by looking at the measurements of Table 2. However, some experts might be inclined to call the red and/or pink line of the LOF algorithm singular outliers as well. See Sect. 5 for a discussion of this aspect.

4.4 Polish Bankruptcy Data

The Polish Bankruptcy data contains financial information of Polish companies and was originally used to model the probability of a bankruptcy [10]. We took the 'five year set' and standardized the ratios. We selected five financial ratios that correspond to major economic indicators of the company: X2 (total liabilities/total assets), X3 (working capital/total assets), X7 (Earnings Before Interest and Taxes/total assets), X9 (sales/total assets), and X10 (equity/total assets). After removing observations with missing values, 5907 observations remain.

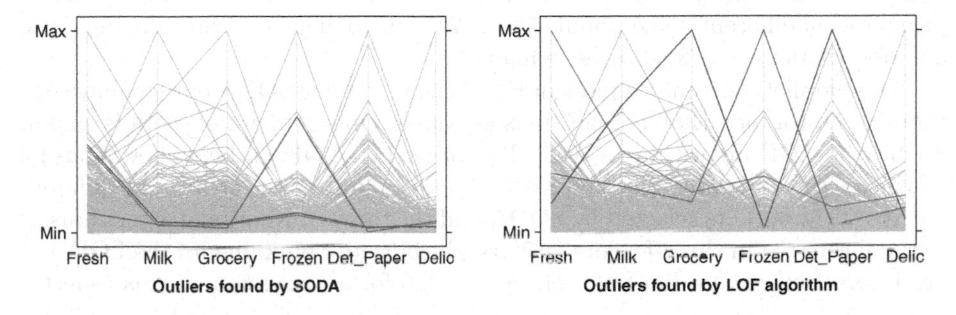

Fig. 5. Parallel coordinates plot for the Wholesale Customers data set with the three largest outliers highlighted. (Color figure online)

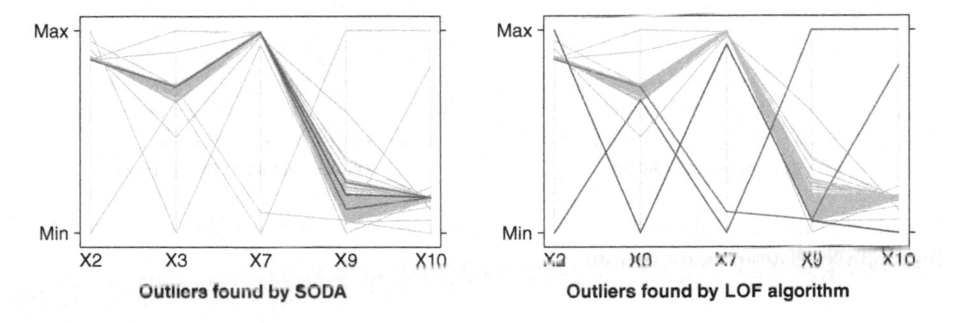

Fig. 6. Parallel coordinates plot for the Polish Bankruptcy data set with the three largest outliers highlighted. (Color figure online)

Comparing the outliers found by SODA and LOF in the parallel coordinates plots of Fig. 6, we see that the SODA outliers show more common behavior than the LOF outliers. It is not clear from the figure whether the SODA outliers have at least one anomalous value. The latter is confirmed by Table 2. The first measurement shows an average LOF value of 1.01 for the SODA outliers, indicating that the observations are not considered outliers without the discriminating feature. Simultaneously, measurement 2 shows that the average LOF score increases by a factor of over 5 when the discriminating feature is added. However, if one is interested in two-feature singular outliers, one may find the blue and pink line of the LOF plot interesting. See Sect. 5 for a discussion on this aspect.

4.5 Algae Data

The data set *Algae*, available via the UCI repository [7], comes from a water quality study where samples were taken from sites on different European rivers of a period of approximately one year. We used the (standardized) concentrations of 8 chemical substances from these samples as features. We exclude the additional measures on different algae populations. After removing rows with missing data, 306 observations and 8 features remain.

The parallel coordinates plots of Fig. 7 show that the SODA outliers have one chemical substance that has an unusual value. One of these outliers is picked up by the LOF algorithm as well. The other two LOF outliers show unusual values for three chemical characteristics. Table 2 confirms the on average different nature of the outliers selected by SODA and the three conventional algorithms. If one is interested in three-feature outliers, the blue and pink lines of the LOF plot are interesting observations as well. See Sect. 5 for a discussion on this aspect.

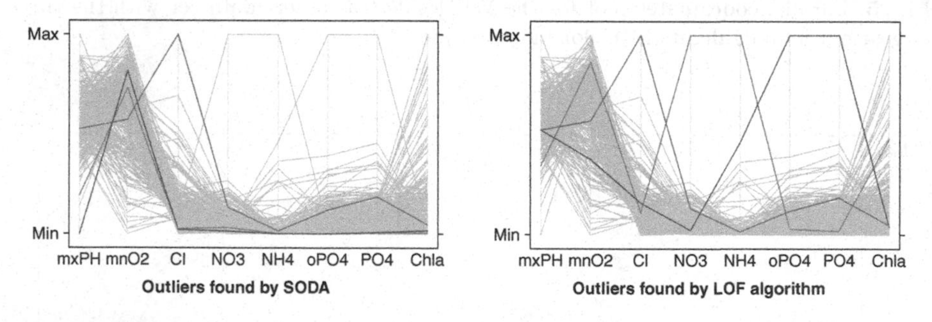

Fig. 7. Parallel coordinates plot for the Algae Data with the three largest outliers highlighted. (Color figure online)

5 Conclusion and Discussion

In this paper, we introduced the concept of a *singular outlier* and pointed to its usefulness in the contexts of fraud detection, data quality and other areas. We introduced an algorithm (SODA) to detect these outliers, based on the Local Euclidean Manhattan Ratio (LEMR). The algorithm has been applied to five publicly available data sets. The parallel coordinates plots as well as the results shown in Table 2 confirm that the SODA algorithm is suited for finding singular outliers.

Although Table 2 points out that the SODA algorithm finds observations that better match the definition of singular outliers when one restricts oneself to *one* discriminating feature, the parallel plots of the latter three data sets show that there might be interesting outliers that have two discriminating features. These observations may not get the highest outlier scores for SODA, despite the fact that the deviations in the discriminating features may be large. When in such a situation, the SODA algorithm can be adapted to focus more on two-feature singular outliers or to take the absolute value of the deviations into account. This might be a fruitful aspect for further research.

Part of such research can be to adjust the Local Euclidean Manhattan Ratio. The Euclidean and the Manhattan distances are based on the L_p norm. Other ratios can be constructed by taking other values for p instead of 2 and 1.

Another interesting path for further research might be to view the detection of singular outliers as an optimization problem where simultaneously attention is given to maximize dissimilarity (discriminating feature) as well as maximizing similarity (non-discriminating features).

Finally, singular outliers might be found with an algorithm with a lower time complexity than SODA. The time complexity of the SODA algorithm is dominated by finding the nearest neighbors of each observation. The nearest neighbors component gives a local flavor to SODA that might be beneficial for data sets that possess local structures (like separate clusters). However, for data sets that do not display this property, comparison of an observation $\mathbf{x}$ to the global center m_{global} might be used. This reduces computational time considerably.

References

1. Abreu, N.G.C.F.M., et al.: Analise do perfil do cliente Recheio e desenvolvimento de um sistema promocional. Ph.D. thesis, ISCTE-IUL (2011)
2. Akbilgic, O., Bozdogan, H., Balaban, M.E.: A novel hybrid RBF neural networks model as a forecaster. Statistics Comput. **24**(3), 365–375 (2014)
3. Breunig, M.M., Kriegel, H.P., Ng, R.T., Sander, J.: LOF: identifying density-based local outliers. In: ACM Sigmod Record, vol. 29, pp. 93–104. ACM (2000)
4. Brunsdon, C., Fotheringham, A., Charlton, M.: An investigation of methods for visualising highly multivariate datasets. Case Studies of Visualization in the Social Sciences, pp. 55–80 (1998)
5. Chandola, V., Banerjee, A., Kumar, V.: Anomaly detection: a survey. ACM Comput. Surv. (CSUR) **41**(3), 15 (2009)
6. Goldstein, M., Uchida, S.: A comparative evaluation of unsupervised anomaly detection algorithms for multivariate data. PLoS ONE **11**(4), e0152173 (2016)
7. Lichman, M.: UCI machine learning repository (2013). http://archive.ics.uci.edu/ml
8. Mardia, K., Kent, J., Bibby, J.: Multivariate statistics (1979)
9. Ramaswamy, S., Rastogi, R., Shim, K.: Efficient algorithms for mining outliers from large data sets. In: ACM Sigmod Record, vol. 29, pp. 427–438. ACM (2000)
10. Zikeba, M., Tomczak, S.K., Tomczak, J.M.: Ensemble boosted trees with synthetic features generation in application to bankruptcy prediction. Exp. Syst. Appl. (2016)

Towards a Semantic Gas Source Localization Under Uncertainty

Javier Monroy(✉), Jose-Raul Ruiz-Sarmiento, Francisco-Angel Moreno, Cipriano Galindo, and Javier Gonzalez-Jimenez

Machine Perception and Intelligent Robotics group (MAPIR), Department of System Engineering and Automation, University of Malaga, Malaga, Spain
jgmonroy@uma.es

Abstract. This work addresses the problem of efficiently and coherently locating a gas source in a domestic environment with a mobile robot, meaning *efficiently* the coverage of the shortest distance as possible and *coherently* the consideration of different gas sources explaining the gas presence. The main contribution is the exploitation, for the first time, of semantic relationships between the gases detected and the objects present in the environment to face this challenging issue. Our proposal also takes into account both the uncertainty inherent in the gas classification and object recognition processes. These uncertainties are combined through a probabilistic Bayesian framework to provide a priority-ordered list of (previously observed) objects to check. Moreover the proximity of the different candidates to the current robot location is also considered by a cost function, which output is used for planning the robot inspection path. We have conducted an initial demonstration of the suitability of our gas source localization approach by simulating this task within domestic environments for a variable number of objects, and comparing it with an greedy approach.

Keywords: Mobile robotics · Semantics · Gas source localization e-nose

1 Introduction

The fusion of different sensing modalities can empower service robots operating in human environments (*e.g.* for elder care at homes or as assistants at offices, airports or hospitals) with new abilities and the possibility to efficiently accomplish complex tasks. With this aim, in this work we focus on the senses of vision and olfaction, and face a challenging task: gas source localization, *i.e.* the finding of the object releasing a particular smell. In this context, *olfaction* is understood as the sensing of volatile chemical substances by means of an electronic nose (e-nose) [1], while *vision* is interpreted as the perception of the environment through a camera capturing light intensity [2,3].

J. Medina et al. (Eds.): IPMU 2018, CCIS 855, pp. 504–516, 2018.
https://doi.org/10.1007/978-3-319-91479-4_42

Given the volatile nature of gases and the complex processes involved in their dispersion (*i.e.* dominated by turbulent flows [4]), after the perception of an unusual gas concentration it is necessary to carry out a search of the object that is releasing it in the environment, process commonly referred as gas source localization (GSL). For the case of domestic environments it includes the location of methane or butane leaks from the home heating system, the presence of smoke, or unpleasant smells coming from spoiled food, the toilet, or the pet sandbox, among others. An efficient localization of these gas sources would permit the robot to act consequently and in time, for instance alerting a human (*e.g.* notifying the presence of smoke from the oven) or suggesting different actions to be carried out (*e.g.* replacing the pet sandbox).

GSL is usually addressed by mimicking animal behaviors through bio-inspired algorithms, assuming the existence of a downwind gas plume (*i.e.* plume tracking) [5,6], or by exploiting other information sources like dispersion models or windflow data [7,8]. However, most of these methods are prone to fail in human-like environments due to the important assumptions they rely on (*e.g.* existence of a gas plume, the predominance of laminar and uniform windflows, or the absence of obstacles in the environment that can interfere with the gas dispersion). Thereby, their success heavily relies on how well the given algorithm adjust itself to the environmental conditions, which determines the way in which gases are dispersed. A way to overcome this issue is to employ artificial vision systems to detect gas source candidates and inspect them, reducing the complexity of the search process. For example, if the e-nose detects an abnormal concentration of a gas classified as smoke, a visually recognized oven is a good candidate to check, while a chair is not. This approach, not being novel, has only been superficially explored under very simple scenarios where the robot exploited knowledge about the source physical characteristics to reduce the locations to search [9]. Yet, what is still needed is a principled way to set the nature of the objects and their possible gas emissions –in other words, their semantics–, from which we can infer what objects in the environment are prone to be the gas source.

Moreover, traditional GSL approaches work, in most cases, with gas classification systems that produce an exact outcome, for example, a detected smell is smoke or not. However, the classification of gases is not extent of uncertainty sources (*e.g.* the cross-sensitivity of gas sensors or the environmental conditions), being mandatory their consideration for a coherent robot operation. For example, an ambiguous gas classification result between smoke and spoiled food (probability of 0.55 vs. 0.45) could end up with the robot only searching for smoke when indeed a dish with fish was forgotten in the kitchen counter. The same holds for the uncertainty inherent to the object recognition process: an object can be recognized as a heater with probability 0.60 or as a fan with 0.40, so it must be also considered.

This work presents, to the best of our knowledge, the first attempt towards a system performing an efficient and coherent gas source localization under uncertainty exploiting semantics. For that, it is built and maintained a semantic representation of the robot environment that provides the GSL task with valuable

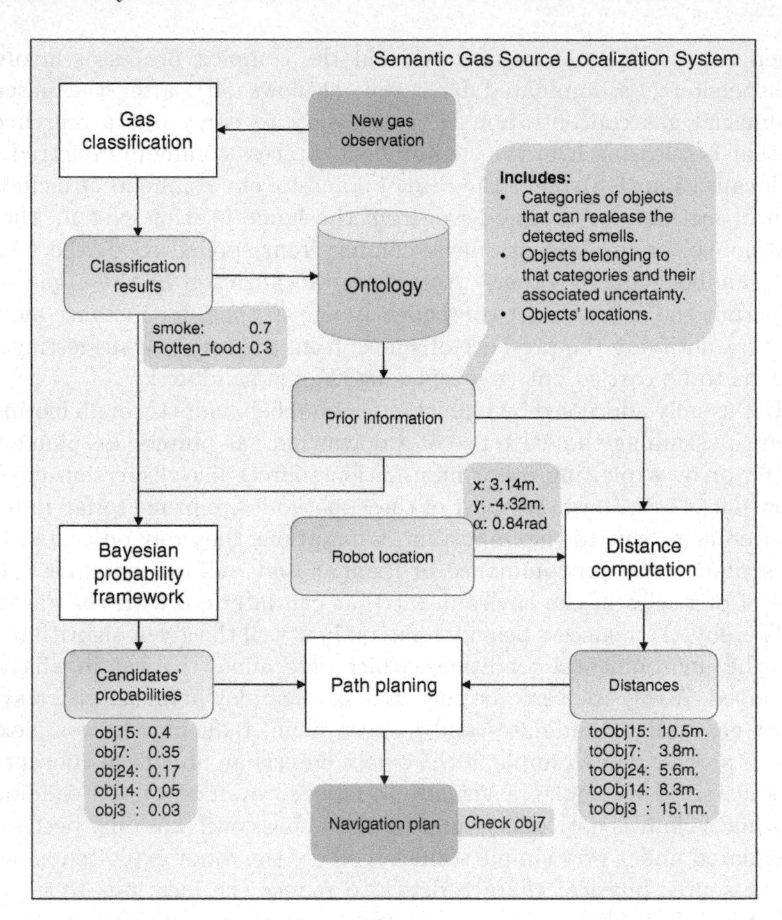

Fig. 1. Overview of the proposed Semantic Gas Source Localization System: from a new gas observation with a detected gas until the generation of the navigation plan for localizating the source. White boxes are processes, while blue shapes are generated/consumed data.

prior information (see Fig. 1). Concretely, an *ontology* [10] is used to encode the semantic knowledge of the domain at hand (e.g. ovens can give off smoke with probability P_a, cocked meal smell with probability P_b, and no smell with P_c), and also to store information about previously perceived objects: their probability of belonging to the considered categories (e.g. heater, cigarette, fish, etc.), and their locations. In this work we assume that the robot workspace has been already visually inspected and a number of objects have been recognized and codified into the ontology. In this way, when a gas emission is perceived and classified as belonging to a number of gas classes with their respective uncertainties, a semantic request is submitted to the ontology which returns: the object categories that can release those gases, and the instances (objects) of that categories already observed in the environment, also with their recognition uncertainty.

A probabilistic Bayesian framework is then in charge of fusing this information and assigning to each object (*i.e.* candidate) a probability of being the gas source. Finally, a cost function is introduced to weight the probability of each candidate by the distances from the current robot location to them, and a path planning module processes its output to provide the navigation plan to be executed by the robot.

A demonstration of the system suitability has been carried out within complex simulated scenarios using GADEN [11]. The obtained results were promising, suggesting that our probabilistic approach is suitable for efficient gas source localization within complex environments, such as domestic ones.

2 Related Work

Gas source localization strategies are many and varied [12]. In this section we focus on two particular approaches: the fusion between the chemical data provided by the e-nose with vision systems in order to boost the GSL task efficiency, and works that consider uncertainty during the search process.

The former approach enables robots to identify candidates from a distance, thus dramatically diminishing the effective search space and greatly enhancing the ability to locate an odor source. It must be noticed that opposed to vision, which is a range sensing modality, most of the gas sensors are point-sampling devices, measuring only the gas that is in contact with them. Despite the notable advantages of considering vision in the GSL task, only very basic algorithms have been proposed so far, most of them relying on strong assumptions about the gas-source shape or color for the visual detection of candidates [9,13]. An exception is the work proposed by Loutfi *et al.* [14], where the authors proposed a symbolic reasoning technique for fusing vision and olfaction. However, focus is placed on object recognition, where gas sensing is only employed for object disambiguation, not to locate the source releasing the volatiles.

Related to works considering some type of uncertainty in the search process, we can highlight some engineered plume-tracing strategies such as infotaxis [7], a gradient-free method exploiting the expected entropy of future samples to guide the robot search towards the gas source, probabilistic approaches based on particle filters [8,15], or strategies based on gas distribution mapping [16]. The latter do not rely on the presence of a plume, neither on strong assumptions about the environmental conditions, however, their limitation resides in the time necessary to sweep the entire environment, and their bad scalability as the environment enlarges.

3 The Semantic Gas Source Localization System

Figure 1 shows an overview of the processes and data involved in the proposed system. In a nutshell, if an unusual gas concentration is detected (e.g. while the robot is exploring the environment or while performing other non gas-related tasks) (see Sect. 3.1), the Semantic Gas Source Localization (SGSL) System is triggered for detecting the object releasing that odor and acting consequently. For that the system performs a semantic query to an ontology to get prior information with different flavors (see Sect. 3.3), which is introduced into a probabilistic framework that yields an ordered list of objects candidates according to their probability of being the source (see Sect. 3.3). This list is the input to a cost function, which is also feed with the distance from the robot current location to the source candidates. A path planning algorithm exploits this function to re-order the candidates list and produce a navigation plan to check them (see Sect. 3.4). For checking if an object candidate is the gas source (process know as validation), the robot will sample the air in the object's proximity, measuring concentration and carrying out a new gas classification. By comparing these values with the ones that triggered the search, the robot is able to discern if the object is or not the gas source it is looking for. The main components of the SGLS system are described next.

3.1 Starting Point: Gas Detection and Classification

In this work we assume that an assistance robot deployed in a home environment is equipped with an e-nose that is sampling the environment on a regular basis. This implies that while the robot is performing its duty tasks (*e.g.* patrolling, assistance, cleaning, etc.), it is also monitoring the gases present in the air. When an abnormal gas concentration level is detected, that is, when the gas concentration observed exceeds a predetermined threshold, the SGSL system triggers the search.

Once the search has been triggered, and in order to determine which objects in the environment are susceptible for releasing the observed gas, we carry out a gas classification. As in many other disciplines, classification corresponds to the process of determining which of a set of classes a new sample belongs. In this work we account for the uncertainty in this process by considering probabilistic classifiers. The output of such classifiers is not a class label, but a set of probabilities representing the belief of the gas observation to belong to each considered gas-class [17,18]. Therefore, any gas classifier giving as output a probability distribution over the set of classes can be employed, *e.g.* Support Vector Machines, Naive Bayes (the one considered in this work), Decision Trees, etc.

3.2 Exploiting Semantic Knowledge: The Ontology

Once a gas has been detected and classified with a certain belief, *e.g.* 0.6 of being smoke and 0.4 rotten food, the first step towards the localization of its source is to obtain valuable prior information to assist the process. With *prior information* in this context we mean: (i) knowledge about the categories of objects that can release such gas, *i.e.* in the case of smoke and rotten food smells, ovens, ashtrays or bins are candidates, and (ii) information regarding the objects already detected in the environment which can belong to that categories. As a reminder, we are assuming that the gas source is between a set of object candidates previously recognized in a visual inspection of the robot workspace.

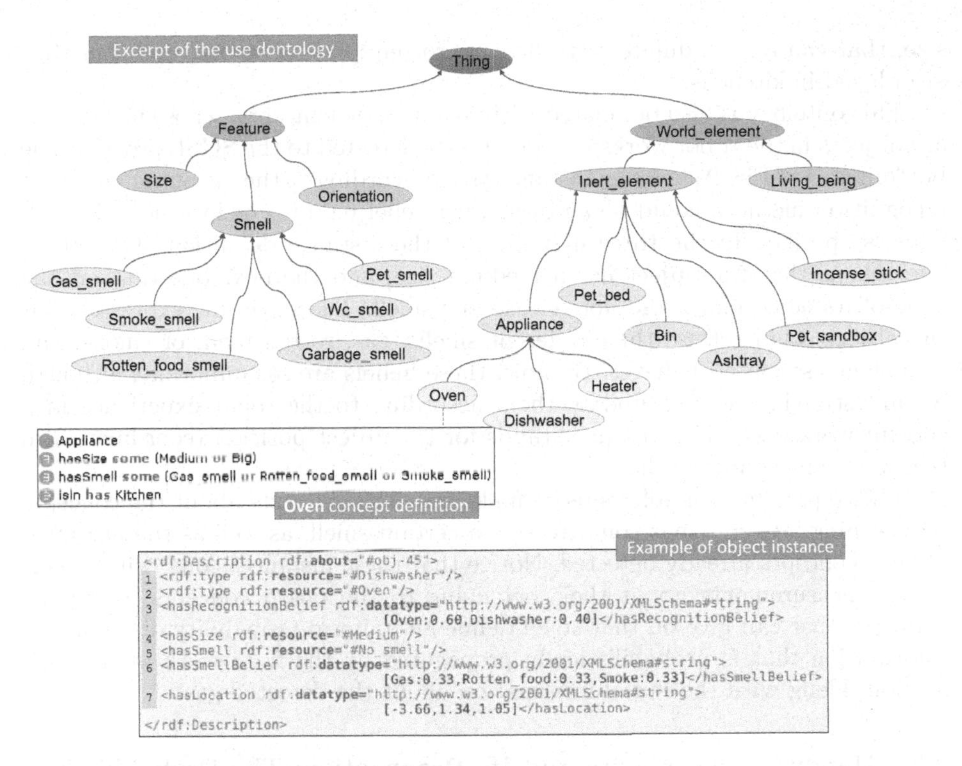

Fig. 2. Excerpt of the ontology used in this work, showing part of the hierarchy of encoded concepts, the definition of the concept `Oven`, and an example of object instance.

The chosen recognition method must be able to provide confidence values about its results, and although this task is simulated in the experiments conducted in this paper, we plan to use Conditional Random Fields (CRFs) [19] given their high recognition rates and proved suitability to this end [20,21].

For codifying the previous information, which is clearly a form of Semantic Knowledge (SK), we have resorted to an ontology [10]. An ontology is a principled way to naturally represent and update SK about a domain of discourse, employing for that a set of concepts arranged hierarchically, properties of that concepts, and instances of them.

As an illustrative example, let us consider an excerpt of the ontology used in this work, shown in Fig. 2. The root concept is `Thing` with two children: `Feature` and `World_element`, the latter establishing the elements that could be found in the robot surroundings and the former their features, *i.e.* `Size`, `Orientation`, and `Smell`. The elements can be `Inert_elements` or `Living_beings`, although in this work we are interested in the first one, which is the parent of concepts like `Oven`, `Astray`, `Dishwasher` or `Bin`. The concepts within this hierarchy are defined by their properties, as it is shown in the same figure for the `Oven` case. From that definition we can retrieve that ovens usually exhibit a medium or big size, that can release different smells: gas, rotten food, or smoke, and that they are placed in kitchens.

This ontology is also populated with instances of concepts, whose in this case are objects in the robot workspace previously detected to the SGSL process. The bottom part of Fig. 2 shows an instance that, according to the output of an object recognition method, could be an oven with belief 0.6 or a dishwasher with 0.4. This is specified in the three first lines of the instance definition. The fourth one tell us that the object has a medium size, and the next one that, at the time of its detection, it did not release any smell. The sixth line expresses that the object could release three different smells: gas, rotten food, or smoke, and also their associated believes. By now, these beliefs are set uniformly, although we are studying how to update them according to the robot experience in a certain workspace. The last line stands for the object position (coordinates) in the environment metric map.

This representation allows us to make semantic requests about the concepts (concerning objects) that could release a certain smell, as well as the instances of that concepts already detected. Notice that these instances come with uncertainty measurements about their belonging to the posed concepts, while the concepts that can give off that smell define an uniform probability distribution, information that is probabilistically propagated by the framework in the next section, along with the initial information about the detected gas.

3.3 Handling Uncertainty and Its Propagation: The Probabilistic Framework

Our probabilistic Bayesian model for uncertainty propagation aims to, given the gas classification results and the prior information from the ontology, provide the probability for each candidate being the source. For that it considers four random variables:

- Z is the gas observation (*i.e.* a measurement of the e-nose (z_g)).
- $G = \{G_i,\ i = 1 : N_G\}$ models the gas class and takes values on the set of N_G possible gases.
- $C = \{C_i,\ i = 1 : N_C\}$ stands for the category of a candidate object, assigning to it a value from the set of N_C categories.
- $S = \{O_i,\ i = 1 : N_O\}$ stands for the gas source, taking values on the set of N_O objects perceived in the environment.

Thus, the probability of a certain candidate object o_i being the gas source, given a gas observation z_g, is modeled as:

$$P(S = o_i|Z = z_g) = \sum_{j=1}^{N_C} P(S = o_i|Z = z_g, C_j)\ P(C_j|Z = z_g) \tag{1}$$

$$P(C_j|Z = z_g) = \sum_{k=1}^{N_G} P(C_j|Z = z_g, G_k)\ P(G_k|Z = z_g)$$

Such source probability is calculated by marginalizing first against the object categories C_j, and second against the gas classes G_k. It allows us to model the probability of each object in the environment of being the gas source as the product of three conditional probability distributions. The first one, $P(S = o_i|Z = z_g, C_j)$, represents the probability of object i being the gas source conditioned on both the gas observation and knowledge about the object category of the gas source (*e.g.* bin, oven, toilet, etc.). Assuming independence with the gas observation given the object category C_j, this probability can be defined as the likelihood of the object belonging to that category (*i.e.* object recognition probabilities), information provided by the ontology (recall line 3 in the bottom part of Fig. 2).

The second probability distribution $P(C_j|Z = z_g, G_k)$, models the likelihood of the source to belong to a certain category C_j conditioned on the gas observation and knowledge of the gas class G_k that has ben released. Again, we can safely assume that this distribution is independent of the gas observation given the gas class, computing its value from the semantic knowledge encoded in the ontology about the object categories that can give off the gas G_k. For example, if $G_k =$ `Smoke` and the defined object categories that can release smoke are `Oven`, `Heater` and `Ashtray`, then: $P(C_{Oven}|G_{Smoke}) = P(C_{Heater}|G_{Smoke}) = P(C_{Ashtray}|G_{Smoke}) = 0.33$, while for the rest of object categories it takes a value of 0, *e.g.* $P(C_{Pet\_sandbox}|G_{Smoke}) = 0$.

Finally, $P(G_k|Z = z_g)$ is interpreted as the probability of the gas release belonging to gas of class G_k conditioned on the gas observation, which corresponds to the output of the probabilistic gas classifier (recall Sect. 3.1). Given the three described probability distributions, the computation of Eq. (1) can be accomplished in short time, enabling a real time operation.

3.4 Giving Coherence to the Localization Process: The Path Planning Algorithm

Once computed the probability of each object in the environment of being the gas source, the robot must plan and inspect the different objects in order to locate the one that is the gas source. For this step we rely on a path planning module that in addition to the referred probabilities also takes into account the distance between the current robot location and the objects. For doing that a cost function is used:

$$\mathcal{L}(o_i) = -\ln\left(P(S = o_i|Z = z_g)\right) distTo(o_i) \tag{2}$$

where $distTo(o_i)$ is the distance between the robot location and the candidate object o_i. This cost function models a trade off between source probability and distance, giving lower values for objects with high probability and/or close to the robot. On each iteration, the path planning module calculates these costs to retrieve the *best* object to check $\hat{o}$ through the optimization:

$$\hat{o} = \underset{o_i,\ 1 \leq i \leq N_O}{arg\min}\ \mathcal{L}(o_i) \tag{3}$$

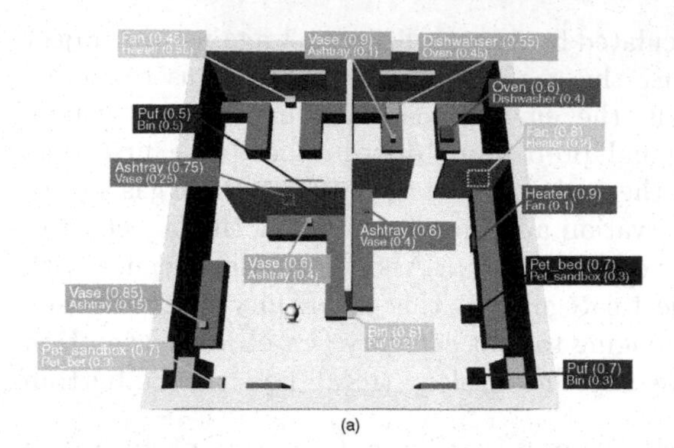

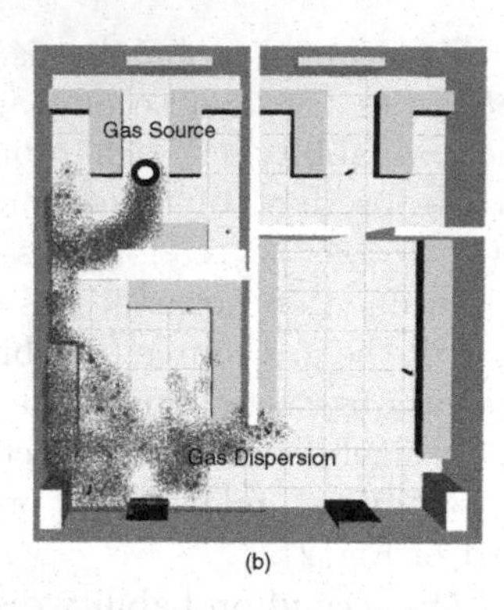

Fig. 3. Experimental setup. (a) 3D simulated environment composed of four rooms and fifteen objects. Objects are shown as 3D colored boxes specifying their location in the environment and their category probabilities. (b) Illustration of a gas dispersion simulation within the environment using GADEN [11]. When the robot is exposed to a gas concentration higher than a set threshold, the search is triggered to locate the source. As can be seen, gas dispersion is chaotic and spreads over multiple rooms, which implies that the robot may be far from the source when the search is triggered.

Once $\hat{o}$ has been calculated, the robot checks if it is the gas source releasing the gas through a process commonly referred as *source validation*. If it is, we are done. If not, the object is removed from the list of candidates, and the optimization in Eq. (3) is carried out again (since the distances from the robot to the remaining candidates have changed), obtaining a new target candidate. Recall that we are assuming that the gas source is among the objects already present in our system, otherwise a more sophisticated search must be implemented for example by performing object recognition along the search process to find new candidates. We will explore that approach in a future work.

4 System Demonstration

This section presents a simulated experiment where a mobile robot equipped with an e-nose must locate a gas emission source in a home environment (see Fig. 3). For this scenario we consider 3 gas classes, namely: `Smoke_smell`, `Gas_smell` and `Rotten_food_smell`, 11 object categories (`Vase`, `Bin`, `Ashtray`, `Oven`, `Heater`, `Dishwasher`, `Fan`, `Puf`, `Incense_stick`, `Pet_sandbox` and `Pet_bed`), and model $P(C_j|Z = z_g, G_k)$ as an uniform probability distribution (see Table 1). Furthermore, we set up fifteen different objects in the environment, which we assume have been previously detected by the robot with the probabilities shown in Fig. 3. All this information is managed by the ontology by means of associations between the objects, categories, gases, the robot and the environment itself.

For demonstration purposes we compare our approach with a deterministic case where there is no uncertainty consideration neither in the gas classification, nor in the object recognition. It must be noticed that this second approach

Table 1. Conditional probabilities of each object category given the gas class being released by the source: $P(C_j|G_k)$. As can be seen, some categories do not release any of the gas classes considered in the experiment ($P(C_j|G_k) = 0$), aspect to be exploited by our system, together with the object recognition uncertainty, to locate the gas source.

Category	Smoke_smell	Gas_smell	Rotten_food_smell
Vase			
Bin			**0.33**
Ashtray	**0.25**		
Oven	**0.25**	**0.5**	**0.33**
Heater	**0.25**	**0.5**	
Dishwasher			**0.33**
Fan			
Puf			
Incense_stick	**0.25**		
Pet_sandbox			
Pet_bed			

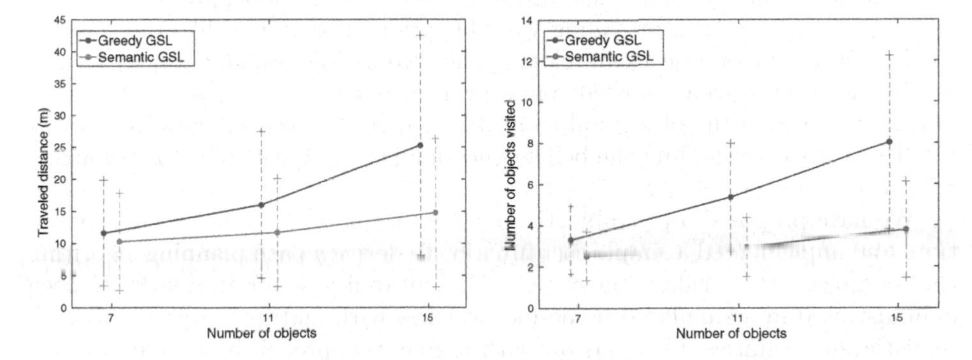

Fig. 4. Traveled distance (left) and number of objects visited (right) during the gas source localization experiments for three different set of objects. In each case, the average ± one standard deviation are plotted. As can be seen our approach improves both magnitudes substantially, specially for a high number of objects.

will fail when the gas or the objects are misclassified (*i.e.* when uncertainty is relevant), being necessary to check all the objects in the environment one by one using only the distance between the robot and the objects to optimize the search. Figure 4 shows the averaged distance traveled by the robot and the number of objects checked before locating the gas source for three setups with different number of objects: 7, 11 and 15. In order to obtain statistically representative results, for each case we run the experiment 1000 times varying (i) the initial robot pose, randomly selecting a pose from within the environment, (ii) the gas source, randomly selecting an object to be the gas source from the list of objects, and (iii) the class of the released gas, generating a gas dispersion in accordance with the types of gases the selected source can emit (see Table 1). As can be seen our approach improves both magnitudes considerably, not only reducing the total distance traveled (which is directly related to the exploration time), but also reduces the number of objects visited before locating the source. The latter is important since the *validation* of a gas source is also an expensive task in terms of time. Furthermore, it can be noticed that the improvement seems to increase with the number of considered objects, something reasonable when comparing with the greedy approach that visits all the objects one by one.

5 Discussion

This work contributes a gas source localization system for mobile robots that aims to find the object releasing a smell efficiently and coherently by exploiting semantic information. On the one hand, it is efficient in the way that selects a set of candidate objects to be the source, and checks them according to their source probability and their distance from the current robot location. On the other hand, its coherence comes from the consideration of the uncertainty coming from both the gas classification and object recognition processes, as well as semantic information providing valuable prior information, like the possible smells that a type of object can release. The system relies on an ontology to naturally encode this prior knowledge in a principled way, and also serves to codify information about the objects already detected in previous explorations of the robot workspace, including the belief concerning their classification as belonging to a certain object category.

We have proposed a probabilistic Bayesian framework to fuse such information, and implemented a simple cost function to derive a path planning algorithm that completes the localization system. The suitability of our approach has been demonstrated in a simulated home-like scenario with multiple objects and with realistic uncertainties. Comparison with a greedy approach based only on distance to the objects has been provided, suggesting that the consideration of semantics and uncertainty represents an interesting approach for tackling this complex problem.

The proposed system has significant room to explore. First of all, experiments in real environments must be carried out in order to find possible limitations and face them. We also plan to replace the simulated object recognition system by one based on Conditional Random Fields. Another certainly interesting point is how to update the beliefs about the smells of objects with the robot experience in a certain environment, which could further improve the search efficiency.

Acknowledgements. This work has been funded by the Spanish Government (project DPI2017-84827-R) and the Andalusia Government (project TEP2012-530). Both financed by European Regional Development's funds (FEDER).

References

1. Sánchez-Garrido, C., Monroy, J., Gonzalez-Jimenez, J.: A configurable smart e-nose for spatio-temporal olfactory analysis. In: IEEE Sensors, pp. 1968–1971 (2014)
2. Forsyth, D., Ponce, J.: Computer Vision: A Modern Approach. Prentice Hall, Upper Saddle River (2011)
3. Falomir, Z.: Qualitative distances and qualitative description of images for indoor scene description and recognition in robotics. AI Commun. **25**(4), 387–389 (2012)
4. Sklavounos, S., Rigas, F.: Validation of turbulence models in heavy gas dispersion over obstacles. J. Hazard. Mater. **108**(1), 9–20 (2004)
5. Marques, L., Almeida, N., De Almeida, A.: Olfactory sensory system for odour-plume tracking and localization. IEEE Sens. **1**, 418–423 (2003)
6. Lochmatter, T., Martinoli, A.: Theoretical analysis of three bio-inspired plume tracking algorithms. In: IEEE International Conference on Robotics and Automation (ICRA), pp. 2661–2668 (2009)
7. Vergassola, M., Villermaux, E., Shraiman, B.I.: 'Infotaxis' as a strategy for searching without gradients. Nature **445**(7126), 406–409 (2007)
8. Li, J.G., Meng, Q.H., Wang, Y., Zeng, M.: Odor source localization using a mobile robot in outdoor airflow environments with a particle filter algorithm. Auton. Robots **30**(3), 281–292 (2011)
9. Ishida, H., Tanaka, H., Taniguchi, H., Moriizumi, T.: Mobile robot navigation using vision and olfaction to search for a gas/odor source. Auton. Robots **20**(3), 231–238 (2006)
10. Uschold, M., Gruninger, M.: Ontologies: principles, methods and applications. Knowl. Eng. Rev. **11**, 93–136 (1996)
11. Monroy, J., Hernandez-Bennetts, V., Fan, H., Lilienthal, A., Gonzalez-Jimenez, J.: GADEN: A 3D gas dispersion simulator for mobile robot olfaction in realistic environments. MDPI Sens. **17**(7:1479), 1–16 (2017)
12. Pearce, T.C., Schiffman, S.S., Nagle, H.T., Gardner, J.W.: Handbook of Machine Olfaction: Electronic Nose Technology. Wiley, Weinheim (2006)
13. Gongora, A., Monroy, J., Gonzalez-Jimenez, J.: Gas source localization strategies for teleoperated mobile robots: an experimental analysis. In: European Conference on Mobile Robotics (ECMR) (2017)
14. Loutfi, A., Coradeschi, S., Karlsson, L., Broxvall, M.: Putting olfaction into action: using an electronic nose on a multi-sensing mobile robot. In: International Conference on Intelligent Robots and Systems (IROS), vol. 1, pp. 337–342 (2004)
15. Neumann, P.P., Hernandez Bennetts, V., Lilienthal, A.J., Bartholmai, M., Schiller, J.H.: Gas source localization with a micro-drone using bio-inspired and particle filter-based algorithms. Adv. Robot. **27**(9), 725–738 (2013)
16. Monroy, J., Blanco, J.L., Gonzalez-Jimenez, J.: Time-variant gas distribution mapping with obstacle information. Auton. Robots **40**(1), 1–16 (2016)
17. Schleif, F.M., Hammer, B., Monroy, J., Gonzalez-Jimenez, J., Blanco, J.L., Biehl, M., Petkov, N.: Odor recognition in robotics applications by discriminative time-series modeling. Pattern Anal. Appl. **19**(1), 207–220 (2016)
18. Bennetts, V.H., Schaffernicht, E., Sese, V.P., Lilienthal, A.J., Trincavelli, M.: A novel approach for gas discrimination in natural environments with open sampling systems. In: IEEE Sensors, pp. 2046–2049 (2014)
19. Ruiz-Sarmiento, J.R., Galindo, C., Gonzalez-Jimenez, J.: A survey on learning approaches for undirected graphical models: application to scene object recognition. Int. J. Approx. Reason. **83**, 434–451 (2017)

20. Ruiz-Sarmiento, J.R., Galindo, C., Gonzalez-Jimenez, J.: Building multiversal semantic maps for mobile robot operation. Knowl.-Based Syst. **119**, 257–272 (2017)
21. Ruiz-Sarmiento, J.R., Guenther, M., Galindo, C., Gonzalez-Jimenez, J., Hertzberg, J.: Online context-based object recognition for mobile robots. In: 17th International Conference on Autonomous Robot Systems and Competition (ICARSC), April 2017

Operational Measurement of Data Quality

Antoon Bronselaer(✉), Joachim Nielandt, Toon Boeckling, and Guy De Tré

Department of Telecommunication and Information Processing, Ghent University, Sint-Pietersnieuwstraat 41, 9000 Gent, Belgium
{antoon.bronselaer,joachim.nielandt,toon.boeckling,guy.detre}@ugent.be

Abstract. In this paper, an alternative view on measurement of data quality is proposed. Current procedures for data quality measurement provide information about the extent to which data misrepresent reality. These procedures are descriptive in the sense that they provide us numerical information about the state of data. In many cases, this information is not sufficient to know whether data is fit for the task it was meant for. To bridge that gap, we propose a procedure that measures the operational characteristics of data. In this paper, we devise such a procedure by measuring the cost it takes to make data fit for use. We lay out the basics of this procedure and then provide more details on two essential components: tasks and transformation functions.

Keywords: Data quality · Measurement · Cost

1 Introduction

Since the proposal of the relational model for databases in 1970, many new areas of research have been developed. One of them is the field of data quality, that emerged around 1995 [25] as a direct consequence of the successful commercialisation of RDBMSs. It was recognized that, due to the versatile possibilities of these storage systems, the term "data quality" had surpassed the simple concept of accuracy. From the early stage, research of data quality has been focusing intensively on the "multi-dimensionality" [23,25,26] of the concept "quality", leading to dimensions like correctness, consistency, currency, reliability and many more. Unfortunately, this focus has not aided much in the development of formally well-defined measurement procedures for data quality. It has been a historical misconception that dimensions necessarily treat different aspects of data quality. For example, when we closely consider the dimensions previously listed, it becomes apparent that they all try to formulate an answer to the same question: "Is data correct or not?". The true difference between these dimensions is that they approach this question in a different manner. With consistency, we consider a set of rules and search for violations of these rules in order to conclude that some data are wrong. With currency and reliability, there is a factor of uncertainty involved with answering the question. If data gets older, the probability that data are still correct may drop whereas with reliability, the uncertainty

J. Medina et al. (Eds.): IPMU 2018, CCIS 855, pp. 517–528, 2018.
https://doi.org/10.1007/978-3-319-91479-4_43

stems from the provider of the data, who might not be trustworthy. Because of the blurred perception of the true nature of dimensions, well understood insights in measurement of data quality only emerged very recently. In this paper, we continue this trend by investigating a formal approach where quality of data is assessed in terms of the cost required to make data fit for use.

The remainder of this paper is structured as follows. In Sect. 2, we provide an overview of the state-of-the-art concerning procedures for measurement of data quality. We thereby focus on procedures with an ordinal nature and show that these can be jointly represented in one framework. In Sect. 3, we introduce a novel procedure for measurement of data quality that fundamentally differs from the existing ones. Whereas existing procedures provide descriptive information about data in light of the real world it describes, the new procedure aims at measuring operational characteristics of data in light of some task(s) that must be completed. We propose to express this operational characteristics in terms of the cost required to make data fit for use. We point out that this cost does not align with utility as it is assumed up front that we want to use data for the task, but cannot do so because there is some quality degradation that holds us back. We detail both the nature of tasks and transformation functions that can make data fit for use. Finally, in Sect. 4, we summarize the most important contributions of this paper.

2 Related Work

Before we introduce a novel approach for measurement of data quality, we first present an overview of current methodologies handling this problem. In this overview, we limit ourselves to those methods that utilize a clear measurement procedure in terms of Representational Measurement Theory [19]. Methods that remain vague as to how numbers are assigned to data items in order to express quality, are left out of the comparison here.

In general, data are created when some observations are made concerning objects or events in the real world, which are registered in a data storage system. As such, the creation of data involves two procedures: an *observation* procedure and a *registration* procedure. In general, it is reasonable to assume that none of these procedures are perfect, which may lead to the situation where the outcome of the registration (the data) is not a true representation of reality. The quality of data reflects the extent to which this is the case.

When trying to devise a measurement procedure for data quality, the first decision to be taken is about the kind of information that must be obtained regarding the quality of data. More specifically, it must be decided how the extent to which data reflects reality must be expressed. Roughly speaking, there are two options. The first option is to accept Boolean information, meaning that we want to know whether data reflects reality, or not. The second option is to require more granular information, which means that we want to obtain some notion on the extent of deviation from reality.

Once this choice is made, we must decide which kind of measurement we wish to perform to obtain information. The type of measurement usually aligns

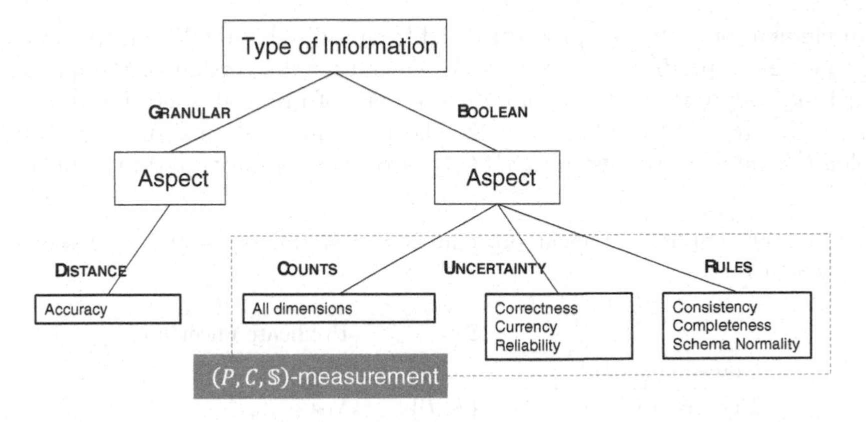

Fig. 1. Different procedures for measurement of Data Quality

with a certain aspect of quality that we want to account for. In case we want granular information, the sole aspect that is considered in literature is a notion of distance between data and reality. The main dimension for which the strategy applies, is accuracy [13] (Fig. 1).

When considering Boolean information, literature is more rich when it comes to its various aspects. The simplest methods adopt a counting procedure where artefacts are counted across a set of data, yielding an absolute scale for data quality. The counting procedure has been somehow formalized in [22], but has been applied in numerous cases [7,8,11,16,21,23]. This procedure is the most simple of measurement procedures and can be applied to virtually any dimension that relies on an observable characteristic of the data. More recently, authors have been considering the case where some characteristics are not easily observed. In that scenario, uncertainty about these characteristics provides a way to express data quality [8,17,18]. Such models are very suitable for measuring, for example, reliability (uncertainty by lack of trust) or currency (uncertainty by age). Usually, these procedures yield a ratio scale, although some uncertainty theories (like possibility theory) might lead to an ordinal scale. Finally, there are also procedures that rely on a set of rules to judge upon the quality of data and these procedures induce ordinal information. The rule-based approach finds its origin in the seminal Fellegi-Holt model, in which two categories of *edit rules* are distinguished [9]. Since their model was published, many forms of edit rules have been proposed. Examples include, among others, functional dependencies [15], conditional functional dependencies [2], denial constraints [5], inclusion dependencies [1] and regular expressions [4].

Interestingly, it has been shown recently that the mentioned aspects of Boolean information can be represented in a single framework [3]. In this framework, one considers a set of predicates P and a capacity function C. The latter is a function that maps the power set of P onto a totally ordered, finite set of quality levels $\mathbb{S}$. The quality of a data item d is then the capacity that corresponds to the set of predicates for which d passes. Quality of data is therefore expressed as an element of $\mathbb{S}$ and the procedure yields an ordinal scale. When the outcome of one or more predicates is uncertain, we can not be certain of the quality of d. In that case, uncertainty about the outcome of predicates can be *propagated* and we obtain a *distribution* over $\mathbb{S}$. The procedures for measurement with a Boolean nature can be cast to a $(P, \mathcal{C}, \mathbb{S})$-structure as illustrated in Table 1.

Table 1. Different measurement procedures with a Boolean nature represented as $(P, \mathcal{C}, \mathbb{S})$-structure.

	P	$\mathbb{S}$	Predicate uncertainty
Counting	$\{p\}$	$\{0, 1\}$	No
Uncertainty	$\{p\}$	$\{F, T\}$	Yes
Edit rules	$\{p_1, ..., p_n\}$	$\{s_1, ..., s_k\}$	No

The fact that all procedures with a Boolean nature can be represented in a common framework, allows us to reason about them in more general terms. More specifically, in the general case, data quality is measured by establishing a suitable structure $(P, \mathcal{C}, \mathbb{S})$ and information comes from either measured differentiations in quality in terms of levels from $\mathbb{S}$, or from an uncertainty model. Essentially, all procedures that can be derived from the $(P, \mathcal{C}, \mathbb{S})$-structure are always *descriptive* with respect to the data in the sense that they provide us numerical information about the correspondence of data and reality. In this paper, we want to introduce a novel way of looking at quality in the sense that we want to measure the extent to which a certain quality loss is stopping us from doing something. The notion of "doing something" is understood here as the completion of a well-defined task. We propose to assess quality by not only looking at the correspondence between data and reality, but also at the inability to use data for this task.

3 Measurement of Data Quality as Inverted Cost

In this section, we step away from descriptive measurements and move in the direction of *operational* measurements. We start from the notion of "fitness for use" installed by Wang and Strong in [26], but we propose a fundamentally different way of measuring "fitness". Traditional techniques measure the quality of data by looking for data artefacts using some knowledge base (i.e., a set of rules or an uncertainty model). Our main argument is that a quantitative assessment of such violations is not always very informative in an operational setting. Imagine a scenario in which there are many observed artefacts in the data, but we have a way (automatically or manually) to efficiently repair these artefacts. In this scenario, the observed issues are not stopping us from using the data and this justifies the question whether we should care about them. Following this line of thought, we propose to assess the quality of data by measuring the "hardness" of preparing the data for completion of a task.

3.1 The Basic Framework

In order to provide a generic theory, we do not initiate from a specific data model, but rather reason about data in simple terms of a set of data items $D \subseteq \mathcal{D}$, where $\mathcal{D}$ is the domain of interest. In order to emphasize the fact that data resources are usually limited, we assume D to be finite. Let us now focus on the problem of measuring quality of elements in D. Rather than measuring descriptive aspects of $d \in D$, we want to measure the extent to which elements from D can serve their purpose. Therefore, we consider a set of tasks $\mathcal{T} = \{t_1, ..., t_n\}$ in such a way that each task $t \in \mathcal{T}$ relies on elements from D. In this setting, we can define a degradation of quality as any *inability* to complete some $t \in \mathcal{T}$. In other words, $d \in D$ is of low quality if, for some reason, we cannot use it to complete some $t \in \mathcal{T}$. In order to present this line of thought more formally, we define a *decision* function and a *transformation* function below.

Definition 1 (Decision function). *Let $\mathcal{D}$ be a universe of discourse for data and $\mathcal{T}$ a set of tasks. A decision function is defined by:*

$$\sigma : \mathcal{D} \times \mathcal{T} \rightarrow \mathbb{B} \tag{1}$$

where $\mathbb{B} = \{T, F\}$ is the set of Boolean truth values. For any $d \in D$ and for any $t \in \mathcal{T}$, $\sigma(d,t) = T$ indicates that task t can be executed with d as input and $\sigma(d,t) = F$ indicates that this is not the case.

Definition 2 (Transformation function). *Let $\mathcal{D}$ be a universe of discourse for data and $\mathcal{T}$ a set of tasks. A transformation function is defined by:*

$$\theta : \mathcal{D} \rightarrow \mathcal{D} \tag{2}$$

In simple terms, a decision function determines whether data $d \in D$ can be used to complete a task t. Note that, for the sake of simplicity, we assume that each task relies on one data item, but this can be generalized easily. If there is no $d \in D$ such that $\sigma(d,t) = T$, we can try to make elements of D fit for use. A transformation function serves to alter $d \in D$ in such a way that $\theta(d)$ is *better fit* for some task $t \in \mathcal{T}$ than d. Note that this does not imply necessarily that $\sigma(\theta(d),t) = T$. The improvement comes from a reduction in cost, as we explain in the following. More precisely, each transformation is assigned a positive cost and therefore comes with a cost function which is defined by:

$$\mathcal{C}_\theta : \mathcal{D} \rightarrow \mathbb{R}_{\geq 0}. \tag{3}$$

The identity transformation is denoted as $\theta_{\mathbb{1}}$ and satisfies $\theta_{\mathbb{1}}(d) = d$, which implies that $\mathcal{C}_{\theta_{\mathbb{1}}}(d) = 0$. With a set of transformations $\Theta = \{\theta_1, ..., \theta_k\}$ at our disposal, the cost of utilizing d for task $t \in \mathcal{T}$ is given by:

$$C(d,t) = \min_{\{\theta \in \Theta | \sigma(\theta(d),t)\}} \mathcal{C}_\theta(d). \tag{4}$$

The above equation allows us to assign a cost to each data item in light of task t for which it ought to be used and with respect to some transformations. If data can not be transformed to complete $t \in \mathcal{T}$, then by convention we say that:

$$C(d, t) = +\infty. \tag{5}$$

A $[0, 1]$-quality indicator, if desired for some purpose like visualisation, can be derived from these costs by means of a quality function.

Definition 3 (Quality function). *Let $D \subseteq \mathcal{D}$ be a set of available data, $\mathcal{T}$ a set of tasks and Θ a set of transformations such that $\theta_{\mathbb{1}} \in \Theta$. A quality function for a task $t \in \mathcal{T}$ is defined by:*

$$Q_t : \mathcal{D} \to [0, 1] : d \mapsto \mathcal{I}(C(d, t)) \tag{6}$$

where $\mathcal{I} : \mathbb{R}_{\geq 0} \to [0, 1]$ is decreasing and satisfies $\mathcal{I}(0) = 1$.

In words, a quality function maps a cost to a number in the unit interval in a decreasing way. Some candidate functions for $\mathcal{I}$ are $\mathcal{I}_1(x) = 10^{-x \cdot \alpha}$, $\mathcal{I}_2(x) = \exp(-x \cdot \alpha)$ and $\mathcal{I}_3(x) = \frac{1}{1 + x \cdot \alpha}$. Hereby, α is a positive real number that plays the role of a shape parameter. Note that $\mathcal{I}_1$ and $\mathcal{I}_2$ are basically variations of the same family. It is easily verified that all of these functions are continuous, decreasing and that they satisfy the boundary condition. We point out a few properties of these functions. For a fixed value of α, we have that:

$$\forall x \in \mathbb{R}_{\geq 0} : \mathcal{I}_1(x) \leq \mathcal{I}_2(x) \leq \mathcal{I}_3(x). \tag{7}$$

For $\alpha = 0$, we have that

$$\forall x \in \mathbb{R}_{\geq 0} : \mathcal{I}_1(x) = \mathcal{I}_2(x) = \mathcal{I}_3(x) = 1 \tag{8}$$

which means that, for the considered quality functions, $\alpha = 0$ models a situation in which the cost to transform data is never perceived as degradation of quality.

In the following, additivity of costs will play an important role, which is why it is interesting that:

$$\mathcal{I}_1(x + y) = \mathcal{I}_1(x) \cdot \mathcal{I}_1(y) \tag{9}$$

from which it follows that:

$$\log(\mathcal{I}_1(x + y)) = \log(\mathcal{I}_1(x)) + \log(\mathcal{I}_1(y)). \tag{10}$$

Additivity of costs is thus transferred to log additivity in terms of quality levels. A similar property holds for $\mathcal{I}_2$ if log is replaced by ln. In the presented framework, we measure the quality of data inverse to the cost we must invest to make those data usable. Hereby, there are two essential aspects we further characterize in the following: (i) the set of tasks and (ii) the set of transformations.

3.2 Characterization of Tasks

By considering a set of tasks, measurement of data quality becomes a procedure that considers data in a specific setting. Therefore, we call this procedure *operational.* The measurement procedures described earlier do not have this property and are therefore called *descriptive.* In [10], the distinction between internal procedures (without context) and external procedures (with context) is made in a similar way. Because of the operational nature, it is tempting to suspect that measurement of quality as an inverse of cost, is somehow equivalent to utility. However, this is not the case. Utility theory deals with preferences and subjectivity and its role in data quality measurement has been treated elsewhere [6–8]. In our approach, the inability to use some $d \in D$ for a task $t \in \mathcal{T}$ is caused only by the fact that d somehow misrepresents reality. Thus, there is a degradation in representation of reality that stops us from executing t.

The assumption that $d \in D$ is required for completion of t, is essential in our reasoning. If we would consider graded completions of t, the problem of devising a measurement procedure for data quality is shifted to a measurement procedure for completion of t. We avoid this problem by considering a Boolean-valued decision function (Definition 1). Note that any system where completion of t is measured on an ordinal scale, can be transferred into the proposed system by decomposing tasks into as many subtasks as there are levels of completion. To illustrate this, suppose there is a task t for which there are three intuitive grades of completion $c_1 \leq c_2 \leq c_3$, then we can consider three subtasks t_1, t_2 and t_3. For each of these subtasks, we can consider a Boolean-valued decision function and the "complex" task is modelled by the set $\{t_1, t_2, t_3\} \subseteq \mathcal{T}$.

In order to get a notion of the applicability of our framework, we present two concrete scenarios in which tasks that rely on data and are hampered by quality degradations: *query resolution* and *predictive modelling.*

Query Resolution. In the first scenario, we consider D to be a database and $\mathcal{T}$ a set of questions that need to be answered. In order to answer a question, a user has to formulate a query q in some query language (e.g., SQL, SPARQL, Cypher, XPath or regular expressions as a degenerate case). Depending on the state of the database, the construction of q may vary from easy to very difficult. There may be different reasons why construction of q is difficult, but as mentioned before, we focus here on misrepresentation of reality. Those misrepresentations may take different forms. One apparent reason is that data are not properly structured, making it difficult to fetch information from D. For example, if a column in a relational database contains many cluttered data, we require extensive use of advanced functions in the SELECT clause in order to complete the query. Another reason might be that certain types of databases are simply not fit for certain types of queries. This is especially the case in the era of NoSQL databases. Document stores are for example very efficient at retrieving all information about a certain object in a single query, but perform bad in queries that involve many relationships. This latter kind of queries is handled very efficiently by graph databases.

Predictive Modelling. In the second scenario, we consider D to be a collection of data and t to be the construction of a predictive model with D as a training set such that there is a certain requirement on the predictive accuracy of the model. As mentioned in [20], application of data mining techniques typically requires a series of additive costs to be invested in preprocessing before the actual training method can be applied. In the context of predictive modelling, the foremost studied data quality problem is completeness, which is usually solved by imputation. However, noise and inconsistencies often pose a problem as well [14]. The scenario of predictive modelling is particularly interesting because, usually, learning methods will have a certain *robustness* against data quality degradations. So it could be that, although we know that D is not a perfect representation of reality, we accept it anyway to complete our task. Consequently, the cost is zero and quality is perceived as perfect. This observation illustrates the operational nature of the approach.

It should be clear from the both scenarios sketched above that the proposal for operational measurements for data quality has no intention of rendering other approaches irrelevant. On the contrary, the cost that we are willing to pay to transform the data will in many cases depend on the outcome of the descriptive measurements listed above. The proposal of cost calculation and operational measurements should be regarded as an intent to better balance the costs and the gains of improving data.

3.3 Characterization of Transformations

In this section, we focus on the set of transformations and their cost functions. We consider again a set of transformations $\Theta = \{\theta_1, ..., \theta_k\}$ and it is assumed that $\theta_{\mathbb{1}} \in \Theta$. We first show that, for a set D and some set Θ, we can derive other data and the entire set of producible data items gives rise to a *partial order*. To see this, we first note that $\theta_{\mathbb{1}}$ has cost zero and is always present. Production is thus reflexive. Second, we argued before that each transformation modifies data in such a way that it becomes *more suitable* for completion of some task. A justification for this is the economic observation that each transformation comes with a non-negative cost. In light of Eq. (4), a transformation that degrades d for all tasks, will never contribute to the minimal cost. Such transformations are therefore irrelevant in our cost calculations and can therefore be ignored. From that however, it follows that:

$$\theta \in \Theta \Rightarrow \theta^{-1} \notin \Theta. \tag{11}$$

In other words, if θ improves quality of d, then the inverse transformation, provided that it exists, will *decrease* quality and can therefore be left out Θ. It follows that production is *anti-symmetric*. Finally, it is often infeasible and even undesirable that Θ contains all necessary transformations directly. Rather, we want to consider compositions of transformations in Θ.

Definition 4 (Composed transformation function). *For a set of transformation functions $\Theta = \{\theta_1, ..., \theta_k\}$, a composed transformation θ^* is a transformation function for which we have:*

$$\theta^* = \theta_{(1)} \circ ... \circ \theta_{(m)} \tag{12}$$

where each $\theta_{(i)} \in \Theta$ and where $\circ$ is the usual functional composition operator.

We can see that any $\theta_i \in \Theta$ can be composed from Θ and that $\theta_{\mathbb{1}}$ is a neutral element of the composition operator $\circ$ because $\theta \circ \theta_{\mathbb{1}} = \theta_{\mathbb{1}} \circ \theta = \theta$. The cost of a composed transformation $\theta^* = \theta_{(1)} \circ ... \circ \theta_{(m)}$ is calculated by:

$$\forall d \in D : \mathcal{C}_{\theta^*}(d) = \sum_{i=0}^{m-1} \mathcal{C}_{\theta_{(m-i)}} (\theta_i^* (d)) \tag{13}$$

where θ_i^* is a composed transformation $\theta_{(m-i+1)} \circ ... \circ \theta_{(m)}$ if $i > 0$ and $\theta_{\mathbb{1}}$ if $i = 0$. From the above, we find that for any $D \subset \mathcal{D}$ and for any Θ, we can consider an acyclic, directed graph G where the set of nodes $N(G)$ satisfies $D \subseteq N(G) \subset \mathcal{D}$, the set of edges $E(G)$ satisfies $E(G) \subseteq \mathcal{D} \times \mathcal{D}$ and

$$\forall (d, d') \in E(G) : d \neq d' \wedge \exists \theta \in \Theta : \theta(d) = d' \tag{14}$$

In this graph representation, a data item $d' \in \mathcal{D}$ satisfies $d \in N(G)$ if there exists some composed transformation that transforms some $d \in D$ into d'. The set of all data we can produce corresponds to those nodes that are reachable from any $d \in D$ and therefore, the total set of data items that can be produced to complete tasks $\mathcal{T}$, takes the structure of a partial order. This is illustrated in Fig. 2, where $D = \{d_1\}$ and $\Theta = \{\theta_1, ..., \theta_4\}$.

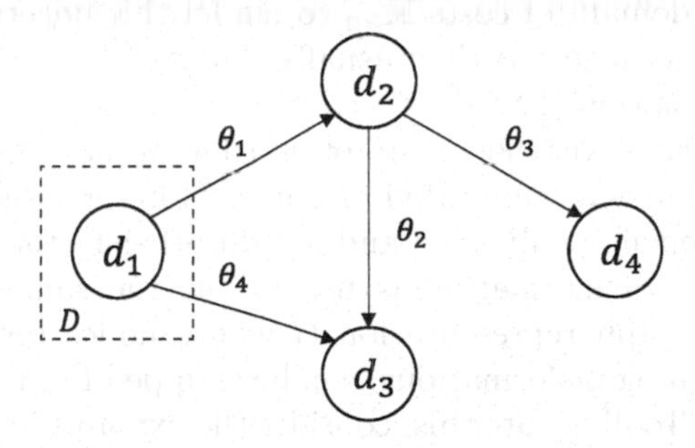

Fig. 2. Example partial order of data items that can be produced from $D = \{d_1\}$ via transformations $\{\theta_1, ..., \theta_4\}$.

From this graph structure, it can be deduced that, in general, we will be able to complete more tasks if $N(G)$ is larger. To obtain this, we could consider more data in D, but this is usually difficult. Instead, we can also make $N(G)$ larger by considering more *atomic* transformations. Transformations are atomic if they can not be logically decomposed into other transformations. If we consider Θ to be as atomic as possible, we can always construct composed transformations and we can produce more data items. To conclude this section, we list some typical characteristics of transformations. A deeper study of these characteristics lies outside the scope of this paper and is deemed as future work.

- **Automation.** We can distinguish between two important categories of transformations: *automated* transformations and *manual* transformations. For automated transformations, we assume that there is some algorithm we can apply to D to improve its quality. Examples of these transformations include standardization, resolution of inconsistencies, repair algorithms, imputation techniques and fusion algorithms. With automated transformations, the main factor to calculate cost is the computational complexity of the algorithm. Automated transformations are always preferable, but not always applicable. By definition, a degradation of quality is a misrepresentation of reality. Hence, some quality improvement steps require knowledge of reality, which usually implies manual supervision. This gives rise to manual transformations, which tend to come with a higher cost, but are sometimes unavoidable.
- **Uncertainty of costs.** So far, we have assumed that the cost of a transformation is deterministic in terms of d. The fact that cost depends on d, is fairly straightforward. For example, in the case of automated transformations, the computational complexity usually depends on d in some way (i.e., linear, polynomial, exponential...). However, if we think of manual transformations, it might not be feasible to assume that the cost of a transformation is fixed for a given d. Experience and work load are two examples that can influence the cost of manual transformations. A more reasonable assumption would be to assume that the cost of a transformation is inherently uncertain. Instead of a deterministic cost, we consider a belief function (e.g., a probability distribution) over the domain of costs $\mathbb{R}_{\geq 0}$ to model this uncertainty. In terms of the graph G, the total cost is then usually calculated in terms of the *expected* cost of a transformation [12, 24].
- **Limited resources.** Another assumption that we have made so far, is that each transformation is an unlimited resource. More precisely, the cost of each transformation depends only on d and not on other factors such as the number of times the transformation is utilized at the same time. Under this assumption, the graph representation G where nodes represent data items and edges represent transformations, can be stripped from all paths that have sub-optimal cost. To illustrate this, consider the example in Fig. 2 and suppose that the cost of applying θ_4 is higher than the sum of the cost of θ_1 and the cost of θ_2. If transformations are unlimited resources, there is no reason why we should ever consider θ_4. However, for manual transformations, the assumption of unlimited resources is infeasible and a more complex cost model that accounts for other factors than d, is required.

4 Conclusion

In this paper, we proposed a fundamentally novel procedure for measurement of data quality. Whereas current procedures aim at measuring aspects of data with respect to the real world it describes, the new procedure measures the cost it takes to make data fit for use for some task. We have introduced the basics of this framework and detailed on the characteristics of tasks and transformation functions.

References

1. Bohannon, P., Fan, W., Flaster, M., Rastoqi, R.: A cost-based model and effective heuristic for repairing constraints by value modification. In: Proceedings of the 2005 ACM SIGMOD International Conference on Management of Data, pp. 143–154 (2005)
2. Bohannon, P., Fan, W., Geerts, F., Jia, X., Kementsietsidis, A.: Conditional functional dependencies for data cleaning. In: Proceedings of the 23rd International Conference on Data Engineering, pp. 746–755 (2007)
3. Bronselaer, A., De Mol, R., De Tré, G.: A measure-theoretic foundation for data quality. IEEE Trans. Fuzzy Syst. **26**(2), 627–639 (2018)
4. Bronselaer, A., Nielandt, J., De Mol, R., De Tré, G.: Ordinal assessment of data consistency based on regular expressions. In: Information Processing and Management of Uncertainty in Knowledge-Based Systems, pp. 317–328 (2016)
5. Chu, X., Ilyas, I., Papotti, P.: Discovering denial constraints. In: Proceedings of the VLDB Endowment, pp. 1498–1509 (2013)
6. Even, A., Shankaranarayanan, G.: Value-driven data quality assessment. In: Proceedings of the International Conference on Information Quality, pp. 265–279 (2005)
7. Even, A., Shankaranarayanan, G.: Understanding impartial versus utility-driven quality assessment in large data-sets. In: Proceedings of the International Conference on Information Quality, pp. 265–279 (2007)
8. Even, A., Shankaranarayanan, G.: Utility-driven assessment of data quality. DATA BASE Adv. Inf. Syst. **38**(2), 75–93 (2007)
9. Fellegi, I., Holt, D.: A systematic approach to automatic edit and imputation. J. Am. Stat. Assoc. **71**(353), 17–35 (1976)
10. Fenton, N.E., Pfleeger, S.L.: Software Metrics: A Rigorous and Practical Approach, 2nd edn. Thomson Publishing, Stamford (1996)
11. Fisher, C.W., Lauria, E.J.M., Matheus, C.C.: An accuracy metric: Percentages, randomness, and probabilities. J. Data Inf. Qual. **1**(3), 16:1–16:21 (2009)
12. Frank, H.: Shortest paths in probabilistic graphs. Oper. Res. **17**(4), 583–599 (1969)
13. Haegemans, T., Snoeck, M., Lemahieu, W.: Towards a precise definition of data accuracy and a justification for its measure. In: Proceedings of the International Conference on Information Quality (ICIQ), pp. 16:1–16:13 (2016)
14. Han, J., Kamber, M., Pei, J.: Data Mining: Concepts and Techniques. Elsevier Science and Technology (2011)
15. Heath, I.: Unacceptable file operations in a relational data base. In: Proceedings of the 1971 ACM SIGFIDET (Now SIGMOD) Workshop on Data Description, Access and Control, pp. 19–33 (1971)

16. Heinrich, B., Kaiser, M., Klier, M.: Does the EU insurance mediation directive help to improve data quality? a metric-based analysis. In: European Conference on Information Systems, pp. 1871–1882 (2008)
17. Heinrich, B., Klier, M.: Metric-based data quality assessment - developing and evaluation a probability-based currency metric. Decis. Support Syst. **72**, 82–96 (2015)
18. Heinrich, B., Klier, M., Kaiser, M.: A procedure to develop metrics for currency and its application in CRM. ACM J. Data Inf. Qual. **1**(1), 5:1–5:28 (2009)
19. Krantz, D., Luce, D., Suppes, P., Tversky, A.: Foundations of Measurement: Additive and Polynomial representations, vol. I. Academic Press, Cambridge (1971)
20. Pipino, L., Kopcso, D.P.: Data mining, dirty data, and costs. In: Ninth International Conference on Information Quality (ICIQ 2004), 5–7 November, pp. 164–169 (2004)
21. Pipino, L., Lee, Y., Wang, R.: Data quality assessment. Commun. ACM **45**(4), 211–218 (2002)
22. Pipino, L.L., Wang, R.Y., Kopcso, D., Rybolt, W.: Developing measurement scales for data-quality dimensions. In: Wang, R.Y., Pierce, E.M., Madnick, S.E., Fisher, C.W. (eds.) Information Quality, chap. 3, pp. 37–51. M.E. Sharpe (2005)
23. Redman, T.: Data Quality for the Information Age. Artech-House, Massachusetts (1996)
24. Sigal, E., Pritsker, A., Solberg, J.: The stochastic shortest route problem. Oper. Res. **28**(5), 1122–1129 (1969)
25. Wang, R., Storey, V., Firth, C.: A framework for analysis of data quality research. IEEE Trans. Knowl. Data Eng. **7**(4), 623–640 (1995)
26. Wang, R., Strong, D.: Beyond accuracy: what data quality means to data consumers. J. Manag. Inf. Syst. **12**(4), 5–34 (1996)

Randomness of Data Quality Artifacts

Toon Boeckling(✉), Antoon Bronselaer(✉), and Guy De Tré(✉)

Department of Telecommunications and Information Processing, Ghent University,
St.-Pietersnieuwstraat 41, 9000 Ghent, Belgium
{toon.boeckling,antoon.bronselaer,guy.detre}@UGent.be

Abstract. Quality of data is often measured by counting artifacts. While this procedure is very simple and applicable to many different types of artifacts like errors, inconsistencies and missing values, counts do not differentiate between different distributions of data artifacts. A possible solution is to add a randomness measure to indicate how randomly data artifacts are distributed. It has been proposed to calculate randomness by means of the Lempel-Ziv complexity algorithm, this approach comes with some demerits. Most importantly, the Lempel-Ziv approach assumes that there is some implicit order among data objects and the measured randomness depends on this order. To overcome this problem, a new method is proposed which measures randomness proportionate to the average amount of bits needed to compress the bit matrix matching the artifacts in a database relation by using unary coding. It is shown that this method has several interesting properties that align the proposed measurement procedure with the intuitive perception of randomness.

Keywords: Data quality · Randomness · Unary codes

1 Introduction

With the enormous increase in data storage capabilities and the use of data in several applications, assessment of the quality of data is increasingly gaining importance. Although it is easy to have an intuitive notion of what *data quality* is, many definitions of the concept exist [1,20,22]. Because of this diverse set of definitions, it is not easy to propose one single measurement procedure [1,2,10,12,21]. However, a general trend that can be seen among the different approaches is that many of them rely on simple counting procedures of basic artifacts. A basic artifact is hereby defined as a low-level degradation of data. The advantages of counting procedures are that they are simple, can be applied in virtually any case and can easily be aggregated. In many approaches, aggregation is implemented by averaging the counts into an error rate [15].

In this paper, basic artifacts are introduced at the level of attribute values in a database relation and the error rate is defined as the relative number of attribute values where an error occurs. This rate is extended by a measure of

J. Medina et al. (Eds.): IPMU 2018, CCIS 855, pp. 529–540, 2018.
https://doi.org/10.1007/978-3-319-91479-4_44

randomness to indicate how randomly data artifacts are distributed, which has an impact on the amount of work to repair the relation. A measure of randomness is already introduced in [8] where it is calculated by using the Lempel-Ziv (LZ) complexity algorithm. Although the concept of adding a randomness measure is useful, calculating a randomness measure with the LZ-complexity algorithm gives rise to some problems. The goal of this paper is first, to illustrate the disadvantages of the LZ approach and after that, to develop a new method to measure randomness by means of binary code compression.

The remainder of the paper is structured as follows. In Sect. 2, it is pointed out that the level of randomness in measured artifacts plays an important role in root cause analysis. Therefore, contributions in the field of data quality measurement and methodologies to calculate randomness as a complexity measure are listed. In Sect. 3, the approach given in [8] is explained in detail and some important drawbacks of the approach are described. After that, in Sect. 4, a novel approach is proposed based on binary code compression and useful properties of the approach are proven. In Sect. 5 some options for future work are listed and finally, the concluding remarks are given in Sect. 6.

2 Related Work

Over the past decades, many measurement procedures in the field of data quality are proposed. Besides that, many attempts have been made to formalize the subjective concept of randomness. Before extending data quality measurement with a randomness value, most important contributions in the field of data quality assessment and randomness are introduced in the following.

Among all definitions of data quality, many of them deal with the multi-dimensional quality model introduced by Wang et al. [27] and Redman [22]. This multi-dimensional view led to the introduction of a broad range of measurement procedures [1,5,6,10,20]. When it comes to fundamental procedures in terms of Representational Measurement Theory [14], most procedures can be expressed in terms of the measurement procedure presented in [2]. In this work, Bronselaer et al. state formally that the highest possible quality can be described by means of a set of predicates. By evaluating and combining those predicates, a given capacity function maps the results to an ordinal scale to indicate a measure of quality.

One of the reasons that measurements of quality are useful, is because they can be used for application in root cause analysis [26,28]. An efficient increase in quality can be guaranteed by eliminating the root causes of artifacts. Originated to increase the quality of software [17], Haegemans et al. recently introduced the idea of root cause analysis in data analysis. In [11], a theoretical framework is proposed to acquire the causes of errors in manually acquired data. Besides that, Haegemans et al. developed a new approach to visually represent objective data quality measurements [9]. The motivation behind this approach is that aggregation of data quality measurements is not enough for root cause analysis and that one needs information about many other aspects to assess data quality. One of the desired properties is that the information should indicate if the distribution of artifacts occurs randomly or not, because this could indicate a common causality.

In order to measure randomness of a sequence, the Minimum Description Length (MDL) principle due to Risannen [23] is considered, which states that the best hypothesis on data is the one with minimal description size. This principle overcomes the non-computability of the Kolmogorov complexity [13,18] by considering codes to describe data. At the same time, it was adopted by Solomonoff [25] and Chaitin [3] in their definitions of randomness and by Lempel et al. in the derivation of the normalized Lempel-Ziv complexity [16]. The motivation to mentally encode a sequence to judge the extent of its randomness is given in [7]. Their judgment is substantiated by the fact that strings of maximum complexity must be patternless and therefore incompressible, otherwise a pattern could have been used to reduce the description length [18]. This insight is adopted in the remainder of the paper to develop a novel measure of randomness.

3 Randomness of Data Artifacts

3.1 Lempel-Ziv Randomness

A simplified form of the procedure introduced in [2] comes down to *counting* items that pass a single test. Because of the simplicity of the procedure, it can be applied to any dimension of quality. However, it has been pointed out in [8] that this measurement procedure does not differentiate between different distributions of artifacts[1] of data. The previous claim is advocated by the observation that data sets with the same error rate may have different error distributions, which, on its turn, has an impact on the complexity of improving the quality. An example is given in Fig. 1 where data artifacts are presented by grey squares and each of the three relations has an error rate of 0.2. In cases *a* and *b*, it takes little effort to fix the artifacts. One may argue that a systematic failure in the procedure that generates data of the fourth column is the root cause of artifacts in case *b*. For these reasons, randomness of artifacts deserves a closer inspection.

The solution given in [8] is to include, besides the error rate, a measure for the randomness of the artifact distribution in a database relation. This measurement is calculated as follows: first, the cells where an artifact occurs are identified and a 0 is assigned to every errorless cell and a 1 is assigned to every cell where at least one error occurs. After that, all the rows of the relation are concatenated into one bit string and finally, the normalized Lempel-Ziv complexity [16] of this bit string is calculated which serves as a measurement value for the randomness. The combination of the error rate and the randomness measure is interpreted as a metric for the accuracy of a database relation.[2] The problem with this approach is that the randomness measure depends on the order

[1] The paper in questions deals with correctness of data, but we aim at a more general approach here.

[2] In [8], Fisher et al. also add a parameter indicating the probability distribution of the errors. This is out of scope of this paper.

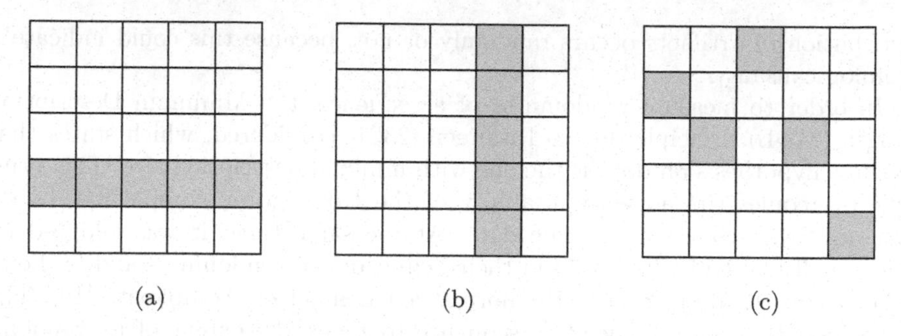

Fig. 1. Three relations with the same error rate (0.2) but a different error distribution

of rows, whereas essentially the "ordering of rows is immaterial" [4]. Therefore, equivalent representations of the same relation, may produce different outcomes of the randomness measurement. A solution to this problem would be to define a specific order on the rows of a relation. An example of such a row order is to map every row of the binary matrix retrieved after the error detection phase to its integer number and sort the rows in natural ordering as illustrated in Fig. 2. By doing this, there is a unique randomness measure for every database relation, but the downside is that a certain structure is still introduced in the bit string by sorting. Besides that, there are certain other properties that the Lempel-Ziv method fails to achieve. For example, it is not possible to derive a relation between the randomness measure and a row/column addition. For those reasons, a different method for measurement of randomness is advocated in the following section.

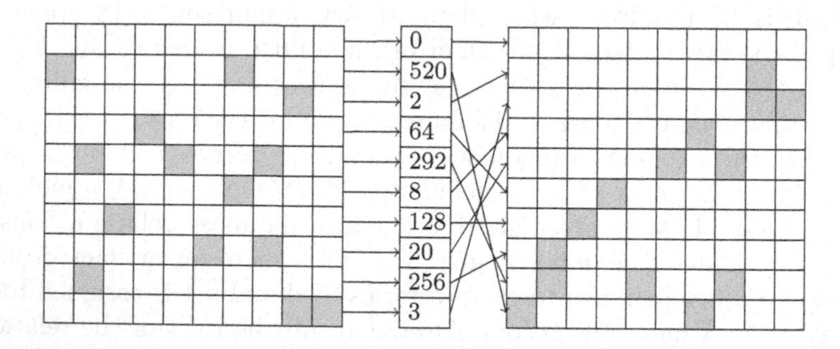

Fig. 2. Reordering of the rows based on their integer value

4 Binary Code Compression

4.1 Method Description

As described in Sect. 3, the Minimum Description Length (MDL) principle [23] is adopted by calculating the normalized Lempel-Ziv complexity of a bit string [16].

However, in the following, a novel randomness measure is introduced that relies on the MDL principle but makes use of *unary codes*. This alternative reveals many interesting properties.

Consider a relational table R with n rows and k columns and assume that a procedure to detect artifacts in the relation already exists. Following the approach in [8], a binary $n \times k$ matrix B can be constructed where $B[i, j] = 1$ indicates that the ith row has an observed artifact for column j. Note that, although an order of both rows and columns is implicitly assumed here, the described approach is independent on the order. This is proved later. In matrix B, each row represents a bit string $\mathbf{b}_i$ of fixed length k. The set of all bit strings of size k is denoted as $\mathbb{B}^k$ and for each $\mathbf{b} \in \mathbb{B}^k$, the number $c(\mathbf{b})$ is defined as the number of times $\mathbf{b}$ occurs in B. Considering the definition of B, each $\mathbf{b} \in \mathbb{B}^k$ corresponds to a certain combination of artifacts in R and the probability that a specific combination appears in B is given by:

$$\Pr[\mathbf{b}] = \frac{c(\mathbf{b})}{n}. \tag{1}$$

If it is assumed that artifacts are random, the distribution over all bit strings occurring in the relation is expected to be uniform. Therefore, a possible approach to measure randomness is to calculate the deviation of the distribution under consideration from the uniform distribution. At first glance, this deviation can be calculated by using a statistical test, but this approach causes some problems. For example, for the Chi-squared test [19], it is difficult to guarantee the sample size assumption as many combinations of artifacts tend to be rare. In addition, if the considered data artifacts are difficult[3] to measure, B is usually based on a small sample from R, yielding a small set of observations. It is also often know upfront that the error distribution will deviate from the uniform distribution, but a notion on which relations deviate more than others is desired.

For those reasons, the MDL principle is considered to measure randomness. In simple words, the observed artifacts in R are deemed random if the set of bit strings represented by B is hard to compress with a certain coding schema. To represent the bit strings in B as efficient as possible, shorter codes are assigned to bit strings with higher probability (or equivalently higher $c(\mathbf{b})$). The method proposed here makes use of unary codes [24], which are a special type of variable-length codes that are *prefix-free* and *self-synchronizing*. Unary coding represents a positive integer n with $n-1$ 1-bits followed by a 0-bit. Therefore, the length of a unary code for the integer n is n. Consider now the set of m unique bit strings observed in B as U and denote this set as $\{\mathbf{u}_1, \ldots, \mathbf{u}_m\}$ where is assumed without loss of generality that $c(\mathbf{u}_i) \geq c(\mathbf{u}_{i+1})$ with $c(\mathbf{u}_i)$ the number of times $\mathbf{u}_i$ appears in B. For every $i \in \{1, \ldots, m\}$, map $\mathbf{u}_i$ to i and encode i to retrieve code word $\mathbf{c}_i$ with unary coding. As a result, the *expected length* of a code word required to encode a bit string from B with unary encoding, is equal to:

$$L(B) = \sum_{i=1}^{|U|} \Pr[\mathbf{u}_i] \cdot i. \tag{2}$$

[3] Determining that data is in error is difficult, while determining that it is missing is easy.

In what follows, randomness of artifacts in R is assessed as $L(B)$. The rationale of this measure is that, in the spirit of the MDL principle, artifacts in R are more random if the corresponding bit matrix B is harder to compress, yielding a higher expected length of codes. Unary codes are used because it turns out that they posses a set of properties that make them very attractive for the intended purpose.

4.2 Properties of $L(B)$

In the following, a series of properties of $L(B)$ is given that align the proposed procedure with the intuitive perception of randomness.

First, it is shown that, for a binary matrix B of a given relation R, the corresponding $L(B)$ has a lower and upper bound. These boundary values can be useful when one would like to compare the randomness of two relations with a different number of rows and/or columns to each other.

Property 1. *The lower bound of $L(B)$, denoted $\underline{L}(B)$, given a binary $n \times k$ matrix B with $n, k \geq 1$ is 1.*

Proof. Given a binary $n \times k$ matrix B with $n, k \geq 1$. To have minimum randomness, all rows in B are equal and only 1 bit is needed to encode this single bit string $\mathbf{u}_1$. Because $\Pr[\mathbf{u}_1] = 1.0$, the following applies:

$$L(B) = \sum_{i=1}^{|U|} \Pr[\mathbf{u}_i] \cdot i = \sum_{i=1}^{|U|} 1.0 \cdot 1 = 1. \tag{3}$$

□

Property 2. *The upper bound of $L(B)$, denoted $\overline{L}(B)$, given a binary $n \times k$ matrix B with $n, k \geq 1$ is*

$$\frac{1}{n} \cdot \left(q \cdot \frac{(2^k) \cdot (2^k + 1)}{2} + \frac{(n \bmod 2^k) \cdot ((n \bmod 2^k) + 1)}{2} \right) \tag{4}$$

with

$$q = \left\lfloor \frac{n}{2^k} \right\rfloor. \tag{5}$$

Proof. Given a binary $n \times k$ binary matrix B with $n, k \geq 1$ and $q = \lfloor \frac{n}{2^k} \rfloor$. In total, when k is given, there are 2^k possible bit strings. Since the goal is to have maximum randomness, the distribution of bit strings appearing in B should be as uniform as possible. Therefore, there are $n \bmod 2^k$ bit strings $\mathbf{u}_1, \ldots, \mathbf{u}_{n \bmod 2^k}$ that appear $(q+1)$ times and $2^k - (n \bmod 2^k)$ bit strings $\mathbf{u}_{(n \bmod 2^k)+1}, \ldots, \mathbf{u}_{2^k}$ that appear q times in B. The expected length of the code used to compress a bit string in binary matrix B is then equal to

$$
\begin{aligned}
L(B) &= \sum_{i=1}^{|U|} \Pr[\mathbf{u}_i] \cdot i \\
&= \frac{(q+1)}{n} \cdot \left(\sum_{i=1}^{n \bmod 2^k} i \right) + \frac{q}{n} \cdot \left(\sum_{i=1+(n \bmod 2^k)}^{2^k} i \right) \\
&= \frac{1}{n} \cdot \left(\frac{(n \bmod 2^k) \cdot ((n \bmod 2^k) + 1)}{2} + q \cdot \left(\frac{2^k \cdot (2^k + 1)}{2} \right) \right)
\end{aligned}
\tag{6}
$$

which is equal to Eq. (4). □

It should be noted that the probability that the measure reaches its upper bound is higher when the number of columns is high. The probability that two rows are the same in a binary $n \times k$ matrix is $1 - (\frac{n!}{2^{kn}} * \binom{2^k}{n})$ which decreases exponentially with k. Therefore, the proposed measure is less useful when a relation has a high number of columns.

Next, it is shown that $L(B)$ is invariant to the order of rows and/or columns of a relation in a database. This is summarized in the following Theorem.

Theorem 1. *Let B and B' be two binary $n \times k$ matrices with $n, k \geq 1$. If B' is a row and/or column permutation of B, then $L(B) = L(B')$.*

Proof. Given a random binary $n \times k$ matrix B with $n, k \geq 1$ for which $L(B)$ is the expected length of a code word needed to compress a bit string of B.

1. Reordering of the rows of B matches the bijective function $X \colon B \longmapsto B'$ with B' consisting of a permutation on the order of the bit strings appearing in B. Therefore the number of times every bit string appears in B' will be the same as the number of times every bit string appears in B and thus $L(B) = L(B')$.
2. Reordering of the columns of B matches the bijective function $X \colon B \longmapsto B'$ with
$$\forall i \in \{1, \ldots, n\} \colon \mathbf{b}'_i = \sigma(\mathbf{b}_i). \tag{7}$$
and σ a permutation function which is the same as the permutation function used to reorder the columns. The result of this is that there are still $m' = m$ unique bit strings $\mathbf{u}_1, \ldots, \mathbf{u}_m$ in B' with
$$\forall i \in \{1, \ldots, m\} \colon c(\mathbf{u}_i) = c(\mathbf{u}'_i). \tag{8}$$
and thus $L(B) = L(B')$.

□

Next, the impact of a changing R on $L(B)$ is studied. From this point of view, there are two important properties that adhere to the intuition of randomness. The first property deals with the change in randomness when a row is added to R. More specifically, if a new row is added to R and the corresponding code for the corresponding new row in B has a size above (resp., below) the previous expected size, then randomness increases (resp., decreases). This result is summarized in the following Theorem.

Theorem 2. *Given a binary $n \times k$ matrix B with $n, k \geq 1$ consisting of m unique bit strings $\mathbf{u}_1, \ldots, \mathbf{u}_m$ and with randomness measure $L(B)$. If a new row is added to B with a code word of length j where $j \in \{1, \ldots, m+1\}$, the randomness will increase (resp., decrease) if and only if $L(B) \leq j$ (resp., $L(B) \geq j$).*

Proof. Given a binary $n \times k$ matrix B with $n, k \geq 1$ consisting of m unique bit strings $\mathbf{u}_1, \ldots, \mathbf{u}_m$ with frequencies $c(\mathbf{u}_1), \ldots, c(\mathbf{u}_m)$ and randomness measure $L(B)$ and a binary $(n+1) \times k$ matrix B' with $n, k \geq 1$ and randomness measure $L(B')$ consisting of n same rows as B. Then

$$\begin{aligned}
& L(B) \leq L(B') \\
\leftrightarrow\ & \frac{1}{n} \cdot \sum_{i=1}^{|U|} i \cdot c(\mathbf{u}_i) \leq \frac{1}{n+1} \cdot \left(\sum_{i=1}^{|U|} i \cdot c(\mathbf{u}_i) + j \right) \\
\leftrightarrow\ & \frac{1}{n \cdot (n+1)} \cdot \sum_{i=1}^{|U|} i \cdot c(\mathbf{u}_i) \leq \frac{j}{n+1} \\
\leftrightarrow\ & \frac{1}{n} \cdot \sum_{i=1}^{|U|} i \cdot c(\mathbf{u}_i) \leq j \leftrightarrow L(B) \leq j
\end{aligned} \tag{9}$$

It should be taken into account that adding a row can change the frequencies and therefore the order of the different bit strings. This happens only when, before the addition, bit strings $\mathbf{u}_i$ and $\mathbf{u}_{i+1}$ appear equally often and a row with bit string $\mathbf{u}_{i+1}$ is added. In this case, both bit strings swap positions in the order and the result still holds. □

Next, the fact that adding a column consisting of all 0-bits or all 1-bits does not change the measure of randomness is shown. This result is summarized in Theorem 3.

Theorem 3. *Given a binary $n \times k$ matrix B with $n, k \geq 1$ consisting of m unique bit strings $\mathbf{u}_1, \ldots, \mathbf{u}_m$. For a binary $n \times (k+1)$ matrix B' derived from B by adding a column of all 0-bits or all 1-bits, $L(B') = L(B)$ applies.*

Proof. Given a binary $n \times k$ matrix B with $n, k \geq 1$ consisting of m unique bit strings $\mathbf{u}_1, \ldots, \mathbf{u}_m$ and randomness measure $L(B)$ and a binary $n \times (k+1)$ matrix B' with $n, k \geq 1$ and randomness measure $L(B')$ where column $k+1$ only consists of 0-bits and the other k columns are the same as in B. The set of unique bit strings U' of B' will contain all the bit strings contained in the set of unique bit string U of B with a 0-bit added at the end. Therefore, $m = m'$ and as a result following from the definition of the randomness measure, $L(B) = L(B')$. □

Theorem 3 is again in line with intuition. If a column is added to R that contains no errors or only errors (e.g., all NULL values), then this new column is systematic when it comes to artifacts. It is expected that it has no impact on the randomness and this is indeed the case with $L(B)$. A corollary of Theorem 3 is that, any column of B in which all bits are equal, has no impact on $L(B)$. Theorem 4 shows that the randomness measure is not influenced by the absolute number of times every bit string appears in B when the relative number of times remains the same. In other words, the measure $L(B)$ is invariant to scaling the counts of bit strings in B with a fixed scale factor.

Theorem 4. *The randomness measure $L(B)$ of a binary $n \times k$ matrix B with $n, k \geq 1$ is scale invariant.*

Proof. Follows immediately by noting that scaling counts does not change frequencies. □

As mentioned before, it is intuitively expected that randomness is maximized when all bit strings in B have equal probability of occurring. Put differently, randomness is maximized if $\Pr[\mathbf{u}]$ has a uniform distribution. It is now fairly easy to see how $L(B)$ behaves in this case. For m unique bit strings in B with equal probability, it applies that:

$$\forall i \in \{1, ..., m\} : \Pr[\mathbf{u}_i] = \frac{1}{m} \tag{10}$$

from which it follows that

$$L(B) = \frac{1}{m} \sum_{i=1}^{m} i = \frac{m+1}{2}. \tag{11}$$

From this result, it is possible to conclude that in case the bit strings in B have a uniform distribution, the expected length of a code word is precisely the average of the longest (length m) and shortest (length 1) code word. This simple expression allows to analyze randomness under the assumption of uniform distributions of (combinations of) artifacts in R. This analysis can be made for a specific scenario that might occur in practice. For example, consider the case of single column artifacts only, meaning that a row contains an artifact in exactly one column or it contains no artifacts at all. In that case, $m = k + 1$ and $L(B) = (k/2) + 1$. Now, it is possible to consider, step by step, more columns that are in error at the same time. For a relation with k columns, the number of possible combinations of $i \leq k$ artifacts is given by $\binom{k}{i}$. When those terms are added for increasing i, the values of $L(B)$ are retrieved in case of uniform distributions, but under certain structural constraints. This analysis gives an idea of the worst-case scenario of quasi-equal probabilities for all $\mathbf{u}_i$. Opposed to that, the best-case scenario for m different bit strings in B is the one where $\mathbf{u}_1$ occurs maximally ($n - m + 1$ times) and all other $\mathbf{u}_i$ occur only once. So what happens in between? To answer that question, the last important result of this paper is summarized in the following Theorem.

Theorem 5. *Let B and B' be two binary $n \times k$ matrices with $n, k \geq 1$ such that B has unique bit strings $\mathbf{u}_1, \ldots, \mathbf{u}_m$ with randomness measure $L(B)$ and B' has unique bit strings $\mathbf{u}'_1, \ldots, \mathbf{u}'_m$ with randomness measure $L(B')$. If $\Pr[\mathbf{u}_i]$ is stochastic dominant over $\Pr[\mathbf{u}'_i]$ then $L(B) > L(B')$.*

Proof. If a probability distribution f is stochastic dominant over another distribution f', then it applies that:

$$\forall j \in \mathbb{N}_{>0} : F(j) \leq F'(j) \tag{12}$$

and there must be some j for which the inequality is strict. Here, F and F' represent the cumulative distributions. If follows that the expected value of f is higher than the expected value of f'. Thus, as a result: $L(B) > L(B')$ □

Theorem 5 has some important consequences. First of all, it shows that strict rankings in randomness can be made even when the set of unique bit strings is different. The reasons for this is because each bit string is translated into a code word and the contribution of that code word to the measured randomness is determined only by its length. So, even with different bit strings in B, the same coding system will be used for both cases and randomness is calculated by looking at the expected length of a code word. This principle can now be used to generate, for fixed m and n, a sequence of distributions $\Pr[\mathbf{u}_i]$ in such a way that each next distribution is dominant over the previous and such that $L(B)$ changes minimally. Therefore, $L(B)$ can be interpreted well.

Let us conclude this section with a brief discussion of the computational complexity of $L(B)$. First, the time complexity of calculating $L(B)$ is $O(n)$ because the occurrence of each $\mathbf{u}_i$ can be counted in a single scan of R. Opposed to that, the complexity of calculating the normalized Lempel-Ziv randomness is $O(n^2)$. The space complexity is upper bounded by m, which is itself upper bounded by 2^k. An important property is that the frequencies of each $\mathbf{u}_i$ can be stored in a histogram, which can be updated incrementally.

5 Future Work

One possible extension of the randomness measure is to give an indication of the location of data artifacts instead of only considering the distribution of the artifacts. Besides that, it can be interesting to identify certain dependencies between the attributes of a relation considering data artifacts. By adding this, it will be much easier to quickly see where most errors are located and what columns are worth to repair with little effort. Besides that, it is the case that the method described in this paper works only when a binary matrix indicating data artifacts is given. Unfortunately, the difficulty of identifying data artifacts depends on the type of artifact. Therefore, it can be useful to distinguish between types of artifacts (e.g. NULL values vs. values that not stroke with reality) or, even more interesting, to introduce a scale of artifacts where a high value implies an unacceptable error in the relation. To calculate the randomness in matrices described above, the proposed measure should be modified to deal with non-binary values.

Finally, one plans to test and evaluate the proposed measure in real-world applications. Therefore, it is necessary to identify artifacts and calculate the measurement on artifact-rich relational databases.

6 Conclusion

In this paper, counting of data artifacts to assess the quality of data is extended by a novel randomness measure to distinguish between systematic and random artifacts in a database relation. The novel measure is calculated as the expected length of a code word required to encode a bit string from B with unary encoding. This novel approach has many advantages compared to the earlier introduced method to calculate a measure for the randomness of data artifacts by means of the Lempel-Ziv complexity algorithm. Besides that, it matches closer with the intuitive perception of randomness.

References

1. Batini, C., Scannapieco, M.: Data Quality: Concepts, Methodologies and Techniques. Springer, Heidelberg (2006). https://doi.org/10.1007/3-540-33173-5
2. Bronselaer, A., De Mol, R., De Tré, G.: A measure-theoretic foundation for data quality. IEEE Trans. Fuzzy Syst. (2017) (published online)
3. Chaitin, G.: Randomness and mathematical proof. Sci. Am. **232**(5), 47–52 (1975)
4. Codd, E.F.: A relational model of data for large shared data banks. Commun. ACM **13**(6), 377–387 (1970)
5. Even, A., Shankaranarayanan, G.: Utility-driven assessment of data quality. ACM SIGMIS Database **38**(2), 75–93 (2007)
6. Even, A., Shankaranarayanan, G.: Dual assessment of data quality in customer databases. J. Data Inf. Qual. **1**(3), 15:1–15:29 (2009)
7. Falk, R., Konold, C.: Making sense of randomness: Implicit encoding as a basis for judgment. Psychol. Rev. **104**(2), 301–318 (1997)
8. Fisher, C.W., Lauria, E.J.M., Matheus, C.C.: An accuracy metric: Percentages, randomness, and probabilities. J. Data Inf. Qual. **1**(3), 16:1–16:21 (2009). http://doi.acm.org/10.1145/1659225.1659229
9. Haegemans, T., Reusens, M., Baesens, B., Lemahieu, W., Snoeck, M.: Towards a visual approach to aggregate data quality measurements. In: Proceedings of the International Conference on Information Quality (Accepted 2017)
10. Haegemans, T., Snoeck, M., Lemahieu, W.: Towards a precise definition of data accuracy and a justification for its measure. In: Proceedings of the International Conference on Information Quality, pp. 16:1–16:13 (2016)
11. Haegemans, T., Snoeck, M., Lemahieu, W., Stumpe, F., Goderis, A.: Towards a theoretical framework to explain root causes of errors in manually acquired data. In: Proceedings of the International Conference on Information Quality, pp. 15:1–15:10 (2016)
12. Heinrich, B., Klier, M.: Metric-based data quality assessment–developing and evaluating a probability-based currency metric. Decis. Support Syst. **72**, 82–96 (2015)
13. Kolmogorov, A.: On tables of random numbers. Theoret. Comput. Sci. **207**(2), 387–395 (1998)
14. Krantz, D.H., Luce, D.R., Suppes, P., Tversky, A.: Foundations of Measurement Volume I: Additive and Polynomial Representations. Academic Press, New York (1971)
15. Lee, Y.W., Pipino, L.L., Funk, J.D., Wang, R.Y.: Journey to Data Quality. MIT Press, Cambridge (2006)

16. Lempel, A., Ziv, J.: On the complexity of finite sequences. IEEE Trans. Inf. Theory **22**(1), 75–81 (1976)
17. Leszak, M., Perry, D.E., Stoll, D.: A case study in root cause defect analysis. In: Proceedings of the 22nd International Conference on Software Engineering, pp. 428–437. IEEE (2000)
18. Li, M., Vitanyi, P.: An Introduction to Kolmogorov Complexity and Its Applications, 2nd edn. Springer, New York (1997). https://doi.org/10.1007/978-0-387-49820-1
19. Pearson, K.: On the criterion that a given system of deviations from the probable in the case of a correlated system of variables is such that can be reasonably supposed to have arisen from random sampling. Phil. Mag. **50**, 157–175 (1900)
20. Pipino, L.L., Lee, Y.W., Wang, R.Y.: Data quality assessment. Commun. ACM **45**(4), 211–218 (2002)
21. Pipino, L.L., Wang, R.Y., Kopcso, D., Rybolt, W.: Developing measurement scales for data-quality dimensions. In: Wang, R.Y., Pierce, E.M., Madnick, S.E., Fisher, C.W. (eds.) Information Quality, chap. 3, pp. 37–51. M.E. Sharpe (2005)
22. Redman, T.C.: Data Quality for the Information Age, 1st edn. Artech House Inc., Norwood (1997)
23. Risannen, J.: Modeling by shortest data description. Automatica **14**, 465–471 (1978)
24. Sayood, K.: Introduction to Data Compression. Morgan Kaufmann Series in Multimedia Information and Systems, 3rd edn. Morgan Kaufmann Publishers Inc., San Francisco (2005)
25. Solomonoff, R.J.: A formal theory of inductive inference. Part i. Inf. Control **7**(1), 1–22 (1964)
26. Wang, R.Y.: A product perspective on total data quality management. Commun. ACM **41**(2), 58–65 (1998)
27. Wang, R.Y., Storey, V.C., Firth, C.P.: A framework for analysis of data quality research. IEEE Trans. Knowl. Data Eng. **7**(4), 623–640 (1995)
28. Wilson, P., Dell, L., Anderson, G.: Root Cause Analysis: A Tool for Total Quality Management. ASQ Quality Press, Milwaukee (1993)

Characterizing Fuzzy y-Models in Multi-adjoint Normal Logic Programming

M. Eugenia Cornejo(✉), David Lobo, and Jesús Medina

Department of Mathematics, University of Cádiz, Cádiz, Spain
{mariaeugenia.cornejo,david.lobo,jesus.medina}@uca.es

Abstract. This paper includes the main notions associated with the syntax and semantics of two interesting paradigms in fuzzy logic programming with default negation: multi-adjoint normal logic programming introduced in [5] and the fuzzy answer set logic programming approach presented in [16]. We will show that fuzzy answer set logic programs can be translated into multi-adjoint normal logic programs, as long as the implication operator used in the former is a residuated implication. Moreover, we will relate the notions of fuzzy y-model and model by means of a characterization theorem which allow us to guarantee the existence of fuzzy y-models of fuzzy answer set logic programs.

Keywords: Multi-adjoint logic programming · Fuzzy model
Negation operator

1 Introduction

Multi-adjoint logic normal programming arises as an extension of multi-adjoint logic programming [14] considering a negation operator in the underlying lattice. This logic programming framework was recently introduced in [5], where a wide study on the syntax and semantics corresponding to this paradigm is carried out. In what regards to the syntax, the most remarkable feature is the use of different implications in the rules of a same multi-adjoint normal logic program and general operators in the bodies of the rules. With respect to the semantics, the developed theory is based on the stable models semantics [8,13]. Note that, when multi-adjoint normal logic programs correspond to some search problem related to a real dataset, the stable models coincide with their possible solutions. Therefore, the results on the existence and unicity of stables models presented in [5] are useful to know both if the logic program is associated with a solvable problem and if only one solution exists. We are interested in applying these

Partially supported by the State Research Agency (AEI) and the European Regional Development Fund (ERDF) project TIN2016-76653-P, and by the research and transfer program of the University of Cádiz.

J. Medina et al. (Eds.): IPMU 2018, CCIS 855, pp. 541–552, 2018.
https://doi.org/10.1007/978-3-319-91479-4_45

results to other logic programming frameworks in which a negation operator is considered. In this paper, we will focus on the fuzzy answer set logic programming framework presented in [16], since its semantics is also based on the notion of stable model.

This logic programming framework [16] was introduced as a combination of two important approaches: answer set programming and fuzzy logic. In recent years, answer set programming [9] has reached a great popularity because it is an useful tool for interesting applications associated with knowledge representation systems and non-monotonic reasoning [6,7,12,13,17,19]. Depending on the considered application, some disadvantages can appear as it is shown in [1,2]. The idea of combining answer set programming with fuzzy logic arises in order to increase the expressive power and the range of potential applications of answer set programming.

This paper will present a first study on the relations between the multi-adjoint logic normal programming and the fuzzy answer set logic programming frameworks given in [5] and [16], respectively, in order to apply the recent existence and unicity results introduced in the first one to the second one. First of all, we will recall the main notions associated with the syntax and semantics of both fuzzy logic programming approaches. In the following, we will present a procedure to translate fuzzy answer set logic programs into multi-adjoint normal logic programs. Considering a particular family of the multi-adjoint normal logic programs obtained in this translation, we will provide the required conditions in order to guarantee the existence of fuzzy y-models of fuzzy answer set logic programs. To reach this goal, a characterization of fuzzy y-models of fuzzy answer set logic programs in terms of models of multi-adjoint normal logic programs will be given. It is important to emphasize that the existence theorem of stable models for multi-adjoint normal logic programs will play a fundamental role in order to ensure the existence of fuzzy y-models. This work will finish with some conclusions and prospects for future work.

2 Multi-adjoint Normal Logic Programming

Multi-adjoint normal logic programming is an logical theory characterized by the use of different implications in the rules of a same logic program, as well as a negation operator and general operators defined on complete lattices in the bodies of the rules. The formal definitions associated with the syntactic structure of multi-adjoint normal logic programming framework are presented below. These definitions are given in detail in [5,15].

Definition 1. *The tuple* $(L, \preceq, \leftarrow_1, \&_1, \dots, \leftarrow_n, \&_n, \neg)$ *is a* multi-adjoint normal lattice *if the following properties are verified:*

1. $(L, \preceq)$ *is a bounded lattice, i.e. it has a bottom* $(\bot)$ *and a top* $(\top)$ *element;*
2. $(\&_i, \leftarrow_i)$ *is an adjoint pair in* $(L, \preceq)$*, for* $i \in \{1, \dots, n\}$*;*
3. $\top \,\&_i\, \vartheta = \vartheta \,\&_i\, \top = \vartheta$*, for all* $\vartheta \in L$ *and* $i \in \{1, \dots, n\}$*;*

4. $\neg\colon L \to L$ *is a negation operator, that is, an order-reversing mapping satisfying the equalities* $\neg(\bot) = \top$ *and* $\neg(\top) = \bot$.

We will define a multi-adjoint normal logic program $\mathbb{P}$ from a multi-adjoint normal lattice together with an additional (symbol of) negation $\sim$. $Lit_{\mathbb{P}}$ will denote the whole set of elements appearing in the rules of $\mathbb{P}$, which will be called literals since they will be either (positive) propositional symbols or negated propositional symbols by $\sim$.

Definition 2. *Let* $(L, \preceq, \leftarrow_1, \&_1, \dots, \leftarrow_n, \&_n, \neg)$ *be a multi-adjoint normal lattice. A* multi-adjoint normal logic program (MANLP) $\mathbb{P}$ *is a finite set of weighted rules of the form:*

$$\langle l \leftarrow_i @[l_1, \dots, l_m, \neg l_{m+1}, \dots, \neg l_n]; \vartheta\rangle$$

where $i \in \{1, \dots, n\}$, $@$ *is an aggregator operator,* ϑ *is an element of* L *and* $l, l_1, \dots, l_n$ *are literals such that* $l_j \neq l_k$, *for all* $j, k \in \{1, \dots, n\}$, *with* $j \neq k$. *The literal* l *is called* head *of the rule,* $@[l_1, \dots, l_m, \neg l_{m+1}, \dots, \neg l_n]$ *is called* body *of the rule and the value* ϑ *is its* weight.

It is important to clarify that the roles played by the negation operators $\neg$ and $\sim$ are different. The truth value of $\neg\,\phi$ can be computed from the truth value of ϕ, while $\sim\phi$ can straightforwardly be inferred from the program. As usual, we will call "default negation" to $\neg$ and "strong negation" to $\sim$. Observe that the strong negation operator used in this paper should not be confused with the well-known notion of involutive negation.

A depth study about the syntax and the semantics of the multi-adjoint normal logic programming framework was given in [3–5]. Specifically, the developed semantics for multi-adjoint normal logic programs is based on the notion of stable model [8] which is closely related to the notion of minimal model. Before introducing the notion of model, we need to include the definition of interpretation and some notational conventions.

Definition 3. *Given a complete lattice* $(L, \preceq)$, *a mapping* $I\colon Lit_{\mathbb{P}} \to L$, *which assigns to every literal appearing in* $Lit_{\mathbb{P}}$ *an element of the lattice* L, *is called* L-interpretation. *The set of all* L*-interpretations is denoted by* $\mathcal{I}_{\mathfrak{L}}$.

It is important to mention that the interpretation of an operator symbol ω under a multi-adjoint normal lattice will be denoted by $\dot{\omega}$. In what regards to the evaluation of a formula $\mathcal{F}$ under an interpretation I, it will be denoted as $\hat{I}(\mathcal{F})$ and it proceeds inductively as usual, until all propositional symbols in $\mathcal{F}$ are reached and evaluated under I.

Taking into account these considerations and following the philosophy used in the semantics of multi-adjoint logic programming [14], we introduce the notions of model and satisfiability as follows.

Definition 4. *Given an interpretation $I \in \mathcal{I}_{\mathfrak{L}}$, we say that:*

(1) A weighted rule $\langle l \leftarrow_i @[l_1, \dots, l_m, \neg l_{m+1}, \dots, \neg l_n]; \vartheta\rangle$ *is* satisfied *by I if and only if* $\vartheta \preceq \hat{I}\,(l \leftarrow_i @[l_1, \dots, l_m, \neg l_{m+1}, \dots, \neg l_n])$.

(2) An L-interpretation $I \in \mathcal{I}_{\mathfrak{L}}$ *is a* model *of a MANLP $\mathbb{P}$ if and only if all weighted rules in $\mathbb{P}$ are satisfied by I.*

Stable models of a normal logic program are related to the minimal models of a monotonic logic program obtained from the original program. For that reason, before introducing the notion of stable model for a MANLP $\mathbb{P}$, a procedure to obtain a positive multi-adjoint logic program from a MANLP is required. Given an L-interpretation I, we will build a positive multi-adjoint program $\mathbb{P}_I$ called *reduct* of $\mathbb{P}$, by substituting each rule $\langle l \leftarrow_i @[l_1, \dots, l_m, \neg l_{m+1}, \dots, \neg l_n]; \vartheta\rangle$ in $\mathbb{P}$ by the rule $\langle l \leftarrow_i @_I[l_1, \dots, l_m]; \vartheta\rangle$ where the operator $\dot{@}_I \colon L^m \to L$ is defined as $\dot{@}_I[\vartheta_1, \dots, \vartheta_m] = \dot{@}[\vartheta_1, \dots, \vartheta_m, \dot{\neg} I(l_{m+1}), \dots, \dot{\neg} I(l_n)]$, for all $\vartheta_1, \dots, \vartheta_m \in L$.

Definition 5. *Given a MANLP $\mathbb{P}$ and an L-interpretation I, we say that I is a* stable model *of $\mathbb{P}$ if and only if I is a minimal model of $\mathbb{P}_I$.*

Below, we will show that each stable model of a MANLP $\mathbb{P}$ is actually a minimal model of $\mathbb{P}$.

Proposition 1. *Any stable model of a MANLP $\mathbb{P}$ is a minimal model of $\mathbb{P}$.*

Requiring the continuity of the operators involved in the rules of MANLPs will guarantee the existence of stable models.

Theorem 1. *Let $(K, \preceq, \leftarrow_1, \&_1, \dots, \leftarrow_n, \&_n, \neg)$ be a multi-adjoint normal lattice, where K is a non-empty convex compact subset of an euclidean space, and $\mathbb{P}$ be a finite MANLP defined on this lattice. If $\&_1, \dots, \&_n$, $\neg$ and the aggregator operators in the body of the rules of $\mathbb{P}$ are continuous operators, then $\mathbb{P}$ has at least a stable model.*

Once the main notions corresponding to the multi-adjoint normal logic programming framework have been recalled, we will continue including the basic notions of the other logical programming framework in which we are interested, that is, fuzzy answer set logic programming.

3 Fuzzy Answer Set Logic Programming

The fuzzy answer set logic programming framework [16] arises as an interesting combination of the concepts of answer set programming and fuzzy logic. The main feature of this logical theory is that elements appearing in the rules of a fuzzy answer set logic program $\mathbb{P}^*$ can be either (positive) propositional symbols, either negated propositional symbols by the negation operator $\sim$ or negated propositional symbols by the negation operator $\neg$. Notice that, the negation operators $\neg$ and $\sim$ behave like the "default negation" and the "strong negation", respectively, described previously in Sect. 2. The set $Lit_{\mathbb{P}^*}$ will collect (positive) the propositional symbols and the negated propositional symbols by $\sim$. We say that the elements belonging to $Lit_{\mathbb{P}^*}$ are called *literals*. The set $ELit_{\mathbb{P}^*}$ will collect the literals and the negated literals by $\neg$. We say that the elements belonging to $ELit_{\mathbb{P}^*}$ are called *extended literals*.

Definition 6. *A fuzzy answer set logic program (FASLP) $\mathbb{P}^*$ is a finite set of rules of the form $l \leftarrow \beta$ such that:*

1. *The head of the rule l is either a literal or the bottom element $\bot \in L$;*
2. *The body of the rule β is a finite set of extended literals;*
3. Constraints *are rules whose head is the bottom element.*
4. Facts *are rules with an empty body.*

Although we have called fuzzy answer set logic programs (FASLPs) to the programs introduced in Definition 6, it is easy to see that the notion of fuzzy answer set is not used in these programs. However, we will mantain this notation in order to indicate that the semantics defined for this kind of programs will be based on the notion of fuzzy answer set.

As far as the fuzzy answer set semantics is concerned, we will firstly state the notion of fuzzy interpretation.

Definition 7. *Given a complete lattice $(L, \preceq)$, a* fuzzy interpretation *is a mapping $I\colon Lit_{\mathbb{P}^*} \to L$ which assigns to every literal appearing in $Lit_{\mathbb{P}^*}$ an element of L.*

We will present the concept of satisfaction function in the following. To do this, we will need a triangular norm (t-norm) $\mathcal{T}\colon L^2 \to L$ and a negation operator $\mathcal{N}\colon L \to L$, which is an order-reversing mapping satisfying that $\mathcal{N}(\bot) = \top$ and $\mathcal{N}(\top) = \bot$, in order to compute the degree of satisfaction of the body of a rule. Specifically, the evaluation of an extended literal $\neg l$ under a given fuzzy interpretation I will be obtained by using the following equality $I(\neg l) = \mathcal{N}(I(l))$.

Moreover, in [16] an specific implication operator $\mathcal{I}\colon L^2 \to L$ will be needed in order to obtain the degree of satisfaction of a rule. That is, the considered implication $\mathcal{I}$ is an order-reversing mapping in the first argument and order-preserving mapping in the second argument; and $\mathcal{I}$ verifies the equalities $\mathcal{I}(\bot, \bot) = \top$ and $\mathcal{I}(\top, x) = x$, for all $x \in L$.

Definition 8. *Let $\mathbb{P}^*$ be a FASLP and I be a fuzzy interpretation. The induced satisfaction function $I_{\models}\colon 2^{ELit_{\mathbb{P}^*}} \cup \mathbb{P}^* \to L$ is defined by:*

$$\begin{aligned}
I_{\models}(\varnothing) &= \top \\
I_{\models}(\{l\} \cup \beta) &= \mathcal{T}(I(l), I_{\models}(\beta)) \\
I_{\models}(\bot \leftarrow \beta) &= \mathcal{I}(I_{\models}(\beta), \bot) \\
I_{\models}(l \leftarrow \beta) &= \mathcal{I}(I_{\models}(\beta), I(l))
\end{aligned}$$

Next, we will use the satisfaction function to develop the notion of fuzzy model. For computing the degree in which a given fuzzy interpretation I is a fuzzy model, we need to consider an aggregator operator $\mathcal{A}$ which combines all the degrees of satisfaction of the rules appearing in $\mathbb{P}^*$ into a single truth value.

Definition 9. *Let $\mathbb{P}^*$ be a FASLP, I be a fuzzy interpretation and $y \in L$. We say that I is a* fuzzy y-model *of $\mathbb{P}^*$ if and only if $y \leq \mathcal{A}(\mathbb{P}^*, I_{\models})$.*

In the following, we will illustrate how the value $\mathcal{A}(\mathbb{P}^*, I_{\models})$ is computed. Given a finite FASLP $\mathbb{P}^*$, we can suppose without loss of generality that the rules appearing in $\mathbb{P}^*$ are denoted as $r_1^*, \ldots, r_m^*$, with $m \in \mathbb{N}$. As we mentioned above, the aggregator operator $\mathcal{A}$ combines all the degrees of satisfaction of the rules in $\mathbb{P}^*$, being these values belonging to L. Consequently, we can say that the aggregator operator assigns to each element in L^n an element in L. Therefore, given an interpretation I such that $I_{\models}(r_i^*) = \alpha_i$, for each $i \in \{1, \ldots, m\}$, with $\alpha_i \in L$, we obtain that $\mathcal{A}(\mathbb{P}^*, I_{\models}) = \mathcal{A}(\alpha_1, \ldots, \alpha_n)$. Notice that, by definition of aggregator operator [10], $\mathcal{A}$ can be any order-preserving operator.

Finally, following the idea presented in [18,20], we will introduce the notion of fuzzy answer set by using unfounded sets. In the classical case, an unfounded set is a set of literals for which there is no motivation to suppose that they have to be true. This fact is due to that these literals either depend on each other, or the rules that can motivate them are not satisfied. As a consequence, an interpretation will be an answer set of a FASLP $\mathbb{P}^*$ if and only if it is a model of $\mathbb{P}^*$ and it does not contain such unfounded sets. In order to translate the classical notion of unfounded set to the fuzzy interpretations environment, we need to introduce the concept of support of a rule.

Given a rule $l \leftarrow \beta \in \mathbb{P}^*$ and a fuzzy interpretation I, the support of the rule $l \leftarrow \beta$ with respect to I is defined as:

$$I_s(l \leftarrow \beta) = \inf\{y \in L \mid \mathcal{I}(I_{\models}(\beta), y) \geq I_{\models}(l \leftarrow \beta)\}$$

Definition 10. *Let $\mathbb{P}^*$ be a FASLP and I be a fuzzy interpretation. A set of literals X is an* unfounded set *with respect to I if and only if for each literal $l \in X$ and each rule $l \leftarrow \beta \in \mathbb{P}^*$, one of the following statements holds:*

1. $\beta \cap X \neq \varnothing$;
2. $I(l) > I_s(l \leftarrow \beta)$;
3. $I_{\models}(\beta) = 0$.

Once the definition of unfounded set has been presented, we are in position to define when a fuzzy interpretation is a fuzzy answer set of a FASLP $\mathbb{P}^*$.

Definition 11. *Given a fuzzy interpretation I and $y \in L$, then:*

(1) I is unfounded-free *if and only if $\{l \in Lit_{\mathbb{P}^*} \mid I(l) > 0\} \cap X = \varnothing$, for every unfounded set X with respect to I.*
(2) I is a fuzzy y-answer set *of a FASLP $\mathbb{P}^*$ if it is an unfounded-free fuzzy y-model of $\mathbb{P}^*$.*

It is convenient to mention that Definitions 9 and 11 have been presented in [16] requiring the x-consistency property. In this paper, we will not consider such extra property because it is not relevant in the translation of the results obtained in the multi-adjoint logic programming framework to the fuzzy answer set logic programming framework.

4 On the Relationship Between Multi-adjoint Normal Logic Programming and Fuzzy Answer Set Logic Programming

In this section, we will study what conditions are required in order to guarantee that a fuzzy answer set logic program can be translated into a multi-adjoint normal logic program. To do this, we need to analyze the similarities and differences corresponding to the syntax and the semantics of both logic programming frameworks. Our main contribution consists in providing a characterization of fuzzy y-models given in fuzzy answer set logic programming by means of models considered in multi-adjoint normal logic programming. This characterization will allow us to apply the results obtained for MANLPs to FASLPs.

First of all, we will see how a FASLP can be translated into a MANLP. Consider a finite fuzzy answer set logic program $\mathbb{P}^*$, that is, a finite set of rules of the form $l \leftarrow \beta$, such that $l \in Lit_{\mathbb{P}^*} \cup \{\bot\}$ and β is a finite set of elements in $ELit_{\mathbb{P}^*}$. On the one hand, observe that β can be seen as a finite union of singletons and, as a consequence, a rule $l \leftarrow \beta$ where $\beta = \{l_1, \dots, l_n\}$ can be rewritten as $l \leftarrow l_1 \cup \dots \cup l_n$. On the other hand, we need to consider a t-norm $\mathcal{T}$, a negation operator $\mathcal{N}$ and an implication operator $\mathcal{I}$ in order to obtain the degree of satisfaction of a rule. This fact is due to that the evaluation of the operator symbols $\cup$, $\leftarrow$ and $\neg$ under a fuzzy interpretation will be $\dot{\cup} = \mathcal{T}$, $\dot{\leftarrow} = \mathcal{I}^{op}$ and $\dot{\neg} = \mathcal{N}$, where $\mathcal{I}^{op}(x, y) = \mathcal{I}(y, x)$, for all $x, y \in L$. If the operator $\mathcal{I}$ is a residuated implication, that is, there exists $\&$ such that $(\&, \mathcal{I}^{op})$ forms an adjoint pair, then we can say that the algebraic structure from which $\mathbb{P}^*$ is defined $(L, \preceq, \&, \mathcal{I}, \neg)$ is a multi-adjoint normal lattice.

Taking into account these considerations and assigning a weight $\vartheta \in L$ to each rule $l \leftarrow \{l_1, \dots, l_n\}$ belonging to $\mathbb{P}^*$, we obtain a program composed of the rules $\langle l \leftarrow @[l_1, \dots, l_n]; \vartheta \rangle$, where $@$ is an aggregator operator symbol interpreted as $\dot{@}\colon L^n \to L$, $\dot{@}[x_1, \dots, x_n] = \mathcal{T}(x_1, \dots, x_n)$, for all $x_1, \dots, x_n \in L$. This program is a MANLP $\mathbb{P}$.

It is important to note that the given FASLP $\mathbb{P}^*$ can contain rules with the bottom element $\bot \in L$ in its head and the translation of these kind of rules $\langle \bot \leftarrow @[l_1, \dots, l_n]; \vartheta \rangle$ cannot be considered in the multi-adjoint approach. In order to deal with this issue, we will introduce an extra literal denoted as *Ct* (simplifying the name "constraint"). The rules $\bot \leftarrow \{l_1, \dots, l_n\}$ in $\mathbb{P}^*$ become into $\langle \mathrm{Ct} \leftarrow @[l_1, \dots, l_n]; \vartheta \rangle$ in the multi-adjoint normal logic program. From the semantical point of view, we will demand that any interpretation I in the multi-adjoint framework satisfies $I(\mathrm{Ct}) = \bot$ so that it also works in the fuzzy answer set programming environment. Observe that, the set $Lit_{\mathbb{P}^*} \cup \{\mathrm{Ct}\}$ in $\mathbb{P}$ is considered to be the set of literals in $\mathbb{P}^*$.

A last remark should be considered in order to check that the rules of the FASLP $\mathbb{P}^*$ are expressed in the same way as the rules of a MANLP given in Definition 2. The elements included in the bodies of the rules of $\mathbb{P}^*$ are extended literals, that is, they can be literals and negated literals by $\neg$. Hence, a rule in the given FALSP $\mathbb{P}^*$ which has been translated into $\langle l \leftarrow @[l_1, \dots, l_n]; \vartheta\rangle$ is actually a rule $\langle l \leftarrow @[p_1, \dots, p_m, \neg p_{m+1}, \dots, \neg p_n]; \vartheta\rangle$, where $p_1, \dots, p_n$ are either (positive) propositional symbols or propositional symbols negated by $\sim$.

After explaining the procedure in order to translate a fuzzy answer set logic program into a multi-adjoint normal logic program, we will state the corresponding notational conventions. Without loss of generality, suppose that the rules of the given finite FASLP $\mathbb{P}^*$ are ordered as $r_1^*, \dots, r_m^*$, with $m \in \mathbb{N}$. Considering $\vartheta_1, \dots, \vartheta_m \in L$, we will define a MANLP which will be denoted as $\mathbb{P}_{\vartheta_1,\dots,\vartheta_m}$ as the set of rules of the form:

$$r_i \equiv \begin{cases} \langle l \leftarrow @[l_1, \dots, l_n]; \vartheta_i\rangle & \text{if} \quad r_i^* \equiv l \leftarrow \{l_1, \dots, l_n\}, l \in Lit_{\mathbb{P}} \\ \langle \mathrm{Ct} \leftarrow @[l_1, \dots, l_n]; \vartheta_i\rangle & \text{if} \quad r_i^* \equiv \bot \leftarrow \{l_1, \dots, l_n\} \end{cases}$$

for each $i \in \{1, \dots, m\}$. From now on, given a fuzzy interpretation $I \colon Lit_{\mathbb{P}^*} \to L$ in the fuzzy answer set logic programming framework, we will define a mapping $I^+ \colon Lit_{\mathbb{P}^*} \cup \{\mathrm{Ct}\} \to L$ such that $I^+(l) = I(l)$ if $l \in Lit_{\mathbb{P}^*}$ and $I^+(\mathrm{Ct}) = \bot$. This mapping I^+ is an L-interpretation in the multi-adjoint logic programming approach on the set of literals $Lit_{\mathbb{P}^*} \cup \{\mathrm{Ct}\}$.

The following theorem shows a characterization of fuzzy y-models in fuzzy answer set programming framework in terms of models in the multi-adjoint approach.

Theorem 2. *Let $\mathbb{P}^*$ be a FASLP and $I \colon Lit_{\mathbb{P}^*} \to L$ be a fuzzy interpretation. Given $y \in L$, consider the following set of MANLPs:*

$$S_y = \{\mathbb{P}_{\vartheta_1,\dots,\vartheta_m} \mid \vartheta_1, \dots, \vartheta_m \in L \text{ and } \mathcal{A}(\vartheta_1, \dots, \vartheta_m) \geq y\}$$

Then, I is a fuzzy y-model of the FASLP $\mathbb{P}^*$ if and only if there exists $\mathbb{P}_{\vartheta_1,\dots,\vartheta_m} \in S_y$ such that I^+ is a model of $\mathbb{P}_{\vartheta_1,\dots,\vartheta_m}$.

In the following, we present an example in order to illustrate Theorem 2.

Example 1. Consider the next FASLP $\mathbb{P}^*$ consisting of the following six rules:

$$\begin{array}{ll} r_1^* : p \leftarrow \{\neg t\} & r_4^* : t \leftarrow \{s\} \\ r_2^* : q \leftarrow \{\neg s\} & r_5^* : t \leftarrow \{\neg p, \neg q\} \\ r_3^* : p \leftarrow \{q, s\} & r_6^* : 0 \leftarrow \{\neg p\} \end{array}$$

The operators involved in the computation of the degree of satisfaction of the previous rules are defined on the complete lattice $([0,1], \leq)$. These operators $\mathcal{T}$, $\mathcal{I} \colon [0,1]^2 \to [0,1]$ and $\mathcal{N} \colon [0,1] \to [0,1]$ are defined, for all $x, y \in [0,1]$, by $\mathcal{T}(x, y) = x * y$; $\mathcal{I}(x, y) = 1$, if $x \leq y$, and $\mathcal{I}(x, y) = y$, otherwise; and $\mathcal{N}(x) = 1 - x$, for all $x, y \in [0, 1]$. Therefore $\mathcal{T}$ is the product t-norm, $\mathcal{I}$ is the Gödel implication and $\mathcal{N}$ is the standard negation [11]. For computing the degree in which a given fuzzy interpretation is a fuzzy model, we will consider the aggregator operator $\mathcal{A}_1\colon [0, 1]^6 \to [0, 1]$ defined by $\mathcal{A}_1(\alpha_1, \ldots, \alpha_6) = \dfrac{\alpha_1 + \cdots + \alpha_6}{6}$, for all $\alpha_1, \ldots, \alpha_6 \in [0, 1]$.

By Theorem 2, we will demonstrate that the fuzzy interpretation given by $I \equiv \{(p, 0.8), (q, 0.4), (s, 0.5), (t, 0.9)\}$ is a fuzzy 0.7-model of $\mathbb{P}^*$. To reach this conclusion, we will see that the mapping I^+ is a model of the multi-adjoint normal program $\mathbb{P}_{1,0.3,1,1,1,0}$ composed of the next five rules:

$$\begin{array}{l}
r_1 : \langle p \leftarrow @[\neg t] \ ; \ 1\rangle \\
r_2 : \langle q \leftarrow @[\neg s] \ ; \ 0.3\rangle \\
r_3 : \langle p \leftarrow @[q, s] \ ; \ 1\rangle \\
r_4 : \langle t \leftarrow @[s] \ ; \ 1\rangle \\
r_5 : \langle t \leftarrow @[\neg p, \neg q] \ ; \ 1\rangle
\end{array}$$

Observe that, the rule r_6 has not been written in the definition of $\mathbb{P}_{1,0.3,1,1,1,0}$ since its weight is equal to zero and therefore, that rule is straightforwardly satisfied. It is easy to see that the program $\mathbb{P}_{1,0.3,1,1,1,0}$ belongs to $S_{0.7}$ since the inequality $\mathcal{A}_1(1, 0.3, 1, 1, 1, 0) = 0.71\widehat{6} \geq 0.7$ holds. Now, we will check that I^+ satisfies the rule r_i, for each $i \in \{1, \ldots, 5\}$. The corresponding computations for these five rules are given below:

$$\begin{array}{l}
r_1 : \hat{I}(p \leftarrow @[\neg t]) = \mathcal{I}(1 - I(t), I(p)) = \mathcal{I}(1 - 0.9, 0.8) = \mathcal{I}(0.1, 0.8) = 1 \geq 1 \\
r_2 : \hat{I}(q \leftarrow @[\neg s]) = \mathcal{I}(1 - I(s), I(q)) = \mathcal{I}(1 - 0.5, 0.4) = \mathcal{I}(0.5, 0.4) = 0.4 \geq 0.3 \\
r_3 : \hat{I}(p \leftarrow @[q, s]) = \mathcal{I}(I(q) * I(s), I(p)) = \mathcal{I}(0.4 * 0.5, 0.8) = \mathcal{I}(0.2, 0.8) = 1 \geq 1 \\
r_4 : \hat{I}(t \leftarrow @[s]) = \mathcal{I}(I(s), I(t)) = \mathcal{I}(0.5, 0.9) = 1 \geq 1 \\
r_5 : \hat{I}(t \leftarrow @[\neg p, \neg q]) = \mathcal{I}((1 - I(p)) * (1 - I(q)), I(t)) = \mathcal{I}(0.2 * 0.6, 0.9) \\
\quad = \mathcal{I}(0.12, 0.9) = 1 \geq 1
\end{array}$$

As a consequence, I^+ satisfies all rules in $\mathbb{P}_{1,0.3,1,1,1,0}$, and thus it is a model of the MANLP $\mathbb{P}_{1,0.3,1,1,1,0}$. Applying Theorem 2, we can ensure that the interpretation I is a fuzzy 0.7-model of $\mathbb{P}^*$.

It is worth to mention that I is not a fuzzy 0.7-model of $\mathbb{P}^*$ when the aggregator operator $\mathcal{A}_2\colon [0, 1]^6 \to [0, 1]$ is given by $\mathcal{A}_2(\alpha_1, \ldots, \alpha_6) = \min\{\alpha_1, \ldots, \alpha_6\}$. Indeed, in this case, I is not a fuzzy y-model of the FASLP $\mathbb{P}^*$ for each $y \in (0, 1]$ since $I_{\models}(r_6^*) = \mathcal{I}(1 - I(p), 0) = \mathcal{I}(0.2, 0) = 0$. □

An interesting consequence of Theorem 2 is the following one. If a FASLP $\mathbb{P}^*$ does not contain literals negated by $\neg$, then it always has at least a fuzzy y-model, for each $y \in L$, when the set S_y is not empty. This proposition is obtained from the results introduced in [14] for multi-adjoint logic programming.

Proposition 2. *Let $\mathbb{P}^*$ be a FASLP without constraints such that there are no elements negated by $\neg$ in the bodies of their rules. Given $y \in L$, consider the following set of MANLPs:*

$$S_y = \{\mathbb{P}_{\vartheta_1,\ldots,\vartheta_m} \mid \vartheta_1, \ldots, \vartheta_m \in L \text{ and } \mathcal{A}(\vartheta_1, \ldots, \vartheta_m) \geq y\}$$

If S_y is a non-empty set then there exists at least a fuzzy y-model of $\mathbb{P}^*$.

It is important to note that the existence of fuzzy y-models in a FASLP $\mathbb{P}^*$ cannot be guaranteed in general, when some literals negated by $\neg$ appear in the bodies of some rules of $\mathbb{P}^*$. Nevertheless, taking into account Proposition 1 and Theorem 2, we can assert that if $I^+ \colon Lit_{\mathbb{P}^*} \cup \{\mathrm{Ct}\} \to L$ is a stable model of $\mathbb{P}_{\vartheta_1,\ldots,\vartheta_m} \in S_y$ with $y \in L$, then I^+ is a model of $\mathbb{P}_{\vartheta_1,\ldots,\vartheta_m} \in S_y$, and thus the restriction of I^+ to the set of literals $Lit_{\mathbb{P}^*}$, which we will denoted by $I^+_{|Lit_{\mathbb{P}^*}}$, is a fuzzy y-model of $\mathbb{P}^*$. Therefore, the results concerning the existence of stable models of MANLPs presented in [5] can be applied in order to ensure the existence of fuzzy y-models. For instance, from Theorem 1 we can deduce that if a FASLP $\mathbb{P}^*$ is defined in a convex and compact subset of an euclidean space and with continuous operators, then it always has fuzzy y-models, whenever the set S_y is non-empty.

Theorem 3. *Let $\mathbb{P}^*$ be a FASLP without constraints and consider the following set of MANLPs:*

$$S_y = \{\mathbb{P}_{\vartheta_1,\ldots,\vartheta_m} \mid \vartheta_1,\ldots,\vartheta_m \in L \text{ and } \mathcal{A}(\vartheta_1,\ldots,\vartheta_m) \geq y\}$$

If S_y is a non-empty set, L is a non-empty convex compact set in an euclidean space and $\mathcal{T}$, $\mathcal{N}$ are continuous operators, then $\mathbb{P}^*$ has at least a fuzzy y-model, with $y \in L$.

The previous theorem will allow us to guarantee the existence of fuzzy-y models for a program built from the FASLP $\mathbb{P}^*$ given in Example 1, when the considered aggregator operator $\mathcal{A}$ is the minimum operator.

Example 2. Coming back to Example 1, we obtained that the fuzzy interpretation $I \equiv \{(p, 0.8), (q, 0.4), (s, 0.5), (t, 0.9)\}$ is not a fuzzy y-model of the FASLP $\mathbb{P}^*$ composed of the rules $r_1^*, r_2^*, r_3^*, r_4^*, r_5^*, r_6^*$, for each $y \in (0, 1]$, when the aggregator operator is the minimum. In order to apply Theorem 3, we will consider the FASLP only composed of the rules $r_1^*, r_2^*, r_3^*, r_4^*, r_5^*$ and we will see that there exist fuzzy y-models for such program considering the minimum. Now, we will check that the hypothesis of Theorem 3 are satisfied.

Notice that, the MANLP $\mathbb{P}_{y,y,y,y,y}$ belongs to S_y for each $y \in [0, 1]$ and, therefore, S_y is a non-empty set. Moreover, $[0, 1]$ is a non-empty convex compact set in the euclidean space $([0, 1], +, *, \mathbb{R})$, where $+, *$ are the usual sum and product defined in $\mathbb{R}$. Since $\mathcal{T}, \mathcal{N}$ are continuous operators, we can apply Theorem 3 which leads us to assert that the program built from $\mathbb{P}^*$ has at least a fuzzy y-model, for each $y \in [0, 1]$.

In particular, given $y = 0.7$, we will see that the fuzzy interpretation given by $J \equiv \{(p, 1), (q, 0.4), (s, 0.9), (t, 0.8)\}$ is a fuzzy 0.7-model with the minimum operator as aggregator operator $\mathcal{A}$. In the following, the computations related to the satisfaction function value for the five rules are given:

$$J_{\models}(r_1^*) = J_{\models}(p \leftarrow \{\neg t\}) = \mathcal{I}(1 - J(t), J(p)) = \mathcal{I}(1 - 0.8, 1) = \mathcal{I}(0.2, 1) = 1$$
$$J_{\models}(r_2^*) = J_{\models}(q \leftarrow \{\neg s\}) = \mathcal{I}(1 - J(s), J(q)) = \mathcal{I}(1 - 0.9, 0.4) = \mathcal{I}(0.1, 0.4) = 1$$
$$J_{\models}(r_3^*) = J_{\models}(p \leftarrow \{q, s\}) = \mathcal{I}(J(q) * J(s), J(p)) = \mathcal{I}(0.4 * 0.9, 1) = \mathcal{I}(0.36, 1) = 1$$
$$J_{\models}(r_4^*) = J_{\models}(t \leftarrow \{s\}) = \mathcal{I}(J(s), J(t)) = \mathcal{I}(0.9, 0.8) = 0.8$$
$$J_{\models}(r_5^*) = J_{\models}(t \leftarrow \{\neg p, \neg q\}) = \mathcal{I}((1 - J(p)) * (1 - J(q)), J(t)) = \mathcal{I}(0 * 0.6, 0.8) = \mathcal{I}(0, 0.8) = 1$$

Therefore, we obtain that $\min\{J_{\models}(r_1^*), \ldots, J_{\models}(r_5^*)\} = 0.8 \geq 0.7$. Consequently, we can conclude that J is a fuzzy 0.7-model of the FASLP whose rules are $r_1^*, r_2^*, r_3^*, r_4^*, r_5^*$. Indeed, we obtain straightforwardly that J is a fuzzy 0.8-model. □

5 Conclusions and Future Work

An initial study on the relationship between multi-adjoint normal logic programming and an interesting fuzzy answer set logic programming framework has been presented. We have shown a procedure which allows us to translate the fuzzy answer set logic programs given in [16] into multi-adjoint normal logic programs, when the implication operator involved in FASLPs is a residuated implication. Moreover, a new literal whose interpretation is the bottom element in the lattice must be considered in the multi-adjoint framework. This fact is due to that FASLPs can contain rules with empty head which must be included in MANLPs. A characterization of fuzzy y-models of fuzzy answer set logic programs in terms of models of multi-adjoint normal logic programs has been introduced. This characterization is very valuable because it leads us apply the existence theorem of stable models in MANLPs and ensure the existence of fuzzy y-models for a FASLP.

As future work, we will continue analyzing more properties corresponding to both logic programming frameworks in order to establish connections between the notions of fuzzy answer set and stable model. Specifically, we are interested in obtaining a characterization of fuzzy answer sets in terms of stable models so that the existence and unicity results for MANLPs [5] can be applied in the considered fuzzy answer set logic programming.

References

1. Balduccini, M., Gelfond, M.: Logic programs with consistency-restoring rules. In: International Symposium on Logical Formalization of Commonsense Reasoning, AAAI 2003 Spring Symposium Series, pp. 9–18 (2003)
2. Brewka, G.: Logic programming with ordered disjunction. In: 18th National Conference on Artificial Intelligence and 14th Conference on Innovative Applications of Artificial Intelligence, pp. 100–105. AAAI (2002)
3. Cornejo, M.E., Lobo, D., Medina, J.: Selecting the coherence notion in multi-adjoint normal logic programming. In: Rojas, I., Joya, G., Catala, A. (eds.) IWANN 2017. LNCS, vol. 10305, pp. 447–457. Springer, Cham (2017). https://doi.org/10.1007/978-3-319-59153-7_39

4. Cornejo, M.E., Lobo, D., Medina, J.: Measuring the incoherent information in multi-adjoint normal logic programs. In: Kacprzyk, J., Szmidt, E., Zadrożny, S., Atanassov, K.T., Krawczak, M. (eds.) IWIFSGN/EUSFLAT -2017. AISC, vol. 641, pp. 521–533. Springer, Cham (2018). https://doi.org/10.1007/978-3-319-66830-7_47
5. Cornejo, M.E., Lobo, D., Medina, J.: Syntax and semantics of multi-adjoint normal logic programming. Fuzzy Sets and Systems, 19 December (2017). https://doi.org/10.1016/j.fss.2017.12.009
6. Dix, J., Kuter, U., Nau, D.: Planning in answer set programming using ordered task decomposition. In: Günter, A., Kruse, R., Neumann, B. (eds.) KI 2003. LNCS (LNAI), vol. 2821, pp. 490–504. Springer, Heidelberg (2003). https://doi.org/10.1007/978-3-540-39451-8_36
7. Eiter, T., Faber, W., Leone, N., Pfeifer, G.: The diagnosis frontend of the dlv system. AI Commun. **12**(1–2), 99–111 (1999)
8. Gelfond, M., Lifschitz, V.: The stable model semantics for logic programming. In: ICLP/SLP 1988, pp. 1070–1080 (1988)
9. Gelfond, M., Lifschitz, V.: Classical negation in logic programs and disjunctive databases. New Gener. Comput. **9**(3), 365–385 (1991)
10. Hájek, P.: Metamathematics of Fuzzy Logic. Trends in Logic. Kluwer Academic, Dordrecht (1998)
11. Klement, E., Mesiar, R., Pap, E.: Triangular Norms. Kluwer Academic, Dordrecht (2000)
12. Madrid, N., Ojeda-Aciego, M.: Measuring inconsistency in fuzzy answer set semantics. IEEE Trans. Fuzzy Syst. **19**(4), 605–622 (2011)
13. Madrid, N., Ojeda-Aciego, M.: On the existence and unicity of stable models in normal residuated logic programs. Int. J. Comput. Math. (2012)
14. Medina, J., Ojeda-Aciego, M., Vojtaš, P.: Multi-adjoint logic programming with continous semantics. In: Eiter, T., Faber, W., Truszczyński, M. (eds.) LPNMR 2001. LNCS (LNAI), vol. 2173, pp. 351–364. Springer, Heidelberg (2001). https://doi.org/10.1007/3-540-45402-0_26
15. Medina, J., Ojeda-Aciego, M., Vojtáš, P.: Similarity-based unification: a multi-adjoint approach. Fuzzy Sets Syst. **146**, 43–62 (2004)
16. Nieuwenborgh, D.V., Cock, M.D., Vermeir, D.: An introduction to fuzzy answer set programming. Ann. Math. Artif. Intell. **50**(3–4), 363–388 (2007)
17. Nogueira, M., Balduccini, M., Gelfond, M., Watson, R., Barry, M.: An a-prolog decision support system for the space shuttle. In: Ramakrishnan, I.V. (ed.) PADL 2001. LNCS, vol. 1990, pp. 169–183. Springer, Heidelberg (2001). https://doi.org/10.1007/3-540-45241-9_12
18. Sacca, D., Zaniolo, C.: Stable models and non-determinism in logic programs with negation. In: Proceedings of the Ninth ACM SIGACT-SIGMOD-SIGART Symposium on Principles of Database Systems, PODS 1990, pp. 205–217. ACM, New York (1990)
19. Soininen, T., Niemelä, I.: Developing a declarative rule language for applications in product configuration. In: Gupta, G. (ed.) PADL 1999. LNCS, vol. 1551, pp. 305–319. Springer, Heidelberg (1998). https://doi.org/10.1007/3-540-49201-1_21
20. Van Gelder, A., Ross, K., Schlipf, J.S.: Unfounded sets and well-founded semantics for general logic programs. In: Proceedings of the Seventh ACM SIGACT-SIGMOD-SIGART Symposium on Principles of Database Systems, PODS 1988, pp. 221–230. ACM, New York (1988)

A New Approach to Hellwig's Method of Data Reduction for Atanassov's Intuitionistic Fuzzy Sets

Eulalia Szmidt[1,2](✉) and Janusz Kacprzyk[1,2]

[1] Systems Research Institute, Polish Academy of Sciences, ul. Newelska 6, 01–447 Warsaw, Poland
{szmidt,kacprzyk}@ibspan.waw.pl

[2] Warsaw School of Information Technology, ul. Newelska 6, 01–447 Warsaw, Poland

Abstract. We propose a new approach to Hellwig's method for the reduction of dimensionality of a data set using Atanassov's intuitionistic fuzzy sets (A-IFSs). We are mainly concerned with the dimension reduction for sets of data represented as the A-IFSs, and provide an illustrative example results which are compared with the results obtained by using the PCA (Principal Component Analysis) method. Remarks on comparisons with some other methods are also mentioned.

Keywords: Data reduction · Hellwig's method · Correlation
Intuitionistic fuzzy sets

1 Introduction

The reduction of dimensionality of a data set is relevant in many fields though a different terminology may be used. For instance, in statistics, the term "variable" is used whereas in machine learning and computer science instead of "variable", the terms "feature" and "attribute" are employed. The idea of the most often used methods lies in exploring relationships among interrelated variables and using some transformations to obtain a smaller yet representative enough new set of the variables describing a data set.

One of the best known and widely employed techniques for the reduction of dimensionality of a data set is Principal Component Analysis (PCA) (Jolliffe [11]) which was first introduced in the early 1900s by Pearson [15], and later developed independently by Hotelling (1933). The method gives reliable results but is complicated from the point of view of calculations for many interrelated variables. Other methods of data reduction have their drawbacks which implies a quest for new methods. This particularly complicates matters when there are crisp, fuzzy, and other "non-standard" types of data representing different specific features of the real systems.

One of the problem faced while constructing a model of a real system is the presence of a lack of knowledge which is crucial for making decisions but at the

J. Medina et al. (Eds.): IPMU 2018, CCIS 855, pp. 553–564, 2018.
https://doi.org/10.1007/978-3-319-91479-4_46

same time difficult to foresee. This problem is amplified when we deal with the behavior of consumers (if they would buy or not a new product), investors (the value of a portfolio and behavior of separate assets), voters (for which candidate they would vote) etc., etc. Atanassov's intuitionistic fuzzy sets (A-IFSs, for short) (Atanassov [1–3]), being an extension of the fuzzy sets (Zadeh [39]), have an inherent possibility to take into account such a lack of knowledge. However, here again, the reliable models can be described by too many variables to efficiently perform simulations. So, we again face the well known problem of the reduction of dimensionality of data. The Principal Component Analysis (PCA) for the A-IFSs (cf. Szmidt and Kacprzyk [35]), Szmidt [19]) gives correct results but, again, it is quite complicated from the point of view of calculations.

In this paper we recall Hellwig's method (Hellwig [9]) for reducing the dimensionality of a linear model (just like in the case of the PCA). Hellwig's method is based, in its original terminology, on the so called capacity of information bearers. The method reduces the data dimensionality by looking for a smaller combination of the variables of a model which is best from some point of view. The best combination of the variables used in a model means here the pointing out of a subset of independent variables with the highest capacity of information. In other words, the chosen independent variables should be strongly correlated with the output of a model (a dependent variable) and weakly correlated among themselves. The method was proposed for crisp data so that we modify it so that it works with data expressed via the A-IFS in which data are described in terms of the membership values, non-membership values, and hesitation margins expressing the lack of knowledge (cf. Sect. 2). Since the A-IFSs become more and more widely applied in diverse fields, exemplified by image processing (cf. Bustince et al. [6,7]), classification of imbalanced and overlapping classes (cf. Szmidt and Kukier [36–38]), group decision making, negotiations, voting and other situations (cf. Szmidt and Kacprzyk [21,23,27,30–32]), the dimensionality reduction of sets of data given as the A-IFSs is of utmost interest, too.

We present an illustrative example showing how Hellwig's method works which makes it possible to see the advantage of the method, i.e., simpler calculations than in other methods. Hellwig [9] emphasizes that for n independent variables there are $2^n - 1$ of the possible subsets of the variables to verify to point out the best subset. However, Hellwig's method of finding an optimal combination of the variables does not require finding the inverse matrices. The illustrative example is a well known example formulated by Quinlan [17]. The example is small but in spite of its size it is not trivial and is a challenge for different learning methods. As we know, Quinlan's optimal set of variables leads to 100% of classification accuracy and we can immediately verify obtained results. We also discuss briefly results obtained by the PCA so to compare necessary steps to receive the solution by using this technique (that has found applications in many fields), and the solution itself. The obtained results are promising both in the sense of finding a proper subset of variables and numerical implementation.

2 A Brief Introduction to Intuitionistic Fuzzy Sets

One of the possible generalizations of a fuzzy set in X (Zadeh [39]) given by

$$A^{'} = \{< x, \mu_{A'}(x) > |x \in X\} \tag{1}$$

where $\mu_{A'}(x) \in [0,1]$ is the membership function of the fuzzy set $A^{'}$, is an A-IFS (Atanassov [1–3]) A is given by

$$A = \{< x, \mu_A(x), \nu_A(x) > |x \in X\} \tag{2}$$

where: $\mu_A : X \to [0,1]$ and $\nu_A : X \to [0,1]$ such that

$$0 \leq \mu_A(x) + \nu_A(x) \leq 1 \tag{3}$$

and $\mu_A(x)$, $\nu_A(x) \in [0,1]$ denote a degree of membership and a degree of non-membership of $x \in A$, respectively. (An approach to the assigning memberships and non-memberships for A-IFSs from data is proposed by Szmidt and Baldwin [20]).

Obviously, each fuzzy set may be represented by the following A-IFS: $A = \{< x, \mu_{A'}(x), 1 - \mu_{A'}(x) > |x \in X\}$.

An additional concept for each A-IFS in X, that is not only an obvious result of (2) and (3) but which is also relevant for applications, we will call (Atanassov [2])

$$\pi_A(x) = 1 - \mu_A(x) - \nu_A(x) \tag{4}$$

a *hesitation margin* of $x \in A$ which expresses a lack of knowledge of whether x belongs to A or not (cf. Atanassov [2]). It is obvious that $0 \leq \pi_A(x) \leq 1$, for each $x \in X$.

The hesitation margin turns out to be important while considering the distances (Szmidt and Kacprzyk [22,24,26], entropy (Szmidt and Kacprzyk [25,28]), similarity (Szmidt and Kacprzyk [29]) for the A-IFSs, etc. i.e., the measures that play a crucial role in virtually all information processing tasks (Szmidt [19]).

The hesitation margin turns out to be relevant for applications – in image processing (cf. Bustince et al. [6,7]), the classification of imbalanced and overlapping classes (cf. Szmidt and Kukier [36–38]), the classification applying intuitionistic fuzzy trees (cf. Bujnowski [5]), group decision making (e.g., [4]), genetic algorithms [16], negotiations, voting and other situations (cf. Szmidt and Kacprzyk papers).

2.1 Correlation Between the A-IFSs

The (degree of) correlation between variables is essential for further analysis. In the case of crisp or even fuzzy data [8,13,18] (in our context, the non-A-IFS-type data), the problem is clear and solved. Unfortunately, the very essence of the A-IFSs, in which 3 degrees (of membership, non-membership and hesitation) characterize information conveyed, the problem of correlation between such types of data is not obvious. An effective and efficient approach, which takes a full advantage of the very essence of the A-IFSs has been proposed by the authors, (cf. Szmidt and Kacprzyk [34,35], Szmidt [19]) in which a new concept of the Pearson's correlation coefficient between two A-IFS-type data was introduced and discussed.

To briefly recall, the correlation coefficient (Pearson's r) between two variables is a measure of the linear relationship between them. The correlation coefficient is 1 in the case of a positive (increasing) linear relationship, -1 in the case of a negative (decreasing) linear relationship, and some value between -1 and 1 in all other cases. The closer the coefficient is to either -1 or 1, the stronger the correlation between the variables.

The same features we demand from a correlation coefficient for two A-IFSs, A and B for which a correlation coefficient should express not only a relative strength but also a positive or negative relationship between A and B (Szmidt and Kacprzyk [34]). Next, all three terms describing an A-IFSs (membership, non-membership values and the hesitation margins) should be taken into account because each of them influences the results (Szmidt and Kacprzyk [34], Szmidt [19]). The above assumptions make the difference between our approach and those from the literature (the arguments for our approach are in (Szmidt and Kacprzyk [34], Szmidt [19]).

Suppose that we have a random sample $x_1, x_2, \ldots, x_n \in X$ with a sequence of paired data $[(\mu_A(x_1), \nu_A(x_1), \pi_A(x_1)), (\mu_B(x_1), \nu_B(x_1), \pi_B(x_1))]$, $[(\mu_A(x_2), \nu_A(x_2), \pi_A(x_2)), (\mu_B(x_2), \nu_B(x_2), \pi_B(x_2))]$, $\ldots$, $[(\mu_A(x_n), \nu_A(x_n), \pi_A(x_n)), (\mu_B(x_n), \nu_B(x_n), \pi_B(x_n))]$ which correspond to the membership values, non-memberships values and hesitation margins of A-IFSs A and B defined on X, then the correlation coefficient $r_{A-IFS}(A, B)$ is given by Definition 1 (Szmidt and Kacprzyk [34]).

Definition 1. The correlation coefficient $r_{A-IFS}(A, B)$ between two A-IFSs, A and B in X, is:

$$r_{A-IFS}(A, B) = \frac{1}{3}(r_1(A, B) + r_2(A, B) + r_3(A, B)) \tag{5}$$

where

$$r_1(A, B) = \frac{\sum_{i=1}^{n}(\mu_A(x_i) - \overline{\mu_A})(\mu_B(x_i) - \overline{\mu_B})}{(\sum_{i=1}^{n}(\mu_A(x_i) - \overline{\mu_A})^2)^{0.5}(\sum_{i=1}^{n}(\mu_B(x_i) - \overline{\mu_B})^2)^{0.5}} \tag{6}$$

$$r_2(A, B) = \frac{\sum_{i=1}^{n}(\nu_A(x_i) - \overline{\nu_A})(\nu_B(x_i) - \overline{\nu_B})}{(\sum_{i=1}^{n}(\nu_A(x_i) - \overline{\nu_A})^2)^{0.5}(\sum_{i=1}^{n}(\nu_B(x_i) - \overline{\nu_B})^2)^{0.5}} \tag{7}$$

$$r_3(A, B) = \frac{\sum_{i=1}^{n}(\pi_A(x_i) - \overline{\pi_A})(\pi_B(x_i) - \overline{\pi_B})}{(\sum_{i=1}^{n}(\pi_A(x_i) - \overline{\pi_A})^2)^{0.5}(\sum_{i=1}^{n}(\pi_B(x_i) - \overline{\pi_B})^2)^{0.5}} \tag{8}$$

where: $\overline{\mu_A} = \frac{1}{n}\sum_{i=1}^{n}\mu_A(x_i)$, $\overline{\mu_B} = \frac{1}{n}\sum_{i=1}^{n}\mu_B(x_i)$, $\overline{\nu_A} = \frac{1}{n}\sum_{i=1}^{n}\nu_A(x_i)$,
$\overline{\nu_B} = \frac{1}{n}\sum_{i=1}^{n}\nu_B(x_i)$, $\overline{\pi_A} = \frac{1}{n}\sum_{i=1}^{n}\pi_A(x_i)$, $\overline{\pi_B} = \frac{1}{n}\sum_{i=1}^{n}\pi_B(x_i)$,

The proposed correlation coefficient (5) depends on two factors: the amount of information expressed by the membership and non-membership degrees (6)–(7), and the reliability of information expressed by the hesitation margins (8).

Remark: Analogously as for the crisp and fuzzy data, $r_{A-IFS}(A,B)$ makes sense for A-IFS variables whose values vary. If, for instance, the temperature is constant and the amount of ice cream sold is the same, then it is impossible to conclude anything about their relationship (as, from the mathematical point of view, we avoid zero in the denominator).

The correlation coefficient $r_{A-IFS}(A,B)$ (5) fulfills the following properties:

1. $r_{A-IFS}(A,B) = r_{A-IFS}(B,A)$
2. If $A = B$ then $r_{A-IFS}(A,B) = 1$
3. $|r_{A-IFS}(A,B)| \leq 1$

The above properties are fulfilled both by the correlation coefficient $r_{A-IFS}(A,B)$ (5) and also by all of its components (6)–(8).

Remark: It is should be emphasized that $r_{A-IFS}(A,B) = 1$ occurs not only for $A = B$ but also in the cases of a perfect linear correlation of the data (the same concerns each component (6)–(8)).

In Szmidt and Kacprzyk [34] there are examples showing that each component may play an important role when considering correlation between A-IFSs. On the other hand, (5) which aggregates (6)–(8)) plays an important role as a bird-eye-view revision of the correlation – in extreme cases, i.e., for the values: −1, 0, and 1 we have exact summarized information about the correlation. For other cases (5) suffers from the same drawbacks each aggregation measure does.

3 Hellwig's Method of Data Reduction for the A-IFSs

Hellwig's method (Hellwig [9]), also called the method of the capacity of information bearers, is a method of variable selection in linear models. The method looks for the best combination of the variables i.e. selects them which means data reduction. The best combination of the variables used in a model means pointing out a subset of a model independent variables with the highest capacity of information. In other words, the chosen independent variables should be strongly correlated with the output of a model (a dependent variable) and weakly correlated among themselves.

The advantage of Hellwig's method lies in a simpler calculations than in other methods reducing the dimensionality of a model. For example, for well known Principal Component Analysis (PCA) method (Jackson [10], Jolliffe [11], [12], Marida et al. [14]) one should find the eigenvectors which is easy only for small matrices. As Hellwig stresses [9], for n independent variables there are $2^n - 1$ of the possible subsets of variables and his method of finding the optimal combination does not require finding the inverse matrices.

To apply Hellwig's method ones needs to use the correlation coefficients, namely,

- a vector of correlation coefficients R_0 between dependent attribute (called as well "feature", "variable" or "predictand") Y and independent attributes ("features", "variables" or "predictors") $X_1, X_2, \ldots, X_n$, and
- a matrix of correlation coefficients R among independent attributes $X_1, X_2, \ldots, X_n$.

In other words, we need a vector $R_o = (r_1, r_2, \ldots, r_n)$ where r_j is a correlation coefficient between X_j and Y, and a symmetric matrix of correlation coefficients R, with the elements $r_{i,j}$ being the correlation coefficients between values X_i and X_j:

$$R = \begin{bmatrix} 1 & r_{12} & \ldots & r_{1n} \\ r_{21} & 1 & \ldots & r_{2n} \\ . & . & . & . \\ r_{n1} & r_{n2} & \ldots & r_{nn} \end{bmatrix}$$

Performing of the Hellwig's method consists of three steps

- calculation of the capacity of an individual information bearer X_j for the k-th combination
$$h_{kj} = \frac{r_j^2}{1 + \sum\limits_{i=1, i \neq j}^{m} |r_{ij}|} \tag{9}$$
where k is number of a combination, $k = 1, 2, \ldots, l$, and j is a number of a variable in the combination, $j = 1, 2, \ldots, m$.
- calculation of the integral capacity of individual information bearers for all combinations
$$H_k = \sum_{j=1}^{m} h_{kj} \tag{10}$$
where $k = 1, 2, \ldots, l$.
- the last step is to find the maximal value among the H_k's.

As an illustration how Hellwig's method works, i.e. how the reduction of the dimensionality of a data set proceeds, we will recall a well known problem formulated by Quinlan [17] but expressed in terms of the A-IFSs. The Quinlan's example, the so-called "Saturday Morning" example, considers the classification with nominal data. This example is small enough and illustrative, yet is a challenge to many classification and machine learning methods. The main idea of solving the example by Quinlan was to select the best attributes (variables) to split the training set (Quinlan used a so-called *Information Gain* which was a dual measure to Shannon's entropy).

We will verify if Hellwig's method gives satisfying results of selecting the best attribute for the problem formulated by Quinlan [17] but expressed in terms of the A-IFSs.

In Quinlan's example [17] (Table 1) we have objects described by attributes. Each attribute represents a feature and takes on discrete, mutually exclusive values. For example, if the objects were "Saturday Mornings" and the classification involved the weather, possible attributes might be [17]:

- **outlook**, with values {sunny, overcast, rain},
- **temperature**, with values {cold, mild, hot},
- **humidity**, with values {high, normal}, and
- **windy**, with values {true, false},

Table 1. The "Saturday Morning" data from [17]

No.	Attributes				Class
	Outlook	Temp	Humidity	Windy	
1	Sunny	Hot	High	False	N
2	Sunny	Hot	High	True	N
3	Overcast	Hot	High	False	P
4	Rain	Mild	High	False	P
5	Rain	Cool	Normal	False	P
6	Rain	Cool	Normal	True	N
7	Overcast	Cool	Normal	True	P
8	Sunny	Mild	High	False	N
9	Sunny	Cool	Normal	False	P
10	Rain	Mild	Normal	False	P
11	Sunny	Mild	Normal	True	P
12	Overcast	Mild	High	True	P
13	Overcast	Hot	Normal	False	P
14	Rain	Mild	High	True	N

The limitation of space does not let us discuss the method of deriving the A-IFS counterpart of Quinlan's example (Table 2) in detail (cf. Szmidt and Kacprzyk [33]) and we only present here the final results.

Next, making use of the A-IFS model (Table 2) and applying (5) we computed the correlation matrices
$R_o = (r_1, r_2, r_3, r_4) = [0.48, 0.45, 0.258, 0.257]$,
and matrix R which values are in Table 3.

As there are 4 variables in our example: Outlook, Humidity, Windy, Temperature (which are abbreviated further as: X_1, X_2, X_3, X_4), it is possible to select

Table 2. The "Saturday Morning" data in terms of the A-IFSs

No.	Attributes				Class
	Outlook	Humidity	Windy	Temperature	
1	(0, 0.33, 0.67)	(0, 0.33, 0.67)	(0.2, 0, 0.8)	(0, 0.33, 0.67)	N
2	(0, 0.33, 0.67)	(0, 0.33, 0.67)	(0, 0.33, 0.67)	(0, 0.33, 0.67)	N
3	(1, 0, 0)	(0, 0.33, 0.67)	(0.2, 0, 0.8)	(0, 0.33, 0.67)	P
4	(0.2, 0.11, 0.69)	(0, 0.33, 0.67)	(0.2, 0, 0.8)	(0, 0, 1)	P
5	(0.2, 0.11, 0.69)	(0.6, 0, 0.4)	(0.2, 0, 0.8)	(0.4, 0.11, 0.49)	P
6	(0.2, 0.11, 0.69)	(0.6, 0, 0.4)	(0, 0.33, 0.67)	(0.4, 0.11, 0.49)	N
7	(1, 0, 0)	(0.6, 0, 0.4)	(0, 0.33, 0.67)	(0.4, 0.11, 0.49)	P
8	(0, 0.33, 0.67)	(0, 0.33, 0.67)	(0.2, 0, 0.8)	(0, 0, 1)	N
9	(0, 0.33, 0.67)	(0.6, 0, 0.4)	(0.2, 0, 0.8)	(0.4, 0.11, 0.49)	P
10	(0.2, 0.11, 0.69)	(0.6, 0, 0.4)	(0.2, 0, 0.8)	(0, 0, 1)	P
11	(0, 0.33, 0.67)	(0.6, 0, 0.4)	(0, 0.33, 0.67)	(0, 0, 1)	P
12	(1, 0, 0)	(0, 0.33, 0.67)	(0, 0.33, 0.67)	(0, 0, 1)	P
13	(1, 0, 0)	(0.6, 0, 0.4)	(0.2, 0, 0.8)	(0, 0.33, 0.67)	P
14	(0.2, 0.11, 0.69)	(0, 0.33, 0.67)	(0, 0.33, 0.67)	(0, 0, 1)	N

Table 3. Evaluation of the correlation coefficients (5) of the "Saturday Morning" data

Attribute	Outlook (X_1)	Humidity (X_2)	Windy (X_3)	Temperature (X_4)
Outlook (X_1)	1	0.05	−0.002	0.06
Humidity (X_2)	0.05	1	0	0.4
Windy (X_3)	−0.002	0	1	−0.14
Temperature (X_4)	0.06	0.4	−0.14	1

from them $2^4 - 1 = 15$ various subsets. If Q_m stands for an element of such arrangement, we get

$Q_1 : \{X_1\}$, $Q_2 : \{X_2\}$, $Q_3 : \{X_3\}$, $Q_4 : \{X_4\}$,
$Q_5 : \{X_1, X_2\}$, $Q_6 : \{X_1, X_3\}$, $Q_7 : \{X_1, X_4\}$, $Q_8 : \{X_2, X_3\}$, $Q_9 : \{X_2, X_4\}$,
$Q_{10} : \{X_3, X_4\}$, $Q_{11} : \{X_1, X_2, X_3\}$, $Q_{12} : \{X_1, X_2, X_4\}$, $Q_{13} : \{X_1, X_3, X_4\}$,
$Q_{14} : \{X_2, X_3, X_4\}$, $Q_{15} : \{X_1, X_2, X_3, X_4\}$

For all the combinations $Q_1 - Q_{15}$ we find from (9)–(10)
$H_1 = h_{11} = r_1^2 = 0.48^2 = 0.23$
$H_2 = h_{22} = r_2^2 = 0.45^2 = 0.2$
$H_3 = h_{33} = r_3^2 = 0.258^2 = 0.067$
$H_4 = h_{44} = r_4^2 = 0.257^2 = 0.066$
$H_5 = h_{51} + h_{52} = \frac{r_1^2}{1+|r_{12}|} + \frac{r_2^2}{1+|r_{12}|} = 0.41$

$H_6 = h_{61} + h_{63} = \frac{r_1^2}{1+|r_{13}|} + \frac{r_3^2}{1+|r_{13}|} = 0.296$

$H_7 = h_{71} + h_{74} = \frac{r_1^2}{1+|r_{14}|} + \frac{r_4^2}{1+|r_{14}|} = 0.279$

$H_8 = h_{82} + h_{83} = \frac{r_2^2}{1+|r_{23}|} + \frac{r_3^2}{1+|r_{23}|} = 0.267$

$H_9 = h_{82} + h_{84} = \frac{r_2^2}{1+|r_{24}|} + \frac{r_4^2}{1+|r_{24}|} = 0.19$

$H_{10} = h_{10,3} + h_{10,4} = \frac{r_3^2}{1+|r_{34}|} + \frac{r_4^2}{1+|r_{24}|} = 0.117$

$H_{11} = h_{11,1} + h_{11,2} + h_{11,3} = \frac{r_1^2}{1+|r_{12}|+|r_{13}|} + \frac{r_2^2}{1+|r_{12}|+|r_{23}|} + \frac{r_3^2}{1+|r_{23}|+|r_{13}|} = 0.476$

$H_{12} = h_{12,1} + h_{12,2} + h_{12,4} = \frac{r_1^2}{1+|r_{12}|+|r_{14}|} + \frac{r_2^2}{1+|r_{12}|+|r_{24}|} + \frac{r_4^2}{1+|r_{14}|+|r_{24}|} = 0.39$

$H_{13} = h_{13,1} + h_{13,3} + h_{13,4} = \frac{r_1^2}{1+|r_{13}|+|r_{14}|} + \frac{r_3^2}{1+|r_{13}|+|r_{34}|} + \frac{r_4^2}{1+|r_{14}|+|r_{34}|} = 0.327$

$H_{14} = h_{14,2} + h_{14,3} + h_{14,4} = \frac{r_2^2}{1+|r_{23}|+|r_{24}|} + \frac{r_3^2}{1+|r_{23}|+|r_{34}|} + \frac{r_4^2}{1+|r_{24}|+|r_{34}|} = 0.245$

$H_{15} = h_{15,1} + h_{15,2} + h_{15,3} + h_{15,4} = \frac{r_1^2}{1+|r_{12}|+|r_{13}|+|r_{14}|} + \frac{r_2^2}{1+|r_{12}|+|r_{23}|+|r_{24}|} + \frac{r_3^2}{1+|r_{13}|+|r_{23}|+|r_{34}|} + \frac{r_4^2}{1+|r_{14}|+|r_{24}|+|r_{34}|} = 0.445$

It is easy to notice that the best combination of the attributes is combination $Q_{11} : \{X_1, X_2, X_3\}$ (Outlook, Humidity, Windy) for which its corresponding value of H_{11} is the highest among all the values of H_l.

It is worth noticing that Quinlan obtained 100% classification accuracy, and the optimal solution (the minimal possible ID3 tree) also involved the same (as pointed out by Hellwig's method) three (of four) attributes.

3.1 Brief Comparison with the Principal Component Analysis (PCA) for the A-IFS Data

Now we will compare the results for the same example but obtained by Principal Component Analysis PCA) as it is one of the best known and widely used linear dimension reduction technique Jackson [10], Jolliffe [11], Marida et al. [14] in the sense of mean-square error.

The Principal Component Analysis (PCA), i.e., the reduction of dimensionality of a data set in which there are lots of interrelated variables, is performed by transforming the source set of data to a new set of uncorrelated variables/features/attributes (the principal components PC) to summarize the features of the original data. The principal components (PCs) are ordered such that the k-th PC has the k-th largest variance among all PCs. The k-th PC points out the direction that maximizes the variation of the projections of the data points such that it is orthogonal to the first $(k-1)$-th PCs. Traditionally, the first few PC are used in data analysis (they capture most of the variation in the original data set).

The steps of PCA for crisp sets (Jolliffe [12], Jackson [10]) are:

- find the correlation matrix,
- find the eigenvectors and eigenvalues of the correlation matrix,
- rearrange the eigenvectors and eigenvalues in the order of decreasing eigenvalues,

- select a subset of the eigenvectors as the basis vectors,
- convert the source data into the new basis.

After performing the above steps adapted to the data expressed via A-IFSs (cf. Szmidt and Kacprzyk [35]), we have noticed that the first three eigenvalues explain most of variability of the data (85%), and summarize the most important features of the data. The obtained result is consistent with the original result Quinlan gives [17] who has indicated that the optimal tree that classifies correctly all data consists of three attributes only (Outlook, Humidity, Windy). The result is just the same as we have obtained while applying Hellwig's method.

Clearly, our example is just for illustration as feature reduction makes sense for large problems (very many features) and then the reduction is usually considerable and very welcome.

4 Conclusions

We presented a novel approach to Hellwig's method for the reduction of data sets for data expressed by the A-IFSs. We used three terms representation of A-IFSs, i.e. taking into account the degree of membership, non-membership and hesitation margin. Such a description turned out important while calculating correlation coefficients in the case of A-IFSs (Szmidt and Kacprzyk [34,35], Szmidt [19]). We hope that the new approach to Hellwig's method for the A-IFS-type data can be important because, on the one hand, the A-IFSs gain a wider and wider importance as a tool for data representation and processing in more and more areas. On the other hand, because of the drawbacks of the existing methods, new techniques in data analysis are constantly looked for. Some methods like the PCA, which is one of the most relevant techniques in data analysis, are difficult to perform for bigger data (complicated calculations). In this context, Hellwig's method with its simpler calculations seems very promising.

References

1. Atanassov, K.: Intuitionistic Fuzzy Sets. VII ITKR Session. Sofia (Centr. Sci.-Techn. Libr. of Bulg. Acad. of Sci., 1697/84) (1983) (in Bulgarian)
2. Atanassov, K.: Intuitionistic Fuzzy Sets: Theory and Applications. Springer, Heidelberg (1999). https://doi.org/10.1007/978-3-7908-1870-3
3. Atanassov, K.T.: On Intuitionistic Fuzzy Sets Theory. Springer, Heidelberg (2012). https://doi.org/10.1007/978-3-642-29127-2
4. Atanassova, V.: Strategies for Decision Making in the Conditions of Intuitionistic Fuzziness. In: International Conference on 8th Fuzzy Days, Dortmund, Germany, pp. 263–269 (2004)
5. Bujnowski, P., Szmidt, E., Kacprzyk, J.: Intuitionistic fuzzy decision trees - a new approach. In: Rutkowski, L., Korytkowski, M., Scherer, R., Tadeusiewicz, R., Zadeh, L.A., Zurada, J.M. (eds.) ICAISC 2014. LNCS (LNAI), vol. 8467, pp. 181–192. Springer, Cham (2014). https://doi.org/10.1007/978-3-319-07173-2_17

6. Bustince, H., Mohedano, V., Barrenechea, E., Pagola, M.: An algorithm for calculating the threshold of an image representing uncertainty through A-IFSs. In: IPMU 2006, pp. 2383–2390 (2006)
7. Bustince, H., Mohedano, V., Barrenechea, E., Pagola, M.: Image thresholding using intuitionistic fuzzy sets. In: Atanassov, K., Kacprzyk, J., Krawczak, M., Szmidt, E. (eds.) Issues in the Representation and Processing of Uncertain and Imprecise Information. Fuzzy Sets, Intuitionistic Fuzzy Sets, Generalized Nets, and Related Topics. EXIT, Warsaw (2005)
8. Fang, Y.-C., Tzeng, Y.-F., Li, S.-X.: A Taguchi PCA fuzzy-based approach for the multi-objective extended optimization of a miniature optical engine. J. Phys. D Appl. Phys. **41**(17), 175–188 (2008)
9. Hellwig, Z.: On the optimal choice of predictors. In: Gostkowski, Z. (ed.) Toward a System of Quantitative Indicators of Components of Human Resources Development, Study VI. UNESCO, Paris (1968)
10. Jackson, J.E.: A User's Guide to Principal Components. Wiley, New York (1991)
11. Jolliffe, I.T.: Principal Component Analysis. Springer, New York (1986). https://doi.org/10.1007/b98835
12. Jolliffe, I.T.: Principal Component Analysis, 2nd edn. Springer, New York (2002). https://doi.org/10.1007/b98835
13. Xia, L., Zhao, C.: The application of PCA-fuzzy probability analysis on risk evaluation of construction schedule of highway. In: IEEE 2010 International Conference on Logistics Systems and Intelligent Management, pp. 1230–1234 (2010)
14. Mardia, K.V., Kent, J.T., Bibby, J.M.: Multivariate Analysis. Probability and Mathematical Statistics. Academic Press, New York (1995)
15. Pearson, K.: On lines and planes of closest fit to systems of points in space. Phil. Mag. **6**(2), 559–572 (1901)
16. Roeva, O., Michalikova, A.: Generalized net model of intuitionistic fuzzy logic control of genetic algorithm parameters. Notes on Intuitionistic Fuzzy Sets **19**(2), 71–76 (2013). ISSN 1310–4926
17. Quinlan, J.R.: Induction of decision trees. Mach. Learn. **1**, 81–106 (1986)
18. Sebzalli, Y.M., Wang, X.Z.: Knowledge discovery from process operational data using PCA and fuzzy clustering. Eng. Appl. Artif. Intell. **14**, 607–616 (2001)
19. Szmidt, E.: Distances and Similarities in Intuitionistic Fuzzy Sets. Springer, Cham (2014). https://doi.org/10.1007/978-3-319-01640-5
20. Szmidt, E., Baldwin, J.: Intuitionistic fuzzy set functions, mass assignment theory, possibility theory and histograms. In: IEEE World Congress on Computational Intelligence 2006, pp. 237–243 (2006)
21. Szmidt, E., Kacprzyk, J.: Remarks on some applications of intuitionistic fuzzy sets in decision making. Notes IFS **2**(3), 22–31 (1996c)
22. Szmidt, E., Kacprzyk, J.: On measuring distances between intuitionistic fuzzy sets. Notes IFS **3**(4), 1–13 (1997)
23. Szmidt, E., Kacprzyk, J.: Group decision making under intuitionistic fuzzy preference relations. In: IPMU 1998, pp. 172–178 (1998)
24. Szmidt, E., Kacprzyk, J.: Distances between intuitionistic fuzzy sets. Fuzzy Sets Syst. **114**(3), 505–518 (2000)
25. Szmidt, E., Kacprzyk, J.: Entropy for intuitionistic fuzzy sets. Fuzzy Sets Syst. **118**(3), 467–477 (2001)
26. Szmidt, E., Kacprzyk, J.: Distances between intuitionistic fuzzy sets: straightforward approaches may not work. In: IEEE IS 2006, pp. 716–721 (2006)

27. Szmidt, E., Kacprzyk, J.: An application of intuitionistic fuzzy set similarity measures to a multi-criteria decision making problem. In: Rutkowski, L., Tadeusiewicz, R., Zadeh, L.A., Żurada, J.M. (eds.) ICAISC 2006. LNCS (LNAI), vol. 4029, pp. 314–323. Springer, Heidelberg (2006). https://doi.org/10.1007/11785231_34
28. Szmidt, E., Kacprzyk, J.: Some problems with entropy measures for the atanassov intuitionistic fuzzy sets. In: Masulli, F., Mitra, S., Pasi, G. (eds.) WILF 2007. LNCS (LNAI), vol. 4578, pp. 291–297. Springer, Heidelberg (2007). https://doi.org/10.1007/978-3-540-73400-0_36
29. Szmidt, E., Kacprzyk, J.: A new similarity measure for intuitionistic fuzzy sets: straightforward approaches may not work. In: 2007 IEEE Conference on Fuzzy Systems, pp. 481–486 (2007a)
30. Szmidt, E., Kacprzyk, J.: A new approach to ranking alternatives expressed via intuitionistic fuzzy sets. In: Ruan, D., et al. (eds.) Computational Intelligence in Decision and Control, pp. 265–270. World Scientific (2008)
31. Szmidt, E., Kacprzyk, J.: Amount of information and its reliability in the ranking of Atanassov's intuitionistic fuzzy alternatives. In: Rakus-Andersson, E., Yager, R., Ichalkaranje, N., Jain, L.C. (eds.) Recent Advances in Decision Making. SCI, vol. 222, pp. 7–19. Springer, Heidelberg (2009). https://doi.org/10.1007/978-3-642-02187-9_2
32. Szmidt, E., Kacprzyk, J.: Ranking of intuitionistic fuzzy alternatives in a multi-criteria decision making problem. In: Proceedings of the conference: NAFIPS 2009, Cincinnati, USA, 14–17 June 2009. IEEE (2009). ISBN 978-1-4244-4577-6
33. Szmidt, E., Kacprzyk, J.: Dealing with typical values via Atanassov's intuitionistic fuzzy sets. Int. J. General Syst. **39**(5), 489–506 (2010)
34. Szmidt, E., Kacprzyk, J.: Correlation of intuitionistic fuzzy sets. In: Hüllermeier, E., Kruse, R., Hoffmann, F. (eds.) IPMU 2010. LNCS (LNAI), vol. 6178, pp. 169–177. Springer, Heidelberg (2010). https://doi.org/10.1007/978-3-642-14049-5_18
35. Szmidt, E., Kacprzyk, J.: A new approach to principal component analysis for intuitionistic fuzzy data sets. In: Greco, S., Bouchon-Meunier, B., Coletti, G., Fedrizzi, M., Matarazzo, B., Yager, R.R. (eds.) IPMU 2012. CCIS, vol. 298, pp. 529–538. Springer, Heidelberg (2012). https://doi.org/10.1007/978-3-642-31715-6_56
36. Szmidt, E., Kukier, M.: Classification of imbalanced and overlapping classes using intuitionistic fuzzy sets. In: IEEE IS 2006, London, pp. 722–727 (2006)
37. Szmidt, E., Kukier, M.: A new approach to classification of imbalanced classes via atanassov's intuitionistic fuzzy sets. In: Wang, H.-F. (ed.) Intelligent Data Analysis: Developing New Methodologies Through Pattern Discovery and Recovery, pp. 85–101. Idea Group (2008)
38. Szmidt, E., Kukier, M.: Atanassov's intuitionistic fuzzy sets in classification of imbalanced and overlapping classes. In: Chountas, P., Petrounias, I., Kacprzyk, J. (eds.) Intelligent Techniques and Tools for Novel System Architectures. SCI, vol. 109, pp. 455–471. Springer, Heidelberg (2008). https://doi.org/10.1007/978-3-540-77623-9_26
39. Zadeh, L.A.: Fuzzy sets. Inf. Control **8**, 338–353 (1965)

Representing Hypoexponential Distributions in Continuous Time Bayesian Networks

Manxia Liu[1,4](✉), Fabio Stella[5], Arjen Hommersom[1,2], and Peter J. F. Lucas[1,3]

[1] Radboud University, ICIS, Nijmegen, The Netherlands
{m.liu,arjenh,peterl}@cs.ru.nl
[2] Open University of the Netherlands, Heerlen, The Netherlands
[3] Leiden University, LIACS, Leiden, The Netherlands
[4] University of Porto, CINTESIS, Porto, Portugal
[5] University of Milano-Bicocca, Milan, Italy
stella@disco.unimib.it

Abstract. Continuous time Bayesian networks offer a compact representation for modeling structured stochastic processes that evolve over continuous time. In these models, the time duration that a variable stays in a state until a transition occurs is assumed to be exponentially distributed. In real-world scenarios, however, this assumption is rarely satisfied, in particular when describing more complex temporal processes. To relax this assumption, we propose an extension to support the modeling of the transitioning time as a hypoexponential distribution by introducing an additional hidden variable. Using such an approach, we also allow CTBNs to obtain *memory*, which is lacking in standard CTBNs. The parameter estimation in the proposed models is transformed into a learning task in their equivalent Markovian models.

Keywords: Continuous time Bayesian networks
Dynamic Bayesian networks · Phase-type distribution · Memory
Hidden variable

1 Introduction

Continuous time Bayesian networks, or CTBNs for short, firstly introduced by Nodelman et al. [1], offer a compact representation for modeling structured stochastic processes that evolve over continuous time. By providing an explicit representation of time, i.e., time acts as a continuous parameter, these models

ML is supported by China Scholarship Council, by a grant from project NanoSTIMA [NORTE-01-0145-FEDER-000016], which was financed by the North Portugal Regional Operational Programme [NORTE 2020], under the PORTUGAL 2020 Partnership Agreement, and through the European Regional Development Fund [ERDF], and by Italian Association for Artificial Intelligence (AI*IA).

J. Medina et al. (Eds.): IPMU 2018, CCIS 855, pp. 565–577, 2018.
https://doi.org/10.1007/978-3-319-91479-4_47

have the advantage of representing a probability distribution over observations that are made at irregularly spaced points in time. The powerful expressiveness of CTBNs to model data in such a form has been demonstrated by numerous early work (see e.g. reliability modeling [2], network intrusion detection [3,4], heart failure modeling [5,6], and gene network construction [7]).

In spite of supporting time irregularity, CTBNs suffer from an important limitation in their expressiveness: the time that a variable stays in a state until transition follows an exponential distribution. This distribution occurs naturally when describing a process where events occur continuously and independently at a constant average rate. In real-world scenarios, however, the assumption is rarely satisfied. In particular, it is inappropriate to describe more complex temporal processes, such as business processes that model interpurchase times [8]. The limitation was firstly described by Nodelman and Horvitz [9]; subsequent work by Gopalratman et al. [10] focuses on Erlang-Coxian distributions to handle time duration. To overcome the limitation, two approaches were proposed by Nodelman et al. [11] to extend CTBNs to phase-type duration distributions, yielding a richer and more flexible distribution. The first approach is to add hidden *states* to the random variables of a CTBN, which is called the *direct approach*. Alternatively, a second and more elegant approach is to add hidden *variables* to a CTBN. From a practical point of view, this approach is attractive, because existing CTBN inference algorithms can be directly applied. For the direct representation, states have to be interpreted as a disjunction of hidden states, which is cumbersome and computationally expensive when using existing software packages. However, the question of how to add the hidden variables to the network structure and what constraints should be imposed on the structure of their parameters was left unresolved [11].

In this paper, we show that the hidden variable approach can be used to represent a large class of duration distributions described by hypoexponential distributions. These distributions significantly generalize the existing exponential distributions. As a second contribution of this paper, we give precise conditions on the CTBN graph and discuss the exact constraints on the parameter structure for representing these distributions. We also show how these models are formally related to the direct representation, which we use for learning the parameters of the model.

The rest of the paper is organized as follows. We start with a motivating example in Sect. 2, followed by a brief summary of CTBNs and hypoexponential distribution in Sect. 3. Then, in Sect. 4, we define hidden continuous time Bayesian networks (HCTBNs). In Sect. 5, we show the relationships between the hidden variable model and the direct models. Subsequently, in Sect. 6, we demonstrate the usefulness of our proposed models by describing non-exponential distribution using HCTBNs and CTBNs, and by modeling dynamics of a medical problem. Finally, the paper is concluded with a brief discussion of possible future work for HCTBNs.

2 Motivating Example

To illustrate the proposed theory, we consider a medical example, viz. factors that influence the cardiac output, i.e., the blood flow to and from the heart. According to the literature, the heart rate, defined in terms of the number of heart beats per minute, has a positive influence on the cardiac output. However, a reduced blood supply, thus oxygen supply, as the result of coronary artery disease, may give rise to a heart attack (myocardial infarction). Consequently, some of the heart muscle fibers will die and the heart may fail to comply with respect to its function as a pump, thus cardiac output will be negatively affected. With regard to the prognosis, (increased) heart rate may be considered a risk factor for myocardial infarction (this is the rationale behind treatment of coronary artery disease patients with beta-blocking drugs, such as propranolol, that decrease heart rate). This causal knowledge is formalized as a directed graph in Fig. 1. Diagnosis of a myocardial infarction is done by examining the shape of the ECG and by determining the levels of troponin (a protein that is released from the dying heart cells) in the blood. In the model we take into account that lab facilities (to determine an ECG and troponin levels in the blood) are not available, as is common in some developing countries. Thus, diagnosing a myocardial infarction in the common way is not an option. As a result, the observations solely consist of the heart rate. With respect to modeling, this also implies hidden causes, such as myocardial infarction, must be taken into consideration when assessing potential causes for reduced cardiac output. More importantly, remembering having a myocardial infarction in the past, which is called *memory*, can alter the evolution of cardiac output in the future. In the remainder of this paper, we propose a method to deal with modeling such hidden causes, in particular to describe the memory behavior of temporal processes that evolve continuously over time.

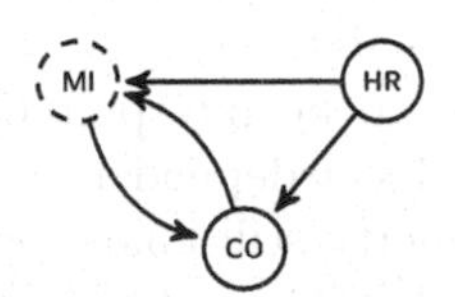

Fig. 1. Causal model for cardiac output: MI = Myocardial infarction; CO = cardiac output, HR = heart rate. The dashed node indicates a hidden cause.

3 Preliminaries

In this section, we will introduce the technical background of continuous time Bayesian networks as originally presented by Nodelman et al. [12] and phase-type distributions. The domain for an n-valued variable X is denoted as $\text{Val}(X) = \{1, 2, \ldots, n\}$ with the notation $X = i$ indicating that variable X has the value i. We also use the notation $\pi(X)$ for the parents of variable X in a given graph.

3.1 Continuous Time Bayesian Networks

A *continuous-time Bayesian network* represents a stochastic process over a structured state space consisting of assignments to a set of local variables. The dynamics of the temporal evolution of the structured state space is described in terms of the evolution of the local variables. Let X be such a *local variable* with finite *domain* $\text{Val}(X) = \{1, 2, \ldots, n\}$, where $i \in \text{Val}(X)$ is called a *state*, and state changes over continuous time. The dynamics of X can be described as a homogeneous Markov process via its *intensity matrix*:

$$Q_X = \begin{pmatrix} q_{11} & q_{12} & \cdots & q_{1n} \\ q_{21} & q_{22} & \cdots & q_{2n} \\ \vdots & \vdots & \ddots & \vdots \\ q_{n1} & q_{n2} & \cdots & q_{nn} \end{pmatrix}$$

where $q_{ii} = -\sum_{j \neq i} q_{ij}$. The time that variable X stays in state i is exponentially distributed with rate $-q_{ii}$ and the expected time is given by $-1/q_{ii}$, and once it transitions from state i, it shifts to state j with probability $-q_{ij}/q_{ii}$.

CTBNs are based on homogeneous continuous time Markov processes, which has the Markov property, also known as a stronger assumption of *memorylessness*. The Markov property states that given the state of the process X at any set of times prior to time t, the distribution of X at time t depends *only* on X at the most recent time prior to time t. It is equivalent to say that given the state of the process X at time s, the distribution of X at any time after s is independent of the entire past of X prior to time s. More formally:

$$P(X_t = j \mid X_{t_1} = k_1, X_{t_2} = k_2, \ldots, X_{t_n} = k_n, X_s = i) = P(X_t = j \mid X_s = i)$$

where $0 < t_1 < t_2 < \cdots < t_n < s < t$.

However, this property has to be interpreted more carefully in CTBNs as these models also express the local dependence of one variable on the others. It is true that the Markov property still holds for CTBNs when conditioned on *all* the local variables in a model; it is not the case when only conditioning on a proper subset of the variables. This is due to the temporal entanglement in CTBNs where time is also considered. Let $\mathbf{X}$ be the variables in a CTBN and $\mathbf{Z}$ be a proper subset of $\mathbf{X}$, i.e., $\mathbf{Z} \subsetneq \mathbf{X}$. When querying the distribution over variables $\mathbf{Z}$ at time t, the distribution over variables $\mathbf{Z}$ at time t is no longer independent from its states at time prior to s given the states of variables $\mathbf{Z}$ at time s. More formally:

$$P(\mathbf{Z}_t = j, \mid \mathbf{Z}_{t_1} = k_1, \mathbf{Z}_{t_2} = k_2, \ldots, \mathbf{Z}_{t_n} = k_n, \mathbf{Z}_s = i) \neq P(\mathbf{Z}_t = j \mid \mathbf{Z}_s = i)$$

This is because the information at time prior to s is propagated to the time t through variables $\mathbf{X} \setminus \mathbf{Z}$. In this paper, we refer such behavior of variables $\mathbf{Z}$ as *memory*. Without loss of generality, we introduce memory to all variables in a given CTBN by adding an additional *hidden* variable. The states of non-hidden variables are dependent on the their states in the past as the hidden variable is always unobservable. In this paper, we restrict ourselves to studying such memory behavior in CTBNs.

3.2 Hypoexponential Distribution

A phase-type distribution is a distribution which describes the time until reaching the absorbing state of a continuous time Markov chain with n transient states and one absorbing state. A phase-type distribution represented by n transient states is said to have order n. This continuous time Markov chain can be described as a state transition diagram. The diagram is a convenient graphical representation in terms of the initial probabilities, i.e., the distribution over the transient states at $t = 0$, the transition rates between the transient states, and the exit rates, i.e., the probability of entering the absorbing state.

Exponential distributions are a special case of phase-type distributions, where the continuous time Markov process has one transient state. The distribution can thus be graphically represented by a state transition diagram with only one state as shown in Fig. 2a. The diagram asserts that the chain enters the first and only transient state 1 with probability one and enters the absorbing state with rate λ. The *hypoexponential distribution*, also known as *generalized Erlang distribution*, is the distribution of the sum of n independent and identically exponentially distributed random variables. The state transition diagram of the Markov chain of n-order hypoexponential distribution is shown in Fig. 2. More details about phase-type distribution can be found in [13].

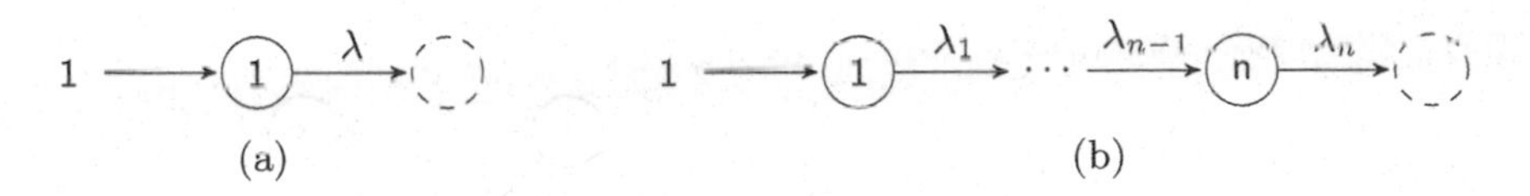

Fig. 2. State transition diagram for exponential distribution as shown in (a) and an n-order hypoexponential distribution as shown in (b). A solid node indicates a transient state and a dashed node indicates an absorbing state.

4 Hidden Continuous Time Bayesian Networks

In this section, we define a new extension of CTBNs, which we call hidden continuous time Bayesian networks, abbreviated to HCTBNs, where there is *only* one variable whose time staying in a state until a transition occurs, given a particular configuration of its parents, is described by a hypoexponential distribution. For other variables, the transition times are exponentially distributed.

4.1 Structure

First, we define the structure associated to an HCTBN, which contains a labeled node X which will correspond to a binary hypoexponential variable, a labeled node H corresponding to a hidden variable that is used to represent the hypoexponential distribution for X, and labeled $\mathbf{Y}$ to exponential variables.

Definition 1 (HCTBN Graph). *An HCTBN graph is a labeled graph defined by a triple $G = (\mathbf{V}, \mathbf{E}, l)$, where $\mathbf{V} = \{X, H\} \cup \mathbf{Y}$ denotes a set of vertices, $\mathbf{E} \subseteq \mathbf{V} \times \mathbf{V}$ a set of arcs on $\mathbf{V}$, and l a label function such that $l(X)$ = hypoexponential, $l(H)$ = hidden and $l(\mathbf{Y})$ = exponential. The following conditions apply to G:*

1. *$H \to X \in \mathbf{E}$ and $X \to H \in \mathbf{E}$;*
2. *For any vertex $Y \in \mathbf{Y}$, $Y \to X \in \mathbf{E}$ iff $Y \to H \in \mathbf{E}$;*
3. *For any vertex $Y \in \mathbf{Y}$, $H \to Y \notin \mathbf{E}$.*

Condition 1 asserts that there is a bidirected edge between vertices X and H. Second, Condition 2 asserts as a parent of vertex X, vertex Y is also a parent of the hidden variable H. Together with Condition 1, it is clear that vertices X and H have the same number of parents. Thus, the number of parameters for H grows exponentially with the number of the parents of vertex X. Third, Condition 1 and 3 state that vertex X is the only child for vertex H.

Example 1. Consider the two simplest HCTBN graphs where we have two vertices X and H. In the first case, we have no other vertices, i.e., $\mathbf{Y} = \emptyset$. In the second case, we have another vertex Y and it is a parent of vertex X, i.e., $\mathbf{Y} = \{Y\}$. The HCTBNs graphs are given in Fig. 3.

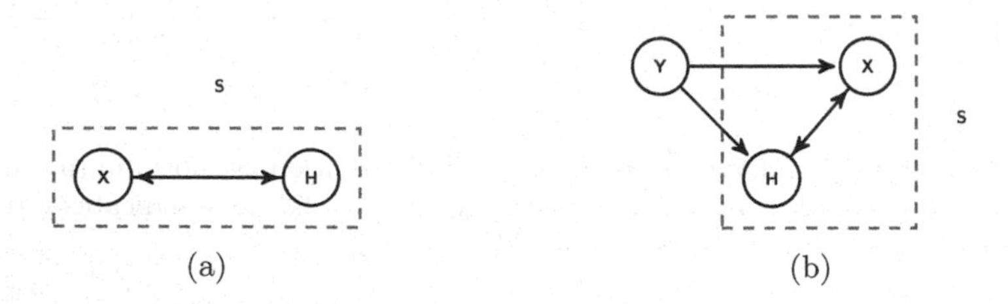

Fig. 3. Two simplest HCTBNs graphs where vertex X has no children: (a) $\mathbf{Y} = \emptyset$ and $\pi(X) = \{H\}$; (b) $\mathbf{Y} = \{Y\}$ and $\pi(X) = \{H, Y\}$.

4.2 Model Definition

Now we give a formal definition of HCTBNs.

Definition 2 (Hidden Continuous Time Bayesian Networks (HCTBNs)). *An n-order hidden continuous time Bayesian network (HCTBN) is a triple $\mathcal{N} = (G, \Lambda, P_0)$ with the graph G as defined in Definition 1. In addition,*

Λ is a set of conditional intensity matrices and P_0 is the initial distribution for the variables associated to the nodes in the graph G with $P_0(X = 1, H = 1) = 1$, and for each configuration $\mathbf{u}$ of the parents $\mathbf{U}$ for variable X, $\mathbf{U} = \pi(X) \setminus \{H\}$, the intensity matrices for variable X and H have the following form:

$$Q_{H|X=1,\mathbf{u}} = \begin{pmatrix} -\lambda_1^{\mathbf{u}} & \lambda_1^{\mathbf{u}} & \dots & 0 & 0 \\ 0 & -\lambda_2^{\mathbf{u}} & \lambda_2^{\mathbf{u}} & \cdots & 0 \\ \vdots & \vdots & \ddots & \vdots & \vdots \\ 0 & 0 & 0 & -\lambda_{n-1}^{\mathbf{u}} & \lambda_{n-1}^{\mathbf{u}} \\ 0 & 0 & 0 & 0 & 0 \end{pmatrix} Q_{H|X=2,\mathbf{u}} = \begin{pmatrix} 0 & 0 & 0 & 0 & 0 \\ \gamma_{n-1}^{\mathbf{u}} & -\gamma_{n-1}^{\mathbf{u}} & \cdots & 0 & 0 \\ 0 & \gamma_{n-2}^{\mathbf{u}} & -\gamma_{n-2}^{\mathbf{u}} & \vdots & 0 \\ \vdots & \vdots & \ddots & \vdots & \vdots \\ 0 & 0 & 0 & \gamma_1^{\mathbf{u}} & -\gamma_1^{\mathbf{u}} \end{pmatrix}$$

$$Q_{X|H=1,\mathbf{u}} = \begin{pmatrix} 0 & 0 \\ \gamma_n^{\mathbf{u}} & -\gamma_n^{\mathbf{u}} \end{pmatrix} Q_{X|H=n,\mathbf{u}} = \begin{pmatrix} -\lambda_n^{\mathbf{u}} & \lambda_n^{\mathbf{u}} \\ 0 & 0 \end{pmatrix} \textit{ If } n \geq 3,\ Q_{X|H=2:n,\mathbf{u}} = \begin{pmatrix} 0 & 0 \\ 0 & 0 \end{pmatrix}$$

The intensity matrices defined in such a form make sure that the time duration distribution for variable X staying in a state is represented by a Markov chain with n transient states.

Example 2. Given the graph where $\mathbf{U} = \emptyset$ as shown in Fig. 3a, and $\lambda_1 = 1, \lambda_2 = 2, \lambda_3 = 3, \gamma_1 = 4, \gamma_2 = 5, \gamma_3 = 6$, we can define a 3-order HCTBN by giving the intensity matrices for variable X and H as below:

$$Q_{H|X=1} = \begin{pmatrix} -1 & 1 & 0 \\ 0 & -2 & 2 \\ 0 & 0 & 0 \end{pmatrix} \quad Q_{H|X=2} = \begin{pmatrix} 0 & 0 & 0 \\ 5 & -5 & 0 \\ 0 & 4 & -4 \end{pmatrix}$$

$$Q_{X|H=1} = \begin{pmatrix} 0 & 0 \\ 6 & -6 \end{pmatrix} \quad Q_{X|H=2} = \begin{pmatrix} 0 & 0 \\ 0 & 0 \end{pmatrix} \quad Q_{X|H=3} = \begin{pmatrix} -3 & 3 \\ 0 & 0 \end{pmatrix}$$

Alternatively, we can view the hypoexponential variable X and the hidden variable H as a whole by amalgamating them into a single variable S, whose state space is the joint state space over X and H. Each state of X now corresponds to a set of instantiations to S. When we amalgamate over the hypoexponential variable X and the hidden variable H, their joint intensity matrix follows a particular structure. The states in the intensity matrix are given by iterating over all the values of X in the ordering before iterating to the next values of H. In this particular case, this gives:

$$Q_{\mathrm{XH}} = \begin{array}{c} \begin{array}{ccccccccccccc} 11 & 21 & 12 & 22 & 13 & 23 & \cdots & 1n-2 & 2n-2 & 1n-1 & 2n-1 & 1n & 2n \end{array} \\ \left(\begin{array}{ccccccccccccc} -\lambda_1 & 0 & \lambda_1 & 0 & 0 & 0 & 0 & 0 & 0 & 0 & 0 & 0 & 0 \\ \gamma_n & -\gamma_n & 0 & 0 & 0 & 0 & 0 & 0 & 0 & 0 & 0 & 0 & 0 \\ 0 & 0 & -\lambda_2 & 0 & \lambda_2 & 0 & 0 & 0 & 0 & 0 & 0 & 0 & 0 \\ 0 & \gamma_{n-1} & 0 & -\gamma_{n-1} & 0 & 0 & 0 & 0 & 0 & 0 & 0 & 0 & 0 \\ \cdots & \cdots & \cdots & \cdots & \cdots & \cdots & \cdots & \cdots & \cdots & \cdots & \cdots & \cdots & \cdots \\ 0 & 0 & 0 & 0 & 0 & 0 & 0 & 0 & 0 & -\lambda_{n-1} & 0 & \lambda_{n-1} & 0 \\ 0 & 0 & 0 & 0 & 0 & 0 & 0 & 0 & \gamma 2 & 0 & -\gamma_2 & 0 & 0 \\ 0 & 0 & 0 & 0 & 0 & 0 & 0 & 0 & 0 & 0 & 0 & -\lambda_n & \lambda_n \\ 0 & 0 & 0 & 0 & 0 & 0 & 0 & 0 & 0 & 0 & \gamma_1 & 0 & -\gamma_1 \end{array}\right) \end{array} \begin{array}{c} \\ 11 \\ 21 \\ 12 \\ 22 \\ \cdots \\ 1n-1 \\ 2n-1 \\ 1n \\ 2n \end{array}$$

Example 3. Consider the HCTBN as given in Example 2. The joint intensity matrix for variable X and H is given as below:

$$Q_{\mathrm{XH}} = \begin{array}{c} \begin{matrix} 11 & 21 & 12 & 22 & 13 & 23 \end{matrix} \\ \begin{pmatrix} -1 & 0 & 1 & 0 & 0 & 0 \\ 6 & -6 & 0 & 0 & 0 & 0 \\ 0 & 0 & -2 & 0 & 2 & 0 \\ 0 & 5 & 0 & -5 & 0 & 0 \\ 0 & 0 & 0 & 0 & -3 & 3 \\ 0 & 0 & 0 & 4 & 0 & -4 \end{pmatrix} \end{array} \begin{matrix} 11 \\ 21 \\ 12 \\ 22 \\ 13 \\ 23 \end{matrix}$$

Proposition 1. *Let X be the hypoexponential variable in an n-order HCTBN. The time that variable X stays in each of its states follows an n-order hypoexponential distribution.*

Given the joint intensity matrix of the hypoexponential variable X and the hidden variable H in an n-order HCTBN, now we can reinterpret the time duration of variable X in terms of the joint state over variables X and H. More specifically, the time of variable X staying in a state is then reinterpreted as the absorbing time of a Markov chain with a sequence of joint states over variable H and X where variable X in the joint states remains in the given state. For example, the time of variable X stays in state 1 is thus viewed as the absorbing time of a Markov chain with a sequence of joint states $11, 12, \ldots, 1n$, where X always stays in state 1 and the final transition in such a chain is the transition from state $1n$ to $2n$. As noted, there is a no explicit absorbing state. It is clear that such a Markov chain describes an n-order hypoexponential distribution. Analogously, we can construct another Markov chain corresponding to state 2 for variable X. Together, we can obtain a single Markov chain that could be graphically represented by a cyclic state transition diagram as shown in Fig. 4a.

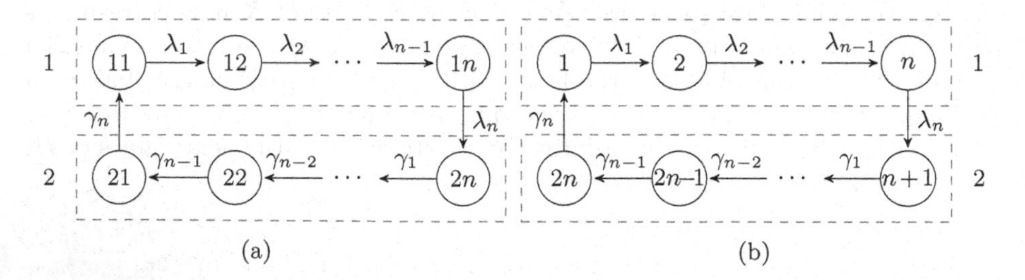

Fig. 4. State transition diagram for joint states over the hypoexponential variable X and hidden variable H in an n-order HCTBN (a) and for states of its extended variable X' in its equivalent Markovian model (b).

5 Equivalent Markovian Models

An important task for any probabilistic graphical models is to estimate parameters from data. In this paper, we transform parameter estimation in HCTBNs into a learning task in their equivalent Markovian models in terms of the same time distribution for the hypoexponential variable. In this section, we devote ourselves to defining such equivalent Markovian models. The introduction of these models only serves as a tool to estimate parameters in HCTBNs.

Definition 3 (Equivalent Markovian Graph). *Let* $G = (\mathbf{V}, \mathbf{E}, l)$, $\mathbf{V} = \{X, H\} \cup \mathbf{Y}$ *be an HCTBN graph. An* equivalent Markovian graph $G' = (\mathbf{V}', \mathbf{E}')$ *is obtained with vertices* $\mathbf{V}' = \{X\} \cup \mathbf{Y}$ *and arcs* $\mathbf{E}' = \mathbf{E} \cap (\mathbf{V}' \times \mathbf{V}')$.

Hence, the graph structure is restricted by excluding the hidden variable H in the graph G' while all other variables remain. However, a different distribution is associated to vertex X in G', in particular the state-space has grown. For example, the equivalent Markovian model graphs associated to HCTBNs introduced in Fig. 3 are shown in Fig. 5.

Fig. 5. Equivalent Markovian graphs associated to HCTBNs as introduced in Fig. 3: (a) $\mathbf{Y} = \pi(X) = \emptyset$; (b) $\mathbf{Y} = \pi(X) = \{Y\}$.

Definition 4 (Equivalent Markovian Models). *Let $\mathcal{N}$ be an n-order HCTBN with intensity matrices Λ. An equivalent Markovian model $\mathcal{M}$ is defined as a triple $\mathcal{M} = (G', \Lambda', P'_0)$ where graph $G' = (\{X\} \cup \mathbf{Y}, E)$ as in Definition 3, Λ' a set of intensity matrices over the vertices of G', and P'_0 the initial distribution with $P'_0(X = 1) = 1$. For any $Y \in \mathbf{Y}$, if $X \notin \pi(Y)$, $Q^{\mathcal{M}}_{Y|\pi(Y)} = Q^{\mathcal{N}}_{Y|\pi(Y)}$; otherwise, $Q^{\mathcal{M}}_{Y|\mathbf{K},X=1:n} = Q^{\mathcal{N}}_{Y|\mathbf{K},X=1}$ and $Q^{\mathcal{M}}_{Y|\mathbf{K},X=n+1:2n} = Q^{\mathcal{N}}_{Y|\mathbf{K},X=2}$, where $\mathbf{K} = \pi(Y) \setminus \{X\}$. Given each configuration $\mathbf{u}$ of parents $\pi(X)$ from $\mathcal{M}$ and joint intensity matrix $Q^{\mathcal{N}}_{XH}$, intensity matrices $Q^{\mathcal{M}}_{X|\pi(X)=\mathbf{u}}$ are defined by re-ordering the states of $Q^{\mathcal{N}}_{XH}$ from current indices $[1, \ldots, 2n]$ to $[1, 3, \ldots, 2n-1, 2n, \ldots, 4, 2]$.*

Definition 4 implies that HCTBNs have the same number of parameters in their equivalent Markovian models.

Example 4. An equivalent Markov model for the HCTBN, as defined in Example 2, with the intensity matrix for variable X is given as below:

$$Q_X = \begin{array}{c} \begin{matrix} 1 & 2 & 3 & 4 & 5 & 6 \end{matrix} \\ \begin{pmatrix} -1 & 1 & 0 & 0 & 0 & 0 \\ 0 & -2 & 2 & 0 & 0 & 0 \\ 0 & 0 & -3 & 3 & 0 & 0 \\ 0 & 0 & 0 & -4 & 4 & 0 \\ 0 & 0 & 0 & 0 & -5 & 5 \\ 6 & 0 & 0 & 0 & 0 & -6 \end{pmatrix} \end{array} \begin{matrix} 1 \\ 2 \\ 3 \\ 4 \\ 5 \\ 6 \end{matrix}$$

Proposition 2. *Let X be the hypoexponential variable in an n-order HCTBN and X' be the extended variable of X in its associated equivalent Markov model. The absorbing time in a Markov chain described by a sequence of states $1, 2, \ldots, n$ of variable X' follows the same distribution as the time distribution of X staying in state 1, and the absorbing time in a Markov chain described by a sequence of states $n+1, n+2, \ldots, 2n$ of variable X' follows the same distribution as the time distribution of X staying in state 2.*

Similar to an HCTBN, we can also construct a state transition diagram for its equivalent Markovian model, as shown in Fig. 4b. The time that X staying in state 1 has an n-order hypoexponential distribution with rates $\lambda_1, \lambda_2, \ldots, \lambda_n$. The same distribution can also be represented by a Markov chain of a sequence of states of variable X', $1, 2, \ldots, n$. The same applies to X staying in state 2.

6 Experiments

In the experiments, we investigate two aspects. First, we investigate whether HCTBNs provide a better approximation than CTBNs when the true temporal processes are governed by a hypoexponential time distribution. Second, we show the usefulness of HCTBNs by modeling a number of factors that influence cardiac output in the medical setting, which was previously introduced in Sect. 2. In this model, an interesting question is how the dynamics of cardiac output are affected by other factors, in particular when a hidden cause is present, i.e., when myocardial infraction is not observed.

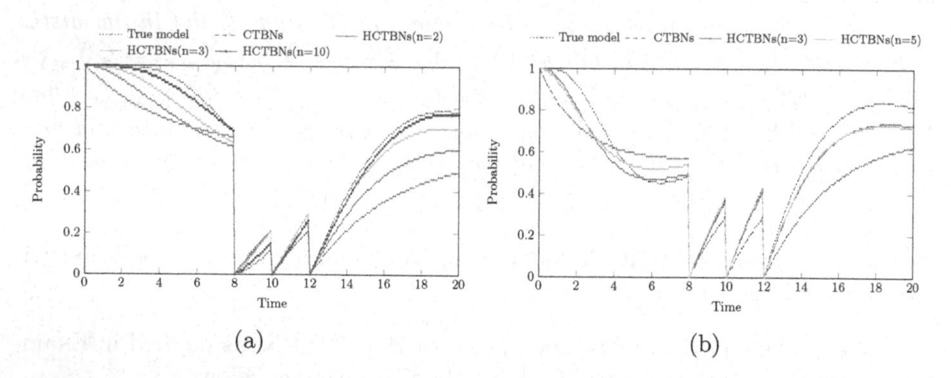

Fig. 6. Probability of X staying at state 1, given evidence $X = 2$ and $Y = 2$ at time 8, 10 and 12. (a): the true process has 10-order hypoexponential distribution and no parents. (b): the true process has 5-order hypoexponential distribution and one parent. The rates in the distribution follow a *Gamma distribution* with rate = 1 and shape = 2. The number of hidden states for the learned HCTBNs is indicated by the number n.

In the experiments, two software packages were mainly used to learn parameters for HCTBNs. The transformation between a given HCTBN and its equivalent Markovian model was implemented by *CTBN-RLE*[1]. We reformulated the parameter estimation task in HTCBNs as one in their equivalent Markovian models, where the EM algorithm is used to approximate a phase-type distribution from data by employing *EMpht*[2]. The EMpht also supports learning parameters from *right censored data*, i.e., a variable staying in a state for at least a given amount of time. A more detailed discussion can be found in [10].

For the first purpose, we generated a number of datasets from temporal processes where the time distribution follows a more complex distribution, rather than simple exponential distribution. In the experiments, a hypoexponential distribution was chosen. With respect to learning parameters for HCTBNs, we also considered the impact of the number of hidden states on the quality of the approximation in learned HCTBNs. The number of hidden states was set to 2, 3 and 10 when the underlying hypoexponential distribution has an order 10, and to 3 and 5 when the distribution has order 5.

For illustrative purpose, we considered learning parameters for a variable with complex time distribution without parents as shown in Fig. 5a and in the presence of one single parent as shown in Fig. 5b. The underlying time distribution was approximated by using the proposed HCTBNs and CTBNs. The dynamics of the hypoexponential variable X in the time interval [0, 20] in learned CTBNs and HCTBNs, as shown in Fig. 6, suggest that HCTBNs have a better approximation of the underlying generalized hypoexponential distribution than CTBNs. It also indicates that other complex distributions may be better approximated using HCTBNs. In addition, we obtained a better approximation using HCTBNs with more hidden states. More importantly, the memory in a given temporal process can be easily captured by HCTBNs, whereas it can not be captured using CTBNs.

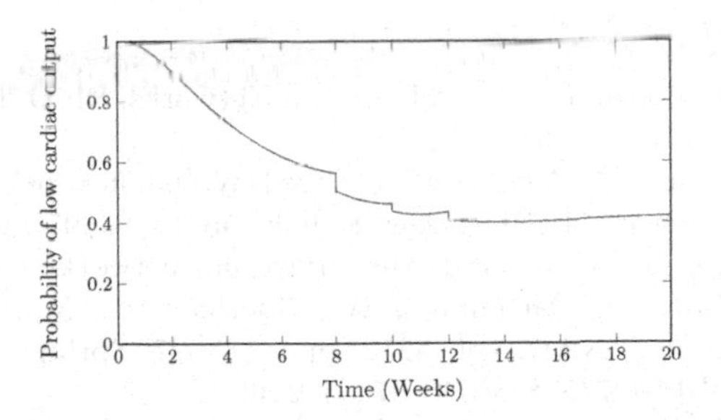

Fig. 7. Probability of having a low cardiac output given the evidence of high heart rate at time 8, 10 and 12.

For the second part of the experiments, we show the usefulness of HCTBNs for the medical example by modeling the dynamics of a patient's cardiac output over time. We computed the probability distribution of cardiac distribution for

[1] http://rlair.cs.ucr.edu/ctbnrle/.

[2] http://home.math.au.dk/asmus/pspapers.html.

a period of 20 weeks. At time 0, the patient has a myocardial infarction and evidence of the patient having high heart rate is given at time 8, 10, 12. Results of this experiment are plotted in Fig. 7. The plot shows that it is less likely for the patient to have a low cardiac output given a high heart rate (see drops at time 8, 10 and 12). The plot also suggests that factors that influence cardiac output cannot be solely explained by heart rate as we have different probabilities of having a low cardiac output at time 8, 10 and 12, even given the same evidence.

7 Conclusions

In this paper, we show that time duration in CTBNs governed by hypoexponential distributions can be modeled by using hidden variables. In addition, we show that the hidden variable also introduces *memory*, which is lacking in standard CTBNs. This memory will make CTBNs better-suited as a modeling tool for more general real-world problems in many domains, such as biology where memory plays a central role. In this paper, we provide a complete formalization of the approach. In addition, experimental results show that HCTBNs indeed can learn this more complex distributions, which was also illustrated by a small medical example.

A limitation of HCTBNs so far is that the observable variables are restricted to two states, as the focus of this paper has been on the introducing a richer time distribution and memory. In future work, we aim to overcome this limitation by supporting multinomial variables. At first glance, the proposed procedure for transforming to equivalent Markovian models can also be applied for multinomial variables but a further careful examination is necessary.

References

1. Nodelman, U.D.: Continuous time Bayesian Networks. Ph.D. thesis, Stanford University (2007)
2. Boudali, H., Dugan, J.B.: A continuous-time Bayesian network reliability modeling, and analysis framework. IEEE Trans. Reliab. **55**(1), 86–97 (2006)
3. Xu, J., Shelton, C.R.: Continuous time Bayesian networks for host level network intrusion detection. In: Daelemans, W., Goethals, B., Morik, K. (eds.) ECML PKDD 2008. LNCS (LNAI), vol. 5212, pp. 613–627. Springer, Heidelberg (2008). https://doi.org/10.1007/978-3-540-87481-2_40
4. Xu, J., Shelton, C.R.: Intrusion detection using continuous time Bayesian networks. J. Artif. Intell. Res. **39**, 745–774 (2010)
5. Gatti, E., Luciani, D., Stella, F.: A continuous time Bayesian network model for cardiogenic heart failure. Flex. Serv, Manuf. J. **24**(4), 496–515 (2012)
6. Liu, M., Hommersom, A., van der Heijden, M., Lucas, P.J.F.: Hybrid time Bayesian networks. Int. J. Approximate Reasoning **80**, 460–474 (2017)
7. Acerbi, E., Viganò, E., Poidinger, M., Mortellaro, A., Zelante, T., Stella, F.: Continuous time Bayesian networks identify Prdm1 as a negative regulator of TH17 cell differentiation in humans. Scientific reports 6 (2016)
8. Angus, I.: An introduction to Erlang B and Erlang C. Telemanagement **187**, 6–8 (2001)

9. Nodelman, U., Horvitz, E.: Continuous time Bayesian networks for inferring users' presence and activities with extensions for modeling and evaluation. Microsoft Research, July–August 2003
10. Gopalratnam, K., Kautz, H., Weld, D.S.: Extending continuous time Bayesian networks. In: Proceedings of the National Conference on Artificial Intelligence, vol. 20, p. 981. AAAI Press, MIT Press; 1999, Menlo Park, Cambridge, London (2005)
11. Nodelman, U., Shelton, C.R., Kollerthu, D.: Expectation maximization and complex duration distributions for continuous time Bayesian networks. In: Proceedings of the Twenty-First International Conference on Uncertainty in Artificial Intelligence, pp. 421–430 (2005)
12. Nodelman, U., Shelton, C.R., Koller, D.: Continuous time Bayesian networks. In: Proceedings of the Eighteenth Conference on Uncertainty in Artificial Intelligence, pp. 378–387 (2002)
13. Verbelen, R.: Phase-type distributions & mixtures of Erlangs (2013)

Sequential Decision Making Under Uncertainty: Ordinal Uninorms vs. the Hurwicz Criterion

Hélène Fargier and Romain Guillaume(✉)

IRIT, CNRS and Université de Toulouse, Toulouse, France
{fargier,guillaum}@irit.fr

Abstract. This paper focuses on sequential decision problems under uncertainty, i.e. sequential problems where no probability distribution on the states that may follow an action is available. New qualitative criteria are proposed that are based on ordinal uninorms, namely R_* and R^*. Like the Hurwicz criterion, the R_* and R^* uninorms arbitrate between pure pessimism and pure optimism, and generalize the Maximin and Maximax criteria. But contrarily to the Hurwicz criterion they are associative, purely ordinal and compatible with *Dynamic Consistency* and *Consequentialism*. This latter important property allow the construction of an optimal strategy in polytime, following an algorithm of Dynamic Programming.

Keywords: Qualitative decision making · Uncertainty
Sequential decision problems

1 Introduction

In a sequential decision problem under uncertainty, a decision maker faces a sequence of decisions, each decision possibly leading to several different states, where further decisions have to be made. A strategy is a conditional plan which assigns a (possibly non deterministic) action to each state were a decision has to be made (also called "decision node"), and each strategy leads to a compound lottery, following Von Neuman and Morgenstern's terminology [17] - roughly, a tree representing the different possible scenarios, and thus the different possible final states that the plan/strategy may reach. The optimal strategy is then the one which maximizes a criterion applied to the resulting compound lottery.

Three assumptions are desired to accept the optimal strategy without discussions on the meaning of optimal strategy. Those assumptions are:

- *Dynamic Consistency*: when reaching a decision node by following an optimal strategy, the best decision at this node is the one that had been considered so when computing this strategy, i.e. prior to applying it.
- *Consequentialism*: the best decision at each step of the problem only depends on potential consequences at this point.

J. Medina et al. (Eds.): IPMU 2018, CCIS 855, pp. 578–590, 2018.
https://doi.org/10.1007/978-3-319-91479-4_48

- *Tree Reduction*: a compound lottery is equivalent to a simple one.

Those three assumptions are linked to the possibility to compute an optimal strategy using an algorithm of dynamic programming [13].

When the problem is pervaded with uncertainty the Hurwicz criterion [7] is often advocated since it generalizes the optimistic maximax and the pessimistic maximin approaches. It makes a "compromise" between these approaches, through the use of a coefficient α of optimism - the Hurwicz value being the linear combination, according to this coefficient, of the two criteria.

Unfortunately, this approach does not suit qualitative, ordinal, utilities: the Hurwicz criterion proceeds to an additive compensation of the min value by the max value. Moreover, the criterion turns out to be incompatible with the above assumptions: it can happen that none of the optimal strategies is dynamically consistent nor consequentialist - as a consequence the optimization of this criterion cannot be carried out using dynamic programming.

In such a situation, a decision maker using the Hurwicz criterion should adopt a resolute choice behavior [2], initially choosing a strategy and never deviating from it later. But many authors insist on the fact that Resolute Choice is not acceptable since a normally behaved decision maker is consequentialist. This leads some of them to use algorithmic approaches based on Veto-process [11] and Ego-dependent process [3] (see also [8,9]).

In the present paper, rather than trying to "repair" the Hurwicz criterion in an algorithmic way, we are looking for new qualitative criteria which can take into account the level optimism/pessimism of the decision maker, like Hurwicz's criterion, and satisfies the three properties stated above (*Dynamic Consistency, Consequentialism* and *Tree Reduction*).

The paper is structured as follows. The next Section presents the Hurwicz criterion, the background on decision trees under pure uncertainty and the principle of dynamic programming. Section 3 then proposes the use of two qualitative uninorms, R^* and R_*, as alternatives to the Hurwicz criterion. Drowning them in the context of sequential decision making, we show in Sect. 4 that R^* and R_* are compatible with *Dynamic Consistency* and *Consequentialism*, and propose to apply an algorithm of dynamic programming to compute an optimal, consequentialist and dynamically consistent strategy. Section 5 eventually summarises the discussion between the two uninorms and the Hurwicz criterion[1].

2 Background

2.1 The Hurwicz Criterion [7]

Let us first consider simple, non sequential decision problems under uncertainty: each decision δ_i is characterized by the multi set of final states $E_{\delta_i} = \{s^i_1, ..., s^i_{m^i}\}$ it can lead to. Given a utility function u capturing the attractiveness of each of these final states, δ_i can be identified with a simple lottery over the utility

[1] The proofs are omitted for the sake of brevity.

levels that may be reached: in decision under uncertainty, where no probability distribution over the consequences of an act is available, a simple lottery is indeed the multiset of the utility levels of the s^i_j, i.e. $L_{\delta_i} = (u^i_1, ..., u^i_{m^i})$ (where $u^i_j = u(s^i_j)$).

A usual way to take the optimism of the decision maker (DM in the following) into account is to use the Hurwicz criterion. The worth of δ_i is then:

$$H(\delta_i) = H(L_{\delta_i}) = (1 - \alpha) \times \min(u^i_1, ..., u^i_{m^i}) + \alpha \times \max(u^i_1, ..., u^i_{m^i}).$$

where $\alpha \in [0, 1]$ is the degree of optimism. H indeed collapses with max aggregation when $\alpha = 1$ (and with the min aggregation when $\alpha = 0$).

2.2 Decision Trees

A convenient language to introduce sequential decision problems is through decision trees [13]. This framework proposes an explicit modeling in a graphical way, representing each possible scenario by a path from the root to the leaves of a tree. Formally, a decision tree $\mathcal{T} = (\mathcal{N}, \mathcal{E})$ is such that $\mathcal{N}$ contains three kinds of nodes (see Fig. 1 for an example):

- $\mathcal{D} = \{d_0, \ldots, d_m\}$ is the set of decision nodes (depicted by rectangles).
- $\mathcal{LN} = \{ln_1, \ldots, ln_k\}$ is the set of leaves, that represent final states in $\mathcal{S} = \{s_1, \ldots, s_k\}$; such states can be evaluated thanks to a utility function: $\forall s_i \in \mathcal{S}$, $u(s_i)$ is the degree of satisfaction of being eventually in state s_i (of reaching node ln_i). For the sake of simplicity we assume, without loss of generality, that only leaf nodes lead to utilities.
- $\mathcal{X} = \{x_1, \ldots, x_n\}$ is the set of chance nodes (depicted by circles).

For any node $n_i \in \mathcal{N}$, $Succ(n_i) \subseteq \mathcal{N}$ denotes the set of its children. In a decision tree, for any decision node d_i, $Succ(d_i) \subseteq \mathcal{X}$: $Succ(d_i)$ is the set of actions that can be chosen when d_i is reached. For any chance node x_i, $Succ(x_i) \subseteq \mathcal{LN} \cup \mathcal{D}$: $Succ(x_i)$ is the set of possible outcomes of action x_i - either a leaf node is observed, or a decision node is reached (and then a new action should be chosen).

The present paper is devoted to *qualitative* decision making under uncertainty; thus:

- the information at chance nodes is a list of potential outcomes - this suits situations of total ignorance, where no probabilistic distribution is available.
- the preference about the final states is purely qualitative (ordinal), i.e. we cannot assume more than a preference order on the consequences (on the leaves of the tree), captured by the satisfaction degrees. The scale $[0, 1]$ is chosen for these degrees, but any ordered set can be used.

Solving a decision tree amounts at building a *strategy*, i.e. a function δ that associates to each decision node d_i an action (i.e. a chance node) in $Succ(d_i)$: $\delta(d_i)$ is the action to be executed when decision node d_i is reached. Let Δ be the set of strategies that can be built for $\mathcal{T}$. We shall also consider the subtree $\mathcal{T}_n$ of

$\mathcal{T}$ rooted at node $n \in \mathcal{T}$, and denote by Δ_n its strategies: they are subtrategies of the strategies of Δ.

Any strategy in Δ can be viewed as a connected subtree of $\mathcal{T}$ where there is exactly one edge (and thus one chance node) left at each decision node - skipping the decision nodes, we get a chance tree or, using von Neuwman and Morgernstern's terminology, a compound lottery[2].

Simple lotteries indeed suit the representation of decisions made at the last step of the tree: $(u_1, ..., u_k)$ is the multiset of the utilities of the leaf nodes $(ln_1, ..., ln_k)$ that may be reached when some decision x is executed. Consider now a decision x made at the penultimate level: it may lead to any of the decision nodes d_i in $Succ(x)$, and thus to any of the simple lotteries $L_i = (u^i_1, ..., u^i_{m^i})$, $d_i \in Succ(x)$ - the substrategy rooted in x defines the compound lottery $(L_i$, s.t. $d_i \in Succ(x))$. The reasoning generalizes for decisions x at any level of the tree, hence the definition of the (possibly multi level) compound lottery L_δ associated to δ.

In order to apply a criterion, e.g. Hurwicz's, a simple lottery is needed. To this extent the *Reduction* of the compound lottery relative to the strategy is computed, which is the simple lottery which gathers all the utilities reached by the inner lotteries. Formally, the reduction of a compound lottery $L = (L_1, ..., L_k)$ composed of lotteries L_i is defined by:

$$Reduction(L) = Reduction(L_1) \cup ... \cup Reduction(L_k) \tag{1}$$

where the reduction of a simple lottery is the simple lottery itself. For instance, if L composed of simple lotteries $(L_1, ..., L_k)$, with $L_i = (u^i_1, ..., u^i_{n^i})$:

$$Reduction(L) = (u^1_1, ..., u^1_{n^1}, ..., u^k_1, ..., u^k_{n^k}) \tag{2}$$

The principle of reduction make the comparison of compound lotteries (and thus of strategies) possible: to compare compound lotteries by some criteria O, simply apply it to their reductions:

$$O(L) = O(Reduction(L)) \tag{3}$$

For instance, considering the Hurwicz criterion, the preference relation over strategies is defined by:

$$\delta \preceq_H \delta' \text{ iff } H(Reduction(L_\delta) \preceq H(Reduction(L_{\delta'})) \tag{4}$$

In all the approaches that follow Eq. (3), and in particular in the approach considered in this paper, *Tree Reduction* is thus obeyed by construction.

Optimality can now be soundly defined, at the global and the local levels:

- $\delta \in \Delta$ is optimal for $\mathcal{T}$ iff $\forall \delta' \in \Delta, O(Reduction(L_\delta)) \succeq O(Reduction(L_{\delta'}))$
- $\delta \in \Delta_n$ is optimal for $\mathcal{T}_n$ iff $\forall \delta' \in \Delta_n, O(Reduction(L_\delta)) \succeq O(Reduction(L_{\delta'}))$

[2] Recall that a simple lottery $L = (u_1, ..., u_k)$ is a multiset of utilities; a compound Lottery $L = (L_1, \ldots, L_k)$ is a multiset of (simple or compound) lotteries.

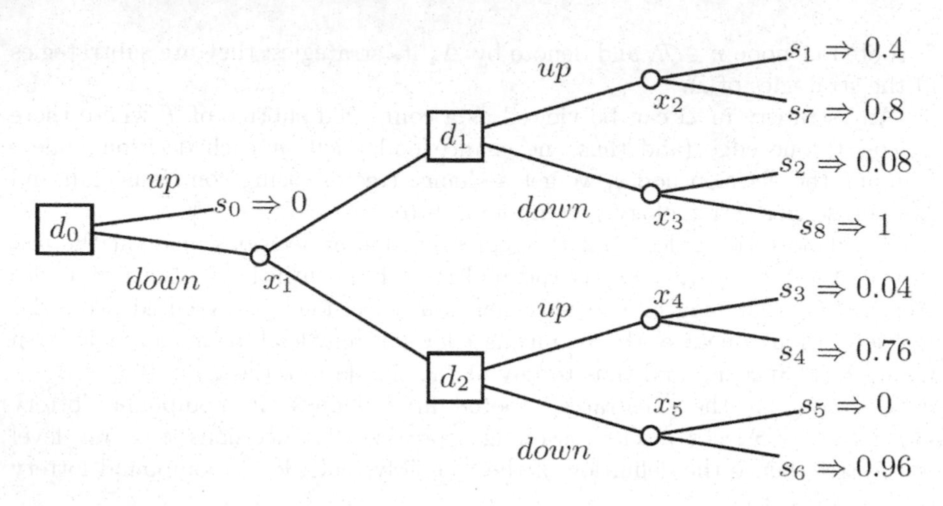

Fig. 1. A decision tree

Let us now consider *Dynamic Consistency.* An optimal strategy δ is said to be dynamically consistent iff for any decision node n, δ_n, the restriction of δ to node n and its descendent, is optimal for the subtree rooted in n. A criterion is said to be compatible with *Dynamic Consistency* if there is always an optimal strategy that is dynamically consistent.

The purely optimist (resp. pessimist) criterion, max (resp. min) is compatible with *Dynamic Consistency* - there always exist an optimal strategy whose substrategies are optimal. Unfortunately, the Hurwicz criterion is not compatible with *Dynamic Consistency.* Let us give a counter example:

Example 1. Consider the decision tree of Fig. 1 and $\alpha = 0.1$; Strategy ($d_0 \leftarrow$ *down*, $d_1 \leftarrow$ *down*, $d_2 \leftarrow$ *up*) is optimal, with a Hurwicz value of $0.1 \cdot 0.04 + 0.9 \cdot 1 = 0.904$; as a matter of fact ($d_0 \leftarrow$ *down*, $d_1 \leftarrow$ *down*, $d_2 \leftarrow$ *down*) has a Hurwicz value of 0.9 and all the strategies with $d_0 \leftarrow$ *up* or $d_1 \leftarrow$ *up* have a lower value. Hence the (only) optimal strategy prescribes "*up*" for d_2. On the other hand, considering the tree rooted in d_2, "*up*" has a H value equal to 0.684, while "*down*" has a H value equal to 0.864 - *up* is not the optimal strategy in this subtree. This counter example shows that Hurwicz is not compatible with *Dynamic Consistency.*

2.3 Dynamic Programming

Consequentialism prescribes that the decision maker selects a plan looking only at the possible futures (regardless of the past or counterfactual history). This is the case when choosing, at each node n, the decision that maximizes O. Hence a consequentialist strategy can be built starting from the anticipated future decisions and rolling back to the present (see Algorithm 1). This is the idea implemented in the algorithm of dynamic programming, which simulates the

Algorithm 1. Dynamic programming

Input: decision tree $\mathcal{T}$ of depth $p > 1$, criterion O
Output: A strategy δ which is optimal for O, its value $O(\delta)$
for $ln \in \mathcal{LN}$ **do**
  $L(ln) = u(ln)$
for $t = p - 1$ *to* 0 **do**
  for $d \in \mathcal{D}_t$ **do**
    `//` $\mathcal{D}_t$ `denotes the decision nodes at depth` t
    for $n \in Succ(d)$ **do**
      $L(n) = \bigcup_{n' \in Succ(n)} L(n')$
    $\delta(d) = argmax_{n \in Succ(d)} O(Reduction(L(n)))$
    $L(d) = L(\delta(d))$
Return $(\delta, O(Reduction(L(d_0))))$

behaviour of such a consequentialist decision maker: the algorithm builds the best strategy by a process of backward induction, optimizing the decisions from the leaves of the tree to its root. Roughly, one can say that a criterion is coherent with *Consequentialism* iff the strategy returned by the algorithm of dynamic programming is optimal according to this criterion.

Unfortunately this is not always the case when optimality is based on the principle of *Tree Reduction*: rolling back the Hurwicz optimization at each node of the tree of Fig. 1 leads to strategy ($d_0 \leftarrow down$, $d_1 \leftarrow down$, $d_2 \leftarrow down$) which is *not* optimal according to Eq. (3).

The correctness of dynamic programming actually relies on an important property, called weak monotonicity:

Definition 1. *A preference criterion over lotteries is said to be weakly monotonic iff whatever L, L' and L'':*

$$O(L) \preceq_O O(L') \Rightarrow O((L, L'')) \preceq O((L', L'')) \tag{5}$$

Proposition 1. *If a criterion O satisfies weak monotonicity then the strategy returned by dynamic programming is optimal according to O.*

By construction, this strategy is dynamically consistent (any of its substrategies is optimal it its subtree), consequentialist and equivalent, according to O, to its reduction.

Corollary 1. *If a criterion O satisfies weak monotonicity then strategy returned by dynamic programming is consequentialist and dynamically consistent.*

3 R_* and R^* as Criteria for Decision Making Under Uncertainty

As we have seen in the previous Section, the Hurwicz criterion which is often advocated for decision making under uncertainty suffers from severe drawbacks, and in particular form its incapacity to satisfy *Dynamic Consistency*. This is regrettable from a prescriptive point of view: when optimizing this criterion, the decision planned for a node is not necessarily the one that would be the best one if the tree rooted at this node were be considered - when reaching this node, a Hurwicz maximizer would be tempted not to follow the plan. That is why we look for alternative qualitative generalizations of the maximax and maximin rules, which, like Hurwicz, allow a balance between pure pessimism and pure optimism.

3.1 The R_* and R^* Uninorms

The uninorm aggregators [18] are generalization of t-norms and t-conorms. These operators allow the identity element (e) to lay anywhere in the unit interval - it is not necessarily equal to zero nor to one, as required by t-norms or t-conorms, respectively.

Definition 2 [18]. *A uninorm R is a mapping $R : [0,1] \times [0,1] \to [0,1]$ having the following properties:*

1. $R(a,b) = R(b,a)$ *(Commutativity)*
2. $R(a,b) \geq R(c,d)$ *if* $a \geq c$ *and* $b \geq d$ *(Monotonicity)*
3. $R(a, R(b,c)) = R(R(a,b),c)$ *(Associativity)*
4. *There exists some element* $e \in [0,1]$, *called the identity element, such that for all* $x \in [0,1]$ $R(x,e) = x$

In this paper we focus on two ordinal uninorms proposed by Yager [18]:

1. $R_* : [0,1]^n \to [0,1]$:
   - $R_*(a_1,...,a_n) = Min(a_1,...,a_n)$ if $Min(a_1,...,a_n) < e$
   - $R_*(a_1,...,a_n) = Max(a_1,...,a_n)$ if $Min(a_1,...,a_n) \geq e$
2. $R^* : [0,1]^n \to [0,1]$:
   - $R^*(a_1,...,a_n) = Min(a_1,...,a_n)$ if $Max(a_1,...,a_n) < e$
   - $R^*(a_1,...,a_n) = Max(a_1,...,a_n)$ if $Max(a_1,...,a_n) \geq e$

R_* specifies that if one of the a_i's is lower than e then the min operator is applied, otherwise max is applied. R^* specifies that if one of the a_i's is greater than e then the max operator is applied, otherwise min is applied. One can see that both R_* and R^* generalize the min and max uninorms, as Hurwicz does (min is recovered when $e = 1$, max when $e = 0$). The identity element e can represent the threshold of optimism (as α for Hurwicz).

R_* and R^* constitute two different ways of generalizing the maximin and maximax criterion, and capture different types of behaviours of the decision maker. In the context of decision making under uncertainty, we propose to interpret $[0,e[$ as an interval of hazards and $[e,1]$ as interval of opportunities:

1. When all the possible utilities lay in the hazardous interval, both R_* and R^* behave in a pessimistic way and evaluate the lottery by its worst outcome.

2. When all the possible utilities lay in the interval of opportunity, both R_* and R^* behave in an optimistic way and evaluate the lottery by its best outcome.
3. When some possible utility belongs to the hazardous interval and others in interval of opportunities, R_* returns a pessimistic value (the worst one) while R^* returns the best, optimistic, one.

Hence, in the simultaneous presence of hazards and opportunities, R_* focuses on the hazards while R^* focuses on the opportunities. In other terms, the comparison of strategies is made as follows:

- R^*: if one of the two strategies may lead to (at least) one opportunity, the DM prefers the strategy with the greatest opportunity. If both lead surely into the interval of hazards, the DM prefers the more robust strategy.
- R_*: if one of the two strategies may lead to (at least) one hazardous utility, the DM prefers the more robust of the strategies. If both are exempt of hazards, the DM prefers the one with the greatest opportunity.

In robust decision making, where performance guarantees are looked for, one will obviously apply the R_* uninorm because of its cautiousness. R^* indeed appears as too adventurous: one single possible opportunity carries the final decision, and this even if all the other utilities lay in the hazard interval. On the contrary, R_* looks for opportunity only when the required level of satisfaction, e, is guaranteed for all the possible outcomes.

Example 2. Let us consider three decisions $\square = (0.55, 0.55)$, $\triangle = (0.7, 0.39)$ and $\bigcirc = (0.9, 0.2)$ (see Fig. 2). In red, on the figure, is represented the zone containing decisions that the DM would like to avoid because too risky when e is set equal to 0.6 ((a). Figure 2 for R_* and (b). Figure 2 for R^*). One can see that if the DM uses R_*, all the solutions are in the red zone hence she/he will select $\square$. Conversely, if the DM uses R^*, decision $\square$ is the only decision in the red zone and $\bigcirc$ will be selected.

Depending on the value $e \in [0, 1]$, the optimal solutions are:

- $\forall e \in [0, 0.2]$ the optimal solution is $\bigcirc$ for both R_* and R^*.
- $\forall e \in]0.2, 0.39]$ for R_*: $\triangle$ and for R^*: $\bigcirc$
- $\forall e \in]0.39, 0.9]$ for R_*: $\square$ and for R^*: $\bigcirc$
- $\forall e \in]0.9, 1]$ the optimal solution is $\square$ for both uninorms.

Notice that $\triangle$ is favoured by R_*, when the degree of guaranteed performance, e, is moderate ($e \leq 0.39$). If a higher degree of performance must be ensured, R_* chooses $\square = (0.55, 0.55)$.

4 R_* and R^* in the Sequential Decision Context

Let us now study the two uninorms in the context of sequential decision. Applying the principle of lottery reduction, we have:

$$\delta \succeq_{R_*} \delta' \text{ iff } R_*(Reduction(\delta)) \succeq R_*(Reduction(\delta')) \quad (6)$$

$$\delta \succeq_{R^*} \delta' \text{ iff } R^*(Reduction(\delta)) \succeq R^*(Reduction(\delta')) \quad (7)$$

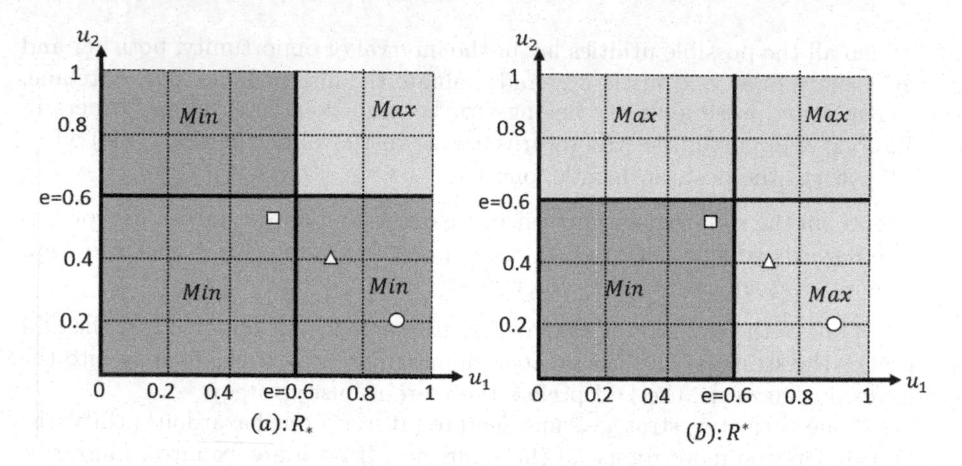

Fig. 2. Illustration of R_* and R^* (Color figure online)

Example 3. Let us go back to the example of Fig. 1 and focus first on criterion R_*. The strategies that decide *down* for d_2 are risky (may reach s_5, which have a utility of 0) and have a R_* equal to 0 whatever the value of e. This is also the case for all the strategies that decide *up* for d_0. Now,

- if $e \in]0, 0.04]$ ($d_0 \leftarrow down$, $d_1 \leftarrow down$, $d_2 \leftarrow up$) is optimal, with a $R_* = 1$.
- if $e \in]0.04, 1]$ there are two optimal strategies, ($d_0 \leftarrow down$, $d_1 \leftarrow up$, $d_2 \leftarrow up$) and ($d_0 \leftarrow down$, $d_1 \leftarrow down$, $d_2 \leftarrow up$), both with $R_* = 0.04$.

It can be checked that any optimal strategy is dynamically consistent. For instance, $R_*(d_2 \leftarrow up)$, which is at least equal to 0.04 (whatever e), is always greater than $R_*(d_2 \leftarrow down)$, which is always equal to 0.

If we consider R^*, both ($d_0 \leftarrow down$, $d_1 \leftarrow down$, $d_2 \leftarrow down$) and ($d_0 \leftarrow down$, $d_1 \leftarrow down$, $d_2 \leftarrow up$) are optimal: their R^* is equal to 1, whatever the value e (and both are dynamically consistent)

Beyond this example, R_* and R^* behave well for sequential problems in the general case; indeed, both are compatible with *Dynamic Consistency* and *Consequentialism*. The reason is that, contrarily to the Hurwicz criterion, they satisfy weak monotonicity:

Proposition 2. *The R_* and R^* satisfies weak monotonicity.*

A direct consequence of Propositions 1 and 2 is that both uninorms can be optimized by dynamic programming (see Algorithm 2).

Theorem 1. *Algorithm 2 computes a strategy optimal w.r.t R^* (resp. R_*) in time polynomial with respect to the size of the decision tree.*

This strategy is thus consequentialist and dynamically consistent; it follows from Theorem 1 that:

Algorithm 2. R^* and R_* under pure uncertainty

Input: decision tree $\mathcal{T}$ of depth $p > 1$, criterion $O \in \{R_*, R^*\}$, optimism coefficient e
Output: A strategy δ which is optimal for O, its value $O(\delta)$

```
for ln ∈ LN do
    V(ln) = u(ln)
for t = p − 1 to 0 do
    for d ∈ D_t do
        // D_t denotes the decision nodes at depth t
        for n ∈ Succ(d) do
            V(n) = O((V(n'), n' ∈ Succ(n)))
        δ(d) = argmax_{n∈Succ(d)} V(n)
        V(d) = V(δ(d))
Return (δ, V(d_0))
```

Corollary 2. *The uninorm R^* and R_* are compatible with* Dynamic Consistency, Consequentialism *and* Tree Reduction.

As already outlined compatibility with *Dynamic Consistency* guarantees that the DM cannot be tempted to deviate from the plan during its execution. Because R_* is consequentialist, the evaluation of a decision can be conservative at some node in the tree (because hazard cannot be excluded) and become optimistic when some safer point is reached (e.g. at node d_1 when $e \leq 0.08$). On the example of Fig. 1, with $e = 0.05$, R_* compares the min values of the two candidate decisions at node d_2, but is optimistic at node d_1: all the outputs that can be reached from d_1 are greater than 0.05, i.e. all the decision are safe when d_1 is reached. Similar examples can be built for R^* (which is nevertheless less in accordance with the intuition, since pessimism is taken into account only when no opportunity is available).

A last algorithmic advantage of R^* and R_* over Hurwicz is that they are associative (like any uninorm). This allows the algorithm of dynamic programming to memorize, for each node, the *value* of the corresponding reduced lottery rather than the lottery itself.

Definition 3. *A criterion O satisfies the decomposition principle iff whatever L, L', $O(L \cup L') = O(O(L), O(L'))$.*

Proposition 3. *R^* and R_* satisfy the decomposition principle.*

Hurwicz, which is not associative, does not satisfy this principle - for instance $H((1,0),(0)) = \alpha^2$ while $H((1,0,0)) = \alpha$.

5 R_* and R^* vs. Hurwicz

Let us now focus on the comparison between the uninorms (and especially of R_*, which has a well founded interpretation in terms of robustness) and Hurwicz's criterion. All are generalization of the maximax and maximin criteria, allow a tuning between optimism and pessimism, and extend to sequential problems through the application of the principle of lottery reduction.

The first remark is that R_* can capture the desiderata of a decision maker who looking for guarantees of performance, the level of performance being represented by e. This kind of requirement cannot be captured by the Hurwicz criterion, unless $\alpha = 0$, i.e. unless Hurwicz collapses with the min (and also collapses with R_* and with R^*, setting $e = 0$).

Moreover, the behaviour of Hurwicz's approach may appear chaotic in its way to move from pessimism to optimism. Consider again Example 2: $\square = (0.55, 0.55)$ and $\bigcirc = (0.2, 0.9)$ are the min optimal and max optimal solutions, respectively. The max (resp. the min) value of $\triangle$ lays between the ones of $\square$ and $\bigcirc$, so $\triangle = (0.39, 0.7)$ appears as an intermediate solution between $\square$ and $\bigcirc$ (see Fig. 2). Nevertheless, $\triangle$ is never optimal for Hurwicz. It can indeed be checked than $H(\square) = 0.55$ whatever α. $H(\triangle) = 0.545$ at $\alpha = 0.5$. When $\alpha \leq 0.5$, $H(\triangle) < 0.55 = H(\square)$; when $\alpha \geq 0.5$ $H(\bigcirc) \geq H(\triangle)$, because $H(\bigcirc)$ increases faster than $H(\triangle)$. Hence a slight variation of α makes Hurwicz jump directly from the pessimistic solution $\square$ to the very optimistic solution $\bigcirc$, without considering $\triangle$, which is Pareto optimal and intermediate between $\square$ and $\bigcirc$.

If we look at the formal properties that may be looked for, the first difference is that the uninorms are purely ordinal. They do not need to assume that the utilities are additive to some extent, while Hurwicz is basically an additive criterion. The second one is their associativity - a basic property that is not satisfied by the Hurwicz's aggregation. Last but not least, R_* and R^* are compatible with *Dynamic Consistency* and *Consequentialism*, while Hurwicz is not.

A first, practical consequence is that a polynomial algorithm of dynamic programming can be designed to find consequentialist and dynamically consistent optimal solutions. *Dynamic Consistency* and *Consequentialism* are also important from a prescriptive point of view. Because the R_* and R^* optimal strategies are dynamically consistent, the DM will never be tempted to deviate from it - we have seen on Example 1 that Hurwicz does not prevent for such deviations.

Consequentialism says that the value of a (sub)strategies only depends on the future consequences - R_* and R_* never care of "parallel", counter factual worlds. As we have seen, Hurwicz is not compatible with this principle: what happens in a world (e.g., in Example 1 in d_2 when *up* is chosen for d_2) may influence the decision in an independent, parallel world (here, in d_1). Indeed, Hurwicz will always prefer $d_1 \leftarrow down$ to $d_1 \leftarrow up$ even in case of a very low - but positive - degree of optimism. This is due to the fact the low value (0.04) for s_3, which is not a descendent of d_1 but of d_2, masks the 0.08 utility of s_2.

Our running example also shows that Hurwicz can be very adventurous even for small positive α's: $(d_0 \leftarrow down, d_1 \leftarrow down, d_2 \leftarrow up)$ might reach a very low utility (0.08) is indeed optimal for Hurwicz as soon as $\alpha > 0$. This strategy will on the contrary be considered as too risky for R_*, unless a low level ($e < 0.08$) of guaranteed performance is looked for.

6 Conclusion

In this paper, we have shown how the R_* and R^* uninorms can be used for decision under uncertainty. They constitute an appealing alternative to Hurwicz's criterion to model the behavior ofa DM who is not purely optimistic nor purely pessimistic: an optimal strategy can be computed in polytime, which satisfies the three natural assumptions of sequential decision making. Moreover, these utilities are purely qualitative; as a perspective, it would be natural to extend them to possibilistic (qualitative) decision trees [14], that allow the expression of some knowledge about the more or less possible consequences of the decisions.

References

1. Ben Amor, N., Fargier, H., Guezguez, W.: Possibilistic sequential decision making. Int. J. Approximate Reasonning **55**(5), 1269–1300 (2014)
2. McClennen, E.F.: Rationality and Dynamic Choice: Foundational Explorations. Cambridge University Press, Cambridge (1990)
3. Dubois, D., Fargier, H., Guillaume, R., Thierry, C.: Deciding under ignorance: in search of meaningful extensions of the Hurwicz criterion to decision trees. In: Grzegorzewski, P., Gagolewski, M., Hryniewicz, O., Gil, M.Á. (eds.) Strengthening Links Between Data Analysis and Soft Computing. AISC, vol. 315, pp. 3–11. Springer, Cham (2015). https://doi.org/10.1007/978-3-319-10765-3_1
4. Gilboa, I.: A combination of expected utility and maxmin decision criteria. J. Math. Psychol. **32**, 405–420 (1988)
5. Grant, S., Kajii, A., Polak, B.: Decomposable choice under uncertainty. J. Econ. Theory **92**(2), 169–197 (2000)
6. Hammond, P.: Consequentialist foundations of expected utility. Theory Decis. **25**, 25–78 (1988)
7. Hurwicz, L.: Optimality Criteria for Decision Making under Ignorance. Cowles Commission Papers 370 (1951)
8. Kikuti, D., Gagliardi Cozman, F., Shirota Filho, R.: Sequential decision making with partially ordered preferences. Artif. Intell. **175**, 1346–1365 (2011)
9. Huntley, N., Troffaes, M.C.M.: An efficient normal form solution to decision trees with lower previsions. In: Dubois, D., Lubiano, M.A., Prade, H., Gil, M.Á., Grzegorzewski, P., Hryniewicz, O. (eds.) Soft Methods for Handling Variability and Imprecision. ASC, vol. 48, pp. 419–426. Springer, Heidelberg (2008). https://doi.org/10.1007/978-3-540-85027-4_50
10. Jaffray, J.-Y.: Linear utility theory for belief functions. Oper. Res. Lett. **82**, 107–112 (1989)
11. Jaffray, J.-Y.: Rational decision making with imprecise probabilities. In: Proceedings of International Symposium on Imprecise Probabilities (ISIPTA 1999), pp. 183–188 (1999)
12. Jeantet, G.: Algorithmes pour la décision séquentielle dans l'incertain: optimisation de l'utilité espérée dépendant du rang et du critère de Hurwicz, Ph.D. thesis, Université Paris VI (2010). (in French)
13. Raiffa, H.: Decision Analysis: Introductory Lectures on Choices Under Uncertainty. Addison-Wesley, Reading (1968)
14. Sabbadin, R., Fargier, H., Lang, J.: Towards qualitative approaches to multi-stage decision making. Int. J. Approximate Reasoning **19**, 441–471 (1998)

15. Savage, L.J.: The Foundations of Statistics. Dover Publications, New York (1972)
16. Shafer, G.: The Art of Causal Conjecture. MIT Press, Cambridge (1996)
17. Von Neumann, J., Morgenstern, O.: Theory of Games and Economic Behavior. Princeton University Press, Princeton (1947)
18. Yager, R., Rybalov, A.: Uninorm aggregation operators. Fuzzy Sets Syst. **80**(1), 111–120 (1996)

Divergence Measures and Approximate Algorithms for Valuation Based Systems

Serafín Moral(✉)

Department of Computer Science and Artificial Intelligence,
University of Granada, Granada, Spain
smc@decsai.ugr.es

Abstract. This paper considers an abstract framework for expressing approximate inference algorithms in valuation based systems. It will provide a definition of a 'more informative' binary relation between representations of information as well as the basic properties of a divergence measure. The approach is illustrated with the cases of probabilistic reasoning (computation of marginal probabilities and most probable explanation) and with inference problems in propositional logic. Examples of divergence measures satisfying the basic properties will be given for these problems. Finally, we will formulate in an abstract way the mean field variational approach and the iterative belief propagation algorithm.

Keywords: Valuation based systems · Inference problems
Approximate algorithms · Divergence measures
Probabilistic Reasoning

1 Introduction

Valuation based systems are a general framework to describe inference algorithms in graphical models [1–4]. The main advantage is that algorithms can be described in an abstract way, in such a way that they can be particularized for different reasoning tasks, as the computation of conditional probabilities [2] or inferences in propositional logic [5,6]. As the problems solved by these algorithms are in general NP-hard, they are exponential and many cases can not be effectively solved. It is for this reason that many approximate algorithms have been devised, in particular for the computation of conditional probabilities. Among them, we can consider Monte Carlo algorithms as Markov chain Monte Carlo [7] or importance sampling [8,9], but also deterministic approaches as mini bucket deletion algorithm [10], variational approaches [11], iterative belief propagation [12,13] or penniless propagation [14]. Very often, it is said that approximate algorithms provide a solution which is 'similar' to the exact one, but without making precise what similar does mean and many basic questions are left unanswered. For example, it is not specified whether the approximation is more or less informative than the exact computation. This is a very important issue, as for example if an approximate algorithm for propositional logic says that p is

J. Medina et al. (Eds.): IPMU 2018, CCIS 855, pp. 591–602, 2018.
https://doi.org/10.1007/978-3-319-91479-4_49

true, it is useful to know whether we can deduce that p is necessarily true or we could have that p is false. In numerical cases, it is also important to know whether we are obtaining an upper bound or a lower bound of the true value. For example, in the problem of computing the most probable explanation [13], if the approximate algorithm provides and upper bound, it can be used as an heuristics for an A^* or a branch and bound search algorithm [15].

Valuation based systems are also useful for an abstract specification of the approximate computation problems and to study the different strategies with independence of the formalism which is used for representing information [4, 6]. In this paper, we will follow this direction by defining a preorder relation between valuations and discriminating between upper and lower approximations. To measure the quality of approximations, we will need and additional divergence measure, as the case of the Kullback and Leibler divergence for probabilities [16], to make precise the meaning of similarity. So, we will provide a set of basic properties for these divergence measures and examples of them for different inference tasks. Finally, to illustrate their use, we will consider the specification of the mean field variational approach in this framework as well as the iterative belief propagation algorithm [12,13] showing as these strategies can be also used in the case of propositional logic.

Section 2 introduces the basics of valuation based systems; Sect. 3 introduces an axiomatic definition of divergence measures and the approximation problems; Sect. 4 considers the approximation of a general valuation by a combination of one-dimensional valuations; and finally Sect. 5 is devoted to the conclusions.

2 Valuation Based Systems

Let $\mathbf{X}$ a finite set of variables. We will assume that each variable $X \in \mathbf{X}$ takes values on a finite set U_X. A generic subset $\mathbf{Y} \subseteq \mathbf{X}$ will take values in the set $U_\mathbf{Y} = \prod_{X \in \mathbf{Y}} U_X$. If $\mathbf{y} \in U_\mathbf{Y}$ and $\mathbf{Z} \subseteq \mathbf{Y}$, we will denote by $\mathbf{y}^{\downarrow \mathbf{Z}}$ the element from $U_\mathbf{Z}$ obtained from $\mathbf{y}$ by removing the coordinates corresponding to variables $X \in \mathbf{Y} \backslash \mathbf{Z}$.

In a valuation based system [1–4] it is assumed that for each $\mathbf{Y} \subseteq \mathbf{X}$ there exists a set of valuations $\mathcal{V}_\mathbf{Y}$. The set of all the valuations is denoted as $\mathcal{V} = \bigcup_{\mathbf{Y} \subseteq \mathbf{X}} \mathcal{V}_\mathbf{Y}$.

If $V \in \mathcal{V}$, the set of variables $\mathbf{Y}$ such that $V \in \mathcal{V}_\mathbf{Y}$ is denoted as $s(V)$. Shenoy and Shafer [1,2]) consider two operations: marginalization and combination. The marginalization is a family of mappings from $\mathcal{V}_\mathbf{Y}$ to $\mathcal{V}_\mathbf{Z}$, where $\mathbf{Z} \subseteq \mathbf{Y}$: if $V \in \mathcal{V}_\mathbf{Y}$ then the marginalization of V to $\mathbf{Z}$ is a valuation from $\mathcal{V}_\mathbf{Z}$ denoted by $V^{\downarrow \mathbf{Z}}$. The combination applies two valuations $V_1 \in \mathcal{V}_\mathbf{Z}, V_2 \in \mathcal{V}_\mathbf{Y}$ into a valuation from $\mathcal{V}_{\mathbf{Z} \cup \mathbf{Y}}$ which is denoted by $V_1 \otimes V_2$. The following basic properties are assumed to be satisfied:

Axiom 1. $\forall V_1, V_2, V_3 \in \mathcal{V}, \quad V_1 \otimes V_2 = V_2 \otimes V_1, \quad (V_1 \otimes V_2) \otimes V_3 = V_1 \otimes (V_2 \otimes V_3)$.
Axiom 2. If $\mathbf{W} \subseteq \mathbf{Z} \subseteq \mathbf{Y}$, and $V \in \mathcal{V}_\mathbf{Y}$, then $(V^{\downarrow \mathbf{Z}})^{\downarrow \mathbf{W}} = V^{\downarrow \mathbf{W}}$.
Axiom 3. If $V_1 \in \mathcal{V}_\mathbf{Y}$, $V_2 \in \mathcal{V}_\mathbf{Z}$, then $(V_1 \otimes V_2)^{\downarrow \mathbf{Y}} = V_1 \otimes V_2^{\downarrow (\mathbf{Y} \cap \mathbf{Z})}$.

In [3] and [4] two additional axioms have been considered:

Axiom 4. *Neutral Element.*- There exists a valuation $V_0 \in \mathcal{V}_{\mathbf{X}}$ such that for any $V \in \mathcal{V}$ with $s(V) = \mathbf{Y}$, then $V_0{}^{\downarrow \mathbf{Y}} \otimes V = V$.

Axiom 5. *Contradiction.*- There exists a valuation $V_c \in \mathcal{V}_{\mathbf{X}}$ such that for any $V \in \mathcal{V}$, then $V_c \otimes V = V_c$.

When the following axiom is also satisfied, we will say that we have an idempotent valuation system or an information algebra [4,6,17]:

Axiom 6. *Idempotence.*- $\forall V \in \mathcal{V}_{\mathbf{Y}}, \forall \mathbf{Z} \subseteq \mathbf{Y}, \quad V^{\downarrow \mathbf{Z}} \otimes V = V$.

Two basic examples will be considered in this paper: probability theory and propositional logic.

Example 1. In probability theory a valuation on $V_{\mathbf{Y}}$ is a mapping (potential) from $U_{\mathbf{Y}}$ to the $\mathbb{R}_0^+$ (the non-negative real numbers). Combination in carried out by multiplication: if $f_1 \in \mathcal{V}_{\mathbf{Y}}$ and $f_2 \in \mathcal{V}_{\mathbf{Z}}$, then

$$f_1 \otimes f_2(\mathbf{u}) = f_1(\mathbf{u}^{\downarrow \mathbf{Y}}).f_2(\mathbf{u}^{\downarrow \mathbf{Z}}), \quad \forall \mathbf{u} \in U_{\mathbf{Y} \cup \mathbf{Z}}.$$

There are two types of marginalization that can be considered depending of the associated inference problem:

- *Sum-marginalization.*- This marginalization is useful for the computation of conditional probabilities $P(x|\mathbf{y})$ when $x \in U_X$ and $\mathbf{y} \in U_{\mathbf{Y}}$ is a set of observations for variables in $\mathbf{Y}$. In this case if $\mathbf{W} \subseteq \mathbf{Z}$, then

$$f^{\downarrow \mathbf{W}}(\mathbf{w}) = \sum_{\mathbf{z}^{\downarrow \mathbf{W}} = \mathbf{w}} f(\mathbf{z}).$$

  We will also assume that two potentials f_1 and f_2 are equivalent if there is $\alpha > 0$ such that $f_1 = \alpha f_2$ (a valuation is an equivalence class).
- *Max-marginalization.*- This marginalization is useful for solving the MPE problem [13]

$$\mathbf{z}_0 = \arg\max_{\mathbf{z} \in \mathbf{Z}} P(\mathbf{z}|\mathbf{y}),$$

  where $\mathbf{Z} = \mathbf{X} \backslash \mathbf{Y}$ is the set of non-observed variables. In this case we will assume that potentials are always valued in $[0,1]$ interval and are not equivalent under multiplication by a positive number. In this case: $f^{\downarrow \mathbf{W}}(\mathbf{w}) = \max\{f(\mathbf{z}) : \mathbf{z}^{\downarrow \mathbf{W}} = \mathbf{w}\}$.

  The neutral element is: $f_0(\mathbf{x}) = 1, \quad \forall \mathbf{x} \in U_{\mathbf{X}}$.
  The potential representing the contradiction is: $f_c(\mathbf{x}) = 0, \quad \forall \mathbf{x} \in U_{\mathbf{X}}$.

Example 2. In the case of propositional logic, the set of variables is identified with a finite set of labels: $\mathbf{X} = \{p, q, r, \ldots\}$. The set U_p will have two values 0 (p is false) and 1 (p is true).

A valuation about $\mathbf{Y} \subseteq \mathbf{X}$ will be a set of clauses F in which all the variables are in $\mathbf{Y}$. Two sets of clauses F_1 and F_2 are said to be equivalent if their set of consequences in propositional logic are the same: $Cons(F_1) = Cons(F_2)$. So, $F_1 = \{p \vee q \vee \neg z, q \vee z\}$ and $F_2 = \{p \vee q, q \vee z\}$ are equivalent.

Combination is union, $F_1 \otimes F_2 = F_1 \cup F_2$. Marginalization of F into a set of variables $\mathbf{Y}$, is the set $Cons(F) \cap \mathcal{L}_{\mathbf{Y}}$, where $\mathcal{L}_{\mathbf{Y}}$ is the set of all the clauses defined over variables $\mathbf{Y}$. An equivalent set of clauses can be obtained by means of a repeated application of Davis-Putnam elimination procedure [5]. If $\mathbf{Y} = \mathbf{Z}\backslash\{p\}$, and $F \in \mathcal{V}_{\mathbf{Z}}$, then $F^{\downarrow \mathbf{Y}}$ can be obtained by computing F_p^+ (clauses containing p), F_p^- (clauses containing $\neg p$), and F_p^0 (the rest of clauses). Then the marginalization is obtained by the union of F_p^0 and the results of the resolution of any clause in F_p^+ with any clause in F_p^-.

The neutral element can be represented by the empty set, and the contradiction by a set containing only the empty clause.

A valuation about variables in $\mathbf{Y}$, can be also represented by a subset $A_F \subseteq U_{\mathbf{Y}}$: $\mathbf{y} \in A_F$ if and only if all the clauses in F are satisfied when the true value of p is $\mathbf{y}^{\downarrow p}$.

A valuation in a set $\mathbf{Z}$, can be extended to a greater set $\mathbf{Y} \supseteq \mathbf{Z}$ by multiplying it by the neutral element: if $V \in \mathcal{V}_{\mathbf{Z}}$ and $\mathbf{Z} \subseteq \mathbf{Y}$, then the *extension* of V to $\mathbf{Y}$ is defined as $V^{\uparrow \mathbf{Y}} = V \otimes V_0^{\mathbf{Y}}$.

It is possible to define a preorder relation in valuation based systems. This preorder is a partial order in the case of idempotent valuations [4,6].

Definition 1. *V_1 is said to be more informative than V_2 ($V_2 \preceq V_1$) if and only if there is a valuation $V \in \mathcal{V}$ such that $V \otimes V_2 = V_1$.*

A consequence of the definition is that if $V_2 \preceq V_1$ then we have that $s(V_2) \subseteq s(V_1)$. It is possible to extend the definition to valuations in different frames, by considering that a valuation is equivalent to all its extensions, but this is not done in this paper.

With this definition, we have that if $V_2 \preceq V_1$, then for any $V \in \mathcal{V}$, $(V_2 \otimes V) \preceq (V_1 \otimes V)$. However, there is a property of ordered valuation algebras [17] that it is not necessarily satisfied: if $V_1 \preceq V_2$, then $V_1^{\downarrow \mathbf{Y}} \preceq V_2^{\downarrow \mathbf{Y}}$, though in all the examples we have considered (probability theory and propositional logic) it is satisfied.

Example 3. In the case of probabilistic valuations we have two different situations depending of whether we have sum-marginalization or the case of max-marginalization. In the former case $f_1 \preceq f_2$ if and only if $Support(f_2) \subseteq Support(f_1)$, where $Support(f)$ is the set of values $\mathbf{y} \in U_{\mathbf{Y}}$ such that $f(\mathbf{y}) > 0$, for $f \in \mathcal{V}_{\mathbf{Y}}$. When $s(f_2) \subset s(f_1)$, a subset A of $U_{s(f_1)}$ is identified with the subset $A \times U_{s(f_2)\backslash s(f_1)}$. In this case, the preorder is not antisymmetrical.

In the case of max-marginalization, the order is $f_1 \preceq f_2$ if and only if $f_1(\mathbf{y}^{\downarrow \mathbf{Z}}) \geq f_2(\mathbf{y})$ for any $\mathbf{y} \in U_{\mathbf{Y}}$ and where $\mathbf{Y} = s(f_1) \subseteq s(f_2) = \mathbf{Z}$. In this case we have a partial order relationship and the supremum, $\sup(\mathcal{R})$, and infimum, $\inf(\mathcal{R})$, always exist for any family of valuations $\mathcal{R} \subseteq \mathcal{V}$. Take into account that $\sup(\mathcal{R})$ is the valuation f given by $f(\mathbf{y}) = \inf_{g \in \mathcal{R}} g(\mathbf{y}^{\downarrow s(g)})$ and that $\inf(\mathcal{R})$ is the valuation f given by $f(\mathbf{y}) = \sup_{g \in \mathcal{R}} g(\mathbf{y}^{\downarrow s(g)})$.

When the order relation is not antisymmetrical, we can define an equivalence relation: $V_1 \sim V_2$ if and only if $V_1 \preceq V_2$ and $V_2 \preceq V_1$. It is important to note that $V_1 \sim V_2$ does not imply that V_1 and V_2 have the same information content. In fact in the case of probabilistic valuations (potentials) under sum-marginalization, we have that $f_1 \sim f_2$ if and only if $Support(f_1) = Support(f_2)$, but the numbers in the potentials can be very different.

In the case of idempotent valuations, the order is always antisymmetrical, and the supremum of two valuations V_1 and V_2 always exists: it is equal to the combination $V_1 \otimes V_2$.

Example 4. In the case of propositional logic we have that $F_1 \preceq F_2$, if and only if $Cons(F_1) \subseteq Cons(F_2)$. In this case, we have that if $\mathcal{R}$ is a family of valuations, then the supremum and infimum always exist and are given by $Sup(\mathcal{R}) = Cons(\bigcup_{F\in\mathcal{R}} F)$ and $Inf(\mathcal{R}) = \bigcap_{F\in\mathcal{R}} Cons(F)$.

3 The Inference Problem, Approximations, and Divergence Measures

The inference problem is the following: given a finite set of valuations $\{V_1, \ldots, V_n\}$ and a variable of interest $X \in \mathbf{X}$, we want to compute $V = (V_1 \otimes V_2 \otimes \cdots \otimes V_n)^{\downarrow X}$.

In the case of probabilistic valuations with sum-marginalization, if $f = V$ is the result of this marginalization, then we can prove that $P(X|\mathbf{y}) \propto f$, if the initial potentials integrate the observations (a value of 0 is assigned to the configurations that are incompatible with the observed values $\mathbf{Y} = \mathbf{y}$). In the case of max-marginalization we have that if $\mathbf{z}_0 = \arg\max_{\mathbf{z}\in\mathbf{Z}} P(\mathbf{z}|\mathbf{y})$, then $\mathbf{z}_0^{\downarrow X} = \arg\max_{x\in U_x} f(x)$, i.e. the value of X in $\mathbf{z}_0$ can be computed using the marginal potential f which is defined for only one variable.

In the case of propositional logic if $F = V$ is the result of the marginalization on p, then if F contains the empty clause or p and $\neg p$, we have that clauses in the initial valuations are inconsistent (there is not a true assignment satisfying all the clauses). In other case, if $F = \{p\}$, then we can deduce p. if $F = \{\neg p\}$, then we can deduce $\neg p$, and if $F = \emptyset$, nothing can be deduced about the true value of p (all the clauses can be satisfied both making p true and making p false). In this case, solving the marginal problem implies the solution of SAT problem which is NP-complete, so not known polynomial algorithm exists. In some cases, it would be necessary to apply approximate algorithms. In general, we want to compute a valuation V' such that V' is 'similar' to the exact marginal V. In the following, we try to make the word 'similar' more precise. A first basic question which is very often neglected is whether we are computing an approximation V' which is less or more informative than V.

In general, the problem of approximating a valuation V can be stated in the following way: we have a subset $\mathcal{V}^* \subset \mathcal{V}$ of simple valuations, then an *upper approximation from* $\mathcal{V}^*$ is a valuation $V' \in \mathcal{V}^*$ such that $V \preceq V'$ and a *lower approximation from* $\mathcal{V}^*$ is a valuation $V' \in \mathcal{V}^*$ such that $V' \preceq V$.

To determine how good is an approximation, we need a divergence measure. In probabilistic valuations with sum-marginalization the most common divergence is the Kullback-Leibler divergence. Assume that f_1 and f_2 are probabilistic potentials such that $s(f_2) = s(f_1) = \mathbf{Y}$ and such that they add 1, i.e. $\sum_{\mathbf{y} \in U_{\mathbf{Y}}} f_1(\mathbf{y}) = \sum_{\mathbf{y} \in U_{\mathbf{Y}}} f_2(\mathbf{y}) = 1$, then

$$KL(f_1, f_2) = \sum_{\mathbf{y} \in U_{\mathbf{Y}}} f_1(\mathbf{y}) \log \left(\frac{f_1(\mathbf{y})}{f_2(\mathbf{y})} \right).$$

When f_1 and f_2 do not add to one, they should be normalized before applying above expression by dividing each value of the potential by its total sum. When f_1 is the contradiction (it is identically equal to 0) then normalization is not possible, and we will assume that $KL(f_1, f_2) = 0$ when f_2 is also the contradiction and $+\infty$ otherwise. If f_2 is the contradiction and f_1 is not the contradiction $KL(f_1, f_2) = +\infty$.

The divergence can be extended to the case of $s(f_2) \subseteq s(f_1)$ by considering $KL(f_1, f_2) = KL(f_1, f_2^{\uparrow s(f_1)})$, i.e. f_2 is first extended to the set of variables of f_1 and then above expression is applied (after normalization when it is necessary).

It is important to remark that $KL(f_1, f_2) = \infty$ when $f_2 \not\preceq f_1$, i.e. when $Support(f_1) \not\subseteq Support(f_2)$, i.e. the divergence is infinite when f_1 is not more informative than f_2. In the following, we define a divergence measure for general valuations, generalizing this and other basic properties.

Definition 2. *A divergence measure in a valuation based systems is a family of mappings, $Di : \mathcal{V}_{\mathbf{Y}} \times \mathcal{V}_{\mathbf{Z}} \rightarrow \mathbb{R}_0^+ \cup \{+\infty\}$, where $\mathbf{Z} \subseteq \mathbf{Y}$ satisfying:*

1. $Di(V_1, V_2) = Di(V_1, V_2^{\uparrow s(V_1)})$.
2. $Di(V_1, V_2) = 0$ *if and only if* $V_1 = V_2^{\uparrow s(V_1)}$.
3. $Di(V_1, V_2) = +\infty$ *when* $V_2 \not\preceq V_1$.
4. *If* $V_3 \preceq V_2 \preceq V_1$, *then there is a valuation* $V_2' \sim V_2$ *with* $Di(V_1, V_2') \leq Di(V_1, V_3)$ *and a valuation* $V_2'' \sim V_2$ *with* $Di(V_2'', V_3) \leq Di(V_1, V_3)$.
5. *If* $V \in \mathcal{V}_{\mathbf{Y}}$, $\mathbf{Z} \subseteq \mathbf{Y}$, *and* $V' \in \mathcal{V}_{\mathbf{Z}}$, *then* $Di(V, V^{\downarrow \mathbf{Z}}) \leq Di(V, V')$, *being this inequality strict when* $V' \neq V^{\downarrow \mathbf{Z}}$.
6. *If* $V_1, V_3 \in \mathcal{V}_{\mathbf{Y}}$, $V_2, V_4 \in \mathcal{V}_{\mathbf{Z}}$, *with* $\mathbf{Y} \cap \mathbf{Z} = \emptyset$, *we have that*
$$Di(V_1 \otimes V_2, V_3 \otimes V_4) = Di(V_1, V_3) + Di(V_2, V_4).$$
7. *If* $s(V_1) = s(V_1')$ *and* $s(V_2) \cap s(V_1) = \emptyset$ *and* $V_1 \otimes V_2 \preceq V$, *then*
$$Di(V, V_1) \leq Di(V, V_1') \Rightarrow Di(V, V_1 \otimes V_2) \leq Di(V, V_1' \otimes V_2).$$

Example 5. It is immediate to prove that $KL(f_1, f_2)$ is a divergence measure for probabilistic pontentials with sum-marginalization. In the case of max-marginalization, KL is not a divergence measure: Property 4 is not satisfied. You only have to consider $U_Y = \{0, 1\}$, and three potentials f_1, f_2, f_3, where

$$f_1(0) = f_1(1) = 1,\ f_2(0) = 1,\ f_2(1) = 0.5,\ f_3(0) = f_3(1) = 0.5$$

We have that $f_1 \preceq f_2 \preceq f_3$, however $KL(f_3, f_1) = 0$ and $KL(f_3, f_2) > 0$, and as in this case the relation $\preceq$ is antisymmetrical, there is not f_2' different from f_2 such that $f_2 \sim f_2'$. A divergence measure in this case can be defined by the following expression when $s(f_1) = s(f_2) = \mathbf{Y}$:

$$Dm(f_1, f_2) = \begin{cases} \log(\sum_{\mathbf{y} \in U_{\mathbf{Y}}} f_2(\mathbf{y})) - \log(\sum_{\mathbf{y} \in U_{\mathbf{Y}}} f_1(\mathbf{y})) & \text{if } f_2 \preceq f_1 \\ +\infty & \text{otherwise} \end{cases}$$

In this expression, it is considered that $Di(f, f) = 0$ when f is the contradiction (it is identically equal to 0). When $s(f_2) \subseteq s(f_1)$, then we consider $Di(f_1, f_2) = Di(f_1, f_2^{\uparrow s(f_1)})$.

In the case of propositional logic, we can define a divergence as follows:

$$Dl(F_1, F_2) = \begin{cases} \log(|A_{F_2}|) - \log(|A_{F_1}|) & \text{if } F_2 \preceq F_1 \\ +\infty & \text{otherwise} \end{cases}$$

where A_{F_i} is the set of true values such that all the clauses of F_i are satisfied (as defined in Example 2), and $|A_{F_i}|$ stands for its cardinal. It is assumed that F_1 and F_2 are defined for the same set of variables. In other case, F_2 is extended to the set of variables of F_1.

With a divergence measure we can state the approximation problems as follows:

- *Lower Approximation:* given a set of simple valuations $\mathcal{V}^*$ and $V \in \mathcal{V}$, to compute $V^* = \arg\min_{V' \in \mathcal{V}^*, s(V')=s(V)} Di(V, V')$.
- *Upper Approximation:* given a set of simple valuations $\mathcal{V}^*$ and $V \in \mathcal{V}$, to compute $V^* = \arg\min_{V' \in \mathcal{V}^*, s(V')=s(V)} Di(V', V)$.

An approximation problem is well defined when this minimum always exists.

Example 6. In the probabilistic case, imagine that $\mathcal{V}^*$ is given by the potentials in each set of variables $\mathbf{Y}$ that are different from zero at most in one point $\mathbf{y}_0 \in U_{\mathbf{Y}}$, i.e. the potentials $f_{\mathbf{y}_0}$ such that $f_{\mathbf{y}_0}(\mathbf{y}) = 0$ if $\mathbf{y} \neq \mathbf{y}_0$. In this case, we are talking about upper approximations as these potentials are very informative (they are the most informative ones among the non contradictory potentials). Given a potential f and an upper approximation $f_{\mathbf{y}_0}$, where $f(\mathbf{y}_0) > 0$, we have that $KL(f_{\mathbf{y}_0}, f) = -\log(f(y_0))$. Therefore, minimizing the KL divergence is equivalent to finding the value $\mathbf{y}_0$ maximizing $f(\mathbf{y}_0)$. In this way, the problem of finding the configuration of maximum probability of a potential (or product of potentials) can be seen as a way of upper approximating it.

A similar approximation can be devised in the propositional logic case. Imagine that $\mathcal{V}^*$ is given by all the set of formulas on variables $\mathbf{Y}$ assigning a true value to any variable in $\mathbf{Y}$, i.e. $F \in \mathcal{V}^*_{\mathbf{Y}}$ if and only if for any $p \in \mathbf{Y}$, we have that $p \in Cons(F)$ or $\neg p \in Cons(F)$, or equivalently $|A_F| \leq 1$. In this way, given F finding $F^* \in \mathcal{V}^*$ such that $Dl(F^*, F)$ is equivalent to finding F^*, such that $F \preceq F^*$ and $|F^*|$ maximum, i.e. finding a true assignment to all the variables in F satisfying all the formulas, when this assignment exists: the FSAT problem.

4 Approximating by a Product of Marginals

Assume that we have a product of valuations $V_1 \otimes \cdots \otimes V_n$ defined on variables $\mathbf{X}$ and we want to approximate this product by a product of valuations, each one of them defined for only one variable $X \in \mathbf{X}$, i.e. the space of approximations is $\mathcal{V}^* = \bigcup_{\mathbf{Y} \subseteq \mathbf{X}} \mathcal{V}^*_{\mathbf{Y}}$ such that for any set of variables $\mathbf{Y}$, the potentials in $\mathcal{V}^*_{\mathbf{Y}}$ are the potentials $\bigotimes_{Y \in \mathbf{Y}} V_Y$, where V_Y is a potential defined for variable Y. This set of potentials $\mathcal{V}^*$ is closed under combination and marginalization and it is a valuation based system. In fact, the two operations are quite simple, if we represent a product $(\bigotimes_{Y \in \mathbf{Y}_1} V_Y)$ by the list of valuations V_Y . For example, the combination is given by the following expression:

$$(\bigotimes_{Y \in \mathbf{Y}_1} V_Y^1) \otimes (\bigotimes_{Y \in \mathbf{Y}_2} V_Y^2) = (\bigotimes_{Y \in \mathbf{Y}_1 \cap \mathbf{Y}_2} (V_Y^1 \otimes V_Y^2)) \otimes (\bigotimes_{Y \in \mathbf{Y}_1 \setminus \mathbf{Y}_2} V_Y^1) \otimes (\bigotimes_{Y \in \mathbf{Y}_2 \setminus \mathbf{Y}_1} V_Y^2).$$

Observe that only one-dimensional combinations are carried out. Marginalization is even simpler. If $\mathbf{Z} \subseteq \mathbf{Y}$, then $(\bigotimes_{Y \in \mathbf{Y}} V_Y)^{\downarrow \mathbf{Z}} = (\bigotimes_{Y \in \mathbf{Z}} V_Y)$. In this case, we only have to remove those valuations defined for variables in $\mathbf{Y} \setminus \mathbf{Z}$.

It is important to remark that once we have approximated $V_1 \otimes \cdots \otimes V_n$ by the product $\bigotimes_{X \in \mathbf{X}} V_X$, computing the marginal valuation to X in the approximate potential is very simple: it is the potential V_X corresponding to this variable.

4.1 Upper Approximation

The upper approximation problem is to find the product $\bigotimes_{X \in \mathbf{X}} V_X$ minimizing the divergence $Di(\bigotimes_{X \in \mathbf{X}} V_X, V_1 \otimes \cdots \otimes V_n)$. This is, in general, a difficult problem and, for example, it implies solving an FNP-complete problem (FSAT) in the propositional logic example. However, in many cases it is efficient to apply an iterative algorithm in which given a current approximation $\bigotimes_{X \in \mathbf{X}} V_X$, we select a variable $X \in \mathbf{X}$, and try to find a valuation V_X^* minimizing the divergence $Di(V_X^* \otimes \left(\bigotimes_{Y \in \mathbf{X}, Y \neq X} V_Y\right), V_1 \otimes \cdots \otimes V_n)$. We will show that this is in some situations a simple computational problem, and in this way, we can devise an iterative algorithm in which, each time a variable X is selected and then the current approximation $\bigotimes_{X \in \mathbf{X}} V_X$ is changed to $V_X^* \otimes \left(\bigotimes_{Y \in \mathbf{X}, Y \neq X} V_Y\right)$ minimizing the divergence.

The efficiency of the local improvement steps is based on the following decomposition property satisfied by some of the divergences defined on valuation based system, as the case of Kullback-Leibler divergence for probabilistic potentials.

Proposition 1. *Divergence KL defined on probabilistic potentials under sum satisfies the following property:*

$$\arg \min_{f \in \mathcal{V}_{\mathbf{Y}}} KL(f.f', f_1.f_2) = \arg \min_{f \in \mathcal{V}_{\mathbf{Y}}} KL(f.f', f_1). \arg \min_{f \in \mathcal{V}_{\mathbf{Y}}} KL(f.f', f_2),$$

for any $f_1, f_2 \in \mathcal{V}$ *and* $f' \in \mathcal{V}_{\mathbf{Z}}$ *with* $\mathbf{Z} \cap \mathbf{Y} = \emptyset$.

Proof. If we compute $\arg\min_{f \in \mathcal{V}_{\mathbf{Y}}} KL(f.f', f^*)$ where $s(f^*) = \mathbf{W}$, we have to minimize in f such that $\sum_{\mathbf{y},\mathbf{z}} f(\mathbf{y}).f'(\mathbf{z}) = 1$ the function

$$\sum_{\mathbf{y} \in \mathbf{Y}, \mathbf{z} \in \mathbf{Z}} f(\mathbf{y}).f'(\mathbf{z}) \log \left(\frac{f(\mathbf{y}).f'(\mathbf{z})}{f^*((\mathbf{y},\mathbf{z})^{\downarrow \mathbf{W}})} \right)$$

where f' is fixed and variables $\mathbf{Z} \cap \mathbf{Y} = \emptyset$. It is not necessary that f^* is defined for all the variables $\mathbf{Y} \cup \mathbf{Z}$.

By applying Lagrange multipliers, we can obtain that this sum is minimized when

$$\sum_{\mathbf{z}} \left(f'(\mathbf{z}) \log \left(\frac{f(\mathbf{y}) f'(\mathbf{z})}{f^*((\mathbf{y},\mathbf{z})^{\downarrow \mathbf{W}})} \right) + f'(\mathbf{z}) \right) + \sum_{\mathbf{z}} (f'(\mathbf{z}).\lambda) = 0,$$

i.e. when $f(\mathbf{y}) \propto e^{\sum_{\mathbf{z}} f'(\mathbf{z}) \log(f^*((\mathbf{y},\mathbf{z})^{\downarrow \mathbf{W}}))}$.

If we call $f^{12} = \arg\min_{f \in \mathcal{V}_{\mathbf{Y}}} KL(f.f', f_1.f_2)$, $f^1 = \arg\min_{f \in \mathcal{V}_{\mathbf{Y}}} KL(f.f', f_1)$, and $f^2 = \arg\min_{f \in \mathcal{V}_{\mathbf{Y}}} Di(f.f', f_2)$, then applying above expression with f^* equal to $f_1.f_2$, f_1, and f_2 respectively, we get:

$$f^{12} \propto e^{\sum_{\mathbf{z}} f'(\mathbf{z}) \log(f_1.f_2((\mathbf{y},\mathbf{z})^{\downarrow \mathbf{W}}))}$$
$$f^1 \propto e^{\sum_{\mathbf{z}} f'(\mathbf{z}) \log(f_1((\mathbf{y},\mathbf{z})^{\downarrow \mathbf{W}}))}$$
$$f^2 \propto e^{\sum_{\mathbf{z}} f'(\mathbf{z}) \log(f_2((\mathbf{y},\mathbf{z})^{\downarrow \mathbf{W}}))}$$

It is immediate to prove $f^{12} = f^1.f^2$, with which we obtain the desired result.

According to this result, to compute f_X^* minimizing $KL(f_X^* \otimes \left(\bigotimes_{Y \in \mathbf{X}, Y \neq X} f_Y\right), f_1 \otimes \cdots \otimes f_n)$ we can compute the potential f_X^i, minimizing the divergence $KL(f_X^i \otimes \left(\bigotimes_{Y \in \mathbf{X}, Y \neq X} f_Y\right), f_i)$ for $i = 1, \ldots, n$ and then making $f_X^* = \prod_{i=1}^n f_X^i$. This is the most important step which makes computationally efficient the local improvement of upper marginal approximations and it is the basis of mean field variational techniques [11]. We can do an additional simplification, taking into account that if $X \notin s(f_i)$, then f_X^i is the neutral potential (constant) and that the minimization of $KL(f_X^i \otimes \left(\bigotimes_{Y \in \mathbf{X}, Y \neq X} f_Y\right), f_i)$ is equivalent to the minimization of $KL(f_X^i \otimes \left(\bigotimes_{Y \in s(f_i), Y \neq X} f_Y\right), f_i)$, i.e. it is only necessary to consider the valuations f_Y where $Y \in s(f_i)$, instead of all the potentials (this is an immediate consequence of Properties 1 and 6 of divergence measures).

In the case that the preorder relation $\preceq$ is antisymmetrical, then property 6, can be rephrased as: if $V_3 \preceq V_2 \preceq V_1$, then $Di(V_1, V_2) \leq Di(V_1, V_3)$ and $Di(V_2, V_3) \leq Di(V_1, V_3)$. If we also assume that for any valuations V_1, V_2 and set of variables $\mathbf{Y}$, with $s(V_2) \subseteq \mathbf{Y} \cup s(V_1)$, then the set $\mathcal{R}^{\mathbf{Y}}_{V_2/V_1} = \{V \in \mathcal{V}_{\mathbf{Y}} : V_2 \preceq V_1 \otimes V\}$ has a minimum element, i.e. a valuation $V^* \in \mathcal{R}^{\mathbf{Y}}_{V_2/V_1}$ such that $V^* \preceq V, \forall V \in \mathcal{R}^{\mathbf{Y}}_{V_2/V_1}$, then the local improvement step, i.e. finding a valuation V_X^* minimizing the divergence $Di(V_X^* \otimes \left(\bigotimes_{Y \in \mathbf{X}, Y \neq X} V_Y\right), V_1 \otimes \cdots \otimes V_n)$ can be found by computing the minimum of the set $\mathcal{R}^X_{(\bigotimes_{Y \in \mathbf{X}, Y \neq X} V_Y)/(V_1 \otimes \cdots \otimes V_n)}$.

The existence of this minimum is satisfied in our two examples of valuations in which $\preceq$ is a partial order. In the case of probabilistic valuations with max-marginalization, the minimum of $\mathcal{R}^{\mathbf{Y}}_{f_2/f_1}$ is given by the potential $f^*(\mathbf{y}) = \min\{1, \min_{\mathbf{z}\in\mathbf{Z}} f_2.f_1^{-1}(\mathbf{y},\mathbf{z})\}$, where $\mathbf{Z} = (s(f_1) \cup s(f_2))\backslash\mathbf{Y}$. In the case of propositional logic potentials the minimum, F^* of $\mathcal{R}^{\mathbf{Y}}_{F_2/F_1}$ is the set of clauses $t \in \mathcal{V}_{\mathbf{Y}}$ such that $t \in Cons(F_2)$ and $t \in Cons(\neg f)$, $\forall f \in F_1$, where $\neg f$ is the set of clauses equivalent to the negation of f.

In the case of idempotent valuations we have a property which is the analogous to Proposition 1 for probabilistic valuations and that allows a decomposition of the computations.

Proposition 2. *If we have an idempotent valuation system in which the minimum of $\mathcal{R}^{\mathbf{Y}}_{V/V'}$ always exists when $s(V) \subseteq s(V') \cup \mathbf{Y}$, then for any V', V_1, V_2 and set of variables $\mathbf{Y}$ with $(s(V_1) \cup s(V_2)) \subseteq \mathbf{Y} \cup s(V')$ we have that*

$$\arg\min \mathcal{R}^{\mathbf{Y}}_{(V_1\otimes V_2)/V'} = \arg\min \mathcal{R}^{\mathbf{Y}}_{V_1/V'} \otimes \arg\min \mathcal{R}^{\mathbf{Y}}_{V_2/V'}$$

Proof. Let us call $V^1 = \arg\min \mathcal{R}^{\mathbf{Y}}_{V_1/V'}$, $V^2 = \arg\min \mathcal{R}^{\mathbf{Y}}_{V_2/V'}$, $V^{12} = \arg\min \mathcal{R}^{\mathbf{Y}}_{(V_1\otimes V_2)/V'}$.

It is clear that if $V^1 \in \mathcal{R}^{\mathbf{Y}}_{V_1/V'}$ and $V^2 \in \mathcal{R}^{\mathbf{Y}}_{V_2/V'}$, then $V_1 \preceq (V^1 \otimes V')$ and $V_2 \preceq (V^2 \otimes V')$, then $(V_1 \otimes V_2) \preceq (V^1 \otimes V' \otimes V^2 \otimes V') = (V^1 \otimes V^2 \otimes V')$, and therefore $V_1 \otimes V_2 \in \mathcal{R}^{\mathbf{Y}}_{(V_1\otimes V_2)/V'}$, and $V^{12} \preceq V^1 \otimes V^2$.

On the other hand, as $V^{12} \in \mathcal{R}^{\mathbf{Y}}_{(V_1\otimes V_2)/V'}$, we have that $(V_1 \otimes V_2) \preceq (V' \otimes V^{12})$ and as $V_1 \preceq (V_1 \otimes V_2)$ we have that $V^{12} \in \mathcal{R}^{\mathbf{Y}}_{V_1/V'}$ and, analogously $V^{12} \in \mathcal{R}^{\mathbf{Y}}_{V_2/V'}$. As a consequence, $V^1 \preceq V^{12}$ and $V^2 \preceq V^{12}$, and $(V^1 \otimes V^2) \preceq V^{12} \otimes V^{12} = V^{12}$.

As, in this case $\preceq$ is antisymmetrical, we have that $V^1 \otimes V^2 = V^{12}$.

A similar decomposition for probabilistic potentials under max-marginalization is not satisfied.

4.2 Lower Approximation

The lower approximation problem is to find the product $\bigotimes_{X\in\mathbf{X}} V_X$ minimizing the divergence $Di(V_1 \otimes \cdots \otimes V_n, \bigotimes_{X\in\mathbf{X}} V_X)$. This is even a more difficult problem than the upper approximation one and we will show that under very general conditions, it is equivalent to solve the marginal problem.

We will say that a valuation system satisfies the *independence consistency* property if and only if for any sets of variables $\mathbf{Y}, \mathbf{Z}$ such that $\mathbf{Y} \cap \mathbf{Z} = \emptyset$, if $V_1 \in \mathcal{V}_{\mathbf{Y}}$ and $V_2 \in \mathcal{V}_{\mathbf{Z}}$ are such that $V_1 \preceq V$ and $V_2 \preceq V$, then $V_1 \otimes V_2 \preceq V$. This property is satisfied in all the cases of idempotent valuations and in probabilistic valuations under sum-marginalization, but not under max-marginalization.

Proposition 3. *Consider $\mathbf{Y}, \mathbf{Z}$ such that $\mathbf{Y} \cap \mathbf{Z} = \emptyset$, $V'' \in \mathcal{V}_{\mathbf{Z}}$, and $V' \in \mathcal{V}_{\mathbf{Y}\cup\mathbf{Z}}$, with $V'' \preceq V'$, then* $\arg\min_{V\in\mathcal{V}_{\mathbf{Y}}} Di(V', V \otimes V'') = V'^{\downarrow\mathbf{Y}}$.

Proof. The proof is an immediate consequence of Properties 5 and 7 of divergence measures.

As a consequence of this proposition, when the independence consistency property is satisfied, the best lower approximation of $(V_1 \otimes \cdots \otimes V_n)$ is given by $\bigotimes_{X \in \mathbf{X}} V_X$ where $V_X = (V_1 \otimes \cdots \otimes V_n)^{\downarrow X}$. In this way, to compute the approximation is equivalent to solve the marginal problem and it makes no sense to use the approximation as a method to compute the marginal problem. A common strategy is to change the global approximation problem to a local approximation one, i.e. instead of approximating $(V_1 \otimes \cdots \otimes V_n)$ by a product of one-dimensional valuations, we try to approximate each V_i by a product $\bigotimes_{X \in s(V_i)} V_X^i$, obtaining a global approximation of $(V_1 \otimes \cdots \otimes V_n)$ by $\bigotimes_{i=1}^n \left(\bigotimes_{X \in s(V_i)} V_X^i\right)$. If we make all the approximations in a fully local way, the results can be poor, so a better procedure should approximate V_j taking into account the other approximations (of V_i for $i \neq j$). For example, by means of a greedy algorithm that tries to compute the best approximation of V_j minimizing $Di(V_1 \otimes \cdots \otimes V_n, \bigotimes_{i=1}^n \left(\bigotimes_{X \in s(V_i)} V_X^i\right))$, where V_X^i is fixed for $i \neq j$. However, this problem is again equivalent to the marginal problem. A further step can be to replace in this divergence each V_i by its approximation (if $i \neq j$). And then to try to minimize an approximate divergence: $Di(V_j \otimes \bigotimes_{i \neq j} \left(\bigotimes_{X \in s(V_i)} V_X^i\right), \bigotimes_{i=1}^n \left(\bigotimes_{X \in s(V_i)} V_X^i\right)$. By Property 6 of divergence measures, this is equivalent to minimize $Di(V_j \otimes (\bigotimes_{X \in s(V_j), i \neq j} V_X^i), \bigotimes_{X \in s(V_j)} V_X^j \otimes (\bigotimes_{X \in s(V_j), i \neq j} V_X^i))$. As this is an approximation problem involving only valuations in variables $s(V_j)$, it is computational feasible. In the case of probabilistic potentials, this is what iterative belief propagation does [12,13]. As in each step the divergence is also approximate and there is not a global measure which is decreasing with the approximations, this approach does not guarantee convergence.

5 Conclusions

In this paper we have considered the abstract framework of valuation based systems to define divergence measures and to state approximation problems. This allows a better understanding of the different strategies and their properties. It also allows to extend the methods designed for one representation of uncertainty to other formalisms. For example, our specification of iterative belief propagation allows to apply it to propositional logic. In the future we plan to extend this framework to schemes where upper and lower approximations are jointly computed in such a way that they cooperate, in the sense that a lower approximation can be used to improve an upper approximation as in importance sampling [8].

Acknowledgements. This research was supported by the Spanish Ministry of Economy and Competitiveness under project TIN2016-77902-C3-2-P and the European Regional Development Fund (FEDER).

References

1. Shafer, G., Shenoy, P.: Local computation in hypertrees. Working Paper N. 201, School of Business, University of Kansas, Lawrence (1988)
2. Shenoy, P., Shafer, G.: Axioms for probability and belief-function propagation. In: Shachter, et al. (ed.) Uncertainty in Artificial Intelligence, vol. 4, pp. 169–198. Elsevier, Amsterdam (1990)
3. Cano, J., Delgado, M., Moral, S.: An axiomatic system for the propagation of uncertainty in directed acyclic networks. Int. J. Approx. Reasoning **8**, 253–280 (1993)
4. Kohlas, J.: Information Algebras: Generic Structures for Inference. Springer, London (2012). https://doi.org/10.1007/978-1-4471-0009-6
5. Kohlas, J., Moral, S.: Propositional information systems. Working Paper N. 96–01, Institute of Informatics, University of Fribourg (1996)
6. Hernández, L., Moral, S.: Mixing exact and importance sampling propagation algorithms in dependence graphs. Int. J. Intell. Syst. **12**, 553–576 (1997)
7. Pearl, J.: Evidential reasoning using stochastic simulation of causal models. Artif. Intell. **32**, 247–257 (1987)
8. Hernández, L., Moral, S., Salmerón, A.: Importance sampling algorithms in belief networks based on aproximate computation. In: Proceedings of Information Processing and Management of Uncertainty in Knowledge-Based Systems Conference (IPMU 1996), vol. 2, Granada, pp. 859–864 (1996)
9. Moral, S., Salmerón, A.: Dynamic importance sampling in Bayesian networks based on probability trees. Int. J. Approx. Reasoning **38**(3), 245–261 (2005)
10. Dechter, R., Rish, I.: Mini-buckets: a general scheme for bounded inference. J. ACM (JACM) **50**(2), 107–153 (2003)
11. Wainwright, M.J., Jordan, M.I., et al.: Graphical models, exponential families, and variational inference. Found. Trends Mach. Learn. **1**(1–2), 1–305 (2008)
12. McEliece, R.J., MacKay, D.J.C., Cheng, J.F.: Turbo decoding as an instance of pearl's "belief propagation" algorithm. IEEE J. Sel. Areas Commun. **16**(2), 140–152 (1998)
13. Darwiche, A.: Modeling and Reasoning with Bayesian Networks. Cambridge University Press, Cambridge (2009)
14. Cano, A., Moral, S., Salmerón, A.: Novel strategies to approximate probability trees in penniless propagation. Int. J. Intell. Syst. **18**(2), 193–203 (2003)
15. Kask, K., Dechter, R.: Branch and bound with mini-bucket heuristics. IJCAI **99**, 426–433 (1999)
16. Kullback, S., Leibler, R.: On information and sufficiency. Ann. Math. Stat. **22**, 76–86 (1951)
17. Kohlas, J., Wilson, N.: Semiring induced valuation algebras: exact and approximate local computation algorithms. Artif. Intell. **172**(11), 1360–1399 (2008)

Topological MI-Groups: Initial Study

Michal Holčapek(✉) and Nicole Škorupová

Institute for Research and Applications of Fuzzy Modelling,
NSC IT4Innovations, University of Ostrava,
30. dubna 22, 701 03 Ostrava 1, Czech Republic
Michal.Holcapek@osu.cz, P17097@student.osu.cz

Abstract. In this paper, we introduce the concept of topological MI-groups, where the MI-group structure, which naturally generalizes the group structure, is enriched by a topology and the respective binary operation and inversion are continuous. To demonstrate that the proposed generalization of topological groups is meaningful, we prove that there are the products of topological MI-groups and the topological quotient MI-groups. The concept of topological MI-group is demonstrated on examples.

Keywords: MI-group · Topological MI-group · Hausdorff metric
Quotient MI-groups

1 Introduction

In practice, data are usually collected by a measurement procedure, which often provides inaccurate results. Therefore, data processing has to admit a measurement uncertainty and use techniques that enable to handle vaguely specified quantities. Fuzzy intervals or stochastic values are typical examples of such quantities. The vaguely specified quantities, however, require novel approaches to computation with them and measurement of their distances, which respect the nature of present vagueness. Because of the failure of group inversion and distributivity law in non–standard arithmetics considered for vaguely specified quantities, the very popular algebraical structure summarizing their fundamental properties became the semimodule structure (a generalization of a vector space). Moreover, the semimodules are often endowed by an appropriate metric, e.g., the extended Hausdorff metric for fuzzy intervals, to form a complete metric space (see [3,17] and references therein). This type of metric spaces then enables us to introduce functions with vaguely specified outputs like intervals, fuzzy intervals or convex sets and investigate the concepts like continuity, measurability or integrability of such functions. The theoretical results are applied, among others, to fuzzy statistic methods, fuzzy differential equations, optimization or image processing [3]. As an example, let us mention fuzzy valued random variables and a strong law of large numbers for this type of random variables (see [13,15–17]) that guarantees the correctness of the application of fuzzy Monte Carlo simulation in risk analysis of construction projects [28].

J. Medina et al. (Eds.): IPMU 2018, CCIS 855, pp. 603–615, 2018.
https://doi.org/10.1007/978-3-319-91479-4_50

As we have mentioned above the non-existence of group inversion forced researches to consider commutative monoids, and therefore, semimodules, as a suitable algebraic structure generalizing the properties of non-standard arithmetics for vaguely specified quantities. Nevertheless, practical applications of vaguely specified quantities and their non-standard arithmetics usually require certain "inverse" quantities, even if they are not the results of the group inversion, i.e., $xx^{-1} \neq 1$ holds in general. To establish the inverse like quantities, the monoid structures are extended by an appropriate unary operation relaxing the properties of the group inverse. A deeper investigation of properties of commutative monoids of intervals and convex bodies that, moreover, satisfy the cancellation law and are extended by a unary operation called the negation has been provided by Markov in [22] (see also [23,24]) and gave rise to a novel algebraic structure standing between monoids and groups.[1] A relaxation of Markov's properties has been proposed by Bica in his algebraic structure for real fuzzy numbers [1,2]. Both above mentioned algebras introduced for particular types of vaguely specified quantities can be included under a unique algebraic structure called the many identities group (MI-group, for short), which is defined as a monoid endowed with an involutive anti-automorphism satisfying the chosen properties of the group inversion. As a consequence of this generalization, the product of an element x and its inversion x^{-1} need not be equal to the identity element 1 and is referred only as a pseudo-identity element. Note that the concept of MI-group has been introduced by Holčapek and Štěpnička in [7,8] (see also [9]) and further developed in [6,10] together with other MI-algebras to describe properties of various approaches to arithmetics of vaguely specified quantities (e.g., stochastic or fuzzy quantities) in a unified way. It is interesting that a submonoid containing pseudoidentities of an MI-group gives rise to a congruence, which generalizes so-called Mareš equivalences introduced and developed in [18–20] and actually interconnects the theory of MI-groups with Mareš theory on fuzzy quantities. A recent deeper investigation of properties of extended monoids of fuzzy intervals in the framework of Mareš theory can be found [25–27].

In [25], Qui et al. enriched the commutative monoid of fuzzy numbers by the extended Hausdorff metric and provided a novel analysis of this algebraic structure from the topological point of view. It should be notated that a similar research on the commutative monoids of fuzzy numbers can be found in [5]. The results obtained in both papers show, among others, that there exists a quotient metric on the quotient monoid, whose equivalence classes are determined by the Mareš equivalence. This non-trivial result extends the knowledge that has been obtained from the research on the metric spaces of fuzzy numbers. Moreover, this type of results belongs rather to the theory of topological (metric) groups than to the theory of metric spaces. Since the considered algebraic structures for fuzzy numbers are particular examples of an MI-group, a natural question arises

---

[1] It should be noted that Markov called this novel structure as *quasimodule*, which is, however, terminologically confusing with the standard denotation, since no scalar operation is consider here.

whether the obtained results can be generalized for MI-groups and a reasonable theory of topological (metric) MI-groups can be developed. Therefore, the aim of this contribution is to provide foundation stones of the possible theory of topological MI-group and demonstrate its meaningfulness by the verification of the existence of two fundamental algebraic constructions – the product of topological MI-groups and quotient topological MI-groups.

The paper has the following structure. The next section presents the summary of definitions and results from the MI-group theory important for the understanding of text. The third section provides basic definitions and results on the topological MI-groups including the product of topological MI-groups and the quotient topological MI-groups. The last section is a conclusion. We assume that the reader has an elementary knowledge in the fuzzy set theory, topological spaces and group theory, otherwise, we refer to [12,14,21].

2 MI-Groups: A Survey

2.1 Basic Definitions

We use the definition of MI-groups considered in [6].

Definition 1. *A triple* $(G, \star, ^{-1})$ *is said to be an* MI-group *if it satisfies*

(G1) $(G, \star)$ *is a monoid with* e_G *denoting the identity element;*
(G2) $^{-1}\colon G \to G$ *is an involutive anti-automorphism, i.e., for any* $x, y \in G$, *it holds*
  (i) $(x \star y)^{-1} = y^{-1} \star x^{-1}$
  (ii) $(x^{-1})^{-1} = x$;
(G3) $x \star (y \star y^{-1}) = (y \star y^{-1}) \star x$ *for any* $x, y \in G$;
(G4) the cancellation law, i.e., for any $x, y, z \in G$, *it holds that*

$$x \star y = x \star z \Rightarrow y = z \quad (\textit{left cancellation law})$$
$$y \star x = z \star x \Rightarrow y = z \quad (\textit{right cancellation law}).$$

An MI-group is said to be abelian or commutative if $x \star y = y \star x$ *holds for any* $x, y \in G$.

Standardly, we write $G = (G, \star, ^{-1})$ and $x \star y = xy$. Let G be an MI-group. We use P_G to denote the least submonoid of the monoid G, which contains the set $\{xx^{-1} \mid x \in G\}$. The elements of P_G are called *pseudo-identities* and P_G. Obviously, if G is a group, then $P_G = \{e_G\}$, and if G is an abelian MI-group, then $P_G = \{xx^{-1} \mid x \in G\}$. Note that P_G contains exactly the symmetric elements of G, i.e., $x = x^{-1}$, if G is a commutative monoid of fuzzy numbers as it was considered in [5,20,27]. Generally, P_G is determined as follows [9].

Lemma 1. *The set* P_G *is determined as*

$$P_G = \{x_1x_1^{-1} \cdots x_nx_n^{-1} \mid n \in \mathbb{N},\, x_1, \ldots, x_n \in G\}.$$

Moreover, $sx = xs$ *for any* $x \in G$ *and* $s \in P_G$.

Recall that fuzzy intervals, sometimes referred as fuzzy numbers, are special fuzzy sets, which are very popular in practice for their bell shaped membership function [4,14]. We use $\mathbb{R}$ to denote the set of real numbers.

Definition 2. *A fuzzy set A on $\mathbb{R}$ is called a* fuzzy interval *if it satisfies the following conditions:*

(i) A *is normal, i.e., there is* $x \in \mathbb{R}$ *such that* $A(x) = 1$,
(ii) A *is convex, i.e.,* $A(\lambda x + (1-\lambda)y) \geq \min\{A(x), A(y)\}$ *for any* $x, y \in \mathbb{R}$ *and* $\lambda \in (0, 1)$,
(iii) A *is upper semi-continuous,*
(iv) A_0 *is compact, where* $A_0 = \mathrm{cl}(\{x \in \mathbb{R} \mid A(x) > 0\})$ *and* cl *denotes the standard topological closure operator of intervals.*

We use $\mathcal{F}$ to denote the set of all fuzzy intervals. A fuzzy interval defined over the set of positive real numbers is called a *positive fuzzy interval.* We use $\mathcal{F}^+$ to denote the set of all positive fuzzy intervals.

Example 1. Let $\mathcal{F}$ be the set of all fuzzy intervals over the real numbers. Then, the triplet $(\mathcal{F}, +, -)$, where $+$, $-$ are defined by the interval operations on α-cuts:

$$\begin{aligned} (A+B)_\alpha &= [x_A^-(\alpha) + x_B^-(\alpha), x_A^+(\alpha) + x_B^+(\alpha)] \\ &= [x_A^-(\alpha), x_A^+(\alpha)] + [x_B^-(\alpha), x_B^+(\alpha)] = A_\alpha + B_\alpha, \\ (-A)_\alpha &= [-x_A^+(\alpha), -x_A^-(\alpha)] = -(A_\alpha) \end{aligned}$$

for any $A = [x_A^-, x_A^+], B = [x_B^-, x_B^+] \in \mathcal{F}$ and $\alpha \in [0, 1]$, is an abelian additive MI-group. Similarly, one can define the abelian multiplicative MI-group $(\mathcal{F}^+, \cdot, ^{-1})$ of positive fuzzy intervals.

An MI-subgroup is a substructure of MI-group which is themselves MI-group.

Definition 3. *Let G be an MI-group, and let H be a non-empty subset of G. If H is itself an MI-group under the product and the inversion of G, then H is said to be an* MI-subgroup *of G. This is denoted by $H \leq G$.*

Theorem 1. *Let $H \subseteq G$ be a non-empty subset. Then H is an MI-subgroup of G iff $e_G \in H$ and $xy^{-1} \in H$.*

A homomorphism of monoids (monoidal part of MI-groups) preserving their inversions is called the homomorphism of MI-groups.

Definition 4. *Let G and H be MI-groups. A map $f \colon G \to H$ is a* homomorphism of MI-groups *if it holds*

(HG1) f *is a monoidal homomorphism of* G *to* H*;*
(HG2) $f(x^{-1}) = f(x)^{-1}$ *for any* $x \in G$.

If f is injective (surjective, bijective), then f is said to be a monomorphism *(*epimorphism, isomorphism*).*

Example 2. A map $\mathrm{Exp}\colon \mathcal{F} \to \mathcal{F}^+$ defined as

$$A = [x_A^-, x_A^+] \longmapsto \mathrm{Exp}(A) = [\exp(x_A^-), \exp(x_A^+)],$$

where $\exp(x_A^-)(\alpha) = \exp(x_A^-(\alpha))$ and $\exp(x_A^+)(\alpha) = \exp(x_A^+(\alpha))$ for $\alpha \in [0,1]$, is a isomorphism of MI-groups.

2.2 Product of MI-Groups

In this part, we show that MI-groups have products. Let $\{X_i, i \in I\}$ be a non-empty family of sets. We use $\prod_{i\in I} X_i$ or simply $\prod X_i$, if no confusion can appear, to denote the Cartesian product of sets. The elements of $\prod X_i$ will be denoted by (x_i). Moreover, we write ι instead of $^{-1}$ to denote the inversion in an MI-group.

Theorem 2. *Let $\{(G_i, \star_i, \iota_i) \mid i \in I\}$ be a non-empty family of MI-groups. Then $(\prod G_i, \star, \iota)$, where $\prod G_i$ is the product of the supports of MI-groups and*

$$(a_i) \star (b_i) = (a_i \star_i b_i),$$
$$(a_i)^\iota = (a_i^{\iota_i}),$$

for any $(a_i), (b_i) \in \prod G_i$, is the product of MI-groups.

Proof: Obvious. □

2.3 Quotient MI-Groups

In this part, we define the quotient MI-groups induced by normal MI-subgroups. For details, we refer to a recent paper [6]. Let us start with the concept of a closed subset in an MI-group G with respect to a submonoid K of P_G.

Definition 5. *Let $(G, \star, ^{-1})$ be an MI-group, and let K be a submonoid of P_G. A non-empty subset H of G is said to be* closed in G w.r.t. K *if $xs \in H$ holds for some $x \in G$ and $s \in K$, then $x \in H$. If H is an arbitrary subset of G, then the* closure of H in G w.r.t. K *is the set*

$$\overline{H}^K = \bigcap\{L \subseteq G \mid H \subseteq L \text{ and } L \text{ is closed in } G \text{ w.r.t. } K\}. \tag{1}$$

The following lemma shows a way how to introduce the closure of H in G w.r.t. K.

Lemma 2. *Let $(G, \star, ^{-1})$ be an MI-group, and let K be a submonoid of P_G. For any $\emptyset \neq H \subseteq G$, we have*

$$\overline{H}^K = \{x \in G \mid \text{ there is } s \in K \text{ such that } xs \in H\}. \tag{2}$$

The concept of the (right, left) congruence modulo H is defined with help of closure of cosets that are formed for the pseudo-identities.

Definition 6. *Let H be an MI-subgroup of an MI-group $(G, \star, ^{-1})$, let $x, y \in G$. We say that x* is right congruent to y modulo H *denoted $x \equiv_r y \pmod H$ if*

$$xy^{-1} \in \overline{yy^{-1}H}^{P_H}. \tag{3}$$

We say that x is left congruent to y modulo H *denoted $x \equiv_l y \pmod H$ if*

$$x^{-1}y \in \overline{xx^{-1}H}^{P_H}. \tag{4}$$

Theorem 3. *Let $H \leq G$.*

(i) Right (resp. left) congruence modulo H is an equivalence relation on G.
(ii) The equivalence class of $x \in G$ under right (resp. left) congruence modulo H is the closure of $Hx = \{hx \mid h \in H\}$ (resp. $xH = \{xh \mid h \in H\}$) in G w.r.t. P_H.
(iii) $|\overline{H}^{P_H}| \leq |\overline{Hx}^{P_H}| \leq |\overline{Hxx^{-1}}^{P_H}|$ and $|\overline{H}^{P_H}| \leq |\overline{xH}^{P_H}| \leq |\overline{xx^{-1}H}^{P_H}|$.

Corollary 1. *Let $H \leq G$.*

(i) G is the union of the closures of the right (left) cosets of H in G.
(ii) Two closures of right (left) cosets of H in G w.r.t. P_H are either disjoint or equal.
(iii) For all $x, y \in G$, $\overline{Hx}^{P_H} = \overline{Hy}^{P_H}$ (resp. $\overline{xH}^{P_H} = \overline{yH}^{P_H}$) iff $xy^{-1} \in \overline{yy^{-1}H}^{P_H}$ (resp. $x^{-1}y \in \overline{xx^{-1}H}^{P_H}$).
(iv) If $\mathcal{R}$ is the set of the closures of all right cosets of H in G w.r.t. P_H and $\mathcal{L}$ is the set of the closures of all left cosets of H in G w.r.t. P_H, then $|\mathcal{R}| = |\mathcal{L}|$.

A normal subgroup H of a group G makes the left and right congruences modulo H (cosets) coincident. A similar idea is applied for the definition of normal MI-subgroup. The following theorem belongs to the folklore in group theory (cf., [11]) and is fundamental for normal MI-subgroups.

Theorem 4. *Let $H \leq G$. Then the following conditions are equivalent:*

(i) the left and right congruence modulo H coincide;
(ii) the closure of each left coset of H in G w.r.t. P_H is the closure of a right coset of H in G w.r.t. P_H;
(iii) $\overline{xH}^{P_H} = \overline{Hx}^{P_H}$ for all $x \in G$;
(iv) $\overline{xHx^{-1}}^{P_H} \subseteq \overline{xx^{-1}H}^{P_H}$ for all $x \in G$;
(v) $\overline{xHx^{-1}}^{P_H} = \overline{xx^{-1}H}^{P_H}$ for all $x \in G$.

The definition of a normal MI-subgroup is as follows.

Definition 7. *An MI-subgroup H of an MI-group G which satisfies the equivalent conditions above is said to be a* normal MI-subgroup of G. *We write $H \lhd G$ if H is normal in G.*

A quotient MI-group induced by a normal MI-subgroup is defined similarly to quotient group (cf. [11]), where the only difference is that the cosets used for groups are replaced now by their closures in G w.r.t. K.

Theorem 5. *If $H \lhd G$, G/H is the set of closures of all left cosets of H in G w.r.t. P_H. Then $G/H = (G/H, \star, ^{-1})$, where $\overline{xH}^{P_H} \star \overline{yH}^{P_H} = \overline{xyH}^{P_H}$ and the inversion $^{-1}$ is given by $\left(\overline{xH}^{P_H}\right)^{-1} = \overline{x^{-1}H}^{P_H}$, is an MI-group called the* quotient MI-group *of G by H.*

3 Topological MI-Groups

3.1 Basic Definition

Topological MI-groups are defined in a similar way as topological groups. We use the following definition, where the continuity of both MI-group operations is considered.

Definition 8. *A quadruplet $(G, \star, ^{-1}, \mathcal{J})$ is said to be a topological MI-group if*

(i) $(G, \star, ^{-1})$ is an MI-group,
(ii) $(G, \mathcal{J})$ is a topological space,
(iii) $\star$ and $^{-1}$ are continuous maps.

Note that $\star$ is a continuous map of the product topological space $(G, \mathcal{J}) \times (G, \mathcal{J})$ to $(G, \mathcal{J})$. Recall that the product topology on $G \times G$ is defined in such a way that the set $\mathcal{B} = \{U \times V \mid U, V \in \mathcal{J}\}$ is its base. An equivalent definition of a topological MI-groups is provided in the following lemma.

Lemma 3. *$(G, \star, ^{-1}, \mathcal{J})$ is a topological MI-group if and only if the map $f: G \times G \to G$ given by $f(x, y) = xy^{-1}$ is a continuous map.*

Sketch of the proof: To prove that f is continuous, one defines two functions $g, h: G \times G \to G \times G$ by $g(x, y) = (x, y^{-1})$ and $h = \star$ and shows that g is continuous. The statement follows from the continuity of g and h and the fact that $f = h \circ g$.

To show that $\star$ and $^{-1}$ are continuous functions, one defines a function $p: G \to G \times G$ by $p(x) = (e_G, x)$, where e_G denotes the identity element in G, and shows that p is continuous. Since $^{-1} = f \circ p$, the inversion $^{-1}$ is continuous. Since $^{-1}$ is continuous, the function g defined above is continuous. The statement follows from $\star = f \circ g$. □

Example 3. Each MI-group $(G, \star, ^{-1})$ endowed with the discrete topology τ_D is a topological MI-group.

Example 4. Let $(\mathcal{F},+,-)$ be the additive MI-group of fuzzy intervals introduced in Example 1, and let us consider the extended Hausdorff metric proposed by Klement et al. in [13]:

$$\varrho_\infty(A,B) = \sup_{0\leq\alpha\leq 1} \max\left\{|x_A^-(\alpha) - x_B^-(\alpha)|, |x_A^+(\alpha) - x_B^+(\alpha)|\right\}.$$

For any $\varepsilon > 0$ and $A \in \mathcal{F}$, we define the ε-ball of A as the set $B_\varepsilon(A) = \{B \mid \varrho_\infty(A,B) < \varepsilon\}$. Let $\mathcal{J}_{\varrho_\infty}$ on $\mathcal{F}$ denote the topology determined by the ε-balls. Then $(\mathcal{F},+,-,\mathcal{J}_{\varrho_\infty})$ is a topological MI-group of fuzzy intervals (induced by the metric ϱ_∞). Similarly, $(\mathcal{F}^+,\cdot,^{-1},\mathcal{J}_{\varrho_\infty})$ is a topological MI-group of positive fuzzy intervals.

A homomorphism of topological MI-groups is defined as a continuous homomorphism of MI-groups.

Definition 9. *A map $f\colon G \to H$ is a continuous homomorphism (simply morphism) of topological MI-groups if f is a homomorphism of MI-groups G and H and simultaneously a continuous map of G to H.*

Example 5. Let us consider the homomorphism $\mathrm{Exp}\colon \mathcal{F} \to \mathcal{F}^+$ of MI-groups introduced in Example 2. From the continuity of the exponential function in the definition of Exp one can show that Exp is a homeomorfism of topological MI-groups $(\mathcal{F},+,-,\mathcal{J}_H)$ and $(\mathcal{F}^+,\cdot,^{-1},\mathcal{J}_H)$. Note that the inverse map from $\mathcal{F}^+$ onto $\mathcal{F}$ can be defined by the logarithmic function, which is also continuous.

Lemma 4. *Let G be a topological MI-group and $a,b \in G$. Then the following maps are all injective continuous maps of G to G:*

(i) $x \mapsto ax$,
(ii) $x \mapsto xb$,
(iii) $x \mapsto x^{-1}$ *(it is a homeomorphism).*

A continuous map $f\colon G \to H$ is said to be *open* (*closed*) if the image of any open (closed) set in G under f is open (closed) in H. If $F, H \subseteq G$, we define

$$FH = \{xy \mid x \in F, y \in H\}. \tag{5}$$

Particularly, we use xF instead of $\{x\}F$. The following lemma is important in the analysis of quotient topological MI-groups.

Lemma 5. *Let $(G,\star,^{-1},\mathcal{J})$ be a topological MI-group, $H \subseteq G$, and $f_a\colon G \to aG$ ($g_a\colon G \to Ga$) is defined by $f_a(x) = ax$ ($g_a(x) = xa$) for $a \in G$, where aG (Ga) is the topological space with the relative topology.*

(i) *f_a (g_a) is open if and only if f_a (g_a) is a homeomorphism of G and aG (Ga) and aG (Ga) is open in G.*
(ii) *f_a (g_a) is closed if and only if f_a (g_a) is a homeomorphism of G and aG (Ga) and aG (Ga) is closed in G.*

(iii) *If f_a (g_a) is open for any $a \in H$ and U is open in G, then HU (UH) is open in G.*
(iv) *If f_a (g_a) is closed for any $a \in H$, where H is finite, and U is closed in G, then HU (UH) is closed in G.*

It should be noted that f_a is always open and closed for topological groups. Indeed, if G is an topological group, then $aG = G$, and the inverse map to f_a is defined by $f_{a^{-1}}$, which is again a continuous homeomorphism. Hence, $f_a(U) = f^{-1}_{a^{-1}}(U)$ is open (closed) in G, whenever U is open (closed) in G. The same statement is not evidently true for MI-groups, since $a^{-1}a \neq e_G$ in general; therefore, $f_{a^{-1}} \circ f_a \neq 1_G$.

Definition 10. *Let $G = (G, \star, ^{-1}, \mathcal{J})$ be a topological MI-group, and let $H \subseteq G$ be a non-empty subset. If H is itself an MI-subgroup under the product and inversion of G, then $(H, \star, ^{-1}, \mathcal{J}_H)$, where $\mathcal{J}_H$ is the relative topology, is a topological MI-subgroup of G.*

It is easy to show (see, e.g., [21]) that each MI-subgroup of a topological MI-group endowed by the relative topology is a topological MI-subgroup.

3.2 Products of Topological MI-Groups

Let us recall that if $\{(G_i, \mathcal{J}_i) \mid i \in I\}$ is a non-empty family of topological spaces, the base of the product topology of $\prod G_i$ is the set $\mathcal{B} \subseteq \{\prod U_i \mid U_i \in \mathcal{J}_i,\, i \in I\}$ such that

$$\prod U_i \in \mathcal{B} \quad \text{iff} \quad U_i \neq G_i \text{ only for a finite number of } i \in I. \tag{6}$$

It is well known that $\pi_i \colon \prod G_i \to G_i$ is a continuous map for any $i \in I$. Moreover, it holds the following statement [12].

Theorem 6. *Let $g \colon H \to \prod G_i$ be a map of a topological space H to the product of topological spaces $\prod G_i$. Then, g is continuous if and only if $\pi_i \circ g$ is continuous for any $i \in I$.*

Now we can introduce the product of topological MI-groups.

Theorem 7. *Let $\{(G_i, \star_i, \iota_i, \mathcal{J}_i) \mid i \in I\}$ be a non-empty family of topological MI-groups. Then, the quadruplet $(\prod G_i, \star, \iota, \mathcal{J})$ such that*

(i) *$(\prod G_i, \star, \iota)$ is the product of MI-groups $\{(G_i, \star_i, \iota_i) \mid i \in I\}$,*
(ii) *$(\prod G_i, \mathcal{J})$ is the product of topological spaces $\{(G_i, \mathcal{J}_i) \mid i \in I\}$*

is the product of topological MI-groups.

Sketch of the proof: To prove the statement, let $f : \prod G_i \times \prod G_i \to \prod G_i$ be defined by $f((x_i), (y_i)) = (x_i \star_i y_i^{\iota_i})$, and put $h_i = \pi_i \circ f$, where π_i denotes the i-th projection. One can simply prove that h_i is continuous for any $i \in I$. The continuity of f follows from Theorem 6, hence, the quadruplet $(\prod G_i, \star, \iota, \mathcal{J})$ is a topological group according to Lemma 3. The proof is finished by the verification that $(\prod G_i, \star, \iota, \mathcal{J})$ is really a product of topological MI-groups. □

3.3 Quotient Topological MI-Groups

Recall that a fundamental concept used in the definition of quotient MI-groups is a closed subset of an MI-group G with respect to a submonoid K of P_G. The following lemma shows an important property of closed subsets.

Lemma 6. *Let $G = (G, \star, ^{-1}, \mathcal{J})$ be a topological MI-group, and let $K \leq P_G$. If U is open in G, then $\overline{U}^K$ is also open in G.*

Sketch of the proof: From Lemma 5, the set $f_s^{-1}(U)$ is open for any open set U, where $f_s : G \to G$ is defined by $f(x) = xs$ for any $s \in K$. If $x \in \overline{U}^K$, then $f_{s_x}(x) \in U$ for a certain $s_x \in K$; therefore, $x \in f_{s_x}^{-1}(U)$. The statement follows from $\overline{U}^K = \bigcup_{x \in \overline{U}^K} f_{s_x}^{-1}(U)$ and the fact that the union of open sets is open. □

Note that a similar statement is not valid for closed sets in G. Let H be an MI-subgroup of a topological MI-group G. Denote G/H the family of closers of all left cosets, i.e., $G/H = \{\overline{xH}^{P_H} \mid x \in G\}$. Recall that G/H becomes an MI-group if H is normal (see Theorem 5). The quotient topology, denoted by $\mathcal{J}_{G/H}$, on G/H is defined as the largest collection of subsets of G/H such that the projection $\pi \colon G \to G/H$ given by

$$\pi(x) = \overline{xH}^{P_H}, \quad x \in G,$$

is a continuous map, i.e., $\pi^{-1}(U) \in \mathcal{J}$ holds for any $U \in \mathcal{J}_{G/H}$. It is well known that each projection $\pi \colon G \to G/H$ is open for the topological groups. The following theorem shows that a similar statement holds also for the topological MI-groups, only we have to assume that $x \mapsto xh$ for $h \in H$ are open maps.

Theorem 8. *Let $H \lhd G$, and let $g_h(x) = xh$ be open for any $h \in H$. Then, the projection $\pi \colon G \to G/H$ is an open map.*

Sketch of the proof: By the definition of the quotient topology, one has to verify that $\pi^{-1}(\pi(U)) \in \mathcal{J}$. From (iii) of Lemma 5 and the assumption on g_h for $h \in H$, it holds that $UH \in \mathcal{J}$. Moreover, $\overline{UH}^{P_H} \in \mathcal{J}$ according to Lemma 6. The proof is finished by the verification of $\pi^{-1}(\pi(U)) = \overline{UH}^{P_H}$. □

Theorem 9. *Let H be a normal MI-subgroup of a topological MI-group G, and let $g_h(x) = xh$ be open for any $h \in H$, and let $g_h(x) = xh$ be open for any $h \in H$. Then $(G/H, \star, ^{-1}, \mathcal{J}_{G/H})$ is a quotient topological MI-group.*

Sketch of the proof: To prove the statement, define $g(x, y) = xy^{-1}$ and consider the following commutative diagram

$$\begin{array}{ccc} G \times G & \xrightarrow{\pi \times \pi} & G \setminus H \times G \setminus H \\ \downarrow g & & \downarrow f \\ G & \xrightarrow{\pi} & G \setminus H, \end{array}$$

where $f(\overline{xH}^{P_H}, \overline{yH}^{P_H}) = \overline{xy^{-1}H}^{P_H}$ and $(\pi \times \pi)(x, y) = (\pi(x), \pi(y))$. Let U be an open neighborhood of $\overline{xy^{-1}H}^{P_H}$ in the quotient topology $\mathcal{J}_{G/H}$. Since g and π are continuous, there exist open neighborhoods V_x of x and V_y of y such that $g(V_x \times V_y) \subseteq \pi^{-1}(U)$; therefore, $(\pi \circ g)(V_x \times V_y) \subseteq U$. Due to Theorem 8, $\pi(V_x)$ and $\pi(V_y)$ are open neighborhoods of $\overline{xH}^{P_H}$ and $\overline{yH}^{P_H}$, respectively; therefore, $(\pi \times \pi)(V_x \times V_y) = \pi(V_x) \times \pi(V_y)$ belongs to the base of the product topology $\mathcal{J}_{G/H}$. From the commutative diagram, it holds that

$$f(\pi(V_x) \times \pi(V_y)) = f \circ (\pi \times \pi)(V_x \times V_y) = (\pi \circ g)(V_x \times V_y)) \subseteq U,$$

which implies that f is a continuous map in a point $(\overline{xH}^{P_H}, \overline{y^{-1}H}^{P_H})$. The statement follows from Theorem 3. □

Corollary 2. *Let H be a normal MI-subgroup of a topological MI-group G, and let $g_y(x) = xy$ be open for any $y \in H$. If H is open, then G/H has the discrete topology.*

4 Conclusions

In this contribution, we introduced topological MI-groups to generalize the concept of topological groups and studied their basic properties. We demonstrated the existence of the product of topological MI-groups and, under certain assumptions, the existence of the quotient topological M-groups. Both constructions indicate that the concept of topological MI-groups is meaningful and a further research in this field should bring interesting and non-trivial results that could enrich the known results from metric spaces over vaguely defined quantities. For example, one open problem is related to the existence of quotient metric MI-groups, where a metric MI-group is an MI-group with a metric such that this MI-group becomes topological with respect to the topology induced by open balls determined by the metric. More precisely, if $(G, \star, ^{-1}, \varrho)$ is a metric MI-group and H is a normal metric MI-subgroup of G, it is not clear how to introduce a quotient metric $\varrho_{G/H}$ on G/H from the metric ϱ on G such that the natural homomorphism π is a quotient map, i.e.,

$$U \in \tau_{G/H} \text{ if and only if } \pi^{-1}(U) \in \tau,$$

where τ and $\tau_{G/H}$ are the sets of open balls determined by the metrics ρ and $\rho_{G/H}$, respectively. The solution of this open problem could integrate the results in [5,25] under a more general theory.

Acknowledgments. This work was supported by the project LQ1602 IT4Innovations excellence in science.

References

1. Bica, A.M.: Categories and algebraic structures for real fuzzy numbers. Pure Math. Appl. **13**(1–2), 63–67 (2003)
2. Bica, A.M.: Algebraic structures for fuzzy numbers from categorial point of view. Soft. Comput. **11**, 1099–1105 (2007)
3. Diamond, P., Kloeden, P.E.: Metric Spaces of Fuzzy Sets: Theory and Applications. World Scientific, River Edge (1994)
4. Dubois, D., Prade, H.: Operations on fuzzy numbers. Int. J. Syst. Sci. **9**, 613–626 (1978)
5. Fechete, D., Fechete, I.: Quotient algebraic structure on the set of fuzzy numbers. Kybernetika **51**(2), 255–257 (2015)
6. Holčapek, M.: On generalized quotient mi-groups. Fuzzy Sets Syst. **326**, 3–23 (2017)
7. Holčapek, M., Štěpnička, M.: Arithmetics of extensional fuzzy numbers - part I: introduction. In: Proceedings of the IEEE International Conference on Fuzzy Systems, Brisbane, pp. 1517–1524 (2012)
8. Holčapek, M., Štěpnička, M.: Arithmetics of extensional fuzzy numbers - part II: algebraic framework. In: Proceedings of the IEEE International Conference on Fuzzy Systems, Brisbane, pp. 1525–1532 (2012)
9. Holčapek, M., Štěpnička, M.: MI-algebras: a new framework for arithmetics of (extensional) fuzzy numbers. Fuzzy Sets Syst. **257**, 102–131 (2014)
10. Holčapek, M., Wrublová, M., Bacovský, M.: Quotient MI-groups. Fuzzy Sets Syst. **283**, 1–25 (2016)
11. Hungerford, T.W.: Algebra, vol. XIX, 502 p. Holt, Rinehart and Winston, Inc. New York (1974)
12. Kelly, J.L.: General Topology. Springer, New York (1975)
13. Klement, E.P., Puri, M.L., Ralescu, D.A.: Limit theorems for fuzzy random variables. In: Proceedings of the Royal Society of London, Series A, vol. 407, pp. 171–182 (1986)
14. Klir, G.J., Yuan, B.: Fuzzy Sets and Fuzzy Logic: Theory and Applications. Prentice Hall, New Jersey (1995)
15. Krätschmer, V.: Limit theorems for fuzzy-random variables. Fuzzy Sets Syst. **126**, 253–263 (2002)
16. Kruse, R.: The strong law of large numbers for fuzzy random variables. Inf. Sci. **28**(3), 233–241 (1982)
17. Li, S., Ogura, Y., Kreinovich, V.: Limit Theorems and Applications of Set-Valued and Fuzzy Set-Valued Random Variables. Springer, Netherlands (2002). https://doi.org/10.1007/978-94-015-9932-0
18. Mareš, M.: Addition of fuzzy quantities: disjunction-conjunction approach. Kybernetika **2**, 104–116 (1989)
19. Mareš, M.: Multiplication of fuzzy quantities. Kybernetika **5**, 337–356 (1992)
20. Mareš, M.: Computation over Fuzzy Quantities. CRC Press, Boca Raton (1994)
21. Markley, N.G.: Topological Groups: An Introduction. Wiley, Hoboken (2010)
22. Markov, S.: On the algebra of intervals and convex bodies. J. Univ. Comput. Sci. **4**(1), 34–47 (1998)

23. Markov, S.: On the algebraic properties of convex bodies and some applications. J. Convex Anal. **7**(1), 129–166 (2000)
24. Markov, S.: Quasilinear space of convex bodies and intervals. J. Comput. Appl. Math. **162**, 93–112 (2004)
25. Qiu, D., Lu, C., Zhang, W., Lan, Y.: Algebraic properties and topological properties of the quotient space of fuzzy numbers based on mareś equivalence relation. Fuzzy Sets Syst. **245**, 63–82 (2014)
26. Qiu, D., Zhang, W., Lu, C.: On fuzzy differential equations in the quotient space of fuzzy numbers. Fuzzy Sets Syst. **295**, 72–98 (2016)
27. Qiu, D., Zhang, W.: Symmetric fuzzy numbers and additive equivalence of fuzzy numbers. Soft. Comput. **17**, 1471–1477 (2013)
28. Sadeghi, N., Fayek, A.R., Pedrycz, W.: Fuzzy Monte Carlo simulation and risk assessment in construction. Comput. Aided Civ. Infrastruct. Eng. **25**, 238–252 (2010)

Virtual Subconcept Drift Detection in Discrete Data Using Probabilistic Graphical Models

Rafael Cabañas[1,2](✉), Andrés Cano[2], Manuel Gómez-Olmedo[2], Andrés R. Masegosa[1], and Serafín Moral[2]

[1] Department of Mathematics, University of Almería, Almería, Spain
{rcabanas,andresmasegosa}@ual.es
[2] Department of Computer Science and Artificial Intelligence, CITIC, University of Granada, Granada, Spain
{acu,mgomez,smc}@decsai.ugr.es

Abstract. A common problem in mining data streams is that the distribution of the data might change over time. This situation, which is known as concept drift, should be detected for ensuring the accuracy of the models. In this paper we propose a method for subconcept drift detection in discrete streaming data using probabilistic graphical models. In particular, our approach is based on the use of conditional linear Gaussian Bayesian networks with latent variables. We demonstrate and analyse the proposed model using synthetic and real data.

Keywords: Concept drift · Data stream · Bayesian networks
Latent variables · Conditional linear Gaussian

1 Introduction

In recent years, the field of mining data streams has received an increasing attention as large amounts of data are continuously being generated (e.g., at financial sector, at social networks, etc.). An important aspect of data streams is that the domain being modelled is often *non-stationary*. In other words, the distribution governing the data changes over time. This situation is known as *concept drift* [6,15,19] and if not carefully taken into account, the result can be a failure to capture and interpret intrinsic properties of the data.

We propose a method for detecting (virtual) subconcept drift in discrete streaming data. This kind of drift affects only to some subspaces of the variables domain. Many approaches might fail in its detection as they usually analyse the whole domain. Our approach is an extension of the one by Borchani et al. [1,4], an approach using the conditional linear Gaussian (CLG) model [7,8] with latent variables. Yet, this previous approach was applied to continuous domains. Thus, we propose transforming the discrete data into continuous. Then, for improving the detection of subconcept drift, our model contains multiple latent variables that are parent of only a few variables in the data.

J. Medina et al. (Eds.): IPMU 2018, CCIS 855, pp. 616–628, 2018.
https://doi.org/10.1007/978-3-319-91479-4_51

The paper is organized as follows. Sections 2 and 3 introduce some basic concepts. Section 4 details our approach for concept drift detection. The empirical analysis is presented in Sect. 5. Finally, the conclusions are given in Sect. 6.

2 Data Streams with Concept Drift

2.1 Definitions and Notation

Let us first introduce the basic notation. We use upper-case letters for random variables and lower-case for their possible values. For example, x is a value of a given variable X. The set of all possible values that a variable X can take is called domain and denoted Ω_X. For the sets of variables and their assignments we use boldface letters, e.g, the set of variables $\boldsymbol{X}$ takes the values in $\boldsymbol{x}$.

In general, a data stream is observed at time-points $t_1, t_2, \ldots$ where $t_j < t_{j+1}$ for all j. At each time point t we have a collection of instances called batch (a.k.a window). For example, if a data stream defined over $\boldsymbol{X}$, then the batch at time point t is $\{\boldsymbol{x}^t[1], \boldsymbol{x}^t[2], \ldots, \boldsymbol{x}^t[N_t]\}$. In classification tasks, data streams are defined over $\boldsymbol{X} \cup C$, where $\boldsymbol{X}$ is the set of *features* or *covariates* and C is the *class variable* or *label*. In this case, a batch is denoted $\{\langle \boldsymbol{x}^t[1], c^t[1] \rangle, \langle \boldsymbol{x}^t[2], c^t[2] \rangle \ldots, \langle \boldsymbol{x}^t[N_t], c^t[N_t] \rangle\}$. We use $P(\boldsymbol{X}, C)$ to denote the joint distribution over the covariates and class labels. In the literature, this is called *concept* [6]. We use $P(C|\boldsymbol{X})$ to denote the conditional distribution over class labels given covariates and $P(\boldsymbol{X})$ to denote the prior probability distribution over covariates. A given distribution $P(\bullet)$ at time-point t is denoted $P_t(\bullet)$.

Data streams are usually non-stationary, which implies changes in the statistical properties of the data stream over time. This is known as *concept drift* [6,15]. More formally, if concept drift is present, it holds that $P_{t_j}(\boldsymbol{X}, C) \neq P_{t_k}(\boldsymbol{X}, C)$ where t_j, and t_k are 2 different time-points. In case of a data stream without a class variable, this is simply $P_{t_j}(\boldsymbol{X}) \neq P_{t_k}(\boldsymbol{X})$. In what follows we shall consider that concept drift only happens across time and not within a set of instances belonging to the same time-point.

2.2 Concept Drift Taxonomy

In the literature, there is much work aiming to characterize concept drift [6,14,18], which can be classified according to different aspects: scope, speed, recurrence, etc. We only describe the different types depending on the scope.

Real concept drift or *class drift* [17] occurs when the likelihood distribution over the class labels given the covariates changes, i.e., it holds that $P_{t_j}(C|\boldsymbol{X}) \neq P_{t_k}(C|\boldsymbol{X})$. As proposed by Webb et al. [18], we will use the term *pure class drift* if also holds that $P_{t_j}(\boldsymbol{X}) = P_{t_k}(\boldsymbol{X})$. On the other hand, *virtual concept drift* or *covariate drift* [17] implies changes in the prior distribution of the covariates, i.e. $P_{t_j}(\boldsymbol{X}) \neq P_{t_k}(\boldsymbol{X})$. Similarly, *pure covariate drift* occurs when the prior distribution of the class remains stable, i.e., $P_{t_j}(C|\boldsymbol{X}) = P_{t_k}(C|\boldsymbol{X})$. Real concept drift can be further divided into two subtypes depending on its scope [11,18].

First, *Subconcept drift* (a.k.a *intersected drift*) occurs when $P(C|\boldsymbol{X})$ changes only for a subspace of $\Omega_{\boldsymbol{X}}$. Otherwise, it is called *full-concept drift*. More formally, subconcept drift can be defined as:

$$P_{t_j}(C|\boldsymbol{X}) \neq P_{t_k}(C|\boldsymbol{X}) \wedge \exists_{\boldsymbol{x}\in\Omega_{\boldsymbol{X}}} \forall_{c\in\Omega_C} P_{t_j}(c|\boldsymbol{x}) = P_{t_k}(c|\boldsymbol{x}) \quad (1)$$

We consider the same idea for virtual concept drift, where the variations in the prior distributions over the covariates affects only to some subspaces of $\Omega_{\boldsymbol{X}}$. Therefore, we can define *virtual subconcept drift* as follows:

$$P_{t_j}(\boldsymbol{X}) \neq P_{t_k}(\boldsymbol{X}) \wedge \exists_{\boldsymbol{x}\in\Omega_{\boldsymbol{X}}} P_{t_j}(\boldsymbol{x}) = P_{t_k}(\boldsymbol{x}) \quad (2)$$

Figure 1 shows virtual concept drift examples across time points t_j and t_k in a data stream over $\{X, C\}$.

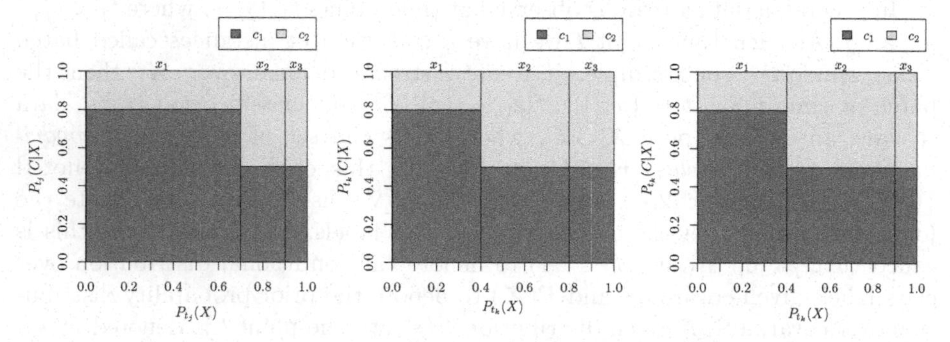

Fig. 1. Concept drift examples across time points t_j and t_k: (left) initial probabilities at t_j, (center) virtual full-concept drift, (right) virtual subconcept drift.

2.3 Detection and Adaptation

In the context of concept drift, we can consider two important tasks: detection and adaptation. Detection methods should be robust to noise and always signal concept drift when it occurs. In recent years, several detection methods have been developed that either monitor the evaluation of performance indicators [19] or compare the distributions on two different data samples [1]. Our approach for concept drift detection belongs to the second group. On the other hand, model adaptation methods aim to keep the model up-to-date [5]. For dealing with subconcept drift, Minku et al. [11] proposed the use of ensembles of classifiers.

3 Bayesian Networks

Our approach for concept drift detection is based on *Bayesian networks* (BNs) [13], which are a class of PGMs representing a joint probability distribution over a finite set of random variables. The nodes represent the variables in the problem being modelled, and the links represent the (conditional) dependencies and independencies among the variables.

Definition 1 (Bayesian network). *A Bayesian network (BN) is a tuple* $\langle \boldsymbol{X}, \mathbf{P}, \mathcal{G} \rangle$ *where:* $\boldsymbol{X}$ *is a set of discrete random variables;* $\mathcal{G}$ *is a DAG where each node represents a variable in* $\boldsymbol{X}$*;* $\mathbf{P}$ *is a set of conditional probability distributions, containing one distribution* $P(X|pa(X))$ *for each* $X \in \boldsymbol{X}$ *where* $pa(X)$ *is the set of parents of* X *according to* $\mathcal{G}$*.*

We will consider conditional linear Gaussian (CLG) BNs [8], which are an extension of BNs allowing discrete and continuous variables. The conditional probability distributions of continuous variables are specified as CLG distributions and discrete variables can only have discrete parents. The conditional distribution of each discrete variable $X_D \in \boldsymbol{X}$ given its parents is a multinomial. On the other hand, the conditional distribution of each continuous variable $Z \in \boldsymbol{X}$ with discrete parents $\boldsymbol{X}_D \subseteq \boldsymbol{X}$ and continuous parents $\boldsymbol{X}_C \subseteq \boldsymbol{X}$, is given by

$$p(z|\boldsymbol{X}_D = \boldsymbol{x}_D, \boldsymbol{X}_C = \boldsymbol{x}_C) = \mathcal{N}(z; \alpha(\boldsymbol{x}_D) + \beta(\boldsymbol{x}_D)^{\mathsf{T}} \boldsymbol{x}_C, \sigma(\boldsymbol{x}_D)), \tag{3}$$

for all $\boldsymbol{x}_D \in \Omega_{\boldsymbol{X}_D}$ and $\boldsymbol{x}_C \in \Omega_{\boldsymbol{X}_C}$. Figure 2 depicts two examples. Note that the BN on the right contains a *latent (i.e., hidden)* variable [12] depicted as a white node. A variable of this kind cannot be directly observed. The rest of the variables are called observed and will be represented with nodes in grey. When we have a set of observable variables, it can happen that the existing relationships among them are very complex, but that this relationships can be highly simplified under the assumption of the existence of one or several hidden variables. Think for example in a Naive Bayes classifier assuming that this is the correct model for the data. If the class variable is not observed, then there are not conditional independence relationships between the attributes (without conditioning to the class). Then, the relationships between these attributes are very complex. These can be simplified if we assume the existence of a hidden variable which is a father of all the attributes (the non-observed class)

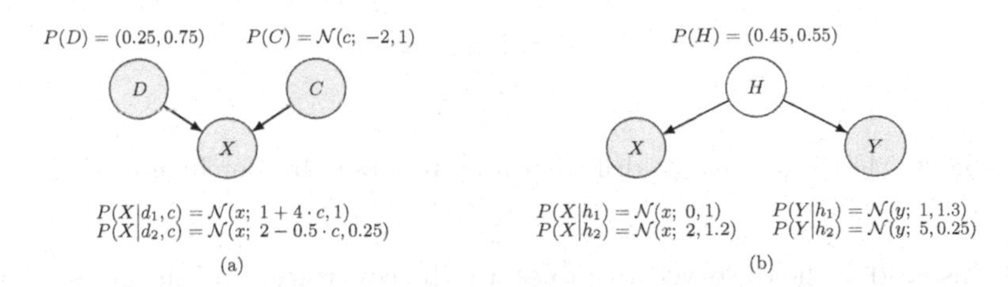

Fig. 2. Example of 2 conditional linear Gaussian BNs.

The BN in Fig. 2b defines a multivariate Gaussian mixture [12, p. 339], which is a latent variable model typically used for unsupervised clustering. Here, the latent variable H represents the clusters or groups while variables X and Y are the data attributes. Note that, if H were continuous, its interpretation would be a variable that summarizes all the data attributes. This is the key idea in the model here proposed for concept drift detection.

In the task of learning BNs from streaming data, the following problem appears. It is not possible to learn the model with the whole data, which might not have been generated yet or it cannot be stored in the memory due to its size. For that reason, some scalable methods for learning BNs from data streams have been developed in the last years, allowing to efficiently update the model when new data is available [2,9,10,20]. With these methods, uniform prior distributions of the latent variables are considered.

4 Concept Drift Detection Using Latent Variables

Herein we address the issue of (virtual) concept drift detection using PGMs with latent variables. First, we give a brief explanation of the method proposed by Borchani et al. [1,4], on which we base our model. Then, we propose a pre-processing algorithm that allows to apply the previous model to discrete data stream. Finally, we explain our model for detecting subconcept drift.

4.1 Full-Concept Drift Detection in Continuous Data

The approach by Borchani et al. [1,4] is defined in the context of classification where $\boldsymbol{X} = \{X_1, X_2, \ldots, X_n\}$ is the set of (continuous) covariates and a discrete class C. Figure 3 shows the BN with plate (a.k.a. plateau) notation [3] proposed by Borchani et al. [1] for modelling concept drift.

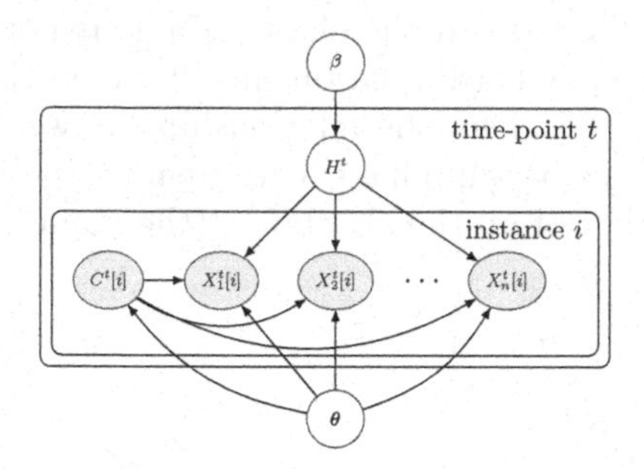

Fig. 3. Model for concept drift detection proposed by Borchani et al. [1].

In this model, the observed variables are the covariates and the class variable. A continuous latent variable H^t is set as a parent of all the nodes in $\boldsymbol{X}$. The nodes labelled with $\boldsymbol{\theta}$ and β represent the parameters, which are shared for all time points and across all instances. As H^t is parent of the observed variables, its values have an influence in the probabilities of the observed variables. So, changes in the value of H^t produce changes in the attribute probabilities. In this way, a concept drift can be associated to variations on H^t. For determining the presence of concept drift, we estimate the posterior distributions of the H^t-variable at each time-point. A variation on its expected value implies that $P(\boldsymbol{X})$ drifts. This is a global detector as all the subspaces of $\Omega_{\boldsymbol{X}}$ are analysed.

4.2 Full-Concept Drift Detection in Discrete Data

Now, we consider how to adapt the previous model to discrete data. In CLG models, discrete variables cannot have continuous parents. Thus, the model in Fig. 3 cannot be directly applied to discrete domains. For that, we propose to apply Algorithm 1 for transforming discrete data over $\boldsymbol{X}$ into equivalent numerical data over $\boldsymbol{X}'$ (that will be considered as continuous). For simplicity, we assume that we have unlabelled data. In case of a class variable, it can be treated as another covariate. The idea is to transform each variable $X \in \boldsymbol{X}$ in a set of numerical variables $\{X^j\}_{j=1}^{|\Omega_X|}$, where $X^j = 1$ if $X = x_j \in \Omega_X$, and 0, otherwise. Note that each new variable $X^j \in \boldsymbol{X}'$ corresponds to a state $x_j \in \Omega_X$.

Algorithm 1. pre-processing

input : $\{\boldsymbol{x}[1], \boldsymbol{x}[2], \ldots, \boldsymbol{x}[N]\}$ (batch of discrete instances over $\boldsymbol{X}$)
output : $\{\boldsymbol{x}'[1], \boldsymbol{x}'[2], \ldots, \boldsymbol{x}'[N]\}$ (batch of continuous instances $\boldsymbol{X}'$)

1: **for** $i \leftarrow 1$ **to** N **do**
2: $\quad \boldsymbol{x}'[i] \leftarrow \emptyset$
3: $\quad$ **for each** $X \in \boldsymbol{X}$ **do**
4: $\quad\quad$ Let $x[i]$ the value of X in the instance $\boldsymbol{x}[i]$
5: $\quad\quad$ **if** X is discrete **then**
6: $\quad\quad\quad$ **for** $j \leftarrow 1$ **to** $|\Omega_X|$ **do**
7: $\quad\quad\quad\quad$ Let x_j the jth state in Ω_X
8: $\quad\quad\quad\quad$ **if** $x[i] = x_j$ **then**
9: $\quad\quad\quad\quad\quad \boldsymbol{x}'[i] \leftarrow \boldsymbol{x}'[i] \cup \{1.0\}$
10: $\quad\quad\quad\quad$ **else**
11: $\quad\quad\quad\quad\quad \boldsymbol{x}'[i] \leftarrow \boldsymbol{x}'[i] \cup \{0.0\}$
12: $\quad\quad\quad\quad$ **end if**
13: $\quad\quad\quad$ **end for**
14: $\quad\quad$ **else**
15: $\quad\quad\quad \boldsymbol{x}'[i] \leftarrow \boldsymbol{x}'[i] \cup \{x[i]\}$
16: $\quad\quad$ **end if**
17: $\quad$ **end for**
18: **end for**
19: **return** $\{\boldsymbol{x}'[1], \boldsymbol{x}'[2], \ldots, \boldsymbol{x}'[N]\}$

For example, let us consider a data stream defined over $\boldsymbol{X} = \{X, Y\}$ with $\Omega_X = \{x_1, x_2, x_3\}$ and $\Omega_Y = \{y_1, y_2, y_3, y_4\}$. Then, the resulting data will be defined over the set of continuous variables $\boldsymbol{X}' = \{X^1, X^2, X^3, Y^1, Y^2, Y^3, Y^4\}$. Table 1 shows an example of this transformation. Once that the data is transformed, we can consider a similar model with latent variables. Figure 4 depicts the proposed BN for detecting concept drift in a data stream over $\boldsymbol{X}$.

Note that the BN is defined over $\boldsymbol{X}'$ instead. Again, the concept drift detection is done by monitoring the H^t-variable. A variation on its expected value implies that $P(\boldsymbol{X}')$ and so does $P(\boldsymbol{X})$. Again, this is a global detector as all the subspaces of $\Omega_{\boldsymbol{X}}$ are analysed.

Table 1. Example of output of Algorithm 1.

X	Y	X^1	X^2	X^3	Y^1	Y^2	Y^3	Y^4
x_1	y_1	1.0	0.0	0.0	1.0	0.0	0.0	0.0
x_3	y_1	0.0	0.0	1.0	1.0	0.0	0.0	0.0
x_2	y_4	0.0	1.0	0.0	0.0	0.0	0.0	1.0
x_1	y_3	1.0	0.0	0.0	0.0	0.0	1.0	0.0

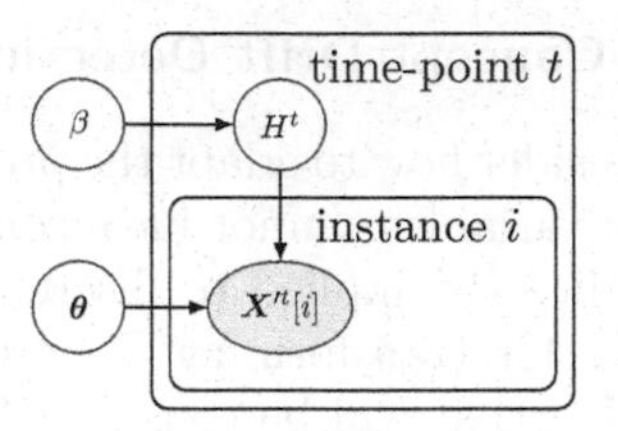

Fig. 4. Model for concept drift detection in a discrete data stream.

4.3 Subconcept Drift Detection in Discrete Data

The previous approaches might fail to detect the presence of subconcept drift: if many subspaces of the variables domains remain stable, the variations in the expected value of the H^t-variable might be too small. To address this problem, we propose a model that detects variations only in some subspaces of the domains.

Initially, the original discrete data over $\boldsymbol{X}$ is transformed using the preprocessing algorithm (and hence we obtain numerical data over $\boldsymbol{X}'$). Then, according to the user preferences, $\boldsymbol{X}'$ is partitioned into the following subsets: $\boldsymbol{S_1}, \boldsymbol{S_2}, \ldots, \boldsymbol{S_N}$ are the subsets of interest while $\boldsymbol{R}$ contains the rest of the variables. Each of these subsets of interest contains the variables corresponding to the subspaces of $\boldsymbol{X}$ that we want to analyse together. For example, for a joint analysis of $P(X = x_i)$ and $P(Y = y_j)$ with $\{X, Y\} \subseteq \boldsymbol{X}$, the variables $X^i \in \boldsymbol{X}'$ and $Y^j \in \boldsymbol{X}'$ will belong to the same subset of interest.

Figure 5 depicts the proposed BN for detecting subconcept drift in data with discrete variables. Unlike previous models, this BN contains multiple H^t-variables, one per each subset of interest. Now the detection can be done independently for each subset: a variation in the expected value of a given H^t_k, implies that the marginal distribution of any of the $X^i \in \boldsymbol{S_k}$ drifts, and so does $P(X = x_i)$.

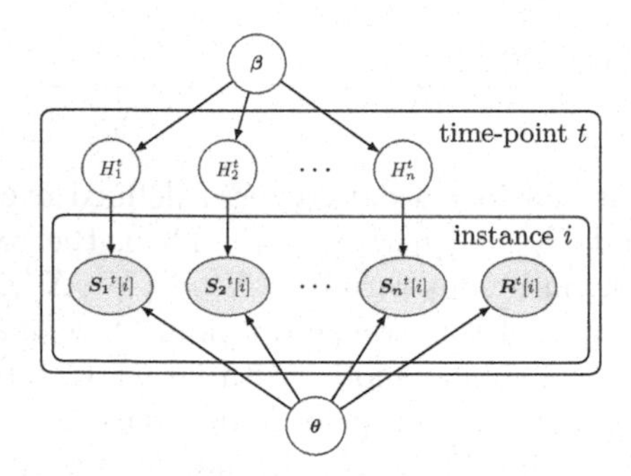

Fig. 5. Model for subconcept drift detection in a data stream with two discrete variables.

In the model presentation, we have assumed that there are no links connecting variables in $\boldsymbol{X}'$. This has been done by simplicity, as there is nothing preventing the existence of these links, taking into account that in that case, we will study the existence of changes in the conditional probabilities of the variable given its parents. For example, we can have a class variable and assume that there is a link from this variable to all the attributes, and in that case we will consider the existence of a concept drift in $P(\boldsymbol{S}_i|C)$.

This model is more appropriate for those cases where subconcept drift occurs: if we suspect that, for a given $\boldsymbol{x} \in \Omega_{\boldsymbol{X}}$ the distribution $P(\boldsymbol{x})$ could remain stable (i.e Eq. (2) holds), then the subset $\boldsymbol{R}$ will contain the corresponding variables in $\boldsymbol{X}'$ associated to the states in $\boldsymbol{x}$.

5 Empirical Validation

Herein we empirically test our approach. We consider a synthetic data stream and another one including information about intrusion detection in a web server. In both cases, Algorithm 1 is applied to each batch in the data stream. Then, for both data streams, the subconcept drift detectors (Sect. 4.3) are compared against the global detectors for full-concept drift (Sect. 4.2). The experimentation was done using the AMIDST Toolbox[1] and all the material for its replication is available at GitHub[2].

5.1 Synthetic Data Stream

When generating the data, at certain time-points, the probabilities used for sampling are changed in order to simulate the presence of concept drift. Here, we consider two discrete variables X and Y with 3 and 4 states respectively. The data set contains a total of 12000 instances sampled from the distributions shown in Table 2. We consider that each time-step contains 1000 instances.

We first apply Algorithm 1 to each batch in the data stream. The result is a stream defined over the set of continuous variables $\{X^1, X^2, X^3, Y^1, Y^2, Y^3, Y^4\}$. Assume that we aim to detect changes in $P(X = x_1)$ or $P(X = x_3)$ on the one hand, and in $P(X = x_2)$, $P(Y = y_2)$ or $P(Y = y_3)$ on the other. Thus, the model in Fig. 5 can be adapted to this scenario by considering the subsets $S1 = \{X^1, X^3\}$, $S2 = \{X^2, Y^2, Y^3\}$ and $R = \{Y^1, Y^4\}$. All in all, the BN modelling a subconcept drift detector is shown in Fig. 6.

The evolution of the expected values of H_1^t and H_2^t are shown in Fig. 7. The changes in the probabilities of interest are proportionally reflected in variations of the latent variables. For example, the most significant variations in H_1^t correspond to the large variations of $P(X = x_1)$ or $P(X = x_3)$ at time points 3,7 and 9. On the other hand, at time point 5, $P(X = x_1)$ barely changes while $P(X = x_3)$ does not vary. This implies a very small variation in H_1^t. If we analyse the evolution of H_2^t, the single significant variation occurs at time point 9,

[1] http://www.amidsttoolbox.com.

[2] https://github.com/PGM-Lab/2018-ipmu-subconcept.

Table 2. Multinomial distributions for sampling the synthetic data. Values shown in bold indicates the variations in the probabilities.

	Time-step t											
	1	2	3	4	5	6	7	8	9	10	11	12
$P(X = x_1)$	0.2	0.2	**0.6**	0.6	**0.8**	0.8	**0.2**	0.2	0.2	0.2	0.2	0.2
$P(X = x_2)$	0.2	0.2	0.2	0.2	**0.0**	0.0	0.0	0.0	**0.5**	0.5	0.5	0.5
$P(X = x_3)$	0.6	0.6	**0.2**	0.2	0.2	0.2	**0.8**	0.8	**0.3**	0.3	0.3	0.3
$P(Y = y_1)$	0.4	0.4	0.4	0.4	0.4	0.4	**0.2**	0.2	**0.0**	0.0	0.0	0.0
$P(Y = y_2)$	0.4	0.4	0.4	0.4	0.4	0.4	0.4	0.4	**0.6**	0.6	0.6	0.6
$P(Y = y_3)$	0.1	0.1	0.1	0.1	0.1	0.1	**0.3**	0.3	0.3	0.3	0.3	0.3
$P(Y = y_4)$	0.1	0.1	0.1	0.1	0.1	0.1	0.1	0.1	0.1	0.1	0.1	0.1

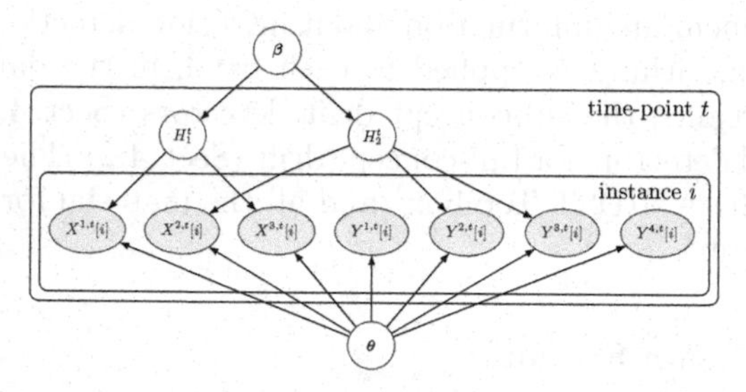

Fig. 6. Proposed model for concept drift detection in the synthetic data stream generated using the distributions given in Table 2.

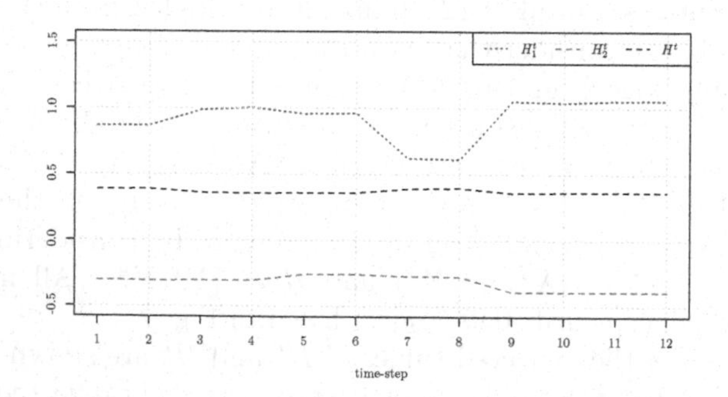

Fig. 7. Results for the synthetic data stream.

which corresponds to a large variation in $P(X = x_2)$. In the probabilities of $P(Y = y_2)$ and $P(Y = y_3)$ there are not drastic variations. The H^t series represents the output of the corresponding full-concept drift detector. We observe that its value remains almost constant, making difficult the concept drift detection.

5.2 Intrusion Detection Data Stream

Now we consider large real-world data. In particular, we use a modified version of the intrusion detection data from KDD Cup 1999 competition [16]. Each instance corresponds to a connection to a web server. It contains 494021 instances with 42 variables. Yet, we only consider the discrete variables V_1, V_2 and V_3 taking 3, 66, 11 states respectively. These variables describe the connection to the server, e.g., we have that $\Omega_{V_1} = \{tcp, udp, icmp\}$. Figure 8 shows the evolution of the distributions of the discrete variables. For simplicity of the display, improbable states in variables with large domains are not shown. In addition, the variables with temporal information in the data stream have been omitted and we consider that each time step is made of 1000 consecutive instances.

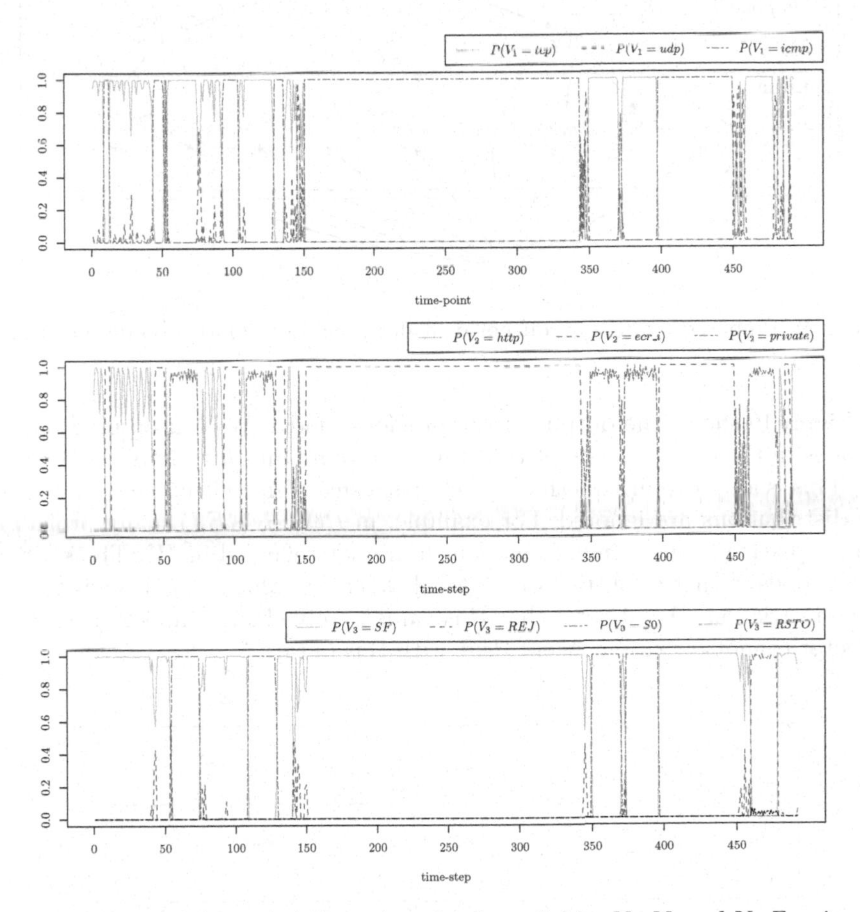

Fig. 8. Evolution of the probability values for the variables V_1, V_2 and V_3. For simplicity, 63 improbable states of V_2 and 7 of V_3 are not shown.

Suppose that we aim to analyse together the probabilities $P(V_1 = tcp)$, $P(V_1 = icmp)$ and $P(V_2 = http)$. The usual traffic in the server are HTTP packages which are sent using TCP protocol. By contrast, during a denegation of service attack, the number of ICMP packages increases. For that reason, it is interesting to concept drift in these 3 states. Suppose that we also aim to analyse the evolution of $P(V_3 = REJ)$ and $P(V_3 = RSTO)$. Taking this into account, and after applying Algorithm 1, we define the subsets of interest $\boldsymbol{S_1} = \{tcp, icmp, http\}$ and $\boldsymbol{S_2} = \{REJ, RSTO\}$, where for simplicity we have represented the variable X^j associated with case x_j of variable X as x_j. The model with these considerations is shown in Fig. 9.

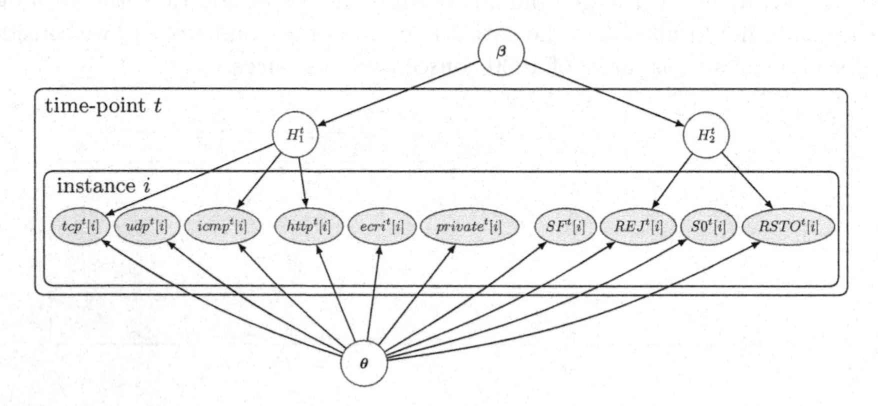

Fig. 9. Proposed model for concept drift detection in the intrusion data stream.

Figure 10 shows the output of the previous model. We can observe that the changes in the distributions of the states $tcp, icmp$ and $http$ imply a change in H_1^t. This is also a robust method which has a smoothing effect: short changes in the distributions are ignored. For example, in $P(V_3 = REJ)$ many probability peaks appear in a few time points which are not reflected in H_2^t. The series H^t corresponds with the output of a global detector, which is not shown due to space restriction. We observe that this value barely changes and hence a global detector cannot capture concept drift with this data.

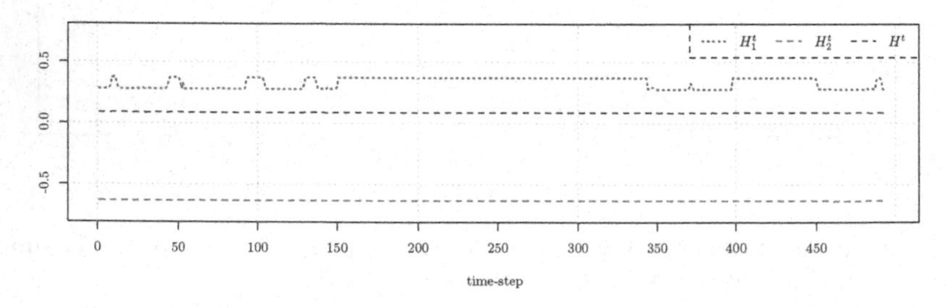

Fig. 10. Results for the intrusion data stream.

6 Conclusions

In this paper we have presented a method for detecting changes in the underlying distributions of discrete streaming data that affect to only some subspaces of the variables domain (i.e. subconcept drift). Our approach, which is based on the use of CLG Bayesian networks, can capture this kind of variations while other global detectors cannot. In the experimental work, we have seen that our method can be applied to large and high dimensional data streams. As future work, our method could be adapted for automatically selecting the subspaces of interest. It could also be extended for detecting subconcept drift in continuous data.

Acknowledgements. Authors have been jointly supported by the Spanish Ministry of Economy and Competitiveness and by the European Regional Development Fund (FEDER) under the projects TIN2013-46638-C3-2-P, TIN2015-74368-JIN, TIN2016-77902-C3-2-P and TIN2016-77902-C3-3-P.

References

1. Borchani, H., et al.: Modeling concept drift: a probabilistic graphical model based approach. In: Fromont, E., De Bie, T., van Leeuwen, M. (eds.) IDA 2015. LNCS, vol. 9385, pp. 72–83. Springer, Cham (2015). https://doi.org/10.1007/978-3-319-24465-5_7
2. Broderick, T., Boyd, N., Wibisono, A., Wilson, A.C., Jordan, M.I.: Streaming variational bayes. In: Advances in Neural Information Processing Systems, pp. 1727–1735 (2013)
3. Buntine, W.L.: Operations for learning with graphical models. JAIR **2**, 159–225 (1994)
4. Cabañas, R., Martínez, A.M., Masegosa, A.R., Ramos-López, D., Samerón, A., Nielsen, T.D., Langseth, H., Madsen, A.L.: Financial data analysis with PGMs using AMIDST. In: 2016 IEEE 16th International Conference on Data Mining Workshops (ICDMW), pp. 1284–1287. IEEE (2016)
5. Gama, J., Castillo, G.: Learning with local drift detection. In: Li, X., Zaïane, O.R., Li, Z. (eds.) ADMA 2006. LNCS (LNAI), vol. 4093, pp. 42–55. Springer, Heidelberg (2006). https://doi.org/10.1007/11811305_4
6. Gama, J., Žliobaitė, I., Bifet, A., Pechenizkiy, M., Bouchachia, A.: A survey on concept drift adaptation. ACM Comput. Surv. (CSUR) **46**(4), 44 (2014)
7. Lauritzen, S.L.: Propagation of probabilities, means, and variances in mixed graphical association models. J. Am. Statist. Assoc. **87**(420), 1098–1108 (1992)
8. Lauritzen, S.L.: Graphical Models. Oxford University Press, Oxford (1996)
9. Masegosa, A.R., Martinez, A.M., Borchani, H.: Probabilistic graphical models on multi-core CPUs using Java 8. IEEE Comput. Intell. Magaz. **11**(2), 41–54 (2016)
10. Masegosa, A.R., Martínez, A.M., Langseth, H., Nielsen, T.D., Salmerón, A., Ramos-López, D., Madsen, A.L.: d-VMP: distributed variational message passing. In: JMLR: Proceedings of the 8th International Conference on Probabilistic Graphical Models, pp. 321–332 (2016)
11. Minku, L.L., White, A.P., Yao, X.: The impact of diversity on online ensemble learning in the presence of concept drift. IEEE Trans. Knowl. Data Eng. **22**(5), 730–742 (2010)

12. Murphy, K.P.: Machine Learning: A Probabilistic Perspective. The MIT Press, Cambridge (2012)
13. Pearl, J.: Bayesian networks: A model of self-activated memory for evidential reasoning. University of California, Computer Science Department (1985)
14. Sayed-Mouchaweh, M.: Learning from Data Streams in Dynamic Environments. Springer, Heidelberg (2016). https://doi.org/10.1007/978-3-319-25667-2
15. Schlimmer, J.C., Granger, R.H.: Incremental learning from noisy data. Mach. Learn. **1**(3), 317–354 (1986)
16. Tavallaee, M., Bagheri, E., Lu, W., Ghorbani, A.A.: A detailed analysis of the KDD CUP 99 data set. In: IEEE Symposium on Computational Intelligence for Security and Defense Applications (CISDA 2009), pp. 1–6. IEEE (2009)
17. Tsymbal, A.: The problem of concept drift: definitions and related work. Computer Science Department, Trinity College Dublin, 106(2) (2004)
18. Webb, G.I., Hyde, R., Cao, H., Nguyen, H.L., Petitjean, F.: Characterizing concept drift. Data Mining Knowl. Disc. **30**(4), 964–994 (2016)
19. Widmer, G., Kubat, M.: Learning in the presence of concept drift and hidden contexts. Mach. Learn. **23**(1), 69–101 (1996)
20. Winn, J.M., Bishop, C.M.: Variational message passing. J. Mach. Learn. Res. **6**, 661–694 (2005)

Opinion Mining in Social Networks for Algerian Dialect

Mehdi Bettiche(✉), Moncef Zakaria Mouffok, and Chahnez Zakaria

Ecole Nationale Supérieure D'Informatique,
BP 68 M Oued Smar 16309, Algiers, Algeria
{cm_bettiche, cm_mouffok, c_zakaria}@esi.dz

Abstract. There has been a significant increase in the volume of Arabic dialect messages on social networks, providing a rich source for opinion mining research. Most research works done on Arabic dialect focus on messages written in Arabic script, with very limited scope on Latin script. In this paper, we are interested in the classification of social networks messages retrieved from Twitter, Facebook and YouTube written in Algerian dialect in Latin script into positive or negative classes using existing opinion mining approaches (lexical-based, machine learning, and hybrid). Also, we apply a regrouping process in the preprocessing step to overcome the issues related to the Algerian dialect such as the orthographic varieties to express the same word. Furthermore, we focus on the hybrid approach which consists in automatically annotating the training corpus with the lexical-based approach and then use the machine learning approach on this corpus for creating the classification model. This approach allows classifying the messages into positive or negative classes, without having to annotate manually a training corpus.

Keywords: Opinion mining · Social networks · Algerian dialect
Classification

1 Introduction

For centuries, researchers have been interested in the study and analysis of human sentiments and opinions. With the growth of social networks, where users give their opinions on several fields and areas, new research works on the field of opinion mining in microblogging have emerged.

Nowadays, the use of social networks such as Twitter has significantly risen, and the data recovered from social networks platforms is being used in many study fields such as economics, politics or social behavior with the various application of text mining. Although the areas of study vary in terms of goals, the dependency on social media has proven to be quite useful as it provides easy access to real time information.

Opinion mining does not only benefit companies in order to enhance their products, but it also helps other fields mentioned before such as politics, where governments could detect and extract the public opinion on government related matters and act accordingly.

J. Medina et al. (Eds.): IPMU 2018, CCIS 855, pp. 629–641, 2018.
https://doi.org/10.1007/978-3-319-91479-4_52

The very first challenge in microblogging opinion mining is language. Social media users tend to use a language, dialect and abbreviations that sometimes differs from a culture to another. Generally, users in the Greater Middle East communicate with Modern Standard Arabic (MSA) and Arabic dialects, which differ from one region to another, because of the various spoken regional dialects of Arabic. Note that the MSA is the only variety of Arabic that is standardized, regulated, taught in schools, and used in written and formal speech. However, in social networks, both MSA and dialectal Arabic are used. In addition to problems related to the use of the Arabic dialect, it's transcription in Latin letters presents additional challenges. This phenomenon is called Arabizi, which is often used in the Maghreb mainly in Tunisian, Moroccan and Algerian dialects where there are no specific or predefined rules in the way of writing the messages. Furthermore, phonetic and orthographic varieties differ for example from Algerian dialects to Moroccan dialects considerably and from one region to another within the same country which means the preprocessing of the data has to be an important part of any Arabic dialect related study.

There are several works on the Arabic dialects in Arabic script. They essentially study the resources building [1–5] and the dialects identification [6–10]. However, in Latin script, there are noticeably less works, in particular in Algerian dialects, which are represented only by the resources building [11]. We noticed that very few works studied the opinion mining in the Arabic dialect written in Latin script. There are for example [12] in the Moroccan dialect and [13] in the Tunisian dialect, but in the Algerian dialect there aren't any studies dealing with opinion mining detection in messages although there is one work on the dialect that focuses essentially on the construction of lexicon [14]. Therefore in our best knowledge there is not yet any opinion mining work done in the Algerian dialect written in Latin script although the dialect is widely used in social networks by the Algerian users. In this paper, we apply the hybrid opinion mining approach to classify social networks messages written in the Algerian dialect in Latin letters on positive or negative classes. In the following, we introduce important works related to our study, then we present our proposed approach, after that we give our test results, and finally, we end with our conclusion and perspectives.

2 Related Works

We can distinguish three main approaches when dealing with researches in the opinion mining in microblogging according to the three classes positive, negative and neutral.

The first approach is the lexical-based approach. In most cases when using this approach, researchers construct a vocabulary of initial opinion words, then use methods to enrich it and finally they classify messages by calculating their score according to the opinion words present in them [15]. Hu and Liu [16] used synonyms and antonyms to identify the semantic orientation of each word in the vocabulary, they achieved an accuracy of 69.3% in detecting opinion sentences and 84.2% in detecting their orientation. In another work, Turney [17] proposed a statistical method using the PMI-Information retrieval (PMI-IR) algorithm which calculates the correlation between a message and the seed words 'excellent' and 'poor' using the search engine 'AltaVista', the accuracy he achieved is 84.2%. Taboada et al. [18], on the other hand, calculated the polarity force of each word of the vocabulary that was mainly composed of adjectives, using the techniques of aggregation and average. They also took into account intensification and negation reporting an accuracy of 90% after testing on several corpora.

The lexical-based approach has the advantage of not requiring a manually annotated training corpus and gives a very precise prediction of semantic orientation but is not suitable for detecting opinions and subjectivity. Indeed, many messages bearing positive or negative opinions are detected as neutral, which is translated by a weak recall. The reason behind this weak recall comes from the fact that many opinion words appearing in the messages and expressing a positive or negative opinion are not present in the vocabulary.

The second approach is the machine learning approach. Pang et al. [19] used the Naïve Bayes (NB), Maximum Entropy (ME) and Support Vector Machine (SVM) classifiers to classify movie reviews into two classes, positive and negative. They took a bag of uni-gram words as features. The SVM classifier gave a better accuracy with 82.9% compared to NB 81.5% and ME 81%. Pak and Paroubek [20] worked on the social network Twitter using Tweets as a corpus. They classify Tweets into three classes: positive, negative or neutral using SVM and Naïve Bayes classifiers. Their approach classifies Tweets into subjective or objective messages first, then the subjective ones into positive and negative using emoticons and POS tags as features. Wang et al. [21] compared between several classifiers: NB, ME, DT, KNN and SVM. They also compared between the Boolean, frequency and TF-IDF representations adding to that comparison the uni-grams and bi-grams features and with several corpora. The best reported result was with the ME classifier scoring 92.62% in accuracy using the frequency representation and the bi-grams. The classifier with the best average score between the different corpora is SVM with 79% accuracy using the TF-IDF representation and the uni-gram features.

The machine learning approach is widely used by researchers because it gives better results and a much better recall than the lexical based approach. Though, it has the disadvantage of needing a manual annotation of the training corpus which is difficult to perform and time consuming when we have a big set of messages.

Finally, the last approach is the hybrid approach which combines the lexical-based approach and the learning-based approach. Zhang et al. [22] used a hybrid approach by automatically annotating the training corpus with the lexical-based approach, and then training the classifier on this corpus which gave an accuracy of 87.7%.

3 Proposed Approach

Our opinion mining approach (see Fig. 1) for the Algerian dialect in Latin script focus on the hybrid approach by combining the lexical based and machine learning based approaches to classify messages in positive or negative. This approach was inspired by [22] although their interest is in entity level sentiment analysis. First we use the lexical based approach to automatically annotate the training corpus that will be used as the input corpus for the machine learning based approach. We compared the results obtained using the hybrid approach with the lexical based and learning based

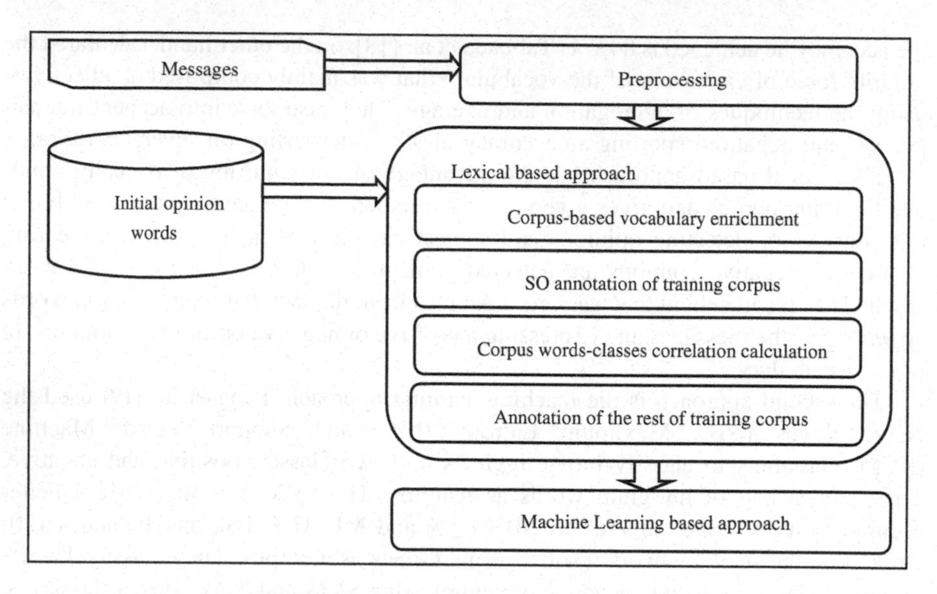

Fig. 1. Global architecture of our approach.

approaches in the case of the Algerian dialect corpus. Figure 1 presents the global architecture of our system.

One of the most important steps is the preprocessing which is performed on the messages collected from the social networks Twitter, Facebook and YouTube. Indeed, the vocabulary needs to be processed and grouped to overcome the multiple spell checking of the same words.

3.1 Preprocessing

The phase of preprocessing consists of applying multiple techniques on the collected corpus messages in order to structure and organize them for further analysis [20].

Like other dialects, Algerian dialect has no orthographic rules, and therefore several spellings can be observed in the corpus for the same word. We can see the following example where a word (which means "everything" in the Algerian dialect) has user using multiple spellings.

Example: kolach, kollach, kolchi, koulach, koulch, kolech etc.

To overcome this problem, we propose to add a regrouping step that is split into two sub-steps: Phonetic regrouping, and similarity regrouping.

Phonetic Regrouping

We used an algorithm called 'soundex' [23] which gives a phonetic code for a given word. This algorithm analyses the word, keeps its first letter in the code, and for the rest of the letters it associates a number according to the table below (Table 1).

Table 1. Soundex code

Code	Letters
1	B, F, P, V
2	C, G, J, K, Q, S, X, Z
3	D, T
4	L
5	M, N
6	R
SKIP	A, E, H, I, O, U, W, Y, H, W, Y

Code = first letter of the word + a number
Example: KOLACH → K42

We assign a phonetic code to each word belonging to the corpus, and then, for the same phonetic code, we collect all the words associated with it.

Example: K42: ['kolach', 'kelach', 'kollach', 'kolchi', 'koulach', 'khlass', 'kolch', 'koulch', 'khalikk', 'kolech', 'khalik', 'koulech', 'khalissa', 'kliké', 'koleche', 'khe-l3oki', 'klacha', 'khlasse', 'khlas', 'kolache', 'koulchi', 'khlak'].

At the end of this step, we have a phonetic dictionary where we associate each group of words having the same phonetic code with their corresponding phonetic code.

Similarity Regrouping

We have noticed that the phonetic regrouping has some weaknesses since several words which have not the same meaning are phonetically regrouped. For example: 'kolach' and 'khlasse' do not have the same meaning. It is therefore necessary to make a second regrouping within each group of words.

We used the similarity regrouping using the Levenshtein distance [24] giving a measure of the similarity between two words. It is equal to the minimum number of characters to be deleted, inserted or replaced to pass from one string to another. We use it to give a percentage of similarity between two words, as follows:

$$\text{Ratio} = 1 - \text{distance}\ (\text{w1},\ \text{w2})/\text{lensum}\ (\text{w1},\ \text{w2})$$

Where:

$$\text{lensum}\ (\text{w1},\ \text{w2}):\ \text{Sum of word}\ (\text{w1 and w2})\ \text{lengths.}$$

Example:

- Ratio (kolach, kollach) = 92%
- Ratio (kolach, khlasse) = 38%.

This test is done for each word in the list of the given dictionary in the phonetic grouping step. Then, we take the most frequent word in the corpus from this list, and we count its similarity to all other words, when the similarity is greater than a certain estimated threshold (that we set at 75% after testing several examples), the words are grouped together, giving the first list that represents the initial regrouping by similarity.

In our previous example the most frequent word is: kolach, after regrouping by similarity we have the following list which accurately contains only the multiple spellings of the word:

kolach: ['kelach', 'kollach', 'kolchi', 'koulach', 'kolch', 'koulch', 'kolech', 'koulech', 'koleche', 'kolache', 'koulchi'].

We do the same thing for the words that remain in the list until it is empty. We then construct a dictionary of similarity where each word of the corpus has its equivalent which is the most frequent word of the group to which it belongs.

In the end we replace each word in the messages preprocessed by their equivalents in the dictionary of similarity.

3.2 Lexical Based Approach

Corpus-Based Vocabulary Enrichment

The lexical based approach relies on a pre-constructed vocabulary. Therefore, we first construct a vocabulary of initial opinion words manually, then we enrich it with the words present in the corpus. To enrich the vocabulary, we must go through several stages. First we create the co-occurrence matrix of the words in the corpus, then we calculate the correlations between the words of the corpus with the PMI, and finally, we enrich the vocabulary by calculating the semantic orientation (SO) of each word of the corpus using the following equation:

$$SO(\mathrm{w}) = \sum_i PMI(w, voc_{pos}[i]) - \sum_i PMI(w, voc_{neg}[i])$$

Where:

- voc_{pos}: set of positive opinion words.
- voc_{neg}: set of negative opinion words.

Then we do a simple comparison:

If $SO(w) > 0$ then the word is assigned to positive opinion words.
If $SO(w) < 0$ then the word is assigned to negative opinion words.

We can do several iterations by taking at each iteration the set of words constructed at the previous iteration.

SO Annotation of Training Corpus

After constructing our vocabulary, the next step is classifying the messages into positive or negative. To do so, we sum the polarity of each word of the message:

$$Polarity(message) = \sum_i Polarity(word_i)$$

$Polarity(word_i)$ can take several values:

Method 1: The basic method would be adding +1 to a message if a positive word is present and −1 if a negative word is present.
Method 2: We take the result given by the semantic orientation (SO) calculated for each word during the construction of the vocabulary.

We used the second method by taking the semantic orientation (SO) of each word to avoid giving the same polarity strength to all words.

Corpus Words-Classes Correlation Calculation
In this approach, many positive and negative messages will not be classified. Then, we add two new steps to classify the rest of the messages.

First, we take all the words of the classified messages as our vocabulary. After that, we calculate their correlation with the positive and negative classes using the PMI to know their polarity.

$$PMI_{pos}(\mathrm{w}) = \log\left(\frac{P_{pos}(w)}{P_{pos}}\right) \; and \; PMI_{neg}(w) = \log\left(\frac{P_{neg}(w)}{P_{neg}}\right)$$

Where:

- P_{pos}: frequency of positive messages.
- P_{neg}: frequency of negative messages.
- $P_{pos}(w)$: frequency of appearance of the word w in positive messages.
- $P_{neg}(w)$: frequency of appearance of the word w in negative messages.

If the PMI between the word and the positive class is positive then the word expresses a positive opinion, otherwise if the PMI between the word and the negative class is positive then the word expresses the negative opinion.

Annotation of the Rest of Training Corpus
In order to classify the messages, we take as a polarity of each word the average PMI between the word and the two positive and negative classes taking into consideration their semantic orientation.

$$\mathrm{PMI}(\mathrm{w}) = P_{pos} * PMI_{pos}(w) + P_{neg} * PMI_{neg}(w)$$
$$Polarity(message) = \sum_{i} \pm PMI(w_i)$$

We can also take the average khi-2 and GI as the value of polarity.

$$khi^2(\mathrm{w}) = \frac{\mathrm{n} \cdot P(w)^2 \cdot (P_{pos}(w) - P_{pos})^2}{\mathrm{P(w)} \cdot (1 - \mathrm{P(w)}) \cdot P_{pos} \cdot (1 - P_{pos})}$$

$$GI(w) = -\left(P_{pos} \cdot \log\left(P_{pos}\right) + P_{neg} \cdot \log\left(P_{neg}\right)\right) + P(w) \cdot \left(P_{pos}(w) \cdot \log\left(P_{pos}(w)\right) + P_{neg}(w) \cdot \log\left(P_{neg}(w)\right)\right) + (1 - P(w)) \cdot \left(\left(1 - P_{pos}(w)\right) \cdot \log\left(1 - P_{pos}(w)\right) + \left(1 - P_{neg}(w)\right) \cdot \log\left(1 - P_{neg}(w)\right)\right)$$

Where:

- $P(w)$: frequency of appearance of the word w in the corpus.

3.3 Machine Learning Based Approach

In this approach, the training corpus must be annotated. Like the lexical based approach, the machine learning based approach relies on a vocabulary which is either constructed using all the words of the corpus after the preprocessing and representation step, or using features selection to select the words which are correlated to positive or negative messages.

Constructing the vocabulary is followed by the classification step. In this step, we use the classifiers: Naïve Bayes (NB), Multinomial Naïve Bayes (MNB), Bernoulli Naïve Bayes (BNB), Decision Tree (DT), K-Nearest Neighbors (KNN), SVM and Artificial Neural Networks (ANN). The goal is to construct a classifying model depending on the annotated training corpus and the vocabulary created, and use this classifying model to classify the messages of test corpus.

4 Tests and Results

First, we tested the machine learning approach and the hybrid approach in a French corpus (see Table 2). For machine learning, we annotated manually the training and the test corpus. We trained the NB, MNB, BNB, DT, KNN, SVM and ANN classifiers on the training corpus (results in Table 3). And then we tested the hybrid approach by

Table 2. French Corpus statistics

French corpus	
Number of messages	2000
Number of words in the vocabulary	3395
Number of messages in training corpus	1200
Number of messages in test corpus	800

Table 3. Machine learning approach results for French corpus

Manually annotation	Classifiers						
	NB	MNB	BNB	DT	KNN	SVM	ANN
Precision	0.734	0.863	0.776	0.694	0.668	0.84	0.835
Recall	0.733	0.826	0.774	0.683	0.657	0.81	0.805
F	0.733	0.825	0.773	0.684	0.658	0.811	0.804
Accuracy	0.733	0.826	0.774	0.683	0.657	0.81	0.805

Table 4. Hybrid approach results for French corpus

Automatic annotation	Classifiers						
	NB	MNB	BNB	DT	KNN	SVM	ANN
Precision	0.68	0.775	0.758	0.704	0.768	0.772	0.767
Recall	0.678	0.774	0.719	0.701	0.766	0.769	0.766
F	0.678	0.774	0707	0.698	0.766	0.768	0.766
Accuracy	0.678	0.774	0.719	0.701	0.766	0.769	0.766

annotating automatically the training corpus with the lexical based approach and training the classifiers on it (results in Table 4). We used the same classifiers.

The goal of the first test is to compare between the real manual annotation and the automatic annotation of the training corpus. Also, French language is used by many Algerian users, and often they combine French words to Algerian words within the same sentence. So, it is interesting to see the French corpus results compare it with the dialectal corpus results.

We used four evaluation metrics, namely precision, recall, F and accuracy. Precision is defined as the fraction of the classified messages that are in the correct class. Recall is defined as the fraction of opinion bearer message that are retrieved by the approach. F characterizes the combined performance of precision and recall. Finally, accuracy is defined as the fraction of the correctly classified messages to all messages.

French corpus is collected from a YouTube video speaking about Islam in France.

The machine learning approach outperforms the hybrid approach. For the MNB classifier, the first approach achieves an F-score of 82.5% whereas the second approach gets 77.4%. This is due to the automatic annotation which has an error rate. However, the hybrid approach gave satisfying results.

Since it is very difficult to find an annotated training corpus in the Algerian dialect and the hybrid approach gave satisfying results, we settled for testing only the lexical-based approach and the hybrid approach for the dialectal corpus. Operating this way helps avoiding the need to manually annotate a corpus for the machine learning approach which is time consuming. For the hybrid approach, we used the same previous classifiers.

Algerian dialectal corpus is collected from YouTube videos about politics in Algeria (Table 5).

Table 5. Algerian dialect corpus statistics

Dialectal corpus	
Number of messages in the corpus	2650
Number of words in the vocabulary before regrouping	13555
Number of words in the vocabulary after regrouping	9274
Number of messages in Training corpus	2250
Number of messages in test corpus	400

4.1 Lexical Based Approach

We tested the approach by taking as a polarity the sentiment orientation (SO) of each word while using multiple iterations of vocabulary enrichment and the presence or not of the regrouping step.

Note that when we do not enrich the vocabulary (number of iteration = 0) we have a very good precision ($\geq 92\%$) but a very low recall ($\leq 48\%$).

The results also show that the more we enrich the vocabulary, the more the precision decreases and the recall grows. This is mainly due to the fact that, when the vocabulary is not enriched, we are sure of the polarity of all the word, therefore having a very good precision.

Furthermore, since the vocabulary is rather poor in words we have a weak recall because several messages bearing opinions are ignored. On the other hand, when the vocabulary is enriched we have a better detection of opinion messages but the error rate increases and thus precision decreases.

In general, the constructed model gave better results when we use the vocabulary enrichment. In fact, the F grows with 18%, for the first iteration (see Table 7). We noticed that when using regrouping, we achieve the highest F score on the first iteration (81.9%), but when we do not use regrouping, the highest F score is seen on the second iteration (78.2%), (see Tables 6 and 7).

Table 6. Lexical based SO classification without regrouping results

Without regrouping	Number of iteration			
	0	1	2	3
Precision	0.924	0.836	0.82	0.812
Recall	0.389	0.723	0.749	0.747
F	0.547	0.775	0.782	0.778
Accuracy	0.389	0.723	0.749	0.747

Table 7. Lexical based SO classification with regrouping results

With regrouping	Number of iteration			
	0	1	2	3
Precision	0.929	0.841	0.824	0.821
Recall	0.48	0.797	0.802	0.801
F	0.634	0.819	0.813	0.811
Accuracy	0.48	0.797	0.802	0.801

Also, the regrouping that we proposed earlier improved the performance of the approach in terms of precision, recall, F and accuracy. The best F achieved without the regrouping is 78.2% in the second iteration, while with the regrouping it's 81.9% in the first iteration. So, this method enhances the results, since we got the highest F score on the first iteration only.

If we go to the third iteration the performance deteriorates for both methods: without regrouping (F = 77.8%) and with regrouping (F = 81.1%) method. Indeed, here, we find that many words which are not opinion-related are detected due to the expansion of the initial set of words.

4.2 Hybrid Approach

We tested the hybrid approach by automatically annotating the training corpus with the lexical based approach. First we used the set of initial opinion words. We then annotated the rest of the ignored messages by taking as the polarity of each word in the corpus, the PMI computed by the correlation between the words and the positive and negative classes.

For this study, we used all the words present in the training corpus as vocabulary and used the TF-IDF representation for the messages. Also, we compared the presence or not of the regrouping step.

The KNN classifier gave better results when we used the regrouping step with 93.7% in terms of F, and then came the ANN and MNB with F = 91.3% for ANN and F = 91% for MNB (see Table 9). However, when regrouping was not used, the results were not as good as previously, with F = 91.2% for KNN, F = 89% for ANN and F = 88.7% for MNB (see Table 8). From the table above, we can witness that this approach gives very good results either in terms of precision, recall, F or accuracy (see Tables 8 and 9). Thus, the hybrid approach gives better results than the lexical based approach in terms of F-Score, with 93.7% for the hybrid approach and 81.9% for the lexical one.

Table 8. Hybrid approach without regrouping results.

Without regrouping	Classifiers						
	NB	MNB	BNB	DT	KNN	SVM	ANN
Precision	0.822	0.893	0.828	0.867	0.918	0.879	0.892
Recall	0.805	0.887	0.747	0.867	0.912	0.877	0.89
F	0.802	0.887	0.731	0.867	0.912	0.877	0.89
Accuracy	0.805	0.887	0.747	0.867	0.912	0.877	0.89

Table 9. Hybrid approach with regrouping results

With regrouping	Classifiers						
	NB	MNB	BNB	DT	KNN	SVM	ANN
Precision	0.847	0.917	0.865	0.867	0.937	0.89	0.915
Recall	0.822	0.91	0.82	0.867	0.937	0.882	0.912
F	0.819	0.91	0.814	0.867	0.937	0.882	0.913
Accuracy	0.822	0.91	0.82	0.867	0.937	0.882	0.912

Overall, this approach presents much better results than the lexical-based approach and is simpler to develop than the machine learning approach since we do not need to manually annotate a training corpus.

Against all odds, the approaches give better results in dialect than in French. In analyzing the messages of our corpus, we noticed that the Algerian dialect corpus is less rich and developed than the French corpus, the positive or negative words are often recurrent, as long as we regroup their different spellings in the regrouping phase.

5 Conclusion/Perspective

In this paper, we presented a hybrid approach in opinion mining. When applied to the Algerian dialect, this approach gave interesting results even though the dialect is not standardized.

Using the lexical based approach to annotate the training corpus, and using that annotated corpus as an input for the machine learning approach gave better results than using the lexical based approach only. We achieved an F-score of 93.7% for the hybrid approach and 81.9% for the lexical one. Also, when working with dialect, we tested a new regrouping step when preprocessing the messages. It consists of performing a phonetic and similarity regrouping, which, when applied to our study was helpful in regrouping the same words with different writings.

Overall, this study on the Algerian dialect enabled us to set new perspective for opinion mining in microblogging, especially with dialectal messages. We can now proceed to other areas of the opinion mining field such as subjectivity analysis, by taking in account for instance a neutral class (for messages that are not classified either as positive or negative) and elaborate on whether those message bear equally positive and negative opinion words or do not contain opinion words at all. Another possible area of interest would be opinion extraction by using for example the opinion vocabulary we constructed to detect or extract messages containing specific opinions.

References

1. Harrat, S., Meftouh, K., Abbas, M., Hidouci, K.W., Smaili, K.: An algerian dialect: Study and resources. Int. J. Adv. Comput. Sci. Appl.-IJACSA **7**(3), 384–396 (2016)
2. Harrat, S., Meftouh, K., Abbas, M., Jamoussi, S., Saad, M., Smaili, K.: Cross-dialectal arabic processing. In: International Conference on Intelligent Text Processing and Computational Linguistics (2015)
3. Bouamor, H., Habash, N., Oflazer, K.: A multidialectal parallel corpus of arabic. In: LREC, pp. 1240–1245 (2014)
4. Cotterell, R., Callison-Burch, C.: A multi-dialect, multi-genre corpus of informal written arabic. In: LREC, pp. 241–245 (2014)
5. Jarrar, M., Habash, N., Alrimawi, F., Akra, D., Zalmout, N.: Curras: an annotated corpus for the Palestinian Arabic dialect. Lang. Resour. Eval. **51**(3), 745–775 (2017)
6. Malmasi, S., Zampieri, M.: Arabic Dialect identification in speech transcripts. In: VarDial, vol. 3, p. 106 (2016)

7. Huang, F.: Improved arabic dialect classification with social media data. In: EMNLP, pp. 2118–2126 (2015)
8. Belgacem, M., Antoniadis, G., Besacier, L.: Automatic identification of arabic dialects. In: LREC (2010)
9. Sadat, F., Kazemi, F., Farzindar, A.: Automatic identification of arabic language varieties and dialects in social media. In: Proceedings of SocialNLP, vol. 22 (2014)
10. Ali, A., Dehak, N., Cardinal, P., Khurana, S., Yella, S.H., Glass, J., Renals, S.: Automatic dialect detection in arabic broadcast speech. arXiv preprint arXiv:1509.06928 (2015)
11. Karima, A., Menacer, M.A., Smaili, K.: CALYOU: a comparable spoken algerian corpus harvested from youtube. In: 18th Annual Conference of the International Communication Association (Interspeech) (2017)
12. Zarra, T., Chiheb, R., Moumen, R., Faizi, R., Afia, A.E.: Topic and sentiment model applied to the colloquial Arabic: a case study of Maghrebi Arabic. In: Proceedings of the 2017 International Conference on Smart Digital Environment, pp. 174–181. ACM (2017)
13. Medhaffar, S., Bougares, F., Esteve, Y., Hadrich-Belguith, L.: Sentiment analysis of tunisian dialects: linguistic resources and experiments. In: Proceedings of the Third Arabic Natural Language Processing Workshop, pp. 55–61 (2017)
14. Guellil, I., Azouaou, F.: Bilingual Lexicon for Algerian Arabic Dialect Treatment in Social Media
15. Ding, X., Liu, B., Yu, P.S.: A holistic lexicon-based approach to opinion mining. In: Proceedings of the 2008 International Conference on Web Search and Data Mining, pp. 231–240. ACM (2008)
16. Hu, M., Liu, B.: Mining and summarizing customer reviews. In: Proceedings of the Tenth ACM SIGKDD International Conference on Knowledge Discovery and Data Mining, pp. 168–177. ACM (2004)
17. Turney, P.D.: Thumbs up or thumbs down?: semantic orientation applied to unsupervised classification of reviews. In: Proceedings of the 40th Annual Meeting on Association for Computational Linguistics, pp. 417–424. Association for Computational Linguistics (2002)
18. Taboada, M., Brooke, J., Tofiloski, M., Voll, K., Stede, M.: Lexicon-based methods for sentiment analysis. Computat. Linguist. **37**(2), 267–307 (2011)
19. Pang, B., Lee, L., Vaithyanathan, S.: Thumbs up?: sentiment classification using machine learning techniques. In: Proceedings of the ACL-02 Conference on Empirical methods in Natural Language Processing, vol. 10, pp. 79–86. Association for Computational Linguistics (2002)
20. Pak, A., Paroubek, P.: Twitter as a corpus for sentiment analysis and opinion mining. In: LREC, vol. 10 (2010)
21. Wang, G., Sun, J., Ma, J., Xu, K., Gu, J.: Sentiment classification: the contribution of ensemble learning. Decis. Support Syst. **57**, 77–93 (2014)
22. Zhang, L., Ghosh, R., Dekhil, M., Hsu, M., Liu, B.: Combining lexicon-based and learning-based methods for twitter sentiment analysis (2011a)
23. Holmes, D., McCabe, M.C.: Improving precision and recall for soundex retrieval. In: International Conference on Information Technology: Coding and Computing, Proceedings, pp. 22–26. IEEE (2002)
24. Levenshtein, V.I.: Binary codes capable of correcting deletions, insertions, and reversals. In : Soviet physics doklad, vol. 10, no. 8, pp. 707–710 (1966)

Using Inductive Rule Learning Techniques to Learn Planning Domains

José Á. Segura-Muros(✉), Raúl Pérez, and Juan Fernández-Olivares

University of Granada, Granada, Spain
{josesegmur,fgr,faro}@decsai.ugr.es

Abstract. When dealing with complex problems, designing a planning domain becomes a hard task that requires time and a skilled expert. This issue can be a problem when trying to model a planning domain intended to work in real-world applications. In order to overcome this problem, domain learning techniques are developed aiming to learn planning domains from existing real-world processes. Domain learning techniques then must face typical problems from this kind of applications such as data incompleteness. In this paper, we extend a classification algorithm developed by our research group, in order to create a highly resistant to incompleteness domain learner. We achieve this by extracting the information contained in a collection of plans and creating datasets, applying cleaning and preprocessing techniques to these datasets and then extracting the hypothesis that model the domain's actions using the classifier. Seeking a first validation of our solution before trying to work with real-world data we test it using a collection of simulated standard planning domains from the International Planning Competition. The results obtained shows that our approach can successfully learn planning actions even with a high degree of incompleteness.

Keywords: Automated planning · Domain learning
Inductive learning · Machine learning · Rule learning

1 Introduction

AI planning systems require the definition of planning domains in order to solve planning problems. Defining handmade planning domains is a task that requires experience and an extensive knowledge of the problems that want to be solved. Also, designing a planning domain able to solve problems that work in a world with a different array of situations is a difficult task because the domain's designer must take into account every single one to allow the domain to handle them. These two factors provoke that build planning domains be a cumbersome task. A way to reduce the time and effort needed to design planning domains is to automate the process as much as possible using planning domain learning techniques. Planning domain learners use plan traces examples and produces as an output a domain able to solve planning problems.

J. Medina et al. (Eds.): IPMU 2018, CCIS 855, pp. 642–656, 2018.
https://doi.org/10.1007/978-3-319-91479-4_53

The solution presented in this paper aims to extend a classification algorithm [1] to learn logical action models from incomplete plan traces. The algorithm selected was NSLV [2] a classification algorithm based on inductive rule learning with fuzzy sets [3]. The extension consist of the addition of preprocessing techniques to convert data obtained from solved planning problems to the data format needed as input to NSLV, as well as postprocessing techniques to build a planning domain from the rules obtained from the classification algorithm. In the early stages of the development of our approach, we choose algorithms that create rule-based models as the core of the learning process because of rule's interpretability and their ability to explicitly show the relationships among the variables involved in the problem. These two characteristics help lessen the problem of translating a classification model to a planning domain. Finally, by considering the domain learning problem as a classification problem, we reduce the problem to find a collection of hypothesis that model the state of the world before and after the execution of each domain's action.

There are several proposed solutions to learn planning domains [4]. In the literature, we can find solutions as different as OBSERVER [5] that monitors executions of expert agents and learns using the information obtained from them, TRAIL [6] that relies on an expert human teacher to guide the learning process or EXPO [7] that starts from an initial incomplete domain and uses plans executions to complete it. New solutions rely on diverse strategies such as LOCM [8] that uses context-free models relating action's parameters, or NLOCM [9] that extends LOCM's strategy in order to learn action's cost. Approaches that can learn from incomplete information are few: ARMS [10] and LAMP [11] are good examples of these solutions. Both generate sets of logic formulas to model the domain's actions and select the best ones using a MAX-SAT solver (ARMS) or a Marvok Net (LAMP).

Most of these solutions can't deal with incompleteness (EXPO, OBSERVER, TRAIL, LOCM, NLOCM) and the few ones that can handle it (ARMS, LAMP) doesn't generate symbolic representations of the actions learned, difficulting the understanding of the results. Another drawback is that although the learned domains obtained with these algorithms are close to the original ones, they usually don't give information about the learned domain's ability to reproduce the plans used during the learning process. The main objective of our research is to use a classifier based on inductive learning resistant to data incompleteness that outputs models easy to understand from a human point of view to learn planning domain's actions models. As secondary objectives, we aim to generate domains that can reproduce correctly the solutions obtained by the original domains.

Our solution has been validated using examples obtained from problems solved using benchmark domains obtained from the international planning community. The noisy and incomplete state's information contained on these examples varies and each one has been used to learn a planning domain. These new learned domains have been compared with the originals ones to measure not only the differences between them but also to measure its capabilities of reproducing the plans used as input on the learning process.

Next section will cover in detail every background concept needed to comprehend how our solution works. In Sect. 3 the NSLV rule learner algorithm will be explained. Then, Sect. 4 will contain the information about the techniques added to extend NSLV. Section 5 will cover our experiments and results. Finally, in Sect. 6 the conclusions drawn from the results will be discussed together with possible improvements of our solution in the near future.

2 Problem Statement

As said earlier, in order to use AP techniques a planning domain is needed along with a planning problem to be solved. The result of a planner is a plan that solves the given problem. A plan is defined as an ordered set of tasks whose execution modify the world until achieving the goals presented in the problem. In AP the world is represented as a conjunction of fluents. A fluent is a statement in the form of $p(arg_1, arg_2, ..., arg_n)$ where p is a logic predicate and arg_x an object of the world. Objects may have a type associated, and those types may have a hierarchical relationship with other types. Each fluent has a value associated: True or False in the case of literal fluents or a numerical value in the case of function fluents. When interpreting the incompleteness of a state two interpretations can be used: the Close World Assumption (CWA) or the Open World Assumption (OWA). CWA interprets the world by considering that unobserved fluents are false. Meanwhile, OWA considers that unobserved fluents are missing, nor true or false, and can't be evaluated. This work follows the OWA interpretation.

A planning domain can be seen as a tuple $<Ont, Act>$ where *Ont* is the ontology of the world, the definition of the predicates and objects of the world, and *Act* is a collection of PDDL actions. In the same way, a PDDL planning action is a tuple $<Header, Pre, Eff>$, where $Header$ is the action's name plus its parameters, Pre the preconditions that must be true to allow the execution of the action and Eff the effects of the action in the world after being executed. An action whose Par, Pre and Eff components are if the parameters are not instantiated with world's objects is called action model. Below, there's the action model of the example action $(boardp1, a1, c1)$ and an example of the execution of the same action over a given state:

```
(:action board :parameters (?p - person ?a - aircraft ?c - city)
 :precondition
    (and
        (at ?p ?c)(at ?a ?c)
    )
 :effect
    (and
        (not(at ?p ?c))(in ?p ?a)
    )
)
```

S_n		S_{n+1}
$(at\ a1\ c1)$	$\longrightarrow$	$(at\ a1\ c1)$
$(at\ p1\ c1)$	(board p1 a1 c1)	$(\neg(at\ p1\ c1))$
$(\neg(in\ a1\ p1))$		$(in\ a1\ p1)$

An ordered set of interleaved states S_x and actions A_x $<S_0, A_0, S_1, A_1, ..., S_n, A_n, S_{n+1}>$ is called Plan Trace (PT). A plan with states S_x interleaved between its actions A_x is called Plan Trace. In a plan trace $<S_0, A_0, S_1, A_1, ..., S_n, A_n, S_{n+1}>$, S_0 is the initial state of the problem solved by the original plan, S_{n+1} is the goal state of that problem and the rest of states are snapshots of the world at a given point during the execution of the plan. Each action has an associated prestate and poststate. The state S_x of an action A_x is the prestate associated with the action and can be seen as the world just before executing the action. In the same way, the state S_{x+1} is the poststate associated with A_x and is the state of the world just after executing the action. Listing 1.1 shows an example of a PT.

S_0: $(at\ p1\ c2) \wedge (= (fuel - level\ a1)\ 100) \wedge (at\ a1\ c2)$
T_0: (board p1 – person a1 – aircraft c2 – city)
S_1: $(at\ p1\ c2) \wedge (= (fuel - level\ a1)\ 100) \wedge (in\ p1\ a1)$
T_1: (fly a1 – aircraft c2 – city c1 – city)
S_2: $(= (fuel - level\ a1)\ 0) \wedge (in\ p1\ a1) \wedge (at\ a1\ c1)$

Listing 1.1. Extract of a PT from a Zeno Travel problem.

The world's states of a PT are usually observed during the execution of a given plan. This can lead to have partially (incomplete) or wrongly (noisy) states observed. Incompleteness occurs when some fluents of the state (or the whole state) are not observed. Noise, on the other hand, is a problem where the value of a fluent is different of the value of the observed fluent. Following the states shown in Listing 1.1 an incomplete version of state S_0 would be $(at\ p1\ c2) \wedge (= (fuel - level\ a1)\ 50) \wedge (at\ a1\ c2)$ an example of noise over the same state.

In order to use the information contained inside a PT in a classification algorithm, datasets must be extracted from the PT's. When translating a PT into a dataset each example is a state from the PT and the attributes correspond with the state's fluents.

Those datasets are size $n * m$ matrices where n is the number of examples of the dataset and m the number of attributes.

$Fluent_1$	$Fluent_2$	...	$Fluent_m$	Class
$Value_{11}$	$Value_{12}$	...	$Value_{1m}$	$Label_1$
$Value_{21}$	$Value_{22}$	...	$Value_{2m}$	$Label_2$
...	...	...	...	...
$Value_{n1}$	$Value_{n2}$	...	$Value_{nm}$	$Label_n$

$Fluent_j$ are the elements which make up the state S_i, and $Value_{ij}$ the values of those fluents. $Value_{ij}$ depends on the $Fluent_j$ type. Literal fluents values can be True or False, while function fluents values are a numerical value. $Label_i$ depends on the problem and the relation of the example i with the rest of the dataset's elements. When representing states with a different number of fluents, the set of all fluents is calculated as the union of the different sets of fluents of each example. If a fluent doesn't appear in an example it's value is set as a Missing Value. Dataset's Missing Values (MV) are treated depending on the world assumption made in the planning domain.

3 NSLV

The algorithm extended was NSLV [2] (New SLaVe) an algorithm of the SLaVe family [12]. NSLV is a classification algorithm based on inductive rule learning with fuzzy sets. NSLV uses the Sequential Covering (SC) strategy described in Algorithm 1, where E is a collection of examples and f a fitness function used to measure the validity of the rules. Other elements of the SC strategy used are:

- **PERFORMANCE.** Measures the difference in the degree of completeness that causes the inclusion of a given rule in the ruleset. In other words, it measures the number of new examples of E explained by the addition of the rule in the collection of previously learned rules.
- **LEARN_ONE_RULE.** Uses a genetic algorithm (GA) to select which tuples $<attribute, value>$ define the antecedent of the rule that best fits a set of examples E. The rule learned must cover at least one example of E. The GA used is a steady state genetic algorithm whose population size maintains constant: each time an element is included in the population the worst element of it is erased.

Starting from an empty ruleset, a new rule is extracted and added to it in each iteration. The examples covered by this new rule are penalized (step 3.b)

Algorithm 1. Sequential covering strategy of NSLV

Input: A set of examples. **Output:** A learned ruleset.

SEQUENTIAL_COVERING (E, f)

1. Learned_rules ← {}
2. Rule ← LEARN_ONE_RULE (E, f)
3. **While** PERFORMANCE (Rule, E) > 0, **Do**
   (a) Learned_rules ← Learned_rules + Rule
   (b) E ← Penalize (Learned_rules, E)
   (c) Rule ← LEARN_ONE_RULE (E, f)
4. **Return** Learned_rules

in order to guide the GA to learn rules that explain new examples. Penalization is realized by marking the examples instead of erasing it from the examples set, helping the algorithm to find a new rule that explains new examples besides previously covered examples. This process ends when the PERFORMANCE of an extracted rule is zero or less.

The rules created by NSLV use a weighted Disjuntive Normal Form (DNF) fuzzy model, following the structure detailed below:

$$\textbf{IF}\, C_1 \text{ and } C_2 \text{ and} \ldots \text{and } C_m \,\textbf{THEN}\, Class \text{ is } B$$

$$\text{with}\,\textbf{weight}\ \ \mathbf{w}$$

where a condition C_i is a sentence X_h is A, with A a fuzzy label (or a set of fuzzy labels) of the domain of the variable X_n. X_n is an element of X the set of antecedent variables of the rule, those antecedent correspond with the attributes of the problem's dataset. Finally, B is the value that represents a class of a particular problem and w a measure of the PERFORMANCE of the rule.

The criterion used to select the best rule is a key element in NSLV. The criterion defined uses a multi-criteria evaluation guided by a Lexicographical Evaluation Function (LEF). The evaluation function's different criteria are ordered by their importance level. This order is essential to assure the rule's accuracy and interpretability level. The criteria defined follows measure the accuracy, consistency and simplicity of the rule.

The most important criterion is the one related to the accuracy. This criterion measures the consistency and completeness of the rule [13] returning the product between them.

On the first hand, rule's completeness degree is measure as:

$$\Lambda(R) = \frac{n^+(R)}{n_{class}}$$

with $n^+(R)$ as the number of examples covered by the rule [14] of a certain class, and n_{class} the number of elements of the examples set of this class. When checking the coverage of an example with MVs, NSLV considers the MV always as correct when evaluating them with a rule's antecedent.

In the other hand, the consistency degree of the rule is measured by considering the possibility of the existence of noise in the rules [13]. Consistency is defined following the next formula:

$$\Gamma(R) = \begin{cases} 1, & \text{if } n^-(R) = 0 \\ \frac{n^+(R) - n^-(R)}{n^+(R)}, & \text{if } n^-(R) < n^+(R) \\ 0, & \text{otherwise} \end{cases}$$

where $n^+(R)$ is the number of examples covered the rule and $n^-(R)$ the uncovered examples [14].

Rule's inference is introduced in the criterion by using n_s as a measure of the success of the rule (number of examples correctly classified) and n_f a measure of the failure of the rule (number of examples incorrectly classified):

$$\Gamma'(R) = \begin{cases} \Gamma(R), & \text{if } (n_s(R) > n_f(R)) \\ 0, & \text{otherwise.} \end{cases}$$

Finally, in order to combine both criterions to ensure a high degree of completeness and (Λ) and consistency (Γ') a new criterion Ψ is defined. The criterion is defined by the next function:

$$\Psi(R) = \Gamma'(R) \times \Lambda(R).$$

The rest of the defined criterions measure the simplicity of the rules. First, svar(R), is the definition of the degree of simplicity of the rule's variables. This measure is the number of irrelevant variables in a rule. Second, sval(R), is the rule's values degree of simplicity. Values degree of simplicity is calculated as the number of understandable assignments. Both criterions are fully defined in [15].

In the end, the multi-criteria evaluation function defined to select the best rules is:

$$fitness(R) = [\Psi(R),\ svar(R),\ sval(R)].$$

As said earlier, the evaluation function follows a lexicographical order. This means that the first criterion in the function guides the selection. In case of a tie between fitness measures, the other two criterions are used sequentially to break the tie.

NSLV can output two kinds of DNF rules: descriptive rules and predictive rules. Predictive rules are rules that contain the minimum information needed to classify an example. While descriptive rules contain the minimum information needed to model a set of examples. In terms of antecedent's information, the difference between these rules is that predictive rules contain only the relevant attributes needed to classify an example, on the other hand, descriptive rules contain every relevant attribute of the example. In this work, we focus in the use of descriptive rules because we need the maximum information in order to model correctly the world's states.

4 PlanMiner-O

As a result of extending NSLV we present the PlanMiner-O planning domain learner. PlanMiner-O takes a collection of plan traces, extracting a dataset for each different action model in it. Then using NSLV, PlanMiner-O generates a set of rules, which once postprocessed form the PDDL planning action. Algorithm 2 shows an overview of PlanMiner-O where PTs is a collection of input plan traces. Other elements of PlanMiner-O are:

- **EXTRACT_INFORMATION.** Generates a dataset from the information stored in PTs. The datasets contain the prestates and poststates of a given *action*.
- **PREPROCESS_DATASET.** Applies various techniques to the dataset in order to increase tolerance to incompleteness of the rule learner.
- **LEARN_RULES.** This step uses the NSLV algorithm to output a set of rules for each class in it, either pre-state or post-state.
- **SELECT_RULES.** By selecting the rule that best fits the prestates and poststates of an *action* PlanMiner-O obtains the information needed to construct an action.

Algorithm 2. Plan Miner algorithm overview.

Input: A collection of Plan Traces.
Output: A set of learned action models.

PlanMiner-O(PTs)

1. ActM ← {}
2. **Foreach** *action* **in** PTs, **Do**
   (a) *Dataset* ← EXTRACT_INFO(*action*, PTs)
   (b) *Dataset* ← PREPROCESS(*Dataset*)
   (c) *rules* ← LEARN_RULES(*Dataset*)
   (d) ActM ← ActM + COMBINE(rules)
3. **Return** ActM

EXTRACT_INFORMATION creates a dataset for each different *action* in PTs. First, selects each pair <*prestate, poststate*> associated with a given *action* across the set of PTs. prestates and poststates of each action that agree with *action* are selected and its schema form is calculated. This is achieved calculating the schema form of the states. State's schema form is calculated by taking each argument $<arg_1, arg_2, ..., arg_n>$ of the action and replacing the i-th argument in every prestate's and poststate's fluent in which it appears with a $Param_i$ token who represents a variable. Finally, every fluent in the state that has not undergone at least one substitution is erased from the state following a criterion of *relevance* [10]. A fluent is relevant if it shares anyone of its parameters with the associated task's parameters. Given an action (*board person*1, *plane*1, *city*2) and an associated state ((*at plane*1 *city*2) ∧ (= (*fuel* − *level plane*1) 100) ∧ (*at person*1 *city*2)) the schema form of the state is: ((*at* $Param_2$ $Param_3$)∧(= (*fuel*−*level* $Param_2$) 100)∧(*at* $Param_1$ $Param_3$)). Datasets' label is *prestate* or *poststate* depending on the relation of the state with the given action.

For each function fluent in the states EXTRACT_INFORMATION adds an extra column to the dataset. The information of this new attribute is computed by selecting the pairs <*prestate*, *poststate*> and calculating the difference between the function's value in *prestate* and *poststate*.

Before beginning the learning process, PREPROCESS_DATASET cleans the dataset by erasing noisy values and missing states. PREPROCESS_DATASET selects a dataset's class and extracts every example of the selected class. Then, it makes a statistical frequency analysis over these examples. This analysis calculates the appearance rate of each tuple <*fluent*, *value*> in the selected examples in comparison with other tuples with the same *fluent*. The analysis ignores those tuples with an MV as value. If the frequency of a tuple is lower than a threshold it value is replaced by an MV in every example it appears. Table 1 shows an example of a frequency table extracted from preselected examples of a dataset. PREPROCESS_DATASET repeats the process with the other dataset's class.

Table 1. Collection of examples and associated frequency table

#	$Attr_1$	$Attr_2$	$Attr_3$	$Attr_4$
1	True	False	False	True
2	True	True	True	True
3	MV	True	False	MV
4	True	True	False	False

Element	Tally	Frequency
$< Attr_1, True >$	3	100%
$< Attr_2, True >$	3	75%
$< Attr_2, False >$	1	25%
$< Attr_3, True >$	1	25%
$< Attr_3, False >$	3	75%
$< Attr_4, True >$	2	66%
$< Attr_4, False >$	1	33%

When PREPROCESS_DATASET has been finished replacing those values that not meet the frequency requisite set, it begins to search for those examples of the dataset whose attributes' values only contains MVs. Any time it founds an example like that, PREPROCESS_DATASET erases it. This ensures that every example in the dataset had at least one attribute with useful information, minimizing future noise problems.

LEARN_RULES takes a *Dataset* and output a set of rules that model the collection of examples of a given class using NSLV. In order to assure the quality of the rules, the dataset is split randomly into two datasets: NSLV create the rules using one of them and test its accuracy with the other. Using a noise-free complete dataset the number of rules created by NSLV is always two: one for each class in the dataset.

SELECT_RULES takes the ruleset created by NSLV and selects the best rule of each class. The rules selected are the ones that cover the maximum number of examples of each dataset's class. Rules with a low coverage of examples are interpreted as rules that model noisy examples and therefore they are ignored. Rule's antecedent contains explicitly the information needed to create a PDDL planning action. In example, given the rule $IF(Attr1 = TRUE) \wedge (Attr2 = FALSE) \Rightarrow C$ its antecedent can be translated directly to the fluents $(attr1) \wedge (not(Attr2))$. Finally, this fluents can be used to create the proper PDDL preconditions or effects using C following a straightforward process:

- Action's preconditions are taken directly from the antecedent of the prestate-class' rule.
- Action's effects are extracted by calculated the difference $\Delta(pre, post)$. $\Delta(pre, post)$ is defined as the set of changes that must be done over a state *pre* in order to make it equal to the state *post*. *pre* is obtained from the prestate rule's antecedent, while *post* is obtained from the poststate rule's antecedent.

Once the whole learning process has finished the rest of the PDDL planning domain is created by simply adding the list of different types' and parametrized fluents extracted from *PT* to it.

5 Experiments and Results

PlanMiner-O was tested using a collection domains from the International Planning Competition IPC. The objective of these experiments is to demonstrate that PlanMiner-O is able to learn planning domains even with high levels of missing states' information. The details of the domains used can be seen in Table 2. From each domain, 200 problems were set. The 80% of these problems were used as train problems and the 20% left as test problems. The experimental process used was defined as follows:

1. Training problems were solved using the original planning domain.
2. For each plan obtained in Step 1, a PT was created.
3. PTs were modified with noise or incompleteness if applicable.
4. A new domain was learned from the collection of PTs.
5. The learned domain and the original one were compared and performance values were calculated.
6. Test problems are solved using the learned domain.
7. New plans generated in Step 6 were validated with the original domain.

In order to ensure the results, Steps 3–7 were repeated 10 times. The final result is the average of the results obtained in Steps 5 and 7. Noise is included in the PTs randomly: A given percentage of fluents' values of the PTs are changed randomly to another valid value. Incompleteness is included following the same philosophy: A given percentage of fluents are selected randomly and erased from the PTs.

Table 2. Benchmark Domains Characteristics(from left to right): domain's number of tasks, domain's number of fluents, average number of tasks in the plans solved, average number of fluents in the plans' states and average CPU time(in seconds) to learn a domain.

Problem	$\lvert tasks \rvert$	$\lvert fluents \rvert$	$\tilde{PL}$	$\tilde{SL}$	$\tilde{CPUt}$
BlocksWorld	4	5	600	500	100
Depots	5	6	236	381	83
DriverLog	6	6	173	169	70
ZenoTravel	5	4	165	95	40
Satellite	5	8	91	178	37
Parking	4	5	57	200	98

PlanMiner-O's Performance is measured using 3 different metrics:

- NSLV rules' accuracy rate.
- Learned domain's error rate.
- Learned domain's validation rate with test problems.

Rule's accuracy is measured as $\frac{classSucc}{classEx}$, where $classSucc$ is the number of examples correctly classified by a given rule and $classEx$ is the total number of examples of the class modelled by that rule in the whole dataset. Then, the ruleset's total accuracy is defined by the following function:

$$Acc = \frac{\sum\limits_{r \in Rules} RAcc(r)}{2}$$

with $Rules$ as the set of rules selected by PlanMiner-O for a single dataset and $RAcc(r)$ the accuracy of a given rule. Total rules' accuracy is calculated as the average of the accuracy of the rules of every action of the domain.

Figure 1 shows that NSLV maintains an accuracy above 90% of success even with certain levels of noise. Taking into account that PlanMiner-O doesn't have implemented any procedure to deal with noise in the input PTs these results show that NSLV is viable to keep working with this kind of problems. In fact, rules accuracy's behaviour is the same that actions error's behaviour when facing incompleteness: results are invariable regardless incompleteness levels of the input information.

The second criterion used to measure the quality of the learned domains is the domain's error rate in comparison with the original domain [11]. Domain's error is defined as $\frac{\sum\limits_{a \in Actions} error(a)}{\lvert Actions \rvert}$ where $Actions$ is the set of Actions of a given domain. Action's error rate is calculated by counting the number of missing or extra fluents in the action's preconditions and effects and dividing it between the number of possible fluents in those preconditions and effects. The number of possible fluents is limited to a finite number of fluents in order to delimit the error rates.

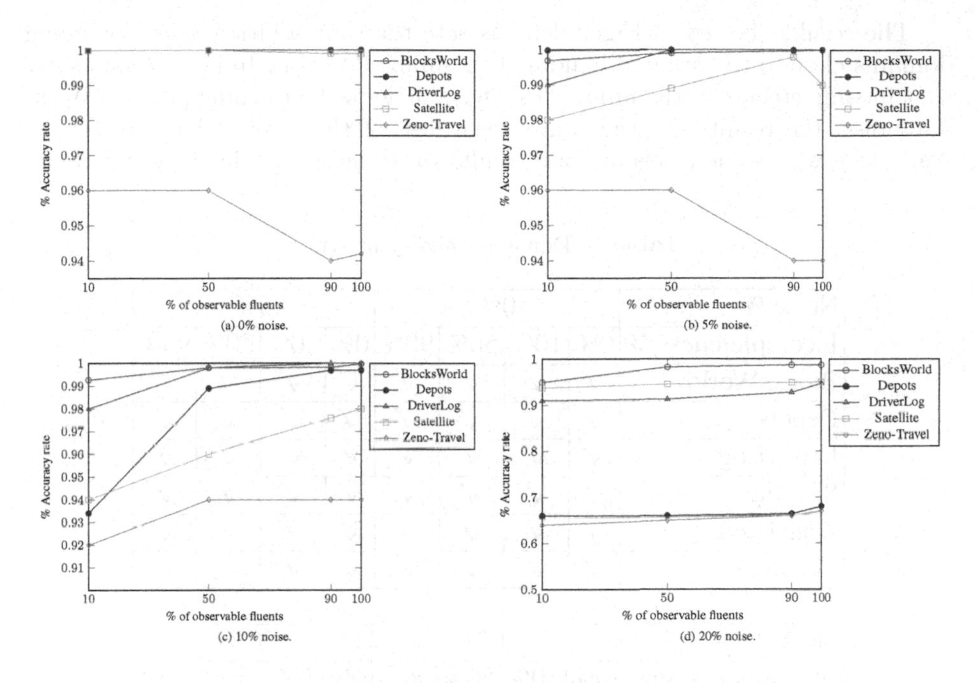

Fig. 1. NSLV error rates.

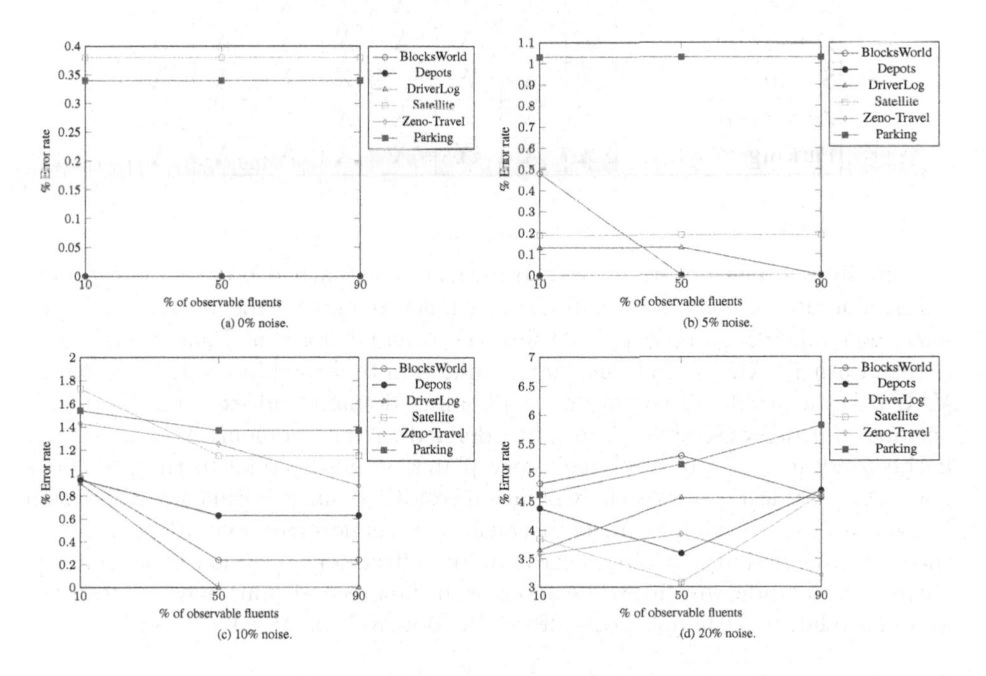

Fig. 2. Learned domains error rates.

The results showed in Fig. 2 demonstrate that our solution learns planning domains close to the original handmade planning domains. In fact, results show that incompleteness in the input PTs affects little to the learning process. Without noise, the results are the same, regardless of the level of incompleteness. And, even with some levels of noise results varies only a 1% in the worst cases.

Table 3. Domains validity matrix.

Noise %	0%				5%			
Incompleteness %	0%	10%	50%	90%	0%	10%	50%	90%
BlocksWorld	✓	✓	✓	✓	✓	✓	✓	✓
Depots	✓	✓	✓	✓	✓	✓	✓	✓
DriverLog	✓	✓	✓	✓	✓	✓	✓	✓
Satellite	✓	✓	✓	✓	✓	✓	✓	✓
ZenoTravel	✓	✓	✓	✓	✓	✓	✓	✓
Parking	✓	✓	✓	✓	✓	✓	✓	✓
Noise %	**10%**				**20%**			
Incompleteness %	**0%**	**10%**	**50%**	**90%**	**0%**	**10%**	**50%**	**90%**
BlocksWorld	✓	✓	✓	✓	X	X	X	X
Depots	X	X	X	X	X	X	X	X
DriverLog	✓	✓	X	X	X	X	X	X
Satellite	X	X	X	X	X	X	X	X
ZenoTravel	X	X	X	X	X	X	X	X
Parking	X	X	X	X	X	X	X	X

Finally, the last criterion used to measure the learned domains is the plan validation rate. A domain is valid if it can generate plans using the test problems and then validating those plans using the original domain. Plan validation is realized using VAL [16], an automatic validation tool used in the IPC. Roughly, VAL takes a problem, a plan and a planning domain and executes the plan's actions in order over the initial state defined in the problem. A plan is valid if the resultant state of applying every plan's action is equal to the problem's goal state. Table 3 contains the validation results of our experiments. Validation is the hardest criterion to meet because of a single error may affect a lot to the domain's validity. A single error in the effects of an action can make the entire action model invalid, while an error in the preconditions may no affect the model's validity. Although both affects the domain's error rate the same.

6 Conclusions and Future Work

In this paper, we have developed PlanMiner-O, an extension of the NSLV algorithm to learn planning domains. The results obtained show that PlanMiner-O is able to learn actions models, even with high levels of incompleteness. The experiments carried out to test our proposal checked not only the error rates from both the domain and NSLV's resultant models but the domain's validity too. Another conclusion extracted from the experiments is that PlanMiner-O can handle some levels of noise, even if there is no procedure implemented to increase noise tolerance. This is achieved thanks to NSLV that is robust enough to deal with these problems.

In the near future, we are going to focus on increase PlanMiner-O's noise tolerance. When increasing noise levels results show an erratic behaviour. As NSLV's accuracy rate modelling states are high enough to fit our expectations, our efforts are going to focus on developing a new procedure able to take the rules learned by NSLV and output a planning domain by combining them instead of selecting the best one. With this new procedure, we aim to fix the erratic behaviour of PlanMiner-O when facing high levels of noise. This will allow us to deal with the problem of create valid domains even with noisy input data.

In order to improve the capabilities of PlanMiner-O we are going to study of to make it deal with numerical values. This addition will increase the research value of PlanMiner-O, opening the door to work with plan traces extracted from more complex problems. These new problems may include the use of action's cost, time constraints or continuous numerical functions. As NSLV is already able to work with this kind of information by discretizing the numerical values using fuzzy labels, we will be going to study how to preprocess the information contained in the plan traces, manipulate it and send it to NSLV.

PlanMiner-O is a concept test of a new family of domains learners we plan to develop. These new domain learners aim to learn HTN planning domains from real-world data. The PlanMiner-O algorithm is the first algorithm that addresses part of this challenge. Before advancing any further in this line of work we developed PlanMiner-O to test our main hypothesis and ideas. Next versions of the PlanMiner domain learners will include new functionalities to deal with data closer to real-world data or learn the hierarchical structures needed in HTN Planning.

Acknowledgements. This research is being developed and partially funded by the Spanish MINECO R&D Project PLAN MINER TIN2015-71618-R.

References

1. Michalski, R.S.: A theory and methodology of inductive learning. In: Michalski, R.S., Carbonell, J.G., Mitchell, T.M. (eds.) Machine Learning. Symbolic Computation, pp. 83–134. Springer, Heidelberg (1983). https://doi.org/10.1007/978-3-662-12405-5_4
2. González, A., Pérez, R.: Improving the genetic algorithm of SLAVE. Mathware Soft Comput. **16**, 59–70 (2009)
3. Zadeh, L.: Fuzzy sets. Inf. Control **8**(3), 338–353 (1965)
4. Jiménez, S., Rosa, T.D.L., Fernández, S., Fernández, F., Borrajo, D.: A review of machine learning for automated planning. Knowl. Eng. Rev. **27**(4), 433–467 (2012)
5. Wang, X.: Learning by observation and practice: an incremental approach for planning operator acquisition. In: Proceedings of the 12th International Conference on Machine Learning, pp. 549–557. Morgan Kaufmann (1995)
6. Benson, S.S.: Learning Action Models for Reactive Autonomous Agents. Ph.D. thesis, Stanford University (1996)
7. Gil, Y.: Learning by experimentation: incremental refinement of incomplete planning domains. In: Cohen, W.W., Hirsh, H. (eds.) Machine Learning Proceedings 1994, pp. 87–95. Morgan Kaufmann, San Francisco (1994)
8. Cresswell, S.N., McCluskey, T.L., West, M.M.: Acquiring planning domain models using LOCM. Knowl. Eng. Rev. **28**(2), 195–213 (2013)
9. Gregory, P., Lindsay, A.: Domain model acquisition in domains with action costs. In: Proceedings of the Twenty-Sixth International Conference on International Conference on Automated Planning and Scheduling, ICAPS 2016, pp. 149–157. AAAI Press (2016)
10. Yang, Q., Wu, K., Jiang, Y.: Learning action models from plan examples using weighted MAX-SAT. Artif. Intell. **171**(2), 107–143 (2007)
11. Zhuo, H.H., Yang, Q., Hu, D.H., Li, L.: Learning complex action models with quantifiers and logical implications. Artif. Intell. **174**(18), 1540–1569 (2010)
12. García, D., González, A., Párez, R.: Overview of the SLAVE learning algorithm: a review of its evolution and prospects. Int. J. Comput. Intell. Syst. **7**(6), 1194–1221 (2014)
13. González, A., Pérez, R.: Completeness and consistency conditions for learning fuzzy rules. Fuzzy Sets Syst. **96**, 37–51 (1998)
14. González, A., Pérez, R.: SLAVE: a genetic learning system based on an iterative approach. IEEE Trans. Fuzzy Syst. **7**(2), 176–191 (1999)
15. Castillo, L.A., González, A., Pérez, R.: Including a simplicity criterion in the selection of the best rule in a genetic fuzzy learning algorithm. Fuzzy Sets Syst. **120**(2), 309–321 (2001)
16. Howey, R., Long, D.: VAL's progress the automatic validation tool for PDDL2.1 used in the international planning competition. In: Proceedings of the ICAPS 2003 Workshop on the Competition: Impact, Organization, Evaluation, Benchmarks, Trento, Italy, pp. 28–37, June 2003

A Proposal for Adaptive Maps

Marina Torres(✉), David A. Pelta, and José L. Verdegay

Models of Decision and Optimization Research Group,
Department of Computer Science and A.I., University of Granada,
18014 Granada, Spain
{marinatorres,dpelta,verdegay}@ugr.es

Abstract. The visualization of specific attributes of the maps is not achieved with standard maps representations or area cartograms. Adaptive Maps can deal with multiple attributes like travel time, quality of the road or tourism interest of the path between two points.

A method to generate and visualize Adaptive Maps is proposed. It departs from a graph with multiple attributes and generates a single measurement matrix that represent the desired distance between points. A Multidimensional Scaling problem on that matrix is solved to finally visualize the adapted map.

To illustrate the proposal, 4 adapted maps are generated and visualized.

Keywords: Adaptive maps · Maps visualization
Multiple attributes · Decision making

1 Introduction

A map is a depiction that emphasizes relationships among elements of some space. In usual geographical maps this relation is just the euclidean distance.

Despite general purpose maps (like tourist city maps) help to understand some features of the depicted area (e.g. the location of sites of interest) they lack of further information about the relationships among the elements on the map (besides the euclidean distance).

However, such information is crucial in some circumstances: for example, people with limited mobility would consider that a destination that looks close in a map is impossible to reach due to certain physical barriers. In other words, not only the euclidean distance is relevant: other features should be considered.

The general concept of modifying maps to represent certain information is related to the so called "area cartograms" where geographic variables are visualized as spatial objects whose size is proportional to certain variable strength. There are several methods to generate cartograms. Rubber maps method is a popular construction method of contiguous area cartograms [10]. Rubber-sheet algorithm uses control points as vertices and map triangles onto corresponding triangles on the stable map [5]. Carto3F algorithm also constructs contiguous

J. Medina et al. (Eds.): IPMU 2018, CCIS 855, pp. 657–666, 2018.
https://doi.org/10.1007/978-3-319-91479-4_54

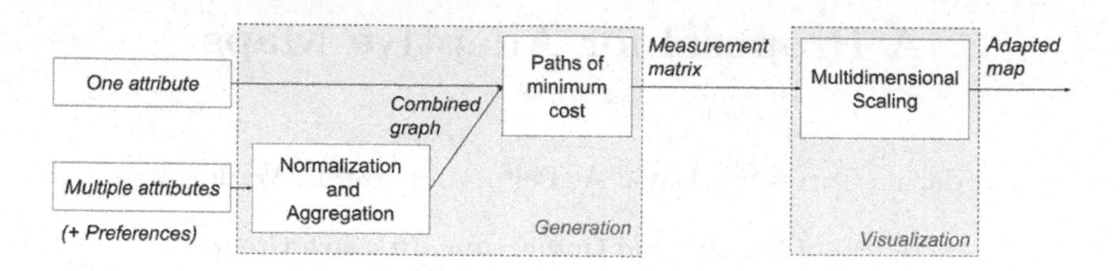

Fig. 1. Workflow diagram of the proposed method.

area cartograms improving the rubber-sheet algorithms in preserving topology [7]. BS Daya Sagar [6] presents a methodology based on mathematical morphology to generate contiguous cartograms that relies on weighted skeletonization by zone of influence. Those cartograms preserves the global shape and local shapes and yields minimal area errors.

The aim of this contribution is to present the idea of "Adaptive Maps" together with a number of procedures to generate and visualize them. Adaptive Maps have a similar purpose as cartograms, in the sense they are a modification of the map to represent certain information. However, as contraposition to the deformed area cartograms, the adapted maps will be used to represent some *measure among points of interest* according to multiple attributes values. Adaptive Maps also represent the information according to the user preferences being possible to weight the attributes in order to give them more of less importance on the desired adapted map.

The paper is organized as follows. In Sect. 2 the concept of Adaptive Map is presented. The generation method is explained in Sect. 2.1 and the visualization of the Adaptive Maps is detailed in Sect. 2.2. We show the application of the proposed method in Sect. 3 giving details on the generation and visualization of adapted maps with one and multiple attributes. The conclusions are detailed in Sect. 4.

2 Adaptive Maps

The idea motivating the concept of Adaptive Maps is that when moving in a city, everything related with distances, travel time, pleasant walk, and so on depends on the eye of the beholder. For example, a cathedral could be just 100 m away which can be perceived as very close, unless you have a mobility problem avoiding you to overcome the stairs that exist in the middle of the way. Or you are carrying a baby in a pushchair and prefer to walk a longer distance that takes you through a pedestrian area instead of the shortest path that goes through a narrow street.

Now the question is: *how can we reflect such information on a map?*

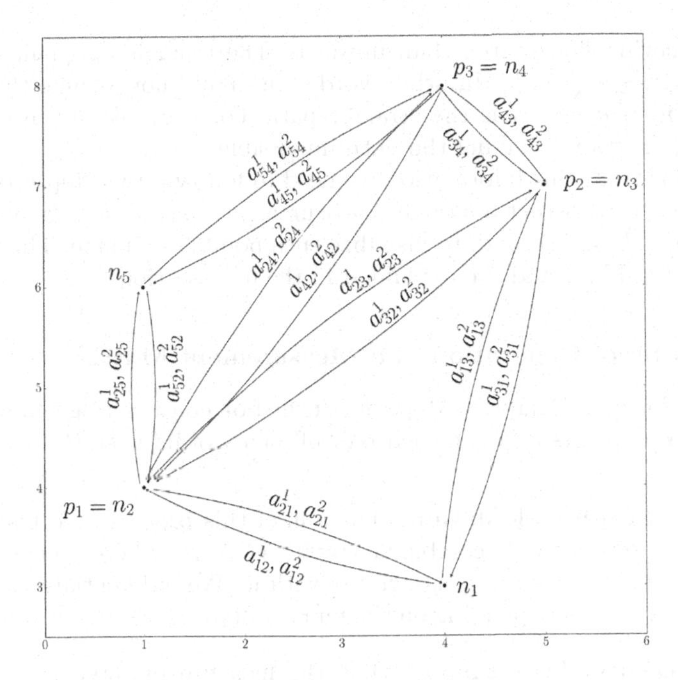

Fig. 2. Elements on the Adaptive Map problem with $|N| = 5$, $|P| = 3$ and $D = 2$.

We start defining an Adaptive Map as a depiction of a measurement, or a combination of measurements, relating elements of interest in some geographical space.

From an operational point of view, the following elements are considered:

1. A number of "layers" $d_1, d_2, \ldots, d_D$ where every d_k with $k = \{1, 2, \ldots, D\}$ stands for a kind of measure (i.e. distance, time, slope, etc.).
2. A directed multigraph $G(N, A)$ where N is a set of nodes and A is a set of labelled arcs. An arc $a_{ij} \in A$ connects two nodes $n_i, n_j \in N$. A multigraph is a graph from which multi-edges are not excluded, but which has no self-loops [4].
3. Every arc a_{ij} has D labels stating some measurements. We denote a_{ij}^k as the value of the measure $d_k \in D$ between nodes $n_i, n_j \in N$.
4. A set $P \subseteq N$, which indicates the points of interest that we want to represent in the adapted map.

The considered elements are depicted in Fig. 2.

A "standard map" is a special case of an adaptive map where $D = 1$, $P = N$ and the measure d_1 is the Euclidean distance.

Please, note that we talk about measures and measurements instead of metrics as we do not require the properties of the latter. We do not assume symmetry: $a_{ij}^1 \neq a_{ji}^1$. For example, let's suppose that d_1 is the travel time by bicycle. Then, consider the case when the path connecting $n_i \rightarrow n_j$ has a slope. It would be faster moving downwards than upwards. The triangle inequality is neither assumed: $a^1ij \nleq a^1_{ik} + a^1_{kj}$. In other words, we should not assume that the best way to go from $n_i \rightarrow n_j$ is the straight path. For example, a physical barrier may exist (e.g. stairs) making the path impossible.

Departing from this information we need to follow a two step process which requires first to construct a sort of "measurement matrix" among every pair of points $p_i, p_j \in P$ and second, to visualize such points in a plane. These steps are summarized in Fig. 1 and are explained in the next sections.

2.1 First Step: Generation of a Measurement Matrix

The generation of an Adaptive Map consists first on constructing a matrix reflecting the measurements between each pair of points of interest P.

Combined Graph Calculation. The aim of this process is to transform the multigraph G into a single combined graph $CG(N, A)$ where every arc $a_{ij} \in A$ will have a single measure c_{ij} associated with it. We differentiate two cases for the calculation of CG depending on the number of measures under consideration:

1. **One measure.** Let's suppose that the measure of interest is d_k. Then, $c_{ij} = a^k_{ij}$. In other words, we just keep the information associated with the measure d_k.
2. **Multiple measures.** In this case, as there are multiple measures to combine, a normalization process should be applied.
   Let's define
   $min_k = \min\{a^k_{ij} \mid a^k_{ij} \in A\}$ and
   $max_k = \max\{a^k_{ij} \mid a^k_{ij} \in A\}$.
   Then, the normalized measure $\hat{a}^k_{ij}$ is calculated:

$$\hat{a}^k_{ij} = \frac{a^k_{ij} - min_k}{max_k - min_k}. \tag{1}$$

   Using the normalized measurements, we calculate a single aggregated cost for each arc as

$$c_{ij} = \sum_{i=1}^{k} w_i \times \hat{a}^k_{ij}, \tag{2}$$

   where we use a set of weights $W = \{w_1, \ldots, w_k\}$ with $\sum_{i=1}^{k} w_i = 1$ and $w_i \in [0, 1]$ to reflect the importance that each measure has for the user.

Measurement Matrix. Departing from the CG graph, we need to calculate a measurement matrix $M^{|P| \times |P|}$ between the points of interest $P \subseteq N$.

This implies searching for the path of minimum cost [2,3] on CG but just among the points of interest in P. The reader should note that even when a direct arc between two points c_{ij}, with $p_i, p_j \in P$ is available, the path of minimum cost should be calculated as the triangle inequality is not assumed.

After these calculations, a measurement matrix M representing the adapted map is obtained. This is shown in Table 1. The value $C(p_i, p_j)$ stands for the minimum cost path joining p_i with p_j.

Table 1. Measurement matrix M. The value $C(p_i, p_j)$ stands for the minimum cost path joining p_i with p_j.

M	p_1	p_2	...	p_i	...	p_n
p_1	0	$C(p_1, p_2)$	...	...	...	$C(p_1, p_n)$
p_2	$C(p_2, p_1)$	0	...	...	...	$C(p_3, p_n)$
...	...	...	...	...	...	
p_i	...	...	...	0	...	
...	...	...	...	...	...	
p_n	$C(p_n, p_1)$	$C(p_n, p_2)$	...	...	...	0

2.2 Visualization

A key aspect in the Adaptive Map concept is visualization.

The problem we need to solve is: find the location in the plane of the points in P in such a way that the distances among them reflects as much as possible the information contained in the measurement matrix M.

When considering path attributes from a non metric space, this representation is not trivial or exact. Because we are looking for approximate solutions for visualizations purposes, we will solve a classical Multidimensional Scaling (classical MDS) problem [1]. MDS problem departs from a matrix known as dissimilarity matrix (M in our case) and calculates the location of the points of interest such that the distances between the points are approximately equal to the values on the matrix. The solution from a classical MDS problem is a coordinate matrix that minimizes a loss function called *strain* given by:

$$Strain(x_1, \ldots, x_n) = \left(\frac{\sum_{i,j} (b_{i,j} - \langle x_i, x_j \rangle)^2}{\sum_{i,j} b_{i,j}^2} \right)^{1/2} \tag{3}$$

where $b_{i,j}$ are coefficients from the matrix $B = -\frac{1}{2} J M^2 J$, with $J = I_n - \frac{1}{N} \mathbb{O}$ a centering matrix and $\mathbb{O}$ a N-by-N matrix of 1.

After solving this problem, the coordinates $x_1, \ldots, x_n \in \Re^2$ are the location of the points $p_1, \ldots, p_n \in P$.

In order to visualize the adapted map, the points of interest P could be easily represented with the locations $x_1, \ldots, x_n$.

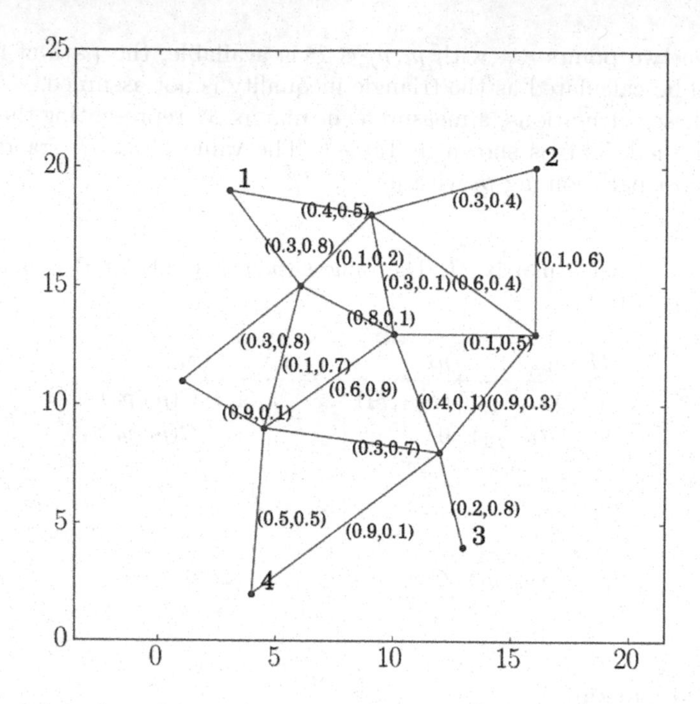

Fig. 3. Original graph with two measures. The measurements are normalized.

3 Example

In this section we will show a detailed example of the construction of an adaptive map, including the generation and visualization steps using 4 cases. For the sake of simplicity we consider an undirected graph (the measurements in the arc are symmetric).

Data. We depart from the graph shown in Fig. 3, where $D = 2$ and there are $|P| = 4$ points of interest located at the positions indicated in Table 2.

Table 2. Original location of the points of interest.

	x	y
p_1	3	19
p_2	16	20
p_3	13	4
p_4	4	2

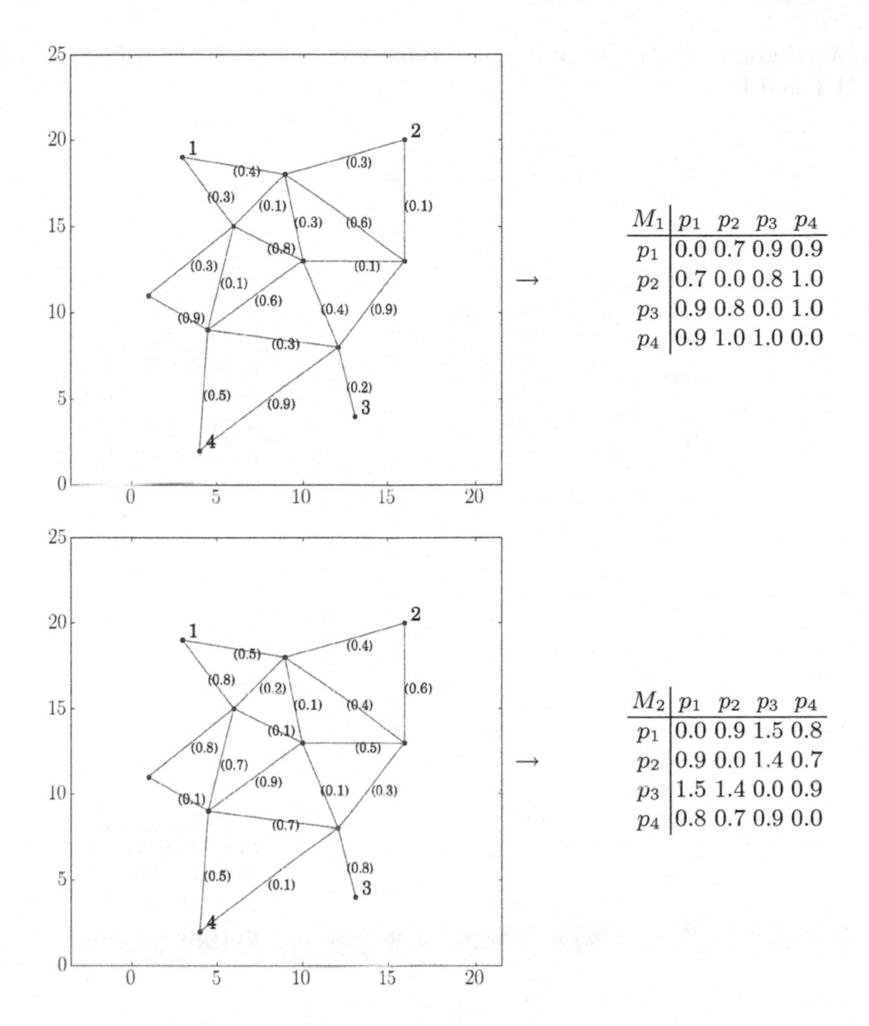

M_1	p_1	p_2	p_3	p_4
p_1	0.0	0.7	0.9	0.9
p_2	0.7	0.0	0.8	1.0
p_3	0.9	0.8	0.0	1.0
p_4	0.9	1.0	1.0	0.0

M_2	p_1	p_2	p_3	p_4
p_1	0.0	0.9	1.5	0.8
p_2	0.9	0.0	1.4	0.7
p_3	1.5	1.4	0.0	0.9
p_4	0.8	0.7	0.9	0.0

Fig. 4. Combined graphs CG_1, CG_2 and their measurement matrices M_1, M_2 obtained from the individual measurements.

Measurement Matrix Calculation. The first step of the generation of an Adaptive Map is the construction of the combined graph. In the case of multiple attributes, the preferences of the user are needed to achieve the required aggregation process, see Sect. 2.1.

We will show 4 different adapted maps. Two of them are constructed from the individual measures, and the other two, using different weights distributions on the user preferences. More specifically, we consider $W1 = \{w_1 = 0.5, w_2 = 0.5\}$ and $W2 = \{w_1 = 0.9, w_2 = 0.1\}$.

The combined graphs for the individual measurements CG_1, CG_2 are shown in Fig. 4 together with the measurement matrices M_1, M_2 that represent the minimum cost of the path between the points of interest. Figure 5 shows the combined graphs CG_3, CG_4 and measurement matrices M_3, M_4 when considering $W1$ and $W2$.

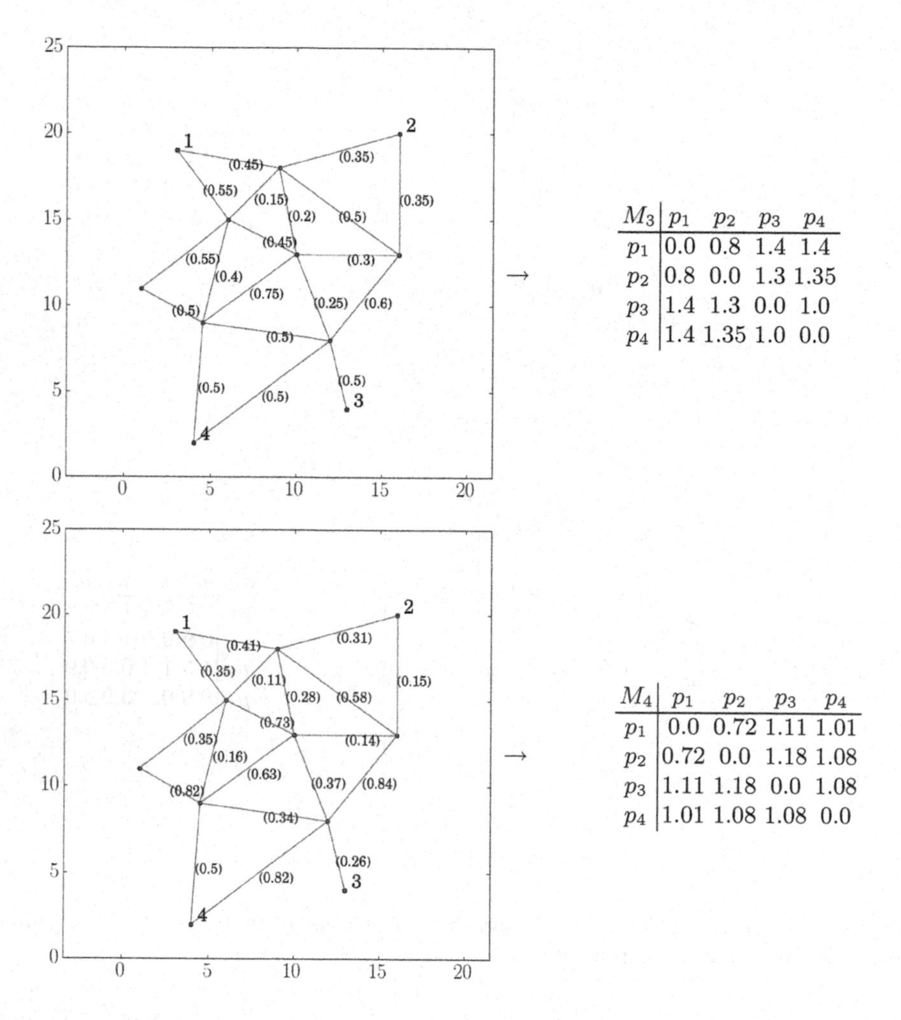

M_3	p_1	p_2	p_3	p_4
p_1	0.0	0.8	1.4	1.4
p_2	0.8	0.0	1.3	1.35
p_3	1.4	1.3	0.0	1.0
p_4	1.4	1.35	1.0	0.0

M_4	p_1	p_2	p_3	p_4
p_1	0.0	0.72	1.11	1.01
p_2	0.72	0.0	1.18	1.08
p_3	1.11	1.18	0.0	1.08
p_4	1.01	1.08	1.08	0.0

Fig. 5. Combined graphs CG_3, CG_4 and their corresponding measurement matrices M_3, M_4 obtained with different weights distributions: (top) $w_1 = 0.5, w_2 = 0.5$ and (bottom) $w_1 = 0.9, w_2 = 0.1$.

Visualization. The visualization process is achieved by solving the MDS problem using the *cmdscale* function [9] on *R* [8]. For each measurement matrix M_i, the solution of the MDS problem determines the location of the points of interest on the final adapted map. The adapted maps obtained are shown in Fig. 6.

If the map is adapted using the individual measurements, as shown in Fig. 6(a), we can observe that the points of interest look closer when considering the first measurement than the second one. In particular, *point 2* is closer to *point 1* and *point 3* using the former metric while the distance between *point 3* and *point 4* looks longer.

If we take into account both measurements simultaneously, as shown in Fig. 6(b), we can observe differences. For example, the distance between *point 1* and *point 4* in *Map 4* is shorter than the one reflected in *Map 3*.

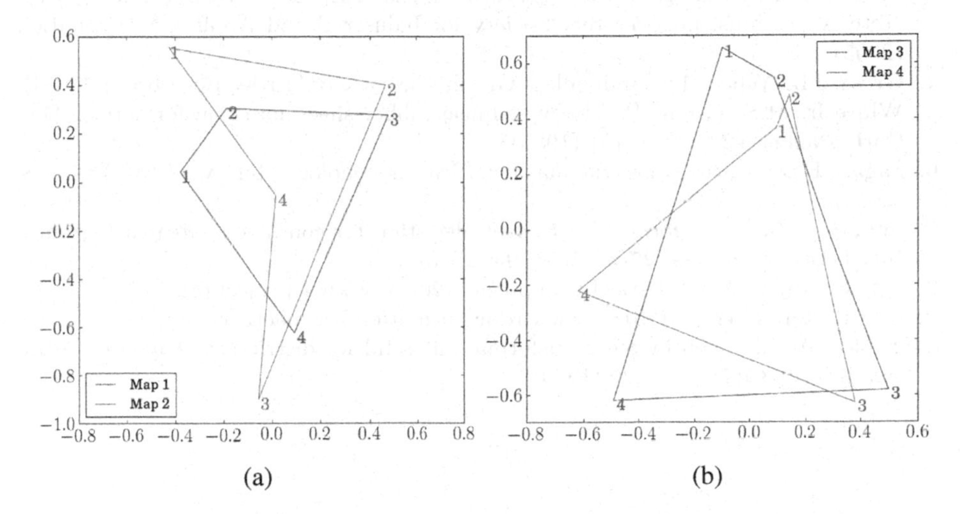

Fig. 6. Visualization of the location of the points in the adapted maps *Map 1*, *Map 2*, *Map 3* and *Map 4* generated with matrices (a) M_1 and M_2, (b) M_3 and M_4.

4 Conclusions

In this paper we introduced the concept of Adaptive Maps, as well as initial procedure to generate and visualize them. Adaptive Maps are useful for representing specific attributes like *walking or cycling time* that standard maps based on Euclidean distance are not able to represent.

The method consists on generating a combined graph and its corresponding measurement matrix that represent the desired distance between points on the adapted map. An approximated visualization of the adapted map is achieved by solving a multidimensional scaling problem. The method is proposed with one or multiple attributes. The examples provided show the suitability and feasibility of the proposal.

Acknowledgments. Authors acknowledge support through projects TIN2014-55024-P and TIN2017-86647-P from the Spanish Ministry of Economy and Competitiveness (including FEDER funds).

M. Torres enjoys a Ph.D. research training staff grant associated with the project TIN2014-55024-P and co-funded by the European Social Fund.

References

1. Cox, T.F., Cox, M.A.: Multidimensional Scaling. CRC Press, Boca Raton (2000)
2. Gallo, G., Pallottino, S.: Shortest path algorithms. Ann. Oper. Res. **13**(1), 1–79 (1988)
3. Goldberg, A.V., Harrelson, C.: Computing the shortest path: a search meets graph theory. In: Proceedings of the Sixteenth Annual ACM-SIAM Symposium on Discrete Algorithms. pp. 156–165. Society for Industrial and Applied Mathematics (2005)
4. Gross, J.L., Yellen, J.: Handbook of Graph Theory. CRC Press, Boca Raton (2004)
5. White Jr., M.S., Griffin, P.: Piecewise linear rubber-sheet map transformation. Am. Cartographer **12**(2), 123–131 (1985)
6. Sagar, B.D.: Cartograms via mathematical morphology. Inf. Vis. **13**(1), 42–58 (2014)
7. Sun, S.: A fast, free-form rubber-sheet algorithm for contiguous area cartograms. Int. J. Geogr. Inf. Sci. **27**(3), 567–593 (2013)
8. The R Project for Statistical Computing (2017). www.r-project.org
9. The R Stats Package (2018). www.rdocumentation.org/packages/stats
10. Tobler, W.R.: A continuous transformation useful for districting. Ann. New York Acad. Sci. **219**(1), 215–220 (1973)

Co-words Analysis of the Last Ten Years of the International Journal of Uncertainty, Fuzziness and Knowledge-Based Systems

Manuel J. Cobo[1(✉)], Wanru Wang[2], Sigifredo Laengle[3], José M. Merigó[3], Dejian Yu[2], and Enrique Herrera-Viedma[4]

[1] Department of Computer Science and Engineering, University of Cádiz, Cádiz, Spain
manueljesus.cobo@uca.es

[2] School of Information, Zhejiang University of Finance and Economics, Hangzhou 310018, Zhejiang, China
wanruwang0401@163.com, yudejian62@126.com

[3] Department Management Control and Information Systems, School of Economics and Business, University of Chile, 8330105 Santiago, Chile
{slaengle,jmerigo}@fen.uchile.cl

[4] Department of Computer Science and Artificial Intelligence, University of Granada, Granada, Spain
viedma@decsai.ugr.es

Abstract. The main aim of this contribution is to develop a co-words analysis of the *International Journal of Uncertainty, Fuzziness and Knowledge-Based Systems* in the last ten years (2008–2017). The software tool SciMAT is employed using an approach that allows us to uncover the main research themes and analyze them according to their performance measures (qualitative and quantitative). An amount of 562 documents were retrieved from the Web of Science. The corpus was divided into two consecutive periods (2008–2012 and 2013–2017). Our key findings are that the most important research themes in the first and second period were devoted with decision making process and its related aspects, techniques and methods.

Keywords: Bibliometric analysis · Science mapping analysis
Co-words analysis

1 Introduction

The *International Journal of Uncertainty, Fuzziness and Knowledge-Based Systems* (IJUFKS) is one of the most important journals in the field of computer science. According to its webpage, it is a forum for research on various methodologies for the management of imprecise, vague, uncertain or incomplete information. In its first year, 1993, there was only two issues with nine articles, and then

J. Medina et al. (Eds.): IPMU 2018, CCIS 855, pp. 667–677, 2018.
https://doi.org/10.1007/978-3-319-91479-4_55

it published four issues each year in 1994 and 1995. Since 1996, it is published bimonthly; and now it has worldwide distribution to researchers, engineers, decision makers, and educators. The latest Journal Citation Reports indicates that IJUFKS had an impact factor of 1.214, and its quartile was Q3 in the Web of Science category of Computer Science, Artificial Intelligence.

So, the main aim of this contribution is to carry out a conceptual science mapping analysis [1–3] of the research conducted by the IJUFKS from 2008 to 2017 (the last ten years). The analysis is developed using SciMAT [4] software tool and partially based in the approach presented in [5].

This article is organized as follows: Sect. 2 introduces the methodology employed in the analysis. In Sect. 3, the dataset is described. In Sect. 4, the science mapping analysis of the IJUFKS is presented. Finally, some conclusions are drawn in Sect. 5.

2 Methodology

Science mapping or bibliometric mapping is a spatial representation of how disciplines, fields, specialties, and documents or authors are related to one another [6]. It has been widely used to show and uncover the hidden key elements (documents, authors, institutions, topics, etc.) in different research fields [7–11].

Science mapping analysis can be carried out with different software tools [3]. Particularly, SciMAT was presented in [4] as a powerful tool that integrates the majority of the advantages of available science mapping software tools [3]. It is an open source software tool that present the following key features:

- It incorporates all the necessary modules to develop all the steps of the science mapping workflow, from data acquisition and preprocessing to the visualization and interpretation of the results.
- It has methods to build the majority of the bibliometric networks, different similarity measures to normalize them and build the maps using clustering algorithms, and different visualization techniques useful for interpreting the output.
- It implements a wide range of preprocessing tools such as detecting duplicate and misspelled items, time slicing, data reduction and network preprocessing.
- It enrich the maps with bibliometric measures based on citation indicators, such as the h-index.

SciMAT was designed according to the science mapping analysis approach presented in [5], combining both performance analysis tools and science mapping tools to analyze a research field and detect and visualize its conceptual subdomains (particular topics/themes or general thematic areas) and its thematic evolution.

Therefore, in this contribution, SciMAT was employed to develop a longitudinal conceptual science mapping analysis [1,3] based on co-words bibliographic networks [2,12]. Thus, the analysis was carried out in three stages:

1. *Detection of the research themes.* In each period of time studied the corresponding research themes are detected by applying a co-word analysis [2] to raw data for all the published documents in the research field, followed by a clustering of keywords to topics/themes [13], which locates keyword networks that are strongly linked to each other and that correspond to centres of interest or to research problems that are the subject of significant interest among researchers. The similarity between the keywords is assessed using the equivalence index [14].
2. *Visualizing research themes and thematic network.* In this phase, the detected themes are visualized by means of two different visualization instruments: strategic diagram [15] and thematic network [5]. Each theme can be characterized by two measures [14]: *centrality* and *density.* Centrality measures the degree of interaction of a network with other networks. On the other hand, density measures the internal strength of the network. Given both measures, a research field can be visualized as a set of research themes, mapped in a two-dimensional strategic diagram (Fig. 1) and classified into four groups:
   (a) Themes in the upper-right quadrant are both well developed and important for the structure of the research field. They are known as the *motor-themes* of the specialty, given that they present strong centrality and high density.
   (b) Themes in the upper-left quadrant have well-developed internal ties but unimportant external ties and so, they are of only marginal importance for the field. These themes are very *specialized and peripheral.*
   (c) Themes in the lower-left quadrant are both weakly developed and marginal. The themes in this quadrant have low density and low centrality and mainly represent either *emerging or disappearing* themes.

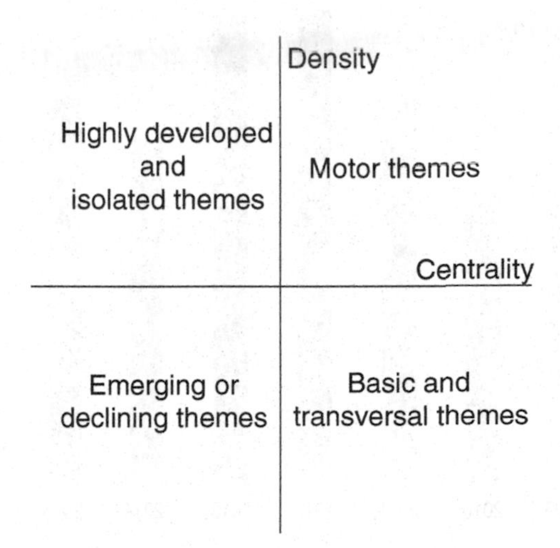

Fig. 1. The strategic diagram.

(d) Themes in the lower-right quadrant are important for a research field but are not developed. This quadrant contains *transversal and general*, basic themes.

3. *Performance analysis*. In this phase, the relative contribution of the research themes to the whole research field is measured (quantitatively and qualitatively) and used to establish the most prominent, most productive and highest-impact subfields. Some of the bibliometric indicators to use are: number of published documents, number of citations, and different types of h-index [16–18].

   For each theme, the performance measure are computed taking into account the documents associated with it. Thus, for instance, the h-index is computed using the citations of the theme's documents.

3 Dataset

In order to carry out the performance and science mapping analysis, the research documents published by the journal IJUFKS during the last ten years must be collected and also, preprocessed.

Since Web of Science (WoS) is the most important bibliographic database, the research documents published by IJUFKS were downloaded from it. The query retrieved a total of 562 documents from 2008 to 2017 (Fig. 2). The corpus was further restricted to articles and reviews. Citations of these documents are also used in this study; they were counted up to 2nd January 2018.

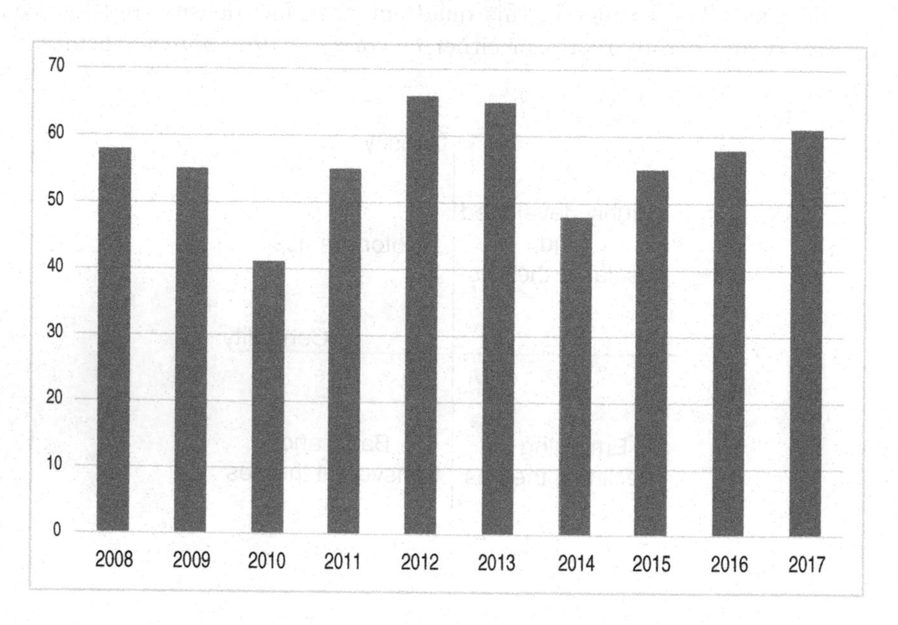

Fig. 2. Distribution of documents retrieved by years.

The raw data was downloaded from WoS as plain text and entered into SciMAT to build the knowledge base for the science mapping analysis. Thus, it contains the bibliographic information stored by WoS for each research document. To improve the data quality, a de-duplicating process was applied (the author's keywords and the Keywords Plus were used as unit of analysis). Words representing the same concept were grouped. Furthermore, some meaningless keywords in this context, such as stop-words or words with a very broad and general meaning, e.g. "MODEL" or "DESIGN", were removed.

Next, using the SciMAT period manager, the corpus was divided in time spans. To avoid data smoothness, the best option would have been to choose one-year periods. However, it was found that not enough data were generated in the span of a single year to obtain good results from science mapping analysis. For this reason, two consecutive periods of five years were established (Fig. 3): 2008–2012 and 2013–2017, with 1,395 and 1,508 keywords, respectively.

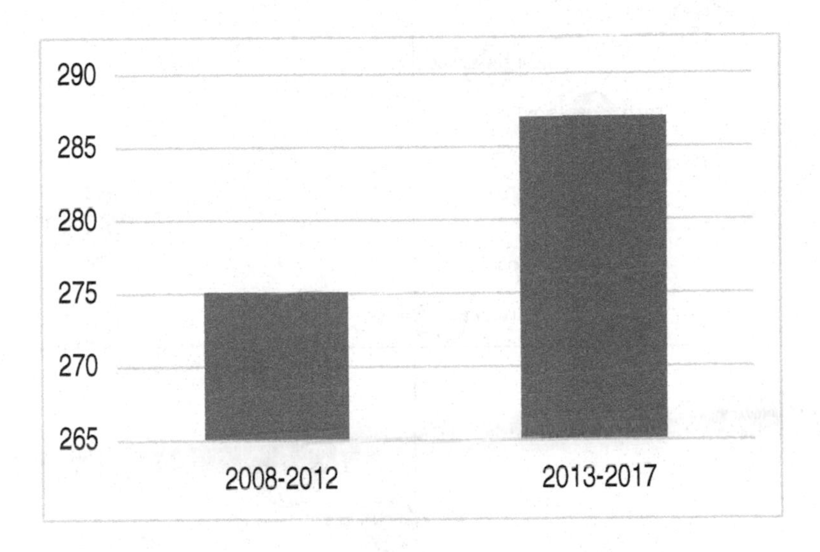

Fig. 3. Distribution of documents retrieved by period.

4 Conceptual Analysis

In order to analyze the most highlighted themes of the IJUFKS, a strategic diagram is shown for each period. In addition, the spheres size is proportional to the number of published documents associated with each research theme.

First Period (2008–2012). According to the strategic diagram shown in Fig. 4, during this period the journal pivoted on fifteen themes, with the following eight major themes (*motor themes* plus *basic themes*): *Fuzzy-Processing-Time*, *Preference-Relations*, *Similarity-Measure*, *Classification*, *Association-Rules*, *Group-Decision-Making*, *Genetic-Algorithm* and *Decision-Making*

The performance measures of the themes are given in Table 1, showing the number of documents, numbers of citations and h–index per theme. According to these performance measures, the following seven themes stand out (more than 100 citations): *Decision-Making*, *Group-Decision-Making*, *Preference-Relations*, *Genetic-Algorithm*, *Similarity-Measure*, *Classification* and *Rough-Set*.

The basic and transversal theme *Decision-Making* gets the highest citations count (more than 1,000) and h-index in this period. It is related with topics such as, fuzzy logic and fuzzy sets, uncertainty, vague set and Dempster Shafer Theory, among others. It plays a central role, and is the basis for other important themes of the journal.

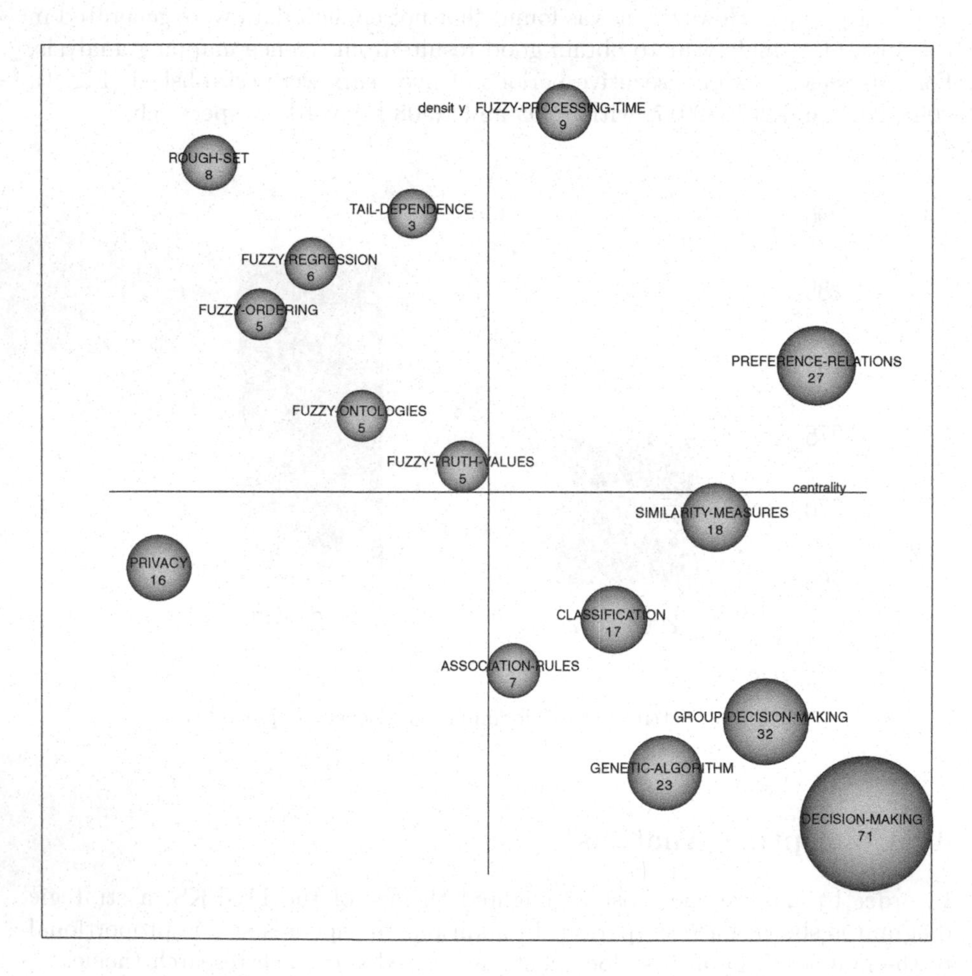

Fig. 4. Strategic diagram for the 2008–2012 period.

Another important basic and transversal theme related with the process of decision making is *Group-Decision-Making* (Fig. 5a), which gets great impact scores. It is devoted with aggregation operators, consensus model, linguistic, etc.

Table 1. Performance of the themes in the 2008–2012 period

Name	Number of documents	Number of citations	h-index
DECISION-MAKING	71	1,023	14
GROUP-DECISION-MAKING	32	695	12
PREFERENCE-RELATIONS	27	908	12
GENETIC-ALGORITHM	23	121	5
SIMILARITY-MEASURES	18	421	7
CLASSIFICATION	17	208	6
PRIVACY	16	48	4
FUZZY-PROCESSING-TIME	9	71	5
ROUGH-SET	8	134	5
ASSOCIATION-RULES	7	48	4
FUZZY-REGRESSION	6	81	4
FUZZY-ORDERING	5	40	4
FUZZY-ONTOLOGIES	5	68	5
FUZZY-TRUTH-VALUES	5	28	3
TAIL-DEPENDENCE	3	35	3

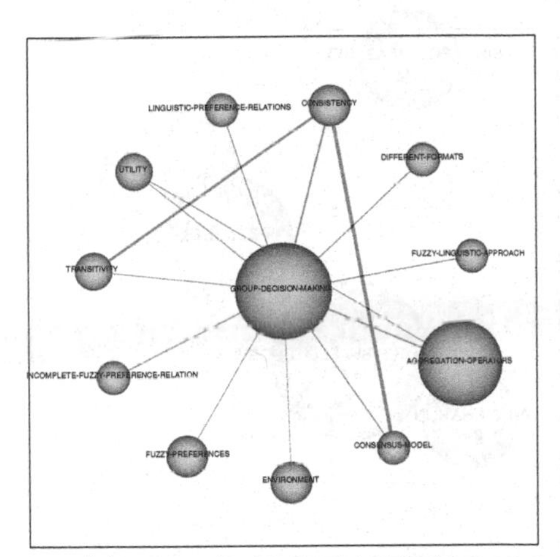

(a) Theme Group Decision Making.

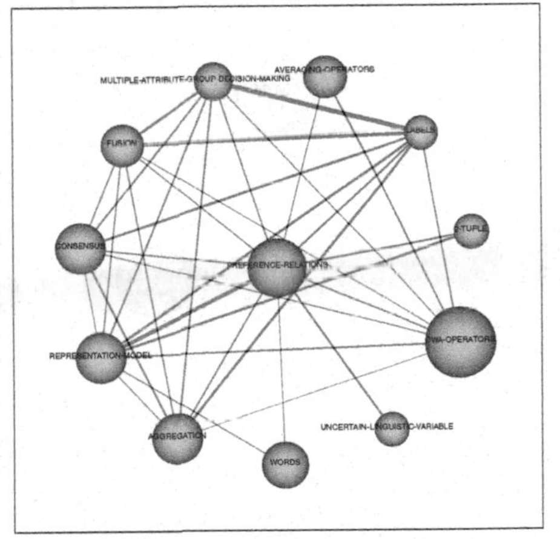

(b) Theme Preference Relations

Fig. 5. Thematic networks for the period 2008–2012.

The motor theme *Preference-Relations* (Fig. 5b) is also related with the decision making process, getting the second position in citations count. It is related with consensus, OWA operators and representation models.

The motor theme *Fuzzy-Processing-Time* gets the highest density in this period, and it is related with due dates.

Finally, the basic and transversal theme *Classification* is devoted with the development of classification algorithms and the related issues, such as, feature selection, decision trees, imbalance datasets, statistical comparison, etc. Moreover, the theme *Genetic-Algorithm* is related with machine learning techniques for optimization.

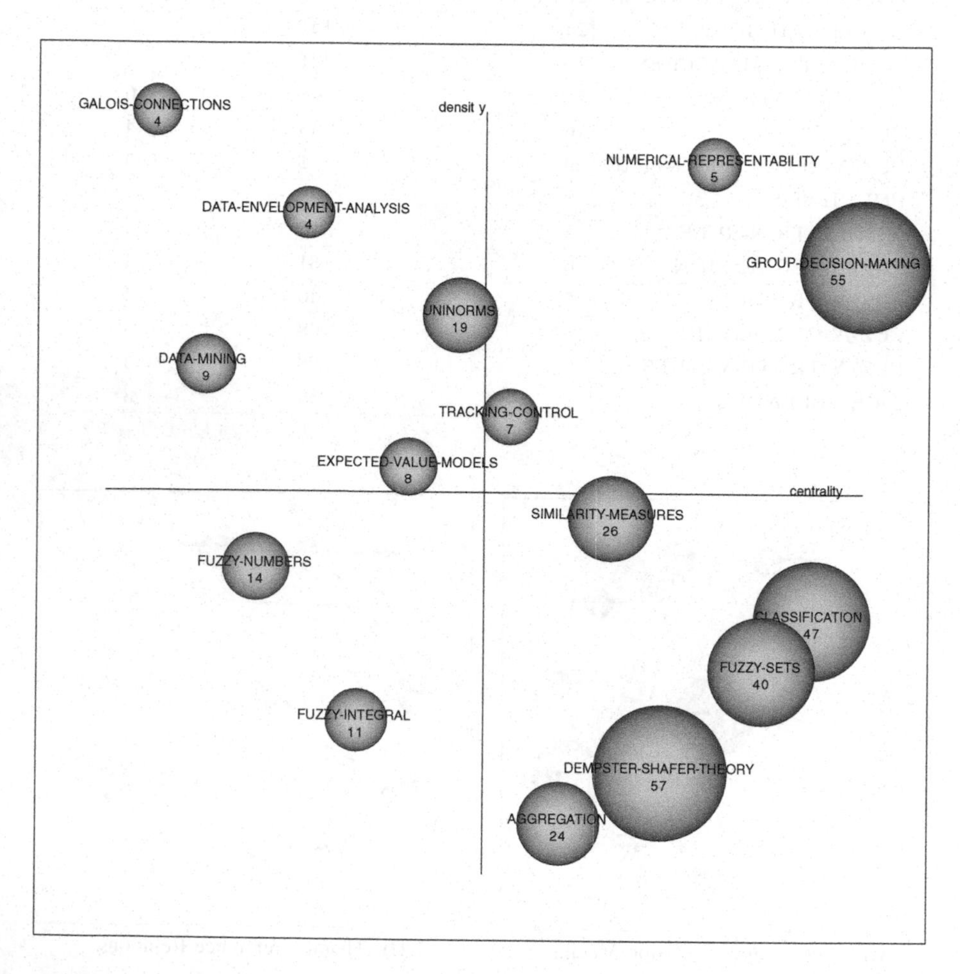

Fig. 6. Strategic diagram for the 2013–2017 period.

Second Period (2013–2017). The research conducted in this period pivots on fifteen themes. According to the strategic diagram shown in Fig. 6, during this period eight themes stand out (*motor themes* plus *basic themes*): *Numerical-Representability*, *Group-Decision-Making*, *Tracking-Control*, *Similarities-Measures*, *Classification*, *Fuzzy-Sets*, *Dempster-Shafer-Theory* and *Aggregation*.

Table 2. Performance of the themes in the 2013–2017 period

Name	Number of documents	Number of citations	h-index
DEMPSTER-SHAFER-THEORY	57	159	7
GROUP-DECISION-MAKING	55	377	11
CLASSIFICATION	47	104	6
FUZZY-SETS	40	49	4
SIMILARITY-MEASURES	26	147	7
AGGREGATION	24	119	7
UNINORMS	19	44	3
FUZZY-NUMBERS	14	24	3
FUZZY-INTEGRAL	11	45	3
DATA-MINING	9	15	2
EXPECTED-VALUE-MODELS	8	43	4
TRACKING-CONTROL	7	7	2
NUMERICAL-REPRESENTABILITY	5	9	2
GALOIS-CONNECTIONS	4	4	2
DATA-ENVELOPMENT-ANALYSIS	4	6	1

According to the performance measures shown in Table 2, five themes could be highlighted (more than 100 citations): *Dempster-Shafer-Theory*, *Group-Decision-Making*, *Classification*, *Similarities-Measures* and *Aggregation*.

The theme *Group-Decision making* (Fig. 7a) evolved from the first period turning into an important motor theme. Furthermore, it gets the highest impact rates. Mostly, it is related with different techniques and tools necessary for the decision making process in group, such as, OWA, preference relations, and distance measure.

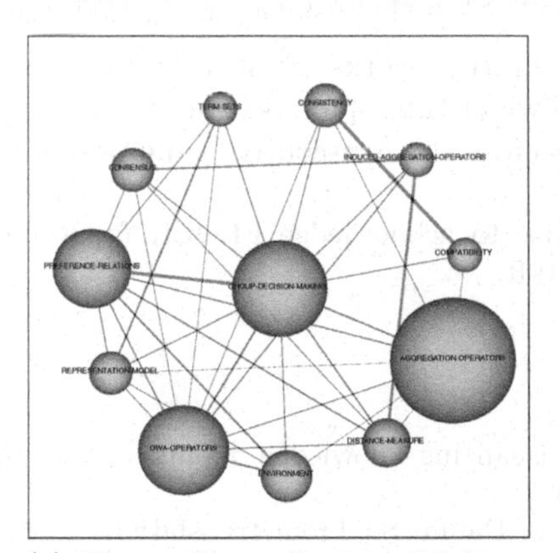

(a) Theme Group Decision Making.

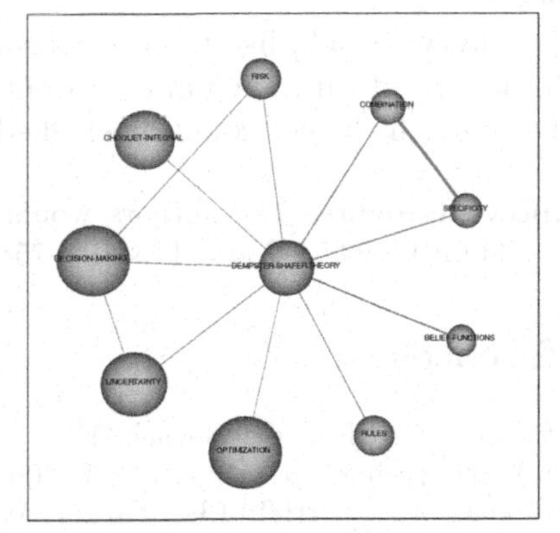

(b) Theme Multi-Attribute-Decision-Making.

Fig. 7. Thematic networks for the period 2013–2017.

The basic and transversal theme *Dempster-Shafer-Theory* (Fig. 7b) gets the second highest impact rates. It is related with the general decision making process, and some techniques such as the Choquet integral.

The theme *Classification* is consolidated in this period, covering a great variety of aspects related with classification algorithms and techniques, such as, pattern classification, accuracy, reduction, rough set, neural networks and genetic algorithm, among others.

Finally, the theme *Aggregation* is devoted to the aspects related with the multicriteria decision making process.

5 Conclusions

In this contribution, a conceptual science mapping analysis of the articles published in the last ten years (2008–2017) by the *International Journal of Uncertainty, Fuzziness and Knowledge-Based Systems* (IJUFKS) has been performed. The analysis was carried out using SciMAT [4].

An amount of 562 documents (articles and reviews) were retrieved. The corpus was split into two consecutive periods of five years length: 2008–2012 and 2013–2017.

In the first period, the themes *Decision-Making*, *Group-Decision-Making*, *Preference-Relations*, *Genetic-Algorithm*, *Similarity-Measure*, *Classification* and *Rough-Set* stand out due to their highest impact rates. It should be point out that the theme *Decision-Making* get more than 1,000 citations. Similarly, in the second period, five themes must be highlighted according to their impact scores: *Dempster-Shafer-Theory*, *Group-Decision-Making*, *Classification*, *Aggregation* and *Similarities-Measures*. It should be mentioned that *Group-Decision-Making* gets two time more citations than the second one. As general conclusion, the themes that get highest impact scores are related with the decision making process.

Finally, we would like to address some future works. First, a global analysis could be carried out taking into account a wider time span. Second, the evolution of the research themes could be studied across the consecutive time periods.

Acknowledgments. The authors would like to acknowledge FEDER funds under grants TIN2013-40658-P and TIN2016-75850-R.

References

1. Börner, K., Chen, C., Boyack, K.W.: Visualizing knowledge domains. Ann. Rev. Inf. Sci. Technol. **37**(1), 179–255 (2003)
2. Callon, M., Courtial, J.P., Turner, W.A., Bauin, S.: From translations to problematic networks: an introduction to co-word analysis. Soc. Sci. Inf. **22**, 191–235 (1983)
3. Cobo, M.J., López-Herrera, A.G., Herrera-Viedma, E., Herrera, F.: Science mapping software tools: review, analysis, and cooperative study among tools. J. Am. Soc. Inf. Sci. Technol. **62**(7), 1382–1402 (2011)

4. Cobo, M.J., López-Herrera, A.G., Herrera-Viedma, E., Herrera, F.: SciMAT: a new science mapping analysis software tool. J. Am. Soci. Inf. Sci. Technol. **63**(8), 1609–1630 (2012)
5. Cobo, M.J., López-Herrera, A.G., Herrera-Viedma, E., Herrera, F.: An approach for detecting, quantifying, and visualizing the evolution of a research field: a practical application to the Fuzzy Sets Theory field. J. Inform. **5**(1), 146–166 (2011)
6. Small, H.: Visualizing science by citation mapping. J. Am. Soc. Inf. Sci. **50**(9), 799–813 (1999)
7. Cobo, M.J., López-Herrera, A.G., Herrera, F., Herrera-Viedma, E.: A Note on the ITS topic evolution in the period 2000–2009 at T-ITS. IEEE Trans. Intell. Transp. Syst. **13**(1), 413–420 (2012)
8. Cobo, M.J., Martínez, M.A., Gutiérrez-Salcedo, M., Fujita, H., Herrera-Viedma, E.: 25 years at knowledge-based systems: a bibliometric analysis. Knowl.-Based Syst. **80**, 3–13 (2015)
9. Martínez, M.A., Cobo, M.J., Herrera, M., Herrera-Viedma, E.: Analyzing the scientific evolution of social work using science mapping. Res. Soc. Work Pract. **5**(2), 257–277 (2015)
10. Murgado-Armenteros, E.M., Gutiérrez-Salcedo, M., Torres-Ruiz, F.J., Cobo, M.J.: Analysing the conceptual evolution of qualitative marketing research through science mapping analysis. Scientometrics **102**(1), 519–557 (2015)
11. Rodriguez-Ledesma, A., Cobo, M.J., Lopez-Pujalte, C., Herrera-Viedma, E.: An overview of animal science research 1945–2011 through science mapping analysis. J. Anim. Breed. Genet. **132**(6), 475–497 (2015)
12. Batagelj, V., Cerinšek, M.: On bibliographic networks. Scientometrics **96**(3), 845–864 (2013)
13. Coulter, N., Monarch, I., Konda, S.: Software engineering as seen through its research literature: a study in co-word analysis. J. Am. Soc. Inf. Sci. **49**(13), 1206–1223 (1998)
14. Callon, M., Courtial, J.P., Laville, F.: Co-word analysis as a tool for describing the network of interactions between basic and technological research: the case of polymer chemsitry. Scientometrics **22**, 153–203 (1991)
15. He, Q.: Knowledge discovery through co-word analysis. Libr. Trends **48**(1), 133–159 (1999)
16. Alonso, S., Cabrerizo, F.J., Herrera-Viedma, E., Herrera, F.: h-Index: A review focused in its variants, computation and standardization for different scientific fields. J. Inform. **3**(4), 273–289 (2009)
17. Hirsch, J.E.: An index to quantify an individual's scientific research output. Proc. Nat. Acad. Sci. **102**(46), 16569–16572 (2005)
18. Martínez, M.A., Herrera, M., López-Gijón, J., Herrera-Viedma, E.: H-Classics: characterizing the concept of citation classics through h-index. Scientometrics **98**(3), 1971–1983 (2014)

The Relationship Between Graphical Representations of Regular Vine Copulas and Polytrees

Diana Carrera[1(✉)], Roberto Santana[1], and Jose A. Lozano[1,2]

[1] Intelligent Systems Group, Department of Computer Science and Artificial Intelligence, University of the Basque Country (UPV/EHU), Manuel de Lardizabal, 1, 20018 Donostia, Gipuzkoa, Spain
dianamaria.carrera@ehu.es

[2] Basque Center for Applied Mathematics, BCAM, 48009 Bilbao, Spain

Abstract. Graphical models (GMs) are powerful statistical tools for modeling the (in)dependencies among random variables. In this paper, we focus on two different types of graphical models: R-vines and polytrees. Regarding the graphical representation of these models, the former uses a sequence of undirected trees with edges representing pairwise dependencies, whereas the latter uses a directed graph without cycles to encode independence relationships among the variables. The research problem we deal with is whether it is possible to build an R-vine that represents the largest number of independencies found in a polytree and vice versa. Two algorithms are proposed to solve this problem. One algorithm is used to induce an R-vine that represents in each tree the largest number of graphical independencies existing in a polytree. The other one builds a polytree that represents all the independencies found in the R-vine. Through simple examples, both procedures are illustrated.

Keywords: Regular vine copulas · Polytrees
(In)dependence relationships · Graphical models

1 Introduction

Graphical models (GMs) [7,13] have been widely used for modeling the dependence structure of multivariate probability distributions through two closely related components: (i) The qualitative component is a graph where nodes correspond to random variables and edges to graphical relationships among them; (ii) The quantitative component is given by a set of local probability distributions that quantify the strength and uncertainty of the (in)dependencies encoded in the graph (or network). According to the type of the graph, directed and undirected, we can distinguish two different GMs: Bayesian networks (BNs) and Markov networks respectively. The interpretation of graphical (in)dependencies is different in directed and undirected graphs.

J. Medina et al. (Eds.): IPMU 2018, CCIS 855, pp. 678–690, 2018.
https://doi.org/10.1007/978-3-319-91479-4_56

In this paper, we focus on two GMs representative of undirected and directed graphs, namely regular vine copulas [4,12] (or simply R-vines) and a subclass of BNs called polytrees [8] respectively.

A copula is a probability distribution function with uniformly distributed margins [14,17]. Copulas allow us to model the dependence structure of multivariate distributions and its margins separately. Despite the generality of the copula-based framework, it turns out that building high-dimensional joint copulas is a difficult problem [1].

Pair copula constructions (PCCs) [11] and their graphical model, called regular vines (R-vines) [3,4], overcome the lack of flexibility of the copula-based modeling in the high-dimensional case. R-vines build multivariate copulas in terms of bivariate copulas (pair-copulas) taking advantage of the fact that the bivariate copulas are more tractable than multidimensional ones. Besides that, bivariate copulas of different families, can be combined in the same decomposition allowing the specification of different types of non-linear dependencies. The qualitative component of R-vines is specified by an R-vine structure (or graph) – a set of nested trees, where the variables are represented by nodes linked by edges, each associated with a pair-copula that captures certain types of pairwise dependence. It is in this sense that we say that R-vine structures encode dependence relationships rather than independencies relationships.

Polytrees (also known as singly connected networks) are directed acyclic graphs (DAGs) where there is no more than one undirected path that connects any two nodes (without undirected cycles). In these graphs, missing edges can represent either conditional independencies or conditional dependencies among random variables.

In general, Bayesian networks, particularly polytrees, have well-studied mathematical properties that have been developed throughout decades. In contrast, R-vines have boomed in the last few years. Previous works have addressed the question of the relationship between directed GMs and R-vines from different perspectives. In [9], a new method for learning the structure of a BN based on PCCs is introduced. In [10], a non-parametric Bayesian belief net as an alternative to a particular subclass of R-vines is introduced. The paper discusses the differences between both models and offers some guidelines on when to use one or the other from a quantitative perspective. In [2], a Bayesian network with pair-copulas is built using PCCs.

However, the problem of verifying whether the graphical independencies found in a polytree can be represented in an R-vine and vice versa has not been answered yet. In this work, we investigate the relationship between the graphical representations of R-vines and polytrees in both directions: (i) Given the graph of a polytree, we want to obtain an R-vine tree-structure that represents the largest number of independencies existing in the starting graph. To this end, a heuristic is proposed that, from the list of independences found in the polytree, performs this task locally, tree-by-tree of the R-vine. (ii) Given an R-vine, we want to build a polytree that represents the largest number of independences existing in the R-vine. Similarly, a heuristic is proposed that, based on the independence list extracted from the R-vine, builds a polytree that represents the independencies existing in the R-vine. These results are useful as they make it clear that properties and algorithms that can be applied to polytrees can be carried over to R-vines, and vice versa.

The paper is organized as follows: In Sects. 2 and 3, we provide the basic concepts as well as a short review of R-vines and polytrees respectively. In Sect. 4, we present the main contribution of this work: two algorithms that induce the graph of a R-vine from the graph of a polytree and vice versa. Section 5 offers a short summary and an outline of future work.

2 Regular Vines

Let $\mathbf{X} = (X_1, \ldots, X_n)$ be an n-dimensional random vector with joint density function $f : \mathbb{R}^n \to [0, \infty)$ and cumulative distribution function $F : \mathbb{R}^n \to [0, 1]$. Furthermore, let $F_i : \mathbb{R} \to [0, 1]$, $i = 1, \ldots, n$ be the corresponding marginal distributions of X_i[1]. Capital letters denote variables and lower letters are their assignments.

A n-dimensional copula C is a multivariate probability distribution function for which the univariate margins are uniform: $C : [0, 1]^n \to [0, 1]$ [14]. Copulas are used to describe the dependence structure among random variables.

The relevance of copulas in probabilistic modeling is given by Sklar's theorem [17], which states that an n-dimensional (multivariate) distribution function F of a random continuous vector $\mathbf{X} = (X_1, \ldots, X_n) \in \mathbb{R}^n$ can be expressed in terms of its marginal distributions $F_i(x_i)$ and a unique copula C. Sklar's theorem for densities is given by

$$f(x_1, \ldots, x_n) = c(F_1(x_1), \ldots, F_n(x_n)) \cdot \prod_{i=1}^{n} f_i(x_i) \tag{1}$$

where f and c denote the density functions corresponding to F and C respectively.

In (1), the copula c can be approximated by an PCC. This decomposition is represented graphically by an R-vine – a sequence of trees, of which each edge corresponds to a pair-copula. An R-vine is a probabilistic graphical model represented as a pair (G, θ). G is the structural part that is composed of a sequence of trees $T_1, T_2, \ldots, T_{n-1}$, where the nodes of T_j are edges in T_{j-1}. Two nodes in T_j (for $j \geq 2$) can only be adjacent if the corresponding edges in the previous tree have a common node (known as proximity condition) [2]. θ contains, for each edge of the trees, a pair-copula and its parameters. The number of edges in an R-vine is $n(n-1)/2$. Figure 1-(right panel) illustrates an R-vine copula c_{12345} for $n = 5$ and its respective factorization.

We define an R-vine formally by following the definition given in [5]: If we denote $T_j = (N_j, E_j)$, the tree of the decomposition at level j, where N_j and E_j

[1] We assume that all multivariate, marginal and conditional distributions are absolutely continuous with corresponding densities.

denote the node and edge sets of the jth tree, the edge $e \in E_j$ joins two vertices of N_j, $X_{k(e)}$ and $X_{l(e)}$, which are determined by the set of indices $k(e)$ and $l(e)$ respectively. Then, in the pair-copula $c_{k(e),l(e)|D(e)}$, the nodes $X_{k(e)}$ and $X_{l(e)}$ are the conditioned nodes, whereas $\mathbf{X}_{D(e)}$, which represents a subvector of $\mathbf{X}$ determined by the indices in $D(e)$, is the conditioning set. Consequently, a regular vine distribution is the distribution of the random vector $\mathbf{X} = (X_1, \ldots, X_n)$ with marginal densities $f_i(x_i)$, $i = 1, \ldots, n$, and where the conditional density of $(X_{k(e)}, X_{l(e)})$ given $\mathbf{X}_{D(e)}$ is specified as $c_{k(e),l(e)|D(e)}$ for the R-vine copula with $n-1$ trees, set of nodes $\mathbf{N} = \{N_1, \ldots N_{n-1}\}$ and set of edges $\mathbf{E} = \{E_1, \ldots E_{n-1}\}$. If the dependence structure of $\mathbf{X}$ is represented by an R-vine copula, then the n-dimensional density $f_{\mathbf{R-vine}}(x_1, \ldots, x_n)$ is given by

$$\underbrace{\prod_{j=1}^{n-1} \prod_{e \in E_j} c_{k(e),l(e)|D(e)} \left(F\left(x_{k(e)} \mid \mathbf{x}_{D(e)}\right), F\left(x_{l(e)} \mid \mathbf{x}_{D(e)}\right)\right)}_{\text{R-vine copula}} \cdot \underbrace{\prod_{i=1}^{n} f_i(x_i)}_{\text{Margins}} \tag{2}$$

In the approximation given in (2), only in the first tree are the pair-copulas unconditional as their arguments are marginal distributions. In the remaining trees, the pair-copulas are conditional as their arguments are conditional distributions. The number of variables in the conditioning set increases in one variable as we go deeper into the R-vine tree-structure: in the second tree, we have first-order conditional copulas; in the third tree, second-order conditional copulas, and so on; that is, in the tree j we have conditional copulas of the order $j-1$.

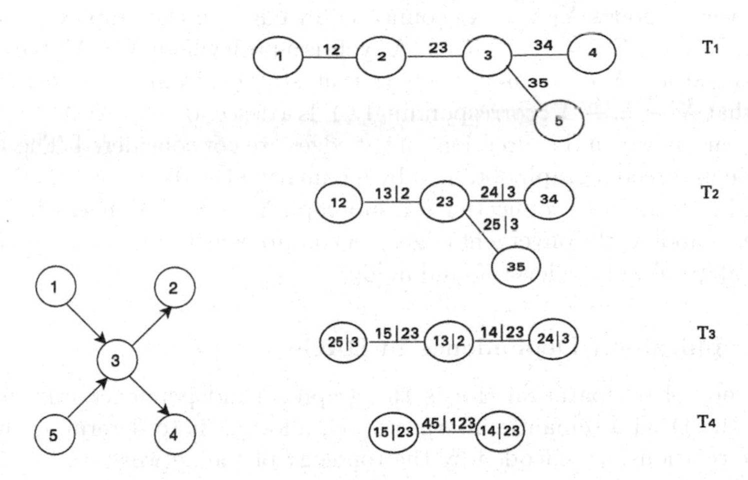

Fig. 1. Example of a polytree (left panel) and an R-vine (right panel) where $n = 5$. The polytree factorization is given as $p(x_1) \cdot p(x_5) \cdot p(x_3 \mid x_1, x_5) \cdot p(x_2 \mid x_3) \cdot p(x_4 \mid x_3)$. The R-vine factorization is given as $\underbrace{c_{12} \cdot c_{23} \cdot c_{34} \cdot c_{45}}_{T_1} \cdot \underbrace{c_{13|2} \cdot c_{24|3} \cdot c_{25|3}}_{T_2} \cdot \underbrace{c_{15|23} \cdot c_{14|23}}_{T_3} \cdot \underbrace{c_{45|123}}_{T_4} \cdot$

2.1 Graphical (In)Dependence in R-vines

We define two types of edges (or links): dashed edges indicate linked nodes are independent and continuous edges indicate linked nodes are dependent with each other.

To provide the R-vine tree-structure with a semantic interpretation in terms of independencies, we define the vine-graphical independence criterion as follows.

Definition 1 (r-separation). *Let $X_{k(e)}$, $X_{l(e)}$, $\mathbf{X}_{D(e)}$ three disjoint subsets in and R-vine. We say that $\mathbf{X}_{D(e)}$ r-separates $X_{k(e)}$ from $X_{l(e)}$ if there is a dashed edge in some tree that joins two vertices, on of them is associated with the vertices in $X_{k(e)}$ the other to the vertices in $X_{l(e)}$, and $\mathbf{X}_{D(e)}$ r-separates nodes $X_{k(e)}$ and $X_{l(e)}$.*

When $\mathbf{X}_{D(e)}$ v-separates $X_{k(e)}$ and $X_{l(e)}$ in G, we write $I\left(X_{k(e)}, X_{l(e)} \mid X_{D(e)}\right)_G$ to indicate that this graphical conditional independence relationship is represented in the graph G. We write $D\left(X_{k(e)}, X_{l(e)} \mid X_{D(e)}\right)_G$ to indicate that $X_{k(e)}$ and $X_{l(e)}$ are conditionally dependent given $\mathbf{X}_{D(e)}$ in the graph G.

3 Polytrees

3.1 Directed Graphs

Let $G = (\mathbf{X}, \mathbf{E})$ be a directed acyclic graph (DAG), consisting of the node set $\mathbf{X}$ and the edge set $\mathbf{E}$. Directed graphs only contain directed edges. A directed edge from node X to node Y is represented as $X \rightarrow Y$. A path from $X = X_1$ to $Y = X_d$ is a sequence of nodes $X_1, \ldots, X_d$ connected by edges in the graph G, where the edge $(X_i, X_{i+1}) \in \mathbf{E}$, $i = 1, \ldots, d-1$. A cycle is a path where $X = Y$. We say that X is an ancestor of Y if we can find a path that, starting from X, reaches the node Y, such that $X \rightarrow \ldots \rightarrow Y$; correspondingly, Y is a descendant of X. An undirected path is a path in which the directions of the edges are not considered. The skeleton of G is the undirected graph obtained by eliminating the directions of edges from G. A head-to-head (h-h) connection is a subgraph $X \rightarrow Z \leftarrow Y$ in which Z is a h-h node (i.e., a node with convergent edges). A comprehensive introduction to graph theory and graphical models is found in [6].

3.2 Graphical (In)Dependence in DAGs

The concept of d-separation [15] is the graphical independence criterion that provides the DAG a semantic interpretation, allowing it to determine the independence relationships encoded by the topology of the network.

Definition 2 (d-separation). *If $\mathbf{X}$, $\mathbf{Y}$ and $\mathbf{Z}$ are three disjoint subsets of nodes in a DAG G, then $\mathbf{Z}$ d-separates $\mathbf{X}$ from $\mathbf{Y}$ or, similarly, $\mathbf{X}$ and $\mathbf{Y}$ are graphically independent given $\mathbf{Z}$ if and only if along any undirected path between any node of $\mathbf{X}$ and any node of $\mathbf{Y}$ there is an intermediate node A such that (i) eitherA is a head-to-head node in the path and neither A nor its descendants are in $\mathbf{Z}$, or (ii) A is not a head-to-head node in the path and it is in $\mathbf{Z}$.*

When $\mathbf{Z}$ d-separates $\mathbf{X}$ and $\mathbf{Y}$ in G, we write $I(\mathbf{X}, \mathbf{Y} \mid \mathbf{Z})_G$ to indicate that the independence relationship is given by the graph G. We write $D(\mathbf{X}, \mathbf{Y} \mid \mathbf{Z})_G$ to indicate that $\mathbf{X}$ and $\mathbf{Y}$ are conditionally dependent given $\mathbf{Z}$ in the graph G.

3.3 Dependence Models

Necessary definitions on dependence models are taken from [6]. The terms of the dependence and the independence models refer exclusively to the qualitative structure of the relationships existing in a set of variables. These models allow us to check which sets of variables are unconditionally or conditionally dependent or independent.

Definition 3 (Dependence Model). *A model M of the variable set $\{X_1, \ldots, X_n\}$ is called a dependence model if it allows to determine whether $I(\mathbf{X}, \mathbf{Y} \mid \mathbf{Z})_M$ is true for all the possible triples of subsets $\mathbf{X}$, $\mathbf{Y}$ and $\mathbf{Z}$.*

Two possible correspondences between a graphical representation G and a dependence model M are I-map and D-map.

Definition 4 (I-map). *The graph G is an I-map of the dependence model M if $I(\mathbf{X}, \mathbf{Y} \mid \mathbf{Z})_G \Rightarrow I(\mathbf{X}, \mathbf{Y} \mid \mathbf{Z})_M$, i.e., if all independence relationships derived from G are verified in M.*

Definition 5 (D-map). *The graph G is a D-map of the dependence model M if $D(\mathbf{X}, \mathbf{Y} \mid \mathbf{Z})_G \Rightarrow D(\mathbf{X}, \mathbf{Y} \mid \mathbf{Z})_M$, i.e., if all dependence relationships derived from G are verified in M.*

An I-map G of M includes some of the independence relationships of M, but not necessarily all of them. An I-map guarantees that the d-separated nodes correspond to independent variables in M, but does not guarantee that connected nodes correspond to dependent variables in M. On the other hand, a D-map G of M includes some of the dependence relationships of M, but not necessarily all of them. A D-map guarantees that connected nodes correspond to dependent variables in M, but does not guarantee that the d-separated nodes correspond to independent variables in M. Empty graphs (the set of edges is empty) and complete graphs (there is an edge between each pair of nodes) are called trivial D-maps and I-maps respectively.

3.4 Polytrees

Bayesian networks (BNs) are GMs based on DAGs. A BN is a pair $(G(\mathbf{X}, \mathbf{E}), P)$, where G is a DAG, $\mathbf{X}$ and $\mathbf{E}$ are the set of variables (or nodes) and the set of directed edges in G respectively, and $P = \{P(X_1 \mid \mathbf{Pa}_1), \cdots, P(X_n \mid \mathbf{Pa}_n)\}$ is a set of n conditional probability distribution functions (one for each variable) where $\mathbf{Pa}_i$ is the set of parents of X_i in G. The set P defines a probability function given by

$$P(\mathbf{X}) = \prod_{i=1}^{n} P(X_i \mid \mathbf{Pa}_i) \tag{3}$$

A particular subclass of BNs are polytrees (of course, (3) also applies to polytrees). In these networks, there is no more than one undirected path that connects any two nodes. Particular types of polytrees include chains: each node has at most one parent and/or only one child, and trees: each node has only one parent. The number of edges in a polytree is $n-1$. Figure 1-(left panel) shows a polytree where $n=5$ and its respective factorization.

4 The Graphical Relationship Between R-vines and Polytrees

This section proposes two methods that induce the graph of an R-vine from the graph of the polytree and vice versa, so that the resulting graph represents the largest number of independencies existing in the other graph.

4.1 From Polytrees to R-vines

We want to obtain an R-vine G_{R-vine} that represents the largest number of independencies found in the polytree G_P. For this purpose, we propose Algorithm 1. To simplify the notation used in Definition 1, we use $X = X_{k(e)}$, $Y = X_{l(e)}$, and $\mathbf{V} = \mathbf{X}_{D(e)}$. Moreover, $|\mathbf{V}| = j-1$ denotes the cardinality of $\mathbf{V}$, and conditioning sets with cardinality $|\mathbf{V}|$ belong to the tree at the level $j = |\mathbf{V}|+1$.

We consider a dependence model M that contains the list of independencies and dependencies represented in the polytree G_P, L_I and L_D respectively. These lists are obtained via the d-separation criterion in Step 1. The elements of these lists have the form $I(X,Y \mid \mathbf{V})$ and $D(X,Y \mid \mathbf{V})$ respectively. Both lists are arranged in ascending order according to $|\mathbf{V}|$.

In Step 2, we obtain the first tree of G_{R-vine}, which is nothing more than the skeleton of the polytree. Notice that T_1 and the skeleton of the polytree have the same edge set, so that both structures represents the same set of unconditional dependencies.

In Steps 3 and 4, the next trees of G_{R-vine} are built inside a *for-loop* that runs over the levels $j = 2, \ldots, n-1$. These trees are maximum weighted spanning trees (MWSTs) [16] that satisfy the R-vine properties. As the edge's weight we use zero for continuous edges and one for dashed edges respectively. Afterward, each relation in L_D of order $j-1$ suggests an edge, which is inserted in the graph if the following two conditions are satisfied: (i) the edge to be inserted does not introduce undirected cycles, which ensures that the resulting R-vine remains singly connected; (ii) the proximity condition holds. If both conditions are met, the boolean function $\varphi(T_j, D_i(X,Y \mid \mathbf{V}))$ is true.

Note that not all independencies found in the polytree can be represented in the R-vine, but those inserted exist in the polytree. On the other hand, we have that all the independence relationships represented in the R-vine exist in the polytree and we denote it as $I(G_{\text{R-vine}}) \subseteq I(G_{\text{P}})$, thus the R-vine is an I-map of the dependence model M obtained from the polytree.

It is worth noting that if the R-vine is a D-vine – a subclass of R-vines where the trees have chain structure – we do not need to select MWSTs since the first tree determines completely the structure of the next trees [1]).

Example 1. Let us illustrate Algorithm 1 based on the polytree of Fig. 2-(left-panel).

Step 1. Obtain $M = \{L_I, L_D\}$ from G_P.

$$L_I = \{I_1(2,4 \mid 3), I_2(1,4 \mid 3), I_3(1,3 \mid 4), I_4(1,4 \mid 2,3)\}$$
$$L_D = \{D_1(1,3 \mid 2), D_2(1,4 \mid 2), D_3(1,2 \mid 3,4), D_4(1,3 \mid 2,4), D_5(2,3 \mid 1,4)\}$$

Step 2. From T_1 (Fig. 2-(right panel)) we obtain the connected nodes: $1-2$, $2-3$, $3-4$.

Step 3. As T_1 is a chain (as in D-vines) we do not need to built an MWST at each level, but only determine if the edges are dashed or continuous. Next we pass to T_2, and from L_I we see that $I_1(2,4 \mid 3)$ may be represented with the dashed edge 23-‑-34. However, the dashed edges 13-‑-34 and 14-‑-34 corresponding to the relationships $I_2(1,4 \mid 3)$ and $I_3(1,3 \mid 4)$, respectively, cannot be inserted in T_2 since the nodes 13 and 14 do not belong to the node set of this tree, which are: 12, 23, and 34.

Step 4. $D_1(1,3 \mid 2)$ can be represented by the continuous edge $12 - 23$ in T_2 without violating the graphical constraints that should hold an R-vine, which implies that $\varphi(T_2, D_1(1,3 \mid 2))$ is true.

Step 5. To build T_3, only I_4 has to be inserted. This can be done by means of the dashed edge 14 | 2-‑-24 | 3.

Notice that the independence relationships represented in this R-vine structure, namely: $[I_1(2,4 \mid 3), I_2(1,4 \mid 2,3)]$, exist in G_P, thus this R-vine is an I-map of the dependence model obtained from the polytree. This way of building the structure of R-vines guarantees that only graphical independencies found in the polytree are inserted in the corresponding R-vine tree. Consequently, all the independencies represented in the R-vine are true in the polytree.

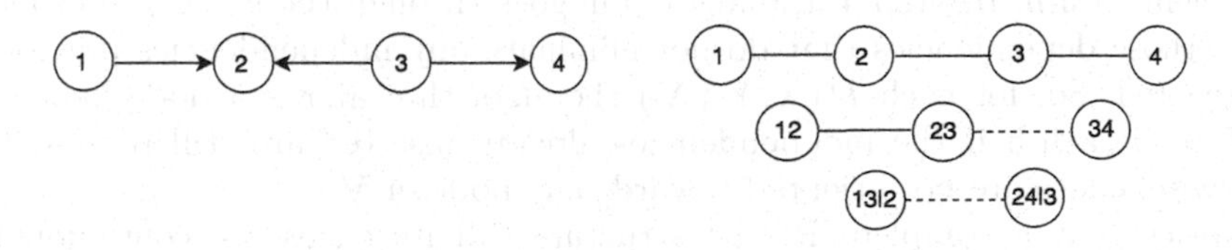

Fig. 2. Illustration of Example 1: (left panel) polytree G_P; (right panel) the resulting R-vine G_{R-vine}.

Algorithm 1. Procedure to build the R-vine tree-structure from the graph of a polytree.

Input: G_P
Output: G_{R-vine}
Step 1 Create the lists L_I and L_D from G_P.
Step 2 Obtain T_1 = skeleton of G_P.
for $j = 2, \ldots, n-1$:
Step 3:
if G_{R-vine} = D-vine:
for each i in $I_i(X, Y \mid V)$ of order $j-1$ in L_I
if $\varphi(T_j, I_i(X, Y \mid \mathbf{V}))$:
Add a dashed edge X--Y in T_j.
else
Build the MWST T_j with dashed edges only.
Step 4:
for i, $D_i(X, Y \mid \mathbf{V})$, in L_D:
if $\varphi(T_j, D_i(X, Y \mid \mathbf{V}))$:
Add a continuous edge $X - Y$ in T_j.

4.2 From R-vines to Polytrees

Similarly to previous section, the idea of this section is to build a polytree G_P that represents the largest numbers of independencies found in an R-vine G_{R-vine}. The heuristic proposed is shown in Algorithm 2.

We consider a denpendence model M that contains the list of independencies and dependencies represented in the R-vine G_{R-vine}, L_I and L_D respectively. Step 1 consists of extracting both lists: L_I is represented by dashed lines and L_D is represented by continuous lines via the v-separation criterion. In Step 2, we obtain the skeleton of G_P that is no other than tree T_1 of G_{R-vine}. To extract L_I and L_D, we can use the procedure given in [6]

Steps 3 and 4 are responsible for determining the direction of the edges of the polytree skeleton. Firstly, the algorithm goes through the list L_I in order to insert the independence relationships. In Step 3, for each $I(X, Y \mid \mathbf{V})$ the edges of the corresponding subgraph $X - \mathbf{V} - Y$ are oriented preventing any node of $\mathbf{V}$ from being a h-h. In Step 4, the algorithm goes through the list L_D in order to insert those dependencies that do not eliminate any independencies previously represented. So, for each $D(X, Y \mid \mathbf{V})$ the algorithm allows a node belonging to $\mathbf{V}$ to be a h-h if the independencies already inserted are still represented; otherwise, edges are not oriented towards any node of $\mathbf{V}$.

Notice that a complete R-vine structure (all its edges are continuous) is a trivial I-map of the dependence model obtained from the polytree. In the opposite case, if all the R-vine's edges are discontinuous or continuous in the first tree, it is a trivial D-map of the dependence model obtained from polytree.

Example 2. Let us illustrate Algorithm 2 based on the R-vine of Fig. 3-(left panel).

Step 1. Obtain $M = \{L_I, L_D\}$ from G_{R-vine}.
$L_I = \{I_1(1,3 \mid 2)\}$
$L_D = \{D_1(2,4 \mid 3), D_2(1,4 \mid 2,3)\}$
Step 2. The skeleton of G_P is the tree T_1 of G_{R-vine}.
Step 3. By representing the independence I_1, the graphs of Fig. 3-(right panel, top) are obtained.
Step 4. To represent $D_1(2,4 \mid 3)$, the node 3 must be considered a head-to-head node. This is done by adding the directed edge $4 \rightarrow 3$. This can be done without adding or removing independencies. However, the relationship D_2 cannot be represented since it can only be inserted as an independence in the polytree. Therefore, the final graph remains as shown in Fig. 3-(right panel, middle).

The resulting polytree, in addition to representing the same independence relationships existing in the R-vine structure, also includes others that are not verified in M (obtained from a polytree). As the independencies of the R-vine are a subset of those of the polytree, we can say that the R-vine is an I-map of M obtained from a polytree, and that the polytree is a D-map of M obtained from an R-vine as all the dependence relationships of the polytree exist in the R-vine. Notice that from the same R-vine, more that one polytree can be obtained, which is illustrated in Fig. 3-(right panel, bottom).

Example 3. Let us illustrate Algorithm 2 based on the R-vine of Fig. 4-(left panel).

Step 1. Obtain $M = \{L_I, L_D\}$ from G_{R-vine}.
$L_I = \{(I_1(2,5 \mid 4), I_2(3,5 \mid 2,4), I_3(1,5 \mid 2,3,4))\}$
$L_D = \{(D_1(1,3 \mid 2), D_2(3,4 \mid 2), D_3(1,4 \mid 2,3))\}$
Step 2. The skeleton of G_P is the tree T_1 of G_{R-vine}.
Step 3. By representing the independence relationships I_1, I_2 and I_3, the graphs of Fig. 4-(middle panel) are obtained. If we take a look at the last graph, we can see that in G_P not only independencies found in G_{R-vine} are represented, but also other independencies that are not visible in L_I. In order to preserve the same independence relationships, in the next step we proceed to insert the conditional dependencies that are in L_D.
Step 4. To represent $D_1(1,3 \mid 2)$, 2 must be considered as head-to-head node, this is done by changing the direction of the edge between the nodes 2 and 3. This can be performed without affecting independence relationships. Analogously, regarding relation D_2, the edge between 2 and 4 is redirected. Nevertheless, D_3 cannot be represented since it would change the existing independencies in the graph. Therefore, the final graph remains as shown in Fig. 4-(right panel).

Algorithm 2. Procedure to build the graph of a polytree from the R-vine tree-structure.

Input: G_{R-vine} consistent
Output: G_P
Step 1 Create the lists L_I and L_D from G_{R-vine}.
for edge e in G_{R-vine}:
  if e is a dashed edge:
    Add $I(X, Y \mid \mathbf{V})$ to L_I.
  else e is a continuous edge:
    Add $D(X, Y \mid \mathbf{V})$ to L_D.
Step 2 Obtain the skeleton of G_P as T_1 of G_{R-vine}.
for $I(X, Y \mid \mathbf{V})$ in L_I:
  if at least one edge between nodes $X, Y, \mathbf{V}$ is an undirected edge:
    Step 3 Orient the edges of the subgraph $X - \mathbf{V} - Y$ without creating a head-to-head node.
for $D(X, Y \mid \mathbf{V})$ in L_D:
  if possible to set some node of $\mathbf{V}$ as a head-to-head node:
    Step 4 Insert the subgraph $X \rightarrow \mathbf{V} \leftarrow Y$.

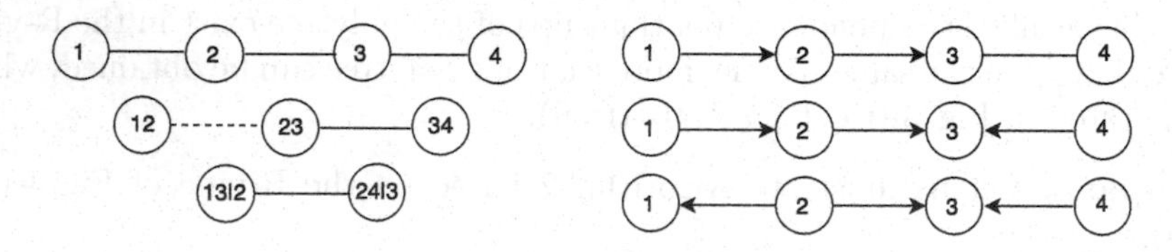

Fig. 3. Illustration of Example 2: (left panel) R-vine G_{R-vine}; (right panel) edge orientation to represent the relationships of $L_I = \{I_1\}$ (top) and $L_D = \{D_1\}$ (middle), and an example of another polytree that can be obtained from the R-vine on the left.

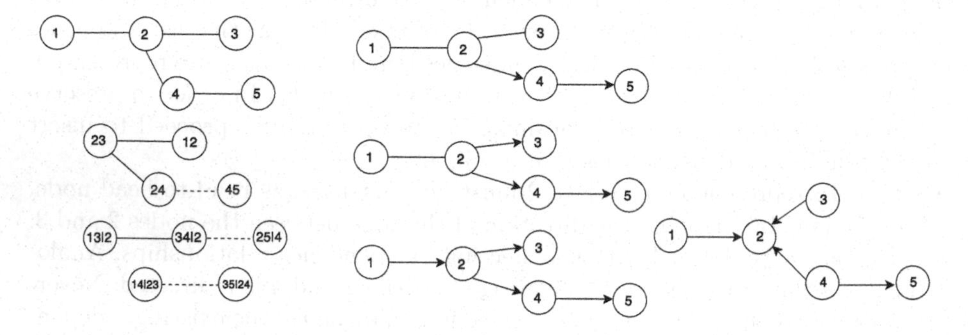

Fig. 4. Illustration of Example 3: (left panel) starting R-vine G_{R-vine}; (middle panel) edge orientation to represent the independence relationships of $L_I = \{I_1, I_2, I_3\}$ (from top to bottom); (right panel) edge orientation to represent the dependence relationships of $L_D = \{D_1, D_2, D_3\}$.

5 Conclusions

In this work, we have studied the connection between the graphical representations of polytrees and R-vines. We have introduced two algorithms for translating between the underlying semantics of polytrees and regular vines from the graphical perspective.

We have shown that we can find an R-vine where all independencies it encodes exist in the polytree, although not all independencies existing in the polytree can be represented in an R-vine. Thus, the R-vine is an I-map of the dependence model obtained from the polytree. On the other hand, given an R-vine, the resulting polytree includes the same independence relationships existing in the R-vine and also others that are not true in the R-vine. As all the dependence relationships inserted in the polytree exist in the R-vine, the polytree is a D-map of the dependence model obtained from the R-vine. An ongoing topic demanding future work is the extension of this study to BNs with undirected cycles.

Acknowledgements. The author would like to thank Dr. Aritz Perez, of Basque Center for Applied Mathematics, BCAM, 48009 Bilbao, Spain, for valuable comments and suggestions. This work is partially supported by the Basque Government (IT609-13 and Elkartek), and Spanish Ministry of Science and Innovation (TIN2016-78365-R). Jose A. Lozano is also supported by BERC 2014–2017 and Elkartek programs (Basque government) and Severo Ochoa Program SEV-2013-0323 (Spanish Ministry of Economy and Competitiveness).

References

1. Aas, K., Czado, C., Frigessi, A., Bakken, H.: Pair-copula constructions of multiple dependence. Insur. Math. Econ. **44**(2), 182–198 (2009)
2. Bauer, A., Czado, C.: Pair-copula Bayesian networks. arXiv:1211.5620 [stat.ME] (2012)
3. Bedford, T., Cooke, R.M.: Probability density decomposition for conditionally dependent random variables modeled by vines. Ann. Math. Artif. Intell. **32**(1), 245–268 (2001)
4. Bedford, T., Cooke, R.M.: Vines - a vew graphical model for dependent random variables. Ann. Stat. **30**(4), 1031–1068 (2002)
5. Brechmann, E.C., Czado, C., Aas, K.: Truncated regular vines in high dimensions with application to financial data. Can. J. Stat. **40**(1), 68–85 (2012)
6. Castillo, E., Gutiérrez, J.M., Hadi, A.S.: Sistemas Expertos y Modelos de Redes Probabilísticas. Monografias de la Academia de Ingeniería (1996)
7. Cowell, G., Dawid, A.P., Lauritzen, S.L., Spiegelhalter, D.J.: Probabilistic Networks and Expert Systems: Exact Computational Methods for Bayesian Networks. Springer, New York (2003). https://doi.org/10.1007/b97670
8. de Campos, L.M.: Independency relationships and learning algorithms for singly connected networks. J. Exper. Theor. Artif. Intell. **10**(4), 511–549 (1998)
9. Haff, I.H., Aas, K., Frigessi, A., Lacal, V.: Structure learning in Bayesian networks using regular vines. Comput. Stat. Data Anal. **101**, 186–208 (2016)
10. Hanea, A.M.: Non-parameteric Bayesian belief nets versus vines. In: Joe, H., Kurowicka, D. (eds.) Dependence Modeling: Vine Copula Handbook, pp. 281–303. World Scientific Publishing (2011)

11. Joe, H.: Families of m-variate distributions with given margins and $m(m-1)/2$ bivariate dependence parameters. In: Rüschendorf, L., Schweizer, B., Taylor, M.D. (eds.) Distributions with Fixed Marginals and Related Topics, pp. 120–141 (1996)
12. Kurowicka, D., Cooke, R.M.: Uncertainty Analysis with High Dimensional Dependence Modelling. Wiley, New York (2006)
13. Lauritzen, L.: Graphical Models. Oxford University Press, Oxford (1996)
14. Nelsen, R.B.: An Introduction to Copulas, 2nd edn. Springer, New York (2006). https://doi.org/10.1007/0-387-28678-0
15. Pearl, J.: Probabilistic Reasoning in Intelligent Systems: Networks of Plausible Inference. Morgan and Kaufmann, San Mateo (1998)
16. Prim, R.C.: Shortest connection networks and some generalizations. Bell Labs Tech. J. **36**(6), 1389–1401 (1957)
17. Sklar, A.: Fonctions de repartition à n dimensions et leurs marges. Publications de l'Institut de Statistique de l'Universite de Paris **8**, 229–231 (1959)

Predicting First-Episode Psychosis Associated with Cannabis Use with Artificial Neural Networks and Deep Learning

Daniel Stamate[1], Wajdi Alghamdi[1(✉)], Daniel Stahl[2], Ida Pu[1], Fionn Murtagh[3], Danielle Belgrave[4], Robin Murray[5], and Marta di Forti[6]

[1] Data Science and Soft Computing Lab, and Department of Computing, Goldsmiths, University of London, London, UK
mapOlwa@gold.ac.uk

[2] Department of Biostatistics and Health Informatics, Institute of Psychiatry, Psychology and Neuroscience, King's College London, London, UK

[3] School of Computing and Engineering, University of Huddersfield, Huddersfield, UK

[4] Microsoft Research Cambridge, Cambridge, UK

[5] Department of Psychosis Studies, Institute of Psychiatry, Psychology and Neuroscience, King's College London, London, UK

[6] MRC Social, Genetic and Developmental Psychiatry Centre, Institute of Psychiatry, Psychology and Neuroscience, King's College London, London, UK

Abstract. In recent years, a number of researches started to investigate the existence of links between cannabis use and psychotic disorder. More recently, artificial neural networks and in particular deep learning have set a revolutionary wave in pattern recognition and machine learning. This study proposes a novel machine learning approach based on neural network and deep learning algorithms, to developing highly accurate predictive models for the onset of first-episode psychosis. Our approach is based also on a novel methodology of optimising and post-processing the predictive models in a computationally intensive framework. A study of the trade-off between the volume of the data and the extent of uncertainty due to missing values, both of which influencing the predictive performance, enhanced this approach. Furthermore, we extended our approach by proposing and encapsulating a novel post-processing k-fold cross-testing method in order to further optimise, and test these models. The results show that the average accuracy in predicting first-episode psychosis achieved by our models in intensive Monte Carlo simulation, is about 89%.

Keywords: First-episode psychosis · Precision medicine · Cannabis use
Prediction modelling · Classification · Neural network · Deep learning
Post-processing · Monte Carlo simulation · Missing data based uncertainty

J. Medina et al. (Eds.): IPMU 2018, CCIS 855, pp. 691–702, 2018.
https://doi.org/10.1007/978-3-319-91479-4_57

1 Introduction

An estimated 183 million people consumed cannabis in 2014 [1] making it the most popular illicit drug in the world. Legalising cannabis, especially in countries such as the Netherlands and Uruguay, and in some states of USA, and the increasing lobby for making cannabis use legal in other countries such as Canada, is an important contributing factor for the popularity of this drug. On the other hand evidence shows that the increase in cannabis consumption is proportionate to the increase in the proportion of people seeking treatment for psychotic disorders [1]. While there is some evidence that consuming cannabis is a risk factor for several types of psychotic disorders [2], the link between these two factors needs to be better quantified.

These days, researchers attempted to understand whether specific patterns of cannabis use such as potency or age are associated with a higher risk of developing psychotic disorders. One study concluded that nearly a quarter of all new psychosis patients in South London (UK) could be associated with the use of high-potency, skunk-like cannabis [3]. Another study [4] estimated that if a person uses cannabis daily for more than six months, then there is a 70% likelihood that this person will suffer from psychotic disorders.

There are few such studies based on risk prediction modelling using advanced machine learning algorithms establishing a link between cannabis use and first-episode psychosis – in fact we are not aware of the existence of other studies apart our recent work [4]. Most studies so far rely only on explanatory research strategies and are mainly based on a number of conventional statistical techniques such as hypotheses formulation and verification via statistical tests, logistic regression modelling, etc. These techniques are well-recognised and used in medical research, but in many situations, they do not match the high potential of machine learning methods. The domain of machine learning has developed at an enormous speed in recent years, with advanced predictive techniques being expanded and improved upon. In particular artificial neural networks and especially deep networks, which are state of the art in prediction, have proven their abilities in many pattern recognition and machine learning applications. One such field of implementation is the domain of medical research [5, 6].

On the one hand, artificial neural networks have been successfully used in understanding the heterogeneous manifestations of asthma [7], diagnosing tuberculosis [8], classifying leukaemia [9], detecting heart conditions in ECG data [10], etc. These studies show that neural networks have been proven to be capable of dealing with complicated medical data such as the ambiguous nature of the ECG signal data, where neural networks show some outstanding results compared to other methods.

On the other hand, recently, deep networks have attracted widespread attention, mainly by defeating alternative machine learning methods such as support vector machines in numerous critical applications such as classifying Alzheimer's disease [11], classifying AD/MCI patients [12], and improving palliative care [13]. While support vector machines are still popular techniques within the machine learning community [4, 14], the family of deep learning techniques are gaining considerable attention [15]. Deep learning methods are types of representation learning methods, which can automatically identify the optimal representation of raw data without requiring prior feature selection.

In this study, we propose a novel machine learning approach based on neural networks and deep learning techniques to develop predictive models for the onset of first-episode psychosis. The dataset that we based our study upon was collected by psychiatry practitioners, and used in previously conducted studies such as [3, 4]. It comprises an extensive set of variables including demographics, drug-related, and several other variables with specific information on the participants' history of cannabis use as seen in Table 1.

Table 1. Cannabis use attributes among other attributes in the analysed dataset

Attribute	Description
lifetime_cannabis_user	Ever used cannabis: yes or no
age_first_cannabis	Age when first used cannabis: 7 to 50
age_first_cannabis_under15	Age less than 15 when first used cannabis: yes, no or never used
age_first_cannabis_under14	Age less than 14 when first used cannabis: yes, no or never used
current_cannabis_user	Current cannabis user: yes or no
cannabis_fqcy	Pattern of cannabis use: never used, only at weekends, or daily
cannabis_measure	Cannabis usage measure: none, hash less than once per week, or hash at weekends, hash daily, skunk less than once per week, or skunk at weekends, skunk daily
cannabis_type	Cannabis type: never used, hash, or skunk
duration	Cannabis use duration: 0 to 41 (months)

Our approach features a gradual control of the limitation of the uncertainty present in the data due to missing values which are usually inherent in clinical datasets due to patients missing appointments, patients not reporting all details, etc. This feature involves considering different thresholds for allowed levels of missingness (per attributes and per records) in the data sets, that we call cutting points, in order to examine how the prediction models' performances may vary with these thresholds. Our approach is based also on a novel methodology of optimising and post-processing the predictive models in a computationally intensive framework. Furthermore, we extended our approach by proposing and encapsulating a novel post-processing k-fold cross-testing method in order to further optimise, and test these models. The results show that the accuracy in predicting first-episode psychosis achieved by our best models in intensive Monte Carlo simulation, falls between 85.13% and 91.54%, with an average of about 89%.

2 Methods

2.1 The Clinical Data

The data used to develop our novel approach to predict the first-episode psychosis is a part of a case-control study at the inpatient units of the South London and Maudsley (SLaM) NHS Foundation Trust in United Kingdom [3]. The clinical data consists of

1106 records, including 489 patients, 370 controls and 247 unlabelled records. Those described as patients were patients of the Trust who at one time presented with first-episode psychosis; controls were healthy people recruited from the local area. Each record refers to a participant in the study and has 255 possible attributes, which were divided into four categories. The first category consists of demographic attributes which represent general features such as gender, race, and level of education. Secondly, drug-related attributes contain information on the use of non-cannabis drugs such as tobacco, stimulants and alcohol. The third category is formed of genetic attributes which were removed from the analysis for the purpose of this study. The final category contains cannabis-related attributes such as the duration of use, initial date of use, frequency, cannabis type, etc. (see Table 1).

2.2 Rationalisation and Refinement

The goal of this stage is to perform a high-level simplification of the dataset, and it embraces several steps. First, records that were missing critical data were removed from the dataset. This included records with missing labels as well as records with missing values on all cannabis-related variables. Secondly, certain variables were removed from the dataset. This primarily involved variables that were deemed to be irrelevant to the study (such as those related to individual IDs of the study participants), and also variables which were outside the scope of the current study (for example, certain gene-related variables). In addition, any numeric predictors that had zero or near-zero variance were dropped. Thirdly, we sought to make the encoding of missing values consistent across the dataset. Prior to this step, values including 66, 99, and −99 all represented cases with missing values – so all such indicators were replaced with a consistent missing value indicator, NA. Fourthly, some variables were re-labelled to provide more intuitive descriptions of the data contained within. Finally, since in multiple situations some variables had a similar meaning, yet there were often missing values for some records in some of these variables, a process of imputation was used to effectively combine the information from related variables into one. For example, two variables described alcohol use but were inconsistently present across the records and presented missing values. These were combined in a way that created one single variable with consistent and as complete as possible values. Such a process was used to generate value-reacher and value-consistent variables related to alcohol use, tobacco use, employment history, and subjects' age.

2.3 A Trade-off Between the Extent of Missing Values and the Dataset Size

A trade-off between the extent of missing values present in the dataset, and the dataset size, needed to be investigated from the point of view of the predictive power of the models that can be built on the dataset. The intuition is that by using a larger subset of the available dataset in the analysis, one would obtain a positive effect on the performance of predictive models (since more data is used to build the models). But this larger subset may also encapsulate more uncertainty due to the presence of more missing values, which usually has a negative effect on the predictive models (even with imputation). Therefore, different cutting points, defined as the thresholds for the percentage of missing values (or level of missingness) allowed in attributes and records, respectively, were considered in order to study the variation of the predictive power of subsets of the dataset. Attributes and records presenting some levels of missingness up to the respective cutting points or thresholds, respectively, were kept in the dataset, and the remaining ones were removed. The considered cutting points for the records were 10%, 20%, …, 100%. For instance 30% in this grid means that we keep in the dataset only the records that have up to 30% missing values (and 100% means practically that all records are kept in the dataset). Moreover, the cutting points for the attributes were identified by first determining the percentage of missing values for each attribute, and then ordering these percentages and splitting them into twenty equal groups. The extreme values in each group formed the cutting points for the attributes.

Overall, these cutting points were applied to the dataset and compared with respect to the performance of single-layer neural network tuned models, in an attempt to determine optimal cutting points which were those for which these models had the highest accuracy. Once these cutting points were determined, they were applied, and a final dataset was thus obtained as the outcome of a trade-off between the extent of missing values present in the dataset, and the dataset size.

How did we exactly proceed to obtain this final dataset? Note that we don't do a full optimisation on all pairs of cutting points for attributes and records to determine this final dataset (because training and tuning neural networks is a computationally expensive procedure), but we just apply a heuristic in our framework. Initially we search for an optimal value among all the attribute cutting points, and we apply it on the dataset. In our case this was 92%. Then, on the resulting dataset, we applied different record cutting points following the grid mentioned above, and we determined the best cutting point, which was 70% in our case. To compare the cutting points and select the best ones, the criterion was the accuracy of the single-layer neural networks which have been tuned on the training set (70% of the data), in a 5-fold cross-validation procedure, on a 10×10 grid for the number of hidden units, and decay values to prevent overfitting with regularisation methods. Random forest imputations of missing values were applied. The models' performances consisting of accuracy and kappa were estimated on the test set (30% of the data).

Figure 1 illustrates the process, in which we observed a decrease in the performance when all the attributes were included or when the cannabis attributes were not present in the obtained dataset.

By applying the determined 92% cutting point for the attributes and 70% cutting point for the records to the original dataset, we obtained 107 attributes and 628 records divided into 360 patients and 268 controls, on which the main phase of predictive modelling with various algorithms was developed, and presented in what follows. We note that the proportion of controls and patients in the final dataset are approximately the same as in the original dataset, so the current dataset is representative.

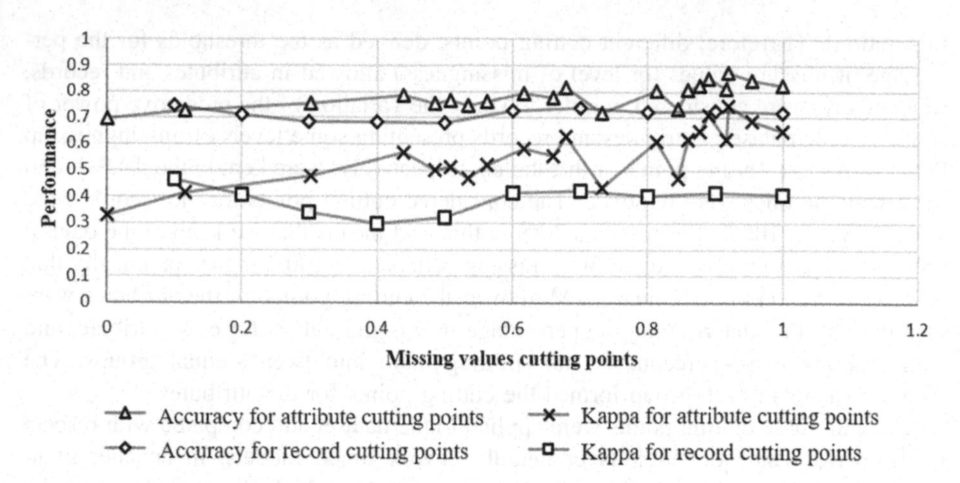

Fig. 1. Model performance for record and attribute cutting points

2.4 Imputation

Missing values' presence in clinical data is rather common due to reasons explained above, and this is the case also of our dataset. The predictive power in the data may depend significantly on the way missing values are treated. While some machine learning algorithms, such as decision trees [16], have the capability to handle missing data outright, most machine learning algorithms do not. In many situations missing values are imputed using a supervised learning technique such as k-Nearest Neighbour (KNN) after suitable scaling to balance the contribution of the numeric attributes. These imputation techniques do not have theoretical formulations but have been much implemented in practice [4, 6]. In this work, we considered different imputations such as the KNN imputation, the tree bagging imputation from the caret package [16], and the random forest imputation from the randomForest package [17]. The last method led to the best results in terms of the performance of the predictive models finally built, although it was more computationally expensive.

2.5 Training and Optimizing (Tuning) Predictive Models

For the purpose of developing optimised predictive models for the first-episode psychosis, the values of the parameters for each of the considered algorithms have been controlled by chosen grids. Predictive models have been fitted, in a 5-fold cross-validation procedure, on each training set after pre-processing techniques were applied on the same training set, and have been tested on each test set. Models based on neural networks with a single-layer, neural networks with multi-hidden-layers, and deep networks, were optimized (tuned) based on maximizing AUC, the area under the ROC curve.

The single-layer neural networks were tuned over 10 values of the size (i.e. the number of hidden units) and 10 values of the decay (i.e. the weight decay), which is the parameter in the penalization method for model regularization to avoid overfitting, similar to the penalization method in ridge regression, based on the L2 norm [16]. The optimal values were 3 and 0.01, respectively. The neural network with multi-hidden layers were tuned over 10 values for each of the 3 hidden layers (i.e. 10 values for the number of hidden units in each layer), and 10 values for the decay. The optimal values were 5, 5, 5 for the 3 layers, and 0.01 for decay, respectively.

As for the deep networks, we employed the H2O's deep learning, which is based on a multi-layer feedforward artificial neural network that is trained with stochastic gradient descent using back-propagation [19]. The deep networks usually contain a large number of hidden layers consisting of neurons with *tanh*, *rectifier*, and *maxout* activation functions. This type of models has many parameters, but it was designed to reduce the number of parameters that the researcher has to specify by applying feature selection and early stopping techniques. We used deep networks with the method of Gedeon [18] to select the best attributes. In our experiments, the early stopping was set to let it stop automatically once the area under the curve AUC does not continue improving, in particular, when AUC does not improve by at least 1% for 10 consecutive scoring events.

Also, a grid optimisation was used with the parameters that need to be tuned such as the activation function, the number and sizes of the hidden layers, the number of epochs, and the 2 parameters corresponding to the L1 and L2 regularisations for preventing overfitting.

The models were tuned over all activation functions, and over 3, 4, …, 25 layers and 30, 35, …, 50 layers. The number of units in each layer had the values 50, 100,…, 250. Also, we used the values 2, 3, 5, and 10 for tuning the number of epochs. Finally, the parameters for the L1 and L2 regularisations were each tuned over the values 10^{-1}, 10^{-2},…, 10^{-10}.

After performing the proposed techniques, the optimal values selected for the deep learning model are *rectifier* as an activation function, 5 epochs, and 8 hidden layers of 200 neurons each. As for the L1 and L2 parameters, the optimal values were 10^{-4} and 10^{-5}, respectively.

2.6 Sampling and Post-processing K-fold Cross-Testing

When there is a priori knowledge of a class imbalance, one direct method to reduce its influence on model training is to select training set samples to have roughly equal event rates [16]. Treating data imbalance usually leads to better predictions models and better trade-off between sensitivity and specificity.

In this study, we considered three sampling approaches to subsample the training data in a manner that mitigates the imbalance problem. The first approach is down-sampling in which we sampled (without replacement) the majority class to be the same size as the minority class. The second method is up-sampling in which we sampled (with replacement) the minority class to be the same size as the majority class. The last approach we used is the synthetic minority over-sampling technique (SMOTE) [20]. SMOTE selects a data point randomly from the minority class, and the K-nearest neighbours to that point are determined and used to generate new synthetic data points by slight alterations to these data points. Five neighbours are used in our analysis. The results show that the up-sampling procedure had no real improvement on AUC or the accuracy performances. Simple down-sampling of the data also had no positive effect on the model performances. However, SMOTE with neural network models has led to an increase in AUC and accuracy.

Figure 2 gives an overall description of the methodology followed here, based on pre-processing, model optimisation, and post-processing. The dataset is randomly split, with stratification, in 60% and 40% parts denoted here by D1 and D2, respectively. D1 is used for training and for optimising the model, as explained in Subsect. 2.5, in a cross-validation fashion, with AUC as optimisation criterion, with and without class balancing. Different pre-processing methods such as missing values imputation and sampling methods that we have explained above, were appropriately integrated into the cross-validation. The optimal model obtained on D1 was then applied to score D2 accounting for the remaining 40% of the dataset. In order to further enhance the model performance, a specially designed post-processing procedure that we introduce here, was applied with the optimised model using D2 dataset. We call it the *k-fold cross-testing method*. In this procedure, we produce k post-processed model variants of the original optimised model. First, we create k stratified folds of D2 dataset. Then, k-1 folds are used to find an alternative probability cut-off on the ROC curve such as the cut-off associated with the largest accuracy. The remaining one-fold is scored with the post-processed model based on the newly found cut-off point. Finally, the whole procedure is repeated until all folds are used for scoring at their turn, then the predictions are integrated, and the model performance is measured on the whole scored dataset D2. We note here as an important remark that in each such iteration of the procedure, the ROC optimisation data (the $k - 1$ folds) and the scored data (the remaining fold) are always distinct, so the data for model post-processing and the data for scoring are always distinct.

2.7 Monte Carlo Simulations

Due to expected potential variations of the predictive models' performance, depending on the datasets for training and testing, but in particular due to the uncertainties introduced by the missing values in the data, we conducted extensive Monte Carlo simulations to study these variations, and the stability of the models. In particular, the simulations for each single-layer neural networks, multi-layers neural networks and deep networks consisted of 2,000 iterations of the procedure included in the bold contour box of Fig. 2. The models' performances consisting of accuracy, sensitivity, specificity, and kappa were evaluated in each iteration. The aggregation of all iterations formed various distributions of the above performance measures. These distributions were visualised using histograms to capture the performance capability and stability of models, as shown in the Results section.

2.8 Hardware and Software

The Monte Carlo simulations that we conducted as explained above are computationally very expensive procedures, therefore a robust framework was required. Parallel processing was performed on a data analytics cluster of 11 servers with Xeon processors and 832 GB fast RAM. The R software was used with a number of packages, including *caret, pROC, e1071, randomForest, ggplot2, plyr, DMwR, AppliedPredictiveModeling, doParallel* and H_2O.

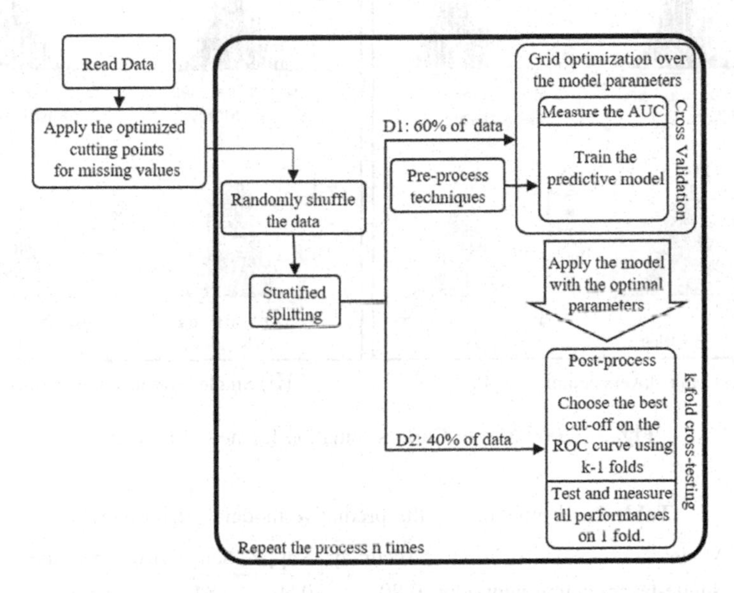

Fig. 2. Summary of the implemented methodology, with k-fold cross-testing method

3 Results

We present here the performances obtained with our approach to predicting first-episode psychosis, investigated with Monte Carlo simulations, as explained above. We should note that, due to lack of space, in this section we only report results regarding models which either are not post-processed, or are post-processed with ROC optimisation based on the largest accuracy cut-off methodology.

The results show that the single-layer neural network scored a mean accuracy of 0.80 (95% CI [0.76, 0.84]) and a mean sensitivity of 0.84 (95% CI [0.76, 0.91]). Also, the multi-layers neural networks achieved a mean accuracy of 0.81 (95% CI [0.77, 0.85]) and a mean sensitivity of 0.85 (95% CI [0.77, 0.92]). Figure 3 shows histogram plots of the Monte Carlo simulations for single and multi-layer neural networks with post-processing and performances evaluated with our k-fold cross-testing method. Results indicate that the difference between single and multi-layer neural networks is not significant regarding the 4 performances (Table 2).

As for deep learning, the results show significantly better performances. Figure 4 illustrates histogram plots of the 2000 Monte Carlo simulations for models based on deep networks without the post-processing (left) and with post-processing (right). The results for the latter show a mean accuracy of 0.89 (95% CI [0.85, 0.92]) and a mean sensitivity of 0.83 (95% CI [0.74, 0.92]).

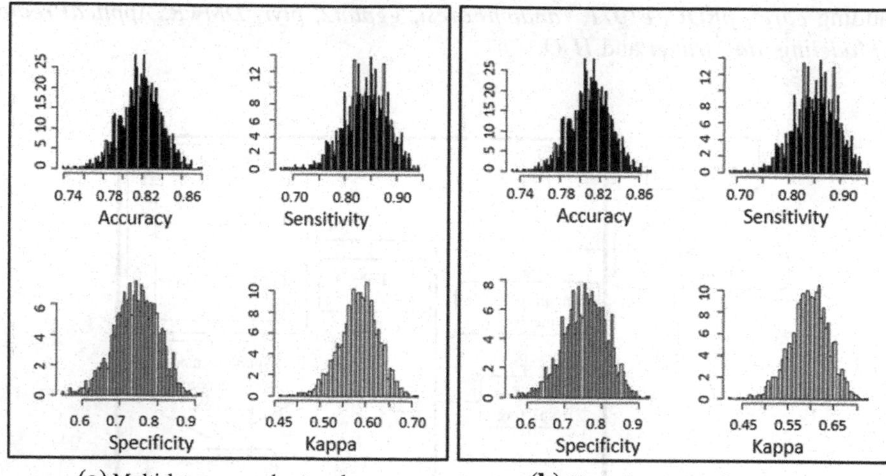

(a) Multi-layers neural networks **(b)** Single layer neural networks

Fig. 3. 2000 Monte Carlo simulation for neural networks.

Table 2. Estimations of the predictive models' performances.

Model	Accuracy	Kappa	Sensitivity	Specificity
Single-layers neural networks	0.80	0.59	0.84	0.74
Multi-layers neural networks	0.81	0.60	0.85	0.75
Deep networks	0.89	0.76	0.83	0.93

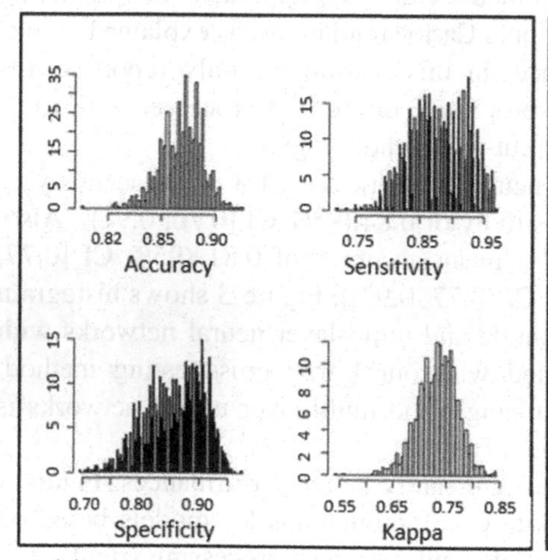

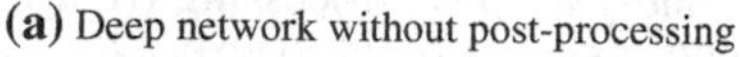

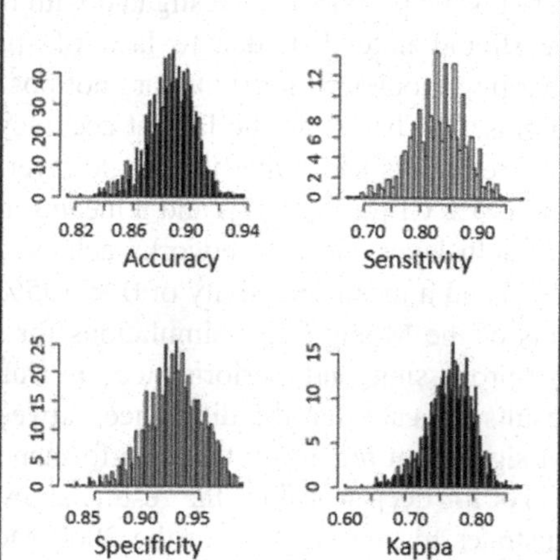

(a) Deep network without post-processing **(b)** Deep network with post-processing

Fig. 4. 2000 Monte Carlo simulation for deep networks.

Overall, we remark a good predictive power and stability of these models, based on an acceptable level of variation of their performance measures evaluated across extensive Monte Carlo experiments. As mentioned before, a significant proportion of this variation may be explained by the uncertainties due to the presence of missing values in the dataset.

4 Conclusion and Directions for Further Work

The aim of this work has been to propose a novel machine learning approach to developing predictive models for the onset of the first-episode psychosis with neural networks and deep learning. To our knowledge, previous studies on the link between cannabis use and first-episode psychosis investigated this highly important relationship via conventional statistical methodologies and techniques and did not tackle the predictability of this condition in relation to the cannabis use. An exception is [4] which is the first study to predict first episode-psychosis using machine learning based on support vector machines, bagged trees, boosted classification trees, eXtreme gradient boosting and random forests. However, the accuracy performances in [4] were slightly under 80%, and as such, under all neural and deep network models' performances achieved in this work.

In this paper, we successfully classified first-episode psychosis from normal control with 89% accuracy using deep learning. This solution proves the high potential of applicability of machine learning, in particular deep learning, in Psychiatry, and enables researchers and doctors to evaluate the risk for and predict first-episode psychosis.

Our approach features a gradual control of the limitation of the uncertainty present in the data by investigating a trade-off between the extent of missing values entailing uncertainty, and the dataset size. Moreover, due to expected potential variations of the predictive models' performance due to the uncertainties entailed by the remaining missing values in the data, we conducted extensive Monte Carlo simulations to study these variations, and the stability of the models.

A potential work direction concerns including genotype data in the study for prediction purposes, and redefining the predictive modelling approach by taking into account the particularities of the newly introduced data, such as the high dimensionality.

References

1. United Nations Office on Drugs and Crime, World Drug Report, (United Nations publication, Sales No. E.16.XI.7) (2016)
2. Radhakrishnan, R., Wilkinson, S., Dsouza, D.: Gone to pot: a review of the association between cannabis and psychosis. Front. Psychiatry **5**, 54 (2014)
3. Di Forti, M., Marconi, A., et al.: Proportion of patients in south London with first-episode psychosis attributable to use of high potency cannabis: a case-control study. Lancet Psychiatry **2**(3), 233–238 (2015)

4. Alghamdi, W., Stamate, D., et al.: A prediction modelling and pattern detection approach for the first-episode psychosis associated to cannabis use. In: IEEE ICMLA, vol. 15, pp. 825–830 (2016)
5. Zhou, H., Tang, J., Zheng, H.: Machine learning for medical applications. Sci. World J. **20**, 1 (2015)
6. Iniesta, R., Stahl, D., McGffin, P.: Machine learning, statistical learning and the future of biological research in psychiatry. Psychol. Med. **46**(12), 2455–2465 (2016)
7. Stamate, D., Cassidy, R., Belgrave, D.: Predictive modelling strategies to understand heterogeneous manifestations of asthma in early life. In: IEEE ICMLA, vol. 16 (2017)
8. Elveren, E., Yumuşak, N.: Tuberculosis disease diagnosis using artificial neural network trained with genetic algorithm. J. Med. Syst. **35**, 329–332 (2011)
9. Adjouadi, M., Ayala, M., et al.: Classification of leukemia blood samples using neural networks. Ann Biomed Engineering **38**(4), 1473–1482 (2010)
10. Yan, Y., Qin, X., et al.: A restricted Boltzmann machine based two-lead electrocardiography classification. In: Proceedings of 12th International Conference on Wearable Implantable Body Sensor Networks (2015)
11. Sarraf, S., Tofighi, G.: Classification of Alzheimer's Disease Using fMRI Data and Deep Learning Convolutional Neural Networks, arXiv:1603.08631 (2016)
12. Li, F., Tran, L., et al.: Robust Deep Learning for Improved Classification of AD/MCI Patients. Int. Workshop on Machine Learning in Medical Imaging, p. 247 (2014)
13. Avati, A., Jung, K., et al.: Improving Palliative Care with Deep Learning (2017)
14. Katrinecz, A., Stamate, D., et al.: Predicting psychosis using the experience sampling method with mobile apps. In: IEEE ICMLA, vol. 16 (2017)
15. Ravì, D., Wong, C., et al.: Deep learning for health informatics. IEEE J. Biomed. Health Inf. **21**(1), 4–21 (2017)
16. Kuhn, M., Johnson, K.: Applied Predictive Modelling. Springer, New York (2013). https://doi.org/10.1007/978-1-4614-6849-3
17. Liaw, A., Wiener, M.: Classification and regression by randomForest. R News **2**(3), 18–22 (2002)
18. Aiello, S., Eckstrand, E., et al.: Machine Learning with R and H2O (2016)
19. Candel, A., Parmar, V., LeDell, E., Arora, A.: Deep Learning with H2O (2015). http://h2o.ai/resources
20. Qazi, N., Raza, K.: Effect of feature selection, smote and under sampling on class imbalance classification. In: 2012 UKSim 14th, pp. 145–150 (2012)

On Cantor's Theorem for Fuzzy Power Sets

Michal Holčapek([✉])

CE IT4Innovations, Institute for Research and Applications of Fuzzy Modeling, University of Ostrava, 30. dubna 22, 701 03 Ostrava, Czech Republic
Michal.Holcapek@osu.cz
http://irafm.osu.cz

Abstract. The aim of the paper is to introduce the concept of fuzzy power set in a universe of sets and investigate its basic properties. We focus here on an analysis of Cantor's theorem for fuzzy sets, which states in the set theory that the cardinality of a set is strictly smaller then the cardinality of its power set. For our investigation of Cantor's theorem we chose two types of equipollency of fuzzy sets, particularly, the binary Cantor's equipollence and its graded version.

Keywords: Cardinal theory · Fuzzy sets · Universe of sets
Fuzzy power sets · Cantor's theorem

1 Introduction

In the elementary set theory, the cardinality of the power set of a set x is strictly greater than the cardinality of the original set x. Symbolically, we write $|x| < |\mathrm{P}(x)|$, where $|x|$ denotes the cardinality of the set x and $|\mathrm{P}(x)|$ the cardinality of the power set of x. This fundamental result is known as Cantor's theorem and has been used to demonstrate that there are sets having cardinality greater than the infinite cardinality of the set of natural numbers. In literature on the set theory, Cantor's theorem is sometimes formulated as there is no function from x onto $\mathrm{P}(x)$ or x is not equipollent $\mathrm{P}(x)$, which is also referred as a more general form of Cantor's theorem. For the purpose of this contribution, we consider the last formulation of Cantor's theorem.

In the standard fuzzy set theory, we can distinguish two concepts: the *power set* and the *fuzzy power set* of a fuzzy set A over a universe of discourse x. The power set of a fuzzy set $A : x \longrightarrow [0,1]$ is the classical set, denoted by $\mathscr{F}(A)$, consisting of all fuzzy subsets of A, where a fuzzy set $B : x \longrightarrow [0,1]$ is a fuzzy subset of A if $B(z) \leq A(z)$ holds for any $z \in x$. A generalization of this concept can be found in the theory of categories under the name of (fuzzy) powerset operator [10,12]. An extension of the power set of a fuzzy set to the fuzzy power set has been proposed by Bandler and Kohout in [1]. In this paper, the fuzzy power set of a fuzzy set A, denoted by $\mathscr{P}(A)$, is defined as a fuzzy set $\mathscr{P}(A) : \mathscr{F}(x) \longrightarrow [0,1]$, where $\mathscr{F}(x)$ is the power set of the fuzzy set x (each

J. Medina et al. (Eds.): IPMU 2018, CCIS 855, pp. 703–714, 2018.
https://doi.org/10.1007/978-3-319-91479-4_58

classical set can be considered as a special fuzzy set), and $\mathscr{P}(A)(B)$ expresses the membership degree in which B belongs to the power set of A, or equivalently, the truth degree of the statement saying that B is a fuzzy subset of A. Using a fuzzy implication operator $\rightarrow$ on $[0,1]$,[1] Bandler and Kohout defined the value of $\mathscr{P}(A)(B)$ as

$$\mathscr{P}(A)(B) = \bigwedge_{z \in x} (B(x) \rightarrow A(x)), \tag{1}$$

where $\bigwedge$ denotes the infimum operation in $[0,1]$. One can see that $B \in \mathscr{F}(A)$ if and only if $\mathscr{P}(A)(B) = 1$. The fuzzy power sets can be extended also for fuzzy sets whose membership degrees are interpreted in more general algebras of truth values. As an example, let us mention the development of lattice-valued set theory provided by Takeuti and Titani in [11] (see also [4]).

In this contribution, we deal with fuzzy sets whose universes of discourse belong to a given universe of sets (e.g., the class of all sets or finite sets; or a set known as a Grothendieck universe). Note that the universe of sets has been introduced in [8] to form a framework for development of fuzzy set theory. The concept of fuzzy power set, which is sound in each universe of sets, has been introduced in [7] and admits only classical (crisp) sets in the universe of discourse of the fuzzy power set. This restriction to crisp sets ensures that each fuzzy power set becomes a fuzzy set in the given universe of sets, which is not true in general, if one admits also fuzzy sets as in the case of Bandler-Kohout definition. A typical example is the fuzzy power set of a fuzzy set over a finite set with the membership degrees interpreted in an infinite algebraic structure of truth values, which does not belong to the universe of all finite sets. For our analysis of Cantor's theorem within the fuzzy set theory, we introduce two types of equipollence for fuzzy sets. The first type of equipollence is a binary class relation on the class of all fuzzy sets in a universe of sets stating that two fuzzy sets have or have not the same cardinality. The second type of equipollence is a graded version of the first type (a fuzzy class relation) and its definition has been proposed in [8] and further developed in [5,6] (see also [7] for finite fuzzy sets).

The main goal of this contribution is to show that Cantor's theorem is valid (valid in a weaker form) for fuzzy sets and proposed fuzzy power sets in each universe of sets if the first (second) type of equipollence is considered.

The paper is structured as follows. The next section introduces basic concepts that are used in the main part of the contribution. The third section is devoted to Cantor's theorem whose validity is verified for Cantor's equipollence. The fourth section provides the proof of Cantor's theorem for graded Cantor's equipolence.

[1] The fuzzy implication operator on $[0,1]$ is often modeled in fuzzy logic as a residuum operation on a complete residuated lattice on $[0,1]$ (see Subsect. 2.1).

2 Preliminaries

2.1 Algebraic Structures of Truth Values

A complete linearly ordered residuated lattice is considered as a structure of membership degrees for fuzzy sets. Recall that a *residuated lattice* is an algebra $\mathbf{L} = \langle L, \wedge, \vee, \otimes, \rightarrow \bot, \top \rangle$ with four binary operations and two constants, for which it holds that

(i) $\langle L, \wedge, \vee, \bot, \top \rangle$ is a bounded lattice, where $\bot$ is the least element and $\top$ is the greatest element of L, respectively,
(ii) $\langle L, \otimes, \top \rangle$ is a commutative monoid,
(iii) the pair $\langle \otimes, \rightarrow \rangle$ forms an adjoint pair, i.e.,

$$a \leq b \rightarrow c \quad \text{if and only if} \quad a \otimes b \leq c \tag{2}$$

holds for each $a, b, c \in L$ ($\leq$ denotes the corresponding lattice ordering).

A residuated lattice is said to be *complete* (*linearly ordered*) if the corresponding lattice $\langle L, \wedge, \vee, \bot, \top \rangle$ is a complete (linearly ordered) lattice. Details and examples of residuated lattices can be found in [2,9].

2.2 Fuzzy Sets in a Universe of Sets

A fuzzy set is usually defined as a function from a fixed non-empty universe of discourse to a set (lattice) of truth values. Nevertheless, the fuzzy set constructions like fuzzy power sets or exponentiation of fuzzy sets requires a system of universes of discourse rather than one fixed universe (cf., [3]). This motivated us to introduce a universe of sets over a complete residuated lattice as a basic framework for our fuzzy set theory [8]. In what follows, we use $x \in y$ to denote that the set x is a member of set y, further, we use $\mathrm{P}(x)$, $\mathscr{D}(f)$ and $\mathscr{R}(f)$ to denote the power set of a set x, the domain and the range of a function f, respectively.

Definition 1. *Let $\mathbf{L}$ be a complete linearly ordered residuated lattice. A universe of sets over $\mathbf{L}$ is a non-empty class $\mathfrak{U}$ of sets in the Zermelo–Fraenkel set theory with the axiom of choice (ZFC) satisfying the following properties:*

(U1) $x \in y$ and $y \in \mathfrak{U}$, then $x \in \mathfrak{U}$,
(U2) $x, y \in \mathfrak{U}$, then $\{x, y\} \in \mathfrak{U}$,
(U3) $x \in \mathfrak{U}$, then $\mathrm{P}(x) \in \mathfrak{U}$,
(U4) $x \in \mathfrak{U}$ and $y_i \in \mathfrak{U}$ for any $i \in x$, then $\bigcup_{i \in x} y_i \in \mathfrak{U}$,
(U5) $x \in \mathfrak{U}$ and $f : x \longrightarrow L$, then $\mathscr{R}(f) \in \mathfrak{U}$,

where L denotes the support of $\mathbf{L}$.

Basic examples of the universes of sets are the classes of all or finite sets. If the ZFC is extended by the axiom admitting the existence of strongly inaccessible cardinals, one can introduce a universe of sets over $\mathbf{L}$ to be a Grothendieck universe.

Definition 2. *Let $\mathfrak{U}$ be a universe of sets over* $\mathbf{L}$. *A function $A : z \longrightarrow L$ (in ZFC) is called a* fuzzy set *in $\mathfrak{U}$ if $z \in \mathfrak{U}$.*

Let $A : z \longrightarrow L$ be a fuzzy set in $\mathfrak{U}$. The domain $z = \mathscr{D}(A)$ is called the *universe of discourse of* A, and the set $\mathscr{S}(A) = \{x \in z \mid A(x) > \bot\}$ the *support of fuzzy set* A. Further, for $\alpha \in L$, the sets $A_\alpha = \{x \in z \mid A(x) \geq \alpha\}$ and $A^\alpha = \{x \in z \mid A(x) = \alpha\}$ are called the *α-cut* and *α-level* of A, respectively. An element $x \in z$ is said to be *negligible* in A whenever $x \notin \mathscr{S}(A)$. A fuzzy set A is said to be *crisp* and referred to a *crisp set* if $A(x) \in \{\bot, \top\}$ for any $x \in z$. The empty function $\emptyset : \emptyset \longrightarrow L$ is called the *empty fuzzy set.* One can see that the empty function as a vacuous fuzzy set is crisp, since the assumption on a crisp set is trivially satisfied. If $x \subseteq y$ are sets in $\mathfrak{U}$, we use χ_x to denote the *characteristic function of x on y*, i.e., $\chi_x : y \longrightarrow L$, which is defined by $\chi_x(z) = \top$ if $z \in x$, and $\chi_x(z) = \bot$, otherwise. A fuzzy set A is a *fuzzy subset* of B in $\mathfrak{U}$ provided that $\mathscr{D}(A) \subseteq \mathscr{D}(B)$ and $A(a) \leq B(a)$ for any $a \in \mathscr{D}(A)$. It is easy to see that $\subseteq$ is a partial ordering on the class $\mathfrak{F}(\mathfrak{U})$ of all fuzzy sets in $\mathfrak{U}$.

We say that two fuzzy sets A and B in $\mathfrak{U}$ are *identical* (symbolically, $A = B$) if $\mathscr{D}(A) = \mathscr{D}(B)$ and $A(a) = B(a)$ for any $a \in \mathscr{D}(A)$. Moreover, A and B are *identical up to negligibility* (symbolically, $A \equiv B$) if $\mathscr{S}(A) = \mathscr{S}(B)$ and $A(a) = B(a)$ for any $a \in \mathscr{S}(A)$. One can observe that the relation "to be identical up to negligibility" is an equivalence on $\mathfrak{F}(\mathfrak{U})$. We use $\mathrm{cls}(A)$ to denote the equivalence class of all fuzzy sets from $\mathfrak{U}$ being identical with A up to negligibility.

2.3 Functions Between Fuzzy Sets

Let $\mathfrak{Fcs}$ and $\mathfrak{Fcs}(x, y)$ denote the class of all functions in $\mathfrak{U}$ and the set of all functions from x to y, respectively. Let $x, y, a, b \in \mathfrak{U}$ such that $a \subseteq x$ and $b \subseteq y$. By the definition, a function $f : x \longrightarrow y$ is a function from a to b if $f(z) \in b$ for any $z \in a$ or

$$\chi_a(z) \leq \chi_b(f(z)) \quad (\text{or } \chi_a(z) \to \chi_b(f(z)) = \top) \tag{3}$$

for any $z \in a$, if we consider the characteristic functions of the sets a and b. Replacing the characteristic functions in condition (3) by fuzzy sets, we obtain a natural definition of a function between fuzzy sets.

Definition 3. *Let $A, B \in \mathfrak{F}(\mathfrak{U})$, and let $f \in \mathfrak{Fcs}$. We say that f is a* function from A to B *(symbolically $f : A \longrightarrow B$) if $f \in \mathfrak{Fcs}(\mathscr{D}(A), \mathscr{D}(B))$ and*

$$A(z) \leq B(f(z)) \quad (\text{or equivalently } A(z) \to B(f(z)) = \top) \tag{4}$$

for any $z \in \mathscr{D}(A)$.

The set of all functions from A to B is denoted by $\mathfrak{Fcfs}(A, B)$. Note that the empty function from the empty fuzzy set to an arbitrary fuzzy set trivially satisfies condition (4) and thus belongs to $\mathfrak{Fcfs}(A, B)$. Obviously, the composition of functions $g \circ f \in \mathfrak{Fcfs}(A, C)$, whenever $f \in \mathfrak{Fcfs}(A, B)$ and $g \in \mathfrak{Fcfs}(B, C)$.

A function $f : x \longrightarrow y$ in $\mathfrak{U}$ is a 1-1 correspondence between x and y if there exists a function $f^{-1} : y \longrightarrow x$ (an inverse function) for which $f^{-1} \circ f = 1_x$ and $f \circ f^{-1} = 1_y$, where 1_x and 1_y denote the identity functions on x and y, respectively. Similarly, we define the 1-1 correspondence between fuzzy sets.

Definition 4. *Let $A, B \in \mathfrak{F}(\mathfrak{U})$, and let $f \in \mathfrak{Fcfs}(A, B)$. We say that $f : A \longrightarrow B$ is a* 1-1 correspondence *(symbolically $f : A \xrightarrow[\text{corr}]{\text{1-1}} B$) if there exists $f^{-1} : B \longrightarrow A$ such that $f^{-1} \circ f = 1_{\mathscr{D}(A)}$ and $f \circ f^{-1} = 1_{\mathscr{D}(B)}$.*

The set of all 1-1 correspondences between fuzzy sets A and B in $\mathfrak{U}$ is denoted by $\mathfrak{Cfs}(A, B)$. Later, we introduce a graded version of 1-1 correspondences that play a fundamental role in the definition of graded equipollence. An equivalent definition in terms of 1-1 and onto functions is the following. Denote $\mathfrak{Fcs}^{\text{1-1}}_{\text{corr}}(x, y)$ the set of all 1-1 correspondences between x and y.

Theorem 1. *Let $A, B \in \mathfrak{F}(\mathfrak{U})$. A function $f : A \longrightarrow B$ is a 1-1 correspondence between fuzzy sets if and only if $f \in \mathfrak{Fcs}^{\text{1-1}}_{\text{corr}}(\mathscr{D}(A), \mathscr{D}(B))$ and $A(a) = B(f(a))$ for any $a \in \mathscr{D}(A)$.*

Proof. ($\Rightarrow$) Let $f : A \longrightarrow B$ be a function such that there exists $f^{-1} : B \longrightarrow A$ such that $f^{-1} \circ f = 1_{\mathscr{D}(A)}$ and $f \circ f^{-1} = 1_{\mathscr{D}(B)}$. Then, f is a 1-1 function of A onto B. Since $A(a) \rightarrow B(f(a)) = \top$ for any $a \in \mathscr{D}(A)$ and simultaneously $B(b) \rightarrow A(f^{-1}(b)) = \top$ for any $b \in \mathscr{D}(B)$, we find that

$$(A(a) \rightarrow B(f(a))) \wedge (B(f(a)) \rightarrow A(a)) = A(a) \leftrightarrow B(f(a)) = \top$$

for any $a \in \mathscr{D}(A)$; therefore, $A(a) = B(f(a))$ for any $a \in \mathscr{D}(A)$.

($\Leftarrow$) Since f is a 1-1 function of $\mathscr{D}(A)$ onto $\mathscr{D}(B)$, there exists $f^{-1} : \mathscr{D}(B) \longrightarrow \mathscr{D}(A)$ such that $f^{-1} \circ f = 1_{\mathscr{D}(A)}$ and $f \circ f^{-1} = 1_{\mathscr{D}(B)}$. To finish the proof, we have to prove that f^{-1} is a function of B to A, i.e., (4) is satisfied for f^{-1}. Let $b \in \mathscr{D}(B)$, and let $a \in \mathscr{D}(A)$ such that $f(a) = b$. Then, we find that

$$B(b) \rightarrow A(f^{-1}(b)) = B(f(a)) \rightarrow A(a) = A(a) \rightarrow A(a) = \top,$$

where we used $A(a) = B(f(a))$. □

Hence, its easy to see that the composition of functions $g \circ f \in \mathfrak{Cfs}(A, C)$, whenever $f \in \mathfrak{Cfs}(A, B)$ and $g \in \mathfrak{Cfs}(B, C)$.

Let $f : x \longrightarrow y$ be a function between sets, and let $z \subseteq x$. The image of z under f is defined by $f^{\rightarrow}(z) := \{b \in y \mid \exists a \in x \;\&\; f(a) = b\}$. The image of a fuzzy set under a function is a straightforward extension of the previous definition and is given by Zadeh's extension principle as follows.

Definition 5. *Let $x, y \in \mathfrak{U}$, and let $f : x \longrightarrow y$ be a function. Let $A : x \longrightarrow L$ be a fuzzy set in $\mathfrak{U}$. The* image of A under f *is denoted by $f^{\rightarrow}(A)$ and defined by*

$$f^{\rightarrow}(A)(b) := \bigvee_{a \in x; f(a) = b} A(a) \tag{5}$$

for any $y \in y$.

2.4 Functions Between Fuzzy Sets in a Certain Degree

Let φ be a formula in fuzzy set theory. Then $[\varphi]$ denotes the truth degree in which the formula φ is true, which is interpreted in the residuated lattice $\mathbf{L}$. For example, the truth degree $[f \in \mathfrak{Fcs}(x, y)]$ expresses how it is true that the function f is a member of the set $\mathfrak{Fcs}(x, y)$. Of course, in this case, the truth degree becomes $\perp$ or $\top$.

Definition 6. *Let $A, B \in \mathfrak{F}(\mathfrak{U})$, and let $f \in \mathfrak{Fcs}$. We say that f is a function of A to B in the degree α provided that*

$$\alpha = [f \in \mathfrak{Fcs}(\mathscr{D}(A), \mathscr{D}(B))] \otimes \bigwedge_{(a,f(a)) \in \mathscr{D}(A) \times \mathscr{D}(B)} (A(a) \rightarrow B(f(a))). \tag{6}$$

By our convention, $[f : A \longrightarrow B]$ denotes the truth degree in which the function f can be considered as a function from A to B. Let us emphasize that if f is not a function from $\mathscr{D}(A)$ to $\mathscr{D}(B)$, then $[f : A \longrightarrow B] = \perp$ even if the infimum value in (6) is greater than $\perp$. Similarly we define the truth degree of a correspondences between fuzzy set.

Definition 7. *Let $A, B \in \mathfrak{F}(\mathfrak{U})$, and let $f \in \mathfrak{Fcs}$. We say that f is approximately a one-to-one correspondence between A and B in the degree α provided that*

$$\alpha = [f \in \mathfrak{Fcs}^{\text{1-1}}_{\text{corr}}(\mathscr{D}(A), \mathscr{D}(B))] \otimes \bigwedge_{(a,f(a)) \in \mathscr{D}(A) \times \mathscr{D}(B)} (A(a) \leftrightarrow B(f(a))). \tag{7}$$

The value $[f : A \xrightarrow[\text{corr}]{\text{1-1}} B]$ denotes the truth degree in which the function f can be considered as a one-to-one correspondence between fuzzy sets A and B.

2.5 Fuzzy Power Sets

As we have mentioned in Introduction, the fuzzy power set for fuzzy sets is considered to be a fuzzy set over the set of appropriate fuzzy sets. Here, we propose an alternative definition that straightforwardly generalizes the classical approach to the power set and it is sound in our fuzzy set theory.

Definition 8. *Let $A \in \mathfrak{F}(\mathfrak{U})$, and $x = \mathrm{P}(\mathscr{D}(A))$. The fuzzy set $\mathscr{P}(A) : x \longrightarrow L$ defined by*

$$\mathscr{P}(A)(y) = \bigwedge_{z \in \mathscr{D}(A)} (\chi_y(z) \rightarrow A(z)) \tag{8}$$

is called the fuzzy power set of A, *where χ_y is characteristic function of y on $\mathscr{D}(A)$.*

One can see that the previous definition copies the Bandler-Kohout definition (1) with the restriction to crisp sets. As a simple consequence of (8), we obtain a simple expression of the membership degrees of fuzzy power set

$$\mathscr{P}(A)(y) = \bigwedge_{z \in y} A(z). \tag{9}$$

The following statement shows that the fuzzy power sets preserve the class equivalence of being identical up to negligibility.

Theorem 2. *$\mathscr{P}(A) \equiv \mathscr{P}(B)$, whenever $A \equiv B$.*

Proof. It can be found in [7]. □

Example 1. Let $\mathbf{L}_{Ł}$ be the Łukasiewicz algebra, and let $A = \{1/a, 0.4/b\}$. Then,

$$\mathscr{P}(A) = \{1/\emptyset, 1/\{a\}, 0.4/\{b\}, 0.4/\{a,b\}\}.$$

Moreover, $\mathscr{P}(\emptyset) = \{1/\emptyset\}$, since $\mathscr{P}(\emptyset)(\emptyset) = \bigwedge\emptyset = 1$.

Example 2. Let $\mathbf{L}$ be a complete residuated lattice on $[0,1]$, let the set of all natural numbers ω belong to $\mathfrak{U}$, and let $A : \omega \longrightarrow L$ be defined by $A(n) = 1/n$. Then, curiously, it holds that $|\mathscr{S}(A)| = |\mathscr{S}(\mathscr{P}(A))|$. Indeed, one can see that $\mathscr{P}(A)$ assigns the zero truth degree to each infinite subset of ω. Hence, we obtain that $x \in \mathscr{S}(\mathscr{P}(A))$ if and only if x is a finite subset of ω. It is well-known that the set of all finite subsets of ω is countable.[2] The statement follows from the fact that the support of A is a countable set.

Theorem 3. *Let $A, B \in \mathfrak{F}(\mathfrak{U})$, and let $f : A \longrightarrow B$ be a function between fuzzy sets. Then, the following diagram commutes*

$$\begin{array}{ccc} A & \xrightarrow{f} & B \\ {\scriptstyle i_A}\downarrow & & \downarrow{\scriptstyle i_B} \\ \mathscr{P}(A) & \xrightarrow[f^{\rightarrow}]{} & \mathscr{P}(B), \end{array}$$

where i_A, i_B are the inclusion functions, i.e., $i_A(a) = \{a\}$ for any $a \in \mathscr{D}(A)$ and similarly i_B, and $f^{\rightarrow}$ is the image function of sets.

Proof. Obviously, $i_A : \mathscr{D}(A) \longrightarrow \mathrm{P}(\mathscr{D}(A))$ given by $i_A(a) = \{a\}$ is a function from A into $\mathscr{P}(A)$, since $A(a) = \mathscr{P}(A)(\{a\})$, and similarly i_B is a function from B into $\mathscr{P}(B)$. Obviously, the diagram commutes. To finish the proof, we show that $f^{\rightarrow}$ is a function from $\mathscr{P}(A)$ to $\mathscr{P}(B)$. If $x \subseteq \mathscr{D}(A)$, then

$$\mathscr{P}(A)(x) = \bigwedge_{a\in x} A(a) \le \bigwedge_{a \in x} B(f(a)) = \bigwedge_{b \in f^{\rightarrow}(x)} B(b) = \mathscr{P}(B)(f^{\rightarrow}(x)),$$

and the proof is finished. □

[2] For example, we can put $\lambda(n) := \{1, \dots, n\}$. Then

$$|\mathscr{S}(\mathscr{P}(A))| = |\bigcup_{n\in\omega} \mathrm{P}(\lambda(n))| \le |\bigcup_{n\in\omega} (\mathrm{P}(\lambda(n)) \times \{n\})| \le |\omega \times \omega| = |\omega|,$$

where $\mathrm{P}(\lambda(n)$ is the power set of $\lambda(n)$ and we used that $|\mathrm{P}(\lambda(n))| < |\omega|$ for any $n \in \omega$.

2.6 Fuzzy Classes

Although the fuzzy sets in $\mathfrak{U}$ are the major objects in our theory, it is useful, similarly to the set theory, to introduce the concept of fuzzy class in $\mathfrak{U}$.

Definition 9. *Let $\mathfrak{U}$ be a universe of sets over* $\mathbf{L}$. *A class function $\mathcal{A} : \mathfrak{Z} \longrightarrow L$ (in ZFC) is called a fuzzy class in $\mathfrak{U}$ if $\mathfrak{Z} \subseteq \mathfrak{U}$.*

Note that each fuzzy set is a fuzzy class because of (U1), but not vice versa. Hence, a fuzzy class $\mathcal{A}$ is said to be *proper* if there is no fuzzy set which is identical to $\mathcal{A}$ up to negligibility (the relation $\equiv$ is extended here to fuzzy classes).

Fuzzy class relations are defined similarly to fuzzy set relations, only fuzzy sets are replaced by fuzzy classes. For the purpose of this paper, we introduce the fuzzy class equivalence and fuzzy class partial ordering.

Definition 10. *A fuzzy class relation $\mathcal{R} : \mathfrak{Z} \times \mathfrak{Z} \longrightarrow L$ is called a fuzzy class equivalence if for any $a, b, c \in \mathfrak{Z}$, it satisfies*

(FE1) $\mathcal{R}(a, a) = \top$,
(FE2) $\mathcal{R}(a, b) = \mathcal{R}(b, a)$,
(FE3) $\mathcal{R}(a, b) \otimes \mathcal{R}(b, c) \leq \mathcal{R}(a, c)$.

3 Cantor's Equipollence

In set theory, two sets are equipollent (equipotent, equivalent, bijective, or have the same cardinality, etc.) if there exists a 1-1 correspondence between them. This definition was proposed by G. Cantor. Formally, the class relation of equipollence denoted by $\sim$ is introduced on the class of all sets as follows:

$$x \sim y \ \text{ iff } \ \exists f : x \xrightarrow[\text{corr}]{\text{1-1}} y. \tag{10}$$

Obviously, the equipollence of sets is a class relation extending the relation to be identical sets. One can see that the substitution of fuzzy sets for the sets in (10) does not reflect the idea that fuzzy sets being identical up to negligibility should be also equipollent. Furthermore, the restriction to particular fuzzy sets in (10) the consistency of our theory is broken as the following simple examples demonstrate.

Example 3. Let $x = \{a, b\}$ and $y = \{c, d, e\}$. Let $A = \chi_x$ and $B = \chi_z$, where $z = \{c, d\} \subset y$. Obviously, the set $\mathfrak{Fcfs}(A, B)$ is empty because there is no 1-1 correspondence between the domains of A and B; hence, $A \not\sim B$. On the other hand, there is a function f such that $f : x \xrightarrow[\text{corr}]{\text{1-1}} z$; therefore, naturally it should be $A \sim B$.

Example 4. Let ω be the set of natural numbers, and assume that $\omega \in \mathfrak{U}$. Let $\mathbf{L}_{Ł}$ be the Łukasiewicz algebra, and let $N, O : \omega \longrightarrow [0, 1]$ be fuzzy sets defined by

$$N(n) = 1 \quad \text{and} \quad O(n) = \begin{cases} 1, \text{ if } n \text{ is an odd number,} \\ 0, \text{ otherwise,} \end{cases}$$

for any $n \in \omega$. Obviously, the set $\mathfrak{Fcfs}(N, O)$ is empty even if there exists a 1-1 correspondence $f : \omega \longrightarrow \omega$; hence, $N \not\sim O$. On the other hand, the sets of odd numbers and natural numbers are equipollent; therefore, it should be $N \sim O$.

To overcome the aforementioned difficulties and simultaneously to accept fuzzy sets that differ up to negligible elements to be identical we propose the following definition of the equipollence of fuzzy sets.

Definition 11. *Let $A, B \in \mathfrak{F}(\mathfrak{U})$. We say that A and B are* Cantor's equipollent *(symbolically $A \overset{c}{\sim} B$) provided that there exist $A' \in \text{cls}(A)$, $B' \in \text{cls}(B)$, and $f \in \mathfrak{Fcs}$ such that $f : A' \xrightarrow[\text{corr}]{\text{1-1}} B'$.*

The following theorem states a necessary and sufficient condition reducing the verification of Cantor's equipollence to two specific fuzzy sets that are identical to the original ones up to negligibility.

Theorem 4. *Let $A, B \in \mathfrak{F}(\mathfrak{U})$, and let $C \in \text{cls}(A)$ and $D \in \text{cls}(B)$ such that $\mathscr{S}(C) = \mathscr{D}(C)$ and $\mathscr{S}(D) = \mathscr{D}(D)$. Then $A \overset{c}{\sim} B$ if and only if there exists $f : C \xrightarrow[\text{corr}]{\text{1-1}} D$.*

Proof. If $A \overset{c}{\sim} B$, then there exist $A' \in \text{cls}(A)$, $B'\text{cls}(B)$ and $f : A' \xrightarrow[\text{corr}]{\text{1-1}} B'$. A simple consequence of Cantor's equipollence, we find that $A'(x) = \bot$ if and only if $B'(f(x)) = \bot$. Hence, f restricted to $\mathscr{S}(A)$ must be a 1-1 correspondence between the supports of A and B such that $A(x) = B(f(x))$ for any $x \in \mathscr{S}(A)$. Therefore, $f \upharpoonright \mathscr{S}(A) : C \xrightarrow[\text{corr}]{\text{1-1}} D$, and the sufficient part is proved. Since the necessary part follows immediately from the definition of Cantor's equipollence, the statement is proved. □

The equipollence of sets is a class equivalence. The same holds for the Cantor's equipollence of fuzzy sets.

Theorem 5. *The class relation $\overset{c}{\sim}$ is a class equivalence on $\mathfrak{U}$.*

The following lemma provides an equivalent definition of Cantor's equipollence of fuzzy sets based on the classical equipollence of α-levels.

Lemma 1. *$A \overset{c}{\sim} B$ if and only if $|A^\alpha| = |B^\alpha|$ for any $\alpha \in L \setminus \{\bot\}$.*

Proof. ($\Rightarrow$) Let $A \overset{c}{\sim} B$. By Theorem 4, we may assume that the domains and the supports of A and B coincide. If $f : A \xrightarrow[\text{corr}]{\text{1-1}} B$ is the respective 1-1 correspondence, then, for any $\alpha \in L \setminus \{\bot\}$, we simply find that $A^\alpha = \emptyset$ if and only if $B^\alpha = \emptyset$; otherwise, we have $f^\rightarrow(A^\alpha) = B^\alpha$. Hence, $|A^\alpha| = |B^\alpha|$ for $A^\alpha = \emptyset$. If $A^\alpha \neq \emptyset$, then $f \upharpoonright A^\alpha : A^\alpha \xrightarrow[\text{corr}]{\text{1-1}} B^\alpha$, which implies that $|A^\alpha| = |B^\alpha|$.

($\Leftarrow$) Let $|A^\alpha| = |B^\alpha|$ for any $\alpha \in L \setminus \{\bot\}$. Since

$$\mathscr{S}(A) = \bigcup_{\alpha \in L \setminus \{\bot\}} A^\alpha$$

and $A(x) = \alpha$ if and only if $x \in A^\alpha$ (note that $A^\alpha \cap A^\beta = \emptyset$ whenever $\alpha \neq \beta$), and similarly for B, we find that a 1-1 correspondence f between A and B can be derived as

$$f := \bigcup_{\alpha \in L \setminus \{\perp\}} f_\alpha,$$

where f_α is an arbitrary 1-1 correspondence of A^α onto B^α for any $\alpha \in L \setminus \{\perp\}$. Note that $f(x) = f_\alpha(x)$ if and only if $x \in A^\alpha$; therefore, $f : \mathscr{S}(A) \xrightarrow[\text{corr}]{\text{1-1}} \mathscr{S}(B)$ such that $A(x) = B(f(x))$; hence, we obtain $A \overset{c}{\sim} B$. □

The following theorem is Cantor's theorem for fuzzy sets based on the equipollence relation $\overset{c}{\sim}$.

Theorem 6 (Cantor's theorem). $A \not\overset{c}{\sim} \mathscr{P}(A)$.

Proof. Let $A \in \mathfrak{F}(\mathfrak{U})$. First, we show that $\mathrm{P}(A^\alpha) \subseteq \mathscr{P}(A)^\alpha$ for any $\alpha \in L \setminus \{\perp\}$. From the fuzzy power set definition, if $y \subseteq A^\alpha$ (including $y = \emptyset$), then $\mathscr{P}(A)(y) = \bigwedge_{x \in y} A(x) = \alpha$; therefore, $y \in \mathscr{P}(A)^\alpha$, which means that $\mathrm{P}(A_\alpha) \subseteq \mathscr{P}(A)_\alpha$. The statement is a simple consequence of Lemma 1 and the fact that $|A^\alpha| < |\mathrm{P}(A^\alpha)| \leq |\mathscr{P}(A)^\alpha|$. □

4 Graded Cantor's Equipollence

We say that fuzzy sets $A, B \in \mathfrak{F}(\mathfrak{U})$ have *cardinal separable supports* if

$$|\mathscr{S}(A)| \leq |\mathscr{D}(B) \setminus \mathscr{S}(B)| \text{ and } |\mathscr{S}(B)| \leq |\mathscr{D}(A) \setminus \mathscr{S}(A)|. \tag{11}$$

In [8], we have introduced the graded version of Cantor's equipollence. The following definition of graded Cantor's equipollence has been presented in [6].

Definition 12. *Let $A, B \in \mathfrak{F}(\mathfrak{U})$, and let $C \in \mathrm{cls}(A)$ and $D \in \mathrm{cls}(B)$ be fuzzy sets that have cardinal separable supports and $|\mathscr{D}(C)| = |\mathscr{D}(D)|$. We say that A and B are Cantor's equipollent in the degree α provided that*

$$\alpha = \bigvee_{f \in \mathfrak{Fcs}(\mathscr{D}(C), \mathscr{D}(D))} [f : C \xrightarrow[\text{corr}]{\text{1-1}} D]. \tag{12}$$

We use $\overset{c}{\approx}$ to denote the fuzzy class relation of being Cantor's equipollent in a certain degree and the value $[A \overset{c}{\approx} B]$ denotes the truth degree in which the fuzzy sets A and B are Cantor's equipollent.

Definition 13. *The fuzzy class relation $\overset{c}{\approx}$ is called the graded Cantor's equipollence of fuzzy sets.*

It is well known that $a \sim b$ implies $\mathrm{P}(a) \sim \mathrm{P}(b)$ in set theory. The following theorem is a natural extension of this statement for fuzzy sets.

Theorem 7. *Let $A, B \in \mathfrak{F}(\mathfrak{U})$. Then,*

$$[A \overset{\mathrm{c}}{\approx} B] \leq [\mathscr{P}(A) \overset{\mathrm{c}}{\approx} \mathscr{P}(B)]. \tag{13}$$

Proof. Let $A, B \in \mathfrak{F}(\mathfrak{U})$. Without lost of generality (due to Theorem 2), we assume that A and B have cardinal separable supports and $|\mathscr{D}(A)| = |\mathscr{D}(B)|$. For $A = B = \emptyset$, the statement is a trivial consequence of Theorem 2. Let $A \neq \emptyset$ or $B \neq \emptyset$. Recall that $\mathscr{D}(\mathscr{P}(A)) = \mathrm{P}(\mathscr{D}(A))$. For any $f \in \mathfrak{Fcs}^{1\text{-}1}_{\mathrm{corr}}(\mathscr{D}(A), \mathscr{D}(B))$, let us define $f^{\rightarrow} : \mathscr{D}(\mathscr{P}(A)) \longrightarrow \mathscr{D}(\mathscr{P}(B))$ by

$$f^{\rightarrow}(y) = \{f(x) \mid x \in y\}, \quad y \in \mathscr{D}(\mathscr{P}(A)). \tag{14}$$

Obviously, $f^{\rightarrow} \in \mathfrak{Fcs}^{1\text{-}1}_{\mathrm{corr}}(\mathscr{D}(\mathscr{P}(A)), \mathscr{D}(\mathscr{P}(B)))$ and

$$\begin{aligned}
[\mathscr{P}(A) \overset{\mathrm{c}}{\approx} \mathscr{P}(B)] &\geq \bigwedge_{y \in \mathscr{D}(\mathscr{P}(A))} (\mathscr{P}(A)(y) \leftrightarrow \mathscr{P}(B)(f^{\rightarrow}(y))) \\
= \bigwedge_{y \in \mathscr{D}(\mathscr{P}(A))} \Big((\bigwedge_{x \in y} A(x)) \leftrightarrow (\bigwedge_{z \in f^{\rightarrow}(y)} B(z)) \Big) &\geq \bigwedge_{y \in \mathscr{D}(\mathscr{P}(A))} \bigwedge_{x \in y} (A(x) \leftrightarrow B(f(x))) \\
&= \bigwedge_{x \in \mathscr{D}(A)} (A(x) \leftrightarrow B(f(x))) = [f : A \xrightarrow[\mathrm{corr}]{1\text{-}1} B].
\end{aligned}$$

Since the previous inequality holds for any $f \in \mathfrak{Fcs}^{1\text{-}1}_{\mathrm{corr}}(\mathscr{D}(A), \mathscr{D}(B))$, we obtain

$$[\mathscr{P}(A) \overset{\mathrm{c}}{\approx} \mathscr{P}(B)] \geq \bigvee_{f \in \mathfrak{Fcs}^{1\text{-}1}_{\mathrm{corr}}(\mathscr{D}(A), \mathscr{D}(B))} [f : A \xrightarrow[\mathrm{corr}]{1\text{-}1} B] = [A \overset{\mathrm{c}}{\approx} B],$$

and the proof is finished. □

One can observe that $A \not\approx \mathscr{P}(A)$ does not imply $[A \overset{\mathrm{c}}{\approx} \mathscr{P}(A)] < \top$. In other words, if there is no 1-1 correspondence between A and $\mathscr{P}(A)$, we cannot immediately exclude that the fuzzy sets A and $\mathscr{P}(A)$ are equipollent in degree $\top$, where $\top$ is a result of the supremum operation in (12). Nevertheless, this claim is true and can be considered as a graded version of Cantor's theorem.

Theorem 8 (Graded version of Cantor's theorem). $[A \overset{\mathrm{c}}{\approx} \mathscr{P}(A)] < \top$.

Since the proof is long we left it out in the paper. The following example demonstrates the graded version of Cantor's theorem on the fuzzy set from Example 2.

Example 5. Assume that $\mathbf{L}$ is the Łukasiewicz algebra, and let $A : \omega \longrightarrow [0, 1]$ be the fuzzy set defined by $A(n) = 1/n$. Since $\mathscr{P}(A)(\emptyset) = 1$, the evaluation of $[A \overset{\mathrm{c}}{\approx} \mathscr{P}(A)]$ is based on one-to-one correspondences f, for which $f(1) = \emptyset$ and $f(2) = \{1\}$ or $f(1) = \{1\}$ and $f(2) = \emptyset$. Obviously, $[f : A \xrightarrow[\mathrm{corr}]{1\text{-}1} \mathscr{P}(A)] = 1/2$, which follows from $A(1) \leftrightarrow \mathscr{P}(A)(\emptyset) = 1 = A(1) \leftrightarrow \mathscr{P}(A)(\{1\})$ and $A(2) \leftrightarrow \mathscr{P}(A)(\emptyset) = 1/2 = A(2) \leftrightarrow \mathscr{P}(A)(\{1\})$. Since there is no one-to-one correspondence in a degree α, which is greater than $1/2$, we obtain $[A \overset{\mathrm{c}}{\approx} \mathscr{P}(A)] = 1/2$.

5 Conclusion

In this contribution, we proposed a novel concept of fuzzy power sets of a fuzzy set defined over the set of crisp subsets of the universe of discourse and analyzed the validity of Cantor's theorem for it with respect to types of equipollences of fuzzy sets. We gave preference to this simpler definition over the Bandler-Kohout concept of fuzzy power set to ensure the soundness of fuzzy set theory, which is built in the framework of a universe of fuzzy sets. Nevertheless, if the Bandler-Kohout fuzzy power sets exist in a universe of sets, a similar analysis can be provided, but this is a subject of our future research.

Acknowledgments. This work was supported by the project LQ1602 IT4Innovations excellence in science.

References

1. Bandler, W., Kohout, L.: Fuzzy power sets and fuzzy implication operators. Fuzzy Sets Syst. **4**, 13–30 (1980)
2. Bělohlávek, R.: Fuzzy Relational Systems: Foundations and Principles. Kluwer Academic Publisher, New York (2002)
3. Gottwald, S.: Set theory for fuzzy sets of higher level. Fuzzy Sets Syst. **2**, 125–151 (1979)
4. Hájek, P., Haníková, Z.: Interpreting lattice-valued set theory in fuzzy set theory. Logic J. IGPL **21**(1), 77–90 (2013)
5. Holčapek, M.: A functional approach to cardinality of finite fuzzy sets. In: Laurent, A., Strauss, O., Bouchon-Meunier, B., Yager, R.R. (eds.) IPMU 2014. CCIS, vol. 443, pp. 234–243. Springer, Cham (2014). https://doi.org/10.1007/978-3-319-08855-6_24
6. Holčapek, M.: Graded dominance and cantor-bernstein equipollence of fuzzy sets. In: Carvalho, J.P., Lesot, M.-J., Kaymak, U., Vieira, S., Bouchon-Meunier, B., Yager, R.R. (eds.) IPMU 2016. CCIS, vol. 611, pp. 510–521. Springer, Cham (2016). https://doi.org/10.1007/978-3-319-40581-0_41
7. Holčapek, M.: A graded approach to cardinal theory of finite fuzzy sets, part I: graded equipollence. Fuzzy Sets Syst. **298**, 158–193 (2017)
8. Holčapek, M., Turčan, M.: Graded equipollence of fuzzy sets. In: Carvalho, J.P., Kaymak, D.U., Sousa, J.M.C. (eds.) Proceedings of IFSA/EUSFLAT 2009, European Soc Fuzzy logic & Technology, Johannes Kepler univ, Austria, pp. 1565–1570 (2009)
9. Novák, V., Perfilieva, I., Močkoř, J.: Mathematical Principles of Fuzzy Logic. Kluwer Academic Publisher, Boston (1999)
10. Rodabaugh, S.E.: Relationship of algebraic theories to powerset theories and fuzzy topological theories for lattice-valued mathematics. Int. J. Math. Math. Sci. **2007**, 1–71 (2007)
11. Takeuti, G., Titani, S.: Fuzzy logic and fuzzy set theory. Arch. Math. Logic **32**, 1–32 (1992)
12. Zhang, Q., Xie, W., Fan, L.: Fuzzy complete lattices. Fuzzy Sets Syst. **160**(16), 2275–2291 (2009)

Clustering of Propositions Equipped with Uncertainty

Marek Z. Reformat[1(✉)], Jesse Xi Chen[1], and Ronald R. Yager[2]

[1] University of Alberta, Edmonton, AB T6G 1H9, Canada
{Marek.Reformat,jesse.chen}@ualberta.ca
[2] Iona College, New Rochelle, NY 10801, USA
yager@panix.com

Abstract. Graph-based data representation formats enable more advanced processing of data that leads to better utilization of information stored and available on the web. Intrinsic high connectedness of such representation provides a means to create methods and techniques that can assimilate new data and build knowledge-like data structures. Such procedures resemble a human-like way of dealing with information.

In the paper, we focus on processing a knowledge graph data. In particular, we propose a simple way of clustering pieces of data that contain levels of uncertainty associated with them. That uncertainty is a result of collecting data from multiple sources. It is due to the fact that information about the same entities occurs a number of times and can be inconsistent. Existence of a number of 'alternative' pieces of data means that we can associate with them different levels of uncertainty. In order to accomplish that, we represent pieces of data from knowledge graphs as propositions with multiple alternatives. Each alternative is associated with an uncertainty value expressing its 'correctness', i.e., a level of confidence that a given alternative represents an accurate piece of information. Those values are generated based on frequency of occurrence and consistency of alternatives. Our method is designed to cluster such propositions. The methodology is presented together with a number of illustrating examples.

Keywords: Knowledge graph · Propositions · Uncertainty · Clustering

1 Introduction

In order to fully utilize web data, i.e., to 'create' data structures reflecting levels of uncertainty in data found on the web and to use that data to extract knowledge, we need techniques that process data and information in a more 'intelligent' way. Existence of multiple sources of information on the web means that the same entities could be described with the same or different statements. Quite often, there is no clarity if we deal with correct or not information. Following a human-like approach, we would like to have an approach that is able

J. Medina et al. (Eds.): IPMU 2018, CCIS 855, pp. 715–726, 2018.
https://doi.org/10.1007/978-3-319-91479-4_59

to determine, based on all found statements and frequency of their occurrence, which pieces of information are correct and what levels of uncertainty we can associate with them.

In our previous work [1–4], we have proposed to use elements of participatory learning [5,6] for a process of aggregating collected pieces of information representing alternatives of statements describing a single entity. The process is able to determine a level of uncertainty associated with each alternative applying elements of possibility theory. The approach is suitable for processing graph-based data format. Each set of alternative statements about an entity is represented as a proposition. Propositions are built based on segments of knowledge graphs in a form of Resource Description Framework (RDF) triples [7].

In this paper, we address a problem of clustering entities represented as sets of proportions with uncertainty. We propose a simple cosine similarity measure based technique for determining similarity between propositions. The process leads to building a similarity matrix for clustering purposes.

2 Knowledge Graph Representation

One of the most important contributions of the Semantic Web concept [8] is the Resource Description Framework (RDF) [7]. This framework is a graph-based format of representing data on the web. The key RDF concept is to represent a piece of data as a triple: `<subject-property-object>`, where the `subject` is an entity being described, the `object` is an entity describing the `subject`, and the `property` is a 'connection' between the `subject` and `object`. For example, *Edmonton is city* is a triple with *Edmonton* as its `subject`, *is* its `property`, and *city* its `object`.

In general, a `subject` of one triple can be an `object` of another triple, and vice versa. The growing presence of knowledge graphs as a data representation format on the web, and Resource Description Framework (RDF) in particular, brings opportunity to develop new ways how data is processed, and what type of information is generated from it.

2.1 RDF Triples as Definitions of Entities

A single RDF triple `<subject-property-object>` can be perceived as a feature of an entity represented by `object`. In other words, an individual triple is a statement about the entity, and multiple triples with the same `subject` constitute a set of statements about the entity. An illustration of this is shown in Fig. 1(a). It is a set of RDF triples – statements – about Edmonton. If we think 'graphically' about it, an RDF-based description of entity resembles a star with an `subject` as its core and tuples `<...-property-object>` as its rays. We will call it an RDF-star.

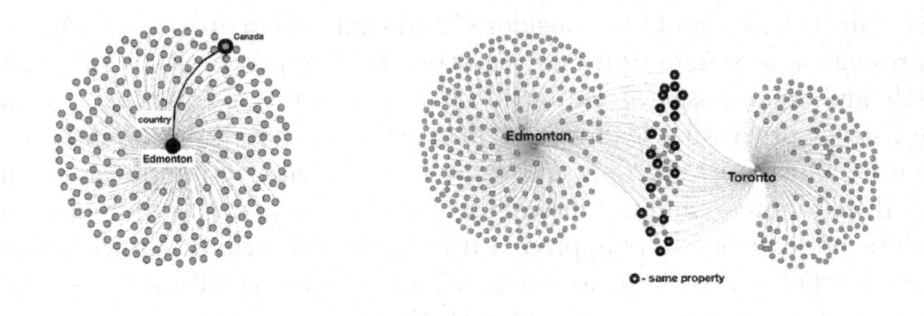

Fig. 1. Resource Description Framework: RDF-based definition of entity – RDF-star (a); evaluation of similarity of RDF-stars (b).

2.2 Similarity of RDF Triples

Evaluation of similarity between RDF-stars is a fundamental process required for clustering. The simplest way of determining similarity between two RDF-stars is to enumerate a number of shared statements/features between both stars. Such a feature-based similarity measure resembles the Jaccard's index [9]. In the case of RDF-stars, it nicely converts into checking how many nodes are shared between two entities. The idea is presented in Fig. 1(b). The entities Edmonton and Toronto share a number of nodes that are connected to the entities representing both cities with the same `property` (black circles in Fig. 1(b)).

2.3 RDF Data Processing

With a growing number of RDF triples on the web – more than 62 billions right now (http://stats.lod2.eu) – processing data represented as RDF triples is gaining attention. There are multiple works focusing on RDF data storage and querying strategies using a specialized query language SPARQL [10]. More and more publications look at handling RDF triples directly.

The work described in [11] looks at the efficient processing of information in RDF data sources for detecting communities. A process of identifying relevant datasets for a specific task or topic is addressed in [12]. A hierarchical clustering algorithm is used for inferring structural summaries to support querying Linked Data sources in [13]. Linked data classification is a subject of the work presented in [14], while an approach for formalizing the hierarchy of concepts from linked data is described in [15].

3 Aggregation and Learning Process – Overview

RDF data contained in repositories represents, in majority of cases, specific pieces of information and statements. Such a vast amount of data can and should be utilized to support learning and knowledge extraction processes. Repeated occurrences of data describing the same entities, missing and/or conflicting pieces of data – all of these should be considered and analyzed in order to create a realistic representation of underlying information. Such representation should reflect strength and weaknesses of data collected from multiple sources, and should enable further processing leading to construction of definitions of entities [16].

Figure 2 illustrates our approach to aggregate data about the same entities from multiple sources, and use these aggregated views of entities for constructing definitions of concepts. The approach uses data in a form of knowledge graphs that are further represented as propositions containing alternatives equipped with levels of uncertainty associated with them.

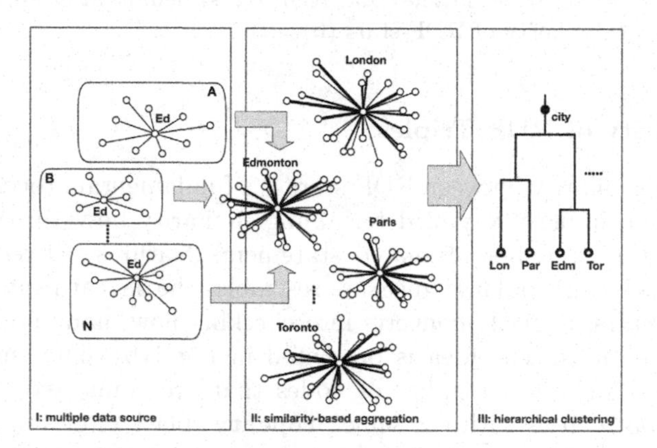

Fig. 2. Aggregation/Learning Process: multiple data sources – $A, B, \ldots, N$ – with descriptions of the same entities (I); aggregation of data about the same entities – *aggregated data about Edmonton, and other cities* (different thickness of lines indicate different levels of uncertainty associated with statements about entities) (II); and clustering of aggregated data – *a snapshot of hierarchy defining* **city**.

As it can be seen, Fig. 2, pieces of data about the same entities – RDF triples – can be found in multiple data repositories. Each repository, no matter where located, can contain the same or different information about the same entities. Therefore, the first step is dedicated to aggregation of information (RDF triples) and creation of 'definitions' of individual entities. It should contain all pieces of information found in all repositories. Also, it should include levels of uncertainty associated with each statement. The uncertainty values reflect levels of significance of statements determined based on the collected data. The work related to that task has been presented in [1–4].

Once the aggregation process is performed, the obtained definitions of entities are clustered. Such an operation leads to creation of *definitions of concepts.* The advantage of using entity definitions with levels of uncertainty in statements would mean building definitions (clusters) of concepts that reflect levels of confidence in statements (RDF triples). In other words, we look for and propose a process that clusters entity definitions when levels of uncertainty associated with different statements are considered and influence a clustering process.

4 RDF Triples and Propositions

An essential component of the proposed process of clustering RDF data/triples is representing a single RDF triple, i.e., a statement about an entity, as a proposition. Eventually, a proposition enables expressing degrees of possibility associated with a number of alternative statements. This section explains the conversion from RDF triples to propositions.

4.1 Propositions in Participatory Learning

In a version of participatory learning adopted for propositional knowledge [6], all statements about an entity are represented as propositions. A single proposition is of a format:

$$P_i : \; V_i \; is \; S_i(x) \tag{1}$$

where a variable V_i is defined over the domain X, S_i is a fuzzy subset of X, and $S_i(x)$ is a degree of possibility that x is a value of V_i. For example, a proposition representing a degree of possibility a that the statement 'Edmonton is located in Canada' is:

$$P_{location} : \; V_{Country} \; is \; S(Canada) = a \tag{2}$$

Another way of representing it (Eq. 1) is:

$$P_i : \; V_i \; is \; \frac{S_i(x)}{x} \tag{3}$$

and then:

$$P_{country} : \; V_{Country} \; is \; \frac{a}{Canada} \tag{4}$$

In general, a proposition could be equipped with a level of certainty [6]:

$$P_i : \; V_i \; is \; S_i(x) \; is \; \alpha_i - certain \tag{5}$$

where α_i represents a degree of certainty in correctness of the statement. According to Zadeh's theory of approximate reasoning [17], we can transform such a propositions into its equivalent non-qualified form:

$$P_i : \; V_i \; is \; S_i(x) \; is \; \alpha_i - certain \Longrightarrow V_i \; is \; F_i(x) \tag{6}$$

where

$$F_i(x) = max(S_i(x), \, 1 - \alpha_i) = S_i(x) \vee (1 - \alpha_i) \tag{7}$$

In a case of multiple x's, a proposition takes the form:

$$P_i : \; V_i \; is \; S_i : \{\frac{a_1}{x_1}, \frac{a_2}{x_2}, \ldots, \frac{a_n}{x_n}\} \tag{8}$$

where $a_1, a_2, \ldots, a_n$ are degrees of possibility that the variable V_i assumes values $x_1, x_2, \ldots, x_n \in X$, respectively.

4.2 RDF Triples as Propositions, Their Aggregation

The ability to perceive an RDF triple as a single statement about an entity leads to a very important observation that becomes a basic idea of 'treating' RDF based data as a set of propositions. We state that each RDF triple is a single proposition defined on the domain of values that can be assigned to the triple's property. For example, the highlighted triple in Fig. 1:

Edmonton – country – Canada

can be expressed as a proposition:

$$P_i : V_{Country}\ is\ S_i : \left\{ \frac{1.0}{Canada}, \frac{0.0}{other\ values\ of\ V_{Country}} \right\} \tag{9}$$

where $V_{Country}$ is a variable, and S_i is a fuzzy subset where *Canada* is associated with 1.0, while all other possible values of $V_{Country}$ have value of 0.0. This means that the Edmonton's country is Canada, and that there is a high certainty that this is a correct statement. Please note, we have added a value *other values of* $V_{Country}$ to indicate that possibility of $V_{Country}$ assuming other values is 0.0. An introduction of such alternative value sets up a stage for 'accepting' other values (see below).

Collecting statements on the web can result in information that could confirm what is already known, i.e., new statements are the same as the ones we know, or they contradict what is known, i.e., new statements could be different. In both cases, we provide a mechanism for aggregating both pieces of data, and determining levels of uncertainty associated with them. The details of the mechanism are presented in [1]. If we have a statement:

$$P_i : V_{Country}\ is\ S_i : \left\{ \frac{1.0}{UK}, \frac{0.0}{other\ values\ of\ V_{Country}} \right\} \tag{10}$$

that should be aggregated with the statement presented in Eq. 9, then the result is:

$$P_i : V_{Country}\ is\ S_i : \left\{ \frac{0.8}{Canada}, \frac{0.6}{UK}, \frac{0.0}{other\ values\ of\ V_{Country}} \right\} \tag{11}$$

The values of 0.8 and 0.6 are exemplary ones. At this time, we would like to emphasis the fact that if we have the following proposition:

$$P_i : V_{Country}\ is\ S_i : \left\{ \frac{1.0}{Canada}, \frac{1.0}{other\ values\ of\ V_{Country}} \right\} \tag{12}$$

then we state that both locations *Canada* and *other values* are equally possible to a degree of 1.0. In other words, this means **'I do not know'**.

Example I. Let us take a look at a little bit more general case. One source of data has the following two propositions:

$$Prop_1^A : V_{Country}\ is\ S_{LP} : \left\{ \frac{a_A}{Canada}, \frac{b_A}{UK}, \frac{c_A}{other\ values} \right\}$$

$$Prop_2^A : V_{TimeZone}\ is\ S_{TZ} : \left\{\frac{x_A}{MST}, \frac{y_A}{other\ values}\right\}$$

while the other two propositions from a different source are:

$$Prop_1^B : V_{Country}\ is\ S_{LP} : \left\{\frac{a_B}{Canada}, \frac{b_B}{US}, \frac{c_B}{other\ values}\right\}$$

$$Prop_2^B : V_{CityPopulation}\ is\ S_{CP} : \left\{\frac{x_B}{1200k}, \frac{y_B}{1300k}, \frac{z_B}{other\ values}\right\}$$

Now we would like to aggregate information from both sources A and B. We obtain:

$$Prop_1 : V_{Country}\ is\ S_{LP} : \left\{\frac{a_1}{Canada}, \frac{a_2}{UK}, \frac{a_3}{US}, \frac{a_4}{other\ values}\right\}$$

$$Prop_2 : V_{TimeZone}\ is\ S_{TZ} : \left\{\frac{x_A}{MST}, \frac{y_A}{other\ values}\right\}$$

$$Prop_3 : V_{CityPopulation}\ is\ S_{CP} : \left\{\frac{x_B}{1200k}, \frac{y_B}{1300k}, \frac{z_B}{other\ values}\right\}$$

The values a_1, a_2, a_3, a_4 are determined using the participatory learning rules adopted for aggregating propositions with uncertainty [1].

4.3 Clustering of RDF-Based Propositions

A process of building categories involves clustering of RDF-stars, and to be more specific, entity definitions represented as sets of $i = 1, \ldots, N$ propositions:

$$\begin{aligned} &Prop_1 :\ V_{Prop_1}\ is\ S_{Prop_1} : \left\{\frac{a_1^1}{x_1^1}, \frac{a_2^1}{x_2^1}, \ldots, \frac{a_p^1}{x_p^1}\right\} \\ &\vdots \\ &Prop_N :\ V_{Prop_N}\ is\ S_{Prop_N} : \left\{\frac{a_1^N}{x_1^N}, \frac{a_2^N}{x_2^N}, \ldots, \frac{a_q^N}{x_q^N}\right\} \end{aligned} \tag{13}$$

where $a_1^i, a_2^i, \ldots, a_n^i$ are degrees of possibility that the variable V_{Prop_i} of the proposition $Prop_i$ assumes alternatives $x_1^i, x_2^i, \ldots, x_n^i \ \in\ X$, respectively. In other words, each definition of entity obtained as a result of the aggregation process (Sect. 4.2), is a set of propositions (Eq. 13).

A clustering process starts with construction of a similarity matrix. Once a set of triples (RDF-stars) is obtained, for example due to a collection process performed by an agent followed by the aggregation process, similarity values should be determined for all pairs of entity definitions. Such similarity matrix is an input to a hierarchical clustering algorithm. The result is a hierarchy of clusters (groups of entity definitions) with the most specific clusters at the bottom, and the most abstract one (one that contains everything) at the top.

The main contribution of this paper is a procedure suitable for determining similarity between propositions with uncertainty – an essential 'ingredient' of building a similarity matrix. Let us explain the idea first. We have two propositions $Prop_i$ and $Prop_j$:

$$Prop_i : V_{Prop_i} \ is \ S_{Prop_i} : \left\{ \frac{a_1^i}{x_1^i}, \frac{a_2^i}{x_2^i}, \frac{a_3^i}{x_3^i} \right\} \tag{14}$$

$$Prop_j : V_{Prop_j} \ is \ S_{Prop_j} : \left\{ \frac{a_1^j}{x_1^j}, \frac{a_2^j}{x_2^j}, \frac{a_3^j}{x_3^j}, \frac{a_4^j}{x_4^j} \right\} \tag{15}$$

and let $x_1^i = x_1^j = x_1$ and $x_3^i = x_4^j = x_2$, while x_2^i, x_2^j, x_3^j are different.

We create a vector based on the union of all different alternatives. So, in our case it will be:

$$v_{Prop_i, Prop_j} = [\ x_1,\ x_2,\ x_2^i,\ x_2^j,\ x_3^j\ ] \tag{16}$$

Based on the vector of alternatives (Eq. 16), we create two vectors with numerical values, where each of them represents a given proposition using its uncertainty values. As the result, we obtain:

$$v^i = [a_1^i,\ a_3^i,\ a_2^i,\ 0.0,\ 0.0\ ] \tag{17}$$

$$v^j = [a_1^j,\ a_4^j,\ 0.0,\ a_2^j,\ a_3^j] \tag{18}$$

In order to determine a degree of similarity between both propositions, we use cosine similarity measure:

$$Sim(V^i, V^j) = \frac{V^i \cdot V^j}{||V^i||_2 \ ||V^j||_2} \tag{19}$$

To generalize, let E^A and E^B be two entities we want to determine similarity of. The entity E^A is defined by a set of RDF triples:

$$E^A = \{< E^A - p_i^A - O_{ik}^A >, where\ i = 1, \ldots, M^A, k = 1, \ldots, N_i^A\} \tag{20}$$

while the entity E^B is of the form:

$$E^B = \{< E^B - p_j^B - O_{jt}^B >, where\ j = 1, \ldots, M^B, t = 1, \ldots, N_j^B\} \tag{21}$$

We generate a set $O^{AB} = O_{ik}^A \cup O_{jt}^B$ for $i = 1, \ldots, M^A, k = 1, \ldots, N_i^A, and\ j = 1, \ldots, M^B, t = 1, \ldots, N_j^B$. This set contains all alternatives of all properties of both entities. The cardinality of O^{AB} is identified as n. Now, we represent both entities as sets of M^A propositions (Eq. 13) describing the entity E^A, and M^B proposition representing the entity E^B. Please note, that a proposition representing a single property contains all its alternatives. Then, two vectors are built: $V^A = [v_1^A, \ldots, v_n^A]$ and $V^B = [v_1^B, \ldots, v_n^B]$ where the values are uncertainty levels associated with alternatives existing in the propositions, or 0.0 in the case a given entity does not have a particular alternative. The final value of similarity is determined using cosine similarity measure (Eq. 19).

Example II. Let us go back to Example I that contains two sets of propositions: $Prop_1^A$ and $Prop_2^A$ describing an entity E^A, and $Prop_1^B$ and $Prop_2^B$ describing an entity E^B. The vector of all alternatives is as follows:

$$\begin{aligned} v = \; & [Canada, UK, US, other\ values_{Loc}, \\ & MST, other\ values_{TZ}, \\ & 1200k, 1300k, other\ values_{CP}] \end{aligned}$$

Based on it, we create two vectors. Each of them describing a single entity:

$$\begin{aligned} v^A = \; & [a_A(Canada), b_A(UK), 0.0(US), c_A(other\ values_C), \\ & x_A(MST), y_A(other\ values_T), \\ & 0.0(1200k), 0.0(1300k), 0.0(other\ values_{CP}] \end{aligned}$$

$$\begin{aligned} v^B = \; & [a_B(Canada), 0.0(UK), b_B(US), c_B(other\ values_C), \\ & 0.0(MST), 0.0(other\ values_T), \\ & x_B(1200k), y_B(1300k), z_B(other\ values_{CP}] \end{aligned}$$

The similarity between both RDF-triples, i.e., sets of propositions, i.e., two vectors v_A and v_B, is calculated using cosine measure (Eq. 19).

5 Clustering of Propositions: Examples

In order to illustrate a clustering process and the influence of uncertainty on the obtained clusters, we include the results of a simple example of utilization of the presented method on six RDF stars/entities downloaded from *dbpedia.org* and representing different cities. Three of the cities are from France: Paris, Marseille, and Lyon; and three from US: Houston, Chicago, and Los Angeles.

5.1 Scenario I

The first experiment focuses on clustering all six cities when a full information about two of them – one from France one from US – become 'more and more' uncertain. The intermediate results of the clustering with increased levels of uncertainty are shown in Fig. 3.

The first dendogram (Fig. 3a) shows the clustering with the 'original', i.e., without uncertainty, information about cities. The second one (Fig. 3b) illustrates changes in clustering – properties of Paris and Houston have a level of uncertainty of 0.4, while the next two dendograms illustrates the results of clustering when levels of uncertainty increased to 0.6 and 1.0 (*do not know* state), respectively.

As expected, both cities that are associated with increased levels of uncertainty are 'drawn' to each other. The distance (Ward's distance) between them shrinks, compare Fig. 3b with Fig. 3c, and with Fig. 3d. At the very end, they create a group by themselves.

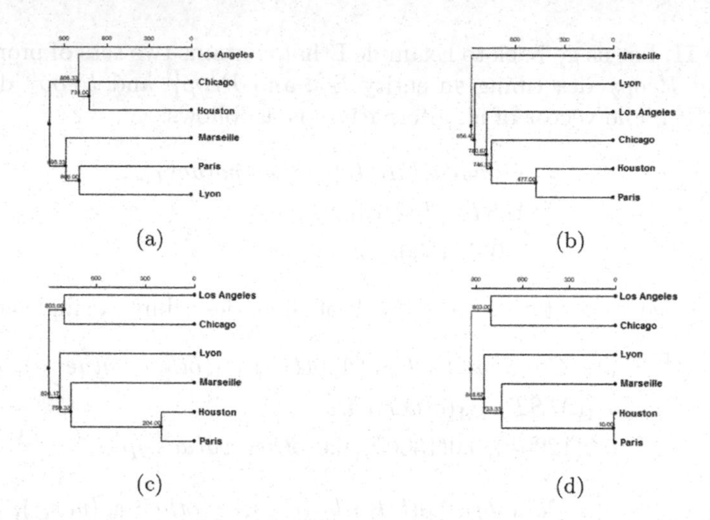

Fig. 3. Dendograms of clustering results: uncertainty for all properties of two cities Paris and Houston gradually increase: – no uncertainty (a); – uncertainty 0.4 (b); – uncertainty 0.6 (c) ; – uncertainty 1.0 (d).

5.2 Scenario II

The second example shows how clustering results change in the case when only a few properties reach the status *do not know.* The experiment also targets two cities: Paris and Houston, Fig. 4.

As before Fig. 4a illustrates the initial – no uncertainty – situation. Figure 4b shows grouping when one property, *type*, exhibits uncertainty of 0.6. As we can see, besides smaller values of distances, the hierarchy stays the same. Situation changes when level of uncertainty of 0.6 'spreads' among top five properties: *type, subject, seeAlso, abstract, externalLink.* The last dendogram show a total degradation of certainty across all properties.

5.3 Scenario III

The last experiment addresses a situation when two pairs of cities have uncertainties associated with different properties. Two sets of cities are used: Paris and Houston; and Lyon and Chicago. The first pair has not definite information for properties *type, subject*, while the second pair for properties *abstract, seeAlso.* Figure 5a show the initial – all is known – situation, while Fig. 5b, the situation *do not know.* As we can see, the distances are smaller when compared with the initial situation. However, they are larger when compared with Figs. 3d and 5d due to the fact that only two properties experiences uncertainty compared with all of them for Scenarios I and II.

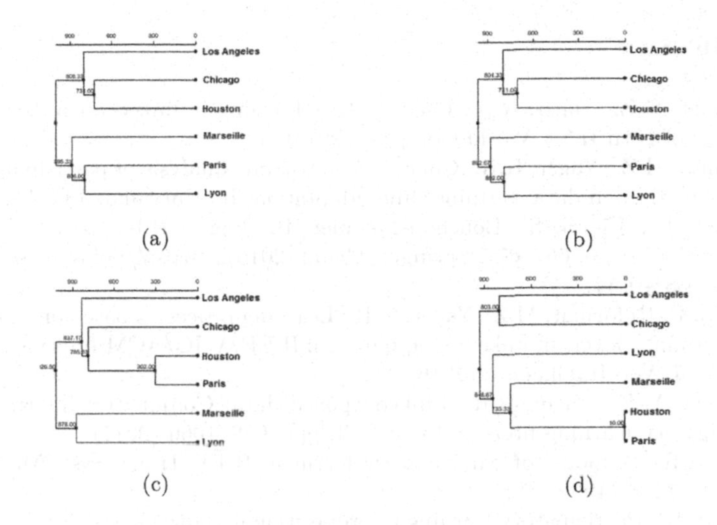

Fig. 4. Dendograms of clustering results: uncertainty for selected properties of two cities Paris and Huston: – no uncertainty (a); – uncertainty 0.4 for one property (b); – uncertainty 0.6 of five top properties (c); – uncertainty 1.0 of five top properties (d).

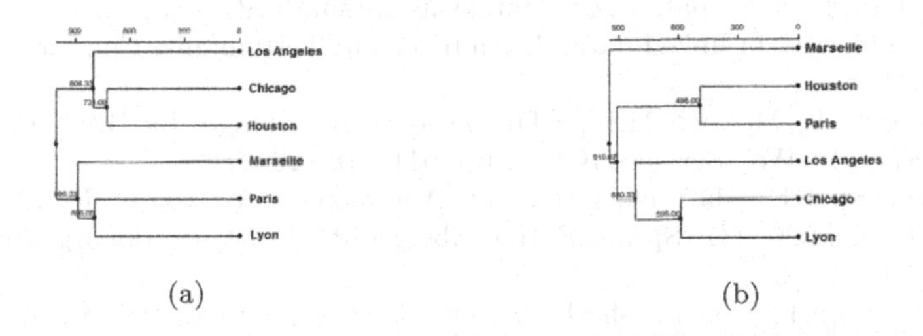

Fig. 5. Dendograms of clustering results: uncertainty for selected properties of two pairs of cities: – no uncertainty (a); – uncertainty 0.6 of five top properties (b).

6 Discussion and Conclusion

The paper introduces a simple methodology for determining similarity between entities that are represented as sets of propositions with uncertainty. Each proposition is a collection of alternative statements describing a single feature of the entity. Further, each alternative is associated with a level of uncertainty expressing a degree of confidence in its correctness. Those propositions are constructed based on RDF triples that are building blocks of knowledge graphs.

We show details of transforming sets of propositions describing two entities into two vectors of uncertainty levels associated with alternative statements. That leads to representing a single entity as a single vector. Once two vectors are created a cosine similarity measure is used to determine similarity between vectors – entities. We illustrate the proposed approach with a number of examples.

References

1. Reformat, M.Z., Yager, R.R.: Participatory learning in linked open data. In: Proceedings of 16th IFSA World Congress (2015)
2. Reformat, M.Z., Yager, R.R., Chen, J.X.: Dynamic analysis of participatory learning in linked open data: certainty and adaptation. In: Carvalho, J.P., Lesot, M.-J., Kaymak, U., Vieira, S., Bouchon-Meunier, B., Yager, R.R. (eds.) IPMU 2016. CCIS, vol. 611, pp. 667–677. Springer, Cham (2016). https://doi.org/10.1007/978-3-319-40581-0_54
3. Chen, J.X., Reformat, M.Z., Yager, R.R.: Learning processes based on data sources with certainty levels in linked open data. In: IEEE/WIC/ACM International Conference on Web Intelligence (2016)
4. Reformat, M.Z., Yager, R.R.: Linked opened data: Conjunctive information and participatory learning process. In: WCCI, pp. 1059–1066 (2017)
5. Yager, R.R.: A model of participatory learning. IEEE Trans. Syst. Man Cybern. **20**, 1229–1234 (1990)
6. Yager, R.R.: Participatory learning of propositional knowledge. IEEE Trans. Fuzzy Sets Syst. **20**, 715–727 (2012)
7. http://www.w3.org/RDF/
8. Berners-Lee, T., Hendler, J., Lassila, O.: The semantic web. Sci. Am. **284**, 29–37 (2001)
9. Zadeh, P.D.H., Reformat, M.Z.: Context-aware similarity assessment within semantic space formed in linked data. J. Ambient Intell. Humaniz. Comput. **4**, 515–532 (2013)
10. Levandoski, J., Mokbel, M.F.: RDF data-centric storage. In: IEEE International Conference on Web Services ICWS, pp. 911–918 (2009)
11. Giannini, S.: RDF data clustering. In: Abramowicz, W. (ed.) BIS 2013. LNBIP, vol. 160, pp. 220–231. Springer, Heidelberg (2013). https://doi.org/10.1007/978-3-642-41687-3_21
12. Lalithsena, S., Hitzler, P., Sheth, A., Jain, P.: Automatic domain identification for linked open data. In: IEEE/WIC/ACM International Conference on Web Intelligence and Intelligent Agent Technology, pp. 205–212 (2013)
13. Christodoulou, K., Paton, N.W., Fernandes, A.A.A.: Structure inference for linked data sources using clustering. In: EDBT/ICDT Workshops, pp. 60–67 (2013)
14. Ferrara, A., Genta, L., Montanelli, S.: Linked data classification: a feature-based approach. In: EDBT/ICDT Workshops, pp. 75–82 (2013)
15. Zong, N., Im, D., Yang, S., Namgoon, H., Kim, H.: Dynamic generation of concepts hierarchies for knowledge discovering in bio-medical linked data sets. In: Proceedings of 6th International Conference on Ubiquitous Information Management and Communication (2012)
16. Chen, J.X., Reformat, M.Z.: Learning categories from linked open data. In: Laurent, A., Strauss, O., Bouchon-Meunier, B., Yager, R.R. (eds.) IPMU 2014. CCIS, vol. 444, pp. 396–405. Springer, Cham (2014). https://doi.org/10.1007/978-3-319-08852-5_41
17. Zadeh, L.: A theory of approximate reasoning. Mach. Intell. **9**, 149–194 (1979)

Robust Lookup Table Controller Based on Piecewise Multi-linear Model for Nonlinear Systems with Parametric Uncertainty

Tadanari Taniguchi[1(✉)] and Michio Sugeno[2]

[1] Tokai University, Hiratsuka, Kanagawa 2591292, Japan
taniguchi@tokai-u.jp
[2] Tokyo Institute of Technology, Yokohama, Kanagawa 2268503, Japan

Abstract. This paper proposes a robust lookup table controller based on piecewise multi-linear model for nonlinear systems with parametric uncertainty. We construct a piecewise multi-linear model of a nonlinear system. The model is a nonlinear approximation and the model can be derived from fuzzy if-then rules with singleton consequents. The piecewise model can be expressed as a lookup table. The model dynamics is described by multi-linear interpolation of the lookup table elements. We design a robust piecewise multi-linear controller for the piecewise model via feedback linearization. The robust piecewise controller can be also expressed as a lookup table. We apply the robust lookup table controller to ball and beam system as a nonlinear system with parametric uncertainty. Examples are shown to confirm the feasibility of our proposals by computer simulations.

Keywords: Lookup table · Robust control · Parametric uncertainty
Feedback linearization · Piecewise model

1 Introduction

A lookup table (LUT) is an array of data that maps input values to output values. Because LUT can reduce computational load and time, it is widely used in various fields. LUT is also widely used in control engineering [1,2], especially industrial fields [3,4]. However many control systems with LUT controllers do not take into account the stability analysis because it is very difficult to analyze the stability.

This paper proposes a robust nonlinear control system represented by LUTs based on piecewise multi-linear (PMLs) models. We construct a PML model of a nonlinear system. The piecewise model is a nonlinear approximation. The model is built on hyper cubes partitioned in state space and is found to be multi-linear [5], so the model has simple nonlinearity. The model has the following features:

J. Medina et al. (Eds.): IPMU 2018, CCIS 855, pp. 727–738, 2018.
https://doi.org/10.1007/978-3-319-91479-4_60

(1) The PML model is derived from fuzzy if-then rules with singleton consequents. (2) It has a general approximation capability for nonlinear systems. (3) It is a piecewise nonlinear model and second simplest after the piecewise linear (PL) model. (4) It is continuous and fully parametric. The stabilizing conditions are represented by bilinear matrix inequalities (BMIs) [6], therefore, it takes long computing time to obtain a stabilizing controller. To overcome these difficulties, we have derived stabilizing conditions [7–9] based on feedback linearization, where [7,9] apply input-output linearization and [8] applies full-state linearization. The control system has the following features: (1) It is not necessary to get the overall model dynamics of an objective plant, but only the vertex values of the plant. (2) These control systems are applicable to a wider class of nonlinear systems than conventional feedback linearization. (3) The piecewise model can be also expressed as an LUT. The internal model dynamics is described by multi-linear interpolation of the LUT elements. Because the piecewise model has an approximation error it is necessary to design a robust controller. We design a robust piecewise controller for the PML model via feedback linearization. The robust piecewise controller can be also expressed as an LUT.

This paper is organized as follows. Section 2 introduces the canonical form of PML models. Section 3 briefly presents ball and beam (BAB) system as a nonlinear system. Section 4 proposes an LUT model based on PML models for BAB system via feedback linearization. Sections 5 and 6 propose LUTs controller and robust controller based on PML models via feedback linearization. Section 7 shows an example demonstrating the feasibility of the proposed methods. Finally, Sect. 8 summarizes conclusions.

2 Canonical Forms of Piecewise Multi-linear Models

2.1 Open-Loop Systems

In this section, we introduce PML models suggested in [5]. We deal with the two-dimensional case without loss of generality. Define vector $d(\sigma,\tau)$ and rectangle $R_{\sigma\tau}$ in two-dimensional space as $d(\sigma,\tau) \equiv (d_1(\sigma), d_2(\tau))^T$,

$$R_{\sigma\tau} \equiv [d_1(\sigma), d_1(\sigma+1)] \times [d_2(\tau), d_2(\tau+1)]. \tag{1}$$

σ and τ are integers: $-\infty < \sigma, \tau < \infty$ where $d_1(\sigma) < d_1(\sigma+1), d_2(\tau) < d_2(\tau+1)$ and $d(0,0) \equiv (d_1(0), d_2(0))^T$. Superscript ^{T} denotes a *transpose* operation.

We consider a two-dimensional nonlinear system: $\dot{x} = f(x)$. For $x = (x_1, x_2) \in R_{\sigma\tau}$, the PML system is expressed as

$$\begin{cases} \dot{x} = f_p(x) = \sum_{i=\sigma}^{\sigma+1} \sum_{j=\tau}^{\tau+1} \omega_1^i(x_1)\omega_2^j(x_2) f(i,j), \\ x = \sum_{i=\sigma}^{\sigma+1} \sum_{j=\tau}^{\tau+1} \omega_1^i(x_1)\omega_2^j(x_2) d(i,j), \end{cases} \tag{2}$$

where $f(i,j)$ is the vertex of nonlinear system $\dot{x} = f(x)$,

$$\begin{cases} \omega_1^{\sigma}(x_1) = (d_1(\sigma+1) - x_1)/(d_1(\sigma+1) - d_1(\sigma)), \\ \omega_1^{\sigma+1}(x_1) = (x_1 - d_1(\sigma))/(d_1(\sigma+1) - d_1(\sigma)), \\ \omega_2^{\tau}(x_2) = (d_2(\tau+1) - x_2)/(d_2(\tau+1) - d_2(\tau)), \\ \omega_2^{\tau+1}(x_2) = (x_2 - d_2(\tau))/(d_2(\tau+1) - d_2(\tau)) \end{cases} \tag{3}$$

and $\omega_1^i(x_1), \omega_2^j(x_2) \in [0,\ 1]$. In the above, we assume $f(0,0) = 0$ and $d(0,0) = 0$ to guarantee $\dot{x} = 0$ for $x = 0$.

A key point in the system is that state variable x is also expressed by a convex combination of $d(i,j)$ for $\omega_1^i(x_1)$ and $\omega_2^j(x_2)$, just as in the case of $\dot{x}$. As seen in Eq. (3), x is located inside $R_{\sigma\tau}$ which is a rectangle: a hypercube in general. That is, the expression of x is polytopic with four vertices $d(i,j)$. The model of $\dot{x} = f(x)$ is built on a rectangle including x in state space, it is also polytopic with four vertices $f(i,j)$. We call this form of the canonical model (2) parametric expression. Figure 1 shows the expression of $f(x)$.

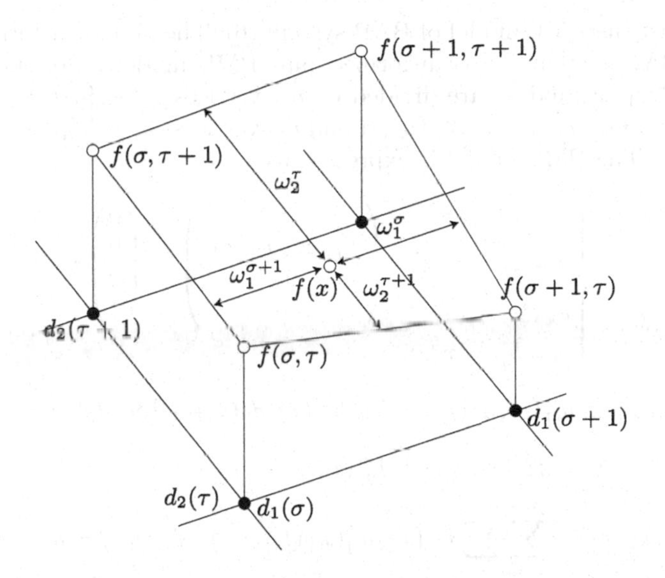

Fig. 1. Piecewise region ($f_{p_1}(x) = \sum_{i=\sigma}^{\sigma+1} \sum_{j=\tau}^{\tau+1} \omega_1^i \omega_2^j f(i,j)$, $x \in R_{\sigma\tau}$)

3 Ball and Beam System

The dynamics of ball and beam (BAB) system [10] is

$$\begin{cases} 0 = \ddot{r} + G\sin\theta - r\dot{\theta}^2, \\ \tau = (Mr^2 + J)\ddot{\theta} + 2Mr\dot{r}\dot{\theta} + MGr\cos\theta \end{cases} \tag{4}$$

where J is the moment of inertia of the beam, M is the mass of the ball and G is the acceleration of the gravity. θ is the angle of the beam and r is the position of the ball. τ is the torque applied to the beam. Using the invertible transformation

$$\tau = 2Mr\dot{r}\dot{\theta} + MGr\cos\theta + (Mr^2 + J)u \tag{5}$$

to define a new input variable u, the system is expressed as

$$\begin{cases} \begin{pmatrix} \dot{x}_1 \\ \dot{x}_2 \\ \dot{x}_3 \\ \dot{x}_4 \end{pmatrix} = \begin{pmatrix} x_2 \\ x_1 x_4^2 - G\sin x_3 \\ x_4 \\ 0 \end{pmatrix} + \begin{pmatrix} 0 \\ 0 \\ 0 \\ 1 \end{pmatrix} u, \\ y = x_1 \end{cases} \tag{6}$$

where $x = (x_1, x_2, x_3, x_4)^T = (r, \dot{r}, \theta, \dot{\theta})$.

4 Lookup Table Model Based on PML Models

We construct the PML model of BAB system (6). The nonlinear terms $x_1 x_4^2$ and $\sin x_3$ of BAB system are transformed into PML model representations. The variables of x_1, x_3 and x_4 are divided by m_1 vertices, $x_1 \in \{d_1(1), \ldots, d_1(m_1)\}$, m_3 vertices, $x_3 \in \{d_3(1), \ldots, d_3(m_3)\}$ and m_4 vertices, $x_4 \in \{d_4(1), \ldots, d_4(m_4)\}$, respectively. The PML model is expressed as

$$\begin{cases} \dot{x} = f_p + g_p u = \begin{pmatrix} x_2 \\ f_{p_2}(x_1, x_3, x_4) \\ x_4 \\ 0 \end{pmatrix} + \begin{pmatrix} 0 \\ 0 \\ 0 \\ 1 \end{pmatrix} u \\ y = h_p = x_1, \end{cases} \tag{7}$$

where $x \in R_{\rho\tau\upsilon} \equiv [d_1(\rho), d_1(\rho+1)] \times [d_3(\tau), d_3(\tau+1)] \times [d_4(\upsilon), d_4(\upsilon+1)]$,

$$f_{p_2}(x_1, x_3, x_4) = f_{p_2}^1(x_1, x_4) + f_{p_2}^2(x_3),$$

$$f_{p_2}^1(x_1, x_4) = \sum_{i=\rho}^{\rho+1} \sum_{\ell=\upsilon}^{\upsilon+1} w_1^i(x_1) w_4^\ell(x_4) f_{s_1}(i, \ell), \;\; f_{s_1}(i, \ell) = d_1(i) d_4(\ell)^2,$$

$$f_{p_2}^2(x_3) = \sum_{k=\tau}^{\tau+1} w_3^k(x_3) f_{s_2}(k), \;\; f_{s_2}(k) = -G\sin d_3(k),$$

$$\omega_1^\rho(x_1) = \frac{(d_1(\rho+1) - x_1)}{(d_1(\rho+1) - d_1(\rho))}, \;\; \omega_1^{\rho+1}(x_1) = \frac{(x_1 - d_1(\rho))}{(d_1(\rho+1) - d_1(\rho))},$$

$$\omega_3^\tau(x_3) = \frac{(d_3(\tau+1) - x_3)}{(d_3(\tau+1) - d_3(\tau))}, \;\; \omega_3^{\tau+1}(x_3) = \frac{(x_3 - d_3(\tau))}{(d_3(\tau+1) - d_3(\tau))},$$

$$\omega_4^\tau(x_4) = \frac{(d_4(\tau+1) - x_4)}{(d_4(\tau+1) - d_4(\tau))}, \;\; \omega_4^{\upsilon+1}(x_4) = \frac{(x_4 - d_4(\upsilon))}{(d_4(\upsilon+1) - d_4(\upsilon))},$$

ρ, τ and υ are integer: $-\infty < \rho, \tau, \upsilon < \infty$, $d_1(\rho) < d_1(\rho+1)$, $d_3(\tau) < d_3(\tau+1)$ and $d_4(\upsilon) < d_4(\upsilon+1)$. The model is found to be fully parametric and its model can be represented by a Lookup table (LUT). Table 1 shows the LUT model with respect to $f_{p_2}(x_1, x_3, x_4)$. The internal model dynamics is described by multi-linear interpolation of the lookup table elements (see Fig. 1).

Note that trigonometric functions of BAB system (6) are smooth functions and are of class C^∞. The PML models are not of class C^∞. In BAB system control, we have to calculate the fourth derivatives of the output y. Therefore the derivative PML models lose some dynamics. In this paper we design a robust piecewise controller as a countermeasure for the approximation error of PML model method.

Table 1. LUT model of $f_{p_2}(x_1, x_3, x_4)$

x_3	x_1	x_4				
		−0.3	−0.15	0	0.15	0.3
$-\pi/8$	−2	3.574	3.709	3.754	3.709	3.574
	−1	3.664	3.732	3.754	3.732	3.664
	0	3.754	3.754	3.754	3.754	3.754
	1	3.844	3.777	3.754	3.777	3.844
	2	3.934	3.799	3.754	3.799	3.934
$-\pi/16$	−2	1.734	1.869	1.914	1.869	1.734
	−1	1.824	1.891	1.914	1.891	1.824
	0	1.914	1.914	1.914	1.914	1.914
	1	2.004	1.936	1.914	1.936	2.004
	2	2.094	1.959	1.914	1.959	2.094
0	−2	−0.180	−0.045	0	−0.045	−0.180
	−1	−0.090	−0.023	0	−0.023	−0.090
	0	0	0	0	0	0
	1	0.090	0.023	0	0.023	0.090
	2	0.180	0.045	0	0.045	0.180
$\pi/16$	−2	−2.094	−1.959	−1.914	−1.959	−2.094
	−1	−2.004	−1.936	−1.914	−1.936	−2.004
	0	−1.914	−1.914	−1.914	−1.914	−1.914
	1	−1.824	−1.891	−1.914	−1.891	−1.824
	2	−1.734	−1.869	−1.914	−1.869	−1.734
$\pi/8$	−2	−3.934	−3.799	−3.754	−3.799	−3.934
	−1	−3.844	−3.777	−3.754	−3.777	−3.844
	0	−3.754	−3.754	−3.754	−3.754	−3.754
	1	−3.664	−3.731	−3.754	−3.732	−3.664
	2	−3.574	−3.709	−3.754	−3.709	−3.574

5 Lookup Table Controller Based on PML Models via Feedback Linearization

We define the output as $y = x_1$ in $x \in R_{\rho\sigma\tau\upsilon}$, the time derivative of y is calculated as $\dot{y} = L_{f_p} h_p = x_2$. The time derivative of y doesn't contain the control inputs u. We calculate the time derivative of $\dot{y}$. We get

$$\ddot{y} = L_{f_p}^2 h_p = f_{p_2}(x_1, x_3, x_4) = \sum_{i=\rho}^{\rho+1} \sum_{\ell=\upsilon}^{\upsilon+1} w_1^i(x_1) w_4^\ell(x_4) f_{s_1}(i, \ell) + \sum_{k=\tau}^{\tau+1} w_3^k(x_3) f_{s_2}(k).$$

The time derivative of $\dot{y}$ also doesn't contain the control inputs u. We continue to calculate the time derivative of $\ddot{y}$. We get

$$y^{(3)} = L_{f_p}^3 h_p + L_{g_p} L_{f_p}^2 h_p u = \frac{\partial f_{p_2}^1(x_1, x_4)}{\partial x_1} x_2 + \frac{\partial f_{p_2}^2(x_3)}{\partial x_3} x_4 + \frac{\partial f_{p_2}^1(x_1, x_4)}{\partial x_4} u. \quad (8)$$

The piecewise controller u derived from (8) can not be defined at $x_1 = 0$ or $x_4 = 0$. Therefore we consider the following approximate feedback linearization [10].

$$y^{(3)} \equiv L_{f_p}^3 h_p = \frac{\partial f_{p_2}^2(x_3)}{\partial x_3} x_4 = \frac{f_{s_2}(\tau+1) - f_{s_2}(\tau)}{d_3(\tau+1) - d_3(\tau)} x_4$$

We continue to calculate the time derivative of $y^{(3)}$. We get

$$y^{(4)} = L_{g_p} L_{f_p}^3 h_p u = \frac{f_{s_2}(\tau+1) - f_{s_2}(\tau)}{d_3(\tau+1) - d_3(\tau)} u$$

The stabilizing controller of (7) is designed as

$$u = \alpha_c(x) + \beta_c(x)\mu \quad (9)$$

where

$$\alpha_c(x) = -\frac{L_{f_p}^4 h_p}{L_{g_p} L_{f_p}^3 h_p} = 0, \ \beta_c(x) = \frac{1}{L_{g_p} L_{f_p}^3 h_p} = \frac{d_3(\tau+1) - d_3(\tau)}{f_{s_2}(\tau+1) - f_{s_2}(\tau)},$$

$\mu = -F\zeta_c$ is the linear controller of the following linear system (10).

$$\begin{cases} \dot{\zeta}_c = A_c \zeta_c + B_c \mu, \\ y = C_c \zeta_c, \end{cases} \quad (10)$$

where $\zeta_c = (h_p, \ L_{f_p} h_p, \ L_{f_p}^2 h_p, \ L_{f_p}^3 h_p)^T$, A_c, B_c and C_c are the matrices of the Brunovsky canonical form. If $f_s(i) \neq f_s(i+1)$ and $d_3(i) \neq d_3(i+1)$, $i = 1, \ldots, m$, there exists a stabilizing controller (9) of BAB system (7) since $\det(L_{g_p} L_{f_p}^3 h_p) \neq 0$. Thus we have to construct the PML model of BAB system such that $f_s(i) \neq f_s(i+1)$ and $d_3(i) \neq d_3(i+1)$, where $i = 1, \ldots, m$ (see Fig. 2).

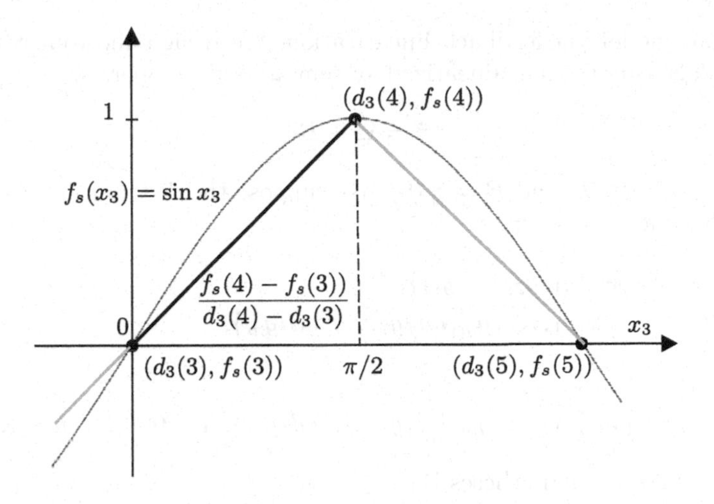

Fig. 2. PML modeling ($d_3(3) = 0$, $d_3(4) = \pi/2$, $d_3(5) = \pi$)

Next we consider the PML model of the torque (5).

$$\tau_p = \tau_{p_1} + \tau_{p_2} + \tau_{p_3} u, \tag{11}$$

where $x \in R^{\rho\sigma\tau\upsilon}$,

$$\tau_{p_1} = \sum_{i=\rho}^{\rho+1} \sum_{j=\sigma}^{\sigma+1} \sum_{\ell=\upsilon}^{\upsilon+1} \omega_1^i(x_1)\omega_2^j(x_2)\omega_4^\ell(x_4)\tau_{s_1}(i,j,\ell),$$

$$\tau_{s_1}(i,j,\ell) = 2Md_1(i)d_2(j)d_4(\ell),$$

$$\tau_{p_2} = \sum_{i=\rho}^{\rho+1} \sum_{k=\tau}^{\tau+1} \omega_1^i(x_1)\omega_3^k(x_3)\tau_{s_2}(i,k), \;\; \tau_{s_2}(i,k) = MGd_1(i)\cos d_3(k),$$

$$\tau_{p_3} = \sum_{i=\rho}^{\rho+1} \omega_1^i(x_1)\tau_{s_3}(i), \;\; \tau_{s_3}(i) = Md_1(i)^2 + J,$$

Finally we get the torque controller applied to the beam when the controller (9) is substituted into the torque controller (11).

The torque controller is found to be fully parametric and its controller can be represented by an LUT. Due to lack of space, the LUT controller at $x_2 = -\pi/16$ is only showed in Table 2a. The internal model dynamics is described by multi-linear interpolation of the LUT elements.

6 Robust Lookup Table Controller Based on PML Models via Feedback Linearization

The PML model is a nonlinear approximation. Therefore it is necessary to design a robust controller. We design a robust piecewise controller for the piecewise multi-linear model via feedback linearization. We design the robust controller from the following tangent linearized system around an operating point.

$$\dot{z} = Az + Bv, \tag{12}$$

where $A = \partial f_p(0)/\partial x$ and $B = g_p(0)$. We suppose the distributions G_0, G_1, ... , G_{n-1} defined as

$$\begin{aligned} G_0 &= \text{span}\{g_1, g_2, \ldots, g_m\}, \\ G_1 &= \text{span}\{g_1, \ldots, g_m, ad_f g_1, \ldots, ad_f g_m\}, \\ &\vdots \\ G_i &= \text{span}\{g_1, \ldots, g_m, ad_f g_1, \ldots, ad_f g_m\},\ i = 0, 1, \ldots, n-1, \end{aligned}$$

satisfy the following hypotheses [11]

(i) Distribution G_i has constant dimension near $x = 0$ for $0 \leq i \leq n-1$
(ii) Distribution G_{n-1} has dimension n.
(iii) Distribution G_i is the involutive for $0 \leq i \leq n-2$.

Consider system and suppose that $f_p(x)$ and $g_p(x)$ satisfy hypotheses (i), (ii) and (iii). Then under the controller

$$u(x, v) = \alpha(x) + \beta(x)\nu \tag{13}$$

and the coordinate transformation vector $z = \zeta(x)$ defined by

$$\alpha(x) = \alpha_c(x) + \beta_c(x)LT^{-1},\ \beta(x) = \beta_c(x)R^{-1},\ \zeta(x) = T^{-1}\zeta_c(x),$$

where

$$L = -L_{g_p}L_{f_p}^3 h_p \left.\frac{\partial \alpha_c}{\partial x}\right|_{x=0},\ T = \left.\frac{\partial \zeta_c}{\partial x}\right|_{x=0},\ R = \frac{1}{L_{g_p}L_{f_p}^3 h_p},$$

the system (7) is transformed into the system (12). A robust linear controller ν is substituted into the controller (13). As discussed in the previous section, substituting the controller (13) to (11) we get the torque controller τ_p.

The torque controller τ_p is also found to be fully parametric and its controller can be represented by an LUT. Due to lack of space, the LUT controller at $x_2 = -\pi/16$ is only showed in Table 2b. The internal model dynamics is described by multi-linear interpolation of the lookup table elements.

7 Simulation Results

We apply the LUT controller in Table 2a and the robust LUT controller in Table 2b to a nominal BAB system (4) and a BAB system with parameter

Table 2. LUT controllers based on PML model ($x_2 = -\pi/16$)

(a) LUT controller

x_3	x_1	x_4: -0.3	-0.15	0	0.15	0.3
$-\pi/8$	-2	-0.792	-0.962	-1.153	-1.364	-1.595
	-1	-0.475	-0.546	-0.621	-0.701	-0.784
	0	0.246	0.199	0.153	0.107	0.061
	1	1.457	1.344	1.234	1.129	1.028
	2	3.248	2.957	2.686	2.435	2.205
$-\pi/16$	-2	-1.327	-1.499	-1.690	-1.900	-2.131
	-1	-0.701	-0.773	-0.848	-0.927	-1.010
	0	0.160	0.114	0.067	0.021	-0.025
	1	1.340	1.227	1.117	1.012	0.911
	2	2.925	2.634	2.364	2.113	1.881
0	-2	-1.781	-1.953	-2.144	-2.354	-2.584
	-1	-0.886	-0.958	-1.033	-1.112	-1.196
	0	0.077	0.030	-0.016	-0.062	-0.108
	1	1.192	1.079	0.970	0.865	0.763
	2	2.546	2.256	1.986	1.735	1.503
$\pi/16$	-2	-2.185	-2.355	-2.546	-2.757	-2.988
	-1	-1.043	-1.114	-1.189	-1.269	-1.352
	0	-0.010	-0.057	-0.103	-0.149	-0.195
	1	1.000	0.887	0.778	0.672	0.571
	2	2.079	1.787	1.516	1.266	1.035
$\pi/8$	-2	-2.489	-2.659	-2.850	-3.061	-3.292
	-1	-1.153	-1.225	-1.300	-1.379	-1.463
	0	-0.094	-0.140	-0.186	-0.232	-0.278
	1	0.778	0.665	0.556	0.450	0.349
	2	1.551	1.260	0.989	0.738	0.508

(b) Robust LUT controller

x_3	x_1	x_4: -0.3	-0.15	0	0.15	0.3
$-\pi/8$	-2	-0.073	-0.331	-0.661	-1.062	-1.518
	-1	0.009	-0.110	-0.246	-0.397	-0.557
	0	0.585	0.509	0.430	0.350	0.272
	1	2.332	2.145	1.967	1.798	1.646
	2	5.924	5.418	4.964	4.566	4.239
$-\pi/16$	-2	-1.481	-1.752	-2.087	-2.484	-2.942
	-1	-0.566	-0.690	-0.827	-0.977	-1.139
	0	0.325	0.247	0.168	0.088	0.009
	1	1.868	1.677	1.497	1.330	1.175
	2	4.735	4.219	3.762	3.366	3.034
0	-2	-2.833	-3.112	-3.447	-3.837	-4.295
	-1	-1.111	-1.238	-1.375	-1.522	-1.683
	0	0.062	-0.018	-0.097	-0.175	-0.254
	1	1.361	1.167	0.987	0.823	0.668
	2	3.458	2.935	2.477	2.089	1.757
$\pi/16$	-2	-4.147	-4.427	-4.758	-5.136	-5.592
	-1	-1.632	-1.760	-1.896	-2.037	-2.198
	0	-0.207	-0.288	-0.367	-0.442	-0.520
	1	0.803	0.607	0.427	0.269	0.117
	2	2.073	1.545	1.091	0.715	0.389
$\pi/8$	-2	-5.322	-5.610	-5.940	-6.311	-6.767
	-1	-2.091	-2.222	-2.357	-2.496	-2.657
	0	-0.465	-0.547	-0.626	-0.699	-0.778
	1	0.232	0.034	-0.145	-0.301	-0.454
	2	0.675	0.139	-0.315	-0.683	-1.010

variation in computer simulations. To construct the PML model of BAB system, the state variables x_1, x_2, x_3 and x_4 of BAB system (4) are divided by the following vertices

$$x_1 \in \{-2,\ -1,\ 0,\ 1,\ 2\},\ x_2 \in \{-1,\ -0.5,\ 0,\ 0.5,\ 1\},$$
$$x_3 \in \{-\pi/8,\ -\pi/16,\ 0,\ \pi/16,\ \pi/8\},\ x_4 \in \{-0.3,\ -0.15,\ 0,\ 0.15,\ 0.3\}.$$

The initial condition is $x(0) = (1,\ 0,\ 0,\ 0)^T$ and the acceleration of the gravity $G = 9.81$ $[m/s^2]$. The nominal values of the system parameters are M is 0.1 [kg] and $J = 0.1$ [kg$\cdot m^2$]. Figure 3 shows that the control responses $x_1, \ldots, x_4$ of the nominal BAB system. In Figs. 4 and 5, we consider the parameter variations with respect to the mass M of the ball. Figures 4 and 5 show the control responses of the BAB systems with parameter variations (the masses of the ball: $M' = 1.1M$ and $M' = 1.2M$). The results show that the feasibility of the proposed robust LUT controller for BAB system with the parameter variations.

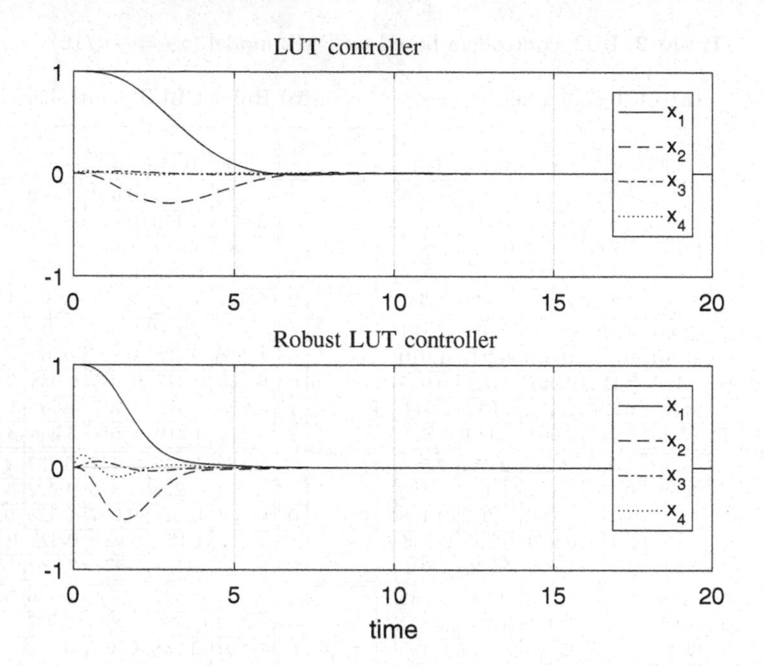

Fig. 3. State responses of nominal BAB system

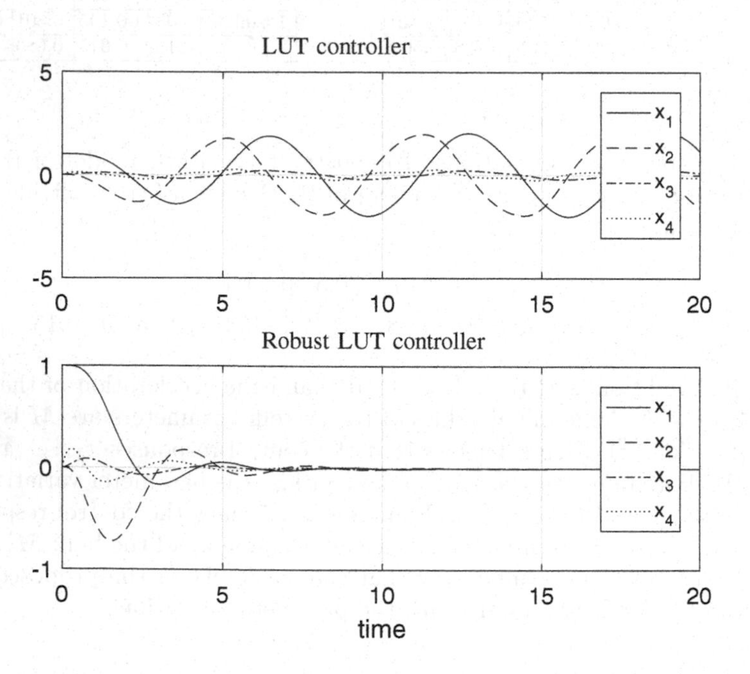

Fig. 4. State responses of BAB system with parameter variation ($M' = 1.1M$)

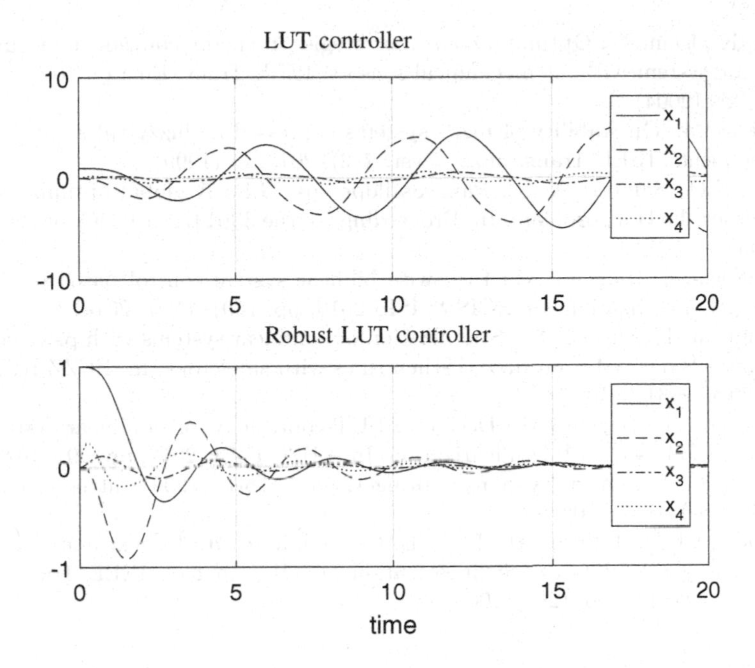

Fig. 5. State responses of BAB system with parameter variation ($M' = 1.2M$)

8 Conclusions

This paper has proposed a robust lookup table controller based on piecewise multi-linear model for nonlinear systems with parametric uncertainty. We have constructed a piecewise multi-linear model of a nonlinear system. The model is a piecewise multi-linear system and a nonlinear approximation. The piecewise model can be expressed as a lookup table. The model dynamics is described by multi-linear interpolation of the lookup table elements. We have designed a robust piecewise multi-linear controller for the piecewise model via feedback linearization. The robust piecewise controller can be also expressed as a lookup table. We have applied the robust lookup table controller to ball and beam system. Examples have been shown to confirm the feasibility of our proposals by computer simulations.

References

1. Ang, K.H., Chong, G., Li, Y.: PID control system analysis, design, and technology. IEEE Trans. Control Syst. Technol. **13**(4), 559–576 (2005)
2. Li, Y.F., Lau, C.C.: Development of fuzzy algorithms for servo systems. IEEE Control Syst. Mag. **9**(3), 65–72 (1989)
3. Wang, Q., Chang, L.: An intelligent maximum power extraction algorithm for inverter-based variable speed wind turbine systems. IEEE Trans. Power Electron. **19**(5), 1242–1249 (2004)

4. Tan, K., Islam, S.: Optimum control strategies in energy conversion of pmsg wind turbine system without mechanical sensors. IEEE Trans. Energy Convers. **19**(2), 392–399 (2004)
5. Sugeno, M.: On stability of fuzzy systems expressed by fuzzy rules with singleton consequents. IEEE Trans. Fuzzy Syst. **7**(2), 201–224 (1999)
6. Goh, K.C., Safonov, M.G., Papavassilopoulos, G.P.: A global optimization approach for the BMI problem. In: Proceedings of the 33rd IEEE CDC, pp. 2009–2014 (1994)
7. Taniguchi, T., Sugeno, M.: Piecewise bilinear system control based on full-state feedback linearization. In: SCIS & ISIS 2010, pp. 1591–1596 (2010)
8. Taniguchi, T., Sugeno, M.: Stabilization of nonlinear systems with piecewise bilinear models derived from fuzzy if-then rules with singletons. In: FUZZ-IEEE 2010, pp. 2926–2931 (2010)
9. Taniguchi, T., Sugeno, M.: Design of LUT-controllers for nonlinear systems with PB models based on I/O linearization. In: FUZZ-IEEE 2012, pp. 997–1022 (2012)
10. Sastry, S.: Nonlinear Systems. Springer, New York (1999). https://doi.org/10.1007/978-1-4757-3108-8
11. Franco, A.L.D., Bourles, H., Pieri, E.R.D., Guillard, H.: Robust nonlinear control associating robust feedback linearization and H_∞ control. IEEE Trans. Autom. Control **51**(7), 1200–1207 (2006)

Credal C4.5 with Refinement of Parameters

Carlos J. Mantas[1(✉)], Joaquín Abellán[1], Javier G. Castellano[1], José R. Cano[2], and Serafín Moral[1]

[1] Department of Computer Science and A.I, University of Granada, Granada, Spain
{cmantas,jabellan,fjgc,smoral}@decsai.ugr.es
[2] Department of Computer Science, University of Jaén, Jaén, Spain
jrcano@ujaen.es

Abstract. Recently, a classification method called Credal C4.5 (CC4.5) has been presented which combines imprecise probabilities and the C4.5 algorithm. The action of the CC4.5 algorithm depends on a parameter s. In previous works, it has been shown that this parameter has relation with the degree of overfitting of the model. The noise level of a data set can influence on the choice of a good value for s. In this paper, it is presented a new method based on the CC4.5 method with a refining of its parameter in the time of training. The new method has an equivalent performance than CC4.5 with the best value of s for each level noise.

Keywords: Classification · Credal sets · Decision trees
Imprecise probabilities

1 Introduction

An algorithm that designs decision trees, called *credal decision trees* (CDTs), has been developed [1] by using Imprecise Dirichlet Model (IDM). In [5,6], the C4.5 algorithm and the theory of CDTs have been related with the presentation of the Credal C4.5 (CC4.5) algorithm, which uses the split criterion called *Imprecise Info-Gain Ratio* (IIGR).

The models CDT and CC4.5, called as *credal trees*, are suitable to classify noisy data sets because decision trees are built by using imprecise probabilities. In this manner, it is supposed that the data sets are not clean.

The parameter s determines the imprecision degree of the probability distributions in the CDTs. According IDM, The probability for each possible value of a variable is within an interval. This interval is wider if the value of s is higher. If the value of s is small then we think that the data set is reliable in order to estimate probabilities. In the previous works, the value $s = 1$ is always used due to computational reasons and the recommendation of the author of the IDM (see [5]). For this reason, the imprecision degree of the probability distributions was constant.

J. Medina et al. (Eds.): IPMU 2018, CCIS 855, pp. 739–747, 2018.
https://doi.org/10.1007/978-3-319-91479-4_61

In a recent work [7], it was analyzed the relation between the noise level of a data set and the action of CC4.5 on this data set with a concrete value for s. It was concluded that it is important to choose correctly the value for s in terms of the noise level of a data set. Usually, it is interesting to use high values of s for data sets with high noise level. However, this is not a rule always correct for all the data sets, as it can be seen in [7]. It can not be determined a value of s for each noise level and for all the data sets. Besides, The estimation of the noise level can be incorrect.

In this paper, it is presented a new method (called cv-CC4.5) where the value of the s parameter for the CC4.5 algorithm is determined by means of a parameter selection procedure by cross-validation [4] for each training data set. An experimental study is finally shown.

2 Credal Decision Trees

The Credal Decision Trees (CDTs) are built by using a split criterion based on imprecise probabilities and uncertainty measures on credal sets. The mathematical basis of this procedure can be described as follows: Let Z be a variable with values in $\{z_1, \ldots, z_k\}$. Let us suppose a probability distribution $p(z_j), j = 1, .., k$ defined for each value z_j from a data set.

Walley's Imprecise Dirichlet Model (IDM) [9] is used to estimate probability intervals from the data set for each value of the variable Z, in the following way:

$$p(z_j) \in \left[\frac{n_{z_j}}{N+s}, \frac{n_{z_j}+s}{N+s} \right], \quad j = 1, .., k;$$

with n_{z_j} as the frequency of the set of values $(Z = z_j)$ in the data set, N the sample size and s a given parameter.

This model produces a type of credal set on the variable Z, $K(Z)$ (see Abellán [2]), defined as

$$K(Z) = \left\{ p \,|\, p(z_j) \in \left[\frac{n_{z_j}}{N+s}, \frac{n_{z_j}+s}{N+s} \right], \quad j = 1, .., k \right\}.$$

On this type of sets (credal sets), uncertainty measures can be applied. The function of the maximum of entropy on the previously defined credal set is used for building CDTs. This function, denoted as H^*, is defined as

$$H^*(K(Z)) = max \left\{ H(p) \,|\, p \in K(Z) \right\},$$

where the function H is the Shannon's entropy function.

An algorithm that calculates the distribution with maximum entropy for $s > 1$ is presented in [7].

3 Credal C4.5

The method to build a Credal C4.5 tree is equivalent to the C4.5 algorithm [8], the difference is that the CC4.5 algorithm estimates the values of the variables with imprecise probabilities and calculates split criteria by using uncertainty measures on credal sets. CC4.5 considers that the training set is not very reliable because it can be affected by class or attribute noise. So, CC4.5 can be considered as a proper method for noisy domains.

CC4.5 is defined by changing the *Info-Gain Ratio* split criterion from C4.5 with the *Imprecise Info-Gain Ratio* (IIGR) split criterion. This criterion is defined as: in a classification problem, let C be the class variable, $\{X_1, \ldots, X_m\}$ the set of features, and X a feature; then

$$IIGR^{\mathcal{D}}(C, X) = \frac{IIG^{\mathcal{D}}(C, X)}{H(X)},$$

where *Imprecise Info-Gain* (IIG) is:

$$IIG^{\mathcal{D}}(C, X) = H^*(K^{\mathcal{D}}(C)) - \sum_i P^{\mathcal{D}}(X = x_i) H^*(K^{\mathcal{D}}(C|X = x_i)),$$

with $K^{\mathcal{D}}(C)$ and $K^{\mathcal{D}}(C|X = x_i)$ are the credal sets obtained by the IDM for the variables C and $(C|X = x_i)$ respectively and for a partition $\mathcal{D}$ of the data set (see Abellán and Moral [1]); $P^{\mathcal{D}}(X = x_i)$ $(i = 1, ..., n)$ is a probability distribution of the credal set $K^{\mathcal{D}}(X)$.

It is chosen the probability distribution $P^{\mathcal{D}}$ from $K^{\mathcal{D}}(X)$ that maximizes the expression $\sum_i P(X = x_i) H(C|X = x_i))$. The procedure for calculating $P^{\mathcal{D}}$ and building Credal C4.5 trees can be found in [5,6].

4 The New cv-CC4.5 Classification Method

The cv-CC4.5 algorithm is an extension of the CC4.5 algorithm where the value of s is established using the training set before using CC4.5. So, this value for s depends on the data set to be classified and it is calculated before applying the algorithm. In previous works about CDTs, the value $s = 1$ was fixed, without observing the data set to be classified.

In the cv-CC4.5 algorithm, the value for s is determined by cross-validation from the training data set. So, s is calculated for each training data set. After that, the CC4.5 algorithm is executed with the calculated value for s.

The k-fold cross-validation [4] is used to find a good value of s for each data set. The original (training) sample is randomly partitioned into k equal sized subsamples. Of the k subsamples, a single subsample is retained as the validation data for testing the built decision tree, and the remaining $(k-1)$ subsamples are used as training data for designing a credal tree by using the CC4.5 algorithm with a concrete value for s. The cross-validation process is then repeated k times (the folds), with each of the k subsamples used exactly once as the validation data. The k results from the folds are averaged to produce an estimation of the goodness of the CC4.5 algorithm with this value of s. 10-fold cross-validation is commonly used.

The steps of the cv-CC4.5 method to obtain a value for the s parameter by k-fold cross-validation from a data set are:

(1) The interval $[0, s_{max}]$ is discretized into q values: s_1, ..., s_q.
(2) For each value $s_i \in \{s_1, ..., s_q\}$ a k-fold cross-validation is carried out to obtain an estimation of the CC4.5 algorithm with $s = s_i$ when the data set is classified.
(3) Finally, the value s_i with the best estimation is selected to be used in the execution of the CC4.5 algorithm.

5 Experimental Analysis

A group of 50 data sets, obtained from the "*UCI repository of machine learning data sets*", are used. They are anneal, arrhythmia, audiology, autos, balance-scale, breast-cancer, wisconsin-breast-cancer, car, cmc, horse-colic, credit-rating, german-credit, dermatology, pima-diabetes, ecoli, Glass, haberman, cleveland-14-heart-disease, hungarian-14-heart-disease, heart-statlog, hepatitis, hypothyroid, ionosphere, iris, kr-vs-kp, letter, liver-disorders, lymphography, mfeat-pixel, nursery, optdigits, page-blocks, pendigits, primary-tumor, segment sick, solar-flare2, sonar, soybean, spambase, spectrometer, splice, Sponge, tae, vehicle, vote, vowel, waveform, wine and zoo.

The C4.5 algorithm from *Weka* software [10], called *J48*, was used for the experimentation. Several methods were added to this software to build Credal C4.5 trees with the same experimental conditions and $s \in \{0.25, 0.5, 1.0, 1.5, 2.0, 3.0\}$ (for $s = 0.0$ CC4.5 and C4.5 are equivalent).

For the cv-CC4.5 algorithm, the selection of the value for s by 10-fold cross-validation was implemented by using the Weka software wrapper method to perform a parameter selection by cross-validation [4]. In the experimentation for cv-CC4.5, the interval [0.0,3.0] was discretized into $q = 7$ values, that is, the discrete set {0.0, 0.5, 1.0, 1.5, 2.0, 2.5, 3.0} was used as possible values for s in the 10-fold cross-validation process. Once selected the best value for s, the CC4.5 algorithm with this value for s was used to compare with the other algorithms.

By using *Weka's* filters, random noise percentages of 0%, 5%, 10%, 20% and 30% were added to the class variable, only in the training data set. Finally, a 10-fold cross validation procedure was repeated 10 times for each data set.

A series of tests were used to compare the methods for a level of significance of $\alpha = 0.05$, the following tests were used: a **Friedman test** to check if all the procedures are equivalents and a pos-hoc **Nemenyi test** to compare all the algorithms to each other (see [3] for more references about the tests).

5.1 Results and Comments

Table 1 presents the average accuracy result for each method and each level of noise. More details are not shown by limitations of space. Table 2 shows

Table 1. Average accuracy result for C4.5, CC4.5 (varying s) and cv-CC4.5 on each level of noise.

Tree	Noise 0%	Noise 5%	Noise 10%	Noise 20%	Noise 30%
C4.5	*82.62*	81.77	80.77	78.20	74.14
$CC4.5_{s=0.25}$	82.45	*81.88*	80.97	78.67	74.89
$CC4.5_{s=0.5}$	82.44	81.87	81.10	79.03	75.41
$CC4.5_{s=1.0}$	82.31	81.85	*81.21*	79.44	76.38
$CC4.5_{s=1.5}$	81.96	81.61	81.03	*79.58*	76.95
$CC4.5_{s=2.0}$	81.54	81.25	80.77	79.55	77.32
$CC4.5_{s=3.0}$	81.00	80.67	80.29	79.35	**77.44**
cv-CC4.5	**82.77**	**82.07**	**81.29**	**79.79**	*77.38*

Table 2. Friedman's ranks of C4.5, CC4.5 (varying s) and cv-CC4.5 on each level of noise

Tree	Noise 0%	Noise 5%	Noise 10%	Noise 20%	Noise 30%
C4.5	*3.75*	4.16	5.32	6.19	6.42
$CC4.5_{s=0.25}$	3.81	*3.81*	4.81	5.82	6.00
$CC4.5_{s=0.5}$	4.01	4.03	3.83	5.24	5.57
$CC4.5_{s=1.0}$	4.00	4.00	**3.61**	4.39	4.91
$CC4.5_{s=1.5}$	4.87	4.58	3.92	*3.54*	3.88
$CC4.5_{s=2.0}$	5.84	5.39	4.98	3.60	**2.96**
$CC4.5_{s=3.0}$	6.12	6.32	5.83	4.00	*2.99*
cv-CC4.5	**3.61**	**3.71**	*3.70*	**3.22**	3.27

Friedman's ranks (numbers in bold fonts are the best result for each noise level, numbers in italic fonts are the second best value). Tables 3, 4, 5, 6 and 7 show the p-values of the Nemenyi test for the methods C4.5, CC4.5 (varying the s parameter) and cv-CC4.5 in the experimentation. In all the cases, Nemenyi procedure rejects the hypotheses that have a p-value$\leq$ 0.001786. The methods in bold fonts are better than its pair about accuracy. It is only shown the pairs of methods that are not equivalent by limitations of space.

From the experimentation, it can be deduced that the new method cv-CC4.5 achieves the best general results. The cv-CC4.5 algorithm is always better than the rest except for 30% of added noise where it is equivalent to the best (CC4.5 with $s = 2.0$). For low levels of added noise, cv-CC4.5 is significantly better than CC4.5 with high values for s. Besides, for these low noise levels, cv-CC4.5 and CC4.5 with low values for s achieves the best results. On the other hand, for high levels of noise, cv-CC4.5 is significantly better than CC4.5 with low values for s.

Table 3. P-values of Nemenyi test for C4.5, CC4.5 (varying s) and cv-CC4.5, without added noise

i	Algorithms	$p-values$
10	$\text{CC4.5}_{s=3.0}$ vs. **cv-CC4.5**	0
9	**C4.5** vs. $\text{CC4.5}_{s=3.0}$	0.000001
8	$\textbf{CC4.5}_{s=0.25}$ vs. $\text{CC4.5}_{s=3.0}$	0.000002
7	$\text{CC4.5}_{s=2.0}$ vs. **cv-CC4.5**	0.000005
6	$\textbf{CC4.5}_{s=1.0}$ vs. $\text{CC4.5}_{s=3.0}$	0.000015
5	$\textbf{CC4.5}_{s=0.5}$ vs. $\text{CC4.5}_{s=3.0}$	0.000017
4	**C4.5** vs. $\text{CC4.5}_{s=2.0}$	0.000020
3	$\textbf{CC4.5}_{s=0.25}$ vs. $\text{CC4.5}_{s=2.0}$	0.000031
2	$\textbf{CC4.5}_{s=1.0}$ vs. $\text{CC4.5}_{s=2.0}$	0.000173
1	$\textbf{CC4.5}_{s=0.5}$ vs. $\text{CC4.5}_{s=2.0}$	0.000187

Table 4. P-values of Nemenyi test for C4.5, CC4.5 (varying s) and cv-CC4.5 (5% of noise)

i	Algorithms	$p-values$
8	$\text{CC4.5}_{s=3.0}$ vs. **cv-CC4.5**	0
7	$\textbf{CC4.5}_{s=0.25}$ vs. $\text{CC4.5}_{s=3.0}$	0
6	$\textbf{CC4.5}_{s=1.0}$ vs. $\text{CC4.5}_{s=3.0}$	0.000002
5	$\textbf{CC4.5}_{s=0.5}$ vs. $\text{CC4.5}_{s=3.0}$	0.000003
4	**C4.5** vs. $\text{CC4.5}_{s=3.0}$	0.000010
3	$\textbf{CC4.5}_{s=1.5}$ vs. $\text{CC4.5}_{s=3.0}$	0.000383
2	$\text{CC4.5}_{s=2.0}$ vs. **cv-CC4.5**	0.000605
1	$\textbf{CC4.5}_{s=0.25}$ vs. $\text{CC4.5}_{s=2.0}$	0.001259

Besides, cv-CC4 and CC4.5 with high values for s are the best for these high noise levels. These facts can be checked with the results of the tests carried out.

The results are analyzed as follows:

- **Average accuracy:** cv-CC4.5 achieves the best result for each added noise level, except for 30% of added noise where it obtains the second best value but close to the best achieved by CC4.5 with $s = 3.0$.
- **Friedman's ranking:** cv-CC4.5 obtains the best Friedman's rank for all the noise levels except for 10% (where the best is CC4.5 with $s = 1.0$) and for 30% (where the best is CC4.5 with $s = 2.0$). In these cases, the ranking of the new method is close to the one of the best method.

Table 5. P-values of Nemenyi test for C4.5, CC4.5 (varying s) and cv-CC4.5 (10% of noise)

i	Algorithms	$p - values$
6	$\mathbf{CC4.5}_{s=1.0}$ vs. $\text{CC4.5}_{s=3.0}$	0.000006
5	$\text{CC4.5}_{s=3.0}$ vs. **cv-CC4.5**	0.000014
4	$\mathbf{CC4.5}_{s=0.5}$ vs. $\text{CC4.5}_{s=3.0}$	0.000045
3	$\mathbf{CC4.5}_{s=1.5}$ vs. $\text{CC4.5}_{s=3.0}$	0.000097
2	C4.5 vs. $\mathbf{CC4.5}_{s=1.0}$	0.000482
1	C4.5 vs. **cv-CC4.5**	0.000944

Table 6. P-values of Nemenyi test for C4.5, CC4.5 (varying s) and cv-CC4.5 (20% of noise)

i	Algorithms	$p - values$
12	C4.5 vs. **cv-CC4.5**	0
11	C4.5 vs. $\mathbf{CC4.5}_{s=1.5}$	0
10	$\text{CC4.5}_{s=0.25}$ vs. **cv-CC4.5**	0
9	C4.5 vs. $\mathbf{CC4.5}_{s=2.0}$	0
8	$\text{CC4.5}_{s=0.25}$ vs. $\mathbf{CC4.5}_{s=1.5}$	0.000003
7	$\text{CC4.5}_{s=0.25}$ vs. $\mathbf{CC4.5}_{s=2.0}$	0.000006
6	C4.5 vs. $\mathbf{CC4.5}_{s=3.0}$	0.000008
5	$\text{CC4.5}_{s=0.5}$ vs. **cv-CC4.5**	0.000037
4	$\text{CC4.5}_{s=0.25}$ vs. $\mathbf{CC4.5}_{s=3.0}$	0.000203
3	C4.5 vs. $\mathbf{CC4.5}_{s=1.0}$	0.000239
2	$\text{CC4.5}_{s=0.5}$ vs. $\mathbf{CC4.5}_{s=1.5}$	0.000520
1	$\text{CC4.5}_{s=0.5}$ vs. $\mathbf{CC4.5}_{s=2.0}$	0.000815

- **Nemenyi test:** The only method that is always significantly better than the worse methods is cv-CC4.5. For low noise levels, the worse methods are CC4.5 with s value in $\{2.0, 3.0\}$; and for high added noise levels, the worse methods are C4.5 and CC4.5 with s values in $\{0.25, 0.5, 1.0\}$.

Table 7. P-values of Nemenyi test for C4.5, CC4.5 (varying s) and cv-CC4.5 (30% of noise)

i	Algorithms	$p - values$
15	C4.5 vs. $\mathbf{CC4.5}_{s=2.0}$	0
14	C4.5 vs. $\mathbf{CC4.5}_{s=3.0}$	0
13	C4.5 vs. **cv-CC4.5**	0
12	$\text{CC4.5}_{s=0.25}$ vs. $\mathbf{CC4.5}_{s=2.0}$	0
11	$\text{CC4.5}_{s=0.25}$ vs. $\mathbf{CC4.5}_{s=3.0}$	0
10	$\text{CC4.5}_{s=0.25}$ vs. **cv-CC4.5**	0
9	$\text{CC4.5}_{s=0.5}$ vs. $\mathbf{CC4.5}_{s=2.0}$	0
8	$\text{CC4.5}_{s=0.5}$ vs. $\mathbf{CC4.5}_{s=3.0}$	0
7	C4.5 vs. $\mathbf{CC4.5}_{s=1.5}$	0
6	$\text{CC4.5}_{s=0.5}$ vs. **cv-CC4.5**	0.000003
5	$\text{CC4.5}_{s=0.25}$ vs. $\mathbf{CC4.5}_{s=1.5}$	0.000015
4	$\text{CC4.5}_{s=1.0}$ vs. $\mathbf{CC4.5}_{s=2.0}$	0.000069
3	$\text{CC4.5}_{s=1.0}$ vs. $\mathbf{CC4.5}_{s=3.0}$	0.000089
2	$\text{CC4.5}_{s=0.5}$ vs. $\mathbf{CC4.5}_{s=1.5}$	0.000561
1	$\text{CC4.5}_{s=1.0}$ vs. **cv-CC4.5**	0.000815

6 Conclusion

In this work, cv-CC4.5 is presented where the problem of finding a good value for s in CC4.5 is minimized. This model cv-CC4.5 is always better or equivalent to the best CC4.5 with distinct values for s when noisy data sets are classified. In this way, an improvement of the method CC4.5 is obtained. By using cv-CC4.5, it is not necessary to indicate a value for s before performing CC4.5, regardless the noise of the data set to be classified.

Acknowledgments. This work has been supported by the Spanish "Ministerio de Economía y Competitividad" and by "Fondo Europeo de Desarrollo Regional" (FEDER) under "Project TEC2015-69496-R".

References

1. Abellán, J., Moral, S.: Building classification trees using the total uncertainty criterion. Int. J. Intell. Syst. **18**(12), 1215–1225 (2003)
2. Abellán, J.: Uncertainty measures on probability intervals from Imprecise Dirichlet model. Int. J. General Syst. **35**(5), 509–528 (2006)
3. Demsar, J.: Statistical comparison of classifiers over multiple data sets. J. Mach. Learn. Res. **7**, 1–30 (2006)
4. Kohavi, R.: Wrappers for performance enhancement and oblivious decision graphs. In: UMI Order No. GAX96-11989, Stanford University (1996)

5. Mantas, C.J., Abellán, J.: Credal-C4.5: decision tree based on imprecise probabilities to classify noisy data. Expert Syst. Appl. **41**(10), 4625–4637 (2014)
6. Mantas, C.J., Abellán, J.: Credal decision trees in noisy domains. In: European Symposium on Artificial Neural Networks. Computational Intelligence and Machine Learning (2014)
7. Mantas, C.J., Abellán, J., Castellano, J.G.: Analysis of Credal-C4.5 for classification in noisy domains. Expert Syst. Appl. **61**, 314–326 (2016)
8. Quinlan, J.R.: Programs for Machine Learning. Morgan Kaufmann series in Machine Learning (1993)
9. Walley, P.: Inferences from multinomial data, learning about a bag of marbles. J. Roy. Stat. Soc. Ser. B **58**, 3–57 (1996)
10. Witten, I.H., Frank, E.: Data Mining, Practical Machine Learning Tools and Techniques, 2nd edn. Morgan Kaufmann, San Francisco (2005)

Tag-Based User Fuzzy Fingerprints for Recommender Systems

André Carvalho, Pável Calado, and Joao Paulo Carvalho(✉)

INESC-ID, Instituto Superior Técnico, Universidade de Lisboa, Lisbon, Portugal
{andre.silva.carvalho,pavel.calado}@tecnico.ulisboa.pt,
joao.carvalho@inesc-id.pt

Abstract. Most Recommender Systems rely exclusively on ratings and are known as Memory-based Collaborative Filtering systems. This is currently dominant approach outside of academia due to the low implementation effort and service maintenance, when compared with more complex Model-based approaches, Traditional Memory-based systems have as their main goal to predict ratings, using similarity metrics to determine similarities between the users' (or items) rating patterns. In this work, we propose a user-based Collaborative Filtering approach based on tags that does not rely on rating prediction, instead leveraging on Fuzzy Fingerprints to create a novel similarity metric. Fuzzy Fingerprints provide a concise and compact representation of users allowing the reduction of the dimensionality usually associated with user-based collaborative filtering. The proposed recommendation strategy combined with the Fuzzy Fingerprint similarity metric is able to outperform our baselines, in the Movielens-1M dataset.

Keywords: Recommender system · Collaborative Filtering
Fuzzy Fingerprint · Tags

1 Introduction

Users of the digital world are overloaded with information [16]. Recommender Systems (RSs) allow us to cope with this, by cataloging a vast list of items, that later can be recommended. Due to their success, RSs can be found in a number of services, providing recommendations for movies, music, news, products, events, services, among others [1]. However, turning state of the art solutions into real-world scenarios is still challenging, mainly due to a large amount of available data and the ensuing scalability issues. For this reason, more traditional approaches, such as Collaborative Filtering (CF) are still the most widely used [18]. Despite its simplicity, CF can provide quite accurate results, thus yielding an advantageous trade-off between engineering effort and user satisfaction.

Memory-based Collaborative Filtering can usually be implemented using one of two different strategies: *user-based* CF, which compares users ratings to determine a neighborhood of similar users; and *item-based* CF, which instead computes item similarities and forms item neighborhoods to produce the rating predictions.

J. Medina et al. (Eds.): IPMU 2018, CCIS 855, pp. 748–758, 2018.
https://doi.org/10.1007/978-3-319-91479-4_62

Over the years, item-based CF has replaced user-based CF, given its better scalability properties [11]. Since the number of users grows over time, and generally at a faster rate than the items, so does the number of similarities, thus posing a scalability problem. Similarities between users also vary more over time than similarities between items, since individual users tend to change their preferences, while the global opinion on a given item tends to remain stable.

In this work, we argue that an effective and efficient user-based CF system can be implemented. To this effect, we use Fuzzy Fingerprints (FFPs) to represent users based on item *tags* and ratings. *Tags*, i.e. short textual labels attached by the users to the items, provide an item description or categorization and are a common resource in current online RSs. They allow us to create a more detailed user representation than traditional CF, in a controlled manner, i.e. by controlling the number of tags used in the FFPs, we can easily fine-tune our system to improve recommendation quality or to speed up the similarity computation. In this work, we mainly focus on obtaining an improved recommendation quality.

Our main contributions are, therefore, (1) a new way to determine relevant items to recommend to users without requiring the computation of rating predictions for user-based CFs, and (2) a novel similarity metric for RSs, using the concept of FFPs [9] to represent users based on tags from rated items. More specifically, we propose to represent users by their low-dimensional Fingerprints, which can then be directly used to determine similarities between them. A similar idea has been previously applied to text authorship identification [9] with success. Our goal is to apply the same principle to RSs using tags from the items rated by each user, to obtain better recommendations. This solution has three major advantages: (1) provides overall better recommendations to users; (2) requires a minimal implementation effort; and (3) its representation of the users is scalable and easily maintainable.

To demonstrate our claims, experiments were performed on a movie dataset providing movies metadata information, allowing the creation of users FFPs.

The remainder of this paper is organized as follows: Sect. 2 contains literature review on similarity metrics for CF; Sect. 3 presents how FFPs can be applied to RSs; Sect. 4 presents an experimental evaluation; finally, in Sect. 5 some conclusions are drawn from the results and directions for future work are proposed.

2 Related Work

Fuzzy systems approaches have been previously used to improve the RS similarity metric [6] focusing exclusively on item-based CF. Our proposal applies concepts of Fuzzy Systems to the problem of user-based Collaborative Filtering. More specifically, we use Fuzzy Fingerprints, in a CF system, to represent users in a more compact way.

CF systems usually rely on the ratings given to items by users to determine similarities between users (or items), through the use of a similarity metric. This allows the creation of neighborhoods of similar users, to predict new ratings.

Traditionally, the similarity is measured using metrics such as Pearson Correlation (PC) or the Cosine similarity (COS) [2]. Nevertheless, many other ways of measuring similarity have been proposed, ranging from simple variations of PC and COS, through the design of more complex functions.

An example is the work of [7], where ratings are combined with a measure of *trust* between users, which is inferred from social information. By introducing the degree of trust between users the authors show that it does improve the overall rating prediction. On a different approach, in [3], the authors propose a combination of the mean squared difference between the user's ratings with the Jaccard coefficient. Through experiments, they demonstrate that results are improved, when compared to traditional CF.

To determine the neighborhood of each user, usually, similarities are computed between the user and all other users, which are then sorted by their degree of similarity and only the top k are kept. In [17] an alternative way to determine neighborhoods is proposed. The authors randomly choose a possible neighbor from the set of all users. This neighbor is kept only if its similarity is above a given threshold. The process is then repeated until a certain amount of neighbors is obtained. Their work has two threshold variables that depend on the data and must be fine-tuned: (1) the minimum similarity for a user to be considered a neighbor; (2) the minimum number of users in the neighborhood.

Combining Recommender systems and tags is not a novel idea [10,13,15]. Tags can help alleviate the so-called *cold-start* and *data sparsity* problems. The cold-start problem occurs when new items, not yet rated by any user, or new users, who have not rated any item yet, cannot receive recommendations since they cannot be compared to other items/users. The *data sparsity* problem is also associated with CF systems since it is common for users and items to have very few ratings, and thus not enough information to produce valuable recommendations [4]. Tags can help address these issues, they only depend on the availability of metadata, for each item. Our RS takes advantage of tags to more accurately represent each user and, therefore, improve the quality of the user similarity computation.

Liu et al. [12] also propose a new similarity metric, which assigns penalties to *bad* similarities, while rewarding *good* similarities. Defining a similarity as good or bad depends on several factors, such as the popularity of the rated items or the similarity of the rating to the other user's ratings.

In [5] a FFP was applied to item-based CF using also movies synopsis to represent items. The FFP results from ratings and synopsis words that are also added as features. A normalization is applied to both ratings and synopsis words, separately, resulting in FFP which combines both. Note that in this work, we are currently creating a user-based CF to represent users with item tags weighted by the ratings, and not represent item using FFP.

The above works show that the selection process of neighbors and the improvement of the similarity measures have a beneficial impact on the overall RS results. This work presents a similarity metric based on FFPs, adapted for user-based CF, using the tags associated to each item, with the main goal of improving the recommendation quality.

3 Tag-Based User Fuzzy Fingerprints for Collaborative Filtering

A Fuzzy Fingerprint (FFP) is a fuzzified ranked vector containing information based on frequencies of occurrence of the elements being encoded [9]. In this Section, we explain how to build and apply a tag-based FFP to represent users in a CF recommender system.

Let N be the total number of tags in the system and let M be the total number of items in the system. Let θ_i represent the set of tags of a given item i: $\theta_i = (t_{1i}, t_{2i}, t_{3i}, \cdots, t_{Ni})$. Any element $t_{ni} \in \theta_i$ can assume the value 1 if the respective tag occurs in the item, or 0 if it does not.

Let r_u be the set of ratings for a given set of items $i_1 \cdots, i_M$, provided by a user u: $r_u = (r_{1u}, r_{2u}, \cdots, r_{Mu})$. We assume, without loss of generality, that $r_{mu} \geq 0$ and that a value of zero means that the user has not yet rated item i_m.

A Fingerprint ϕ_u is built by counting, for user u, the number of occurrences of each tag in the items rated by u, multiplied by the respective item's rating, i.e. $\phi_u = (c_{1u}, c_{2u}, \cdots, c_{Nu})$, where:

$$c_{nu} = \sum_{\forall i=1}^{M} t_{ni} \times r_{iu} \tag{1}$$

The rationale behind Eq. (1) is that tags from items a user has rated higher should also get a higher importance in the Fingerprint. The next step consists in ordering ϕ_u according to c_{nu} and keeping only the k highest values. The Fingerprint size k is a parameter of the system and can be optimized offline.

To illustrate the previous procedure, let $r_u = (5, 2, 4)$ for items a, b, and c. Assume there are only 5 tags and let $\theta_a = (1, 0, 0, 1, 1)$, $\theta_b = (0, 1, 0, 0, 1)$, and $\theta_c = (0, 0, 1, 1, 0\}$. Assuming that $k = 4$, the resulting Fingerprint ϕ_u will be $(c_{4u} = 9, c_{5u} = 7, c_{1u} = 5, c_{3u} = 4)$.

The Fingerprint ϕ_u is, therefore, an *ordered set* of tags. The rank of each tag reflects its importance in representing the user. This Fingerprint still needs to be transformed into a Fuzzy Fingerprint. The fuzzification of the Fingerprint leverages the importance of the order (and not of the frequency) to distinguish between users. The FFP of user u, Φ_u, is obtained by fuzzifying the rank (the position in the Fingerprint) of each tag.

The choice of the fuzzifying function can affect the obtained results [8,9]. Here, we have tested the linear approach, shown in Eq. 2, where p_{u_j} is the rank of tag t_n within ϕ_u (starting with $t = 0$).

$$\mu_{linear}(\mathrm{p}_{t_n}) = \frac{k - \mathrm{p}_{t_n}}{k} \tag{2}$$

Preliminary experiments indicate that using other fuzzifying functions does not significantly improve or degrade the quality of the results in this approach.

After the fuzzification step, we can now define the FFP Φ_u as:

$$\Phi_u = \{(t_n, \mu(\mathrm{p}_{t_n})), \forall t_n \in \phi_u\} \tag{3}$$

The FFP is, therefore, a ranked set of tags, each of which is associated with a membership value, built based on the description of the items rated by the user.

Once the FFP for each user is determined, it is possible to compute similarities between users.

Consider Φ_u and Φ_j the FFPs of users u and j. The FFP similarity between users u and j is defined as:

$$sim_{FFP} = (\Phi_u, \Phi_j) = \sum_{t_n \in U_i \cap U_j} \frac{\min(\Phi_u(t_n), \Phi_j(t_n))}{k} \tag{4}$$

where $\Phi_x(t_u)$ denotes the membership value associated to tag t_n in Φ_x. Note that the use of k in this equation as a normalization factor is only needed to facilitate development and parameter optimization. It can be omitted during system operation when computing similarities, largely improving computational efficiency.

The recommendation process of the proposed RS does not rely upon rating predictions as in traditional Collaborative Filtering (see Sect. 4). Instead, it identifies the user's nearest neighbors (according to Eq. 4) and uses the items seen and liked by them to extrapolate possible items to recommend to the user.

The RS starts by computing which users are the nearest neighbors of user u, based on the FFP similarity metric. Users are considered neighbors if the similarity is greater than a defined threshold $sim_{\text{threshold}}$.

We consider that any item rated highly by a neighbor (e.g., 4 or 5 on a 0–5 scale) and rated higher than that neighbor's item rating average, is recommendable to the user.

The final step in the recommendation process consists in getting the difference between the rating of the recommendable item, the average rating given to that item by the neighbor, and multiplying it by the similarity between the user and the neighbor. This allows to create a ranking of recommendable items.

4 Evaluation

To assert the effectiveness of the proposed RS experiments were performed using a movie dataset. Precision, Recall, and F1-score are used as evaluation metrics.

The similarity metrics used as baselines for comparison are the traditional Pearson Correlation (PC) and Cosine similarity (COS). In addition, we also include the Jaccard Mean Squared Difference (JMSD) [3], an improvement on previous metrics that offers a high rating prediction accuracy, while using a lower number of neighbors. Finally, a similarity metric, that uses FFPs [6] yet is only applicable to traditional item-based CF and which only relies upon ratings to compute similarities. We refer to this baseline [6] throughout the rest of this document as FFP_{rating}. While the FFP proposed in this document will be referenced as FFP_{tags}. All similarity metrics baselines use both user-based and item-based, except FFP_{rating} that is only applicable to item-based CF.

Pearson Correlation coefficient has been widely used since it is simple to implement, intuitive, and provides good quality results [3]. PC is defined in Eq. 5, where I is the set items both user u and j rated.

$$sim_{PC}(u,j) = \frac{\sum_{i \in I}(r_{u,i} - \bar{r}_u) \times (r_{j,i} - \bar{r}_j)}{\sqrt{\sum_{i \in I}(r_{u,i} - \bar{r}_u)^2} \times \sqrt{\sum_{i \in I}(r_{j,i} - \bar{r}_j)^2}} \tag{5}$$

The resulting similarity will be in within the interval $[-1, 1]$, where -1 corresponds to an inverse correlation, $+1$ to a positive correlation, and values near zero show that no linear correlation exists between the two users.

Another often used similarity measure is the Cosine similarity, as defined in Eq. 6. COS will yield a value between 0 and 1, where 0 corresponds to no similarity between u and j and 1 to exactly proportional ratings between both users.

$$sim_{COS}(u,j) = \frac{\sum_{i \in I} r_{u,i} \times r_{j,i}}{\sqrt{\sum_{i \in I} r_{u,i}^2} \times \sqrt{\sum_{i \in I} r_{j,i}^2}} \tag{6}$$

The idea behind Jaccard Mean Squared Difference (JMSD) is to combine the Jaccard coefficient, which captures the number of ratings in common between users, with the Mean Square Difference (MSD) of those ratings, resulting in Eq. 7:

$$sim_{JMSD}(u,j) = Jaccard(u,j) \times \big(1 - MSD(u,j)\big); \tag{7}$$

where $Jaccard$ and MSD are defined as:

$$Jaccard(i,j) = \frac{|I_u \cap I_j|}{|I_u \cup I_j|} \qquad MSD(i,j) = \frac{\sum_{i \in I}(r_{u,i} - r_{j,i})^2}{|I|} \tag{8}$$

where I_s is the set of items rated by user s.

The FFP_{rating} metric uses an approach that is totally different to the one proposed in this work: each item has its own FFP and the recommendation is based exclusively on ratings. The user's ratings constitute the item Fingerprint and ratings are sorted taking into consideration the total amount of ratings from each user.

We now explain how a traditional CF computes rating predictions. Let $\hat{r}_{ui}$ be the *predicted* rating that a given user u would assign to item i. We start by computing the *neighborhood* N_u, of user u, i.e. the set of n users in the database that are more similar to u, using a similarity function. The value of $\hat{r}_{ui}$ is defined as:

$$\hat{r}_{ui} = \bar{r}_u + \frac{\sum_{v \in N_u} \text{sim}(u,v) \times (r_{vi} - \bar{r}_v)}{\sum_{v \in N_v} \text{sim}(u,v)} \tag{9}$$

where r_{vi} is the rating assigned by user v to item i, $\bar{r}_x$ is the average of all ratings assigned to user x. A traditional CF system usually performs these predictions for a large set of items and returns those with the highest rating predictions, as recommendations.

An evaluation was conducted using MovieLens-1M (ML-1M) dataset, from the movie domain. By using Dbpedia[1], Tags and other meta-data, regarding each movie, were collected. In this work, we focus exclusively on Tag information.

The ML-1M dataset has 1 million ratings, 6040 users, 3706 items, a sparsity of 95.53% and has an average of 125 ratings per user.

The evaluation process was performed through 5-fold cross-validation, using RiVal [14], a framework to make RSs evaluation fair process, completely separating the recommendation task of a RS from the Evaluation of the recommendations.

We define any item with rating greater than or equal to 4 as a relevant (i.e. should be recommended) to the user.

Precision can be computed using Eq. 10 and Recall using Eq. 11. In this work, we do not set a threshold for a maximum number of recommendations i.e. the RS can recommend as many relevant items to a user as possible. Even though we calculate the F1-score (Eq. 12.), we support the idea that Precision is a far better indication for a good RS, as long as Recall is within a range that allows the retrieval of a sufficient number of relevant items (in the tested cases, all approaches fulfill the Recall criteria).

$$PR = \frac{\#TruePositives}{\#TruePositives + \#FalsePositives} \tag{10}$$

$$RC = \frac{\#TruePositives}{\#TruePositives + \#FalseNegatives} \tag{11}$$

$$F1 = 2 \times \frac{PR \times RC}{PR + RC} \tag{12}$$

We start by comparing the similarity distribution using our similarity metric and the baselines, this allows us to determine the best $sim_{\text{threshold}}$ when selecting the neighborhood. We then vary the number of neighbors used by the FFP_{tags} over different sizes of k. This allows us to determine not only the best k for the FFP_{tags} but also the most adequate number of neighbors to use. Finally, we present a summary table with baselines and how do they perform in comparison to the proposed RS.

Figure 1 shows the similarity distributions. By analyzing Fig. 1d, we notice that the average similarity is around 0.2. This provides a good indicator to experiment different $sim_{\text{threshold}}$ around 0.2. Experimentally, we determined that using 0.25 provides good results, for this dataset.

Figure 2 compares different sizes for the FFP and for each size we vary the number of neighbors used by the RS. According to the $F1 - measure$ the best results are obtained using k equal to 200. Knowing that, on average, each user has 637 tags associated to rated movies, the proposed FFP similarity metric uses only 31% of existing tags, being able to correctly select relevant tags to represent each user.

[1] Dbpedia: http://www.dbpedia.org.

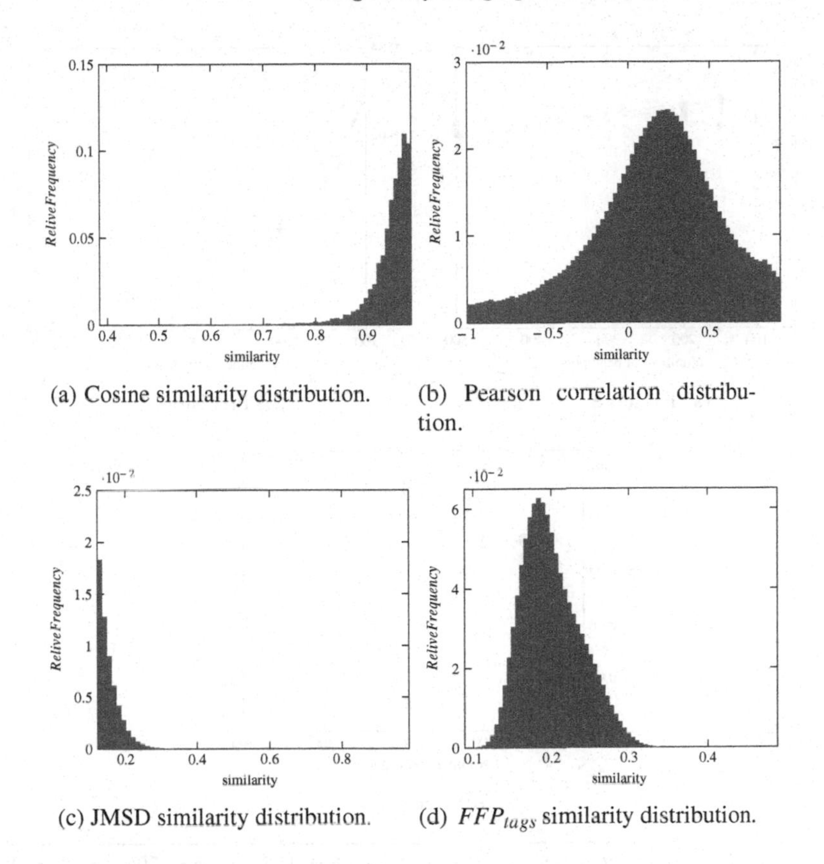

(a) Cosine similarity distribution.

(b) Pearson correlation distribution.

(c) JMSD similarity distribution.

(d) FFP_{tags} similarity distribution.

Fig. 1. Histograms show the similarity distribution of different similarity metrics when applied to user similarity computation. FFP_{tags} uses $k = 200$ tags to represent a user FFP.

Table 1 shows how the different tested approaches perform. The proposed Tag-user based FFP performs better overall than any other approach, even when compared to the state-of-the-art JMSD, although the improvement is not significant.

An interesting result is how much better the proposed approach is when compared to other previously proposed user-based approaches, thus opening the door to further developments in user-based RS. It should be noted that item-based approaches have been thoroughly used in the past and have been highly optimized. Yet user-based approaches are also viable. For example, it is very easy to enrich the FFP using data other than simple tags, from movie descriptions to a user's favorite actors, directors or genres.

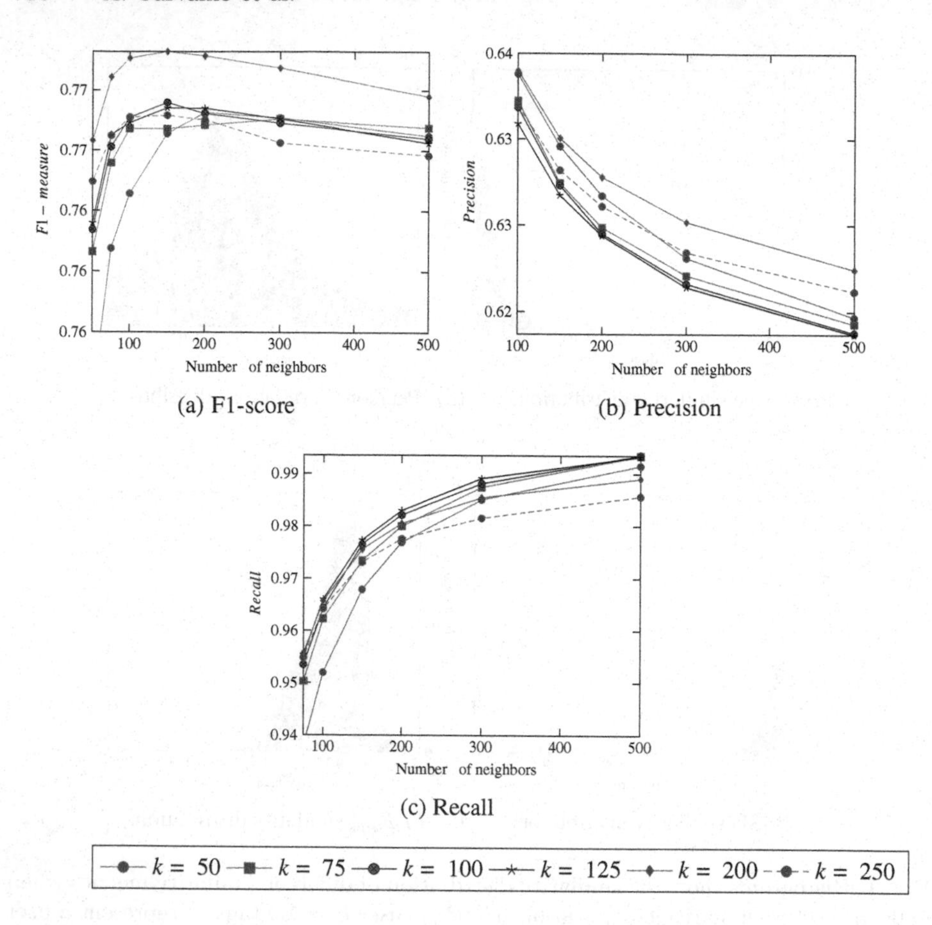

Fig. 2. Comparison between different sizes of the FFP, while varying the number of neighbors used.

Table 1. Summary results in which FFP_{tags} (using $k = 200$) combined with the proposed recommendation algorithm is compared with several baselines using item-based and user-based CF.

Similarity Metric	Approach	Num. neighbors	F1-score	Precision	Recall
FFP_{tags}	User-based	150	**0.76929**	**0.63504**	0.97554
COS	Item-based	50	0.76622	0.62112	**0.99978**
PC	Item-based	75	0.76621	0.62115	0.99969
JMSD	Item-based	20	0.76623	0.62112	0.99980
$FPP_{ratings}$	Item-based	20	0.76623	0.62113	0.99979
COS	User-based	200	0.42356	0.26869	0.99989
PC	User-based	100	0.42338	0.26854	0.99990
JMSD	User-based	100	0.42356	0.26869	0.99989

5 Conclusion

In this work, we have applied the concept of Fuzzy Fingerprints to user-based Collaborative Filtering and represented users based on tags according to the items they rated. FFPs are used to create a new concise user representation that improves the F1-score and Precision of an RS. The best result for the proposed approach was obtained for $k = 200$. In this dataset, each user has on average 637 tags, which shows that the FFPs are able to reduce the problem complexity while still improving recommendation quality.

We have experimentally compared our proposal to two traditional similarity measures, Pearson Correlation and Cosine similarity, and a state-of-the-art similarity metrics such as Jaccard Mean Squared Difference.

Results show that FFPs are a promising approach since they can be applied with success in recommendation tasks. In fact, using FFPs we are able to represent a user using, on average, 68% less features. In addition, and even though we do not address such issue in this paper, FFP similarity is a much more computationally efficient process than any of the other similarity measures. This can be arguably enough to compensate for the fact that there are usually much more users than items in RS, as we will try to show in a future work.

Future work includes more extensive parameter optimization, enriching the FFP with other features, and improving the last step of the recommendation algorithm by using more sophisticated ways to aggregate the influence of each neighbor.

Acknowledgments. This work was supported by national funds through Fundação para a Ciência e a Tecnologia (FCT) with reference UID/CEC/50021/2013, by project GoLocal (ref. CMUPERI/TIC/0046/2014) and co-financed by the University of Lisbon and INESC-ID.

References

1. Adomavicius, G., Tuzhilin, A.: Toward the next generation of recommender systems: a survey of the state-of-the-art and possible extensions. IEEE Trans. Knowl. Data Eng. **17**(6), 734–749 (2005). https://doi.org/10.1109/TKDE.2005.99
2. Bobadilla, J., Ortega, F., Hernando, A., Gutiérrez, A.: Recommender systems survey. Knowl.-Based Syst. **46**, 109–132 (2013). https://doi.org/10.1016/j.knosys.2013.03.012. http://www.sciencedirect.com/science/article/pii/S0950705113001044
3. Bobadilla, J., Serradilla, F., Bernal, J.: A new collaborative filtering metric that improves the behavior of recommender systems. Knowl.-Based Syst. **23**(6), 520–528 (2010). https://doi.org/10.1016/j.knosys.2010.03.009. http://www.sciencedirect.com/science/article/pii/S0950705110000444
4. Bogdanov, D., Haro, M., Fuhrmann, F., Xambó, A., Gómez, E., Herrera, P.: Semantic audio content-based music recommendation and visualization based on user preference examples. Inf. Process. Manage. **49**(1), 13–33 (2013). https://doi.org/10.1016/j.ipm.2012.06.004. http://www.sciencedirect.com/science/article/pii/S0306457312000763

5. Carvalho, A., Calado, P., Carvalho, J.P.: Combining ratings and item descriptions in recommendation systems using fuzzy fingerprints. In: 2017 IEEE International Conference on Fuzzy Systems (FUZZ-IEEE), pp. 1–6 (2017). https://doi.org/10.1109/FUZZ-IEEE.2017.8015604
6. Carvalho, A., Calado, P., Carvalho, J.P.: Fuzzy fingerprints for item-based collaborative filtering. In: Kacprzyk, J., Szmidt, E., Zadrożny, S., Atanassov, K.T., Krawczak, M. (eds.) IWIFSGN/EUSFLAT -2017. AISC, vol. 641, pp. 419–430. Springer, Cham (2018). https://doi.org/10.1007/978-3-319-66830-7_38
7. Chen, S., Luo, T., Liu, W., Xu, Y.: Incorporating similarity and trust for collaborative filtering. In: Sixth International Conference on Fuzzy Systems and Knowledge Discovery, FSKD 2009, vol. 2, pp. 487–493 (2009). https://doi.org/10.1109/FSKD.2009.720
8. Dimiev, V.: Fuzzifying functions. Fuzzy Sets Syst. **33**(1), 47–58 (1989). https://doi.org/10.1016/0165-0114(89)90216-9. http://www.sciencedirect.com/science/article/pii/0165011489902169
9. Homem, N., Carvalho, J.P.: Authorship identification and author fuzzy 'fingerprints'. In: Annual Meeting of the North American Fuzzy Information Processing Society (NAFIPS), pp. 1–6 (2011). https://doi.org/10.1109/NAFIPS.2011.5751998
10. Kim, H.N., Ji, A.T., Ha, I., Jo, G.S.: Collaborative filtering based on collaborative tagging for enhancing the quality of recommendation. Electron. Commer. Res. Appl. **9**(1), 73–83 (2010). https://doi.org/10.1016/j.elerap.2009.08.004. http://www.sciencedirect.com/science/article/pii/S1567422309000544, Special Issue: Social Networks and Web 2.0
11. Linden, G., Smith, B., York, J.: Amazon.com recommendations: item-to-item collaborative filtering. IEEE Internet Comput. **7**(1), 76–80 (2003). https://doi.org/10.1109/MIC.2003.1167344
12. Liu, H., Hu, Z., Mian, A., Tian, H., Zhu, X.: A new user similarity model to improve the accuracy of collaborative filtering. Knowl.-Based Syst. **56**, 156–166 (2014). https://doi.org/10.1016/j.knosys.2013.11.006. http://www.sciencedirect.com/science/article/pii/S0950705113003560
13. Osmanli, O., Toroslu, I.: Using tag similarity in SVD-based recommendation systems. In: 2011 5th International Conference on Application of Information and Communication Technologies (AICT), pp. 1–4 (2011). https://doi.org/10.1109/ICAICT.2011.6111034
14. Said, A., Bellogín, A.: RiVal: a toolkit to foster reproducibility in recommender system evaluation. In: RecSys 2014 Proceedings of the 8th ACM Conference on Recommender Systems, pp. 371–372. ACM Press (2014). https://doi.org/10.1145/2645710.2645712, http://dl.acm.org/citation.cfm?doid=2645710.2645712
15. Smith, G.: Tagging: People-Powered Metadata for the Social Web, 1st edn. New Riders Publishing, Thousand Oaks (2007)
16. Tsai, C.F., Hung, C.: Cluster ensembles in collaborative filtering recommendation. Appl. Soft Comput. **12**(4), 1417–1425 (2012). https://doi.org/10.1016/j.asoc.2011.11.016. http://www.sciencedirect.com/science/article/pii/S1568494611004583
17. Wibowo, A.T., Rahmawati, A.: Naive random neighbor selection for memory based collaborative filtering. In: 2015 International Seminar on Intelligent Technology and Its Applications (ISITIA), pp. 351–356 (2015). https://doi.org/10.1109/ISITIA.2015.7220005
18. Ye, T., Bickson, D., Ampazis, N., Benczur, A.: LSRS'15: Workshop on large-scale recommender systems. In: Proceedings of the 9th ACM Conference on Recommender Systems, RecSys 2015, pp. 349–350. ACM, New York (2015). https://doi.org/10.1145/2792838.2798715, http://doi.acm.org/10.1145/2792838.2798715

Author Index

Printed in the United States
By Bookmasters

Printed in the United States
By Bookmasters